STRUCTURE AND EVOLUTION OF ACTIVE GALACTIC NUCLEI

ASTROPHYSICS AND SPACE SCIENCE LIBRARY

A SERIES OF BOOKS ON THE RECENT DEVELOPMENTS
OF SPACE SCIENCE AND OF GENERAL GEOPHYSICS AND ASTROPHYSICS
PUBLISHED IN CONNECTION WITH THE JOURNAL
SPACE SCIENCE REVIEWS

Editorial Board

R.L.F. BOYD, *University College, London, England*

W. B. BURTON, *Sterrewacht, Leiden, The Netherlands*

L. GOLDBERG, *Kitt Peak National Observatory, Tucson, Ariz., U.S.A.*

C. DE JAGER, *University of Utrecht, The Netherlands*

J. KLECZEK, *Czechoslovak Academy of Sciences, Ondřejov, Czechoslovakia*

Z. KOPAL, *University of Manchester, England*

R. LÜST, *European Space Agency, Paris, France*

L. I. SEDOV, *Academy of Sciences of the U.S.S.R., Moscow, U.S.S.R.*

Z. ŠVESTKA, *Laboratory for Space Research, Utrecht, The Netherlands*

VOLUME 121
PROCEEDINGS

TABLE OF CONTENTS

PREFACE   xv

WELCOME ADDRESSES   xvii

LIST OF PARTICIPANTS   xxi

INVITED LECTURES:

EVOLUTION OF THE LUMINOSITY FUNCTION OF QUASARS
         M. Schmidt   3

IRAS OBSERVATIONS OF ACTIVE GALAXIES
         G. Neugebauer, B.T. Soifer, M. Rowan-Robinson   11

THE PROPERTIES OF "DWARF" SEYFERT NUCLEI IN NEARBY GALAXIES
         A. V. Filippenko, W. L. W. Sargent   21

CONTINUITY IN THE OBSERVED PROPERTIES OF QSOs, HIGH-REDSHIFT EMISSION-LINE RADIO GALAXIES, BL LAC OBJECTS, N SYSTEMS AND SEYFERT GALAXIES, AND POSSIBLE INTERPRETATIONS
         G. Burbidge   47

SPECTRAL PROPERTIES OF BLAZARS OBSERVED AT ULTRAVIOLET AND X-RAY FREQUENCIES
         A. Treves, G. Ghisellini, L. Maraschi, E. G. Tanzi   63

OPTICAL VARIABILITY IN QUASARS
         S. Cristiani   81

INTERNAL DYNAMICS AND FORMATION OF EMISSION CLOUDS IN ACTIVE GALACTIC NUCLEI
        I. Shlosman, P. A. Vitello, G. Shaviv   95

ON THE PAPALOIZOU AND PRINGLE INSTABILITY
        M. A. Abramowicz, O. M. Blaes, J. Lu   113

THE BIRTH OF AGNs AND QUASARS VIA THE COLLAPSE OF DENSE STAR CLUSTERS TO SUPERMASSIVE BLACK HOLES
        S. L. Shapiro   129

THE POSSIBLE MECHANISM OF THE FORMATION OF THE HARD SPECTRUM OF ACTIVE GALACTIC NUCLEI
        I. D. Novikov, B. E. Stern   149

THE PROTOGALACTIC CONNECTION
        J. Silk, R. F. G. Wyse   173

SEYFERT GALAXIES IN THE UNIVERSE
        D. E. Osterbrock   193

EVOLUTIONARY CONNECTION OF SEYFERT GALAXIES AND QUASARS
        D. W. Weedman   215

EVOLUTIONARY CONTINUITY OF THE OPTICAL AGNs
        A. Cavaliere, E. Giallongo, F. Vagnetti   231

SURVEYS OF LOCAL AGNs
        P. Véron   253

DISCOVERY OF NARROW AND VARIABLE LINES IN THE ULTRAVIOLET SPECTRUM OF THE SEYFERT GALAXY NGC 4151, AND AN OUTLINE OF OUR PREVIOUS RESULTS
        M. H. Ulrich, A. Altamore, A. Boksenberg, G. E. Bromage, J. Clavel, A. Elvius, M. V. Penston, G. C. Perola, M. A. J. Snyders   275

STRUCTURE AND EVOLUTION OF ACTIVE GALACTIC NUCLEI

INTERNATIONAL MEETING HELD IN TRIESTE, ITALY,
APRIL 10–13, 1985

Edited by

G. GIURICIN
Dipartimento di Astronomia, Università degli Studi di Trieste, Italy

F. MARDIROSSIAN
Osservatorio Astronomico di Trieste, Italy

M. MEZZETTI
Dipartimento di Astronomia, Università degli Studi di Trieste, Italy

and

M. RAMELLA
Dipartimento di Astronomia, Università degli Studi di Trieste, Italy

D. REIDEL PUBLISHING COMPANY

A MEMBER OF THE KLUWER  ACADEMIC PUBLISHERS GROUP

DORDRECHT / BOSTON / LANCASTER / TOKYO

Library of Congress Cataloging in Publication Data

Main entry under title:

Structure and evolution of active galactic nuclei.

(Astrophysics and space science library; v. 121)
Sponsored by the European Physical Society, Astronomy Division.
Includes indexes.
1. Galaxies–Congresses. 2. Galaxies–Evolution–Congresses. 3. Astrophysics–Congresses. I. Giuricin, G. (Giuliano) II. European Physical Society. Astronomy and Astrophysics Division. III. Series.
QB856.S77   1985   523.1'12   85-25675
ISBN   90-277-2155-6

Published by D. Reidel Publishing Company,
P.O. Box 17, 3300 AA Dordrecht, Holland.

Sold and distributed in the U.S.A. and Canada
by Kluwer Academic Publishers,
190 Old Derby Street, Hingham, MA 02043, U.S.A.

In all other countries, sold and distributed
by Kluwer Academic Publishers Group,
P.O. Box 322, 3300 AH Dordrecht, Holland.

All Rights Reserved
© 1986 by D. Reidel Publishing Company, Dordrecht, Holland
No part of the material protected by this copyright notice may be reproduced or
utilized in any form or by any means, electronic or mechanical
including photocopying, recording or by any information storage and
retrieval system, without written permission from the copyright owner

Printed in The Netherlands

TABLE OF CONTENTS

LARGE SCALE IONIZED GAS IN RADIO GALAXIES AND QUASARS
R. A. E. Fosbury   297

HOST GALAXIES OF QUASARS
J. W. Fried   309

OBSERVATIONS OF FAINT QUASARS: SHAPE AND EVOLUTION OF THE LUMINOSITY FUNCTION
D. C. Koo   317

A NEW COMPLETE SAMPLE OF FAINT, OPTICALLY SELECTED QUASARS: COMPARISON WITH PREVIOUS RESULTS
B. Marano, G. Zamorani, V. Zitelli   339

EVOLUTION OF THE LUMINOSITY FUNCTION OF EXTRAGALACTIC OBJECTS
V. Petrosian   353

SPECTRAL EVOLUTION IN YOUNG ACTIVE GALACTIC NUCLEI
E. Boldt, D. Leiter   383

QUASAR EVOLUTION AND LUMINOSITY FUNCTION: AN X-RAY PERSPECTIVE
Y. Avni   397

PROPERTIES OF THE SHARP METAL-RICH ABSORPTION LINES OBSERVED IN QSO SPECTRA
J. Bergeron   421

RADIO GALAXY POPULATIONS: A PROGRESS REPORT
H. Van Der Laan, P. Katgert, M. J. A. Oort   437

SOME THEORETICAL ASPECTS OF AGNs
M. J. Rees   447

POSTER PAPERS:

MEGAMASERS IN NUCLEI OF GALAXIES

W. A. Baan   461

QUASAR CANDIDATES IN THE FIELD OF S.A. 94 (2h53m+0° 20')
C. Barbieri, S. Cristiani   465

IMPROVEMENT OF THE HUBBLE DIAGRAM FOR QSOs LEADING TOWARDS Qo
J. E. Beckman, M. Kidger   471

LINERS: THE LOW LUMINOSITY END OF AGN
L. Binette   475

OBSERVATIONS OF ACTIVE NUCLEI AT CALAR ALTO OBSERVATORY
K. Birkle, P. Rafanelli, U. Thiele   479

FREQUENCY DEPENDENT POLARIZATION IN BLAZARS
C.-I. Björnsson   483

CCD PHOTOMETRY OF A SAMPLE OF SEYFERT GALAXIES: A PROGRESS REPORT
C. Bonoli, F. Bortoletto, F. Bonoli, F. Delpino, V. Zitelli   487

A NEW FAINT QSO SURVEY
B. J. Boyle, R. Fong, T. Shanks, B. A. Peterson   491

THE RELATIVE FREQUENCY OF COMPACT EXTRAGALACTIC OBJECTS
P. Brosche   499

AN IDEALIZED MODEL FOR DUST ACCRETION IN ACTIVE GALACTIC NUCLEI
J. Buitrago, E. Mediavilla   503

RAM PRESSURE CONFINEMENT OF THE EMISSION LINE GAS AND THE ASSOCIATED RADIO EMISSION IN AGNs
T. J. Carroll   509

TABLE OF CONTENTS

ULTRAVIOLET SPECTRA OF 3 SEYFERT 1 GALAXIES WITH OPTICAL FE II EMISSION LINES
F.-Z. Cheng, C. A. Grady, P. L. Selvelli — 513

DISTRIBUTION OF QUASARS AND THE FORMATION OF LARGE SCALE STRUCTURE IN THE UNIVERSE
Y. Chu, L. Fang — 517

RELATIVISTIC PARTICLE STREAMS IN AGN
C. S. Coleman — 521

THE BEHAVIOUR OF THE CIV EMISSION LINE STRENGTH IN VARIABLE SEYFERT 1 GALAXIES
L. Colina, W. Wamsteker — 525

THE LOW EXCITATION LINES IN QUASARS AND AGN IN THE FRAME WORK OF PHOTOIONIZATION MODELS
S. Collin-Souffrin — 531

COSMOLOGICAL EVOLUTION OF OPTICALLY SELECTED QSOs
L. Danese, G. De Zotti, A. Franceschini — 537

THE SMALL WIDE ANGLE TAIL NGC4874 IN THE CENTER OF COMA CLUSTER
L. Feretti, G. Giovannini — 541

COLLIMATED WINDS FROM ACTIVE GALACTIC NUCLEI
A. Ferrari, E. Trussoni, R. Rosner, K. Tsinganos — 543

COMPARISON BETWEEN OPTICALLY SELECTED AND X-RAY SELECTED ACTIVE GALACTIC NUCLEI
A. Franceschini, I. M. Gioia, T. Maccacaro — 547

GROUPS AROUND BRIGHT SEYFERT NUCLEI
K. J. Fricke, W. Kollatschny, H. H. Loose — 551

STATISTICS OF BRIGHT GALAXIES AND CLUSTER MORPHOLOGY
G. Giuricin, F. Mardirossian, M. Mezzetti ... 555

FIRST RESULTS FROM A CATALOG OF DATA CONCERNING AGN-FWHM AND FWZI VERSUS L_o FOR CIV AND $H\beta$
M. Joly ... 559

COSMOLOGICAL EVOLUTION IN THE EXTENT OF DOUBLE RADIO GALAXIES
V. K. Kapahi ... 565

THE SPECTRAL INDEX-FLUX DENSITY RELATION AND THE COSMOLOGICAL EVOLUTION OF RADIO SOURCES
V. K. Kapahi, V. K. Kulkarni ... 569

EFFECT OF DYNAMICAL FRICTION ON THE ESCAPE OF A SUPERMASSIVE BLACK HOLE EJECTED FROM THE CENTER OF A GALAXY
R. C. Kapoor ... 573

CLASSIFICATION OF OPTICAL JETS IN GALAXIES
W. C. Keel ... 579

CERENKOV LINE RADIATION - A NEW INTERPRETATION OF THE EMISSION LINES OF QUASARS
T. Kiang, J. H. You ... 583

AN INTERPRETATION OF THE LIGHTCURVE OF BL LAC
M. R. Kidger, J. E. Beckman ... 587

THE I.A.C./Q.M.C. CATALOGUE OF QUASAR MULTIBAND SPECTRA
M. R. Kidger, J. E. Beckman ... 591

THE OPTICAL VARIABILITY OF 3C 345
M. R. Kidger, J. E. Beckman ... 601

DOUBLE NUCLEUS GALAXIES
W. Kollatschny, K. J. Fricke, J. Hellwig ... 605

TABLE OF CONTENTS

A TEST FOR THE GRAVITATIONAL LENS HYPOTHESIS IN THE MOST LUMINOUS QUASAR : S5 0014+81
H. Kühr 611

THE PROPERTIES OF X-RAY SELECTED BL LAC OBJECTS
T. Maccacaro, I. M. Gioia, D. Maccagni, J. T. Stocke 615

KISO SURVEYS OF ULTRAVIOLET-EXCESS OBJECTS
H. Maehara, T. Noguchi, M. Kondo, N. Miyauchi-Isobe, B. Takase 619

STEEP SPECTRUM RADIO SOURCES SHOWING LOW FREQUENCY VARIABILITY
F. Mantovani, I. Browne, R. Fanti, A. Ficarra, T. Muxlow, L. Padrielli, J. Romney ... 623

THE LUMINOSITY FUNCTION OF QUASARS AND LOW LUMINOSITY ACTIVE GALACTIC NUCLEI
H. L. Marshall 627

NEAR INFRARED SURFACE PHOTOMETRY OF NGC1068
E. Mediavilla Gradolph, C. Sanchez Magro 633

ACTIVE GALAXIES IN HIGH DENSITY ENVIRONMENTS
T. K. Menon, P. Hickson 637

TOWARDS THE LUMINOSITY FUNCTION OF SEYFERT GALAXY NUCLEI
E. J.A. Meurs 641

DISTANCE OF THE FAR GALAXIES AND SCATTERING OF THE PHOTONS IN THE SPACE
M. Missana 645

THE OPTICAL LUMINOSITY OF THE INVISIBLE NUCLEUS OF M 82
P. Notni 649

FLUX VARIATIONS AND STRUCTURAL CHANGES IN

EXTRAGALACTIC RADIO SOURCES
L. Padrielli, M. F. Aller, H. D. Aller, N. Bartel, C. Fanti, R. Fanti, A. Ficarra, L. Gregorini, F. Mantovani, L. Matveenko, G. D. Nicolson, J. D. Romney, K. W. Weiler   653

THE MAGNETIC MONOPOLES CONTENT OF GALACTIC NUCLEI, QUASARS, STARS AND PLANETS
Q. Peng, Z. Li, D. Wang   659

A MONOPOLE MODEL FOR GALACTIC NUCLEI
Q. Peng, D. Wang, Z. Li   663

AN EVOLUTIONARY LINK BETWEEN SEYFERT I AND II GALAXIES?
E. Pérez, M. V. Penston   669

THE POSSIBLE DETERMINATION OF TWO NEW BL LAC REDSHIFTS
M. Persic, P. Salucci   675

MAGN; MILDLY ACTIVE GALACTIC NUCLEI
P. Pismis   679

ON THE DUPLICITY OF SEYFERT GALAXY NGC 1275 NUCLEUS
I. Pronik, L. Metik   683

PCD SPECTROSCOPY OF MKN 463 A DOUBLE NUCLEUS SEYFERT-2 GALAXY
P. Rafanelli, S. di Serego Alighieri   689

NARROW BAND PHOTOMETRY AND THE REDDENING OF CYG A
K. D. Rakos, N. Fiala   693

EVOLUTION OF POWERFUL AGN AND THE COLLAPSE OF RICH CLUSTERS
N. Roos   697

LOW-LUMINOSITY ACTIVE NUCLEI IN NEARBY ELLIPTICAL GALAXIES
E. M. Sadler   701

TABLE OF CONTENTS

KINEMATICS IN THE NARROW LINE REGION OF NGC 4151 - EVIDENCE FOR A MERGER?
    H. Schulz   705

T COR B, A "SEYFERT 1 NUCLEUS" AT 1200±100 PARSECS?
    P. L. Selvelli, J. Clavel, A. Cassatella, M. Hack   709

INTERACTION AND EMISSION MORPHOLOGY
    N. A. Sharp   713

FLOPPY DISCS: A RECIPE FOR THE "OBSERVERS' DREAM" MODEL
    M. D. Smith, D. J. Raine   717

THE LEO INTERGALACTIC NEUTRAL HYDROGEN CLOUD
    Y. Terzian, S. E. Schneider, E. E. Salpeter   723

EMISSION LINES INDICATING A UNIVERSAL L/M RATIO FOR THE CENTRAL ENGINE IN AGN ?
    A. Wandel   727

THE EXPLANATION OF THE POSITRON-ELECTRON ANNINIHILATION LINE AT OUR GALACTIC CENTER UNDER THE MODEL OF GALACTIC NUCLEI WITH MONOPOLES
    D. Wang, Q. Peng, Z. Li   737

INDEX OF NAMES   741

INDEX OF SUBJECTS   757

INDEX OF OBJECTS   761

LIST OF CONTRIBUTORS   765

PREFACE

The structure and the evolution of the active galactic nuclei is a rapidly developing subject, and one of the leading research fields in modern astrophysics. It thus fully justifies the meeting held in Trieste (Italy) from April 10th to 13th, 1985.

The invited lectures - as well as the poster papers - covered the topics mentioned in the title of the conference both from an observational and from a theoretical point of view.

The meeting was hosted by the International Centre for Theoretical Physics of Trieste and was attended by about 150 participants from 25 nations, who presented 27 invited lectures and 62 poster papers.

The Scientific Organizing Committee consisted of Profs. M. Abramowicz (co-Chairman), A. Cavaliere (co-Chairman), B. Coppi, N. Dallaporta, M. Hack, F. Pacini, L. Radicati di Brozolo, D.W. Sciama, G. Setti and A. Treves. We acted as the four members of the Local Organizing Committee. Moreover, we are pleased to thank the Chairmen of the Sessions (Profs. J. Beckman, F. Bertola, N. Dallaporta, M. Hack, L. Maraschi, D. Sciama, M. Shutz) for their invaluable help.

The meeting was sponsored by the European Physical Society (Astronomy Division). It was organized and supported by the Dipartimento di Astronomia, Universita' degli Studi, Trieste; the Dipartimento di Fisica, Universita' degli Studi, Milano; the Consiglio Nazionale delle Ricerche; the Gruppo Nazionale di Astronomia; the International Centre for Theoretical Physics, Trieste; the International School for Advanced Studies, Trieste; the Osservatorio Astronomico, Trieste; and the Scuola Normale

Superiore, Pisa. It was also supported by the Azienda Autonoma di Turismo e Soggiorno di Trieste e la sua Riviera; the Banco di Napoli, Trieste; the Banco di Sicilia, Trieste; the Comune di Trieste; and the Regione Autonoma a Statuto Speciale Friuli Venezia-Giulia.

We are very grateful for all the support we received. In particular we wish to thank Prof. P. Fusaroli, Magnifico Rettore of the University of Trieste, for his heartly welcome, and Prof. P. Budinich (Director of the International School for Advanced Studies of Trieste), Prof. M. Hack (Director of the Department of Astronomy of the University of Trieste and of the Trieste Astronomical Observatory), and Prof. A. Salam (Director of the International Centre for Theoretical Physics of Trieste) and the staffs of their Institutions for their help, which greatly contributed to the success of the meeting.

Finally, we are pleased to thank the Authors of the Poster Papers who followed our editorial suggestions and compressed their texts to four pages only.

The invited lecture by Elihu Boldt and Darryl Leiter was presented by Amri Wandel since both the authors were unable to attend the meeting. The text of the answers given in the course of the discussion was, however, revised by the authors.

Unfortunately, the text of the invited lecture by Dr. E.S. Phinney has not arrived.

All questions and answers have been typed from the written versions we have received from questioners and authors. In this way we have been able to collect more than 70% of the discussions.

The Participants are grieved at the news of the death of Prof. I.S. Shklowsky, who was planning to attend this Trieste Meeting.

Giuliano Giuricin
Fabio Mardirossian
Marino Mezzetti
Massimo Ramella
Editors

ADDRESS OF PROFESSOR PAOLO FUSAROLI, MAGNIFICO RETTORE DELL'UNIVERSITA DEGLI STUDI DI TRIESTE

Authorities, ladies and gentlemen, distinguished colleagues and dearest friends,

I am glad and I feel highly honoured to express my warmest welcome and that of Trieste University, to the eminent foreign and italian scholars who, by their presence, honour - in this magnificent hall of the International Centre for Theoretical Physics - the international congress on the "Structure and Evolution of Active Galactic Nuclei" sponsored by the European Physical Society, Astronomy Division, and supported by the Department of Astronomy of the University of Trieste, by the National Research Council, by the International Centre for Theoretical Physics of Trieste, by the International School for Advanced Studies of Trieste, by the Astronomical Observatory of Trieste, and to express to all the organizers, to the Scientific Organizing Committee, the distinguished colleagues Prof. Abramowicz, Cavaliere, Coppi, Dallaporta, Hack, Pacini, Radicati di Brozolo, Sciama, Setti and Treves and to the Local Organizing Committee, colleagues Prof. Giuricin, Mardirossian, Mezzetti and Ramella, my feelings of the strongest and most affectionate appreciation for having realized, in our town of Trieste, a congress of so high a scientific level.

This congress is rewarded by the presence of many qualified researchers to whom I give my especial welcome - scholars coming from all European Countries, from North and South America, from Australia, China, Thailand, Japan, India, Russia and Israel.

The holding of this important meeting in Trieste makes it possible both for the city, and for our University, to reaffirm its historical vocation which is in harmony with such an international, cultural and scientific conference.

I must be allowed to express particular thanks and my warmest gratitude to my dearest colleague and friend Margherita Hack, head of the new Department of Astronomy, whose prestigious and assiduous research and scientific activity honours the international role of Trieste and her University: a role to which we have devoted, for some time now, our best energies, in collaboration with other scientific Institutions of the town, convinced as we are that the common language of science is the best instrument for pursuing the great aim of civil progress and world peace.

Expressing, therefore, my best wishes that you may achieve the results you are aiming at, I open the International Congress on the "Structure and Evolution of Active Galactic Nuclei".

ADDRESS OF PROFESSOR ABDUS SALAM, DIRECTOR OF THE INTERNATIONAL CENTRE FOR THEORETICAL PHYSICS, TRIESTE

Illustrious Rector, Professor Hack, Professor Sciama,

I have been asked by the local organizers of this scientifically illustrious Conference to say a few words of welcome on behalf of the International Centre for Theoretical Physics. I am most happy to take this opportunity of welcoming all of you, and particularly those among this distinguished gathering who have been associated, during the course of the years, with the programmes of the Centre. Among them is Professor Maarten Schmidt, who at the 1968 Symposium which we held to review the situation in theoretical physics, gave us a beautiful exposition of the state of the art in the subject of quasars at that time. Our old friend, Geoffrey Burbidge, was also there at that meeting. Among those who have visited the Centre are Professors Novikov, Rees, Shapiro, Silk and, of course, Professor Abramowicz and our Italian colleagues, Professors Cavaliere, Coppi, Dallaporta, Pacini, Radicati di Brozolo, Setti, Treves, Giuricin, Mardirossian, Mezzetti and Ramella.

Ever since the emergence of the gauge unification ideas in particle physics and cosmology, galaxies and their nucleation are now active areas of research in the new discipline of particle-astrophysics. I am naturally anxious to have this opportunity to hear of the new developments in this area.

I have been asked to say a few words about the Centre itself. The Centre was founded in 1964 by the International Atomic Energy Agency (IAEA) for internatinal collaboration in theoretical physics in general and for looking after the needs of physicists from developing countries in particular. We hold seminars, courses, workshops and research sessions throughout the year. Some 2.400 physiscists, half of them from developing countries and half from the rich countries visit the Centre every year. We are supported by generous grants from the Government of Italy, as well as, of course, from IAEA and UNESCO. The University of Trieste and the town and region of Trieste, whose Mayor is present today, have taken the Centre to their hearts and for their innumerable courtesies, we are very grateful.

Let me, in conclusion, express once again our deepest appreciation to the organizers. I wish the partecipants in the meeting a very happy period of stay in Trieste and at the Centre.

LIST OF PARTICIPANTS

ABRAMOWICZ, M., International School for Advanced Studies, Strada Costiera 11, I-34100 Trieste, Italy
AMES, S., Radioastronomy Institute, Auf dem Hugel 71, 5300 Bonn, FRG
AVNI, Y., Harvard-Smithsonian Center for Astrophysics, Cambridge, Massachusetts 02138, USA and Weizmann Institute of Science Rehovot 76100, Israel
BAAN, W.A., Arecibo-Observatory, Cornell University, P.O.BOX 995, Arecibo, Puerto Rico 00613
BASSANI, L., Istituto TESRE, Via Castagnoli 1, 40126 Bologna, Italy
BECKMAN, J.E.; Instituto de Astrofisica de Canarias, Universidad de La Laguna, Tenerife, Spain
BERGERON, J.,Institut d'Astrophysique, 98bis boulevard Arago, F-75014 Paris, France
BERTOLA, F., Istituto Astronomico, Universita' di Padova, Vicolo dell'Osservatorio 5, 35100 Padova, Italy
BINETTE, L., European Southern Observatory, D-8046 Garching bei Muenchen, FRG
BJORNSSON, C.I., Nordita, Blegdamsvej 17, DK-2100 Copenhagen 0, Denmarksics, University of Durham, South Rd.,
BONOLI, C., Osservatorio Astronomico e Istituto di Astronomia, 35100 Padova, Italy
BOYLE, B.J., Department of Physics, University of Durham, Science Laboratories, South Road, Durham, UK
BROSCHE, P., Observatorium Hoher List, Universitaets-Sternwarte Bonn, D-5568 Daun, FRG
BUITRAGO, J., Instituto de Astrofisica de Canarias, University of La Laguna, Tenerife, Spain
BURBIDGE, G., Center for Astrophysics and Space Sciences University of California, San Diego La Jolla, California 92093, and Kitt Peak National Observatory, Tucson, Arizona, USA
CALVANI, M., Istituto di Astronomia, Vicolo dell'Osservatorio 5, 3100 PADOVA,
CARROLL, T.J., Department of Astrophysics, University of Oxford, Oxford, UK
CAVALIERE, A., Dipartimento di Fisica, Universita' Roma 2, via O. Raimondo, 00173 Roma, Italy
CHU, Y., Center for Astrophysics, University of Science and Technology of China, Hefei, Anhui, China
CLAVEL, J., Observatoire de Meudon, 92190 Meudon, France,

and ESA Villafranca Tracking Station, Ap.do 54065, Madrid, Spain
COLEMAN, C.S., Department of Astrophysics, South Parks Road, Oxford OX1 3RQ, England
COLINA, L., Space Astrophysics Division, SSD-ESTEC, Noordwijk, Holland
COLLIN-SOUFFRIN, S., Observatoire de Paris-Meudon, 92195 Meudon Principal Cedex, France
COLPI, S., Dipartimento di Fisica, Universita' di Milano, Italy
CRISTIANI, S., European Southern Observatory, Casilla 19001, Santiago 19, Chile
DALLAPORTA, N., International School for Advanced Sudies, Strada Costiera 11, I-34100 Trieste, Italy
DE RUITER, H., Istituto di Radioastronomia, via Irnerio 46, 35100 Bologna, Italy
DE ZOTTI, G., Istituto di Astronomia, Vicolo dell'Osservatorio 5, I-35122 Padova, Italy
DRINKWATER M.J., Institute of Astronomy, The Observatories, Madingley Road, Cambridge, CB3 0DS, UK
DUSCHL, W.J., Max-Planck-**Institut fuer** Astrophysik, Karl-Schwarzschild Str. 1, D-8046 Garching bei Muenchen, FRG
FALOMO, R., Osservatorio Astrofisico di Asiago, Asiago, Vicenza, Italy
FANTI, C., Istituto di Radioastronomia, via Irnerio 46, 40126 Bologna, Italy
FANTI, R., Istituto di Radioastronomia, via Irnerio 46, 40126 Bologna, Italy
FERETTI, L., Istituto di Radioastronomia, via Irnerio 46, 40126 Bologna, Italy
FERRARI, A., Istituto di Fisica Generale dell'Universita', Torino, Italy
FILIPPENKO, A.V., Department of Astronomy, University of California, Berkeley CA 94720, USA
FOSBURY, R.A.E., Space Telescope European Coordinating Facility, European Southern Observatory, Karl-Schwarzschild-Str. 2, D-8046 Garching bei Muenchen, FRG
FRANCESCHINI, A., Istituto di Astronomia, Vicolo dell'Osservatorio 5, I-35122 Padova, Italy
FRICKE, K.J., Universitaets-Sternwarte, Geismarlandstr. 11, 3400 Gottingen, FRG
FRIED, J.W., **Max-Planck**-Institut fuer Astronomie, D-6900 Heidelberg 1, FRG
GIALLONGO, E., Istituto Astronomico, via Lancisi 29, 00161 Roma, Italy
GIOMMI, P., ESOC-Exosat, Robert Brosche Strasse 5, 61000 Darmastad, FRG
GIOIA, I.M., Center for Astrophysics, 60 Garden st., Cambridge MA 02138, USA also from Istituto di

LIST OF PARTICIPANTS

GIURICIN, G., Osservatorio Astronomico, via G.B. Tiepolo 11, I-34131 Trieste, Italy
Radioastronomia, CNR, Bologna, Italy
GREGORINI, L., Istituto di Radioastronomia, via Irnerio 46, 40126 Bologna, Italy
HACK, M., Osservatorio Astronomico, via Tiepolo 11, I-34131 Trieste, Italy
HICKSON, P. Department of Geophysics and Astronomy, University of British Columbia, 2219 Main Mall, Vancouver, B.C., V6T 1W5, Canada
JOLY, M., Observatoire de Paris-Meudon, 92195-Meudon Principal Cedex, France
KAPAHI, V.K., Tata Insitute of Fundamental Research, P.O. Box 1234, Bangalore 560 012, India
KAPOOR, R.C., Indian Institute of Astrophysics, Bangalore 560034, India
KATGERT, P., Leiden Observatory, P.O. Box 9513, 2300 RA, Leiden, The Nederland
KEEL, W.C., Kitt Peak National Observatory, National Optical Astronomy Observatories, P.O. Box 26732, Tucson, AZ 85726, USA
KIANG T., Dunsink Observatory, Dublin, Ireland
KIDGER, M., Instituto de Astrofisica de Canarias, Universidad de La Laguna, Tenerife, Spain
KINMAN, T.D., Kitt Peack National Observatory, 950 N.Cherry Ave. P.O. Box 26732, Tucson, Arizona 85726, USA
KOLLATSCHNY, W., Universitaets-Sternwarte, Geismarlandstr. 11, 3400 Gottingen, FRG
KOO, D.C., Space Telescope Science Institute, 3700 San Martin Drive, Baltimore, Maryland 21218, USA
KUHR, H., Max-Planck-Institut fuer Astronomie, Koenigstuhl, D-6900 Heidelberg 1, FRG
LIVIO, M., Departement of Physics, Tedmion, Haifa 32000, Israel
MACCACARO, T., Center for Astrophysics, 60 Garden st., Cambridge MA 02138 USA also from Istituto di Radioastronomia, CNR, Bologna, Italy
MACCAGNI, D., Istituto di Fisica Cosmica, Milano, Italy
MAEHARA, H., Tokyo Astronomical Observatory, Osawa, Mitaka, Tokyo 181, Japan
MANTOVANI, F., Istituto di Radioastronomia, via Irnerio 46, 40126 Bologna, Italy
MADAU, P. International School for Advanced Studies, Strada Costiera 11, I-34100 Trieste, Italy
MARANO, B., Dipartimento di Astronomia, via Zamboni 33, 40126 Bologna, Italy
MARASCHI, L., Dipartimento di Fisica, Universita' di Milano, Italy
MARDIROSSIAN, F., Department of Astronomy, University of Trieste, via G.B. Tiepolo 11, I-34131, Trieste, Italy

MARSHALL, H.L., Space Telescope Science Institute, 3700 San Martin Drive, Baltimore, MD 21218, USA
MEDIAVILLA, GRADOLPH, E. Instituto de Astrofisica de Canarias, University of de La Laguna, Tenerife, Spain
MENON, T.K., Department of Geophysics and Astronomy, University of British Columbia, 2219 Main Mall, Vancouver, B.C. V6T 1W5, Canada
MEURS, E.J.A., Institute of Astronomy, Madingley Road, Cambridge, UK
MEZZETTI, M., Osservatorio Astronomico via G.B. Tiepolo 11, I 34131, Trieste, Italy
MICHALEC, A., Observatorium Astronomierne UJ, Ul. Orla 171, 30-244 Krakow, Poland
MISSANA, M., Osservatorio Astronomico di Brera, via Brera 28, 20121 Milano, Italy
MORGANTI, R., Dipartimento di Astronomia, via Zamboni 33, 40126 Bologna, Italy
MUXLOW, T., Nuffield Radio Astronomy Labs., Jodrell Bank, UK
NANNI, D., INFN Laboratori Nazionali di Frascati, Casella Postale 13, 00044 Frascati, Roma, Italy
NEUGEBAUER, G., Division of Physics, Matematics and Astronomy, California Institute of Technology, Pasadena, California 91125, USA
NOTNI, P., Zentralinstitut fuer Astrophysik der AdW der DDR, 15 Potsdam, DRG
NOVIKOV, I., Space Research Institute, Academy of Sciences of USSR, Profsoyuznaja 84/32, Moscow 117810, USSR
OSTERBROCK, D.E., Lick Observatory, Board of Studies in Astronomy and Astrophysics, University of California, Santa Cruz, CA 95064
PADRIELLI, L., Istituto di Radioastronomia, via Irnerio 46, 40126 Bologna, Italy
PENG, Q., Kapteyn Laboratorium, Box 800-9700 AV, Groningen, Holland
PEREZ, E., Royal Greenwich Observatory, Herstmonceux Castle, Hailsham, Easr Sussex BN27 1RP, UK
PEROLA, C., Istituto di Astronomia, Universita' di Roma, via Lancisi 29, 00161 Roma, Italy
PERRY, J.J., Institute of Astronomy, Madingley Road, Cambridge, UK
PETROSIAN, V., Center for Space Science and Astrophysics, Stanford University, Stanford, CA 94305, USA
PFLEIDERER, J., Institut fuer Astronomie, Innsbruck, Austria
PFLEIDERER, M., Institut fuer Astronomie, Innsbruck, Austria
PHILLIPS, M.M., Observatorio Interamericano De Cerro Tololo, Casilla 603, La Serena, Chile
PHINNEY, E.S., Institute for Advanced Studies, Princeton NJ, USA
PIRO, L., Istituto TESRE, via dei Castagnoli 1, 40126 Bologna, Italy

LIST OF PARTICIPANTS

PISMIS, P., Instituto de Astrofisica de Canarias, Universidad de La Laguna, Tenerife, Spain - on leave from - Instituto de Astronomia, Universidad Nacional Autonoma de Mexico, Apartado Postal 70-264, Ciudad Universitaria, 04510 Mexico, D.F., Mexico

PRONIK, I., Crimean Astrophysical Observatory, USSR

RAFANELLI, P., Istituto di Astronomia dell'Universita' di Padova, Vicolo dell'Osservatorio 5, 35122 Padova, Italy

RAKOS, K.D., Institut fuer Astronomie, Tuerkenschanzstrasse 17, Wien, Austria

RAMELLA, M., Osservatorio Astronomico, via G.B. Tiepolo 11, I-34131 Trieste, Italy

RAY, T.P., Physics Dept., University College Dublin, Belfield, Dublin 4, Ireland

REES, M.J., Institute of Astronomy, Medingley Road, Cambridge, UK

ROOS, N., Sterrewacht Leiden, P.O. Box 9513, 2300 RA Leiden, The Netherlands

SADLER, E.M., European Southern Observatory, Karl-Schwarzschild-Str. 2, D-8046 Garching bei Muenchen, FRG

SALAM, A., International Center for Theoretical Physics, Strada Costiera 11, 34100 Trieste, Italy

SCHMIDT, M., Palomar Observatory, California Institute of Technology, Pasadena, CA 91125, USA

SCHUTZ, B., Department of Applied Mathematics and Astronomy, University College, Cardiff, CF1 IXL, UK

SCIAMA, D.W., International School for Advanced Studies, Stada Costiera 11, I-34100 Trieste, Italy and Department of Astrophysics, University of Oxford, South Parks Road, Oxford OX1 3RQ, UK

SECCO, L., Istituto di Astronomia, Universita' di Padova, Vicolo dell' Osservatorio 5, 35122 Padova, Italy

SELVELLI, P., Osservatorio Astronomico di Trieste, via G.B. Tiepolo 11, I-34131 Trieste, Italy

SHAPIRO, S.L., Cornell University, Center for Radiophysics and Space Research, Space Sciences Building, Ithaca, New York 14853, USA

SHARP, N.A., Kitt Peak National Observatory, National Optical Astronomy Observatories, P.O. Box 26732, Tucson, AZ85726, USA

SHAVER, P.A., European Southern Observatory, Karl-Schwarzschild-Str. 2, D-8046 Garching bei Muenchen, FRG

SHAVIV, G., Department of Physics, Technion-Israel Institute of Technology, Haifa, Israel 32000

SILK, J., Department of Astronomy, University of California, Berkeley, CA 94720, USA

SMITH, M.D., Department of Astronomy, University of

Leicester, Leicester LEI 7RH, UK
STEPHEN, J.B., Istituto TESRE, via Castagnoli 1, 40126 Bologna, Italy
STIRPE, G.M., Sterrewacht, Postbus 9513, 2300 RA, Leiden, The Netherland
STOCKENHUBER, H., Institut fuer Astronomie, Tuerkenschanzstrasse 17, 1180 Wien, Austria
SVENSSON, R., NORDITA, Blegdamsvej 17, DK-2100, Copenhagen, Denmark
TADHUNTER, C., Royal Greenwich Observatory, Herstmonceux, Hailsham, E. Sussex, UK
TERZIAN, Y., NAIC and Department of Astronomy, Cornell University, Ithaca, New York 14853, USA
TREVES, A., Dipartimento di Fisica, Universita' di Milano, Italy
TREVESE, D., Osservatorio Astronomico di Roma, Viale del Parco Mellini 84, 00100 Roma, Italy
ULRICH, M.H., European Southern Observatory, 8046 Garching bei Muenchen, FRG
VAGNETTI, F., Istituto Astronomico, via Lancisi 29, 00161 Roma, Italy
VAN DER LAAN, H., Sterrewacht Leiden, Postbus 9513, 2300 RA Leiden, Nederland
VAN HEERDE, G.M., Sterrewacht Leiden, Postbus 9513, 2300 RA Leiden, Nederland
VAN WOERDEN, D.H., Kapteyn Laboratorium, der Rijksuniversiteit te Groningen, Landleven 12, Postbus 800, 97000 AV Groningen, Nederland
VETTOLANI, G., Istituto di Radioastronomia, via Irnerio 46, 40126 Bologna, Italy
VERON, M., Observatoire de Medoun, 92190 Medoun, France
VERON, P., Observatoire de Haute Provence, 04870 Saint Michel l'Observatoire, France
VIGNATO, A., INFN. Laboratori Nazionali di Frascati, Casella Postale 13, 00044 Frascati, Roma, Italy
WANDEL, A., Astronomy Program, University of Maryland, College Park, MD-20742, USA
WEEDMAN, D.W., Department of Astronomy, The Pennsylvania State University, 525 Davey Laboratory, University Park, PA 16802, USA
WILKES, B.J., Smithsonian Astrophysical Observatory, 60 Garden Street, Cambridge, MA 02138, USA
YEE, H.K.C., Departement de Physique, Universite de Montreal, C.P. 6128, Succ. A, Montreal, Canada
ZAMORANI, G., Istituto di Radioastronomia, via Irnerio 46, 40126 Bologna, Italy
ZIEBA, S., Observatorium Astronomierne UJ, Ul. Orla 171, 30-244, Krakow, Poland
ZITELLI, V., Dipartimento di Astronomia, via Irnerio 46, 40126 Bologna, Italy

INVITED LECTURES

EVOLUTION OF THE LUMINOSITY FUNCTION OF QUASARS

Maarten Schmidt
Palomar Observatory
California Institute of Technology
Pasadena, CA 91125, USA

ABSTRACT. The luminosity function of quasars shows a strong dependence on redshift. We discuss alternate descriptions of this evolution. Density evolution is assumption-free if the density variation with redshift is also allowed to be a function of absolute magnitude. Luminosity evolution and, in particular, pure luminosity evolution make a priori assumptions about the shape of the luminosity function. We illustrate the derivation of density evolution and its strong dependence on optical luminosity. We discuss work by Schmidt and Green (1985) on X-ray counts of quasars and the need for negative X-ray luminosity evolution to explain the counts and the low average redshift of X-ray quasars. As a consequence the quasar contribution to the 2 keV X-ray background is only around 8-13%. The evolution of active galactic nuclei of lower optical luminosity is tightly constrained by the observed X-ray background.

1. THE ROLE OF THE LUMINOSITY FUNCTION

The luminosity function of quasars represents the distribution of quasar properties such as optical luminosity, X-ray luminositiy, radio luminosity, etc., as a function of redshift or cosmic epoch. It may also depend on such properties as spectral index, morphology, etc., but these have hardly been explored yet.

The luminosity function is the underlying basic property that determines the counts of optical, X-ray, or radio counts, the redshift distribution of samples selected by optical, X-ray, and/or radio flux, etc.

The evolution of the luminosity function is represented by the dependence on redshift or cosmic time. It is therefore a statistical property of the ensemble of quasars, and not to be confused with the evolution of individual quasars.

Individual quasars may be born at different times and exhibit a light history $L(t)$ after birth that may be different for different quasars. The distribution of luminosities of quasars at a given epoch, each evolving separately, constitutes the luminosity function. The relation between the individual light history $L(t)$ and the luminosity function is complex. It appears unlikely that we can learn much about individual quasar evolution from the evolution of the quasar luminosity function, unless we make assumptions.

One set of assumptions is that quasars were all born at the same time and that the shape of their light history was the same for all of them. In that case, the luminosity function is constant in total number and shape, and it shifts only in the luminosity coordinate. This is the case of pure luminosity evolution, discussed below.

2. DERIVATION OF THE LUMINOSITY FUNCTION

2.1. Complete Coverage of Hubble Diagram

If we had complete knowledge of quasars in the $(m, \log z)$ Hubble diagram, then the distribution of apparent magnitudes m in a strip of redshift z to $z + \Delta z$ is related to the luminosity function $\Phi(M,z)$, by

$$n(m) = \Delta V(z) \cdot \Phi(M,z) , \qquad (1)$$

where $\Delta V(z)$ is the co-moving volume corresponding to Δz and

$$M = m + 5 - 5 \log D , \qquad (2)$$

where D is the luminosity distance which depends both on the cosmological model and the spectral index of the continuum.

For a redshift around 2 or 2.5, there is considerable low-dispersion slitless spectroscopic survey material. The luminosity function so determined has been discussed by Gaston (1983). The dispersion among different surveys is rather large. There is a suspicion that Schmidt objective prism surveys yield substantially smaller numbers than 4-meter grism surveys (Osmer 1982).

2.2. Incomplete Coverage of Hubble Diagram

If surveys cover the Hubble diagram incompletely, as is the case, then assumptions have to be made and it is these assumptions that generate discussion. In Figure 1 we illustrate the luminosity function $\Phi(L,z)$ at a given redshift z, and the local luminosity function $\Phi(L,0)$.

2.2.1. Density evolution - Define a density function

$$\rho(L,z) = \Phi(L,z)/\Phi(L,0) . \qquad (3)$$

In the case of density evolution, the relation between the two luminosity functions of Figure 1 is made at given L, i.e.,

$$\Phi(L,z) = \rho(L,z) \Phi(L,0) . \qquad (4)$$

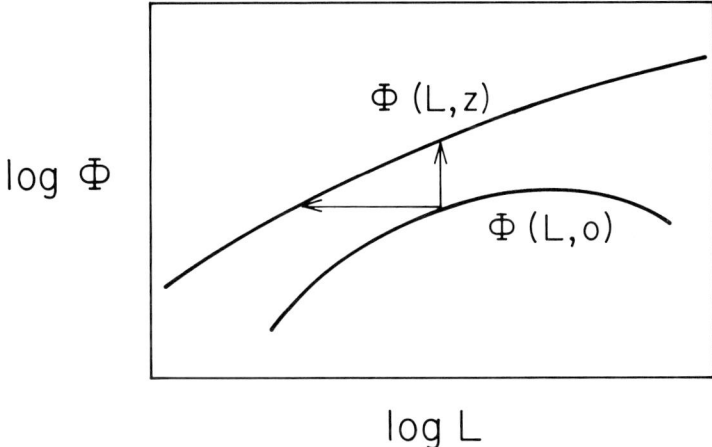

Figure 1. Illustration of luminosity functions Φ at redshift z and at z = 0. Density evolution relates the functions at given L(vertical arrow), and luminosity evolution at given Φ(horizontal arrow).

Initially, when information about $\Phi(L,z)$ was extremely scarce, I assumed that $\rho(L,z) = \rho(z)$, i.e., that the density variation is independent of the luminosity. This is the case of pure density evolution.

Eventually, it became clear that the density increase with redshift does vary with luminosity (Schmidt and Green 1983). We then did use $\rho(L,z)$ and, to distinguish it from the case of pure density evolution, we referred to this case as "luminosity-dependent density evolution." This term may have suggested the introduction of a particular assumption about evolution, but it is not: $\rho(L,z)$ is simply defined by eq. (3), and hence eq. (4), which follows trivially from eq. (3), does not introduce any assumptions about the z-dependent part of the luminosity function.

2.2.2. Luminosity evolution - In this case, the two luminosity functions of Figure 1 are related by the horizontal arrow, i.e.,

$$\Phi(L,z) = \Phi\{L \cdot f(L,z), 0\} . \tag{5}$$

As shown by the example in Figure 1, a horizontal connection from one curve to the other is not always possible and, if possible, may not be unique.

If $f(L,z) = f(z)$, i.e., independent of luminosity, then a constant shift in log L translates one luminosity function into the other. This is the case of pure luminosity evolution. As discussed in Section 1, this case results if all quasars formed at the same time and had a common light history log L(t).

As long as we have only fragmentary understanding of what is going on in quasars, let alone how their properties vary with cosmic timme, I see no good reason to start by making the strong assumptions inherent in pure luminosity evolution. I prefer to describe the evolution of the luminosity function in terms of density evolution, involving the density function $\rho(L,z)$ whose definition introduces no assumptions. If pure luminosity evolution were the correct description, then we should eventually find that the luminosity functions derived at different redshifts are identical except for a shift in the log L coordinate.

3. DETERMINATION OF DENSITY FUNCTION AND LOCAL LUMINOSITY FUNCTION

Since available samples cover the Hubble diagram very poorly, there is in practice a need to make an assumption about the dependence of the density function on redshift. Usually, the density function is taken to be a power law $(1+z)^p$ or an exponential function $\exp(c\tau)$, where τ is the look-back time expressed in the age of the universe.

If the samples available are constrained by more than one flux limit (say, optical and radio), then the V/V_{max} method (Schmidt 1968), provides a test whether the objects are drawn from a population with a uniform distribution in space. If not, the value of p or c can be derived by trial and error through the V/V_{max} test. If several samples are available, the generalized parameter V_e/V_a (Avni and Bahcall 1980) can be used instead of V/V_{max}.

If we have two or more samples, constrained by a flux limit at one frequency (say, optical, or X-ray), then we proceed as follows. For each source in sample A, we derive the maximum redshift z_{max} to which it is observable within the sample limits. Now let

$$V'_{max} = \int_0^{z_{max}} \rho \frac{dV(z)}{dz} dz , \qquad (6)$$

where $dV(z)$ is the co-moving volume element. The contribution of this one source to the local luminosity function is $1/V'_{max,A}$. If there are $N(A)$ sources in sample A, then the local luminosity function is the sum of each of their $1/V'_{max,A}$ contributions.

If $V'_{max,B}$ is the maximum volume accessible to each of the sources in sample A within the limits of sample B, then the number of sources in sample B is expected to be

$$N(B) = \sum_{N(A)} \frac{V'_{max,B}}{V'_{max,A}} . \qquad (7)$$

The density parameter p or c is determined by trial and error such that the predicted number N(B) agrees with the observed number.

Except for the parametrization of the density function, the above method yields a direct determination of the local luminosity function. There is no need to parametrize the local luminosity function. If there is a desire to do so, it would be desirable to first follow the direct derivation described above: the luminosity function so derived can then be used as a guide to a suitable functional representation.

4. EVOLUTION OF OPTICAL LUMINOSITY FUNCTION

There exist only a few samples of spectroscopically confirmed quasars with measured optical magnitudes, complete to a given limiting magnitude over a given sky area. The discussion in this section is based on Schmidt and Green (1983). They use eq. (7) to predict the content of four fainter samples from that of the Palomar Bright Quasar Survey (BQS).

As an example of evidence for evolution, the predicted number of quasars in the high-redshift 4-meter grism surveys falls short by a factor of 200 compared to the observations, if $\rho(L,z) = 1$. This constitutes probably the most impressive evidence for evolution of the quasar luminosity function. Assuming that the co-moving space density of quasars varies as $\exp(c\tau)$, where τ is the light-travel time expressed in the age of the universe, the observed 4-meter grism number can be shown to require around $c = 17$.

Next, using this value of c, one predicts a number density of quasars with $B < 21$ and $z < 3$ of around 1000 per square degree, whereas the observed surface density is around 50 per dquare degree. The main difference between the large-redshift 4-meter grism quasars and those with $B < 21$, $z < 3$ is in absolute luminosity: the former are around $M(B) = -27$ while the latter range down in luminosity to $M(B) = -23$. Accordingly, Schmidt and Green (1983) proposed that the density evolution depends on absolute luminosity. Specifically, they assumed that c is a linear function of absolute magnitude $M(B)$. For details, the reader is referred to the article, in particular to Table 5 and Figure 4, which illustrates the luminosity function at different redshifts.

5. EVOLUTION OF X-RAY LUMINOSITY FUNCTION

Estimates of the 2 keV X-ray background contributed by quasars range from 20% (Setti and Woltjer 1982) to 79% (Maccacaro, Gioia, and Stocke 1985). In the absence of any evolution, the contribution of active galactic nuclei (AGNs) of lower luminosity ($M(B) > -23$), BL Lacs, clusters of galaxies and galaxies is around 45% (Schmidt and Green 1985). This suggests that the X-ray background may impose constraints on the evolution of the luminosity function of quasars. As we will see below, this is not actually the case.

Many of the same Palomar Bright Quasar Survey objects used to establish the optical luminosity function have been observed with the Einstein Observatory (Tananbaum et al. 1985). Using these objects we can predict the quasar content of the Medium Sensitivity Surveys (Gioia et al. 1984) through eq. (7), assuming the density function for optical quasars found from the BQS (Schmidt and Green 1983). This exercise predicts four times too many quasars, with a mean redshift 50%

larger than actually observed in the Medium Sensitivity Surveys (Schmidt and Green 1985).

To explain this discrepancy, we postulate X-ray luminosity evolution of assumed shape $(1+z)^\gamma$ and find that, depending on spectral index, a value of γ = -1.9 to -1.2 is required to reproduce the number of quasars, and approximately the mean redshift, observed in the Medium Sensitivity Surveys (Schmidt and Green 1985, model HH5).

This negative X-ray luminosity evolution leads to a rather low predicted X-ray background contribution for quasars of 13% to 8%, again depending on the X-ray spectral index. This includes only luminous quasars with M(B) < -23. Apparently, then, the X-ray background does not constrain the evolution of the luminosity function of these quasars.

Schmidt and Green (1985) show that AGNs of lower luminosity (M(B) > -23) contribute about 29% to the X-ray background if they show no evolution. For these objects, the background severely constrains the evolution: even for density evolution as mild as $\exp(2\tau)$ the background produced by these objects together with quasars, BL Lac objects, clusters of galaxies and galaxies exceeds 100%.

We conclude that while quasars contribute a minor fraction of the X-ray background, the contribution by lower luminosity AGNs is so large that any substantial density evolution for these objects can be excluded. This conclusion is consistent with the luminosity-dependent density evolution deduced by Schmidt and Green (1983), which predicts that for objects of M(B) > -23 there should be little or no evolution.

This research was supported in part by the National Science Foundation under grants AST-8111754 and AST-8314134.

REFERENCES

Avni, Y., and Bahcall, J. N. 1980, Astrophys. J., **235**, 694.
Gaston, B. 1983, Astrophys. J., **272**, 411.
Gioia, I. M., Maccacaro, T., Schild, R. E., Stocke, J. T., Liebert, J. W., Danziger, I. J., Kunth, D., and Lub, J. 1984, Astrophys. J., **283**, 495.
Osmer, P. S. 1982, Astrophys. J., **253**, 28.
Schmidt, M. 1968, Astrophys. J., **151**, 393.
Schmidt, M., and Green, R. F. 1983, Astrophys. J., **269**, 352.
Schmidt, M., and Green, R. F. 1985, in preparation.
Tananbaum, H., Avni, Y., Green, R. F., Schmidt, M., and Zamorani, G. 1985, in preparation.

DISCUSSION

SCIAMA: It is my impression that one can significantly constrain the evolution of quasars by requiring that they do not produce too many ionising ultra-violet photons which would be incident on galaxies, Lyman α clouds etc. Have you looked into this?

SCHMIDT: I believe that in connection with the Lyman α clouds, Sargent and Weymann and Wolff have isolated the redshift magnitude distribution of quasars, as based on counts or evolution models, in their evaluation of the UV radiation field.

MARSHALL: When examining the Z-L plane and including the faint Braccesi sample (Marshall et al. 1984) then I find that LDDE is necessary for an exponential parametrization but that a power law form can be fit with a constant index independent of luminosity (for $B<20, z<2.2$). How are your results dependent on the exponential form?

SCHMIDT: The parametrization of the z-dependent part of the luminosity function is only needed as long as the Hubble diagram is poorly covered by surveys. As soon as new surveys are in conflict with the parametric form, it should be replaced. But in any case LDDE (= luminosity-dependent density evolution) is unrelated to the parametrization of the z-dependent part od the luminosity function.

PHYNNEY: Do I understand correctly that you have found evidence that L_x/L_{opt} decreases with redshift over and above the inverse correlation of L_x/L_{opt} with L_{opt} which is automatically included in your analysis. And is it possible that this is due to a correlation of X-ray luminosity with UV excess? (e.g. Usher finds higher number counts from variability studies than found by prism or a UV excess).

SCHMIDT: 1) Our derivation of quasar X-ray counts indeed includes implicitly any correlation of L_x/L_{opt} with L_{opt}. Our evidence for a decrease pf L_x/L_{opt} for increasing redshift means that pass of the correlation of L_x/L_{opt} with L_{opt}, seen from plotting these quantities for individual quasars, has to be attribuited to a correlation of L_x/L_{opt} with redshift. 2) Part of the correlation of L_x/L_{opt} with L_{opt} or z might be due to the UV excess, i.e., to a difference of the spectral index of the optical and X-ray continua.

PETROSIAN: I am not certain how many parameters are used for the fitting of counts to the background X-ray radiation. In particular did you assume a redshift cut-off for the high luminosity objects (quasars)?

SCHMIDT: I assumed for all objects discussed a redshift cut-off at z=3. Direct observations exist only for quasars where we (Schmidt, Schneider and Gunn) now find evidence for such a cut-off, at least for quasars with magnitudes 20 and fainter.

KOO: Have you tried to see if the application of mainly evolution luminosity with slight density <u>decrease</u> with time would be consistent with both the optical and X-ray data, assuming <u>no change</u> of the shade of the luminosity function? Most of the total counts or energy comes from the lowest luminosity objects (for realistic luminosity functions), so that even strong luminosity evolution is allowable if accompanied by relatively slight overall density decrease with lookback time.

SCHMIDT: I imagine that you are proposing to describe the evolution of the quasar luminosity function in terms of a combination of pure luminosity evolution and pure density evolution. We have not tried this particular scheme.

MARSHALL: The BF sample, if taken as an X-ray survey, is consistent with the deep surveys of Murray et al. and reaches the same approximate limits in X-ray flux. Shouldn't the results or predictions be essentially similar for these two surveys?

SCHMIDT: I guess I do not agree: the quasar surface density in the BF down to log f(2keV)=-30.8 is as large or larger than that in the HSS to log f(2heV)=-31.15, effectively. On the basis of the general slope of quasar X-ray source counts, one would expect the BF to show one-third of the surface density of the HSS.

IRAS OBSERVATIONS OF ACTIVE GALAXIES

G. Neugebauer, B. T. Soifer

Division of Physics, Mathematics and Astronomy
California Institute of Technology
Pasadena, California 91125
U.S.A.

M. Rowan-Robinson

Department of Applied Mathematics
Queen Mary College
Mile End Road
London, E1 4NS
England

ABSTRACT: The IRAS survey gives an unbiased view of the infrared properties of the active galaxies. Seyfert galaxies occupy much the same area in color-color plots as do normal infrared bright galaxies, but extend the range towards flatter 60-25 μm slopes. Statistically the Seyfert 1 galaxies can be distinguished from the Seyfert 2 galaxies, lying predominantly closer to the area with constant slopes between 25 and 100 μm. The infrared measurements of the Seyfert galaxies cannot distinguish between the emission mechanisms in these objects although they agree with the currently popular ideas; they do provide a measure of the total luminosity of the Seyferts. The quasar's position in the color-color diagrams continue the trend of the Seyferts. The quasar 3C48 is shown to be exceptional among the radio loud quasars in that it has a high infrared luminosity which dominates the power output of the quasar and is most likely associated with the underlying host galaxy.

The survey by the *Infrared Astronomical Satellite* (IRAS) provides an unbiased way in which to understand the infrared properties of active nuclei. In this review, we present a preliminary summary of the gross infrared properties of the active galaxies as seen by IRAS (Neugebauer *et al*. 1984; *The Explanatory Supplement to the IRAS Catalogs and Atlases* 1985).

In order to set the context for the active galaxies seen by IRAS, the spectral indexes of a sample of infrared "bright" galaxies seen in the survey proper are shown in Figure 1. The spectral indexes α are defined such that the flux density f_ν is proportional to ν^α where ν is the frequency. By "bright" is meant a flux density limit of 5 Jy at 60 μm with detections at 25 and 100 μm as well. The sample represents about 3% of the $\sim$ 20,000 galaxies detected by IRAS with Galactic latitude $|b| > 30°$. No attempt was made to isolate sources that were, in fact, Galactic, nor was there any attempt to take account that some of the galaxies were extended. Both of these effects probably do not change the

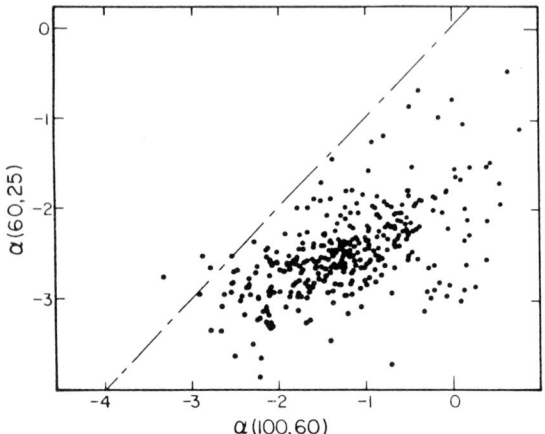

Figure 1: A plot is given of the spectral indexes between 25 and 60 μm ($\alpha(60,25)$) and those between 60 and 100 μm ($\alpha(100,60)$) for the bright infrared galaxies, i.e., sources with $f_\nu(60\ \mu m) > 5$ Jy, found in the IRAS survey. The spectral indexes and sample are defined in the text. The dashed line represents equal spectral indexes. The uncertainty in each coordinate is about 0.3.

statistical conclusions of the study. The 60 to 100 μm color temperatures lie between 30 and 60 K. The sample, which will be discussed in detail by Soifer *et al.* (1985), probably is representative of the population of galaxies seen by IRAS, although significant selection effects, primarily having to do with the requirement that the galaxy be observed at 25 and 100 μm, are present. The sample includes examples of nearby normal galaxies, such as M33, and more distant infrared active galaxies such as Arp 220, as well as several active galactic nuclei as exemplified by MKN 231.

Figure 2 shows the location in the same plot and scale as Figure 1 of the Seyfert galaxies listed by Veron-Cetty and Veron (1983) and detected in the survey. Although the distribution of Seyfert galaxies covers the same area as do the bright galaxies, the distribution extends towards flatter 25 to 60 μm slopes.

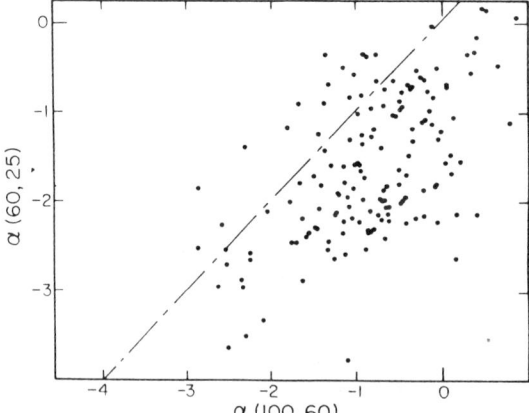

Figure 2: The same plot and on the same scale as Figure 1 is shown, but for the Seyfert galaxies in the Veron-Cetty and Veron Catalog (1983).

The IRAS properties of the Seyfert galaxies, as derived from a preliminary look at the IRAS catalog, are discussed by Miley, Neugebauer and Soifer (1985) who studied the properties of all Seyfert galaxies in the Markarian and NGC catalogs. One hundred and sixteen out of a possible 186 Seyfert galaxies are included in the study by Miley et al.,(1985). About 50 % of all Seyfert galaxies have 60 μm luminosities in excess of 10^{10} $L_\odot$, and the mean 60 μm luminosity increases with optical B luminosity. The luminosity functions of Seyfert 1 and Seyfert 2 galaxies are quite similar. It is possible, however, to statistically separate the two types of galaxies in color-color plots as shown in Figure 3.

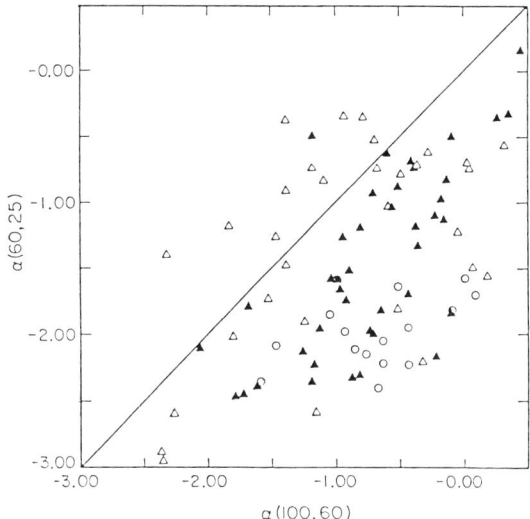

Figure 3: The same plot as in Figure 2 is shown, but for Seyfert galaxies listed in the Markarian and NGC catalogs; from Miley et al. (1985). The Seyfert galaxies are separated according to type. Seyfert 1 galaxies are denoted by open triangles, Seyfert 2 galaxies by filled triangles and galaxies with HII region nuclear regions by circles.

Figure 3, from Miley et al. (1985), is similar to Figure 2, but the Seyfert 1 and Seyfert 2 galaxies are given different designations. It is seen that the area above the line of constant spectral index is preferentially, but not exclusively, filled with Seyfert 1 galaxies. As a further base of comparison, galaxies listed in Veron-Cetty and Veron (1983) as having nuclear emission-line spectra characteristic of HII regions are also included in this plot. It is seen these lie preferentially in the same area as the "bright" galaxies. Although the infrared measurements provide a measure of the total bolometric luminosity of the Seyfert galaxies, they do not discriminate between the physical processes involved. They are, however, consistent with the current prejudices that the infrared emission of Seyfert 2 galaxies is dominated by starburst activity, while the infrared emission of Seyfert 1 galaxies is dominated by non-thermal emission.

Figure 4 shows the distribution of all sources with $|b| > 45°$ and detections at 25, 60 and 100 μm, on the same plot and scale as Figures 1 and 2. Although this plot is subject to many selection effects, it includes many sources which

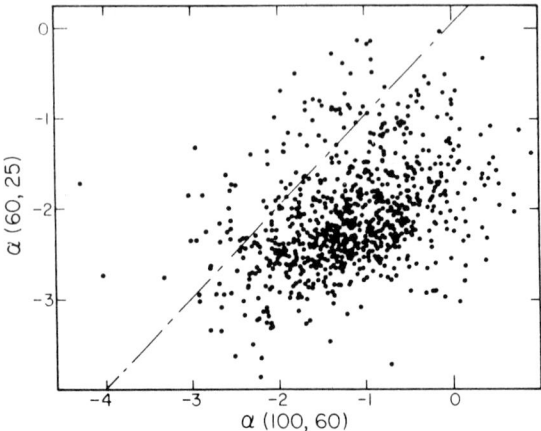

Figure 4: The same plot and scale as Figures 1 and 2 are shown, but for all sources with $|b| > 45°$ and detections at 25, 60 and 100 μm.

have a larger ratio of 25 to 60 μm flux than does Figure 1, i.e., more positive spectral indexes $\alpha(60,25)$. It is seen that there are many sources in the upper left hand corner, i.e., in the area where the slopes are steeper from 60 to 25 μm than from 100 to 60 μm. To the extent that this area is the primary location of Seyfert 1 galaxies, see Figure 3, it is clear that the IRAS survey is an ideal searching ground for Seyfert 1 galaxies; see also de Grijp, *et al.* (1985).

Figure 5 shows the distribution of those sources defined by Veron-Cetty and

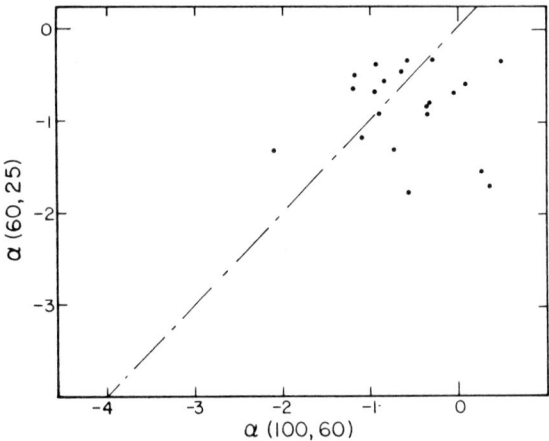

Figure 5: The same plot and scale as Figures 1, 2 and 4 but for those sources identified as quasars by Veron-Cetty and Veron(1983). The three quasars with the lowest values of $\alpha(60,25)$ are 3C48, MKN 231 and MKN 1014.

Veron (1983) to be quasars which were detected in the IRAS survey with good quality fluxes in at least one band. For all but six of these, pointed observations were used to provide sufficient signal to noise ratios in the different wavelength bands to provide reasonable measures of the slopes. For those six with no pointed measurements, the survey observations were coadded in order to increase the signal to noise ratios. The parent sample included all quasars from the catalog of Veron-Cetty and Veron who "arbitrarily defined a quasar as a starlike object, or object with a starlike nucleus, brighter than absolute (visual) magnitude -23." In addition to the detections whose spectral indexes are shown in Figure 5, there was an approximately equal number of IRAS sources with positional coincidences with quasars, but which were either poor quality IRAS detections or which were masked or confused by the presence of a nearby galaxy. The luminosities in the infrared range from 3×10^{10} to 10^{14} $L_\odot$.

Although the IRAS survey lasted only nine and one half months, the survey scanned approximately 70 % of the sky at times separated by about 6 months. Thus there are some data available on potential variability of the sources. In addition, several of the well known active galaxy nuclei were monitored as part of the pointed observation program. A study of the variability of these sources has not been finished at this time, but the effects on Figure 5 should be small compared to the uncertainties.

It is seen that the quasars cluster around the line of equal spectral indexes more closely than do either the galaxies as a whole or the Seyfert galaxies. Three quasars by the luminosity definition of Veron-Cetty and Veron (1983) -- 3C48, MKN 231 and MKN 1014 -- lie well below the line of constant spectral index. Of these three, two -- MKN 231 and MKN 1014 -- have also been classified as Seyfert 1 galaxies. The three lie in the area which is steeper in $\alpha(60,25)$ and flatter in $\alpha(100,60)$ than the constant spectral index line; i.e., they have a bump at 60 μm. The energy distribution of 3C48, shown in Figure 6, shows that the

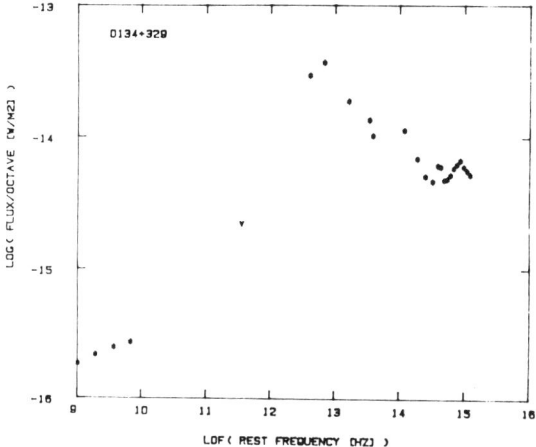

Figure 6: The continuum energy distribution of 3C48 is given. The bump near 60 μm (frequency ~ 5×10^{12} Hz), which is ascribed to thermal reradiation of heated dust grains, is clearly seen. The near-infrared points were obtained at the Palomar Observatory. The radio points are from Kellermann, Pauliny-Toth and Williams (1969) while the millimeter limit is from Ennis, Neugebauer, and Werner (1982) and the visual observations are from Neugebauer et al. (1979).

infrared emission does dominate the power output of this quasar. The luminosity in the infrared is $\sim 5 \times 10^{12}\ L_\odot$, about six times that in the visible. Its observed color temperature between 60 and 100 μm is 55 K.

Figure 7 shows the continuum energy distributions of 3C273 and 3C345, two radio loud, variable quasars. The energy distribution of 3C345 is a result of a

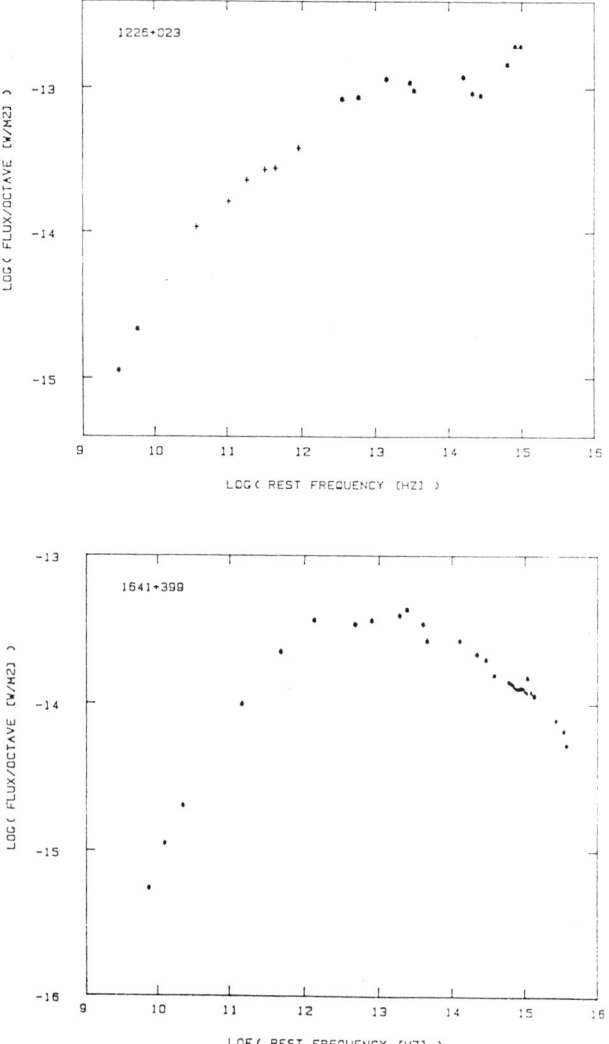

Figure 7: The continuum energy distributions for 3C273 and 3C345 are shown on the same scales. The points marked with (+) in 3C273 are from Clegg et al. (1983), and were obtained in 1982, April; the difference from a smooth curve probably reflects variability in the source. The measurements of 3C345, including the IRAS and near infrared points, were obtained in the 1983 April-May time period (Bregman et al., 1985).

study by Bregman *et al.* (1985) and represents observations taken within two months during 1983 April and May. The observations of 3C273 were taken over a period of several years and the discontinuities are probably the result of variability. The continuum distribution of 3C48 is obviously different from those of the other radio loud quasars. Furthermore, there is no evidence in the radio structure of 3C48 for the presence of a milli arc-second source characteristic of flat spectrum radio sources, and there is no sign of strong variability in either the radio or near infrared continuua. Unlike 3C273, 3C345 and other radio loud quasars found in the survey, the infrared emission in 3C48 therefore most likely does not come from an extension of the radio emission.

It should be emphasized that the radio loud quasars generally show much more radio emission relative to their infrared than do normal galaxies. Figure 8 shows the ratio of the radio luminosity to that the infrared luminosity for the quasars and for a sample of normal galaxies studied by Helou (1985). The samples include only those objects with measured emission both in the radio and infrared, so that the sample does not contain the bulk of the radio quiet quasars. The figure is intended only to show that the "radio loud" quasars are, indeed, very radio loud relative to normal galaxies. No statistical inferences should be drawn from the figure.

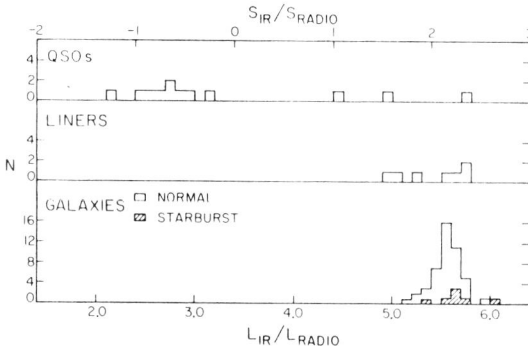

Figure 8: Histograms of the ratio of the radio luminosity to that in the infrared is given for the different types of objects listed. The ratios for the galaxies are from Helou, 1985. Only those quasars with measurements in both the radio and infrared have been counted, so, e.g., radio quiet quasars are under represented.

Before discussing the infrared emission of 3C48 further, it is of interest to ask how common an infrared peak is in the quasars. Of the 22 quasars from the sample of Veron-Cetty and Veron (1983), eight in this preliminary study show definite maxima between 25 and 100 μm. Examples of the energy distributions of two are shown in Figure 9. Most, but not all of the objects with an infrared peak are also defined to be Seyfert galaxies. Thus while infrared emission is not uncommon, it does certainly not universally dominate the emission from quasars.

The origin of the infrared emission of 3C48 is discussed by Neugebauer and Soifer (1985). They conclude that the most likely source of the emission is thermal reradiation from heated dust. In the rest system of the quasar, the

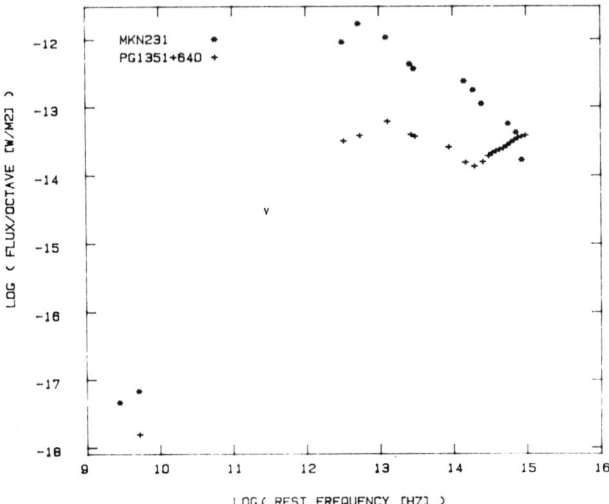

Figure 9: The continuum energy distributions for PG 1351+640 and MKN231 are shown. For 1351+640, the millimeter limit is from Ennis et al. (1982), the radio point is from Condon et al. (1981) and the visual observations are from Neugebauer et al. (1979). For MKN231 the visual and radio points are from references given in Veron-Cetty and Veron (1983); the near infrared points are from Palomar Observatory.

representative temperature of the grains, assuming a (wavelength)$^{-1}$ emissivity dependence, is ~ 57 K. The mass in dust inferred from the observations is 6×10^7 $M_\odot$. If the grains were heated by a single luminosity source, the grains must be ~ 2 kpc from the source. All of these numbers place the origin of the radiation well outside the classical broad and narrow emission line regions of the quasar. It must be acknowledged, however, that 3C48 is well known to have a narrow line emission region which extends asymmetrically out some 40 kpc from the nucleus (Sandage and Miller 1966, Wampler et al., 1975). The direct association of the infrared emission with this region is excluded because the mass derived from the infrared measurements again far exceeds that obtained from the line measurements.

We conclude that the heated dust is most probably located in an extremely luminous host galaxy surrounding 3C48. The 3C48 system is well known to contain a highly luminous galaxy (Gehren et al., 1984; Boroson and Oke (1982); Fried (1986). If this host galaxy is the location of the infrared emission, the relative infrared and visual properties would be similar to those of Arp 220 (Soifer et al., 1984). If the dust is heated by a single central luminosity source, a special geometry must be invoked since there is no evidence of extinction within the broad line region. If the dust is heated by a set of luminosity sources spread throughout the galaxy, the most likely explanation is that an enormous amount of star formation is occurring in the galaxy. In either case, the infrared observations of 3C48 might thus represent the harbingers of a new type of phenomena associated with quasars.

ACKNOWLEDGEMENTS

We thank all the members of the IRAS team who made these observations possible. We especially thank George Helou for providing us with the radio-to-infrared luminosities. The *Infrared Astronomical Satellite* was developed and is operated by the Netherlands Agency for Aerospace Programs (NIVR), the US National Aeronautics and Space Administration (NASA), and the UK Science and Engineering Research Council (SERC).

REFERENCES

Boroson, T. A., and Oke, J.B. 1982, *Nature*, **296**, 397.

Bregman, J. *et al.*, 1985, *Ap. J.*, submitted.

Clegg, P.E., Gear, W.K., Ade, P.A.R., Robson, E.I., Smith, M.G., Nolt, I.G., Radostitz, J.V., Glaccum, W., Harper, D.A., and Low, F.J. 1983, *Ap. J.*, **273**, 58.

Condon, J.J., O'Dell, S.L., Puschell, J.J., and Stein, W.A. 1981, *Ap. J.*, **246**, 624.

Ennis, D.J., Neugebauer, G., and Werner, M. 1982, *Ap. J.*, **262**, 460

The Explanatory Supplement to the IRAS Catalogs and Atlases 1985, edited by Beichman, C.A., Neugebauer, G., Habing, H.J., Clegg, P.E., and Chester, T.J. (Washington D.C., U.S. Government Printing Office).

deGrijp, M.H.K., Miley, G.K., Lub, J., deJong, T., 1985, *Nature*, in press

Fried, J. 1986, this volume

Gehren, T., Fried, J., Wehinger, P.A., and Wyckoff, S. 1984, *Ap. J.*, **278**, 11.

Helou, G. 1985, in preparation

Kellermann, K.I., Pauliny-Toth, I.I.K., and Williams P.J.S. 1969, *Ap. J.*, **157**, 1.

Miley, G.K., Neugebauer, G., and Soifer, B.T. 1985 *Ap. J. (Letters)*, in press.

Neugebauer, G., Oke, J.B., Becklin, E.E., and Matthews, K. 1979, *Ap. J.*, **230**, 79.

Neugebauer, G. *et al.* 1984, *Ap.J. (Letters)*, **278**, L1.

Neugebauer, G., and Soifer, B.T. 1985, *Ap.J. (Letters)*, submitted.

Sandage, A.R. and Miller, W.G. 1966, *Ap. J.*, **144**, 1238.

Soifer, B.T. *et al.* 1985, in preparation.

Soifer, B.T., Helou, G., Lonsdale, C.J., Neugebauer, G., Hacking, P., Houck, J.R., Low, F.J., Rice, W., and Rowan-Robinson, M. 1984, *Ap. J.*, **283**, L1.

Veron-Cetty, M.P. and Veron, P. 1984, *A Catalogue of Quasars and Active Nuclei*, (Munich: European Southern Observatory).

Wampler, E.J., Robinson, L.B., Burbidge, E.M., and Baldwin, A.J. 1975, *Ap. J.*, **198**, L49

THE PROPERTIES OF "DWARF" SEYFERT NUCLEI IN NEARBY GALAXIES*

Alexei V. Filippenko
Department of Astronomy, University of California
Berkeley, CA 94720 USA

Wallace L. W. Sargent
Palomar Observatory, California Institute of Technology
Pasadena, CA 91125 USA

ABSTRACT. Accurate knowledge of the luminosity function of active galactic nuclei over a wide range of absolute magnitudes is a prerequisite to understanding their structure and evolution. Here we describe preliminary results from a large survey of bright, nearby galaxies which is designed to quantify the *faint* end of the luminosity function. Many high-quality spectra are shown to emphasize the great variety of physical conditions in the nuclei and to illustrate the methods used in a detailed analysis of the data. The characteristics of Seyfert galaxies are detected in a surprisingly large number of objects, although at much fainter levels than usually reported for classical AGNs. In particular, broad Hα emission is visible in at least 10% of the nuclei, and the relative intensities of narrow lines are often indicative of photoionization by a nonstellar continuum. A very wide range of densities is found in the narrow-line regions of certain low-ionization nuclei, just as in many QSOs and Seyfert 1 galaxies. These results suggest that intrinsically faint ("dwarf"), but nevertheless active, nuclei are much more common than previously believed, and that they may have evolved from luminous quasars.

1. INTRODUCTION

It is now generally thought that distant QSOs reside in the nuclei of galaxies, deriving their power by physical processes similar to those in the intrinsically fainter Seyfert galaxies and related objects. This hypothesis rests largely on the continuity which exists between the observed properties of different types of active galactic nuclei (AGNs). The spectral characteristics of type 1 Seyferts, for example, strongly resemble those of optically selected QSOs, and there is substantial overlap in their absolute luminosities if cosmological redshifts are assumed (Weedman 1976, 1977). Moreover, the morphologies and colors of extended "fuzz" around relatively nearby quasars are similar to those of galaxies (Hutchings *et al.* 1982; Malkan, Margon, and Chanan 1984; Malkan 1984). In some cases, absorption lines from normal stars have even been detected in the nebulosity (Boroson and Oke 1982; Boroson, Oke, and Green 1982; Balick and Heckman 1983).

The least luminous "classical" Seyfert 1 nucleus is that of NGC 4051 (Véron 1979), with $M_V \approx -15.6$ ($H_0 = 50$ km s^{-1} Mpc^{-1}). For comparison, the nucleus of NGC 4151 is more than three magnitudes brighter, while Markarian Seyfert nuclei typically have $-22 \lesssim M_V \lesssim -20$. In recent years, however, a number of galactic nuclei with features

---
*Research based on observations made at Palomar Observatory, Caltech.

similar to, but much weaker than, those in classical AGNs have been found (Heckman 1980; Keel 1983; Osterbrock 1981; Phillips, Charles, and Baldwin 1983; Shuder 1980; Stauffer 1982; Véron et al. 1980). A particularly striking object is M81, whose nucleus has $M_V \approx -10$ to -11; as shown by Peimbert and Torres-Peimbert (1981) and by Shuder and Osterbrock (1981), Hα exhibits a component whose full width at zero-intensity (FWZI) is 4000–5000 km s^{-1}, substantially broader than the forbidden lines. M81 is therefore a low-luminosity ("dwarf") type 1 Seyfert galaxy. The discovery of X-ray emission from its nucleus (Elvis and Van Speybroeck 1982) supports this classification.

Of fundamental importance in many studies of AGNs is the local luminosity function $\Phi(L,0)$, which gives the present ($z = 0$) number density of active nuclei having intrinsic luminosity L, for all values of L. The behavior of $\Phi(L,z)$ as a function of redshift z can be used to derive the manner in which AGNs evolve with time (e.g., Schmidt and Green 1983; Maccacaro, Gioia, and Stocke 1984; Marshall 1985), and hence to explore possible relationships between QSOs in the early universe and galaxies at the currect epoch. These problems can best be addressed if the *faint* end of $\Phi(L,z)$ is as accurately known as the bright end. It is particularly crucial to determine how widespread low-luminosity AGNs really are when contemplating the nature of the soft X-ray (2–10 keV) background: Véron (1979) showed that down to $M_V = -18$ the luminosity function is of the form $\Phi(L,0) \propto L^{-3}$, and that Seyfert nuclei can easily account for the observed X-rays if this extends to $M_V = -16.5$ (see also Elvis, Soltan, and Keel 1984). Removal of starlight from the host galaxy, which was not generally done in earlier surveys (Huchra and Sargent 1973; Meurs and Wilson 1984) but has recently been stressed by Cheng et al. (1985), is clearly necessary to accurately study the faintest nuclei.

We are therefore conducting an extensive investigation of a statistically complete sample of bright, nearby galaxies in order to quantify the AGN luminosity function at the faint end. Optical spectra of exceptionally high quality are being used to analyze the emission lines and deduce the physical characteristics of the nuclei. The following questions are especially interesting:

a) Is there a lower limit to the AGN/QSO phenomenon? Can its presence or absence tell us anything about the basic physical mechanisms involved?

b) What is the contribution of faint Seyfert nuclei to the soft X-ray background?

c) What is the distribution of host galaxies among different Hubble types and luminosities? Are many of the galaxies interacting with others?

d) Are there significant correlations between low-level optical activity and the observed radio properties?

e) Do the widths of emission lines correlate with atomic parameters such as ionization potential and critical density? Is there a relationship between line width and stellar velocity dispersion?

f) What differences exist in the emission-line properties of galaxies in the Virgo cluster, in small groups, and in the field?

g) If AGNs are powered by accretion of matter onto a black hole, are faint Seyferts generally "starving" for fuel or do they simply lack extremely massive black holes?

h) Did most mildly active galaxies evolve from the distant, luminous QSOs, or do they represent a separate population of objects?

In this paper we will describe the survey and some of our initial results. Since the observations and analysis are still far from being completed, the conclusions will necessarily be preliminary and of limited scope. Much more thorough discussions will be published elsewhere.

2. OBSERVATIONS AND REDUCTIONS

Our goal is to observe every galaxy north of the celestial equator whose apparent blue magnitude (B_T) is ≤ 12.5 (de Vaucouleurs, de Vaucouleurs, and Corwin 1976; Sandage and Tammann 1981). There are roughly 500 such objects, but initially we concentrated on classical Seyferts, Heckman's (1980) "low-ionization nuclear emission-line regions" (LINERs), and other galaxies which seemed particularly interesting. The results for these 75 galaxies, some of which have $\delta < 0°$ and/or $B_T > 12.5$, were described by Filippenko and Sargent (1985, hereafter Paper I). Spectra of ~ 80 additional objects have subsequently been taken, but not yet published. Here we will present a few old spectra in greater detail, as well as some of the new observations.

Data are obtained with the Double Spectrograph (Oke and Gunn 1982) at the Cassegrain focus of the Hale 5.08 m telescope at Palomar Observatory. A long slit whose width is $2''$ makes a radial cut through the nucleus of each galaxy, and a dichroic filter directly behind the slit reflects blue light ($\lambda \lesssim 5500$ Å) to one camera and transmits red light ($\lambda \gtrsim 5500$ Å) to the other. Two-dimensional spectra are recorded on CCD detectors whose dynamic range, linearity, and temporal stability are crucial to the success of the project. Typical integration times are 20–30 minutes.

The full width at half-maximum (FWHM) of unresolved features is 4.0–4.5 Å in the blue camera and 2.0–2.6 Å in the red, while the corresponding wavelength ranges are $\sim \lambda\lambda$ 4230–5100 and $\sim \lambda\lambda$ 6200–6880. A significant number of emission lines can therefore be studied, including Hγ, [O III] λ4363, He II λ4686, Hβ, [O III] λ5007, [O I] λ6300, Hα, [N II] λ6583, [S II] λ6716, and [S II] λ6731. Moreover, the high signal-to-noise ratios ($S/N \approx 100/1$ in the nucleus) and excellent spatial resolution (0.58 arcsec/pixel for the red spectra) allow thorough studies to be made of the ionization structure, velocity dispersion, abundances, and kinematics as a function of radial distance from the nucleus.

Standard procedures are used to reduce the spectra (Paper I). Cosmic rays, a bias level, geometric distortions, and the background sky are removed from each data frame. Next, one-dimensional spectra of the nucleus (generally the central $2'' \times 4.1''$) and other relevant regions are extracted. The wavelength scale is then established by comparison with spectra of inert gases and Fe. Finally, the data are flux-calibrated with bright secondary standards (Oke and Gunn 1983) and corrected for atmospheric extinction. Although an attempt is made to eliminate telluric absorption features such as the B band (~ 6860 Å) by comparison with the intrinsically smooth continua of the standard stars, a weak line at ~ 6280 Å (primarily due to O_2) was overlooked during the reduction of spectra shown in this paper.

Various computer programs are used interactively to analyze the resulting one-dimensional spectra. As described in the next section, an important step is the accurate removal of underlying absorption lines from the integrated stellar population in each nucleus. This affects both the relative intensities and profiles of the emission lines, and in some cases reveals weak emission previously not visible (see Filippenko and Halpern 1984, hereafter FH; Filippenko 1985).

3. A POTPOURRI OF SPECTRA

We now discuss and illustrate a representative sample of galaxies, concentrating on the red spectral region since it has thus far been analyzed more completely than the blue region. Although the detection of broad Hα emission is hampered by the adjacent [N II] lines, in most cases it is nevertheless easier to see than the corresponding component of

Hβ, which is at best one-third the strength of Hα and can be much weaker if significant reddening is present.

3.1. Brief Overview

Figure 1 shows typical spectra of eight galactic nuclei, obtained in February 1984 (Paper I). The S/N ratio is so high that most features are real. A wide range of relative intensities and profiles is visible in the emission lines, which are generally resolved (as can be seen by comparison with the neon-argon spectrum). The broad component of Hα in NGC 4151 is prominent, even though the active nucleus was in a "low" state at the time and resembled that of type 2 Seyfert galaxies in spectra having lower S/N ratios (Ulrich et al. 1985; Perez and Fosbury 1986). NGC 4235 is a known type 1 Seyfert galaxy possibly located in the Virgo cluster (Abell, Eastmond, and Jenner 1978); asymmetries are present in both the narrow and broad emission lines, but the dip near 6495 Å is attributed to Ca + Fe absorption. These and other absorption lines are very noticeable in objects such as NGC 4192 and NGC 4258. NGC 4278 and NGC 4258 exhibit the characteristics of Heckman's (1980) LINERs (strong [O I] λ6300 and [S II] $\lambda\lambda$6716, 6731); as discussed later, they also contain a broad component of Hα emission similar to, but far weaker than, that in classical type 1 Seyferts. The nuclei of NGC 4274 and NGC 4303 obviously contain

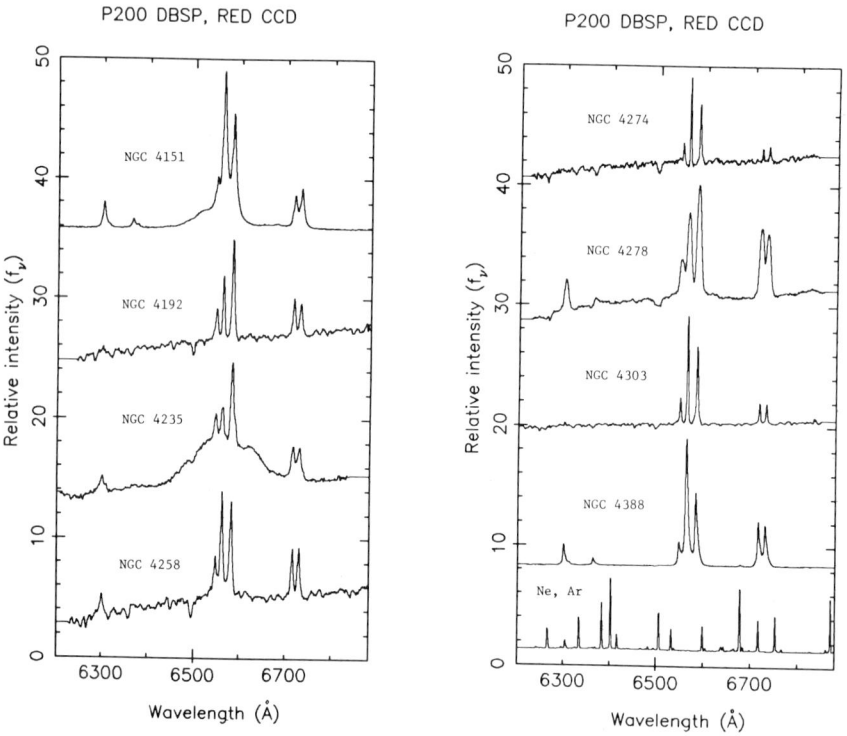

Figure 1: Spectra of eight representative galactic nuclei (2″ × 4.1″ entrance aperture), and a comparison arc, are shown with different ordinate offsets and multiplicative scaling factors. In each case the redshift has been removed. Short, horizontal line segments indicate the absence of data at the extreme blue and/or red ends.

extensive regions of star formation, while NGC 4388 is an X-ray galaxy whose narrow lines have relative strengths indicative of photoionization by a nonstellar continuum.

A better idea of the quality of the data may be gained by comparing the two spectra of NGC 2681 in Figure 2. These were obtained during separate integrations on different locations of the CCD, but even minor discrepancies are minimal. The nucleus of NGC 2681 displays several relatively strong emission lines (equivalent width of [N II] is ~ 7 Å) superposed on a rich absorption-line spectrum. Hα emission, however, is only barely visible because the underlying absorption is so deep, and it would not be detectable in spectra of substantially lower S/N ratio or resolution. No broad component of Hα appears to be present, but the narrow-line ratios resemble those in LINERs and hence suggest low-level Seyfert activity (Section 4).

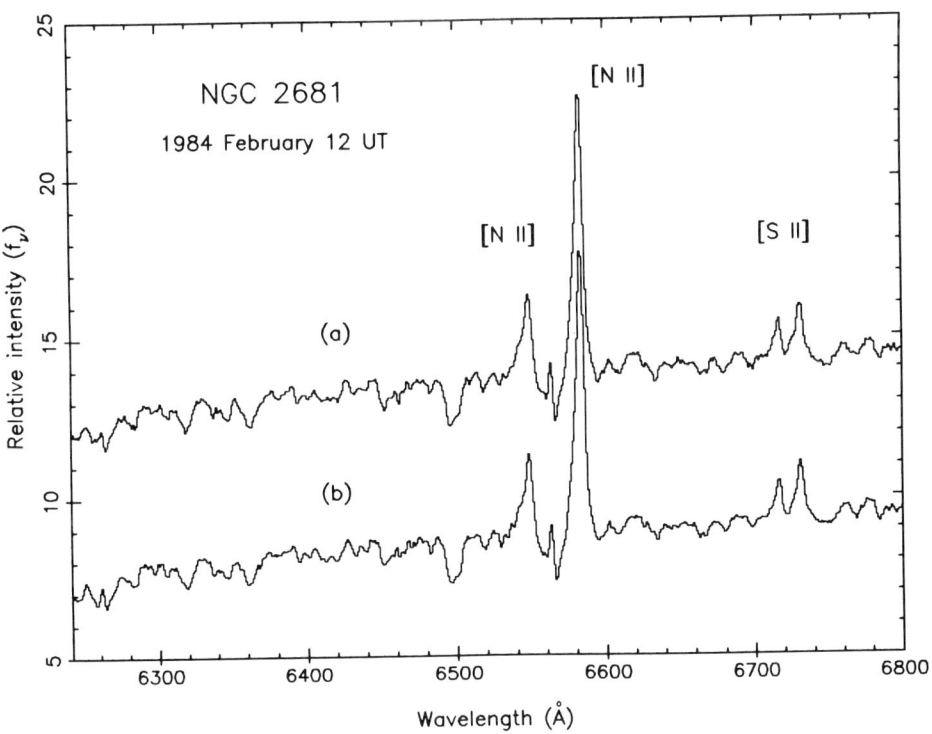

Figure 2: Two independent spectra of NGC 2681 demonstrate that almost all observed features are real. The intensity scale (relative flux density per unit frequency interval) is linear and refers to spectrum (a), while (b) is offset by −5 units. Note the weak Hα emission inside the fairly deep Hα absorption line.

3.2. Type 1 Seyfert Galaxies

An object whose classification is beyond doubt is shown in Figure 3. The spectrum of NGC 4639 exhibits very strong, broad Hα emission (FWZI ≈ 8600 km s^{-1}, FWHM ≈ 3700 km s^{-1}) in addition to the usual narrow lines, yet this bright (B_T = 12.2) SBb galaxy is not listed in any current catalogs of AGNs (e.g., Véron-Cetty and Véron 1984). It is puzzling that the Seyfert 1 characteristics were not noticed in the surveys done by Ford, Rubin, and Roberts (1971) and by Sandage (1978); perhaps the Hα line was markedly weaker in the past. Also, broad Hβ is faint, and many surveys are more sensitive near ~ 5000 Å than near ~ 6500 Å. Considerable reddening is probably present in, or along the line of sight to, the broad-line region. NGC 4639 is a member of the Virgo cluster, which until recently was thought to contain no Seyfert galaxies (Phillips and Malin 1982; Abell, Eastmond, and Jenner 1978).

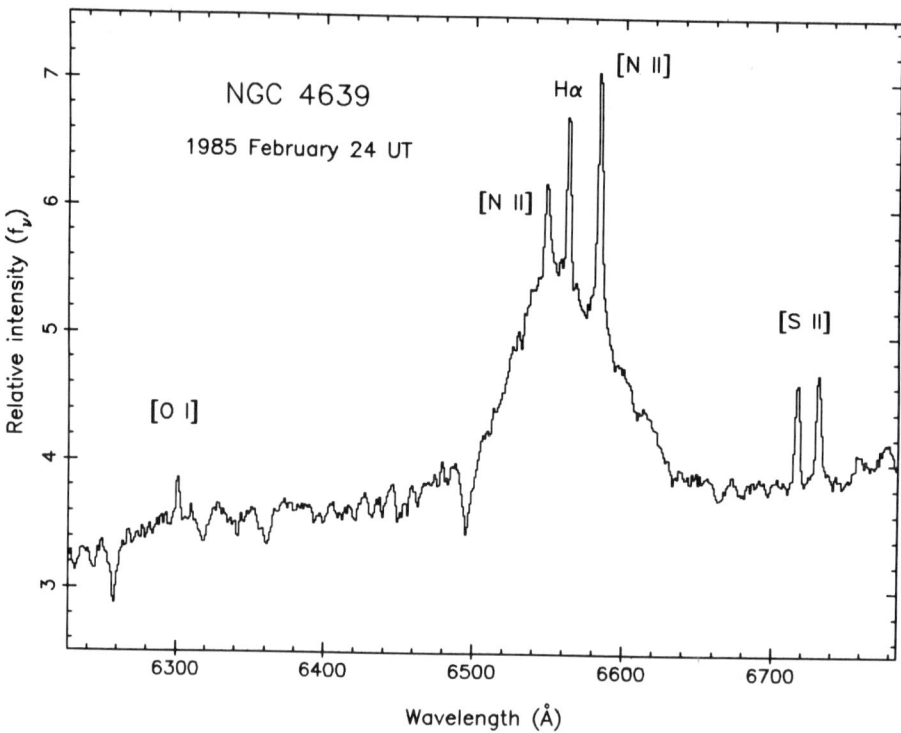

Figure 3: Red spectrum of NGC 4639, a newly discovered Seyfert 1 galaxy in the Virgo cluster. The broad component of Hβ corresponding to that at Hα is very weak, indicating that considerable reddening is probably present. Strong absorption lines from an old stellar population are visible.

Although the great strength of Hα in NGC 4639 is atypical for galaxies in our survey, several other objects exhibit easily visible wings on either side of the {Hα + [N II]} blend. The broad Hα emission in NGC 3998 (Fig. 4), for example, has FWZI ≳ 8000 km s^{-1}; it was initially detected by Heckman (1980), and confirmed several times (Blackman, Wilson, and Ward 1983; Keel 1983). [O I] λ6300 (FWHM ≈ 700 km s^{-1}) is broader than each of the [S II] lines (FWHM ≈ 400 km s^{-1}) and has considerably stronger wings than a Gaussian profile. These properties provide valuable clues to the physical conditions in the narrow-line region (NLR), as described in Section 4.

Subtle, but nevertheless perceptible, broad components of Hα are present in many of the galaxies analyzed in Paper I. A particularly significant case is M87, the famous Virgo A radio source which is associated with an optical jet and possibly harbors a supermassive black hole in its nucleus (Sargent el al. 1978; Young et al. 1978). Figure 5 shows that not only are the forbidden lines quite broad (FWHM ≈ 1000 km s^{-1}) compared with those in most galactic nuclei, but that very faint extensions are visible

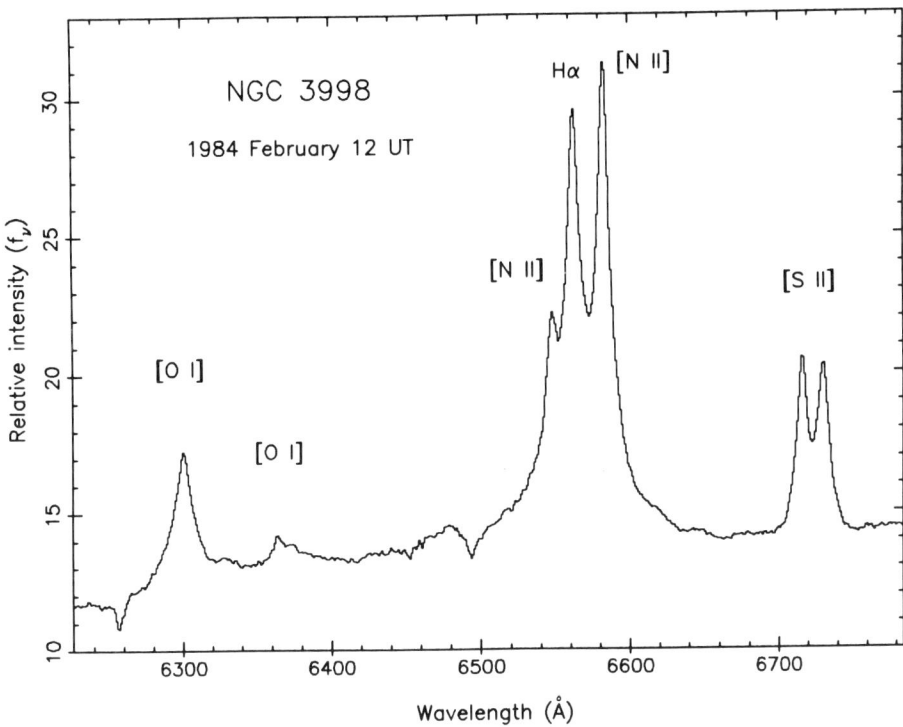

Figure 4: Spectrum of NGC 3998, which clearly shows broad Hα emission as well as differences in the widths of narrow forbidden lines. The absorption at ~ 6495 Å is due to Ca and Fe, while that at ~ 6258 Å is mostly atmospheric O_2.

redward of [N II] λ6583 and blueward of [N II] λ6548. Similar wings are absent in the profiles of [S II], [O III] (in the blue spectrum), and other lines, so they are likely to be associated with Hα (FWZI $\gtrsim$ 6500 km s^{-1}) instead of [N II]. Thus, M87 is a low-luminosity type 1 Seyfert galaxy, albeit a very peculiar one. (Note that we generally list the FWZI of the broad Hα, rather than its FWHM or full width at quarter-maximum, because the narrow Hα and [N II] lines introduce large errors into measurements of the ill-determined broad profile at these other locations.)

3.3. Contamination by Starlight

Most of the spectra illustrated so far show substantial contributions from stars. Sometimes the salient features of emission lines are still easily visible, such as in NGC 3998, but usually removal of the underlying absorption lines is necessary to facilitate accurate measurement of faint emission lines. This can be accomplished by subtracting the "template" spectrum of a different galactic nucleus whose integrated stellar population is similar to that of the object being analyzed, but which lacks emission lines. Another technique is to use as a template the spectrum of the *same* galaxy, but many arc seconds away from its nucleus, where there are often no emission lines. In both cases corrections to the template must frequently be made because of differences in metallicity, velocity dispersion, and reddening (see FH).

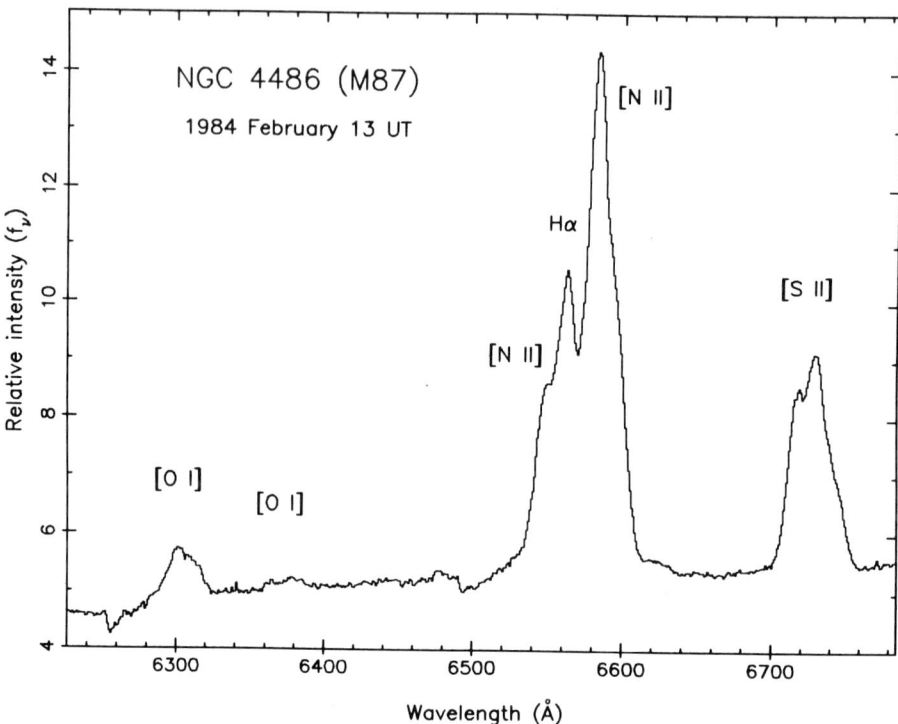

Figure 5: Emission lines in the nucleus of M87 are strong, relatively broad, and asymmetrical. The {Hα + [N II]} blend exhibits weak wings that are probably due to a very broad component of Hα emission similar to, but much fainter than, that in "classical" type 1 Seyfert galaxies.

3.3.1. *Normal, old population.* Figure 6 illustrates the first of these methods for the galaxy NGC 2841. This object has strong, narrow emission lines of Hα, [N II], and [S II] (spectrum a); moreover, a broad, very low-contrast component of Hα emission also seems to be present at first glance. The nucleus of NGC 3115 (Fig. 6b), however, shows a similar weak bump roughly centered on Hα, and the absence of *narrow* emission lines or any other signs of activity in this galaxy suggests that the feature is produced by shallow absorption lines in the neighboring continuum. Indeed, TiO bands at $\sim \lambda 6700$ in the M0 giant 69 ν Gem (Fig. 6d) obviously give rise to a similar continuum shape, while early-K giants (Fig. 6e), which account for a major portion of the total luminosity, dilute the strength of the depression. Spectrum 6f of an A3 star demonstrates that no substantial oscillations in the continuum were introduced by the reduction procedure.

The net result of subtracting the spectrum of NGC 3115 from that of NGC 2841 is shown in Figure 6c. It is clear that broad Hα emission is almost certainly absent, and that the narrow Hα line is stronger than before. The intensity ratio of the [S II] $\lambda 6716$ and [S II] $\lambda 6731$ lines, which is a good indicator of the electron density (n_e) if $n_e \approx 1000$ cm^{-3}, has changed due to the removal of Ca I $\lambda 6718$ absorption. Finally, although [O I] λ6300 was not even suspected in the original spectrum, it is now visible at the correct wavelength and can be used along with other important diagnostics to determine the dominant ionization mechanism in the clouds of gas (Baldwin, Phillips, and Terlevich 1981). Thus, accurate subtraction of the stellar component is crucial in a proper analysis of NGC 2841.

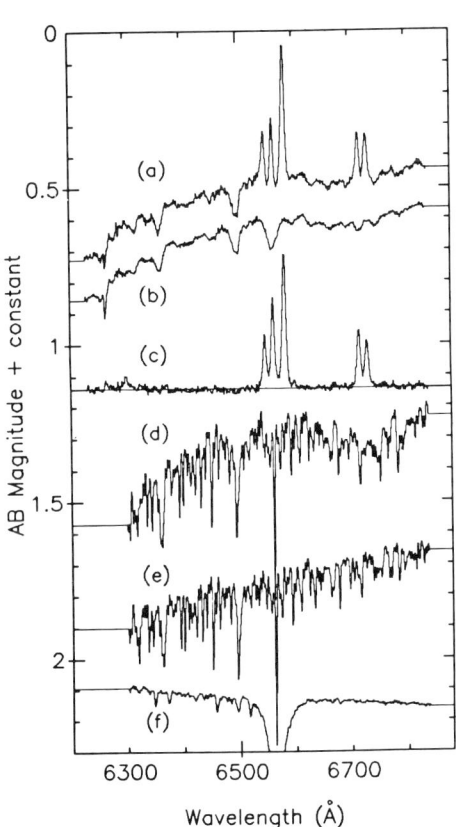

Figure 6: (a) Red spectrum of NGC 2841. A faint, very broad component of Hα emission appears to be present, but the continuum of the absorption-line template NGC 3115 (b) shows a similar shape. Subtraction of NGC 3115 from NGC 2841 (c) confirms that there is *no* broad Hα in NGC 2841. Moreover, the relative intensities of the narrow lines have been altered. Spectra (d), (e), and (f) are of an M0 giant, a K2 giant, and an A3 dwarf, respectively (see text). The data are plotted on a logarithmic (magnitude) scale.

3.3.2. *Young stars.* The situation is more complicated for galactic nuclei with peculiar stellar properties. In some objects, for instance, there is ample evidence for a significant contribution from young stars, and it is then very difficult to find appropriate absorption-line templates because of the emission lines usually associated with such stars. Furthermore, the strength of the Balmer absorption is quite sensitive to the mass function in the nucleus. Thus, removal of the underlying starlight can perhaps be best accomplished by spectral synthesis techniques (Keel 1983), but even these have severe limitations.

A good example of a galaxy whose nucleus has experienced a relatively recent episode of star formation, and whose continuum is therefore not well matched by templates such as NGC 3115, is NGC 404 (Fig. 7). Deep Balmer absorption lines are visible in blue spectra ($\lambda\lambda$ 3600–5200), but Hα is obliterated by the corresponding emission line. Although hot, massive stars are known to be present and the emission lines are very narrow (Paper I), it is interesting that the *relative strengths* of the lines resemble those in Heckman's (1980) *LINERs* rather than in H II regions. Terlevich and Melnick (1985) hypothesize that galactic nuclei with vigorous bursts of star formation may evolve into type 2 Seyferts and eventually into "Blue LINERs" by passing through a stage during which extremely hot ($T > 10^5$ K) stars, which they call "Warmers," ionize the surrounding gas; as the hot stars age, the ionization level decreases and the gas emits a LINER spectrum instead of a high-excitation Seyfert spectrum. Nuclei having relatively small starbursts, on the other hand, are thought to evolve directly into "Red LINERs." NGC 404 may be a prototypical "Blue LINER," but for a variety of reasons we consider it unlikely that the emission-line spectra of most LINERs and type 2 Seyfert galaxies are produced in this manner.

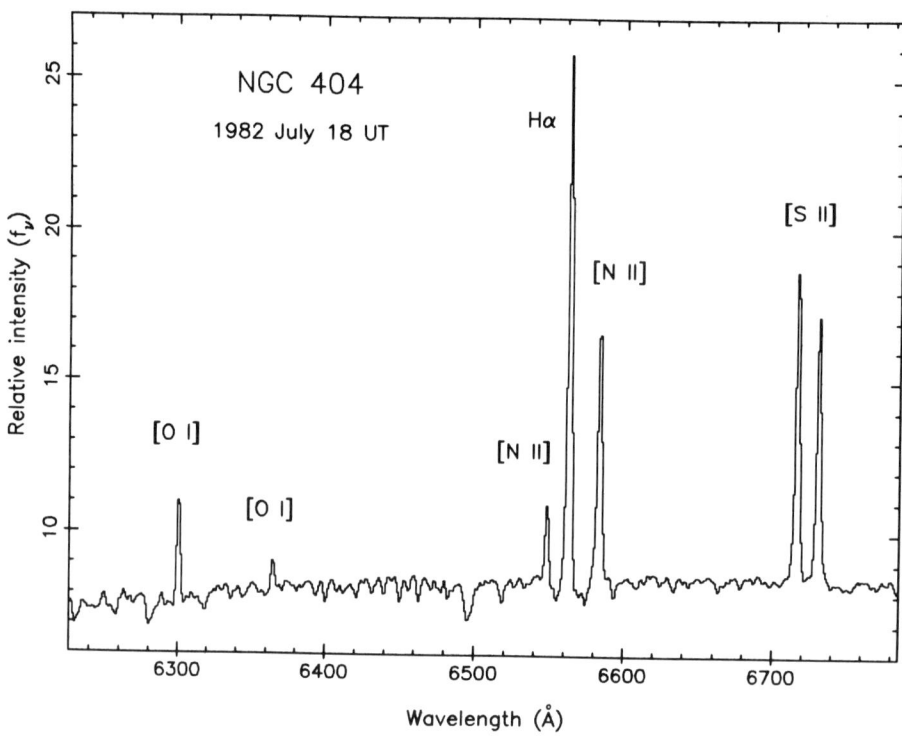

Figure 7: Red spectrum of NGC 404, a peculiar LINER whose emission lines are remarkably narrow. The blue spectrum contains prominent, high-order Balmer absorption lines, indicating the presence of young, massive stars. Suitable template galaxies are difficult to find; perhaps an adequate subtraction of the continuum may be obtained through spectral synthesis.

3.4. The Elusive Broad Hα Emission

We have seen that subtraction of the stellar continuum greatly simplifies measurement of narrow emission lines and increases the accuracy of the results. Even so, the detection of a very weak, broad component of Hα emission is by no means straightforward. The [N II] λλ6548, 6583 doublet is often prominent, sometimes surpassing the strength of the narrow Hα line, and the intrinsic widths produce severe blending of the profiles. Hence, one must carefully search for extended wings on either side of the {Hα + [N II]} blend, as was done by Stauffer (1982), Keel (1983), and others.

3.4.1. *Gaussian profiles.* If the isolated emission lines such as [O III] λ5007 are well represented by Gaussians, a fruitful procedure is to look for positive residuals by decomposing the blend into a set of three such components. This is illustrated for the X-ray LINER NGC 3884 (Reichert el al. 1982) in Figure 8a. High-frequency oscillations in the residuals (8b) indicate that the narrow lines are not exactly symmetrical Gaussians, but they average out to zero and cause no problems. Excess emission at the extreme red and blue portions of the blend, on the other hand, is of greater significance; similar wings are not present in the [S II] doublet and in uncontaminated emission lines throughout the spectrum, so the [N II] lines are unlikely to have them as well. Instead, we propose that these wings are due to a component of Hα emission which is substantially broader than the narrow lines, just as in type 1 Seyfert galaxies. The simplest, though not necessarily most realistic, profile to adopt is a Gaussian, so in Figure 8c we include a fourth component when calculating the best least-squares fit. The new residuals (8d) still show the high-frequency oscillations attributed to the narrow lines, but the extended wings are now absent. Of course, the inclusion of additional free parameters should produce a better fit irrespective of their physical significance, but in this case the use of a fourth

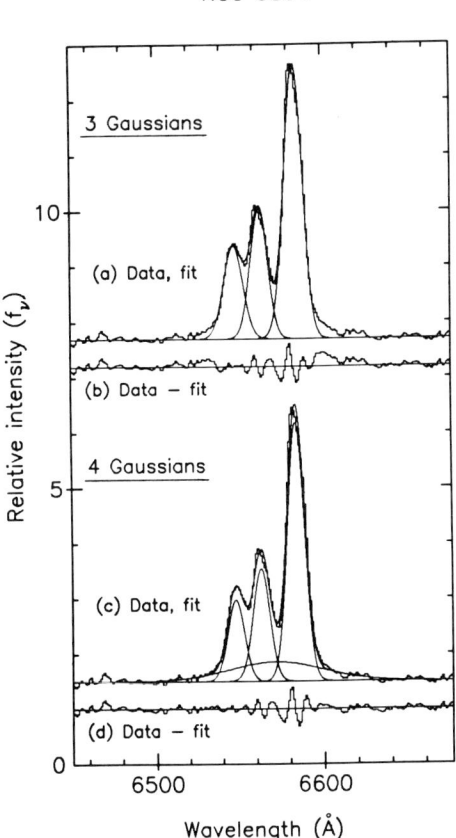

Figure 8: Gaussian decomposition of the {Hα + [N II]} blend in NGC 3884. Three Gaussians are fit to the data in (a); the residuals (b) have almost no net average, except at the extreme ends where significant positive deviations are visible. These wings are unlikely to be caused by [N II], since other emission lines lack them. A least-squares fit with four Gaussians (c), however, leaves only the oscillatory residuals (d); a very weak, broad component of Hα emission is therefore probably present.

component is justified and leads to reasonable results; $I([\text{N II}]\ \lambda 6583)/I([\text{N II}]\ \lambda 6548)$ ≈ 3.0, for example, as predicted by atomic theory. The same is true for NGC 4278, NGC 4258 (Fig. 1), and many additional galaxies, including the classical LINER NGC 1052 (Heckman 1980). In fact, our analysis shows that at least 19, and possibly up to 28, of the 75 objects initially surveyed (Paper I) have a broad component of Hα. Further study of these 28 candidates should improve the accuracy of the measured fluxes and eliminate the galaxies which are not type 1 Seyferts by weeding out (Weed man 1977) those with no broad Hα emission.

It is important to realize that a large fraction of the apparent width of the narrow lines in NGC 3884 and other objects is clearly due to global rotation of the galaxies. Their two-dimensional spectra exhibit steep rotation curves of large amplitude which in some cases (e.g., NGC 5005) amount to a velocity difference of 300–350 km s^{-1} over only $1''$–$2''$ of the nuclear region. Orientation of the $2'' \times 4.1''$ rectangular aperture along the rotation axis will simplify the detection of broad Hα in these objects by minimizing the contribution of the narrow lines.

3.4.2. *Non-Gaussian profiles.* The decomposition procedure outlined above is not justified for galaxies whose nuclei contain asymmetrical or non-Gaussian low-ionization emission lines, as shown in Figure 9. The [S II] and [N II] profiles in NGC 2639 obviously have

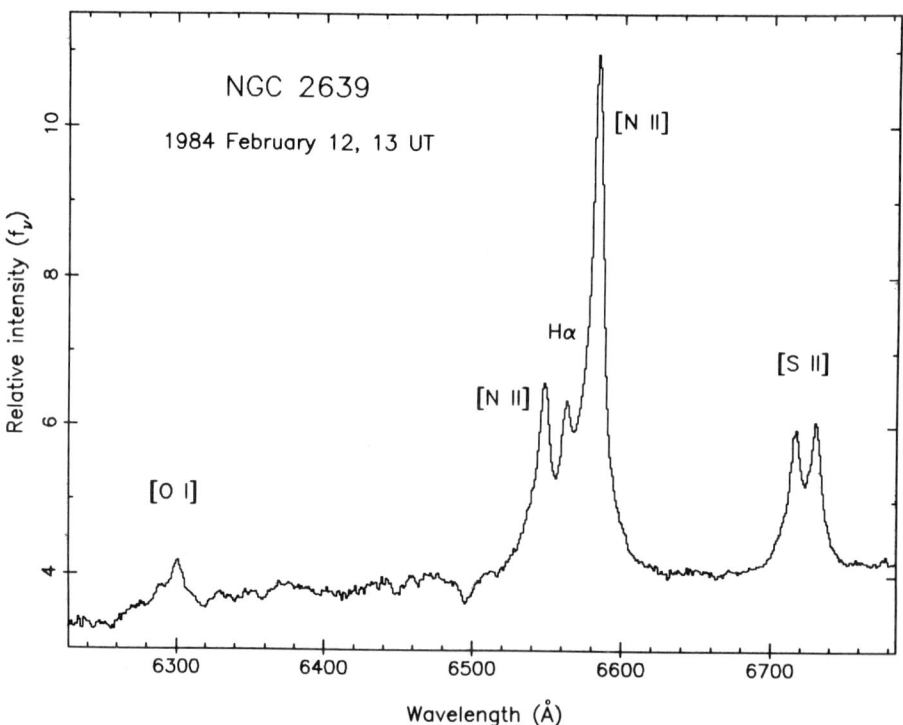

Figure 9: Red spectrum of NGC 2639, a galaxy whose emission lines are far from Gaussian. Isolated lines, or synthetic profiles constructed from the least contaminated portions of the [S II] lines, must be used in a proper analysis of the {Hα + [N II]} blend. Broad Hα emission is present, but not to the extent reported previously.

extended wings, and the derived strength of broad Hα is significantly overestimated if the {Hα + [N II]} blend is decomposed with Gaussians. This danger is especially prevalent in data of low resolution or poor S/N ratio, since the intrinsic asymmetries are not apparent and Gaussians are often simply *assumed* to represent the line profiles. Instead, one must deconvolve the blend either by using isolated emission lines (such as [O III] λ5007) from other regions of the spectrum or by constructing a synthetic narrow-line profile from the blue side of [S II] λ6716 and the red side of [S II] λ6731. This is a difficult and somewhat subjective procedure if the lines are blended, especially when a great range of widths and profiles characterizes the emission-line spectrum (FH; Filippenko 1985; De Robertis and Osterbrock 1984). Analysis of NGC 2639 indicates that broad Hα is present, but it is not nearly as strong as might be inferred from previous descriptions (Keel 1983; Huchra, Wyatt, and Davis 1982).

A more striking example of how asymmetrical profiles can affect the measurement of broad Hα emission is provided by NGC 4388. Figures 1 and 10 show that the forbidden lines in this type 2 Seyfert galaxy have stronger, more extended red wings than blue wings, which is an exception to the general trend in AGNs (Heckman *et al.* 1981; Whittle 1985*a,b*). Casual inspection of the spectra shows that much or even most of the blending of the [N II] and Hα lines is produced by these asymmetrical wings. In data of lower resolution (Stauffer 1982) the asymmetries are not visible, and Gaussian decomposition of the blend indicates that a strong, broad, redshifted component of Hα is necessary

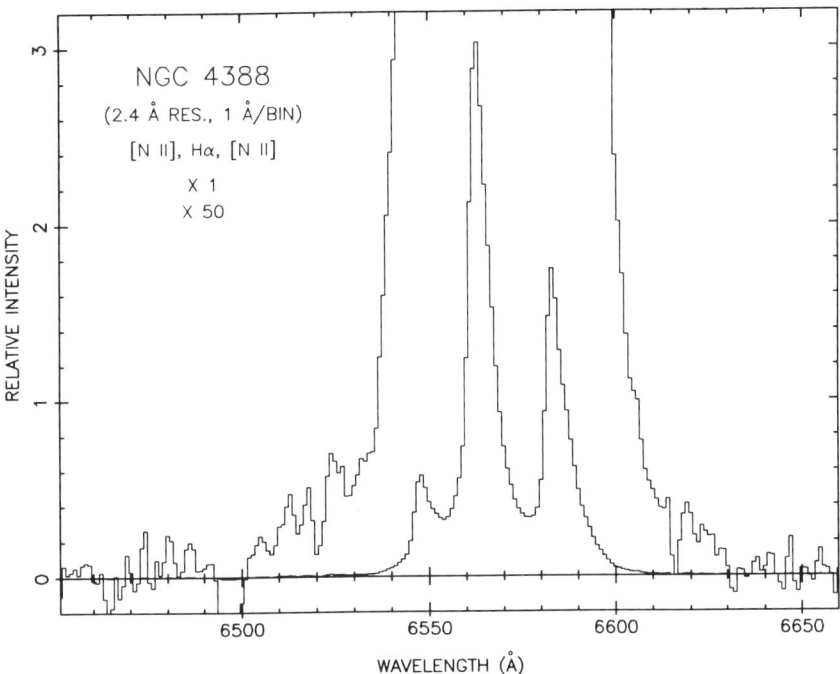

Figure 10: The {Hα + [N II]} blend in NGC 4388 is shown with two intensity scales differing by a factor of 50. Emission lines have very extended wings (especially on the red side) which produce severe blending. Gaussian decomposition of low-resolution data incorrectly yields a broad, redshifted component of Hα which is *much* more intense than that visible in the expanded plot.

to provide an adequate fit by filling the "valleys" between [N II] and Hα. Note that the *expanded* (×50) plot in Figure 10 does in fact reveal broad Hα emission, but it is exceedingly faint relative to the narrow lines and has an equivalent width of only ~ 8 Å, far weaker than reported by Stauffer (1982). This detection supports the idea that all narrow-line X-ray galaxies are actually type 1 Seyferts (Véron *et al.* 1980), some of whose nuclei are highly obscured.

3.5. Faint Seyferts Hidden by H II Regions

Many of the galaxies included in our survey are of type Sbc or later, and often exhibit prominent emission lines from H II regions near the nucleus. NGC 4303 (Fig. 11) is one such object; narrow lines (FWHM $\lesssim$ 100 km s^{-1}) are visible all along the slit as well as in the nucleus itself. Close examination of the line profiles, on the other hand, shows that a broader, "triangular" base is present in some lines, especially after the underlying absorption lines have been removed. The FWZI of [N II] λ6583, for example, is ~ 800 Å, and its intensity is greater than that of the corresponding Hα component. Likewise, the base of [O III] λ5007 appears to be substantially stronger than that of Hβ, suggesting that an extremely weak type 2 Seyfert nucleus (Baldwin, Phillips, and Terlevich 1981) is being hidden by prominent H II regions in NGC 4303. This method of analyzing line

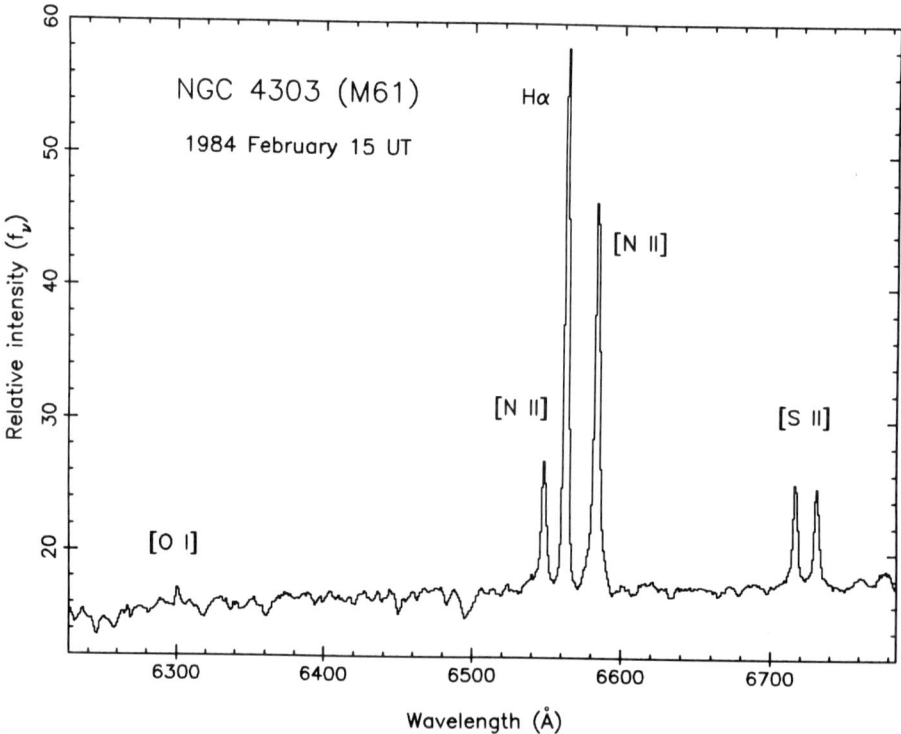

Figure 11: Spectrum of NGC 4303, whose nucleus is dominated by H II regions. Careful inspection of the narrow [N II] λ6583 line, however, reveals a distinct triangular base. The relative intensities of this "broad" [N II] and the corresponding components of other lines are similar to those in type 2 Seyfert galaxies.

profiles to find components produced by both nonstellar and stellar activity in a given nucleus has previously been described by Véron et al. (1981). To fully exploit it, data of exceptionally high *spatial* resolution as well as high spectral resolution and S/N ratio are necessary, since a smaller fraction of the light will come from H II regions if spectra are obtained through progressively smaller apertures. The Space Telescope will clearly be invaluable for such studies.

3.6. Radial Ionization Gradients

Another method for determining whether ionized gas near the nucleus of a galaxy is indicative of nonthermal activity is to analyze the radial gradients of emission-line intensity ratios, as was first done in a systematic manner for M51 by Rose and Searle (1982). It was shown that the gradients can reasonably be explained by geometrical dilution of a power-law ionizing continuum of index ~ -1 emanating from the center of the galaxy, and that other possible excitation mechanisms (e.g, heating by shocks, ionization by OB stars) are much less probable. We have been conducting similar studies of the most interesting galaxies for which we have long-slit spectra, and in some cases the preliminary results are relatively unambiguous. NGC 6951, for example, exhibits intensity ratios and radial gradients very similar to those in M51, and the interpretation is basically the same as that proposed by Rose and Searle (1982). The electron density is measured from the [S II] doublet, and a photoionization model is subsequently used to explore the consequences of different input parameters such as power-law slope and cloud covering fraction. Thus far, the best fit to the data indicates that gas in the nuclear region is photoionized by a nonstellar continuum whose relative prominence decreases rapidly with radius. Farther out the line intensity ratios are typical of H II regions, as are the small line widths.

4. DISCUSSION

The spectra in the preceding section illustrate the wide range of properties found in the nuclei of nearby galaxies. One of the most important from our point of view is the presence of broad Hα emission, albeit at a faint level, in a significant fraction of objects, including over 40% of the LINERs studied by Heckman (1980). This provides a direct link to the classical Seyferts and QSOs, and strongly suggests that similar physical processes occur in all these objects. Although the initial part of our survey was directed specifically at those galaxies which were the best candidates for low-luminosity AGNs, "dwarf" Seyfert nuclei will still probably account for at least 10% of the complete sample.

4.1. Line Profiles in LINERs

Another interesting property of many galaxies in our sample is that various *narrow* lines sometimes have different widths and profiles. This was mentioned in connection with NGC 3998 (Section 3.2), and is seen more clearly in high-quality spectra of M81 (Fig. 12). Not only does M81 have a prominent component of broad Hα emission (FWZI ≈ 5000 km s^{-1}), but the [O I] $\lambda 6300$ profile exhibits wings of intermediate width which are definitely absent in the [S II] doublet. Differences in the profiles are generally not so extreme, but [O I] $\lambda 6300$ is noticeably broader than each of the [S II] $\lambda\lambda 6716, 6731$ lines in over 30% of Heckman's (1980) LINERs.

Recent studies (FH; Filippenko 1985) have demonstrated that this phenomenon provides an important clue to the physical conditions in LINERs. The spectrum of NGC 7213 (FH), for example, shows blatant differences in the [O I] and [S II] profiles just

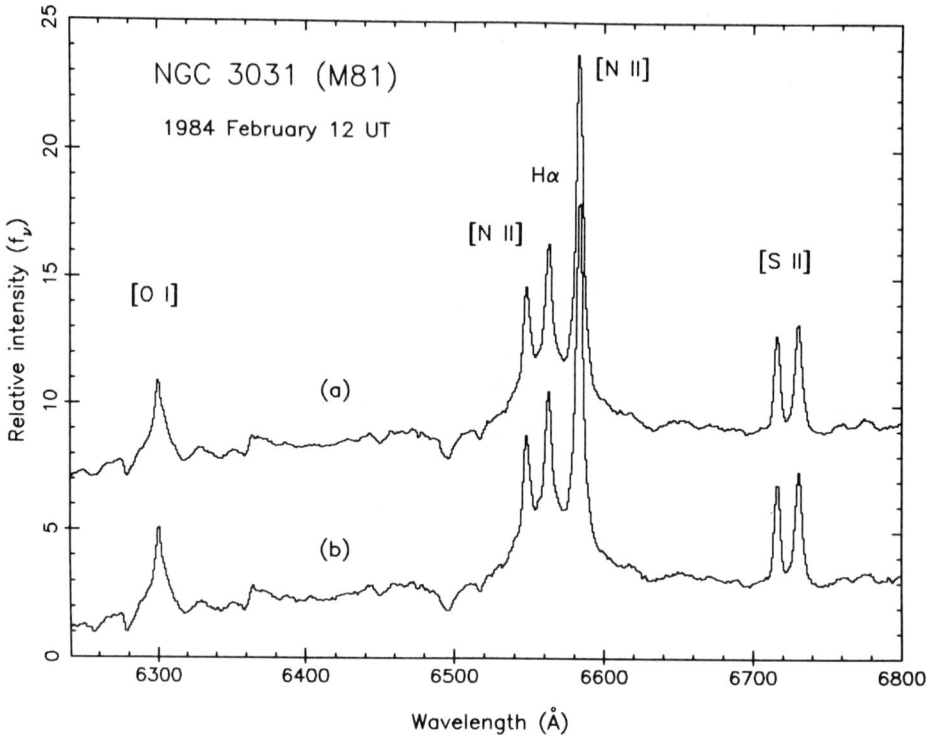

Figure 12: Two separate spectra of M81 (*b* offset by −6 units), showing the high quality of the data. Broad Hα is obvious, but weaker than in classical Seyfert 1 galaxies. Large differences in the profiles of narrow lines, particularly [O I] λ6300 and the [S II] doublet, are also visible.

like those in M81. Even more remarkable is the fact that different transitions of the O^+ ion (and others as well) have dissimilar widths (Fig. 13). In general, lines associated with high critical densities for collisional deexcitation are broader than those with low values of n_e(crit) (Fig. 14). Since the gas emissivity per unit mass is proportional to n_e below n_e(crit), and is roughly independent of n_e above n_e(crit), a given line will receive its largest contribution from clouds whose density is near the corresponding value of n_e(crit) if a large range of densities is present among clouds of comparable mass. Dense clouds ($n_e \approx 10^6$–10^7 cm^{-3}) therefore have the greatest bulk motions, and the simple model of FH shows that they probably reside closer to the nucleus than gas of low density ($n_e \approx 10^3$ cm^{-3}). Similar results were found for MR 2251−178 (a nearby QSO), Pictor A (an N galaxy), and the classical LINER PKS 1718−649 (Fig. 15).

4.2. LINERs: Shock Heated or Photoionized?

These results are of special relevance for determining the ionization mechanism in LINERs. Although the overall spectra of LINERs strongly resemble those of supernova remnants, evidence supporting the hypothesis that the gas is predominantly photoionized by a nonstellar continuum, rather than heated by shocks, has rapidly been accumulating.

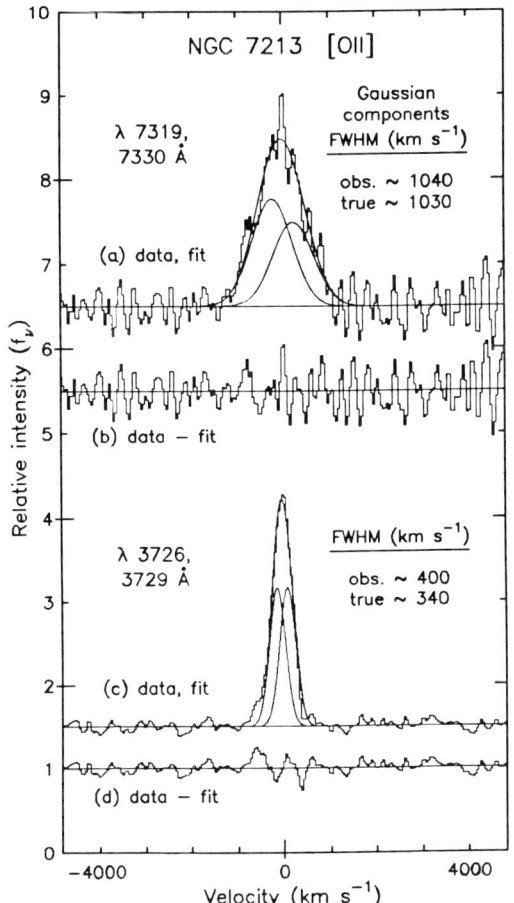

Figure 13: Gaussian decomposition of the [O II] blends in NGC 7213, a type 1 Seyfert with many features of LINERs. The auroral lines ($\lambda\lambda 7319$, 7330), associated with high critical densities, are much broader than the nebular (low-density) pair but have comparable strength. Dense ($n_e \gtrsim 10^6$ cm^{-3}), rapidly-moving clouds ($v \approx 1000$ km s^{-1}) contribute predominantly to the auroral [O II] doublet, while [O II] $\lambda 3727$ is produced by slow clouds ($v \approx 300$ km s^{-1}) of far lower density ($n_e \approx 10^3$ cm^{-3}).

Models by Ferland and Netzer (1983), Halpern and Steiner (1983), and Binette (1985) show that the emission lines in LINERs and classical AGNs form a continuous sequence in which the main variable is the ratio of ionizing photons to nucleons at the face of a cloud in the NLR. This "ionization parameter" (U) is large ($\sim 10^{-2}$) in QSOs and luminous type 1 Seyfert galaxies, but drops to $\sim 10^{-3.5}$ in LINERs, so the absence of an obvious nonstellar continuum in the latter objects is not surprising. One of the main problems with this interpretation, however, is that the electron temperature (T_e) derived from the relative intensities of [O III] $\lambda 4363$ and [O III] $\lambda 5007$ may be too high ($T_e \gtrsim 25000$ K) in a few of the most accurately measured LINERs, including the prototype NGC 1052 (Koski and Osterbrock 1976; Fosbury *et al.* 1978; but see also Keel and Miller 1983, and Rose and Tripicco 1984). Such high values of T_e indicate collisional excitation of the gas, as in shock models. In this case LINERs and Seyfert galaxies might be totally unrelated classes, and LINERs should not be included in the luminosity function of AGNs.

It is important to emphasize that the high-T_e problem exists only if the low densities ($n_e \approx 10^3$ cm^{-3}) derived from the red [S II] doublet also apply to the O^{++} region. A *range* of densities in the narrow-line clouds would remove the problem if $n_e \gtrsim 10^6$ cm^{-3} in the densest clouds. But recall that this is exactly what is implied by the different line profiles

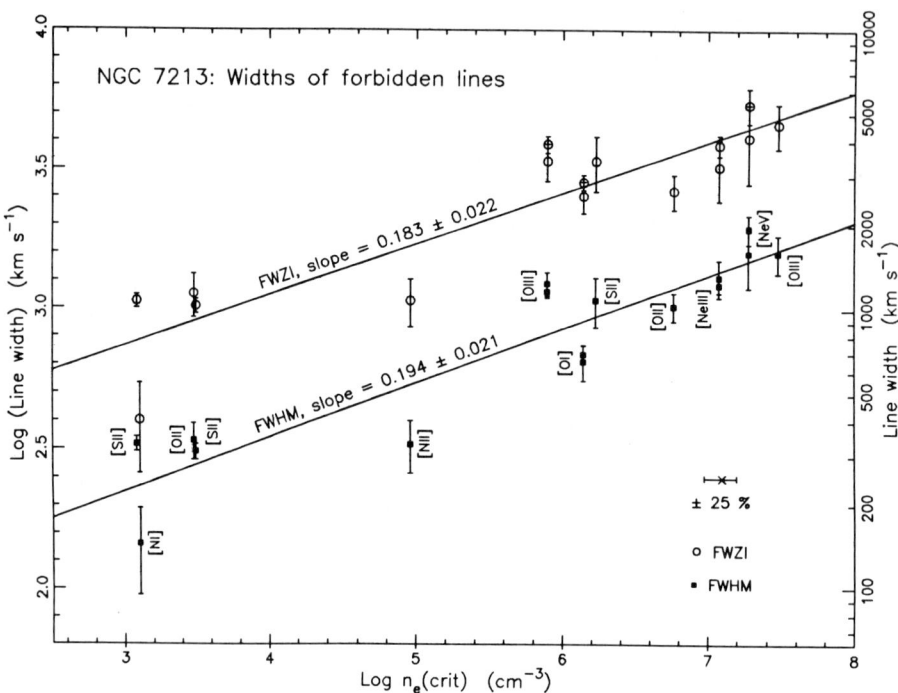

Figure 14: Widths of emission lines in NGC 7213 are plotted against the corresponding critical densities, revealing an excellent correlation. Since $n_e(\text{crit})$ is the density at which collisional deexcitation occurs as frequently as radiative decay, a given line can be a "tracer" of gas with $n_e \approx n_e(\text{crit})$ if many clouds of comparable mass but spanning a wide range of densities are present. High-density clouds in NGC 7213 clearly have greater bulk motions than those of lower density.

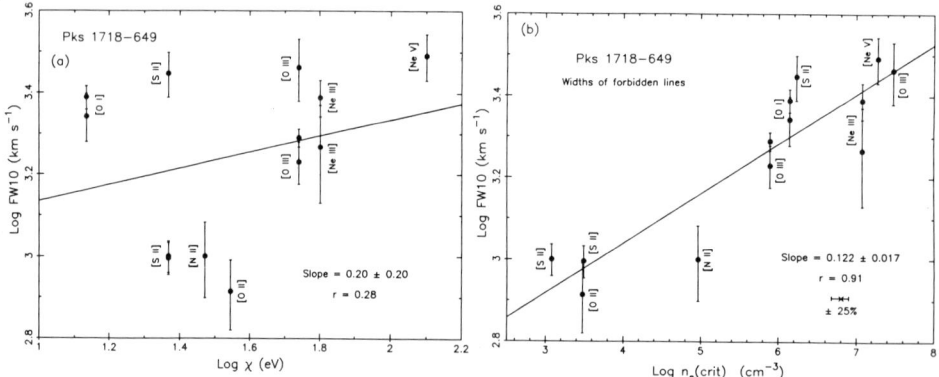

Figure 15: The full width at 10%-intensity for various emission lines in the classical LINER PKS 1718–649 is plotted against (*a*) the ionization potential χ of the emitting species, and (*b*) the appropriate critical density. Line width and χ are essentially uncorrelated, while the correlation between width and $n_e(\text{crit})$ is excellent. Optically thick clouds having densities between 10^3 cm^{-3} and 10^7 cm^{-3} must be present in the NLR.

in many LINERs! Moreover, in NGC 7213 and Pictor A there is compelling evidence that high densities, not high temperatures, produce the anomalously strong [O III] $\lambda4363$; these objects have prominent nonstellar continua and are strong X-ray sources, so there can be no doubt that photoionization is the dominant excitation mechanism. In fact, Figure 16 shows that [O III] $\lambda4363$ is so intense in NGC 7213 that it cannot be produced at *any* plausible temperature unless $n_e \gtrsim 10^{5.5}$ cm^{-3}, just as in certain QSOs and broad-line radio galaxies (Baldwin 1975; Osterbrock, Koski, and Phillips 1976; Neugebauer et al. 1976). Finally, the inclusion of a large density range in photoionization models alleviates other, more minor, problems that commonly characterize models which consider only low densities for all the clouds (Péquignot 1984). The very strong [O I] $\lambda6300$ line, for example, is easily explained by the presence of dense ($n_e \approx 10^6$ cm^{-3}), optically thick clouds, in large regions of which only $\sim 10\%$ of the hydrogen is ionized by penetrating X-rays.

Note that a range of densities may be present in the NLR even if all emission lines have roughly the *same* widths, as would be the case for clouds orbiting in a logarithmic potential (see below) or flowing ballistically away from the nucleus. Thus, it is likely that the emission lines in most LINERs can be explained in terms of photoionization by a nonstellar continuum, even if they do not exhibit significant differences in width.

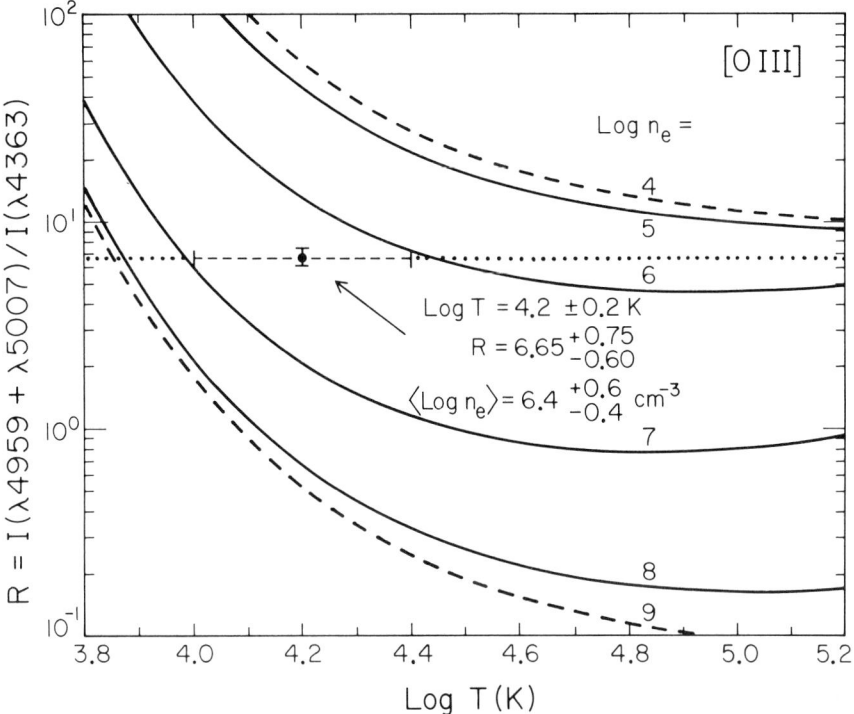

Figure 16: The ratio of nebular to auroral [O III] emission is strongly dependent on electron temperature (labeled T) and density (n_e), although it is primarily an indicator of T at the densities normally associated with the NLR of AGNs ($n_e \lesssim 10^4$ cm^{-3}). In NGC 7213, however, the measured value $R = 6.65$ is *inconsistent* with $n_e \lesssim 10^{5.5}$ cm^{-3} for the O^{++} zone. At $T \approx 16000$ K the average density is $\sim 10^{6.4}$ cm^{-3}, but much higher values ($n_e \gtrsim 10^7$ cm^{-3}) also exist since [O III] $\lambda4363$ is broader than [O III] $\lambda5007$ (Fig. 14).

4.3. Probing the Nuclear Potential

The quantitative correlation between line width and $n_e(\text{crit})$ can be interpreted with reasonable success in a number of dynamical models of the NLR, but perhaps the simplest is to assume that clouds move along Keplerian orbits in a gravitational potential dominated by the central engine (Carroll and Kwan 1983; FH). If the ionization parameter is independent of distance from the nucleus, then $v \propto n^{1/4}$. The observed relation in NGC 7213 is $v \propto n^{1/5}$, which is not too different. Some objects, on the other hand, exhibit a weaker dependence of v on n, and in many cases all of the emission lines have equal widths to first order. If radiation pressure, gas drag, and other forces are assumed to be negligible, the line widths can be used as a probe of the gravitational potential by generalizing the arguments discussed above. A logarithmic potential ($\rho \propto r^{-2}$), for example, produces no variation in line widths, while one that is intermediate between $\ln r$ and $1/r$ should yield $v \propto n^\beta$, where $0 \lesssim \beta \lesssim 0.25$. Hence, the line widths may be used to study the relative dominance of the central object and the stars near the nucleus, although a complete treatment must be much more complicated than outlined here.

5. CONCLUSIONS

Preliminary results from an extensive spectroscopic survey show that the characteristics of Seyfert galaxies and QSOs are present at a faint level in a significant fraction of nearby galactic nuclei. In particular, an Hα emission line which is broader than the forbidden lines has been detected in $\sim 30\%$ of the initial, heavily biased sample of objects (Paper I). New data suggest that $\gtrsim 10\%$ of all galaxies exhibit this feature. Many of the remaining objects may have a broad-line region which is simply too faint to be seen among the narrow emission lines and stellar continuum in the currently available spectra. When finished, this survey should provide a fairly complete description of the luminosity function of AGNs at the faint end $(-17 \lesssim M_V \lesssim -11)$.

It is significant that over 40% of Heckman's (1980) LINERs, in which shock heating was originally thought to be the dominant excitation mechanism, show signs of broad Hα emission. This supports the more recent hypothesis that gas in LINERs is actually *photoionized* by nonstellar radiation similar to, but much weaker than, that in classical AGNs. Furthermore, analysis of the profiles of forbidden lines provides direct evidence for a wide *range* of densities ($n_e \approx 10^3$–10^6 cm^{-3}) in the narrow-line clouds of many LINERs, improving the agreement between observed relative intensities and the predictions of photoionization models.

Although Seyfert nuclei and LINERs are found predominantly in early-type spiral galaxies, there are indications of nonstellar activity in late-type spirals as well, but it can easily be masked by strong emission lines from H II regions. Broad Hα has even been detected in some elliptical galaxies, such as M87. Complicated methods must often be used to isolate the line from adjacent features, and the underlying starlight is a serious contaminant which must be removed with care.

These results demonstrate that low-level Seyfert activity is a common property of nearby galaxies. Comparisons with theoretical models indicate that the physical mechanisms involved may be fundamentally similar to those in AGNs of higher luminosity. On the other hand, it is still not known whether most of the nuclei are faint because the putative black hole is orders of magnitude less massive than in typical QSOs or because the accretion rate of matter is very low. Perhaps this question will soon be partially answered by ground-based studies, but the Space Telescope, with its high spatial resolution, will undoubtedly provide more definitive conclusions. It seems likely, however, that at least some of these "dwarf" Seyfert galaxies were luminous quasars in the distant past, at a time when the gas supply near the central "monster" was much more plentiful than it is today.

ACKNOWLEDGMENTS

We thank the Time Allocation Committee of Palomar Observatory for providing a generous number of nights on the Hale reflector. Many technicians and telescope operators were of great help during the acquisition of data. John Biretta provided one of the main computer programs used in the subsequent analysis. Figures 1, 6, 8, 10, 13, 14, 15, and 16 have previously been published (FH; Filippenko 1985; Paper I), and are reproduced here courtesy of *The Astrophysical Journal*. This research was supported by NSF grant AST 82-16544 to W. L. W. S., as well as by fellowships to A. V. F. from the Fannie and John Hertz Foundation and from the Miller Institute for Basic Research in Science (U.C. Berkeley). A. V. F. is also grateful to the organizing committee of this conference for financial assistance.

REFERENCES

Abell, G. O., Eastmond, T. S., and Jenner, D. C. 1978, *Ap. J. (Letters)*, **221**, L1.
Baldwin, J. A. 1975, *Ap. J.*, **201**, 26.
Baldwin, J. A., Phillips, M. M., and Terlevich, R. 1981, *Pub. A.S.P.*, **93**, 5.
Balick, B., and Heckman, T. M. 1983, *Ap. J. (Letters)*, **265**, L1.
Binette, L. 1985, *Astr. Ap.*, **143**, 334.
Blackman, C. P., Wilson, A. S., and Ward, M. J. 1983, *M.N.R.A.S.*, **202**, 1001.
Boroson, T. A., and Oke, J. B. 1982, *Nature*, **296**, 397.
Boroson, T. A., Oke, J. B., and Green, R. F. 1982, *Ap. J.*, **263**, 32.
Carroll, T. J., and Kwan, J. 1983, *Ap. J.*, **274**, 113.
Cheng, F.-Z., Danese, L., De Zotti, G., and Franceschini, A. 1985, *M.N.R.A.S.*, **212**, 857.
De Robertis, M. M., and Osterbrock, D. E. 1984, *Ap. J.*, **286**, 171.
de Vaucouleurs, G., de Vaucouleurs, A., and Corwin, H. G., Jr. 1976, *Second Reference Catalogue of Bright Galaxies* (Austin: University of Texas Press).
Elvis, M., Soltan, A., and Keel, W. C. 1984, *Ap. J.*, **283**, 479.
Elvis, M., and Van Speybroeck, L. 1982, *Ap. J. (Letters)*, **257**, L51.
Ferland, G. J., and Netzer, H. 1983, *Ap. J.*, **264**, 105.
Filippenko, A. V. 1985, *Ap. J.*, **289**, 475.
Filippenko, A. V., and Halpern, J. P. 1984, *Ap. J.*, **285**, 458 (FH).
Filippenko, A. V., and Sargent, W. L. W. 1985, *Ap. J. Suppl.*, **57**, 503 (Paper I).
Ford, W. K., Rubin, V. C., and Roberts, M. S. 1971, *A. J.*, **76**, 22.
Fosbury, R. A. E., Mebold, U., Goss, W. M., and Dopita, M. A. 1978, *M.N.R.A.S.*, **183**, 549.
Halpern, J. P., and Steiner, J. E. 1983, *Ap. J. (Letters)*, **269**, L37.
Heckman, T. M. 1980, *Astr. Ap.*, **87**, 152.
Heckman, T. M., Miley, G. K., van Breugel, W. J. M., and Butcher, H. R. 1981, *Ap. J.*, **247**, 403.
Huchra, J., and Sargent, W. L. W. 1973, *Ap. J.*, **186**, 433.
Huchra, J. P., Wyatt, W. F., and Davis, M. 1982, *A. J.*, **87**, 1628.

Hutchings, J. B., Crampton, D., Campbell, B., Gower, A. C., and Morris, S. C. 1982, *Ap. J.*, **262**, 48.
Keel, W. C. 1983, *Ap. J.*, **269**, 466.
Keel, W. C., and Miller, J. S. 1983, *Ap. J. (Letters)*, **266**, L89.
Koski, A. T., and Osterbrock, D. E. 1976, *Ap. J. (Letters)*, **203**, L49.
Maccacaro, T., Gioia, I. M., and Stocke, J. T. 1984, *Ap. J.*, **283**, 486.
Malkan, M. A. 1984, *Ap. J.*, **287**, 555.
Malkan, M. A., Margon, B., and Chanan, G. A. 1984, *Ap. J.*, **280**, 66.
Marshall, H. L. 1985, *Ap. J.*, in press.
Meurs, E. J. A., and Wilson, A. S. 1984, *Astr. Ap.*, **136**, 206.
Neugebauer, G., Becklin, E. E., Oke, J. B., and Searle, L. 1976, *Ap. J.*, **205**, 29.
Oke, J. B., and Gunn, J. E. 1982, *Pub. A.S.P.*, **94**, 586.
Oke, J. B., and Gunn, J. E. 1983, *Ap. J.*, **266**, 713.
Osterbrock, D. E. 1981, *Ap. J.*, **249**, 462.
Osterbrock, D. E., Koski, A. T., and Phillips, M. M. 1976, *Ap. J.*, **206**, 898.
Peimbert, M., and Torres-Peimbert, S. 1981, *Ap. J.*, **245**, 845.
Péquignot, D. 1984, *Astr. Ap.*, **131**, 159.
Perez, E., and Fosbury, R. A. E. 1986, these proceedings.
Phillips, M. M., Charles, P. A., and Baldwin, J. A. 1983, *Ap. J.*, **266**, 485.
Phillips, M. M., and Malin, D. F. 1982, *M.N.R.A.S.*, **199**, 905.
Reichert, G. A., Mason, K. O., Thorstensen, J. R., and Bowyer, S. 1982, *Ap. J.*, **260**, 437.
Rose, J. A., and Searle, L. 1982, *Ap. J.*, **253**, 556.
Rose, J. A., and Tripicco, M. J. 1984, *Ap. J.*, **285**, 55.
Sandage, A. 1978, *A. J.*, **83**, 904.
Sandage, A., and Tammann, G. A. 1981, *A Revised Shapley-Ames Catalog of Bright Galaxies* (Washington, DC: Carnegie Institution of Washington).
Sargent, W. L. W., Young, P. J., Boksenberg, A., Shortridge, K., Lynds, C. R., and Hartwick, F. D. A. 1978, *Ap. J.*, **221**, 731.
Schmidt, M., and Green, R. F. 1983, *Ap. J.*, **269**, 352.
Shuder, J. M. 1980, *Ap. J.*, **240**, 32.
Shuder, J. M., and Osterbrock, D. E. 1981, *Ap. J.*, **250**, 55.
Stauffer, J. R. 1982, *Ap. J.*, **262**, 66.
Terlevich, R., and Melnick, J. 1985, *M.N.R.A.S.*, **213**, 841.
Ulrich, M.-H., et al. 1985, *Nature*, **313**, 747.
Véron, P. 1979, *Astr. Ap.*, **78**, 46.
Véron, P., Lindblad, P. O., Zuiderwijk, E. J., Véron, M. P., and Adam, G. 1980, *Astr. Ap.*, **87**, 245.
Véron, P., Véron, M.-P., Bergeron, J., and Zuiderwijk, E. J. 1981, *Astr. Ap.*, **97**, 71.
Véron-Cetty, M.-P., and Véron, P. 1984, *ESO Scientific Report*, No. 1.
Weedman, D. W. 1976, *Ap. J.*, **208**, 30.
Weedman, D. W. 1977, *Ann. Rev. Astr. Ap.*, **15**, 69.
Whittle, M. 1985a, *M.N.R.A.S.*, **213**, 1.
Whittle, M. 1985b, *M.N.R.A.S.*, **213**, 33.
Young, P. J., Westphal, J. A., Kristian, J., Wilson, C. P., and Landauer, F. P. 1978, *Ap. J.*, **221**, 721.

DISCUSSION

SHAPIRO: The distinction between "low-luminosity" versus "high-luminosity" AGNs <u>may</u> be explained by gas-starved versus gas-rich accretion onto a black hole, as you suggest. Alternatively, it may be explained by accretion onto a low-mass versus high-mass black hole. Can your observations distinguish between these two possibilities? For example, can you measure velocity dispersions from your line profiles and use the Keplerian law to determine the underlying black hole mass?

FILIPPENKO: As you say, low-luminosity AGNs may simply be galaxies which have low-mass black holes, and I didn't mean to imply that they necessarily have massime holes and low accretion rates. Unfortunately, the <u>narrow</u> lines are produced in such a large volume taht normal stars contribute substantially to the potential, so it will be difficult to determine the mass of the black hole itself from them. The <u>broad</u> components seem much more promising, however, and I am currently working on this problem.

WANDEL: In light of your two-Gaussian fit to Hα and other lines, would you suggest a two-region scheme (one of high velocities and high density, and the other contributing the narrow lines with low velocity and density), or a continuous variation between these two extremes, extended in radius? If the clouds are pressure-supported, the wide range in critical density implies a wide range in the radius of the emission-line region.

FILIPPENKO: The observations are usually not good enough to discriminate between these two possibilities. In most objects it is difficult enough to even <u>detect</u> the broad Hα, and its profile is impossible to measure quantitatively due to interference from the [NII] line, narrow Hα, and starlight. Some galaxies, however, do seem to show a more-or-less continuous range in density and velocity. It is almost certainly true that in all cases the different components come from clouds at different distances from the exact nucleus.

OSTERBROCK: You have excellent data on NGC 1052, with considerably better signal-to-noise ratio than Koski and I had in 1976. As you know, Keel and Miller have also worked on this galaxy more recently. The very weak broad components of Hα emission you find in M87 and NGC 1052 are very interesting, and indeed suggest, as you said, that they have Seyfert-like properties. Can you say anything about the widths of the narrow emission lines in NGC 1052, after correcting for the integrated stellar absorption-line spectrum, and about the relative strength of [OIII] $\lambda 4363$?

FILIPPENKO: In a 2"x4".1 aperture (PA=30°) centered on the nucleus, [OI] $\lambda 6300$ has a FWHM of 700-800 km/s while the red [NII] and [SII] lines have FWHM 400-500 km/s. Since the rotation curve at this PA exhibits a sharp gradient across the nucleus, significant portions of the quoted widths are of rotational origin. Relative differences in the <u>intrinsic</u> widths of different lines are therefore probably even larger. The blue spectrum hasn't been analyzed yet, so I can't say anything about [OIII] $\lambda 4363$.

PISMIS: If the phenomenon giving rise to the activity in a nucleus is not isotropic (say it is bipolar, or any other directional one), we would expect the emission line widths to depend on the geometry of the outgoing matter. For example, if the orientation of the ejecta in close to normal to the line of sight we would observe narrow emission lines. Such galaxies - narrow line galaxies - may then not be different physically form other nuclei with wider emission lines.

FILIPPENKO: It is true that geometry must play an important role in such models, but the number of free parameters is often so large that one can almost always devise a model that explains the observed properties of a <u>given</u> object. The model may fail, on the other hand, when all the data for a large sample of objects are considered. For example, the difference between type I and type 2 Seyfert galaxies cannot simply be due to the angle between our line of sight and some sort of "broad-line jet", since differences in the large-scale (kpc) radio properties would remain unexplained.

SCHUTZ: If these nuclei are simply QSOs that have run out of the fuel available to them in their galaxies, then one might expect that in interacting galaxies some refueling might occur. Have you looked at nearby interacting galaxies with similar precision?

FILIPPENKO: Others (such as Keel and Dahari) have, but I haven't. My impression is that interacting galaxies are slightly more likely to show activity, but certaily most local nuclei of this type are not preferentially in interacting galaxies.

VAN DER LAAN: I have a comment and a question. Your beautiful results demonstrate the great importance of spectral dynamic range. In radio _and_ optical astronomy, improved sensitivity and spectral resolution yield optimal information only to the extent that instrumental stability and data handling techniques provide maximal dynamic range. That is now being achieved. How does your estimated proper density of "mildly active nuclei" at the present epoch compare with the proper density upper limit at $Z \simeq 3$ estimated for quasars by Schmidt?

FILIPPENKO: My initial subsample of 75 galaxies was chosen to include many classical Seyferts and the best candidates for low-luminosity active nuclei. Because of this extreme bias, I have resisted the temptation to make comparisons until a greater degree of uniformity is achieved in the survey.

CONTINUITY IN THE OBSERVED PROPERTIES OF QSOs, HIGH-REDSHIFT EMISSION-LINE RADIO GALAXIES, BL LAC OBJECTS, N SYSTEMS AND SEYFERT GALAXIES, AND POSSIBLE INTERPRETATIONS

G. Burbidge
Center for Astrophysics and Space Sciences
University of California, San Diego
La Jolla, California 92093, and
Kitt Peak National Observatory[+]
Tucson, Arizona

ABSTRACT

The evidence for continuity between the observed properties of QSOs and the active nuclei of galaxies of various types is reviewed. It is concluded that the continuity argument is compatible with many of the observations, provided (i) that the idea of directed bulk relativistic motions in many objects which show apparent superlight velocities or very large rapid flux variations is accepted, and (ii) that all of the evidence for the physical associations of QSOs and galaxies, and pairs of QSOs, with very different redshifts, is ignored. Since bulk motions with $v \geq 0.99c$ are very difficult to maintain in the presence of any appreciable amounts of diffuse gas, and since much of the evidence for non-cosmological redshifts is compelling, a new approach is outlined. It is proposed that many QSOs and BL Lac objects are gravitationally stable coherent objects which are ejected from galaxies comparatively closeby. They radiate assymetrically, in a cone directed away from their direction of motion. They must be ejected from galaxies with velocities which range from very small values for QSOs with low redshifts which appear to lie in galaxies, to values in the range 0.1c-0.9c for the higher redshift QSOs. Some BL Lac objects may be embedded in galaxies, while some may be coming towards us.

It is suggested that models of this type which imply reduced distances and lower luminosities for the QSOs can be used to explain the full range of phenomena seen. Physical models need to be worked out, since the only mechanism which has so far been proposed to eject coherent objects from galaxies is the slingshot mechanism which appears to give rather low velocities of ejection.

[+]Operated by the Association of Universities for Research in Astronomy Inc., under contract with the National Science Foundation.

INTRODUCTION

More than twenty years ago Burbidge, Burbidge, and Sandage (1963) using observational evidence in radio and optical wavelengths, made the case for the idea that violent events occur in the nuclei of many normal galaxies. At that time the arguments were based on mass motions in Seyfert galaxies, the evidence for ejection resulting from detailed studies of velocity fields in comparatively nearby galaxies like M82 and NGC 1275, on radio synchrotron radiation from powerful radio galaxies, and on unique objects like the jet in M87. The powerful radio sources already showed us that very large amounts of energy, up to conservatively 10^{60}-10^{62} ergs, were released in violent events, and while a number of different scenarios were under discussion, it became clear (in the discussion at the Texas meeting in 1963 and at the Solvay Conference in 1964) that the energy released must either be gravitational in origin or be associated with creation. There was clearly no way of reducing the energy requirements of the most powerful sources other than by supposing that they were not at the distances implied by their redshifts, and this, for radio galaxies, appeared unthinkable.

The discovery of the QSOs and the rapid identification of considerable numbers of them in the 1960s, together with the similarity of their emission line spectra with those of Seyfert nuclei and N galaxies, led Sandage (1973) to suppose that N systems could be made up of QSOs embedded in galaxies. The argument that the weakest (least luminous) objects are the low redshift Seyfert nuclei while the most luminous and most energetic are the high redshift QSOs became obvious. This continuity argument is attractive, and is the majority view at present.

Implicit in the idea is that all of the violent events take place in galaxies of stars, as is undoubtedly the case in the bright classical Seyfert galaxies. Thus, the continuity argument requires that extended images corresponding to the galaxies be found in low redshift QSOs, with the angular sizes of the extensions decreasing with redshifts. More than this, however, it is necessary to show that these extended images are due to starlight. Kristian (1973) argued that there was an inverse correlation between the angular diameters of the fuzz around QSOs and the redshifts, which supported the hypothesis. However, in no case were any spectroscopic data available.

CONTINUITY BETWEEN QSOs AND SEYFERT GALAXIES

Early attempts were made to study the fuzz around a few QSOs. In the cases of 3C 48 (Wampler et al., 1975), 3C 249.1 (Richstone and Oke, 1977), and 4C 37.43 (Stockton, 1976), narrow emission lines at the same redshift as the QSOs were detected showing that at least hot gas was present. In the last few years strenuous efforts have been made to detect and study the fuzz around QSOs with the continuity argument assumed.

Three techniques have been used. The first is direct imaging. Gehren et al. (1984) have looked at 17 QSOs, 16 with z in the range 0.044-0.543, and one with z=0.838. They concluded that all of the low redshift QSOs had extended images comparable with galaxies. Hutchings et al. (1984) looked at 75 QSOs with $z \leq 0.63$ and concluded that all but 7 had fuzz. This then was claimed to be compatible with their being galaxies. However, 30% of these objects have $z \leq 0.1$ and are probably not QSOs at all. Malkan et al. (1984) looked at 24 QSOs selected as x-ray sources. They found that 15 with values of z up to 0.4 have fuzz, but all 7 with $z > 0.5$ are unresolved. Malkan (1984) then looked at 23 low redshift QSOs which are strong radio sources. Seventeen are resolved, and Malkan concluded that, provided we assume that the extended images are due to starlight, the radio-quiet QSOs are embedded in spiral galaxies while those which are radio sources are embedded in bright elliptical galaxies.

The claim is then made (cf Malkan 1984) that all QSOs with $z < 0.4$ have extended images which are galaxies.

Is this claim borne out by the spectroscopic evidence?

The second technique is to study the fuzz spectroscopically. Spectroscopy of the fuzz around 0241 + 622 (z = 0.044) has been attempted by Romanishin et al. (1984). While a galaxy may possibly be present the result is hardly convincing. In the most detailed investigation, Boroson and Oke (1984) studied eight QSOs, 3C 48, 3C 249.1, 3C 273, 4C 37.43, 3C 323.1, PKS 2141+174, 4C 31.63 and 4C 11.72, some of which had been studied earlier. The only QSO which unambiguously shows an early-type absorption spectrum in the fuzz is 3C 48 (Boroson & Oke, 1982). All of the others either show a blue continuum with strong emission lines, or a red continuum with weak or no emission lines. MacKenty and Stockton (1984) found stellar absorption features in MK 1014 (z=0.1631), an object which Osterbrock and Dehari (1983) called either a QSO or a Seyfert galaxy. Balick and Heckmann (1983) studied four QSOs, one of which was 3C 48. In the core of 0351+026 (z=0.036) they found one absorption feature, in 1059+730 (z=0.089) they found an early type absorption spectrum, and in 0845+378 (z=0.307) they found questionable absorption features.

Thus, it cannot be said that stellar absorption lines expected in the outer parts of spiral galaxies have been found in many cases. In the only two cases for which there is good evidence, 3C 48, and 1059+730, the absorption is to be attributed to early type stars. The question must also be asked as to whether or not we can clearly draw a line between Seyfert galaxies and QSOs with small redshifts. Practically all of the QSOs which have had this fuzz studied spectroscopically have small enough redshifts so that some might classify them as Seyfert galaxies and not QSOs. This leads us to ask the question, what do we know about the stellar properties of Seyfert galaxies outside their nuclear regions? In fact, it appears that very little work has been done on this problem. It may be that early type spectra are the rule rather than the exception in Seyfert galaxies but we do not know. A detailed study of the stellar population in the outer part of a large sample of Seyfert galaxies where the spiral structure can clearly be seen is required.

The third approach has been to study the faint galaxies lying at small angular distances from the QSOs. Several studies have been made of the faint galaxies near to mostly low redshift QSOs and in a number of cases it has been found that the redshifts of the faint galaxies are approximately the same as those of the QSOs. This type of work by Stockton (1978) and more recently by Heckman et al. (1984), and Yee and Green (1984), has led to very strong statements concerning the cosmological nature of the redshifts, particularly by Heckman et al. However, the work of Yee and Green, who have looked at a range of z between 0.05 and 2.05 and who find that low redshift QSOs have more galaxies nearby than those in control fields $\sim 1^\circ$ away, is the most convincing.

All of these results do to some considerable extent support the continuity argument though the spectroscopy of the fuzz gives the weakest result.

Entirely against this simpler picture is the evidence obtained over several years of the association of QSOs with large redshifts with very bright galaxies with very small redshifts (cz < 3000 km sec^{-1}). We shall discuss this in detail later.

Of particular interest here is the recent discovery by Huchra et al. (1985) of a galaxy in the Zwicky Catalogue, 2237+0305, which has a redshift z=0.0347, but a nucleus which is a genuine QSO with z=1.697. Here we have a <u>high</u> redshift QSO with fuzz around it which is a genuine galaxy with a low redshift. This is not at all what is expected on the conventional continuity picture, since at z=1.7 the galaxy would not be visible. This must either be interpreted as further evidence for local QSOs ejected by galaxies (Burbidge 1985) or, less likely, a rather remarkable gravitational lens effect (Huchra et al. 1985).

CONTINUITY AND THE BL LAC OBJECTS

The BL Lac objects were originally defined as non-thermal sources which show rapid flux variability at radio frequencies, and which show no emission lines characteristic of QSOs in their spectra. It is now known that these objects may be variable in optical and x-ray frequencies also. In practice there is considerable ambiguity about which objects reasonably fall into this category, in part because they do not form a homogeneous group, and in part because there has been a tendency by some to identify almost any star-like radio source which does not show an emission line QSO spectrum as a BL Lac object.

Some well known BL Lac objects have extended images, while others appear to be star-like in appearance. Those with extended images which have been studied spectroscopically and appear to consist of a powerful variable nonthermal source embedded in a galaxy include: BL Lac (z=0.069), Mk 421 (z=0.030), Mk 501 (z=0.034), Ap Libra (z=0.049), OT 546 (z=0.055), 3C 371 (z=0.050), and PKS 0521-36 (z=0.055). 3C 371 was originally listed as a normal radio galaxy. It has been concluded that all of these BL Lac objects are embedded in bright ellipticals. The emission lines seen from the fuzz are weak and narrow.

There are also some BL Lac objects which were originally classified as QSOs, but which have undergone outbursts in which their emission line spectra have disappeared, or have become extremely weak. The prototype is 3C 446 (z=1.404). There is also a group of objects which show the variations associated with the BL Lac phenomenon but in which only one emission line can be seen, leaving the redshift or blueshift ambiguous. (In view of what we shall discuss later, the possibility that these objects are coming towards us should not be discounted.) These objects, and also those which show absorption redshifts, but no emission lines, tend to be morphologically indistinguishable from QSOs. A classical example of this latter type is AO 0235+164 which shows two absorption redshifts (z=0.524 and z=0.852) but no emission lines. This object has been detected in absorption in the 21-cm line at the lower redshift and these 21-cm absorption lines have varied in time. This is one of the most difficult objects to interpret. There is nebulosity adjacent to, but not symmetrically placed about the BL Lac object, and this nebulosity has an emission line redshift of 0.524. The conventional explanation requires that there is a galaxy at z=0.524 responsible both for the emission line redshift and the (variable) absorption line redshift, and that this galaxy is accidentally aligned with a background galaxy in which the BL Lac object is embedded. This hypothetical galaxy must have a cosmological redshift > 0.852. Unless this galaxy gives rise itself to the absorption at z=0.852, there must also be an intervening cloud aligned with the two galaxies to give the second absorption line system.

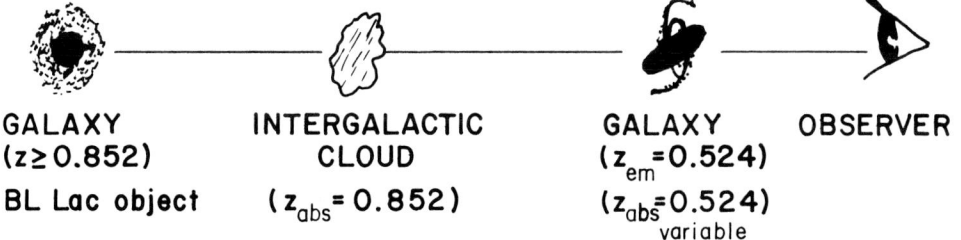

GALAXY	INTERGALACTIC	GALAXY	OBSERVER
(z ≥ 0.852)	CLOUD	(z_{em}=0.524)	
BL Lac object	(z_{abs}= 0.852)	(z_{abs}=0.524) variable	

This, to put it mildly, is a highly unlikely configuration.

We can briefly summarize the continuity arguments as they relate to BL Lac objects. These objects are clearly related to galaxies, since some of them, at least, lie in elliptical galaxies, though the evidence is confined to those at very small redshifts. Some of them are also clearly related to QSOs and can transform back and forth depending on the degree of the outbursts.

CONTINUITY AND JETS

Jets, implying directivity and the channeling of energy, are frequently seen in radio sources (Bridle and Perley, 1984). Their existence in both QSOs and radio galaxies implies a continuity in their properties at least for the subset of QSOs that are strong radio sources. Optical jets are much rarer than radio jets but they are seen in M87 and in 3C 273. Several other objects are now known to show optical jets; PKS 0521-36 is a radio galaxy (N system) (z=0.055) with a BL Lac object embedded in it (Ulrich 1981, Danziger et al. 1983). Keel (1985) has studied the optical jet investigated by Danzinger et al., and has concluded that it is similar to the jet in M87. In addition to this, the radio galaxies 3C 31 (z=0.017), and 3C 66B (z=0.0215) also show optical jets. It is of some interest that 3C 66B is very close to 3C 66A which is another BL Lac object.

Continuity in the scales of the jets in the sense that the largest and the most powerful come from the most energetic sources is not established. In radio frequencies, even under the cosmological redshifts assumption, the energies in jets in radio galaxies (at comparatively low redshifts) and QSOs (at high redshifts) are comparable.

CONTINUITY AND APPARENT SUPERLIGHT VELOCITIES

Among the QSOs and the active nuclei of galaxies there is a subset of objects which show variations in radio flux and the small scale structure of the components. The common explanation which has been put forward for this is that we are seeing bulk relativistic motions. The classical QSOs in which the so-called superlight velocities have been detected through changes in angular sizes of radio structures include 3C 273, 3C 345, 3C 279, 3C 454.3, 3C 179, 4C 39.25 and NRAO 140 (Kellermann, 1984, Cohen, 1984). Also the peculiar Seyfert galaxy 3C 120 (z=0.033) and BL Lac (z=0.069) show the same effect. The changes in angular size converted into length scales by using the cosmological redshifts show that in x years the structure has moved, or changed in size by y light years where $y/x=\gamma>1$. For different objects γ can range up to about 20 (for 3C 279). Thus, the apparent speed is many times c. For 3C 120, $\gamma = 4$, (Walker et al. 1982).

Since $\gamma = (1-\beta^2)^{-\frac{1}{2}}$ where the velocity = βc, this explanation requires extreme relativistic velocities. For $\gamma = 20$, $\beta = 0.995$, and for $\gamma = 4$, $\beta = 0.968$.

Rapid flux variability which sets a size limit for the radiating source of cτ, where τ is the time over which the variability has occurred, shows that objects at their cosmological redshifts, have exceedingly small angular dimensions and hence very large surface brightnesses. Early in the investigations synchrotron calculations for such cases led to very strong magnetic fields, exceedingly high energy densities of radiation and particles, and the corresponding Compton catastrophe. The only seemingly viable way to avoid these problems was to either bring the objects closer, i.e. accept the concept of non-cosmological redshifts, or argue that these sources are expanding relativistically with bulk motions comparable to those required for the super-luminal radio sources. The latter solution which was proposed originally in 1966 by Rees, is despite its difficulties, the one most preferred at present.

While I am not aware of any objects of small redshift which require that we invoke bulk relativistic motion to explain the flux variations, the cases of 3C 120 and BL Lac together with the QSOs provide a continuity argument for superluminal motions in radio sources.

CONTINUITY BETWEEN RADIO GALAXIES AND QSOs

Originally it was the very powerful energetics of radio sources which led to the realization that energies of the order of 10^{60} ergs or greater were being generated in the form of relativistic particles. When the QSOs were discovered shortly after this, their very high luminosities combined with assumed lifetimes of the order of $10^6 - 10^7$ years led to comparable energies, so that it could be argued that since the objects had similar energy contents there was continuity, and that the phenomena were related.

However, with the realization that the galaxies and QSOs were morphologically different, and that the bright radio galaxies preferentially lie in rich clusters, and the QSOs do not, and that the majority of the QSOs are not powerful radio sources, the differences outweighed these similarities, so that the continuity arguments were weakened.

Now the argument may have changed again. The conviction of many astronomers that QSOs lie in galaxies, might suggest that the only difference is one of the possibly different evolutionary histories of spirals (containing QSOs) and ellipticals.

More important is the fact that Spinrad and his colleagues have begun to show that many of the very faint 3C radio galaxies are objects of high redshift (3C 256 has $z=1.82$) with strong emission lines in their spectra. While there are some differences between the spectra of QSOs and of these "galaxies," the range in redshifts is comparable, and the only real observational difference is that these objects are fainter and more diffuse than QSOs.

SUMMARY OF CONTINUITY ARGUMENTS

We have described in some detail the evidence for continuity in the observed properties of low luminosity and high luminosity objects assuming continuity in the nature of the redshifts. If all of the active centers lie in galaxies, the model of increasing energetic events ranging from the classical Seyferts to the higher luminosity QSOs is plausible. This model also requires that a fraction of all of these objects have outbursts which give rise to highly directed relativistic bulk motions with values of $\beta \geq 0.99$. This scheme also necessarily implies that there is a much larger population of objects which are violently variable and are emitting radiation in other directions, i.e. not only must it be argued that the compact radio sources are beamed, but that outbursts in optical, infrared and x-ray luminosities of the variable objects are also highly directional.

What else does acceptance of the continuity argument entail?

(a) It requires that galaxies underlie all QSOs and BL Lac objects. While the evidence for extensions around low-redshift QSOs and some BL Lac objects is strong, the spectroscopic evidence demonstrating that we are looking in general at galaxies of stars is very much weaker for the QSOs. The best case for a stellar absorption spectrum is in the fuzz associated with 3C 48. Here the spectrum is not that of a normal stellar population, but of a much younger population (A type).

(b) Continuity requires that bulk relativistic motions are commonplace, and this to my mind is a real difficulty. First, there is no independent evidence anywhere in astrophysics for such high velocities of bulk material. The highest velocities which can be directly measured with some confidence are velocities of some thousands of km/sec, i.e., 0.01-0.03c in displacements in Seyfert nuclei, and velocities of the order 0.1-0.2c in some broad-absorption-line QSOs. We also have the remarkable case of SS 433 (Margon 1984) where there is unequivocal evidence of velocities of 0.26c. While the existence of highly relativistic jets has been widely promoted, based on the apparent existence of small scale superluminal motions, de Young (1984) for example, has cast doubt on the existence of relativistic jets in extended radio sources.

As was pointed out by Jones and Burbidge (1973) highly relativistic bulk motions will almost instantaneously be slowed down by interaction with gas of only moderate density in the source. It is inconceivable to me that such low density gas is not present in the path of the ejected cloud.

(c) The continuity argument, however appealing it may appear, requires us to discount all of the evidence that some QSOs with high redshifts are ejected from comparatively nearby galaxies. Very good evidence that MK 205 (z=0.070) and NGC 4319 are physically connected has recently been shown by Sulentic (1983) following a long history of debate about this connection. The apparent association of many QSOs and bright galaxies found by Arp and others have been shown to be statistically significant (Burbidge 1979, 1981). Two examples of at least three QSOs very close to bright galaxies are now known - NGC 1073 with 3 QSOs with very different redshifts within 2' of its center (Arp and Sulentic 1979; Burbidge et al. 1979), and NGC 3842, with 3 QSOs within 90" of its center (Arp 1984a; Arp 1984b), and it is very hard indeed to attribute these to accident. More recently the accidental discovery of 2E 0104.2 + 3153, a large redshift QSO only 10" from a low redshift elliptical galaxy (Stocke et al. 1984) and the accidental discovery of a QSO with z=1.7 lying $\leq$ 0.3" from the center of the galaxy 2237 + 0305 (Huchra et al. 1985) both provide strong evidence for physical association (Burbidge 1985).

In addition to the evidence that QSOs with large redshifts are, in a number of cases, associated with comparatively nearby galaxies with small redshifts, we also have the curious situation concerning close pairs of QSOs. According to the conventional wisdom, close pairs if they have precisely the same redshifts are evidence for gravitational lensing. If they are not quite so close they are members of superclusters, and if they are close together and have different redshifts they are assumed to be accidental configurations and are ignored. Burbidge, Narlikar and Hewitt (1985) have recently made a study of all pairs of QSOs in the literature with angular separations of 2' or less. Thirty such pairs are now known. If we restrict ourselves to those pairs for which one member is a radio source, and they are both no fainter than 18.5 (V), six pairs are known. Taking into account the search procedures, Burbidge et al. have concluded that only 0.7 pairs would be expected by chance. The Poisson probability of obtaining six successes with a mean of 0.7 is 1.3×10^{-4}. Thus, it appears highly probable that the majority of the pairs with different redshifts are physically associated.

A NEW APPROACH

How can we reconcile all of the observational evidence? On the one hand we have the continuity arguments, and on the other the evidence that galaxies and QSOs with very different redshifts are associated. For a new approach the key assumption that I shall make is that extremely relativistic bulk motions do not take place. Thus, we shall suppose that none of the objects which appear to show superluminal motions are at the distances given by their redshifts. We will assume that they are local QSOs ejected from galaxies lying within 100-200 Mpc, as are the QSOs apparently associated with bright galaxies.

We shall suppose that most QSOs are coherent self-gravitating objects ejected from the nuclei of galaxies sometimes at high speeds. They are galactic cannonballs.

To avoid the well known redshift-blueshift problem we assume that the line spectrum arises only in the tail of the object, i.e., we assume that we are only seeing the QSOs moving away from us. This idea was first proposed by Strittmatter (in Burbidge & Burbidge 1967) to explain the absence of blue shifted QSOs. More recently Hoyle (1980) and Narlikar and Subramanian (1983) have taken up this hypothesis. For only redshifts to be seen the semi-vertical angle of the backward emitting cone must be

$$\Theta = \cos^{-1}\left\{\frac{1-(1-\beta^2)^2}{\beta}\right\}.$$

One of the first points to be made concerning this scenario is that the ejected objects do not need to have anything like the very high speeds required for relativistic bulk motion in the conventional picture. Since the redshifts z for the ejected objects are related to β by the relation

$$\beta = \frac{(1+z)^2 - 1}{(1+z)^2 + 1}$$

values of β between about 0.1 and 0.9 will cover the full range of redshifts seen in QSOs. In fact, the well known drop off in redshifts beyond $z \simeq 2.5$ may well be understood in this picture as due to the difficulty of accelerating coherent objects to high values of β.

If QSOs are in general coherent objects moving away from us with the radiation directed down their tails, what do these objects look like when they are coming towards us. One obvious possibility is that they are seen as BL Lac objects. However, since BL Lac objects are rare compared with QSOs, this idea cannot be correct if the front and back hemispheres have comparable luminosities. It is more likely that the objects when seen coming towards are intrinsically much fainter than the typical QSOs, and thus they have not been detected. What then are the BL Lac objects?

Those objects which appear between outbursts to be QSOs with emission line spectra, e.g. 3C 446, must be objects which are moving away from us with flares in the continuum which appear from time to time. If they show a series of absorption redshifts, then these must be due to gas ejected in the tail of the QSO which then slows down and cools by interaction with the intergalactic medium. A similar argument will explain all of the absorption in QSOs. It is due to gas which is ejected from the objects in a series of discrete events, each shell having a characteristic redshift which will decrease with time as the gas slows down. Many hundreds of shells are indicated in some objects to explain the $L\alpha$, $L\beta$ systems. In the conventional picture these latter systems are attributed to a hitherto unknown component in the universe. In this picture all of the absorption ultimately originates in the QSOs.

What about the BL Lac objects and QSOs which appear to lie in galaxies. For all of the BL Lac objects in this category the evidence for galaxies appears strong, and the galaxies look normal. We have no evidence concerning the actual velocities of the BL Lac objects in these galaxies since no lines are seen. Thus, they may be objects ejected from the galaxy which move towards us nearly along the line of sight, or they may be parts of the galaxies at rest with respect to the galaxies. The critical issue to my mind is whether or not the distance based on the redshift of the galaxy leads to apparent superluminal motion. If it does, then I would argue that such an object must have been ejected towards us from the galaxy so that it lies much closer to us than the galaxy. A specific case of some interest is 3C 66A, which may well have been ejected from the radio galaxy 3C 66B. In the case of 3C 120 which shows apparent superluminal motion with a redshift for the galaxy of 0.033, we must suppose that the compact radio source in the BL Lac nucleus has been ejected towards us. The same argument must apply to BL Lac itself.

For the QSOs lying apparently in galaxies we have a more complex problem. In the case of 2237 + 0305, the QSO must have been ejected from the galaxy in a direction away from us. In the cases of the QSOs which lie close to faint galaxies with nearly the same redshifts, we must conclude that they are only moving very slowly away from these galaxies with velocities $\sim 10^3$ km sec^{-1}. A test of this would be to measure accurately the redshifts of the "galaxies" in which they appear to be embedded. If our hypothesis is correct we would not expect the redshifts from the stars, if they are there, to agree exactly with the QSO redshifts. For QSOs embedded in galaxies at exactly the same redshifts it would be necessary to argue that they lie practically at rest with respect to the parent systems. However, any differences should be in the sense that the QSO redshift is slightly greater than the galaxy redshift.

If this model is correct, no objects of this type should show apparent superluminal motion. Also, none of the QSOs which are apparently associated with nearby bright galaxies should show apparent superluminal motion when they are assumed to lie at the distances of the bright galaxies.

This schematic model has a number of attractive features. The model suggests that the large redshift QSOs are not at great distances but are objects ejected from comparatively nearby galaxies. Small redshift QSOs apparently embedded in galaxies may indeed lie in those galaxies at cosmological distances, or only have very small relative velocities compared with the velocities given by the redshifts. Critical cases would be any QSOs which could be definitely shown to lie in galaxies at the same redshifts and could be shown to have superluminal motions.

In this model the absorption in QSOs is all intrinsic to the objects and is due to ejected gas.

The model is incomplete in the sense that we do not yet have a theory apart from the slingshot mechanism (Saslaw et al. 1974) to explain how the objects are ejected from the nuclei, nor do we understand why the objects are radiating assymetrically. However, the existence of very highly relativistic bulk motions which are required in the normal picture are also unexplained in terms of a realistic physical model.

How does this model look from an energetic standpoint? Very high luminosities are no longer indicated, but the typical luminosity of a QSO may be 10^{42}-10^{43} erg sec^{-1} rather than values of 10^{47}-10^{48} ergs sec^{-1} which are normally used for highly energetic objects. If some low redshift QSOs really lie in galaxies at cosmological distances then the tail of the luminosity distribution may extend to higher luminosities than 10^{43} erg sec^{-1}. In this picture the anomalous QSO 3C 273 is naturally assumed from its position to have been ejected from a galaxy in the Virgo cluster.

There is a large amount of kinetic energy associated with these rapidly moving objects. However, we have no estimate of their masses. For typical velocities a few tenths of the speed of light, and for masses in the range 10^3-10^5 (a pure guess) their kinetic energies will lie in the range 10^{56}-10^{58} ergs.

ACKNOWLEDGEMENTS

I wish to thank A. Hewitt for much help in the preparation of this paper, and also research support from the National Science Foundation through grant NSG AST84-17650.

REFERENCES

Arp, H.C. and Sulentic, J.W., 1979, Ap.J., 229, 496.
Arp, H.C., 1984a, Ap.J., 283, 59.
Arp, H.C., 1984b, Astron. Ap., 139, 240.
Balick, B. and Heckman, T.M., 1983, Ap.J. (Letters), 265, L1.
Boroson, T.A. and Oke, J.B., 1982, Nature, 296, 397.
Boroson, T.A. and Oke, J.B., 1984, Ap.J., 281, 535.
Bridle, A. and Perley, R., 1984, Ann. Rev. Astron. Ap., 22, 319.
Burbidge, E.M., Junkkarinen, V., and Koski, A.T., 1979, Ap.J. (Letters), 233, L97.
Burbidge, G.R., Burbidge, E.M., and Sandage, A.R., 1963, Rev. Mod. Phys., 35, 947.
Burbidge, G.R. and Burbidge, E.M., 1967, Quasi-Stellar Objects, (W.H. Freeman & Co., San Francisco).
Burbidge, G.R., 1979, Nature, 282, 451.
Burbidge, G.R., 1981, Annals, N.Y. Acad. Sci., 375, 123.
Burbidge, G., 1985, A.J., in press.
Burbidge, G., Narlikar, J.V., and Hewitt, A., 1985, submitted to Nature.
Cohen, M.H. and Unwin, S.C., 1984, I.A.U. Symp. 110, VLBI and Compact Radio Sources, Eds. R. Fanti, K. Kellermann, G. Setti, (D. Reidel: Dordrecht), p. 95.
Danziger, I.J., Bergeron, J., Fosbury, R.A.E., Maraschi, L., Tanzi, E.G., and Treves, A., 1983, M.N.R.A.S., 203, 565.
DeYoung, D., 1984, Science, 225, 677.
Gehren, T., Fried, J., Wehinger, P.A., and Wyckoff, S., 1984, Ap.J., 278, 11.
Heckman, T.M., Bothun, G.D., Balick, B., and Smith, E.P., 1984, A.J., 89, 958.
Hoyle, F., 1980, preprint.
Huchra, J., Gorenstein, M., Kent, S., Shapiro, I., Smith, G., Horine, E., and Perley, R., 1985, A.J., in press.
Hutchings, J.B., Crampton, D., and Campbell, B., 1984, Ap.J., 280, 41.
Jones, T.W. and Burbidge, G.R., 1973, Ap.J., 186, 791.
Keel, W., 1985, preprint.
Kellermann, K.I., 1984, preprint.
Kristian, J., 1973, Ap.J. (Letters), 179, L61.
MacKenty, J.W. and Stockton, A., 1984, Ap.J., 283, 64.
Malkan, M.A., 1984, Ap.J., 287, 555.
Malkan, M.A., Margon, B., and Chanan, G.A., 1984, Ap.J., 280, 66.
Margon, B., 1984, Ann. Rev. Astron. Ap., 22, 507.
Osterbrock, D.E. and Dehari, O., 1983, Ap.J., 273, 478.
Richstone, D.O. and Oke, J.B., 1977, Ap.J., 213, 8.
Romanishin, W., Ford, H., Ciardullo, R., and Margon, B., 1984, Ap.J., 277, 487.
Sandage, A., 1973, Ap.J., 180, 687.
Saslaw, W.C., Valtonen, M.J., and Aarseth, S.J., 1974, Ap.J., 190, 253.
Stocke, J.T., Liebert, J., Schild, R., Gioia, I., and Maccacaro, T., 1984, Ap.J., 277, 43.

Stockton, A., 1976, Ap.J. (Letters), 205, L113.
Stockton, A., 1978, Ap.J., 223, 747.
Subramanian, K. and Narlikar, J.V., Quasars and Gravitational Lenses, Proc. Liege Conf., June, 1983.
Sulentic, J.W., 1983, Ap.J. (Letters), 265, L49.
Ulrich, M.H., 1981. Astron. Ap., 103, L1.
Walker, R.C., Seielstad, G.A., Simon, R.S., Unwin, S.C., Cohen, M.H., Pearson, T.J., and Linfield, R.P., 1982, Ap.J., 257, 56.
Wampler, E.J., Robinson, L.B., Burbidge, E.M., and Baldwin, J.A., 1975, Ap.J. (Letters), 198, L49.
Yee, H.K.C. and Green, R.F., 1984, Ap.J., 280, 79.

DISCUSSION

FILIPPENKO: You mentioned that galaxies close to QSOs and having the same redshift are simply systems in which the galaxy ejected a QSO with a low relative velocity. But Heckman et al. (1984) recently showed that 95% of all galaxies with projected distances of less than 50 Kpc from QSOs have the same redshift as the neighboring QSOs. How can this be consistent with your interpretation?

BURBIDGE: It must simply mean that low velocities of ejection with values comparable to the random motion between galaxies occur. The redshifts of the QSOs are never <u>exactly</u> equal to those of the adjacent galaxies. To reiterate, if we treat the evidence for physical association between galaxies with very different redshifts from the QSOs as firm as I do; and if we treat the other evidence as equally firm as you do, then the the only model left, if the spectral shifts are Doppler shifts is the one I have proposed.

SCHMIDT: Geoffrey, you will not be surprised that I bring up a problem for your STAR WARS scenario based on quasar statistics: the steep counts of quasars in Euclidian space require a non-uniform space distribution

BURBIDGE: As always I respect your arguments. However I am attempting to explain evidence of a different kind which I feel is compelling and this leads me in the direction I described. It is largely a matter of taste as to which evidence is given the highest credibility.

VAN DER LAAN: Let me underline dr. Burbidge's scepticism on the propagation of particle beams and of gaseous strems. In powerful radio galaxies very straight channels for energy transport extend over hundreds of kpc's. So many hazards of instabilities threaten beams, and streams described hydrodynamically, that the fantastically effective long distance transport observed is not credibly explained theoretically. Moreover, in complex sources related to orbiting double galaxies and to fast cluster members (3C 449 and 3C129 for example) ballistic trajectories of relativistic particle sources explain the morphology convincingly, as shown by Icke. Both sets of phenomena call for the formation of high speed (but non-relativistic) compact clouds which travel in ballistic orbits for up to 10^8 years, converting kinetic energy to particle energy along the way and at the end. Theoreticians need to construct plausible cloud-cannons.

FILIPPENKO: CCD images under excellent seeing conditionts have recently shown that Huchra's QSO in the low-redshift galaxy is composed of two images separated by about 1".4. Can you comment on this? Also, what will you say if the spectra turn out to be identical in both redshift and overall appearance?

BURBIDGE: If this is true it gives support to the gravitational lens hypothesis. This would also be the case if the spectra of two components were identical. However, I have become very suspicious of observers who first make interpretations and then find evidence to support them. In any case there is much evidence for non-cosmological redshifts which has remained good for many years.

SPECTRAL PROPERTIES OF BLAZARS OBSERVED AT ULTRAVIOLET AND X-RAY FREQUENCIES

A.Treves[1], G.Ghisellini[2], L.Maraschi[1] and E.G.Tanzi[3]
1) Dipartimento di Fisica, Università di Milano, Italy
2) International School for Advanced Studies, Trieste, Italy
3) Istituto di Fisica Cosmica CNR, Milano, Italy

ABSTRACT. The large majority of all Blazars observed in the far ultraviolet (26 objects) are considered. Broad band spectral data are collected choosing measurements close in time and corrected for contamination from the surrounding galaxy. It is found that the overall spectrum of this class of objects steepens substantially between radio and infrared and between infrared and ultraviolet frequencies but not between ultraviolet and X-ray frequencies. The spectral slopes in different bands are correlated from X-ray to IR frequencies. The 6 X-ray selected objects contained in the sample show spectra which are flatter than average in all bands. The average X-ray luminosity of the whole sample and that of the X-ray selected group are the same. It is suggested that relativistic beaming may enhance the radio but not the X-ray emission of radio selected blazars.

1. INTRODUCTION

A considerable number of Blazars (26) have been observed in the far ultraviolet (1200-3000 A) by means of the IUE satellite. This body of data, directly accessible through the satellite archives, and in great part already published, is sufficiently large and homogeneous as to allow a study of the ultraviolet spectral properties of these objects as a class. In particular the ultraviolet band has, with respect to the optical one, the important advantage of a lesser contamination from the surrounding galaxy, which enables a reliable determination of the non thermal continuum. The sensitivity threshold of IUE corresponds to $m_B < 16$. For these bright objects the spectral information from radio to X-ray frequencies is practically complete and in many cases quasi simultaneous observations in different frequency bands are available. It is therefore possible to collect reliable broad band spectral parameters for the whole sample (section 2).
The sample includes 6 objects which appear in the X-ray catalogue of the HEAO1 A1 observations of Wood et al. (1984), at a count level (>7 c/s) and at galactic latitude, b >20°, for which the catalogue is more than 90% complete. They are : 0323+022, 0548-322, 1101+384 (MK

421), 1218+304, 1652+398 (MK 501), 2155-302. With the only exception of MKN 421 the objects are contained also in the "complete" HEAO1 A-2 X-ray survey of Piccinotti et al (1983). These 6 objects will be called X-ray selected in that they have, or could have been, discovered from X-ray observations, had they not been known in advance from the observations at other wavelengths.

In section 3 the distributions of spectral indices in various bands are given and the correlations between them are examined. The monochromatic luminosity distributions at different frequencies are then considered (section 4). In particular it appears that the X-ray selected objects have notably different spectral properties with respect to the whole group, but a similar average X-ray luminosity.

The overall spectral behaviour of BL Lacs is discussed in section 5, first comparing with QSOs and then considering in particular the significance of the different behaviour of radio and X-ray selected objects. An interpretation of the data in terms of emission models is given in Section 6. The main results are summarized in section 7.

Preliminary results on the subjects discussed in this paper were presented in Maraschi et al. 1983b, 1984.

2. BLAZARS OBSERVED IN THE FAR ULTRAVIOLET

BL Lac objects, Optically Violently Variables (OVVs) and Highly Polarized Quasars (HPQs) are usually grouped together under the denomination of Blazars, which eliminates the somewhat ambiguous issue of the strength of the emission lines as a classification criterion.

As a main reference list of Blazars we have considered the one of Angel and Stockman (1980). To this, highly polarized quasars from the work of Moore and Stockman (1984) have been added, plus a small number of newly discovered BL Lac objects. The resulting list of objects has been checked with the IUE archives up to Dec.83. UV spectra have been found for the 26 objects reported in Table 1, plus for other 5, which have not yet been included in the analysis. These are PKS 0736+017, 1510-089, 1641+399 (3C 345) 1845+797 (3C.390.3) and NGC 1275.

In most cases the analysis of the ultraviolet spectra found in the literature was used in the compilation. Direct analysis performed in several cases with the criteria mentioned below did not yield significant differences. In the cases of 0829+046, 1308+32, 1514-24, for which no published information could be found, the spectra were retrieved from the archives and analyzed with the procedure described in Maraschi et al, 1985b. Spectral indices were obtained by best fitting the spectrum with a single power law in the two wavelength ranges of sensitivity of IUE (SW, 1200-1950 A, and LW, 1900-3000 A), whenever both were available. In the case of 0716+71 only spectral indices for the two individual ranges were published. The spectra were therefore reanalyzed to give the combined spectral index.

Table 1 lists the objects in order of right ascension together with one of the catalogue designations.

Column 2 gives the redshifts of the objects. Those based either on a single line or on the size of the associated nebulosity are followed by a star. Those based on absorption lines by presumably intergalactic matter are indicated with ⩾.

In column 3 the 5 GHz flux is reported, choosing in order of priority: a) measurements simultaneous with those in other bands; b) measurements made with the VLA by Weiler and Johnston (1980) or other authors; c) other available measurements.

In columns 4, 6 and 9 the infrared (2μ) optical (5500 A) and ultraviolet (2500 A) fluxes are given in mJy, choosing, when available, measurements simultaneous to those in other bands. In the other cases preference was given to measurements obtained at epochs near to those of the ultraviolet and X-ray observations. The reported fluxes are dereddened with the extinction values reported in Column 11. Spectral indices derived from separate power law fits in each of the three bands are also reported (columns 5,8,10). In the optical and IR range spectral indices and fluxes corrected for the galactic contribution were chosen, whenever available from the literature. In column 11 the 2 KeV flux is given in μJy, as deduced in all cases, but for 1156+295, from Einstein observations. References are given in each column.

3. AVERAGE SPECTRAL ENERGY DISTRIBUTION OF BLAZARS

The distributions of spectral indices in the IR, optical and UV bands of the 26 Blazars observed with IUE are shown in Fig.1a, b, c and the mean values are given in Table 2, together with the standard deviation of the mean. Broad band spectral indices, derived from flux ratios at widely different frequencies, are also considered. These are α_{RU}, connecting the radio (5 GHz) and the UV band (2500 A), α_{UX} connecting the UV (2500A) and X-ray (2 KeV) bands and α_{RX} connecting directly the two extreme frequecies. They are defined as follows:

$$\alpha_{12} = - \frac{\lg (F_1/F_2)}{\lg (\nu_1/\nu_2)}$$

where F_1 and F_2 are the monochromatic fluxes at the rest frequencies ν_1, ν_2, K-corrected using

$$F_1 = F_1^{obs} (1+z)^{\alpha-1}$$

where α is the spectral index in the appropriate band. For the radio and X-ray bands we assume $\alpha_R=0$, $\alpha_X = 1$, respectively, while in the ultraviolet we use the quoted value of α_{UV}. When the redshift z and/or α_{UV} are not available, we use the corresponding average values, namely 0.4 for the redshift and 1.4 for α_{UV}.

The distributions of the broad band indices are shown in Fig.1d, e, f and the corresponding mean values are given in Table 2.

From Fig.1 and Table 2 it appears that the overall spectrum steepens with increasing frequency up to 5×10^{14} Hz. A large change in slope occurs between radio and IR frequencies (compare for instance

Table 1 Observational data on Blazars

Coord.		z	Ref	F_r	Ref	F_K	Ref	α_{IR}	Ref	F_V	Ref
0215+015		⩾1.686	24	0.4	3	14.7	1	0.96	1	0.3	7
0219+428	3C66A	0.444*	47	0.52	77	13.1	85	0.7	85	3.6	85
0235+164	AO	⩾0.852	58b	1.95	77	20	16	1.7	1	3.1	55
0323+022	PKS	-	-	0.04	18	-	-	-	-	0.92	18
0521-365	PKS	0.0554	63	9.3	33	9	17	1.5	17	1	17
0537-441	PKS	0.894	54	4	32	10	1	1.21	1	1.6	21
0548-322	PKS	0.069	77	0.23	16	6.5	16	0.3	78	1.7	78
0716+71		-	-	0.65	4	-	-	-	-	2.4	4
0735+178	PKS	⩾0.424	14	2.1	12	28.5	12	0.9	12	6.9	12
0754+10	OI090.4	-	-	0.81	85	28	85	1.2	85	4	85
0829+046	OJ049	-	-	0.47	77	12.7	16	1.1	16	1.4	3
0851+202	OJ287	0.306*	47	2.2	82	12.6	82	0.75	82	4	82
1101+384	MK421	0.0308	69	0.56	77	63	45	0.92	45	17.8	45
1133+704	MK180	0.046	47	0.2	76	-	-	-	-	27	49
1156+295	TON599	0.729	60	1.4	27	24.1	27	1.25	80	5.07	27
1215+303	ON325	-	-	0.47	84	10.9	84	0.7	84	3.4	84
1218+304		0.13*	79	0.05	16	46	35	0.3	79	1.2	35
1308+32	B2	0.996	47	2.62	77	12	16	0.9	50	1.7	50
1418+54	OQ530	-	-	1.15	84	10.6	84	.1	84	2.4	84
1514-24	ApLiB	0.049	29	2.42	77	27	28	0.85	29	2.9	79b
1652+398	MK501	0.0337	69	1.25	49	13.5	65	0.65	49	3.3	49
1727+502	IZ187	0.0554	52	0.21	11	40	52	0.87	11	1.22	11
1807+698	3C371	0.05	59	2	83	23	59	0.7	83	5.6	83
2155-304	PKS	0.118	9	0.26	2	31.2	26	0.5	26	17.2	46
2200+420	BL Lac	0.069	47	3.27	77	90	16	1.3	29	5.9	11b
2223-05	3C446	1.404	13	4	11b	22	11b	1.3	11b	3.9	11b

Table 1 Observational data on Blazars (continued)

Coord.		α_O	Ref	F_{UV}	Ref	α_{UV}	Ref	F_X	Ref	A_V	Ref
0215+015		1.8	23	0.7	6	-	-	0.052	61	-	-
0219+428	3C66A	1.3	85	1.27	85	1.58	85	0.08	36	0.15	85
0235+164	AO	2.8	55	0.01	66	-	-	0.17	53	0.1	55
0323+022	PKS	-	-	0.31	75	1.5	75	2.7	18	-	-
0521-365	PKS	1.5	17	1.5	17	1.5	17	0.44	61	-	-
0537-441	PKS	1.6	21	0.8	42	1.5	42	0.15	42	0.2	42
0548-322	PKS	0.3	78	0.4	72	0.84	72	2.15	72	0.13	78
0716+71		-	-	0.17	-	1	-	0.11	4	-	-
0735+178	PKS	1.27	12	2.84	12	1.43	12	0.17	12	0.15	12
0754+10	OI090.4	1.1	85	1	85	2.1	85	0.085	36	0.25	85
0829+046	OJ049	2.2	30	0.6	-	2.1	-	0.095	36	-	-
0851+202	OJ287	1	82	0.72	82	1.48	82	0.65	82	0.07	82
1101+384	MK421	0.9	45	6	70	0.97	70	4.3	62	-	-
1133+704	MK180	0.7	49	0.58	49	0.73	49	2.82	49	-	-
1156+295	TON599	1.5	80	0.81	27	1.7	27	0.075	45b	-	-
1215+303	ON325	2	84	1.02	84	1	84	0.5	84	-	-
1218+304		0.9	79	0.59	73	0.64	73	3.42	62	-	-
1308+32	B2	1.5	50	0.5	-	0.7	-	0.14	36	-	-
1418+54	OQ530	1.8	84	0.21	84	1.83	84	-	-	-	-
1514-24	ApLiB	1.5	79b	1.9	-	1.7	-	0.22	61	0.16	79b
1652+398	MK501	0.65	49	1.3	49	0.65	49	5.88	49	-	-
1727+502	IZ187	0.87	11	0.33	11	1	11	1	11	0.08	11
1807+698	3C371	1.5	83	1.73	83	1.74	83	0.3	83	0.13	59
2155-304	PKS	0.9	46	16	74	0.72	74	4	74	-	-
2200+420	BL Lac	2.3	19	1.5	11b	3	11b	0.53	61	1	11b
2223-05	3C446	1.3	11b	0.86	22	2.9	22	0.46	11b	0.05	22

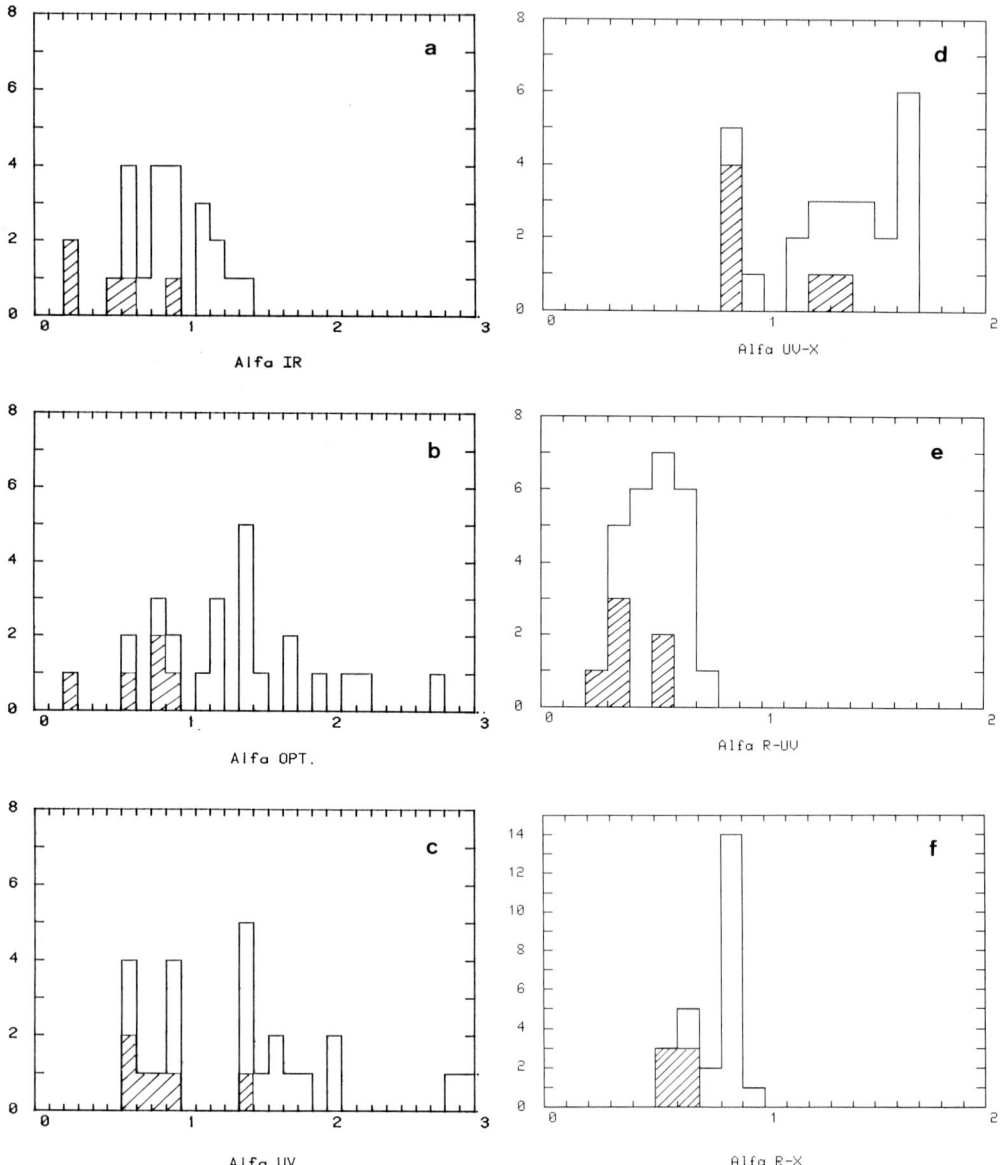

Fig. 1. Distribution of spectral indices. a) infrared index Alfa IR;
b) optical index Alfa OPT; c) Ultraviolet index Alfa UV;
d) Ultraviolet (2500 A) to X-ray (2 keV) index Alfa UV-X;
e) Radio (5 GHz) to Ultraviolet (2500 A) index Alfa R-UV;
f) Radio (5 GHz) to X-ray (2 keV) index Alfa R-X. Shaded
areas correspond to X-ray selected objects.

$<\alpha_{RU}>=0.51+0.02$ with $<\alpha_{IR}>=0.9+0.07$), however more data at millimiter and submillimiter frequencies would be needed to better constrain the spectral shape in this frequency domain. The few available observations (Ennis et al 1982, Landau et al 1983) indicate that the change should occur between 3×10^{11} and 3×10^{13} Hz.

A significant spectral steepening or "break" occurs on average between the IR and the optical bands with $\Delta <\alpha>=0.47+0.13$. Above 5×10^{14} Hz the steepening is halted : the slopes of the optical and UV bands are equal, within the uncertainties. The average spectral index in the UV band and that obtained connecting the UV to X ray frequencies do not differ significantly, i.e. on average the X-ray flux is compatible with being the extrapolation of the UV one.

Another interesting point, which emerges from the data collected in Table 1, is that in general the spectral shape in different bands is correlated, i.e. if an object is flat (steep) in a given band it tends to have an overall flat (steep) spectrum. In Fig.2 the correlations of α_{UV} vs α_{IR} and of α_{UX} vs α_{UV} are shown. Both are significant at better than 10^{-3} chance probability. However no correlation appears between α_{RU} and α_{UX} (Fig.2c).

In Fig.1 the values referring to X-ray selected objects are indicated by shaded areas and in Table 2 the average values of the spectral indices for this group of 6 objects are given separately. It has already been noted that these objects are radio deficient (Chanan et al 1982, Stocke et al 1985). It is very remarkable that in each spectral band as well as in their overall spectrum, these objects appear to be significantly flatter than the whole population. They consistently show up at the flat end of the spectral indices distributions, so that the weakness of the radio emission appears to

TABLE 2. Mean Spectral Indices and Luminosities of BL Lacs

	All (26)	X-ray selec.(6)
α_{IR}	0.91 ± 0.07	0.54 ± 0.12
α_O	1.38 ± 0.12	0.73 ± 0.12
α_{UV}	1.43 ± 0.13	0.89 ± 0.13
α_{RU}	0.51 ± 0.02	0.39 ± 0.05
α_{UX}	1.30 ± 0.06	1.00 ± 0.10
α_{RX}	0.77 ± 0.03	0.59 ± 0.03
Log L_R	33.08 ± 0.31	31.67 ± 0.14
Log L_{UV}	30.35 ± 0.28	29.53 ± 0.39
Log L_X	26.91 ± 0.21	26.82 ± 0.24
α_{UV}^{eff}	$1.09 ^{+0.06}_{-0.04}$	$0.92 ^{+0.06}_{-0.04}$

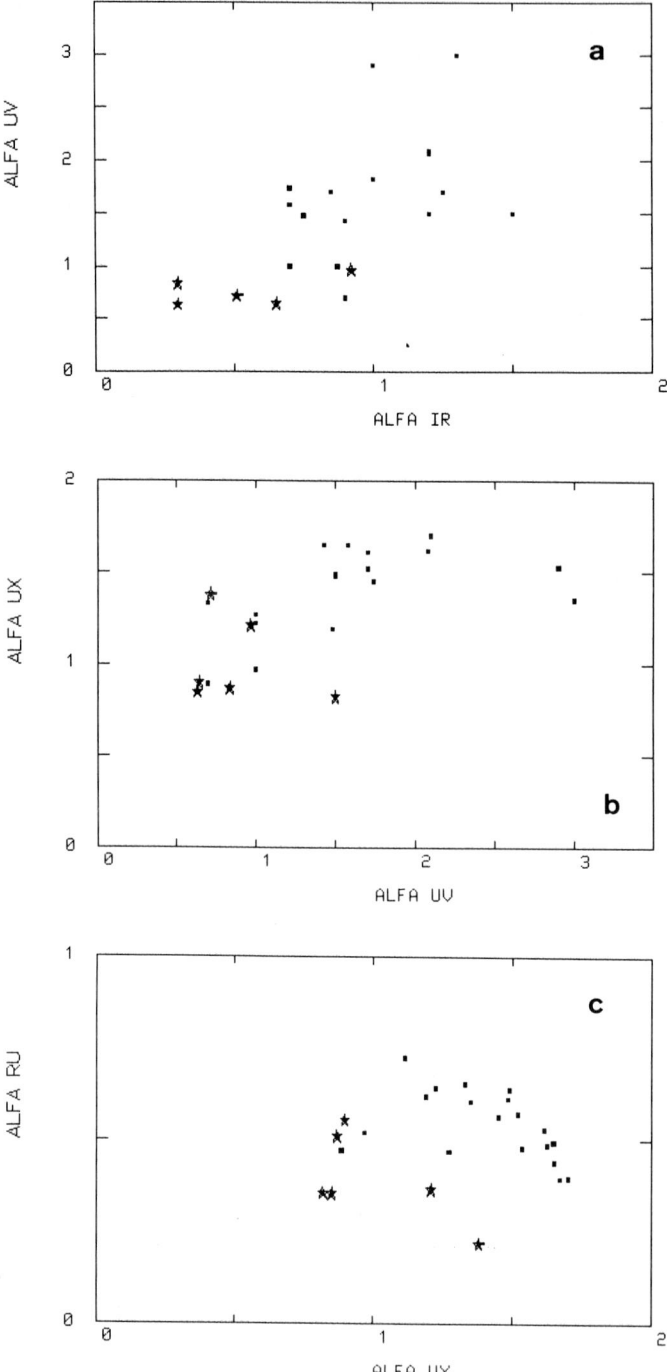

Fig. 2. Correlations between spectral indices (stars represent X-ray selected objects. a) UV vs. IR spectral index; b) UV to X-ray vs. UV spectral index; c) Radio to UV vs. UV to X-ray spectral index.

be related to the flatness of their broad band energy distribution. In fact although no general correlation appears between α_{RU} and α_{UX}, for the subgroup of X-ray selected objects the average α_{UX} and α_{RU} are lower than the averages for the whole group. However for non X-ray selected objects an anticorrelation between α_{RU} and α_{UX} seems to be present.

4. REDSHIFT AND LUMINOSITY DISTRIBUTIONS

The red-shift distribution of the objects is shown in fig.3. it appears to be bimodal with a large deficiency of intermediate red shifts, $0.1 \leqslant z \leqslant 0.4$, which may be due to observational difficulties either in finding objects or in determining red-shifts in this range, in the absence of emission lines. The X-ray selected objects belong to the low red-shift group.

For the objects with known red-shift it is possible to study the luminosity distribution. Ho=50 and $q_o=1/2$ were assumed and the fluxes were K-corrected as described in the previous section. The distributions of the Radio (5 GHz), Ultraviolet (2500 A) and X-ray (2 Kev) monochromatic luminosities are shown in Figs 4a,b,c. It is clear from the figures and from the (logarithmic) averages reported in Table 2 that X-ray selected objects have lower luminosity than the entire sample at radio and UV frequencies, but are indistinguishable from the rest on the basis of the X-ray luminosity.

In Fig.5a and b the composite spectral index α_{UX} is plotted as a function of L_{UV} and L_X respectively. A significant correlation is found with L_{UV}, while none is present with L_X. The result is consistent with the indications from the luminosity distributions. It shows that, in absolute intensity units, the spectra of all objects

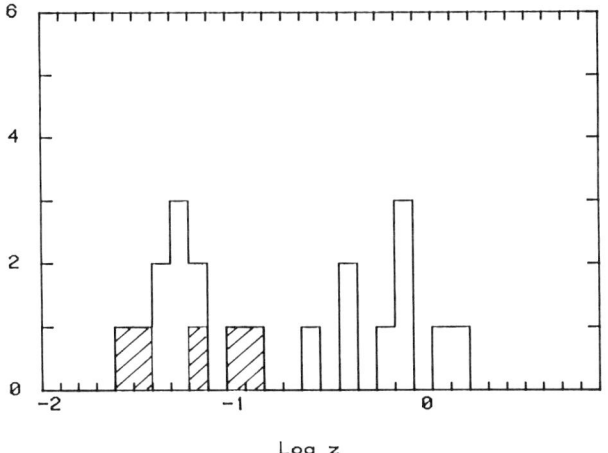

Fig.3. Red-shift distribution of Blazars. The shaded areas correspond to X-ray selected objects.

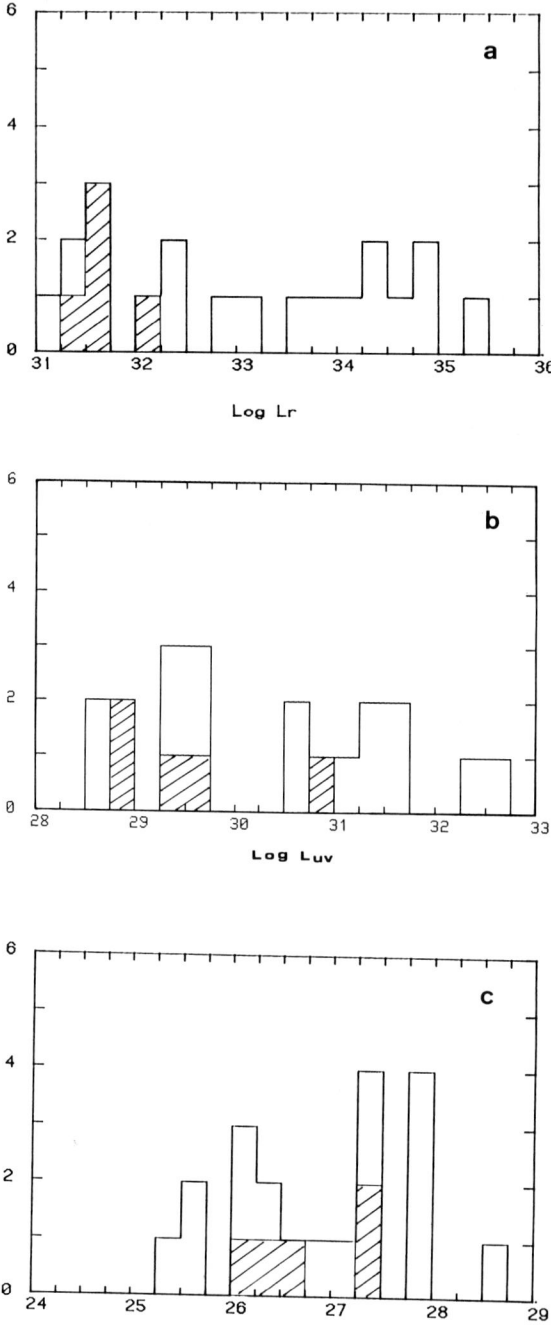

Fig. 4. Distribution of monochromatic luminosity at radio (a), ultraviolet (b) and X-ray (c) frequencies. Shaded areas represent X-ray selected objects.

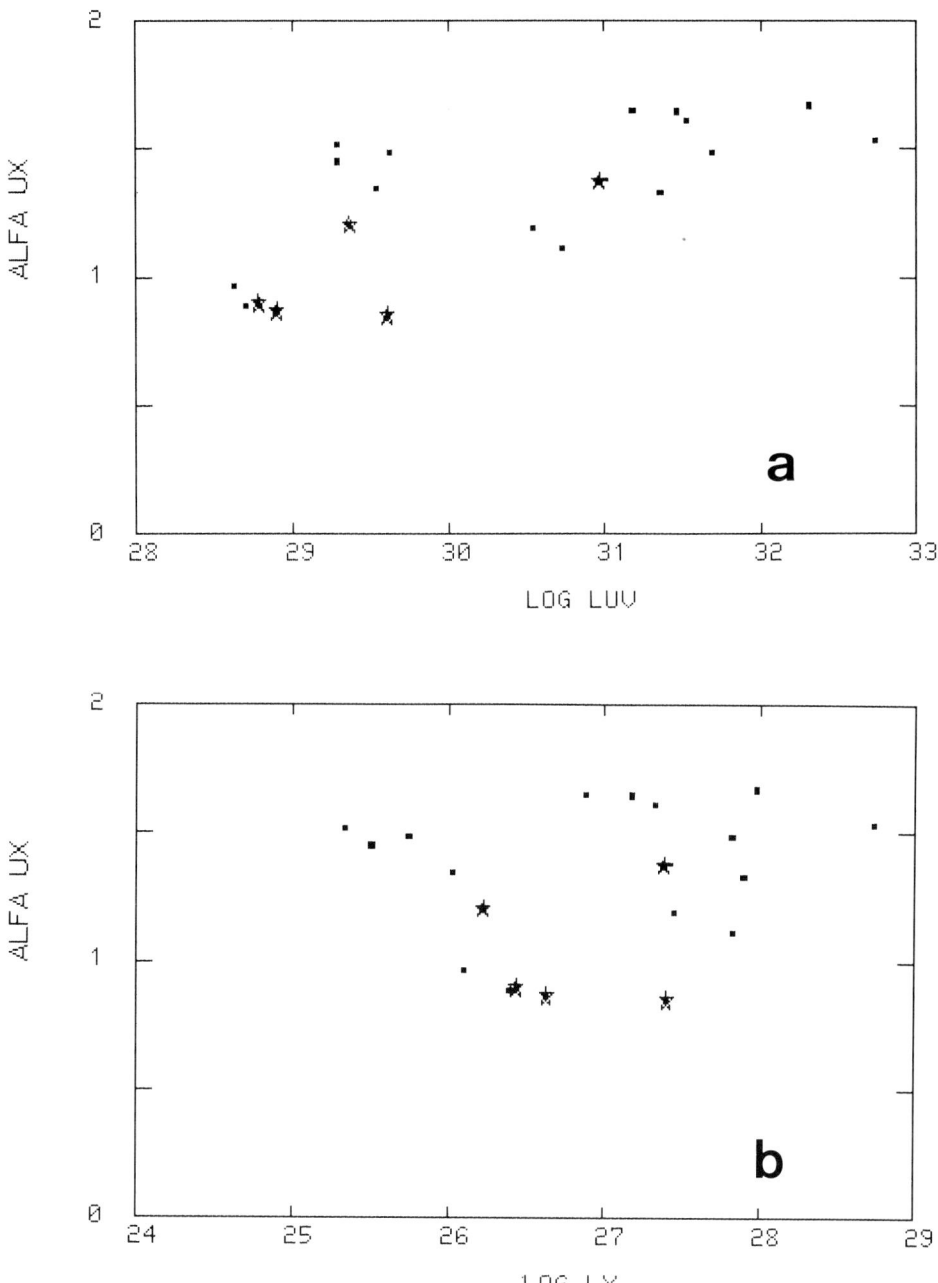

Fig. 5. Ultraviolet to X-ray spectral index vs. monochromatic ultraviolet luminosity (a), and vs. monochromatic X-ray luminosity (b). Stars represent X-ray selected objects.

superpose in the X-ray region (i.e. L_X has relatively low dispersion). As a consequence the objects with lower spectral indices are weaker at lower frequencies (UV and radio).

5. COMPARISON WITH QUASARS

The spectral properties of Blazars described above can be summarized as follows. The spectral slope increases from radio to X-ray frequencies, the major changes being between the radio and IR frequencies, and between the IR and UV. The X-ray flux is on average consistent with an extrapolation from the UV. Moreover there is a strong correlation between the IR, UV and UV to X-ray spectral indices, in the sense that, if the spectrum is flat (steep) in a given band, it tends to be globally flat (steep). These spectral properties can be interestingly compared with those of QSOs. The most remarkable difference, already noted by us (e.g. Maraschi et al, 1983b), is the shape of the optical-UV spectrum. While for our sample of 26 objects we find $<\alpha_{UV}>= 1.43+0.13$ (see table 2), Richstone and Schmidt (1980), in their study of ~ 100 QSOs in the optical band, find a much harder spectral shape with $<\alpha_{2500}>=0.6+0.05$.

On the other hand, comparing the spectral indices deduced from the interpolation of the UV to the X-rays, one finds steeper values for QSOs than for BL Lacs. For this scope we refer to the paper of Zamorani et al (1981) who considered a sample of QSOs distinguishing between radio loud and radio quiet objects. They define an effective spectral index α_{UX}^{eff}, using the average X-ray to optical flux ratio:

$$\alpha_{UX}^{eff} = - \frac{\lg <F_{2Kev}/F_{2500}>}{2.6}$$

Computing the same quantity with the present data we find for our sample $\alpha_{UX}^{eff}=1.1+0.05$ while Zamorani et al (1981) give $\alpha_{UX}^{eff}=1.27+0.03$ for Radio Loud Quasars and $\alpha_{UX}^{eff}=1.46+0.06$ for Radio Quiet Quasars.

In the infrared we have for Blazars $<\alpha_{IR}>=0.9+0.1$ while Ennis et al (1982) for quasars give $<\alpha_{IR}>=1.3+0.1$. It appears therefore that the broad band energy distribution of Quasars is marginally steeper in the infrared, substantially flatter in the optical and steeper again in the UV to X-ray range, with respect to the much smoother energy distribution of Blazars. On the whole, the comparative picture of the spectral characteristics of the two classes is consistent with the suggestion that in QSOs a spectral component is present, peaking in the UV range, which is absent in BL Lacs. This component could be responsible for the different degree of variability and of polarization of the two classes of objects and if it could be removed, the underlying Quasar continuum connecting the IR to the X-rays, may be essentially similar to that of BL Lac objects. The interpretation of the 3000 A bump in terms of thermal emission from an accretion disk e.g. (Malkan and Sargent 1982, Malkan 1983) is consistent with the above suggestion.

The global flatness of X-ray selected Blazars is another point which requires some comment. Examining the monochromatic luminosity distributions one finds that the X-ray selected group appears underluminous from UV to radio frequencies, but does not differ from the entire sample in X-ray luminosity.

Given that the X-ray luminosity is not a discriminating parameter, in the sense that the X-ray luminosity distribution is the same for the entire sample and for this subsample, it follows that X-ray "complete" surveys should yield informations on the entire Blazar population. On the other hand, it is clear from the previous section that X-ray selected BL Lacs have very homogeneous spectral properties, which we shall call of type 1, distinct from those of the rest of the sample (type 2), which is essentially radio selected. We therefore tentatively conclude that type 1 BL Lac Objects (with flat spectra and weak radio emission) are the most frequent. This is to say that their space density for a given X-ray luminosity is higher than that of type 2 BL Lac Objects of comparable X-ray luminosity, which are characterized by steep spectrum, and strong radio emission. The radio selection, on the basis of which most of the BL Lacs have been discovered, obviously favours objects with high radio luminosity, i.e. with steep overall spectrum (type 2). This inference is consistent with the fact that the average redshift of X-ray selected objects is smaller than that of the radio selected ones.

6. IMPLICATION FOR EMISSION MODELS

There exists a practically unanimous consensus that the electromagnetic spectrum of BL Lac objects derives at low frequencies from synchrotron radiation while at high frequencies there is an important contribution from Compton scattering of synchrotron photons off the relativistic electrons. The limit between the two regimes is usually set between ultraviolet and X-ray frequencies. The models differ in the specification of the spatial structure of the source, (homogenous or inhomogeneous), and in the assumptions about the bulk motion of the emitting relativistic plasma. In particular the hypothesis that the Blazar phenomenon is due to relativistic enhancement of the emitted radiation in the observer's direction associated with relativistic bulk motion, (relativistic beaming) has been under discussion since it was first proposed by Blandford and Rees (1978).

In our opinion the most important indication from the data discussed above is that the average X-ray luminosity of objects with widely different broad band spectrum is essentially the same.
This on one hand suggests that the mechanism responsible for the X-ray emission is somehow primary, and on the other hand requires to understand what causes the differentiation in the broad band spectra.

As mentioned above, at about 2 KeV, the synchrotron and Compton mechanisms may both be of importance. The correlation between α_{UX} and α_{UV} indicates a common mechanism active in UV and X-rays, favouring synchrotron radiation as the dominant source of soft X-rays. On the other hand the indication of an anticorrelation between α_{RU} and α_{UX} within the subgroups of X-ray selected and radio selected objects suggests that in the stronger radioemitting sources a Compton contribution may be present in the X-ray band.

A possible answer to the question of the broad band spectrum is simply that the electron spectra are intrinsically different in different sources. However, since this assumption seems somewhat ad hoc, we would like to discuss a different possibility.

Let us consider jet models, in which plasma flows in an elongated region defined by external boundary conditions (e.g. Marscher 1980, Konigl 1981). Reynolds (1982) has shown that the Lorentz factor Γ of the bulk motion of the plasma can increase along the jet on a scale which may be 100 times the scale of the injection region r_c. Let us make the basic assumption that synchrotron X-rays are generated in the inner part of the source, while lower frequencies are produced at larger radii (see Ghisellini et al.,1985) for a discussion of models of this type). With this hypothesis X-rays would be emitted from plasma with low Γ , i.e. quasi isotropically, while lower frequencies would be increasingly boosted along the jet axis. The observed spectrum would depend on the orientation of the jet, with lower frequencies enhanced in the case of alignment. Since this is the less probable case, steep overall spectra (type 2) would be less frequent than flat ones (type 1). One could thus account for the fact that X-ray selection, which relies on quasi isotropic emission normally yields type 1 objects, while radio selection obviously favours type 2 objects.

Although this model is only qualitative at the present stage it has some interesting implications. One is that rather small values of Γ are indicated, $\Gamma \simeq 5-3$, in order to avoid a too large depression of the radio flux for the misaligned observer. (In the simplest model, the flux ratio at 0° and 90° from the jet goes as $\Gamma^{6+2\alpha}$). Another implication is that type 1 objects should be less variable at lower frequencies ($\nu < 10^{15}$Hz) than type 2 objects, since for the first class the Doppler factor

$$\delta = \left[\Gamma (1-\beta\cos\theta)\right]^{-1}$$

could not be larger than 1 and no time contraction should be present.

A detailed discussion of the observed Compton flux in the two cases is complex and beyond the scope of this paper. However we would like to mention the following. In the rest frame of the plasma the Compton flux from intermediate or outer regions of the jet may be larger than that from the inner regions (e.g.Ghisellini et al, 1985). For the aligned observer this Compton emission will be amplified by the same factor as the synchrotron emission from the same region. Therefore in some cases a Compton contribution may be of importance at X-ray frequencies in type 2 objects.

Since this may derive from an intrinsically large region, the variability time scale, even corrected for the Doppler factor may be larger than that associated with the region emitting synchrotron

X-rays. This kind of situation may account for the cases in which X-ray variability occurs on a time scale larger or comparable to that of lower frequencies (e.g. 0735+178, Bregman et al. 1982).

7. SUMMARY AND CONCLUSIONS

The spectral data on Blazars indicate a rather uniform and coherent spectral phenomenology.

1) The average broad band spectrum steepens substantially between radio and infrared and between infrared and ultraviolet frequencies, but not between ultraviolet and X-ray frequencies.

2) The spectral slopes in different bands are correlated i.e. a flat slope in the UV or optical or IR bands implies a globally flat overall spectrum from IR to X-ray frequencies. This statement does not in general extend to the radio range.

3) X-ray selected BL Lac objects are flatter than the average of the sample at all frequencies. For these objects as a subgroup also the radio to UV spectral index is lower than average, which implies that their radio emission is low compared to their UV and X-ray emissions.

4) The X-ray luminosity of the subgroup of X-ray selected objects is the same as that of the whole sample. The fact that X-ray selection favours flat spectrum objects is therefore interpreted as due to the higher space density of this type of objects.

5) It is suggested that the less frequent, steep spectrum objects, which are found by radio selection, have jets closely aligned with the observer's line of sight. Only the radio emission should be enhanced by relativistic beaming, while the X-ray emission, should derive from a region of the jet where the bulk Lorentz factor of the plasma is close to 1.

8. REFERENCES

1) Allen, D.A., Ward, M.J., Hyland, A.R., 1982, MNRAS 199, 969
2) Aller, M.F., Aller, H.D. and Hodge, P.E., 1982, in "Extragalactic Radio Source", P.335, Eds.Heeshen D.S. and Wade C.M.
3) Angel, J.R.P., Stockman, H.S., 1980, Ann.Rev.A.A., 18, 321
4) Biermann, P., et al., 1981, Ap.J. 247, L53
5) Biermann, P., et al., 1982, Ap.J. 252, L1
6) Blades, J.C., et al., 1982, ESA SP 176, 538
7) Blades, J.C., et al., 1982, MNRAS 200, 1091
8) Blandford, R., and Rees, M.J., 1978, in Pittsburgh Conference on BL LAC objects, 1978, Eds. Wolfe A.M.
9) Bowyer, S., et al., 1984, Ap.J. 278, L103
10) Bregman, J.N., et al., 1985, Ap.J., 288, 32
11) Bregman, J.N., et al., 1982, Ap.J., 253, 19
11bis) Bregman, J.N., et al., 1982, ESA SP 176, 589
12) Bregman, J.N., et al., 1984, Ap.J., 276, 454
13) Burbidge, G.R., et al., 1977, Ap.J. Suppl.S. 33, 113

14) Carswell, R.F., et al., 1974, Ap.J., 190, L123
15) Chanan, G.A., et al., 1982, Ap.J. 261, L31
16) Cruz-Gonzales, I. and Huchra, J.P., 1984, An.J. 89, 441
17) Danziger, I.J. et al., 1983, MNRAS, 203, 565
18) Doxsey, R., et al., 1983, Ap.J. 264, L43
19) Ennis, D.J., Neugebauer, G., Werner, M., 1982, Ap.J. 262, 451
20) Ennis, D.J., Neugebauer, G., Werner, M., 1982, Ap.J. 262, 460
21) Falomo, R., 1985, private communication
22) Garilli, B., and Tagliaferri, G., 1985, submitted to Ap.J.
23) Gaskell, C.M., 1978, B.A.A.S. 10, 662
24) Gaskell, C.M., 1982, Ap.J. 252, 447
25) Ghisellini, G., Maraschi, L. and Treves, A., 1985, Astron. Astroph., in press
26) Glass, I.S., 1981, MNRAS 194, 795
27) Glassgold, A.E., et al., 1983, Ap.J., 274, 101
28) Hewckman, T.M., et al., 1983, Ap.J., 272, 400
29) Impey, C.D., et al., 1982, MNRAS, 200, 19
30) Kinman, T.D., 1976, Ap.J., 87, 859
31) Konigl, A.P., 1981, Ap.J., 243, 700
32) Komesaroff, M.M., et al., 1984, MNRAS, 208, 409
33) Kuhr, H., et al., 1981, Astron.Astroph.Suppl.S. 45, 367
34) Landau, R., et al., 1983, Ap.J. 269, 68
35) Ledden, J.E., et al., 1981, Ap.J., 243, 47
36) Madejski, G.M., Schwartz, D.A., 1983, Ap.J. 275, 467
37) Malkan, M.A., 1983, Ap.J. 268, 582
38) Malkan, M.A. and Sargent, L.W., 1982, Ap.J. 254, 22
39) Maraschi, L. et al., 1983a, Astron.Astroph., 125, 117
40) Maraschi, L., Tanzi, E.G. and Treves, A., 1983b, 24th Liege Coll., p.437
41) Maraschi, L., Tanzi, E.G., and Treves, A., 1984, Adv.Space Res. 3, 167
42) Maraschi, L., et al., 1985a, Ap.J., 294,
43) Maraschi, L., et al., 1985b, preprint
44) Marscher, A.P., 1980, Ap.J., 235, 386
45) Maza, J., Martin, P.G., Angel, J.R.P., 1978, Ap.J., 224, 374
45bis) McHardy, I., 1984, 18th ESLAB Symp.on X-ray Astronomy (Reidel), 559.
46) Miller, H.R., McAlister, H.A., 1983, Ap.J., 272, 26
47) Miller, J.S., French, H.B., Hawley, S.A., 1978, Pittsburg Conf. on Bl Lac Objects, A.M. Wolfe Eds.
48) Moore, R.L., Stockman, H.S., 1984, Ap.J., 279, 465
49) Mufson, S.L., et al., 1984, Ap.J., 285, 571
50) O'Dell, S.L., et al., 1978, Ap.J.Suppl., 38, 267
51) O'Dell, S.L., Pushell, J.J., Stein, W.A., 1977, Ap.J. 213, 351
52) Oke, J.B., 1978, Ap.J. 219, L97
53) Owen, F.N., Helfand, D.J., Spangler, S.R., 1981, Ap.J. 250, L55
54) Peterson, B.A., et al., 1976, Ap.J., 207, L5
55) Pica, A.J., et al., 1980, Ap.J., 236, 84
56) Piccinotti, G., et al., 1982, Ap.J. 253, 485
57) Reynolds, S.P., 1982, Ap.J., 256, 13
58) Richstone, D.O. and Schmidt, M., 1980, Ap.J., 235, 361
58bis) Riecke, G.H., et al., 1976, Nature 260, 754

59) Sandage, A., 1966, Ap.J. 145, 1
60) Schmidt, M., 1975, Ap.J. 185, 253
61) Schwartz, D.A. and Ku, W.H.M., 1983, Ap.J., 266, 459
62) Schwartz, D.A., et al., 1979, Ap.J., 299, L53
63) Searle, L. and Bolton, J.G., 1968, Ap.J., 154, L101
64) Smith, H.E., Spinrad, H., and Smith, E.O., 1976, PASP 88, 621
65) Snijders, M.A.J., et al., 1979, MNRAS, 189, 873
66) Snijders, M.A.J., et al., 1982, MNRAS, 201, 801
67) Stein, W.A., O'Dell, S.L., Strittmatter, P.A., 1976, Ann.Rev.Astron.Astroph., 173
68) Stoke, J.T., et al., 1985, preprint
69) Ulrich, M.H., et al., 1975, Ap.J., 198, 261
70) Ulrich, M.H., et al., 1984, Ap.J., 276, 466
71) Urry, C.M. and Mushotzky, R.F., 1982, Ap.J., 253, 38,

72) Urry, C.M., et al., 1982, Ap.J., 261, 12
73) Urry, C.M., et al., 1984, preprint
74) Urry, C.M., 1984, Ph.D. thesis
75) Urry, C.M., et al., 1985, private communication
76) Wardle, J.F.C., Moore, R.L., Angel, J.R.P., 1984, Ap.J. 279, 93
77) Weiler, K.W. and Johnston, K.J., 1980, MNRAS, 190, 269
78) Weistrop, D., Smith, B.A. and Reitsema, H.J., 1979, Ap.J.233,504
79) Weistrop, D., et al., 1981, Ap.J., 233, 504
79bis) Westerlund, B.E., Wlerick, G. and Garnier, R., 1982, Astron.Astroph., 105, 284
80) Wills, B.J., et al., 1983, Ap.J., 274, 62
81) Wood, K.S., et al., 1984, Ap.J.Suppl. 56, 507
82) Worral, D.M., et al., 1982, Ap.J., 261, 403
83) Worral, D.M., et al., 1984a, Ap.J., 278, 521
84) Worral, D.M., et al., 1984b, Ap.J., 284, 512
85) Worral, D.M., et al., 1984c, Ap.J., 286, 711
86) Zamorani, G., et al., 1981, Ap.J., 245, 357

DISCUSSION

WILKES: I didn't understand from your data why you concluded that the type 1, flat Blazars have a higher spatial density that type 2.

TREVES: The argument is based on two points: 1) The two types of Blazars are indistinguishible on the basis of the X-ray luminosity - 2) Complete X-ray surveys select type 1 objects (flat spectrum).

PHYNNEY: Is there extended radio structure around your X-ray selected BL Lacs? Moore et al. 1981 and Antonucci and Ulvestad 1985 found such structure on scales around most low-luiminosity BL Lacs, consistent with the hypothesis that they are beamed radio galaxies. Could the differences between your "type 1" and "type 2" radio galaxies be that they are respectively beamed radio galaxies (as Seyferts) and beamed quasars, respectively? This would explain the relative space densities and the difference in Lx/Lopt.

TREVES: Your suggestion seems reasonable, but I don't have elements to strengthen it.

OPTICAL VARIABILITY IN QUASARS

S. Cristiani
European Southern Observatory
Casilla 19001
Santiago 19
Chile

1. INTRODUCTION

Since the early days, variability represented one of the most important tools for the understanding of the physical processes involved in Active Galactic Nuclei.
  The very first evidence of violent activity in the nuclei of some galaxies was provided by the observations of Seyfert (1943) on a number of galaxies showing broad emission lines whose widths corresponded to motions of the ionized gas of several thousand kilometres per second.
  Later on, studies of radio galaxies showed that those objects are place of intense activity. The observations, interpreted in terms of synchrotron acceleration, implied a considerable amount of energy in relativistic particles and magnetic fields (of the order of 10^{60} ergs). Morphological evidences (e.g. the jet in M 87 and the so-called "classical" radio structure, with two lobes and a source of small angular size coincident with the optical nucleus) suggested also that the source of this activity had to be searched in the nucleus and gave origin to the "problem of the energy supply" (Burbidge 1958). This problem became dramatic when quasars were discovered.
  The identification by Schmidt (1963) of the radio source 3C 273 with a stellar object at $z = 0.158$ was already surprising for its absolute magnitude (about -27) deriving from the cosmological interpretation of its redshift. But the observation that quasars can be variable provided the puzzling information that the enormous energy output comes from a compact region, by means of the thumb rule: source dimensions $<$ c·dt. Analysis of QSO light curves shows variations often on timescales of one year and, in some cases, of a few days, implying dimensions $<$ 1 pc for the central source. Very simple computations indicate that, in this situation, the underlying energy source must be gravitational in nature (Cavaliere 1981) as modelled by Zeldovich and Novikov and by Salpeter as early as 1964.
  But the wealth of information for theoreticians coming from studies of variability in quasars is not limited to these simple results. I will recall here the phenomenon of superluminal velocities (see Cohen and Unwin 1982) observed in some radio sources. Scheuer (1984) reviews

various explanations. Among Christmas trees, relativistic beaming and gravitational lensing, not to mention the non-cosmical redshift hypothesis, I understand the second is, at present, the most fashionable, in its original version or in the combination with Christmas trees. On a similar basis, the discovery of low-frequency radio variability has led to a revision of the "Classical" interpretation of incoherent synchrotron radiation from an expanding plasmoid (Fanti and Salvati 1980).

2. OPTICAL SAMPLES

In the following part I will deal in more detail with optical variability and the relative problems.

The optical domain offers the largest quantity of data about quasar variability and is potentially more interesting, for instance, than the radio waveband, since it is probing (as can be inferred from the variation timescales) regions closer to the central engine. The possibility to classify quasar light curves is of the greatest importance for a number of models predicting some preferential timescales as well as correlations between absolute magnitude and amplitude of the variations, between amplitude and redshift, etc. The difficulties in this respect are apparent when we look at figure 1. Some quasars, like 3C 351 or 3C 380, show regular variations of moderate amplitude. Others, like 3C 345 or 3C 446, have sudden bursts of large amplitude. Finally, cases of flares superimposed on longer term variations, as in PHL 658, are not excluded, or cases which can be described as "flaring down" of the object, as OQ 530. These real differences probably reflect, as for variable stars, different physical mechanisms and render very difficult the use of simple statistical parameters.

The data illustrated in figure 1 are a result of a long-term project undertaken at the Asiago Observatory since 1967 in order to study the optical variability of QSOs. Material has been collected and published for 96 objects (see Barbieri et al. 1983 and references therein). The criteria of selection were the apparent magnitude and the presence in the Asiago Survey of Supernovae plates; no selection was made on the basis of redshift, radio emission or other properties. The results of this survey can be compared with other surveys (Pica and Smith 1983, Bonoli et al. 1979, Netzer and Sheffer 1983) to investigate some of the basic properties of quasar variability.

2.1. Optical Versus Radio Properties

One of the most investigated problems in the variability of AGN is the existence of a relationship between optical and radio behaviour. Penston and Cannon (1968) found a definite connection between flat radio spectrum and the class of highly variable quasars. In table 1, some results of the Asiago survey are summarized.

Although the number of radio-quiet QSOs is too small to draw a definite conclusion, it is clear that radio-loud QSOs are more variable than radio-quiet QSOs. Ratios, however, are contained around a factor of 2 and, surely, the Asiago survey cannot confirm the difference of a

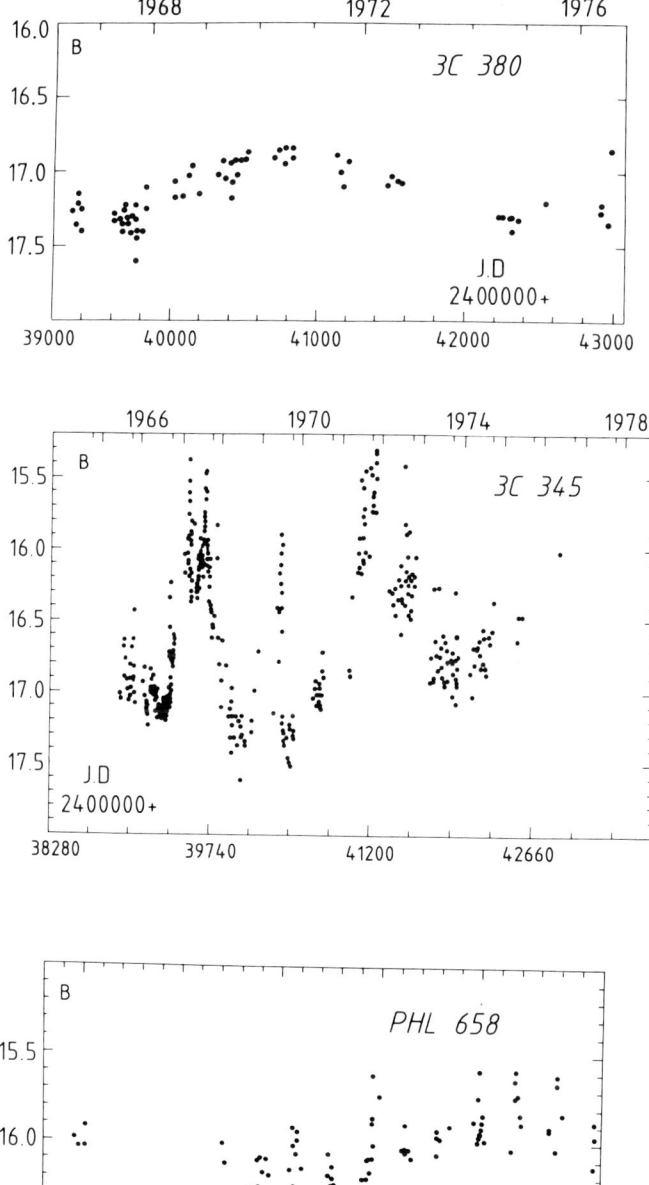

Figure 1: Light curves of three quasars.

Table 1

	Radio-Loud	Radio-Quiet
Total	75	21
OVV	18 (24%)	2 (10%)
OVV+V	36 (40%)	5 (24%)

factor of about 10 found by Bonoli et al. (1979). It was also investigated whether some relationship exists between optical and radio variability for radio-loud objects, as shown in table 2.

Table 2

	OVV+V	Non Variable
Radio-Loud	47	28
Radio Variable	20 (43%)	3 (11%)

The different behaviour is rather convincingly established: radio-loud QSOs tend to be variable at both radio and optical wavelengths.

2.2. Variability Versus Absolute Magnitude

Bonoli et al. (1979) and Netzer and Sheffer (1983) found no relation between absolute luminosity and variability. Figure 2 represents the distribution of the Asiago sample in the plane M, Dm (for a standard Friedman universe with $q_0 = .5$ and $H_0 = 50$ km/s/Mpc and neglecting the K correction).

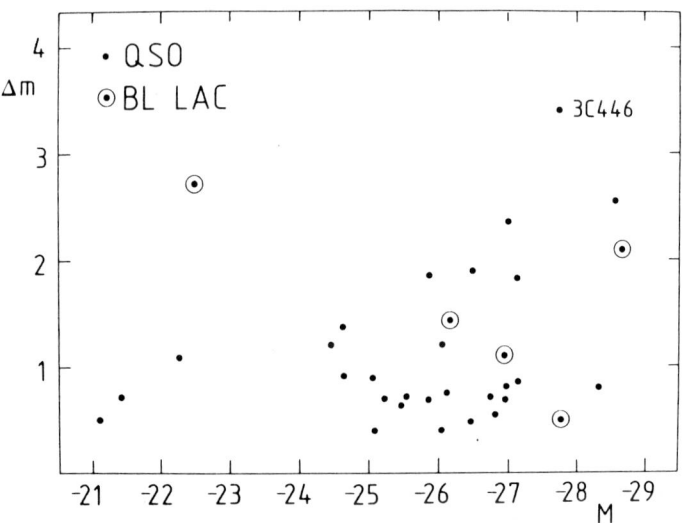

Figure 2: The M, Δm plane for variable AGNs in the Asiago sample.

Separately considered, QSOs show a slight appearance of a trend, in the sense that the greates Δm's have been observed in the most luminous objects. However, this trend can result from a connection with the apparent magnitude of the type discussed in the next paragraph. The conclusion is that the amplitude of the variation could be only weakly, if at all, correlated with the absolute magnitude. Pica and Smith (1983) find in their sample a slight anticorrelation between absolute magnitude and variability. All those results together point against the theories assuming that quasars are composed of multiple subunits of random flaring objects. In these models, in fact, the most luminous quasars exhibit the smallest variability, with a strong anticorrelation.

2.3. Variability Versus Redshift

Figure 3 shows the distribution of objects in redshift. The percentage of variable quasars drops with increasing z. A similar result is shown in the sample of Pica and Smith, while no significative correlation between variability and redshift is found by Bonoli et al. (1979) and by Netzer and Sheffer (1983). A time expansion effect is not the only responsible for the observed trend: the main dependence is determined by the apparent magnitude as shown in figure 3. The result for the Asiago and Rosemary Hill surveys can be understood in terms of the biases included in those two samples. Since these samples are not complete, among the bright objects of low redshift the percentage of radio-loud, variable quasars tend to be higher than in a complete sample. Furthermore, the uneven time coverage is favourable for detection of variability in objects with brighter apparent magnitude, for which poorer quality plates are still usable and S/N ratio is larger.

For a better assessment of the last two points, complete samples as the one of Bonoli et al. (1979) and the one by Netzer and Sheffer (1983) are necessary. For several reasons, the amount and the quality of available data are not sufficient at present to settle the question. A more refined survey with better photometric calibration and smaller photometric errors and, possibly, a larger number of objects, is needed. In this respect, a work in progress by Hawkins, Cristiani, Véron-Cetty, Véron and Braccesi on the Braccesi field using the facilities of the Cosmos machine of ROE could provide valuable information.

3. Particular Objects

Incomplete samples have been shown, due to the presence of biases, to be partially unsatisfactory to study some statistical properties of the AGN class. Monitoring of particular objects is, on the other hand, extremely important to understand some special behaviour which, in turn, can shed light on the whole subject of variability.

An interesting and debated point to consider, for instance, is the existence of periodicities in the light curves of quasars. As shown before, the photometric variations of those sources are, in general, erratic. Nevertheless, sometimes it is possible to single out some kind of regularity. Two cases can be used as examples: 3C 345 and 3C 446,

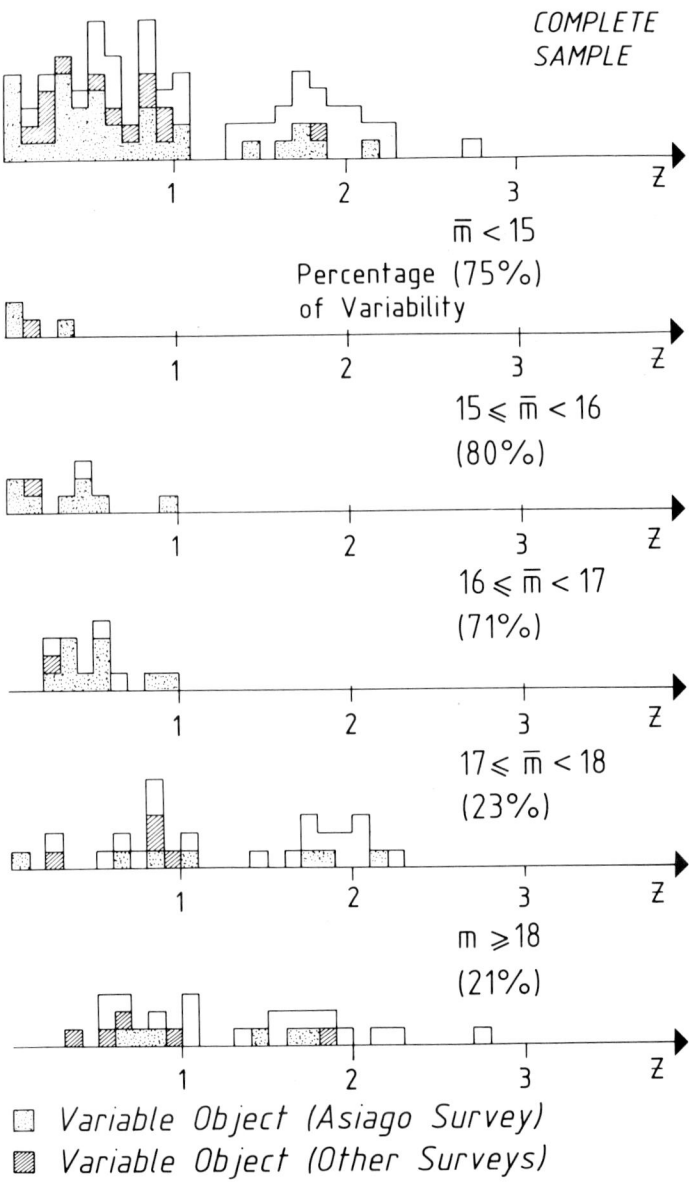

Figure 3: Variability versus redshift.

3C 345, already shown in Figure 1 has been suggested on the basis of a Deeming-type analysis, to show periodicities of 1600 and 800 days (Barbieri et al. 1977). For 3C 446 recurrent outbursts at a distance of 1540 days are indicated (Figure 4).

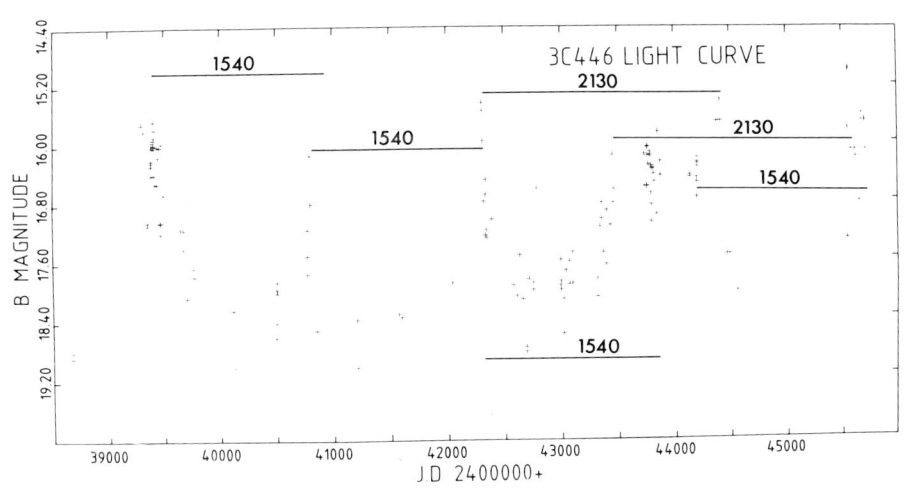

Figure 4: Light curve of 3C 446.

The data collected on 3C 446 (Barbieri et al. 1985) are of extreme interest. In the optical, this object shows activity at several levels, down to significant intraday variations, setting a typical value of 1 light day for the emitting region. An even tighter upper limit can be derived from polarization data. Stockman and Angel (1978) observed a rotation of 90 degrees in the polarization angle in about 7.5 hours, implying that the polarized flux is emitted in a region contained in a radius of 8.10^{14} cm. Spectroscopic data indicate that emission lines and continuum do not vary in the same way. The C III] 1909 Å line, for instance, is prominent in the quiescent state, but seems not to vary while the continuum level is increasing.

The radio behaviour of 3C 446 is also of extreme interest. At 408 MHz it does show only marginal variability (Fanti et al., 1981), while at higher frequencies strong variations have been recorded. The longest series of data in the literature is that at 2.8cm published by Medd et al. (1972) and by Andrew et al. (1978). During an interval of 3797 days, the radio flux underwent an almost sinusoidal fluctuation with a period of 2400 days. There is no apparent correlation of the radio curve with the optical data and a simple time shift between the two is excluded. The amplitude of the radio variations is much smaller than the optical one, and no evidence of significant short time scale changes can be found, indicating a considerably larger size for the region emitting the 2.8cm radiation, which can be estimated to be of the order of a few light years.

In the X-ray band 3C 446 has been monitored with the Einstein Observatory for short timescales variations, at 100s, 1000s, 5000s and 1d, with a negative result (Zamorani et al., 1984). On longer timescales, however, this object does show variability also at these frequencies, with an increase of a factor 0.5 in about six months.

A case similar to 3C 446 is represented by PKS 0537-441. This object has been classified as a quasar of redshift 0.894 on the basis of spectra taken by Peterson et al. (1976). It has been observed to vary with large amplitude in the optical, radio and IR.

The emission lines C III] 1909 Å and Mg II 2798 Å observed by Peterson et al. (1976) are sometimes invisible against the continuum.

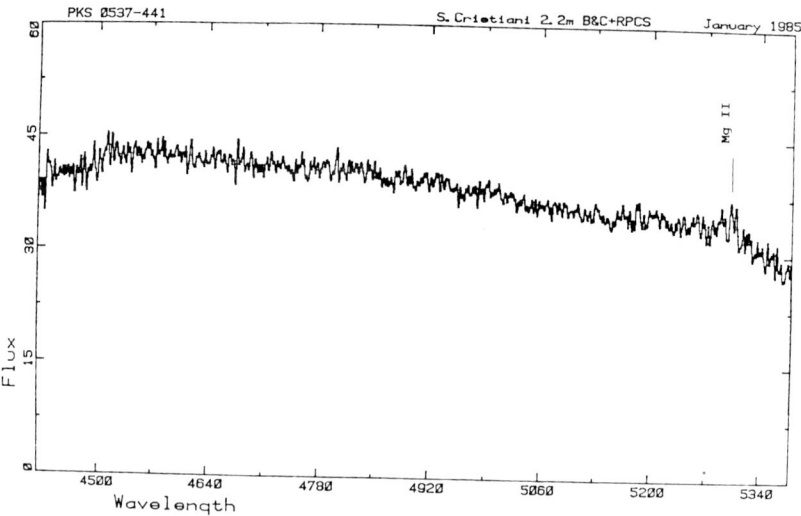

Figure 5: A spectrum of PKS 0537-441. Expected position of Mg II 2798 Å line is indicated.

In Figure 5 it is shown a spectrum of the object I took on January 10, 1985 at the ESO-MPI 2.2m telescope with a relatively high resolution (~ 1 Å) and a good S/N ratio. Only a marginal evidence (E.W.<2 Å) of the Mg II line is present. The picture for PKS 0537 appears to be less simple in this respect than in the case of 3C 446. When the spectrum was taken, in fact, the object had a "medium-low" continuum level (m_v= 15.6). A simple scheme as constant flux lines superimposed on a variable continuum may not hold for PKS 0537-441. Unfortunately, the lack of published data about the flux in the Mg II line in other situations prevents from any further quantitative discussion.

January 1985 was however a period of considerable activity for PKS 0537-441. Rapid changes were taking place, as proved by the increase of about 0.4 magnitudes in the V band from January 10 to January

12 and by further observations of Tanzi et al. (1985) in the IR, UV and X-ray bands. The overall variability of PKS 0537-441 appears extremely significative. If we put together the data collected by Maraschi et al. (1985) with those of Tanzi et al. (1985), we obtain what is shown in Figure 6.

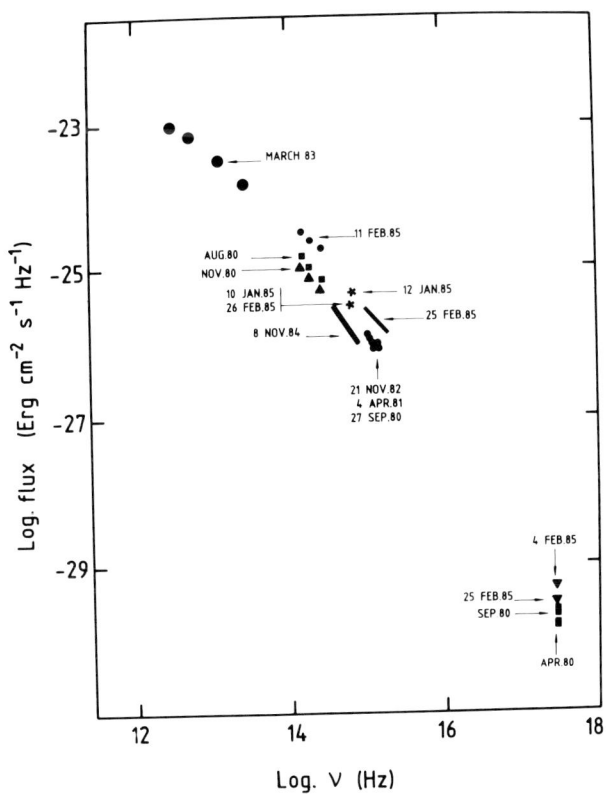

Figure 6: Overall spectrum of PKS 0537-441. Abscissa is logarithm of frequency (Hz), ordinate logarithm of flux in erg cm^{-2} s^{-1} Hz^{-1}.

From those data a global increase of flux, in all the observed bands, is apparent for the 'burst' taking place about the end of January 1985. This remarkable behaviour seems to be usual for PKS 0537-441: a similar event has probably occurred around September 1980 (Maraschi et al., 1985).

What can we learn from those two examples?

The picture emerging from the data of 3C 446 is in many respects a 'classical' one, with a core of about 10^{15} cm size emitting the highly polarized optical continuum, a region originating the X-ray emission at $\simeq 10^{18}$ cm, probably overlapping the radio emitting region, while the size of the region of the broad emission lines is (lower limit) about 10^{20} cm. Ultimately, the increase in the continuum radiation should cause an increase in line emission with a delay determined by the distance of the BLR from the central source. To measure this delay (10 yr is the order of magnitude expected from the current models) would provide additional confidence in the models themselves.

One of the puzzling problems in comparing observations in different wavelength ranges is the difficulty in revealing correlations between different bands of the electromagnetic spectrum. This may simply reflect the structural complexity of the compact sources and the different dependence of the emission at various frequencies on radiative losses and particle acceleration. Even if all the radiation is emitted with the same mechanism, a given spatial component is not necessarily dominating at every frequency, except perhaps in the major outburst. This seems indeed to be the case of PKS 0537-441 for the activity phase of January 1985. The overall increases of emission from the radio to the X-ray band is extremely suggestive when interpreted in terms of a Sincro Self Compton process of electrons and photons producing the radio or infrared-optical emission.

It is apparent how variability data on 3C 446 and PKS 0537-441 provide extremely useful information for the study of the structure and phenomenology of Active Galactic Nuclei. However, the question can be raised, whether these objects are truly representative of any AGN class. Their properties seem, in fact, intermediate between the typical quasar properties (prominent broad emission lines) and the BL Lac properties (high percentage of linear polarization, featureless continuum), and the picture derived for them may not hold for other objects of these two classes. Monitoring of these and a number of other objects is therefore continuing.

To conclude, I would like to emphasize again the lines of research which, in my opinion, should be pursued in the future: statistical analysis of complete samples of quasars and multifrequency monitoring of selected objects. Such observations are potentially suitable to disclose the 'Pandora box' of variability. It is a kind of work requiring considerable patience and coordinating efforts, but the wealth of puzzling problems and unanswered questions contained in the box, awaiting to be 'dispersed and play havoc among mankind', is certainly worth the travail.

Acknowledgements. I am grateful to C. Barbieri, L. Maraschi, E. Tanzi and A. Treves for providing a large quantity of unpublished data and useful suggestions.

References

Andrew, B. H., Macleod, J. M., Harvey, G. A., Medd, W. J., 1978: Astron. J. 83, 863.
Barbieri, C., Cristiani, S., Nardon, G., Romano, G., 1983: XXIV Liege Astrophysical Colloquium, p. 443.

Barbieri, C., Cristiani, S., Omizzolo, A., Mardon, G., Romano, G., 1985: Astron. Astrophys. 142, 316.
Barbieri, C., Romano, G., di Serego, S., Zambon, M., 1977: Astron. Astrophys. 59, 419.
Bonoli, F., Braccesi, A., Federici, L., Zitelli, V., Formiggini, L., 1979: Astron. Astrophys. Suppl. 35, 391.
Burbidge, G. R., 1958: Paris Symposium of Radio Astronomy, ed. R. N. Bracewell (Stanford Univ. Press), p. 541.
Cavaliere, A., 1981: Plasma Astrophysics, Course and Workshop held in Varenna, p. 97.
Cohen, M. H., Unwin, S. C., 1982: Extragalactic Radio Sources , IAU Symposium no. 97, p. 345.
Fanti, C., Fanti, R., Mantovani, F., Padrielli, L., Weiler, K. V., 1981: Astron. Astrophys. Suppl. 45, 61.
Fanti, R., Salvati, M., 1980: Proc. European Regional Meeting in Astronomy, Variability in Stars and Galaxies, p. C.3.1.
Maraschi, L., Schwartz, D. A., Tanzi, E. G., Treves, A., 1985: Astrophys. J. in press.
Medd, W. J., Andrew, B. H., Harvey, G. A., Locke, J. L., 1972: Mem. R. astr. Soc. 177, 109.
Miller, J. S., French, H. B., 1978: Pittsburgh Conference on BL Lac Objects, ed. A. M. Wolfe, p. 228.
Metzer, H., Sheffer, Y., 1983: Mon. Not. R. Astron. Soc. 203, 935.
Penston, M. V., Cannon, R. D., 1968: Royal Obs. Bull. 159, 85.
Peterson, B. A., Jauncey, D. L., Wright, A. F., Condon, J. J., 1976: Astrophys. J. 207, L5.
Pica, A. J., Smith, A. G., 1983: Astrophys. J. 272, 11.
Salpeter, E. E., 1964: Astrophys. J. 140, 796.
Scheuer, P. A. G., 1984: IAU Symp. no. 110, p. 197.
Schmidt, M., 1963: Nature 197, 1040.
Seyfert, C. K., 1943: Astrophys. J. 97, 28.
Stockman, H. S., Angel, R. P., 1978: Astrophys. J. 210, L67.
Tanzi, E. G., Barr, P., Bouchet, P., Chiappetti, L., Cristiani, S., Danziger, J., Gionni, P., Maraschi, L., Treves, A., Wamsteker, W., in prep.
Wilkes, B. J., Wright, A. E., Jauncey, D. L., Peterson, B. A., 1983: Proc. Astron. Soc. Austr. 5, 2.
Zamorani, G., Gionni, P., Maccacaro, T., Tananbaum, H., 1984: Astrophys. J. 278, 28.
Zeldovich, Y. B., Novikov, I. D., 1964: Dokl. Acad. Nauk. 155, 1033.

DISCUSSION

KIDGER: A study of 3C 345 by John Beckman and myself has suggested that, since the three major bursts between 1967 and 1975, which are well fitted by 800 and 1600 d periods, there is no real evidence of subsequent behaviour following these periods. For example, the bursts predicted in 1976 and 1980, do not seem to have occurred and that periodic behaviour has been replaced more by a "trend" in the light curve.

CRISTIANI: I think it is worth pointing out that Fourier analysis does not represent adequately all the characteristics of light curves. As stressed in Barbieri et al. (1985), a simple superposition of sinusoidal waves with constant phases usually cannot reproduce the narrowness of the peaks and is incapable of following precisely the overall pattern of the curve. Furthermore, some of the major outbursts are not single events but instead very active phases with several short term luminosity fluctuations. In such a complicate framework the results of a Deeming-type analysis should be regarded only as as indication of some preferential time scales.

BECKMAN: Although there appears to be only a weak correlation between the amplitude of QSO variability and the absolute magnitude (which is, of course, begging the question about one determines the latter), there does seem to be evidence for a stronger correlation between the power frequency spectrum and absolute luminosity. Do you have any comment on this correlation?

CRISTIANI: There are a number of models, beaming for instance, in which this correlation could be explained. However, in my opinion, the evidence derived from the present data is not very significative.

WILKES: The spectrum you displayed from Wilkes et al. (1983) is the sum of two spectra taken by Peterson et al. (1976) which showed the lines Mg II and C III]. One other spectrum was also taken by them, which showed no emission lines, similar to your recent data.

CRISTIANI: As a comment to this comment, I would like to emphasize that PKS 0537-441 looks under many respects like a

BL Lac Object turning from time to time into a quasar, rather than the reverse as in the case of 3C 446.

INTERNAL DYNAMICS AND FORMATION OF EMISSION CLOUDS IN ACTIVE GALACTIC
NUCLEI

Isaac Shlosman
Department of Physics
University of Florida
Gainesville, Florida, U.S.A.

Peter A. Vitello
Science Application Inc.
McLean, VA 22102, U.S.A.

and

Giora Shaviv
Department of Physics
Technion-Israel Institute of Technology
Haifa, Israel 32000

ABSTRACT. We show that cool UV accretion disks, if present in active galactic nuclei, can be sources of line driven winds. Continuum sub-critical disks are shown to possess optically thin continuum winds which are accelerated by radiation pressure in lines.
    Results of numerical simulation of outflows from disks around supermassive black holes are presented. The necessary condition for wind initiation is the existence of an appropriate circumstance in which the absorption lines coincide with the radiation peak. Such a situation is shown to arise in the shielded region of the disk, the shield being provided by the disk atmosphere.
    The winds are characterized by steep velocity gradients and high asymptotic velocities $v/c \sim 1/30$, as observed in broad emission-line regions of QSO's and Seyfert galaxies, and are thermally unstable as they heat up. This results in a two-phase equilibrium: cold clouds embedded in a hot, Compton-heated background. The hot wind provides the pressure confinement for the dense and massive clouds.

I. INTRODUCTION: ACCRETION DISKS AND DYNAMICS OF AGNs

    The morphology of the continuum radiation is frequently considered as the defining characteristics of AGNs. Most of the AGNs also display broad emission-like spectra (BELS). At least 10% of all intermediate to high-redshift QSO's and few Seyfert galaxies show strong absorption features in their spectra, some of which have a P-Cygni profile. The broad-absorption lines (BALS) imply the existence of extensive outflow of matter. It is plausible that the phenomenon of BALS is more general and the detectability of these regions in any particular QSO or Seyfert depends on the viewing angle.

We adopt here the picture that the energy source of these objects is accretion onto a single central compact object, very probably a supermassive black hole (SBH).

The nature of the accretion processes far away from the SBH depends on the particular boundary conditions. Spherically-symmetric accretion can be obtained only if the angular momentum of the captured matter is very low and can be neglected.

The other frequently observed situation is the redistribution of the angular momentum by viscous dissipation, i.e. formation of accretion disk. In any case, the "dragging of inertial frames" by a rapidly rotating black hole will probably lead to disk accretion for $R/R_s < 10^2$ (Bardeen and Petterson, 1974). In our notation R is the radial distance from the black hole and R_s is its Schwarzschild radius.

The existence of an accretion disk induces a non-spherical geometry. If, for example, an outflow starts close to the plane of the disk it will be very anisotropic. The radiation emitted by the accretion disk is also anisotropic.

The observations of collimated radio-jets imply that at least some of the processes are anisotropic.

The anisotropy of the radiation emerging from the disk and the wind affect the ionization parameter Ξ defined by Krolik, McKee and Tarter (1981) as

$$\Xi = \frac{(P_{rad})_{ion}}{P_{gas}} \tag{1}$$

where $(P_{rad})_{ION}$ is the pressure of the ionizing radiation (between 1 and 1000 ryd) and P_{gas} is gas thermal pressure in the region where the BEL is formed. The value of Ξ is thus inferred from the observed BEL intensity ratios, and will be discussed later. The dimensions of BELRs, r_{BELR} follow from

$$r_{BELR} = \left(\frac{L_{ion}}{4 \pi c \, \Xi \, P_{gas}} \right)^{1/2} \tag{2}$$

where L_{ION} is the ionizing luminosity of AGN.

The above definition of r_{BELR} is based on the assumption of isotropically emitted radiation and spherically symmetric gas distribution. Both assumptions are ad hoc.

The emerging picture is therefore of an outflow from an accretion disk around a SBH. We direct our attention here to the following questions:
(a) What kind of instability (if any) is responsible for initiating an outflow (wind) from a subcritical disk?
(b) What is the parameter space in which a wind forms and to what extent the wind influences the dynamics of the BEL region.

(c) Can the observed two phase character of the BEL region be a natural outcome of the outflow from the accretion disk?

We show that certain parts of the subcritical disk are unstable to radiation pressure in the line driven outflow. Emission clouds form in the wind as a result of two phase instability. We estimate the cloud parameters and discuss their fate.

II. LINE DRIVEN WIND IN ACCRETION DISKS

The excess of UV radiation from AGNs can be attributed to the presence of relatively cold $10^4 - 10^{5\circ}$K matter. This matter is probably distributed in accretion disks (Malkan and Sargent, 1982) and numerous emission-line clouds (for review, see Davidson and Netzer, 1979, and references therein). This naturally suggests that the line opacity of disk atmosphere, combined with the ultraviolet radiation, will produce radiation pressure, which may be sufficiently high to cause line-driven winds.

We suggest therefore that accretion disks in AGNs can be sources of winds accelerated by means of radiation pressure in the lines.

Let's concentrate on the radiation dominated innermost region (RDR) of an accretion disk around a super massive black hole, say $R \lesssim 10^{3.5} R_s$ (where R_s is the Schwarzschild radius). We find that typically the luminosity emerging from this region is close to the local Eddington luminosity calculated on the basis of electron scattering, say $L/L_E > 0.1$. However $L/L_E < 1$ and the matter is gravitationally bound. The height of the atmosphere above the symmetry plane Z=0 is

$$Z_1 \simeq 3 \times 10^{15} \left(\frac{\dot{m}}{\dot{M}_E}\right)\left(\frac{M_{BH}}{10^9 M_\odot}\right)(1 - L/L_E)^{-1} \qquad (3)$$

where $\dot{M}_E = 3 \times 10^{10} M_\odot/yr$ is the Eddington accretion luminosity, M_{BH} is the mass of the black hole in solar masses. Optically thick accretion disks around super massive BHs appear to be relatively cold in contrast with disks around stellar mass BHs. Under such conditions hydrogen is completely ionized while several metals (C, N, O etc.) are partly ionized. The metals provide substantial line opacity. Although electron scattering opacity in the disk atmosphere dominates over the Rosseland mean one, the line opacity may be greater at some wavelengths.

As is well known absorption of radiation in resonance lines of abundant elements gives rise to winds in early type stars. Our basic claim is that a similar situation exists in our case. The radiation force in the lines is sufficient to accelerate the matter to escape velocities.

The origin of the wind can be understood in the following way: In the case of static atmosphere we have for the monochromatic optical depth

$$\tau_\ell = \int_{Z_a}^{\infty} \varkappa_\ell \, \rho \, dz \qquad (4)$$

while in a differently accelerating atmosphere we have

$$\tau_\ell = \varkappa_\ell \, \rho \, v_{th} \left| \frac{dv}{dz} \right|^{-1} \tag{5}$$

Here v_{th} is the thermal velocity of the absorbing ion, $\varkappa_\ell$ is the properly normalized line absorption coefficient, z is the vertical coordinate, v the gas velocity and $z_a = z$ (τ(continuum)=1) is the base of the atmosphere.

The above expressions yield similar results in the case of static atmospheres namely $\tau_\ell \to \infty$. However, the actual optical depth in the line is reduced drastically if velocity gradients are present. The reduction in the number of absorbers is due to the Doppler shift in the absorption lines.

The radiation force due to electron scattering opacity is given by

$$F_{es} = \frac{\varkappa_{es} \, \mathcal{F}(R)}{c} \tag{6}$$

where $\varkappa_{es} = 0.2(1+x) \, cm^2 g^{-1}$ is the Thomson scattering opacity and $F(R)$ is the frequency integrated radiative flux of the disk.

The net force per unit volume $F_{\ell,tot}$ due to absorption lines is given by

$$F_{\ell,tot} = \sum_\ell \frac{\mathcal{F}(R, \nu_\ell)}{c} \varkappa_\ell \, \rho \, \frac{(1 - e^{-\tau_\ell})}{\tau_\ell} \tag{7}$$

where ν_ℓ is the local frequency of the line-center, $\varkappa_\ell$ is the line absorption coefficient normalized by the line profile function in the absence of stimulated emission, namely,

$$\varkappa_\ell \, \rho = \frac{\pi e^2}{m_e c} f_\ell \, n_\ell \tag{8}$$

where f_ℓ and n_ℓ are the oscillator strength and number density of the absorbing ions in the wind.

We note that a decrease of the optical depth of the line frequency increases the line radiation force on the outflowing species. The line acceleration continues till the non-linear effects saturate. At this point the wind reaches the asymptotic velocity.

The wind is therefore a result of having cool matter over hot radiation. In hot matter there are too few lines and in cold radiation there are not enough photons for the acceleration.

III NUMERICAL SIMULATION OF WINDS FROM ACCRETION DISKS

The theory of line-driven stellar winds was initiated by Lucy and Solomon (1970) and later expanded by Castor, Abbot and Klein (1975, hereafter CAK).

We adapt the CAK theory to the present case by:
(a) allowing for non-LTE effects
(b) coupling the hydrodynamics with the ionization balance equations.

The true problem of simulating the radiation field and the mass flow generated by an accretion disk is clearly a two-dimensional one. In view of the exploratory nature of our model and the complications involved in true two-dimensional calculations, we choose to approximate the 2D flow by a 1D one with a geometrical correction factor

$$f(z, R) = 1 + \frac{(z-z_0)^2}{R^2} \tag{9}$$

While sufficiently close to the disk (low z) the wind possesses a plane parallel symmetry. However, further out the flow streamlines diverge due to centrifugal effects, transverse gradients of the gas thermal pressure and the variation of the local gravitational acceleration with z. The geometrical factor f provides a continuous change of the geometry from a plane one to a spherical one.

The radiation field at (R,z) close to the disk can be approximated by a superposition of the local flux generated by the disk and hard X-ray flux from the central source. The atmosphere of the disk is optically thick in the radial direction. Consequently, the relative contribution of the central X-ray source to the flux at some point (R,z) varies with height. The contribution vanishes at the base of the wind and increases with z.

The local flux is assumed to be a Planck spectrum modified by electron scattering.

The X-ray flux from the central source is attenuated by

$$\xi \cong e^{-\tau(R, z)} \tag{10}$$

where $\tau(R,z)$ is the optical depth to X-rays in the radial direction. At the base of the wind $\xi \ll 1$ while $\xi \to 1$ at the top of the atmosphere.

The degree of anisotropy of the central X-ray source is unknown, we assume therefore the radiation force due to electron scattering of the central X-ray photons to be isotropic and sub-Eddington.

Since the wind ascension time is much shorter than the disk orbital period we neglect the rotation of the disk.

A steady state flow is assumed and the governing equations are:
Continuity:

$$\rho v_z f = \dot{M} \tag{11}$$

Momentum conservation:

$$(\sigma_z^2 - \sigma_s^2)\frac{d\sigma_z}{dz} = \frac{\sigma_z}{z}\left[\frac{2\sigma_s^2 z^2}{f R_o^2}\right.$$

$$\left. - z\frac{d\sigma_s^2}{dz} - \frac{GM_{BH} z^2}{f^{3/2} R_o^3}[1 - \Gamma_{es} - \Gamma_{\ell}]\right] \quad (12)$$

and energy conservation:

$$\frac{dT}{dz} = \frac{2}{3}\frac{1}{\sigma_z}\left(\frac{H - \Lambda}{k_B n} - T\left[\frac{d\sigma_z}{dz} + \frac{2z\sigma_z}{f R_o^2}\right]\right) \quad (13)$$

Here $\dot{M}$ is the mass outflow rate per unit of disk area. Γ_{es} and Γ_{ℓ} are the ratio of the electron scattering and lines radiation forces respectively to the z-component of the gravitation. v_s is the sound velocity. H and Λ are the heating and cooling rates per unit volume.

We note that since the radiation forces depend on the velocity gradient the momentum equation is not linear and contains the velocity gradient in an implicit form. A unique hydrodynamical solution is determined by the condition of continuity of the acceleration through the critical point.

The local ionization and thermal balance in the line acceleration phase are affected only by the black body photons and includes all ions of H, He, C, N and O with relative solar abundances.

The assumption of local ionization and thermal balance is justified by the facts that the time scales for ionization/recombination and heating/cooling are much shorter than the hydrodynamical time scale. This condition is valid for number densities above 10^7 cm^{-3} and is satisfied by all our wind models.

Fifteen resonance lines in the wavelength range of 500-1600 Å are included in the calculations of the line force. Doublets are treated as single lines with an effective oscillator strength equal to the sum of the individual components. Inasmuch as the outflow region is optically thin to continuum absorption and to electron scattering of the local disk radiation, LTE is not valid and the ionization equilibrium is calculated assuming steady state. Also, photon production and backscattering in the wind are neglected.

The following processes are taken into account in the thermal balance: Compton heating, free-free absorption heating, photoionization and resonance line heating. The cooling processes are: inverse Compton, free-free, dielectronic radiative recombination and collisionally excited line emission cooling.

The wind structure can be characterized by three distinct regions:
(a) line accelerated outflow with large velocity gradients, plane parallel symmetry and isothermal temperature distribution.

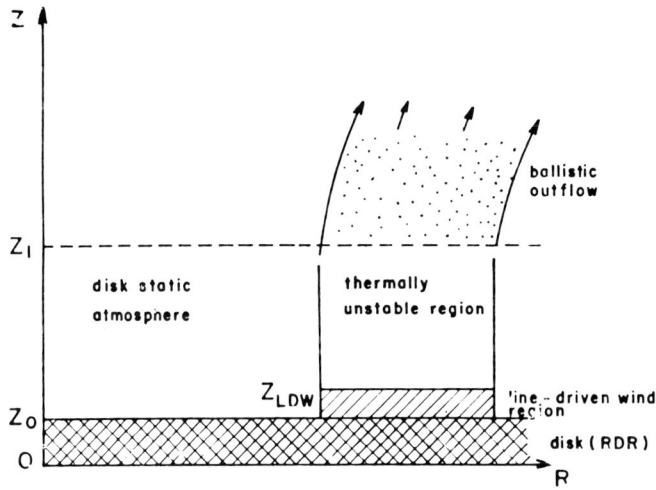

Fig. 1 The outflow from accretion disk (not to scale)
The flow is symmetric with respect to Z and R axes.

(b) inertial flow, almost constant velocity, plane-parallel symmetry, positive temperature gradient and thermally unstable flow.
(c) two component ballistic flow. Conical and slowly decelerating isothermal wind.

The self consistent hydrodynamical calculations were performed only for the first region. Single particle approximation is used in the two other regions. The justification for this treatment is provided in the next section.

The details of the disk model are not crucial because the wind originates from the atmosphere and we use only global parameters which do not vary from one disk model to the other.

The numerical calculations show that a line driven wind (LDW) originates from the disk in the radial region

$$R_o = 3.4 \times 10^{16} \, \alpha^{2/15} \left(\frac{\dot{m}}{\dot{m}_E}\right)^{2/15} \left(\frac{M_{BH}}{10^9 M_\odot}\right)^{-2/15} \text{cm} \quad (14)$$

with radial width of $\delta R_o \simeq 10^{16}$ cm, where the effective temperature is in the range $3 \times 10^4 - 5 \times 10^4$ °K.

The maximal velocity of the wind is reached at $z_{LDW} \simeq 1.1\, z_o$. At this point the velocity gradient becomes small and the Sobolev approximation for the radiative force calculation breaks down.

The maximal velocity of the line-driven wind was found to scale with the local escape velocity from the disk $v_{esc} = (GM_{BH}/R)^{1/2}(1 - \Gamma_{es})^{1/2}$ being slightly higher than v_{esc}. The maximal velocity v_{LDW}/v_{esc} changes as $R^{-1/2}$ above the ring where the wind occurs, so that the ratio v_{LDW}/v_{esc} remains approximately constant (the CAK theory for spherical stellar winds yields a similar scaling). Note, that the binding energy of the matter in a Keplerian disk is smaller by a factor of 2 compared to the non-rotating case. The wind density at z_{LDW} scales as $n_{LDW} \sim R^{-2}$. This is to be expected, as $n_{LDW} v_{LDW}$ must decrease as $\sim R^{-3}$ at least, because of the variation with R of the local disk flux.

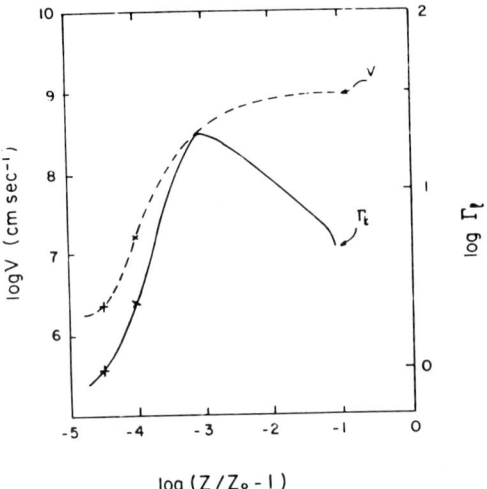

Fig. 2. The velocity (dashed line) and the line force profiles for the line-driven wind. The crosses show the position of the sonic point and X-es show the critical point in the flow. See Table 1 for physical parameters.

The physical properties of the line driven wind from an α-disk model are summarized in Table 1. These results are taken as initial conditions for the further wind development. The profile of the radiation force (and velocity) are shown in Fig. 2. The electron temperature is almost constant.

In the second phase $z > z_{LDW}$ the wind propagates at essentially constant velocity, (plane-parallel symmetry is still valid) and becomes thermally unstable.

	Values at the sonic point $\approx z_0$	Values at $z_{LDW} \approx 1.1 z_0$
n_9 cm^{-3}	100	0.2
v_9 cm sec^{-1}	2×10^{-3}	0.9
T_4 °K	3.3	3.4
$\dot{M}_{WIND}$ (M$_\odot$/yr)	10	10

Table 1. The sonic and asymptotic line-driven wind (LDW) properties at $R_0 = 3.4 \times 10^{16}$ cm. The effective temperature of radiation is $T_{eff} = 4 \times 10^4$ °K. $\tau_{es} = 0.8$.

IV THERMAL INSTABILITIES AND FORMATION OF CLOUDS IN THE WIND

The resonance lines in the line-driven wind play a double role: thermal and dynamical. First, the resonance lines act as a thermostat in keeping the temperatures relatively low $10^4 - 10^5$ °K. Second, the lines provide the momentum absorption medium, the existence of which is required to drive the plane parallel gas off the disk. The radiative acceleration of the LDW's from our accretion disk models is extremely efficient and as a consequence the scale height is small compared to the scale height of the relevant atmospheric layers. The wind reaches the velocity of escape while still in the atmosphere. This fact is important because the disk atmosphere provides the necessary shielding from the hard X-ray flux. If it were not for the shielding the wind would be completely ionized and heated on a very short time scale.

The temperature of the wind in the photosphere is determined in the shielded region essentially by the electron scattering modified black body radiation. For $z > z_1$, above the shielded region, the temperature is determined by the equilibrium between compton heating and inverse compton cooling. The effect of other factors, like attenuation, atmospheric density distribution, etc. is small.

The compton equilibrium temperature depends only on the X-ray spectrum. Using the Krolik, McKee and Tarter's (1981, hereafter KMT) composite spectrum results in compton temperature of $T_8 \approx 1$. Since the wind has a small optical depth it will not affect the X-ray spectrum.

The gradual heating of the wind will not affect the plane-parallel character of the ascending wind, nor will it influence its dynamics, as long as the wind stays supersonic. Thus in the post LDW phase it can be described as inertial flow with constant density and velocity but increasing pressure. The atmospheric region over which the inertial flow takes place is $\Delta z_{16} \sim 1$.

As long as the main heating and cooling processes in the wind are photoionization and radiative recombination, the gas is thermally stable. Following Field (1965), the necessary condition for the onset of an isobaric thermal instability in gas is that the heating rate at constant pressure must increase with temperature faster than the cooling rate, i.e.

$$\left[\frac{\partial}{\partial T} (H - \Lambda) \right]_P > 0 \qquad (15)$$

where H and Λ are the dominant heating and cooling rates of the gas. The dominant cooling in the wind changes at $T \sim 10^{6.5}$ °K. For $10^{6.5} < T < 10^8$ °K, when the gas is fully ionized, Compton heating by the central X-ray photons becomes dominant (and is $\sim T$ at constant pressure) while free-free provides the main cooling (being $\sim T^{1/2}$ at constant pressure), because it is more effective than inverse Compton cooling. As it follows from inequality (15), small perturbations will grow rapidly under such conditions in the initially smooth flow, which separates into two phases. The regions of enhanced density, the first phase, will begin to cool, initially by free-free emission (and will therefore be compressed by the rising external pressure) and later on by line cooling, while the second phase, the smooth background will heat up.

The final state consists of a two-component flow. The low temperature - high density one is stabilized at $T_{c4} \sim 1$, as the gas starts to recombine and the radiative cooling shuts off (from now on the subscripts c and h will refer to cold and hot phases respectively). The high temperature - low density phase is stabilized at $T_{h8} \sim 1$ for standard AGN spectra.

We consider next the timescales of relevant processes, which characterize the linear stage of the thermal instability and define the final parameters of cooling condensations (clouds). Their stability is discussed in the subsequent section.

The upper limit on the timescale for cloud formation is the wind crossing time of the disk atmosphere:

$$\tau_{cross} \approx 10^7 \, \Delta Z_{16} \, V_9^{-1} \; sec. \qquad (16)$$

Let λ be the initial size of the perturbation. In order to cool at approximately pressure equilibrium with the surroundings, the sound crossing time $\tau_s \simeq 10^6 \lambda_{14} T_{c8}^{-1/2}$ sec of the condensation must be shorter than the free-free cooling time

$$\tau_{cool} = 2 \times 10^6 \, T_{c8}^{1/2} \, n_9^{-1} \, \text{sec.} \tag{17}$$

As the condensation contracts

$$\tau_s \sim T_c^{-1/6}$$

$$\tau_{cool} \sim T_c^{3/2}$$

and eventually the cooling time will become shorter than τ_s. The cooling will proceed isochorically ($n_c \sim$ const) and dynamic effects will become important, namely a collapse of the cloud.

The collapse can be described approximately by

$$\rho_c \ddot{\lambda} \approx \nabla P_h \tag{18}$$

Here $\ddot{\lambda} \approx \lambda/(\Delta t)^2$ is the acceleration, ∇p_h is the gradient of the external pressure across the cloud and ρ_c is the cloud mass-density. The approximate analytical solution of the equation yields the following estimate for the largest perturbation which can collapse in time τ_{cross}

$$\lambda_{o,max} \approx V_{SH} \left(\frac{T_{cf}}{T_{hf}} \right)^{1/6} \quad \tau_{cross} \simeq 2 \times 10^{14} \, T_{h8}^{1/2} \, \Delta Z_{16} \, V_9^{-1} \tag{19}$$

cm

where subscript f is for "final", and we used $T_{cf}/T_{hf} \approx 10^{-4}$.

Perturbation with $\lambda < \lambda_{min}$ will heat up and damp out by thermal conduction. By equating the rate of free-free cooling to the rate of conductive heating of the perturbation we find the limiting wavelength to be

$$\lambda_{min} = 9 \times 10^{12} \, T_{h,8}^{3/2} \, n_9^{-1} \quad \text{cm} \tag{20}$$

The final sizes of the clouds can be estimated assuming their contraction takes place in all three dimensions, i.e. $n_h \lambda_0^3 \sim n_c \lambda_f^3$:

$$4 \times 10^{11} \, \text{cm} \lesssim \lambda_f \lesssim 10^{13} \, \text{cm} \tag{21}$$

The clouds will contract by a factor ~ 20 in radius, their internal number density will approach $n_{c13} \sim n_{h9} T_{h8}/T_{c4}$, their mass-range being $M_c \simeq 10^{24} - 10^{28}$ g, which is high compared to the values usually quoted in the literature.

Evidently, the radiation pressure in the contracting clouds is comparable to the thermal gas pressure in the vicinity of the accretion disk. The estimate for an α-disk model gives $P_{rad}/P_{gas} < 10$, where an upper limit is obtained for an unattenuated X-ray flux. Two processes will affect the growth of perturbations under such conditions, the Compton drag and the trapping of radiation. An important question is: can the thermal instability be suppressed in the radiation dominated matter?

It was argued (Shlosman, Vitello and Shaviv 1984) that for every P_{rad}/P_{gas} ratio there is a maximal instability length which still permits the condensation to develop. It follows that the contraction of the largest cloud will be affected by Compton drag if subject to unshielded X-ray flux. However, for cases with attenuation factor $\xi = 0.3 - 0.5$, the drag is already reduced below its critical value and will not prevent the collapse. Such values of ξ seem a rather modest requirement for the disk atmosphere.

The second process which must be considered is the radiation trapping due to electron scattering and the resonance line opacities. For the most massive and opaque clouds the diffusion time τ_{leak} of the trapped photons is small compared to the characteristic time for the cloud contraction namely

$$\frac{\tau_{leak}}{\tau_{cross}} \simeq 7 \times 10^{-4} \, T_{h,8}^{1/2} \, \tau_{es} \ll 1 \qquad (22)$$

implying that electron scattering opacity τ_{es} will not affect the growth of perturbations in our case.

Large-scale velocity gradients discussed in the previous section, i.e. dynamical effects in the contracting clouds, will reduce substantially the leakage time for the resonance line opacity trapped photons, by systematically blue-shifting them (Rees and Ostriker, 1977, and references therein). Large densities ($n_c > 10^{11}$ cm^{-3}) encountered in the cooling clouds will enhance the collisional de-excitation of the resonance excited levels (Davidson and Netzer, 1979). An additional important point to be stressed here is that the clouds which form under the present scenario may be somewhat hotter than the usual emission-line clouds, say at $T_{c4} \simeq 3 - 4$ instead of $T_{c4} \simeq 1 - 2$, which would greatly affect their visual appearance and reduce the density of the trapped resonance photons.

It is difficult to estimate the initial amount of angular momentum of the collapsing cloud. The disk and its atmosphere are highly turbulent and the velocity field chaotic on all relevant scales up to $\sim Z_0$. In some cases, the angular momentum of the cloud will affect the latest stages of the collapse, possibly leading to cloud fragmentation. Part of the cloud internal rotation will be dissipated by the bulk motion of the fragments.

The picture of a thermal instability in the wind close to the disk is attractive mainly because of the relatively high densities encountered, which result in short timescales in spite of the high-luminosity central X-ray source.

V. ACCRETION DISKS IN AGNs: A CLUE TO BEL REGIONS?

Clouds formed as a result of thermal instability are in pressure equilibrium with the confining medium. The non-attenuated X-ray flux restricts the existence of the cool phase to a relatively narrow range of the ionization parameter (KMT)

$$0.3 \lesssim \Xi \lesssim 10. \qquad (23)$$

Assuming $L_{ion}/L \sim 0.1$ (Oke et al. 1970) we estimate $\Xi \approx 1.6/P_{17}$ in the wind region for the α-disk model. Here $P_{17} \equiv n\,T/10^{17}$. Thus, the two phases will coexist for $0.1 < n_{c,13} < 5$.

The amount of matter which condenses into clouds cannot be estimated without detailed calculations and knowledge of the attenuation factor $\xi(z)$. A too efficient cloud formation reduces the density in the hot gas and the pressure support for the cool phase. Consequently the clouds expand on a short sound crossing time scale τ_s, increase their ionization parameter over the critical one given above, evaporate and increase the external pressure. Even under such conditions a large fraction (e.g. $\sim 1/2$) of the gas in the wind can exist as cold clouds.

As we shall see below, the only dynamical influence of radiation on the hot and cold components of the wind at $z > z_{LDW}$, is to modify the effective mass of the BH to $M_{BH}(1 - \Gamma_{es})$. Our approach, therefore, will be to specify the flow streamlines by single particle trajectories in 3D-space (ballistic approach, see also Icke, 1977). The flow is restarted at $z = z_{LDW}$ (where the calculation of the line-driven wind was terminated), its initial velocity consists of the z-component, taken to be the maximal velocity of the line-driven wind, and the azimuthal component in the disk plane, given by Keplerian rotation.

The trajectory of a single particle that escapes from a rotating disk around BH and has an initial velocity directed along the rotation axis is an opening helix and lies on a conical surface. The trajectory becomes radial in less than one orbital time, $2 \times 10^8 m_9 (R/R_o)^{3/2}$ sec, where R_o is a typical distance of the wind base from the SBH.

Rotation can be neglected for $R/R_o > 5-10$ where R is the radial coordinate along the streamline. The outflow occurs over a vast region between the two radial conical surfaces forming a constant angle with the z-axis $(v_{LDW}/v_{esc})^{-1} \approx$ const. Far enough, when the flow is already radial, we can assume constant velocity and approximate the density profile by $\sim n_{LDW}(r/R_o)^2$. This expression is correct to within a factor of 2, because of the implicitly assumed constant velocity. Although the wind is highly supersonic with initial Mach number ~ 10, the internal pressure results in a small additional (~ 10 percent) and constant widening of the cone with $v_{sh} \approx 10^8 T_{h8}^{1/2}$ cm sec^{-1}, as it is not collimated by an external pressure. Thus on the large scale $\sim 10^{17} - 10^{18}$ cm the symmetry of the wind from an accretion disk strongly differs from a spherical one and has a bi-conical character. Using the terminology adopted in the literature for the extragalactic jets (e.g. Readhead, Cohen and Blandford, 1978), the high-velocity outflow from an accretion disk resembles an unconfined free jet.

The large scale Kelvin-Helmholtz instabilities accompanying the propagation of jets confined by an external medium, are not important in the latter case.

What will be the effect of the radiative flux on the newly formed clouds? The resonance lines are very optically thick for clouds with column densities $n_c \lambda_f > 10^{17}$ cm^2. Being optically thick also in the $L\alpha$ continuum and having column densities n_c in excess of the critical column density

$$N_{CRIT} \simeq 10^{23} L_{47}^{-1} m_9^{-1} \text{ cm}^{-2} \tag{24}$$

the clouds cannot be accelerated by means of radiation pressure in hydrogen continuum. The merit of the present model is that the clouds with high column densities are formed with the flow bulk velocity and thus are expected to be hydrodynamically stable. As the external pressure decreases the clouds expand ($N_c \sim (R_0/R)^{4/3}$). Clouds formed with $40 < N_{c23} < 10^3$, well above N_{crit}, are subject to radiative acceleration by hydrogen continuum further outward at $6 \times 10^{17} \lesssim r \lesssim 6 \times 10^{18}$ cm, where the densities decrease to $10^7 - 6 \times 10^9$ cm^{-3}.

The question whether the clouds can be stable against the pressure of internally generated resonance line photons, mainly $L\alpha$ photons, is an open one. The pressure of the $L\alpha$ photons depends strongly on the cloud temperature, which is expected to be somewhat higher in the vicinity of the disk, compared to its "classical" value $T_{c,4} = 2$. The overall hydrogen emissivity is expected to be reduced in the very high density ($N_c > 10^{10}$ cm^{-3}) clouds, making them very difficult to observe.

Iron lines and Balmer continuum can be responsible for cloud cooling, although no detailed calculation has been performed yet. The high densities encountered in clouds will affect the line ratios even for constant ionization parameter. In fact it is difficult to understand how any model of outflowing BEL clouds can avoid such high densities. It seems that the clouds formed in the high velocity wind close to the disk are destroyed mainly by evaporation.

The classical evaporation rate applies whenever the saturation parameter $\sigma_0 = 1.3 \cdot 10^{-2} T_{h8}^2 (n_{h9} \lambda_{f13})^{-1} (r/R_0)^{4/3} \simeq 0.03 - 1$ (KMT; Balbus and McKee, 1982). For $1 < \sigma_0 < 100$ the evaporation is saturated. Radiative losses fully compensate the conductive heating of clouds with $\sigma_0 \lesssim 0.03$. This means that clouds formed within λ_f-range given above and with $0.013 < \sigma_0 < 0.3$ start to dissipate on the classical evaporation timescale:

$$\tau_{class} \simeq 1.2 \times 10^{10} n_{c13} \lambda_{f13}^2 T_{h,8}^{-5/2} (r/R_0)^{-2/3} \text{ sec} \tag{25}$$

taking into account the cloud expansion when moving outward in pressure equilibrium with the hot medium $(n_c \sim r^{-2}, \lambda \sim r^{2/3})$. This timescale is applicable until the evaporation parameter σ_0 becomes of order unity at $r_{class} \simeq 26 R_0 (n_{h9} \lambda_{f13})^{3/4} T_{h8}^{-3/2}$ cm. For $r > r_{class}$ the characteristic timescale for saturated evaporation is given by KMT:

$$\tau_{satur} \simeq 6.1 \times 10^5 N_{c23}^{7/6} P_{17}^{-1} (r/r_{class})^{4/9} \text{ sec} \qquad (26)$$

which must be compared with the flow crossing time of BELR:

$$\tau_{BELR} \simeq 8.8 \times 10^8 v_9^{-1} (r/r_{class}) \text{ sec} \qquad (27)$$

Figure 3 shows all mentioned timescales. Evidently, the largest clouds have the capability to cross BELR.

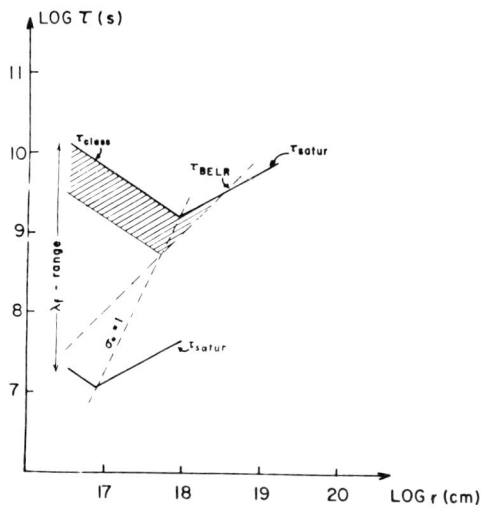

Fig. 3   Comparison of relevant timescales for cloud evolution. The shadowed region restricts the clouds that survive BELR crossing.

VI. SUMMARY AND CONCLUSIONS

We have shown that the cool accretion disks, if present in AGNs can be sources of LDWs. Subcritical disks are thus capable of producing winds which are optically thin in the continuum and are accelerated by radiation pressure in UV-resonance lines.

The necessary conditions for initiating a wind is that the local radiative flux peaks in the energy range of the existing lines. The base of the wind must therefore be shielded from the central region, which is assumed to emit a power law flux. The disk atmosphere can provide such shielding.

The winds are characterized by steep velocity gradients and asymptotic velocities in the range of 10^9 cm/sec, generally in the same bulk part of the observed velocities in BEL regions of QSOs and Sy galaxies. The winds become thermally unstable upon heating. The onset of the instability is apparently sufficiently fast to involve a wide range of wavelengths.

The so formed clouds are distinguished from usual BEL clouds by the high density which causes a strong suppression of hydrogen emissivity.

REFERENCES

1. Balbus S.A. and McKee C.F. 1982 Ap. J. 252, 529.
2. Bardeen J. and Petterson J.A. 1974 Ap. J. Lett. 195, L65.
3. Castor J.I., Abbot D.C. and Klein R.I. 1975 Ap. J. 195, 157.
4. Davidson K. and Netzer H. 1979, Rev. Mod. Phys. 51, 715.
5. Field G.B. 1965 Ap. J. 142, 531.
6. Hutchings J.B. and Gower A.C. 1984 preprint.
7. Icke V. 1977 Nature 266, 699.
8. Krolik J.H., McKee C.F. and Tarter C.B. 1981 Ap. J. 249, 422.
9. Lucy L.B. and Solomon P.M. 1970 Ap. J. 159, 879.
10. Malkan M.A. and Sargent W.L.W. 1982 Ap. J. 254, 22.
11. Oke J.B., Neugebauer G. and Backlin E.E. 1970 Ap. J. 159, 341.
12. Readhead A.C.S., Cohen M.H. and Blandford R.D. 1978 Nature 272, 131.
13. Rees M.J. and Ostriker J.P. 1977 Ap. J. 179, 541.
14. Shakura N.I. and Sunyaev R.A. 1973 Astr. Ap. 24. 337.
15. Shlosman I., Vitello P.A. and Shaviv G. 1984 Ap. J. (in press).
    Shlosman I., Vitello P.A. and Shaviv G. 1985 (in preparation).
16. Wilson A.S., Baldwin J.A. and Ulvestad J.A. 1984 preprint.

DISCUSSION

WANDEL: What is the scale height of the vertical wind to diverge into radial flow? If it is of the order of the radius of the disk ($\sim 3 \times 10^{16}$ cm) as suggested by your figure, it would contradict the observational evidence for radial flow in the BELR, which in Seyfert 1 galaxies (e.g. in NGC 4151) is of the order of a few $\times 10^{16}$ cm.

SHAVIV: First, the flow is conical. Details are given in the paper. Secondly, all scales are of the order of few $\times 10^{16}$ cm. As said in my talk, the change from plane symmetry to spherical (conical) flow is rapid, namely over scales $\simeq 10^{16}$ cm. The figure is not to scale and hence misleading. As stressed in my talk, if you start spherically symmetric, the total mass loss is prohibitively high. Our model circumvents this difficulty by starting from the disk with plane symmetry.

CARROLL: Since your wind accelerates so rapidly, all clouds will be born with the terminal wind velocity. This produces a rectangular emission line profile shape, unlike the profile wing shapes observed. Also, no possibility exists for producing line profile differences since all clouds have the same spacial velocities regardless of their physical conditions and location.

SHAVIV: The clouds form after the acceleration phase. We do not predict any line shapes. The prediction of line profiles must take into account (a) the whole geometry which is a conical flow, so you expect blue shifts and red shifts, and (b) the wind originates from a "region" in the disk and hence you expect a spread in the velocities.

ON THE PAPALOIZOU AND PRINGLE INSTABILITY

Marek A. Abramowicz, Omer M. Blaes and Jufu Lu

International School for Advanced Studies
Strada Costiera, 11
34014 Trieste, Italy

ABSTRACT. The recent discovery of Papaloizou and Pringle that fluid rings are dynamically unstable against global, nonaxisymmetric perturbations if of great importance to the theory of accretion disks. However, before these disks can be proved to be stable or unstable, substantial theoretical progress along the lines scketched in this lecture must be achieved.

1. INTRODUCTION

Thick accretion disks orbiting supermassive black holes form an important basis in the current theoretical discussion of quasars and other active galactic nuclei (Rees 1984). However, the viability of this model was recently questioned by the important theoretical discovery of Papaloizou and Pringle (1984, 1985). They demonstrated that fluid rings, or tori, orbiting central objects are dynamically unstable to global, adiabatic, nonaxisymmetric perturbations.

Fluid tori are not accretion systems and they differ from accretion disks, both thin and thick, in a number of fundamental ways globally. However, in some of their corresponding regions, both tori and thick disks are locally very similar. Is this local similarity relevant to the global instability found by Papaloizou and Pringle? Can one argue that because rings are unstable, thick accretion disks are unstable as well?

In our opinion the answer to the first question is most probably yes, but the answer to the second one must be no.

Thick accretion disks are, in a sense, half way between nonaccretion tori and spherical accretion, sharing the properties of both these flows. Moncrief (1980) has proved that adiabatic, spherically symmetric accretion onto a Schwartzschild black hole is stable with respect to any infinitesimally small perturbation. It is the accretion itself which makes the flow stable by dragging some of the perturbations into the black hole and not allowing the others to grow. Is this result relevant to accretion with angular momentum? Can one argue that thick accretion disks are stable for at least small angular momenta?

Again: most probably <u>yes</u> to the first question, no to the second.

Rings are unstable and spherical accretion is stable, but the question is are accretion disks stable? This is a complex problem, involving the study of nonstationary, 3D, relativistic flows in the curved spacetime of a rotating black hole. One can easily agree with Blandford, Jaroszyński and Kumar (1985) who emphasize the role of numerical methods in this context. (See e.g. Hawley, Smarr and Wilson 1984 for a description of 2D methods of this type.) However, analytic methods can be very useful because of their ability in concentrating on the most important, general aspects of the problem. This gives a wider perspective and deeper understanding. But which are the most important, general aspects of the problem? This is precisely what we do not know yet, as we still do not understand the physical nature of the Papaloizou and Pringle instability, despite interesting suggestions which have been made by Papaloizou and Pringle themselves, Goldreich and Narayan 1985, and Blandford, Jaroszyński and Kumar 1985.

In this lecture we give a review of a few theorems and speculations which, in our opinion, are relevant to the question <u>is the Papaloizou and Pringle instability important for thick accretion disks</u>? We shall try to base our resoning on physical or geometrical intuition rather than on complicated mathematics, which can be found elsewhere (Blaes 1985, Abramowicz, Lu and Livio 1985).

2. THEORY: DIFFERENCES AND SIMILARITIES BETWEEN THEORETICAL MODELS OF ADIABATIC ACCRETION

Fig. 1 shows the three basic models of adiabatic accretion, which will often be used in our discussion of the Papaloizou and Pringle instability. in all three of them it is assumed that the specific angular momentum is constant.

The first panel shows the equatorial planes of the three flows. Note the global, <u>topological</u>, differences and similarities:

ON THE PAPALOIZOU AND PRINGLE INSTABILITY 115

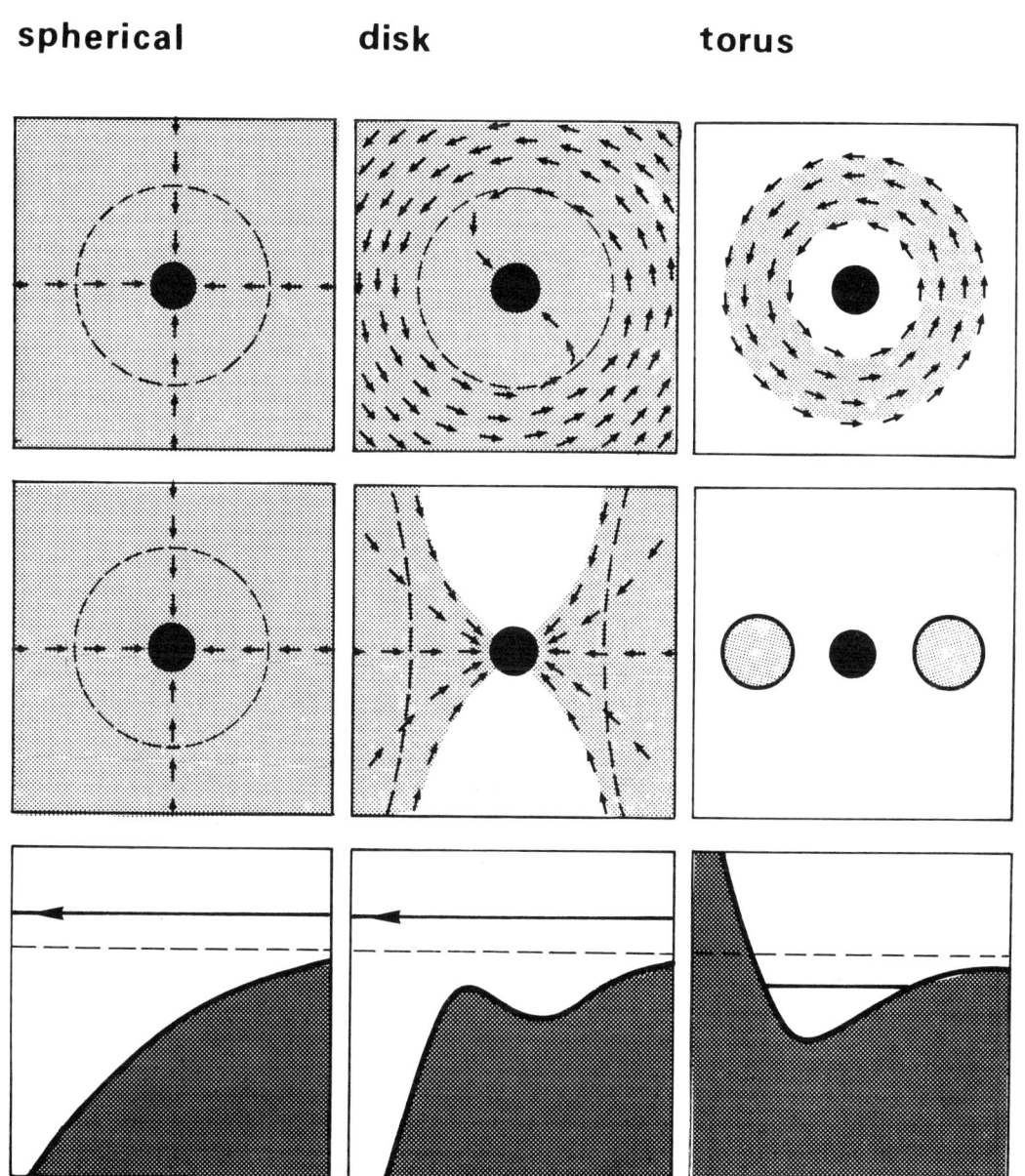

Figure 1. The three models for accretion

	Spherical	Disk	Torus
[1] Unbounded fluid distribution?	yes	yes	no
[2] Sound horizon* present?	yes	yes	no
[3] Flow lines open?	yes	yes	no

Note that the equatorial plane spherical and disk models are identical and they differ from the ring model.

The second panel shows the meridional section (i.e. the plane which contains the symmetry axis) of the flow. This time we have:

	Spherical	Disk	Torus
[4] Sound horizon present?	yes	yes	no
[5] Flow lines open?	yes	yes	no
[6] Free surface present? (Does p=0 somewhere?)	no	yes	yes
[7] Extremal points of the pressure present? (Does grad p=0 somewhere?)	no	yes	yes
[8] Supersonic motion at the free surface present?	no	yes	yes

The third panel shows the effective potential (gravitational and centrifugal) for radial motion on the equatorial plane. The most important points to note are:

	Spherical	Disk	Torus
[9] Is the flow locked in the potential well	no	no	yes
[10] Is the potential barrier at the inner edge of the flow infinite?	no	no	yes

The fact that there is no infinite potential barrier when the centre accretes fluid with constant angular momentum in the case of the disk is

---

*   If there are two different regions in the flow (the inner and the outer) such that it is impossible to send sound signals from the inner to the outer region, the boundary between them is called the sound horizon. The existence of the sound horizon is an important global property of the flow.

connected with general relativistic effects. In the Newtonian case the centrifugal force $\sim \ell^2/r^3$ blows up, and so does the potential when a fluid with ℓ=const goes to the centre, r=0. (We use cylindrical coordinates r, z, ϕ).

Let us now consider (after Abramowicz, Lu and Livio 1985) a model of quasi radial accretion which is confined to a solid angle ω and in which the θ component of the velocity (in spherical coordinates) vanishes. The specific angular momentum ℓ, the specific energy e and the specific accretion rate $\dot{m} = \dot{M}/\omega$ are constant in the space and time. The condition for the regularity of the transonic solution $F(\ell, e, m) = 0$ reduces the number of independent, global flow parameters to two. We use e and ℓ as the parameters and assume e > 0 as only then can flow start from infinity. Then the location of the sonic point r_s and the accretion rate are both functions of e and ℓ. The behaviour of $r_s = r_s(\ell)$ and $\dot{m} = \dot{m}(\ell)$ for a given value of $e < 0.02c^2$ is shown in Fig. 2. One can clearly

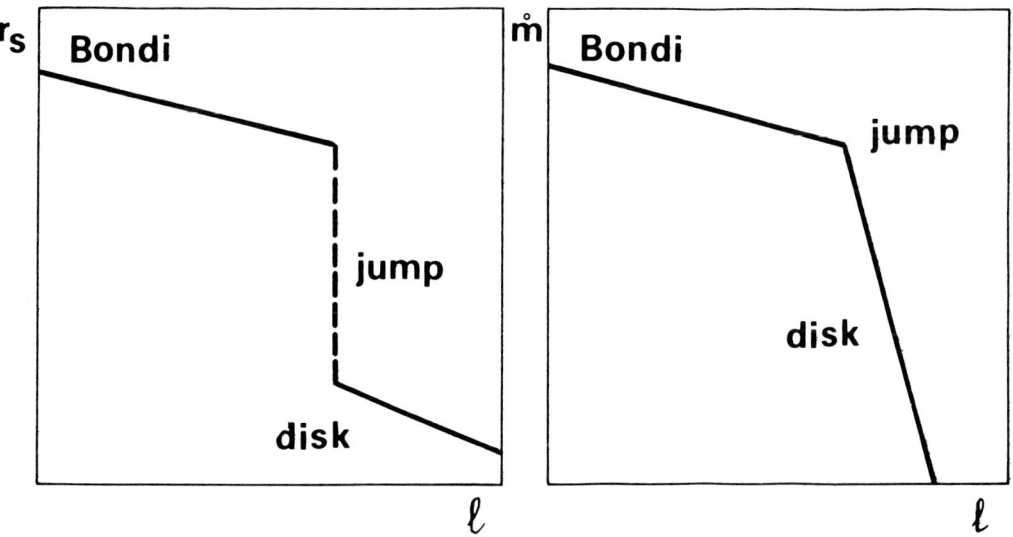

Figure 2. <u>Non uniqueness of the transonic accretion</u>
Location of the sonic point depends discontinuously

define two regimes of accretion. The first one bears a strong resemblance to the classical <u>Bondi</u> solution. The sonic point lies very far from the balck hole and in the transonic part the flow properties are not affected by relativistic or rotational effects. The second, or <u>disk</u>, regime is relativistic: it has no Newtonian analogy. The sonic point lies very close to the hole and the rotation is one of the dominant features in the transonic part of the flow. For $e < 0.02c^2$ the transition between these two regimes is discontinuous: the position of the sonic point experiences a jump when the angular momentum continuously increases.

The reason for such a behaviour should be apparent from Fig. 3 where the integral curves for the accretion velocity are shown. The physically acceptable ones (corresponding to a transonic solution) are distinguished by broken or dotted lines.

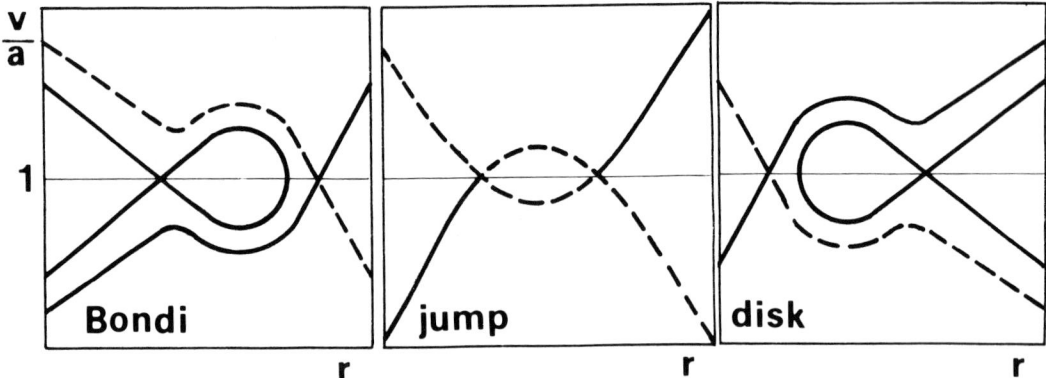

Figure 3. Integral curves for transonic accretion

When the assumption of quasi radial flow is relaxed a hysteresis type behaviour is possible. It was suggested that the real accretion flow can sometimes oscillate quasi-periodically between the "Bondi" and "disk" states (Abramowicz, Lu and Livio 1985); see Fig. 4.

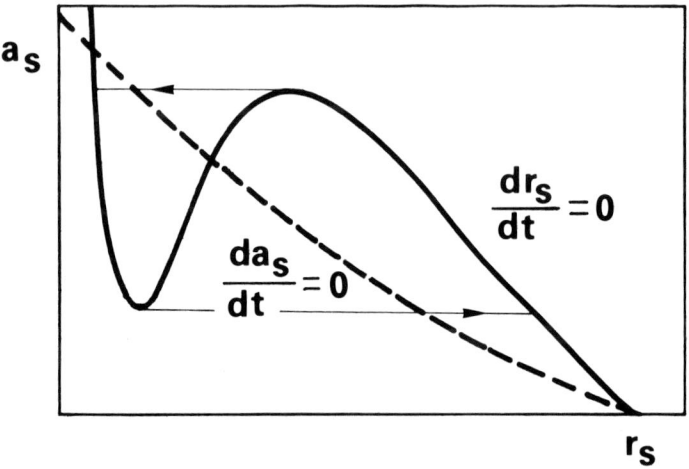

Figure 4. Limit cycle for non stationary transonic accretion

3. THEORY: NON-ACCRETION TORI ARE UNSTABLE

For simplicity we assume that the torus is polytropic and rotates with constant specific angular momentum $\ell = \Omega r^2 = $ const. We restrict our discussion to the case of <u>normal modes</u> which depend on time t and azimuthal angle ϕ through the factor

$$\exp[i(m\phi + \sigma t)] \tag{3.1}$$

with m being the integer azimuthal wavenumber and σ the frequency of the mode. Whem $\text{Im}(\sigma) < 0$ the mode is unstable.

Let us consider a particular meridional cross-section of the torus. The equipressure surfaces $p(r,z) = $ const. form a family of closed curves centred on the point $r = r_c$, $z = 0$ where the pressure has a maximum. Let r_T be the mean cross-section radius of the torus, i.e. the radius of the curve $p(r,z) = 0$. We define the parameter

$$\beta \equiv \frac{r_T}{r_c}. \tag{3.2}$$

when $\beta \ll 1$ the torus is <u>slender</u>. In the limit $\beta = 0$ all the equipressure surfaces shrink into the two central points. However, we can still speak about the structure of the torus in this limit by adopting a new coordinate η such that:

$$p(r,z) = p_c(1-\eta^2)^{n+1}, \tag{3.3}$$

with n being the polytropic index and the subscript c referring to the centre.

Clearly, $\eta = 0$ at the centre and $\eta = 1$ at the surface, independently of the value of β. Although the cross section radius goes to zero together with β, the linear scale is at the same time expanded by the factor $1/\beta$, so the structure of the torus in the $\beta = 0$ limit is not singular in the η coordinate. This is shown in Fig. 5. Note that in this limit all the equipressure surfaces are perfect, concentric circles! In addition not only the angular momentum is constant, but also the angular velocity as $\Omega = \Omega_c + \theta(\beta)$. This means that no geometrical or dynamical property of the torus depends on θ in the $\beta = 0$ limit. Twisting the torus along the lines $\eta = $ const. by <u>any</u> value of θ will not change the torus structure. The fact that the torus in its equilibrium state does not depend on t, ϕ, and θ enables one to write the normal mode solution to the perturbation equation (with the appropriate boundary conditions at $\eta = 0$ and $\eta = 1$) in the form (Blaes 1985b):

$$A(\eta) \exp[i(K\theta + m\phi + \sigma t)] \tag{3.4}$$

where $A(\eta)$ are known functions (related to the Jacobi polynomials). The eigenfunctions of the pulsation equation $W(\eta, \theta, \phi) = A_j(\eta) \exp[i(K\theta + m\phi)]$ form a complete orthonormal set. The corresponding eigenvalues, or frequencies, are

$$\sigma_{(o)} = -\Omega_c m \pm [\frac{1}{n}(2j^2 + 2jn + 2j|K| + n|K|]^{\frac{1}{2}} \Omega_c \tag{3.5}$$

Here the index (0) denotes the $\beta = 0$ limit, and the physical interpretations of m, K and j are

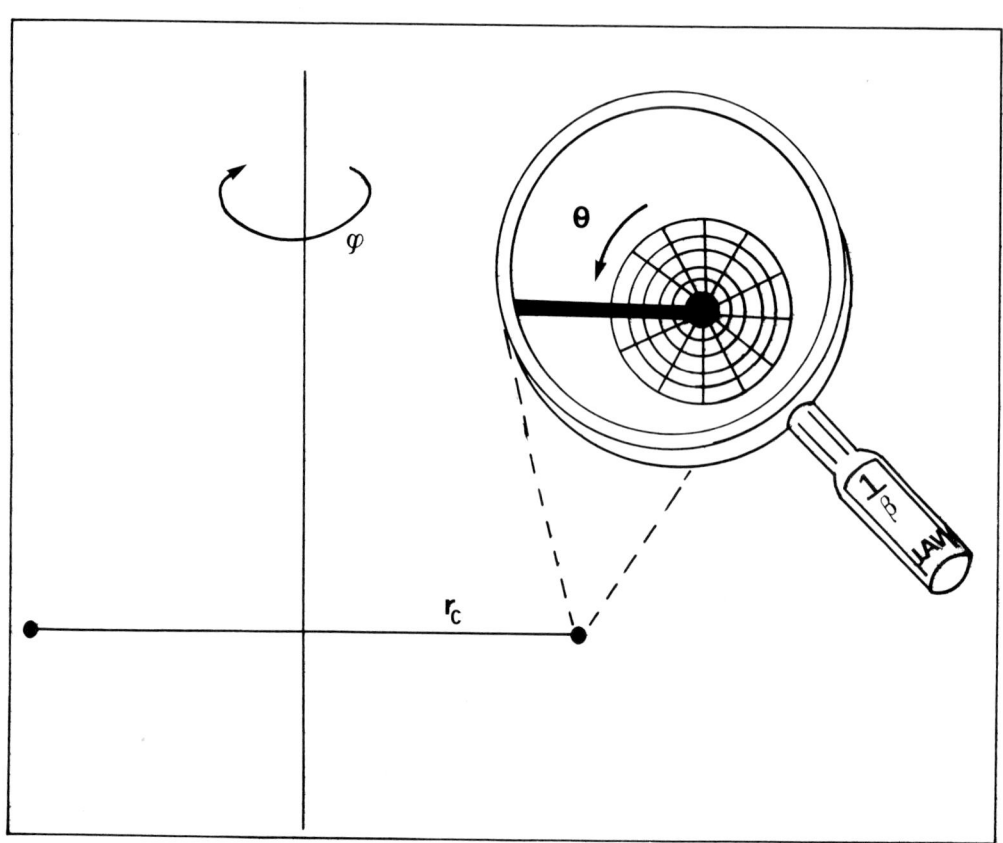

Figure 5. <u>The slender torus limit</u>

m = azimuthal wavenumber (an integer)
 $2|m|$ gives the number of nodes in the ϕ direction
K = cross-section wavenumber (an integer)
 $2|K|$ gives the number of nodes in the θ direction
j = radial wavenumber (a non-negative integer)
 j gives the number of nodes in the η direction.

It appears that all the eigenvalues are real and therefore stable. However the mode for which j = 0 = K:

$$\sigma_{(o)} \equiv \sigma^{CR}_{(o)} = -\Omega_c m \qquad (3.6)$$

is only <u>marginally stable</u>. Such a mode corotates with the basic flow and all the unstable modes arise from it when the slender, constant specific angular momentum, constant entropy torus is slightly changed to be infinitesimally nonslender, to have infinitesimally small gradient of angular momentum, etc. To see this, let us expand the frequency σ with respect to $\beta \ll 1$. For the corotating mode the result is (Blaes 1985b)

$$\sigma = \sigma_{(o)} + \beta\sigma_{(1)} = -\Omega_c m \pm i\Omega_c m \left[\frac{3}{2(n+1)}\right]^{1/2} \qquad (3.7)$$

and this shows instability, as the mode with the minus sign has $\text{Im}(\sigma) < 0$. All other, non- corotating, modes remain <u>stable</u> when $0 \neq \beta \ll 1$. The e-folding time of the unstable mode is of the order of the dynamical time $\tau = 1/|\text{Im}(\sigma)| \sim (1/\Omega_c)(1/m\beta)$. For any nonzero β one can find m big enough (when $m > 1/\beta$) to make $\tau = (1/\Omega_c)$, i.e. dynamical growth of the unstable mode. This is precisely the Papaloizou and Pringle instability.

4. THEORY: SPHERICAL ACCRETION IS STABLE

Moncrief (1980) studied the stability of adiabatic, potential (i.e. vorticity free) accretion onto black holes. He noticed that if the entropy and vorticity perturbations are of bounded extent for a given initial moment $t = t_o$ then, after a finite time, they will be dragged by the flow into the hole and disappear. Thus, not only in the unperturbed flow can the velocity be derived from the potential $v_i = \nabla_i \psi$, but also in the perturbed flow it is $\delta v_i = \nabla_i (\delta\psi)$. We shall denote $\delta\psi \equiv X$.

The accretion flow has a sound horizon which is located outside the event horizon of the black hole. Only the region outside the sound horizon is relevant for the stability. In particular, if the norm of the velocity perturbation $N = \|\delta \vec{v}\|^2$ is bounded by its initial value in the subsonic region, the flow is stable. This statement does not depend on a normal mode analysis.

Moncrief has proved that

$$\frac{dN}{dt} = \dot{M} \int A^2 (\frac{x}{t})^2 dS - \int B^2 (\frac{\partial x}{\partial t})(\frac{\partial x}{\partial T}) \frac{V_T}{V_S} dS - C^2 (\frac{\partial x}{\partial t})(\frac{\partial x}{\partial \phi}) \frac{V_\psi}{V_S} dS \quad (4.1)$$

Here $\dot{M}$ is the accretion rate, A^2, B^2, C^2 some positive functions, T indicates a coordinate tangent to the sound horizon (and orthogonal to ϕ), V_S in the sound velocity, and all the integrals are taken along the sound horizon. In the case of purely radial, spherical accretion onto a Schwarzschild balck hole we have $V_\phi = 0$ and $V_T = 0$. Therefore:

$$\frac{dN}{dt} = \dot{M} \int A^2 (\frac{\partial x}{\partial t})^2 dS \quad \begin{array}{l} < 0 \text{ for accretion } (\dot{M} < 0) \\ > 0 \text{ for wind } \quad (\dot{M} > 0) \end{array} \quad (4.2)$$

Note that, if $dN/dt < 0$ then the norm of the velocity perturbation is bounded by its initial value $N(t_o)$ and the flow is <u>stable</u>.

Obviously, it is the accretion itself which stabilizes the flow by sweeping away the perturbation beyond the sound horizon and making dN/dt negative.

A similar situation takes place when a viscous fluid moves in between two inclined plates (Landau and Lifshitz, 1959). The "accretion" type symmetric flow, with converging fluid lines and a sink, is stable for any inclination angle $\alpha < \pi$ and any Reynolds number R, but the "wind" type flow, i.e. fluid lines diverging from a source, is stable only if $R < R_{max} (\alpha)$ with $R_{max} (\pi) = 0$.

A substantial modification is needed to extende the Moncrief proof to the case of the Bondi type conical accretion discussed in the second section. Although in this case angular momentum can be assumed to be infinitesimally small i.e. $V_\phi \sim 0$, the existence of the supersonic accretion flow on the free surface causes the sound horizon to be highly nonspherical and therefore the component V_T of the accretion velocity is not small. It is not possible to tell if $dN/dt < 0$ in this case without solving the pulsation equation.

Moncrief argues that in the case of accretion of rotating matter the energy density of the perturbations becomes indefinite inside the region between the sound horizon and a <u>stationary limit for the sound metric</u>, which plays a similar role to the ergosphere in the case of a rotating black hole (Fig. 6) The occurrence of this region signals, as it does in Kerr spacetime, the possibility of superradiant scattering. This possibility was not explored and its significance to the accretion process is not known.

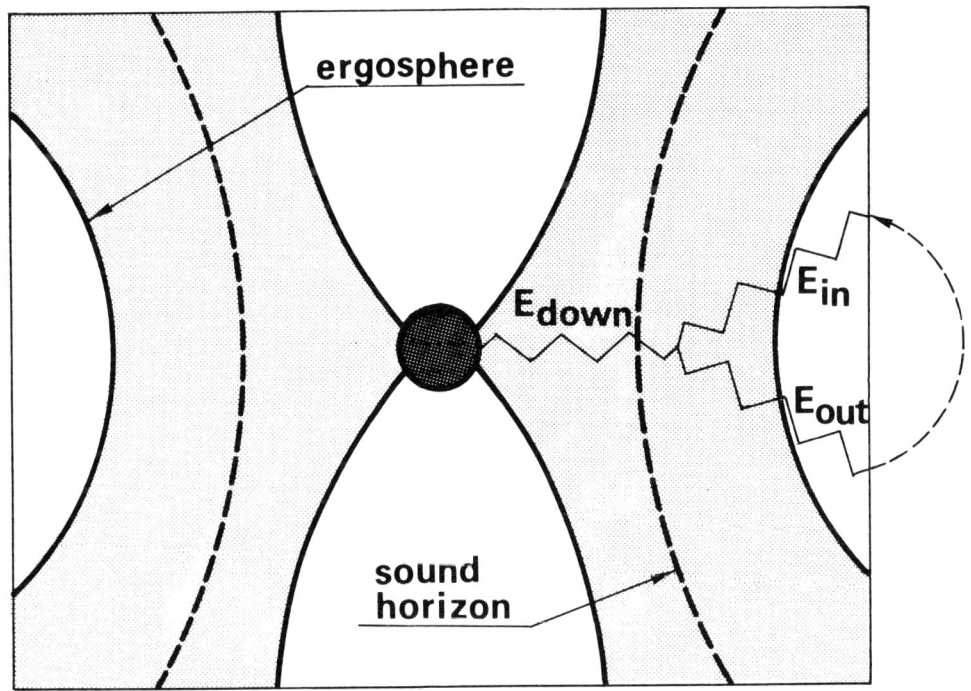

Figure 6. The sound horizon and superradiant scattering

5. SPECULATIONS: CAN FLUID TORI BE STABILIZED?

Several problems can be discussed in terms of the general theory of stability of axially symmetric, stationary flows (see Schutz 1983 for a compact presentation of the theory and detailed references). Other problems are quite particular to the fluid tori only.

(a) Is the corotating normal mode the only unstable mode of the torus?

The answer is no. Blaes (1985, unpublished) found a general analytic solution which gives all the possible pulsations of the slender torus. In addition to the normal modes, there exist modes with time dependence $\sim t^N (N = 1, 2, 3)$. No solution for $\beta \neq 0$ is known at the moment, but it seems that a perturbation method, similar to that used in the case of the normal modes, can be applied to get the most general solution also in the $0 \neq \beta \ll 1$ case. Although preliminary results seem to indicate no other unstable modes when $\beta \ll 1$, such additional modes may exist for finite β because two different stable modes can merge at some value of β and then give rise to an unstable mode. This possibility cannot be studied by perturbation techniques (as σ is not analytic in β there) so a numerical approach is necessary.

(b) Can non-linear effects stabilize the tori?

We do not know the answer to this question. Schutz (1983) remarked: "Although much can be learned from linear perturbation theory, a linear instability is simply a signal that one has to do more work! Once the instability grows into the non-linear regime, it may provide to be relatively harmless, or to produce only small changes in the background flow".

The fact that, at least for slender tori, the Papaloizou and Pringle instability is oscillatory (Blaes 1985a),

$$\frac{\text{Re}(\sigma)}{\text{Im}(\sigma)} \gg 1 \tag{5.1}$$

makes it possible to employ the methods of resonant wave interactions to study the non-linear growth of the instability. There is a possibility that the instability will saturate on a finite amplitude and induce turbulent viscosity. In Blandford's (1984) opinion: "The cores of thick disks may become supported as much by turbulent motion as by centrifugal forces. This offers the exciting prospect of furnishing a believable prescription for the viscosity in accretion disks in general. Unfortunately, it is not yet clear if dynamical instabilities will always be present".

(c) What is the physical nature of the Papaloizou and Pringle instability?

No general sufficient condition for instability is known in the case of nonaxisymmetric perturbations. Nevetheless, several authors have already made several important and interesting suggestions. Papaloizou and Pringle themselves made a few of them:

(1) Constant specific angular momentum tori made of incompressible fluid are stable and finite compressibility had a destabilizing effect.

(2) The instability occurs when there are shearing supersonic motions in the fluid. This always happens for tori as the sound velocity

on the free surface is zero, but the rotation velocity is finite. <u>The same is true for the accretion flows.</u>

(3) The instability occurs when the gradient of the ratio of vorticity to density changes sign. This always happens for tori with non-constant specific angular momentum.

Point (1) is irrelevant for real accretion flows as they are always made of compressible fluids. Points (2) and (3) provide a necessary condition for the instability: either (2) or (3) must occur.

Note that in the case of potential (zero vorticity) spherical accretion, which is stable, neither (2) nor (3) are fulfilled, but in the case of accretion disks at least one of them (point (2)) is.

Using a geometrical optics approximation Goldreich and Narayan (1985) demonstrated that:

(4) A good reflecting surface at either the inner or outer edge of the disk (torus) is a crucial ingredient for the Papaloizou and Pringle instability. (The "reflecting surface" could be, for example, the density gradient, potential barrier, etc.)

It would not be possible to compare this with the analytic solution presented earlier, because of the different method used (one must first re-express the analytic solution in terms of geometrical optics, we plan to do this in the future). However, it is clear that the Goldreich and Narayan result is very important to the stability problem of accretion flows: if there is a good reflecting boundary, the modes can carry energy from the region of negative energy density (inside the corotation resonance) to the region of positive energy density (outside the corotation resonance). This may resemble the possibility of superradiant scattering discussed earlier (Fig. 6). Is it possible that the inward accretion flow is just large enough to limit the amplitude of these modes to a value that gives sufficient angular momentum transport to maintain the flow?

6. CONCLUSION

Are thick accretion disks dead after the discovery of the Papaloizou and Pringle instbaility? We think they are not, or to put this more accurately, there is no way to know the answer to this question before investigating several problems. In view of the global differences between accretion flows and tori one must:

(1) <u>Investigate the global linear stability of adiabatic accretion flows onto black holes.</u>

Suppose the result would be that the Papaloizou and Pringle instability is present. What then? The obvious thing to do will be:

(2) Check whether the instability saturates at a finite amplitude in the non-linear regime.

If the answer is no, then before claiming that thick accretion disks are not viable as a model of active galactic nuclei one must do another thing:

(3) Consider the non-stationary model of accretion (e.g. quasi periodic changes between the "Bondi" and the "disk" state). Is it stable?

The alternative approach would be to make a large number of sophisticated numerical experiments on non-stationary, asymmetric accretion flows onto a black hole.

Although it is of course impossible to foresee the result of these theoretical works, one should note that Nature, maybe, knows the answer very well: in the only case where real accretion disks are actually observed, the instabilities found by Papaloizou and Pringle are absent. Accretion disks in systems similar to U Geminorum (Smak 1984) are in a steady, low viscosity (i.e. also low turbulence) state during thousands of orbital periods!

REFERENCES

Abramowicz, M.A., Lu, J. and Livio, M. 1985, Proceedings of Marcel Grossmann Conference

Blaes, O.M. 1985a, Mon. Not. R. Astr. Soc., 212, 37P

Blaes, O.M. 1985b, Mon. Not. R. Astro. Soc., in press

Blandford, R.D. 1984, preprint

Blandford, R.D., Jaroszyński, M. and Kumar, S. 1985, Mon. Not. R. Astr. Soc., submitted for publication

Hawley, J., Smarr, L., and Wilson, J.R. 1984 Ap. J., 277, 296

Moncrief, V. 1980 Ap. J., 235, 1038

Papaloizou, J.C.B. and Pringle, J.E. 1984 Mon. Not. R. Astr. Soc. 208, 721

Goldreich, P. and Narayan, R. 1985, Mon. Not. R. Astr. Soc., 213, 7P

Papaloizou, J.C.B., and Pringle, J.E. 1985, Mon. Not. R. Astr. Soc., 213, 199

Rees, M.J. 1984, Ann. Rev. Astron. Astrophys., 22, 471

Schutz, B.F. 1983 in Lecture in Applied Mathematics, 20, 99

Smak, J 1984 Pub. A.S.P., 96, 5

DISCUSSION

REES: 1) If non-linear effects quench these instabilities, do you think it may be possible to estimate the effective viscosity, or to specify the rotation law? 2) Would you agree that the most serious consequences of these instabilities are likely to be for the optically thick tori with thermal photospheres, because these require, for consistency, a large column density, and thus a very low effective viscosity?

ABRAMOWICZ: 1) The possibility exists, but unfortunately a nonlinear analysis is extremely difficult. 2) If the only way to stop these instabilities is through the non-linear regime, then I agree.

SCHUTZ: You mentioned that we ought to study adiabatic accretion, but might viscosity have a significant effect on such instabilities?

ABRAMOWICZ: Certainly in the cases of high-m modes, even small viscosity will matter, but may be not for low m.

SMITH: I would like to ask a question that may be irrelevant. Since the slender torus is neutrally stable to θ-rotation, could it not attain vortex ring properties i.e. take of vertically, such as observed smoke rings?

ABRAMOWICZ: The θ-symmetry makes the infinitely slender torus neutrally stable to θ-displacements, just as the axisymmetry makes it neutrally stable to azimuthal displacements. However, this does not mean that it is neutrally stable to an induced θ-velocity field, and therefore I do not think that a vortex ring is a seriuos physical possibility.

PHINNEY: You mentioned in your introduction, but not in the talk, the possibility that the presence of a critical surface ("sound horizon") in real accretion disks, but not in fluid tori, is a possible stabilizing mechanism -it provides an absorbing boundary for unstable sound waves. Could you elaborate, and do you take the possibility seriously?

ABRAMOWICZ: [This elaboration will be found in the text].

THE BIRTH OF AGNs AND QUASARS VIA THE COLLAPSE OF DENSE STAR CLUSTERS
TO SUPERMASSIVE BLACK HOLES

Stuart L. Shapiro
Cornell University
Center for Radiophysics and Space Research
Space Sciences Building
Ithaca, New York 14853

ABSTRACT. We consider the fate of a dense cluster of compact stars – neutron stars or stellar-mass black holes – embedded in the nucleus of an active galaxy. The combined effects of two-body dynamical relaxation (the "gravothermal catastrophe") and star-star collisions and coalescence can drive such a cluster to a relativistic state in a Hubble time. When the core becomes sufficiently relativistic, the cluster inevitably undergoes catastrophic gravitational collapse to a supermassive black hole. This process leads naturally to the birth of supermassive black holes of the "right size" to explain quasars and AGNs: $10^6 \lesssim M/M_\odot \lesssim 10^9$.

1. INTRODUCTION AND MOTIVATION

There is a wealth of observational evidence, albeit circumstantial, that supermassive black holes reside in dense stellar systems (see Table 1). Although some of energetic phenomena observed in AGNs and quasars can be explained by models without black holes, the supermassive black hole picture does furnish a plausible and all-encompassing interpretation of the data.

If we <u>assume</u> that supermassive black holes are the central engines that <u>power</u> AGNs and quasars, simple considerations then allow us to estimate their masses. The observed total luminosities of these sources are typically in the range

$$10^{44} \lesssim L(\text{erg s}^{-1}) \lesssim 10^{47}. \tag{1}$$

Now the maximum luminosity that a black hole of mass M can generate via gas accretion is roughly the Eddington luminosity,

$$L_{Edd} \sim 10^{38} (M/M_\odot) \text{ erg s}^{-1}. \tag{2}$$

Requiring that $L \lesssim L_{Edd}$ then establishes the (minimum) mass range for supermassive black holes in AGNs and quasars:

$$M/M_\odot \gtrsim 10^6 - 10^9 \qquad (3)$$

This is the mass range that is frequently cited in black hole models of these sources.

TABLE 1 CIRCUMSTANTIAL EVIDENCE FOR SUPERMASSIVE BLACK HOLES IN STELLAR SYSTEMS

Phenomenon	Examples	References
optical light cusps	M87	Young et. al. (1978)
velocity dispersion rise	M87	Sargent et. al. (1978)
broad emission lines and rapid time variability	AGNs; quasars (e.g. NGC 4151)	Strittmatter and Williams (1976); Ulrich et. al. (1984)
energetic activity: radio jets; x-rays; nonthermal emission; high total luminosities	AGNs; quasars	Begelman et. al. (1984) Rees (1984)

If the black hole hypothesis is in fact correct, an essential question remains: how and under what circumstances did such a supermassive black hole form? This is the main question we shall attempt to answer below. Our discussion is patterned after Shapiro and Teukolsky (1985a).

Martin Rees and his collaborators have sketched various routes by which a supermassive black hole may form in a dense stellar system (see Rees 1978; Begelman and Rees 1978; Rees 1984). Figure 1 summarizes some of the possibilities. They involve gas dynamical and/or stellar dynamical processes leading to the formation and ultimate collapse of a supermassive star or dense star cluster of compact stars. Rees makes the important point that the general tendency for binding energy per unit mass to increase in dense stellar systems makes the formation of black holes via gravitational collapse almost inevitable.

One of the first routes by which a supermassive black hole might form in a dense stellar system was mapped out twenty years ago by Zel'dovich and Podurets (1965) (hereafter ZP). It is sketched on the far right of Figure 1. ZP argued that the coupled effects of secular core collapse, i.e. the "gravothermal catastrophe" (which they referred to as "stellar evaporation") and star-star collisions and coalescence would ultimately drive the core of a dense star cluster to a

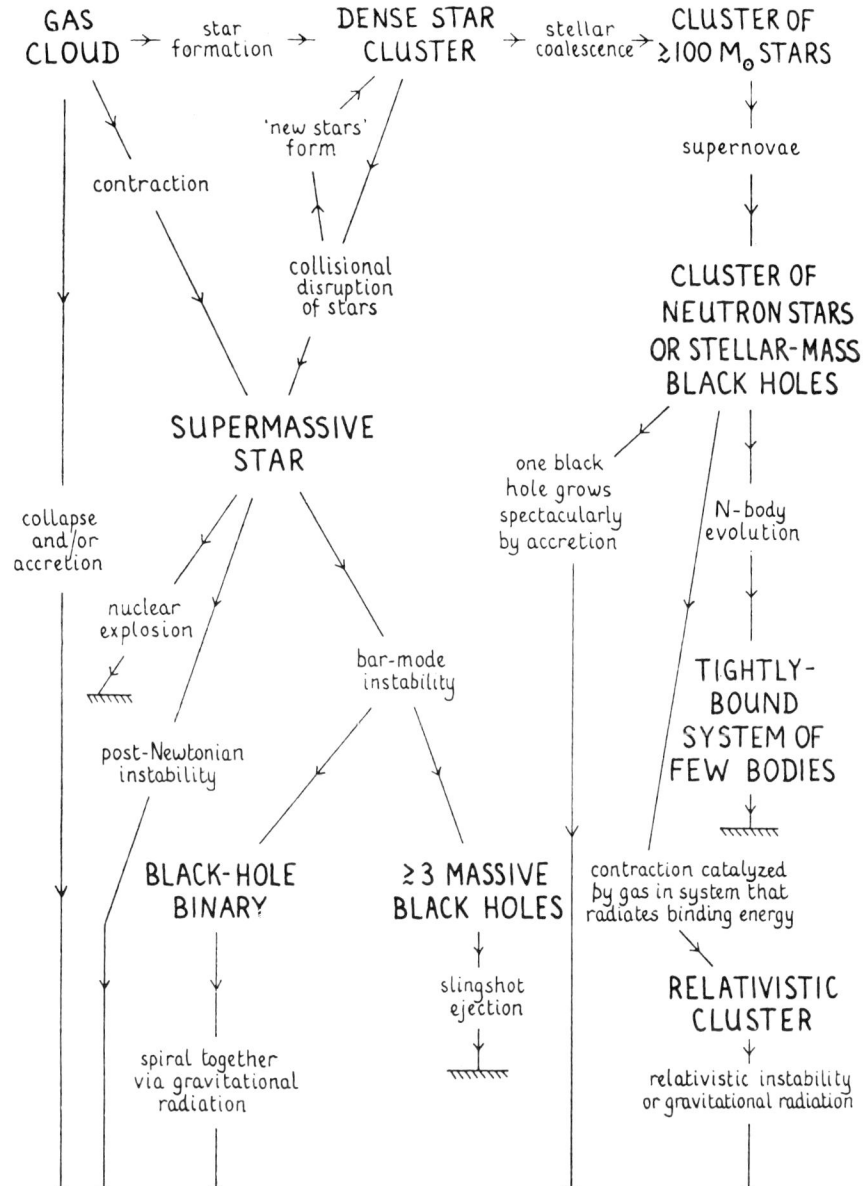

FIGURE 1. Possible routes for runaway evolution in active galactic nuclei leading to the formation of a supermassive black hole. [From Rees 1978].

relativistic state. They then speculated that when such a stellar core became sufficiently relativistic, the cluster would become unstable due to general relativity and would undergo catastrophic collapse to a supermassive black hole in a <u>dynamical</u> timescale, t_d. The original discussion of ZP focussed on a <u>dense</u> star cluster of stellar-mass black holes (but should apply equally well to a cluster of neutron stars.) They thus argued that such a cluster would inevitably collapse to a supermassive black hole around which remnant stars would orbit.

We (Shapiro and Teukolsky 1985a) have re-examined the scenario of ZP in light of two recent computational developments reviewed in Section 2 below. Our reanalysis suggests that, with some modification, the ZP scenario ought to be regarded as a prime candidate for the path by which supermassive black holes form in dense galactic nuclei.

2. RECENT COMPUTATIONAL ADVANCES

2.1. The 'Gravothermal Catastrophe'

The 'gravothermal catastrophe' has only recently been elucidated by detailed 1+1 and 2+1 dimensional Fokker-Planck calculations that follow core collapse for many e-foldings (Cohn 1979, 1980; Marchant and Shapiro 1980; Duncan and Shapiro 1982). Here the gravothermal catastrophe refers to the process by which the cumulative effect of two-body, small-angle, gravitational (Coulomb) scattering causes the core of a self-gravitating, large-N, Newtonian star cluster to undergo <u>secular</u> collapse on a relaxation timescale, $t_r \gg t_d$. The recent <u>Newtonian</u> calculations show that advanced core collapse in spherical clusters is <u>not</u> described by the simple "evaporation model", derived by Ambartsumian (1938) and Spitzer (1940) and assumed by ZP. Nevertheless, as these early analytic models suggested and as the recent, quasi-analytic model of Lynden-Bell and Eggleton (1980) more accurately predicts, core collapse does proceed homologously in a dynamically evolving star cluster.

The homologous nature of core collapse is dramatically illustrated in Figure 2 taken from Cohn (1980). The density profile of an isotropic, equilibrium star cluster evolving secularly according to the 1+1 dimensional Fokker-Planck equation is plotted at different times. As time increases, the central (core) density grows while the core radius shrinks. However, apart from an overall scale factor, the profile maintains a self-similar shape ($\rho(r) \propto r^{-2.23}$ throughout the stellar halo outside the core). Runaway core collapse proceeds homologously, a fact we shall exploit below.

The 'catastrophe' encountered during core collapse is simply that, after evolving for a total time $\tau = 330\ t_r$, the central star density n_c, central velocity dispersion v_c, central potential ϕ_c, and central redshift z_c of a star cluster all blow-up to infinity! On the other hand, the core radius r_c and the total number of core stars N_c simultaneously collapse to zero. Thus, gravitational encounters in a large-N, Newtonian star cluster inevitably drive the core to a relativistic state in a finite time.

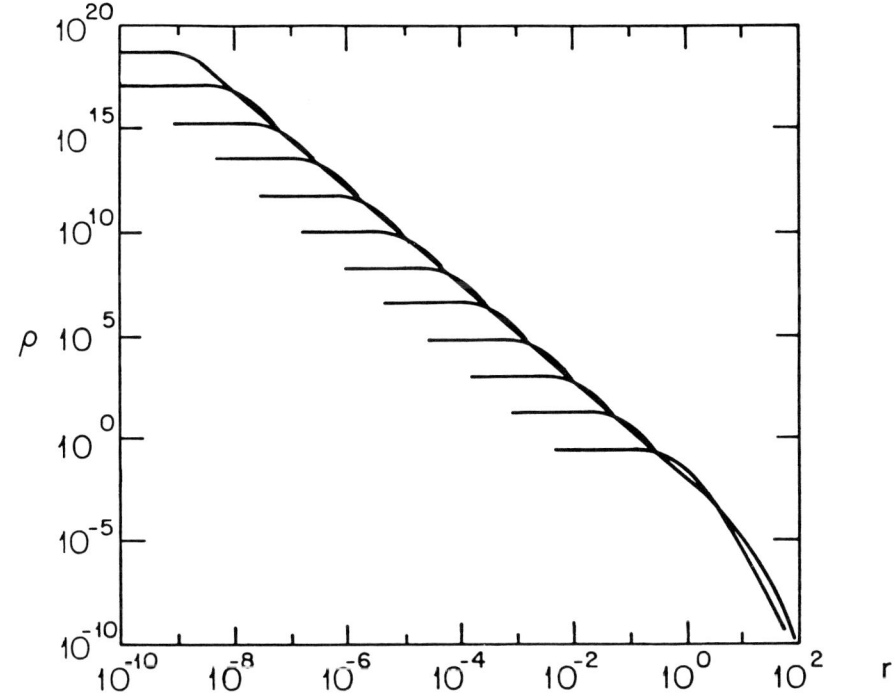

FIGURE 2. Evolution of the density profile of a Newtonian cluster via the 'gravothermal catastrophe'. The central density increases in time while the radius of the central core decreases. The time interval between the first two epochs is roughly 80% of the time required to complete core collapse. [From Cohn 1980].

2.2 Relativistic Stellar Dynamics on the Computer

For the first time, it is now possible to numerically integrate the full Einstein equations for the dynamical evolution of a spherical, collisionless configuration in general relativity (Shapiro and Teukolsky 1985 b,c). These Vlasov integrations enable one to follow on the computer the evolution of a relativistic star cluster on dynamical timescales, even during epochs characterized by total gravitational collapse leading to the formation of a supermassive black hole. The formation and growth of the black hole can be followed remarkably accurately without the appearance of numerical or physical singularities.

Although the computations to date are restricted to spherical systems, the gravitational fields can be arbitrarily strong, the particle (e.g. star) velocities arbitrarily close to c (the speed of light) and the particle density and velocity profiles arbitrarily shaped. The method of integration is described elsewhere (Shapiro and

Teukolsky 1985b). Suffice it to say that the scheme combines the
techniques of <u>numerical relativity</u> with the tools of direct N-body
<u>particle simulations</u>. The matter obeys the general relativistic
Vlasov equation, which is solved by particle simulation. The gravitational (metric) field satisfies the ADM 3+1 equations, which are
solved by finite difference techniques. The particles move on geodesics in a momentarily "fixed" background field. At regular time
intervals they are binned on a spatial lattice. The matter densities
and currents are are then determined on the lattice sites and used as
source terms for the field equations. The gravitational fields are
then updated by solving the field equations and the particles are
evolved to the next time slice. And so forth until the integrations
terminate.

The resulting code has been applied to a number of interesting
nonlinear dynamical problems (Shapiro and Teukolsky 1985c), including
the stability of equilibrium star clusters, binding energy criteria
for stability, the catastrophic collapse of star clusters to black
holes and relativistic 'violent relaxation'.

Most significantly, these fully relativistic numerical integrations provide dramatic confirmation of the original speculations of ZP
that star clusters become relativistically unstable at sufficiently
high central redshift, $z_c \gtrsim 0.5$ (see Figures 3-5). While this instability had already been demonstrated rigorously in perturbation
theory (Ipser and Thorne 1968; Ipser 1969a,b; Fackerell 1970), the
numerical integrations can follow its nonlinear growth and determine
the ultimate fate of unstable clusters. In particular, the integrations show quite generally that clusters of sufficiently high z_c do
undergo catastrophic collapse to a black hole on a dynamical timescale. They further reveal that <u>even in the case of extremely centrally condensed configurations with extensive Newtonian halos, an
appreciable fraction of the total mass ultimately collapses to a
central black hole in a few crossing times</u>, provided the central core
is sufficiently relativistic. This result is due to the cascading
nature of collapse in collisionless configurations, a collective
phenomenon predicted by ZP and called "avalanche-type contraction" by
them.

These recent Fokker-Planck and general relativistic dynamical
calculations provide fresh impetus for reconsidering the ZP supermassive black hole formation scenario. We do so below.

3. SECULAR CORE COLLAPSE: THE HOMOLOGICAL EQUATIONS

Consider now a very dense, Newtonian cluster of compact stars –
neutron stars or stellar-mass black holes – embedded in a dynamically
evolved galactic nucleus. Such a cluster is the likely outcome of
evolution in a dense ($\gtrsim 10^8$ stars p_c^{-3}) but otherwise normal, galactic
nucleus initially composed of main sequence stars (see Begelman and
Rees 1978 and references therein). The homological evolution of the
core of a cluster of compact stars is approximately governed by three
ordinary differential equations which account for the combined effects

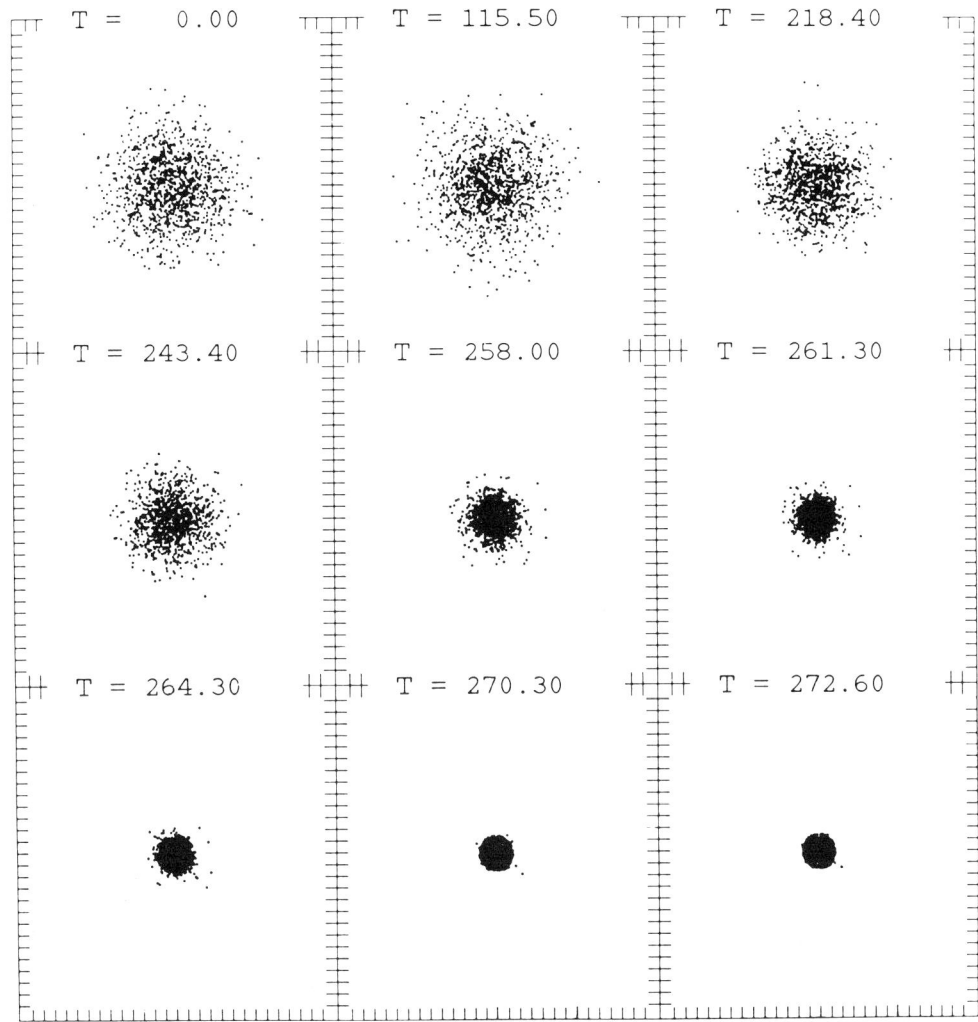

FIGURE 3. Fully relativistic numerical simulation of the gravitational collapse of an unstable relativistic star cluster initially in equilibrium. At t=0 the cluster has a radius $R/M = 9.16$, $z_c = 0.517$ and satisfies a truncated Maxwell-Boltzmann (isothermal) equilibrium profile. When the dynamical integrations were terminated virtually all of the mass was inside the black hole event horizon. Snapshots of the collapsing configuration are shown at various times. Time and spatial gridpoints are in units of M, the arbitrary total mass of the cluster. [From Shapiro and Teukolsky 1985c].

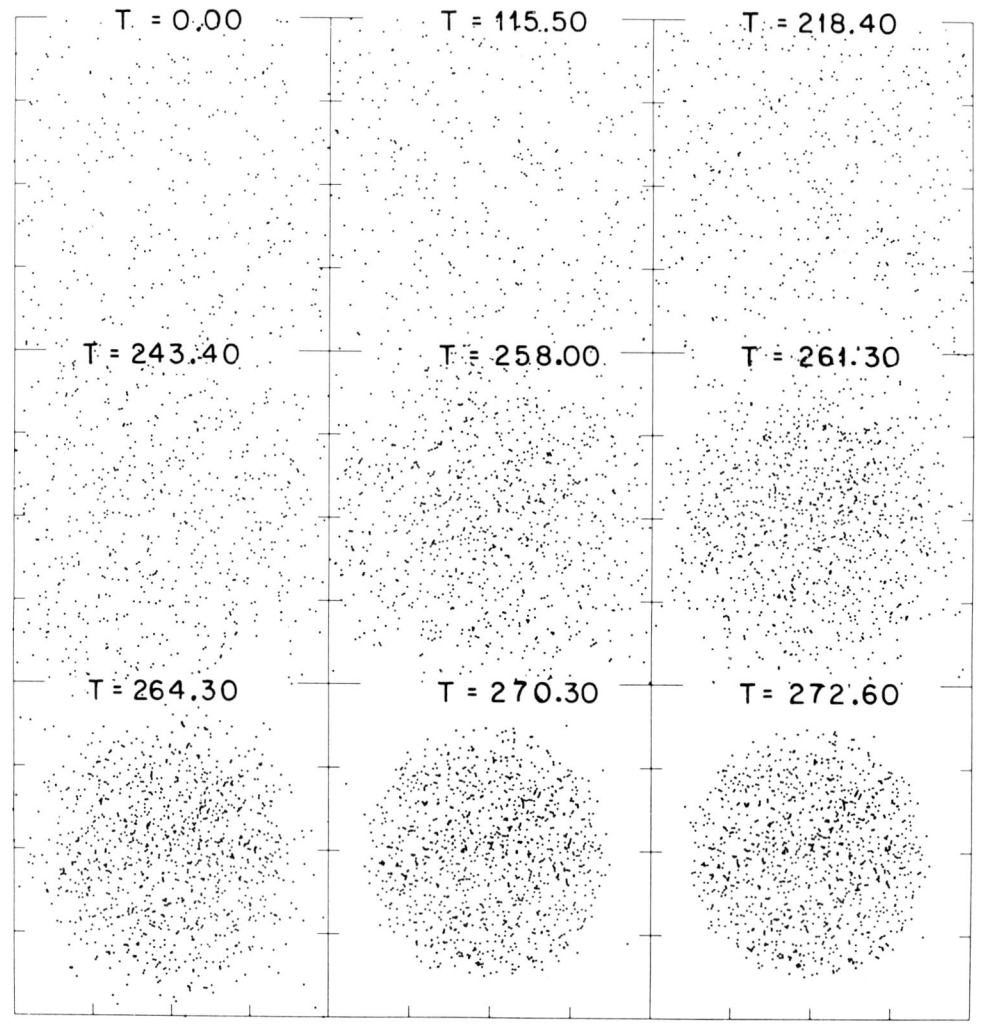

FIGURE 4. Blowup of the central regions shown in Figure 3. Note that virtually all of the mass lies inside the event horizon at $r_S/M = 2$ at late times. The cluster surface approaches an asymptotic limit near $r_S/M = 1.5$. [From Shapiro and Teukolsky 1985c].

THE BIRTH OF AGNs AND QUASARS

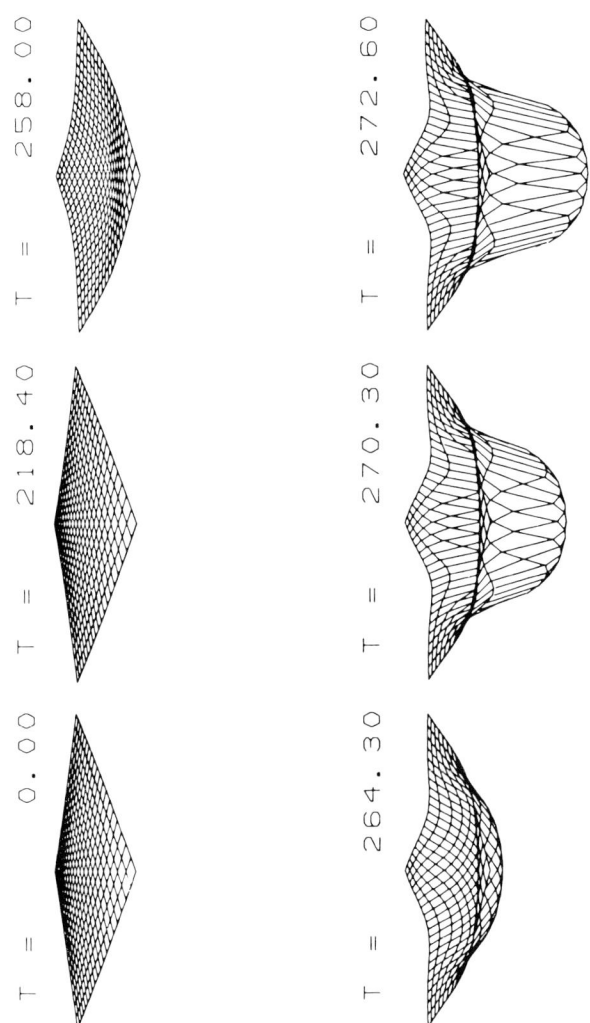

FIGURE 5. Contour plot of the gravitational "potential" for the collapse shown in Figure 3. Here the log of the lapse function α is plotted as a function of areal radius r_s on selected time slices. In Newtonian theory the lapse is related to the gravitational potential ϕ according to $\alpha = 1+\phi$. Collapse to a black hole is characterized by exponential "collapse of the lapse" in the (maximal) gauge chosen here.

of gravitational scattering and star collisions and coalescence. With $G = c = 1$, the equations are:

$$\dot{E}_c/E_c = -\alpha_1/t_r + 1/t_{coll}, \qquad (4)$$

$$\dot{N}_c/N_c = -\alpha_2/t_r - 1/t_{coll}, \qquad (5)$$

$$\dot{m}_c/m_c = 1/t_{coll}, \qquad (6)$$

(cf. Lightman and Shapiro 1978). Here $N_c(t)$ is the number of core stars, $m_c(t)$ is the mean stellar mass, and $E_c(t)$ is the characterstic core binding energy. These quantities are related to the characteristic core density n_c, r.m.s. velocity v_c, and radius r_c according to the virial theorem

$$E_c \approx 1/2\, N_c m_c v_c^2 \approx 1/4(N_c m_c)^2/r_c, \qquad (7)$$

and the relations

$$n_c \approx N_c/(4\pi r_c/3) \approx 2 \times 10^9 \text{ pc } (N_{c,8}^{-2} v_{c,3}^6 m_{c,*}^{-3}),$$

$$r_c \approx 0.2 \text{ pc}(N_{c,8} v_{c,3}^{-2} m_{c,*}), \qquad (8)$$

where $N_{c,8} \equiv N_c/10^8$, $v_{c,3} \equiv v_c/10^3 \text{ km s}^{-1}$ and $m_{c,*} \equiv m_c/M_\odot$. The central relaxation timescale is given by Spitzer and Hart (1971):

$$t_r \approx v_c^3/\lfloor (3/2)^{1/2} 4\pi m_c^2 n_c \ln(0.4 N_c)\rfloor$$

$$\approx 0.8 \times 10^8 \text{ yr}(N_{c,8}^2 v_{c,3}^{-3} m_{c,*} \Lambda_8^{-1}), \qquad (9)$$

where $\Lambda_8 \equiv \ln(0.4 N_c)/\ln(0.4 \times 10^8)$. The corresponding dynamical timescale $t_d \ll t_r$ is

$$t_d \approx r_c/v_c \approx 200 \text{ yr}(N_{c,8} v_{c,3}^{-3} m_{c,*}). \qquad (10)$$

The central two-body collision timescale is

$$t_{coll} \approx \frac{1}{n_c \sigma_{coll} v_\infty} \approx 6 \times 10^{13} \text{ yr}(N_{c,8}^2 v_{c,3}^{-5} m_{c,*}), \qquad (11)$$

where $v_\infty \approx \sqrt{2} v_c$ is the typical asymptotic relative velocity and σ_{coll} is the collision cross-section. For any compact object with radius

R < 4m, the critical specific angular momentum for test particle capture is $\ell = 4m$, implying

$$\sigma_{coll} \approx \pi\tilde{\ell}^2/v_\infty^2 = 16\pi m^2/v_\infty^2. \tag{12}$$

We employ equation (12) as an approximation to σ_{coll} for both neutron stars and black holes. (Note that a $1.4 M_\odot$ neutron star constructed from, e.g., a moderately stiff TNI equation of state has $R \approx 5.1m$ and $\tilde{\ell} \approx 4.1m$.)

The constants α_1 and α_2 appearing in equations (4) and (5) may be fixed by matching to the detailed Fokker-Planck calculations for advanced core collapse in the absence of collisions (i.e. $t_{coll} = \infty$). These calculations (e.g. Cohn 1980) give $\alpha_1 = 8.72 \times 10^{-4}$ and $\alpha_2 = 1.24 \times 10^{-3}$ (hence $\alpha \equiv \alpha_1/\alpha_2 = 0.701$). We remark that "stellar evaporation" theory gives $\alpha = \alpha_1 = 0$ and $\alpha_2 = 7.4 \times 10^{-3}$, which results in quite different evolution.

It is useful to parametrize the cluster core by z_c, the central redshift, which is related to v_c according to

$$z_c \approx \phi_c \approx 14 v_c^2/3 \approx 5 \times 10^{-5} v_{c,3}^2 \tag{4'}$$

at late times. Equation (4) may then be replaced by

$$\dot{z}_c/z_c = (\alpha_2 - \alpha_1)/t_r + 1/t_{coll} \tag{14}$$

For systems of interest, $t_{coll} > t_r$ initially (cf. eqns. 9 and 11) and star collisions are dynamically unimportant. Eventually, when the core collapses sufficiently, the inequality reverses and star collisions and coalescence dominate the evolution. Accordingly, the evolution of a Newtonian cluster of compact stars is characterized by two distinct epochs: (1) an initial long, low-redshift ($z_c < z_{col} \sim 10^{-2}$) "point-mass" epoch during which the core undergoes secular collapse via the gravothermal catastrophe, and (2) a later short, high-redshift ($z_c \gtrsim z_{coll}$) "finite-radius" epoch during which compact star collisions and coalescence dominate the core evolution. During the brief collision-coalescence period, a cluster core of neutron stars undergoes an internal "phase change" to a core of stellar-mass black holes: when such collisions become dynamically important, the velocity dispersion in the core v_c is less than the escape speed from the neutron star surface. Consequently, $1.4 M_\odot$ neutron stars collide at freefall velocity, which results in coalescence and little mass-loss (Seidl and Cameron 1972; Gilden and Shapiro 1984). Upon cooling by rapid neutrino emission (timescale ~ minutes), the coalesced neutron star collapses to form a ~ $2.8 M_\odot$ black hole, which will later collide with another compact star, and so forth.

During both epochs the number of core stars decreases while the central velocity dispersion and redshift increase. Ultimately the core becomes relativistic ($z_c = z_{crit} \sim 0.5$), at which point we know that it undergoes catastrophic gravitational collapse to a black hole. (cf. Section 2.2). The hole rapidly grows outward beyond the core, eventually swallowing a significant fraction of the total mass of the compact star cluster in a few mean dynamical timescales via the "avalanche" instability.

The subsequent dynamical evolution of a dense galactic nucleus with a central supermassive black hole has been studied by many authors (Hills 1975; Rees 1978; Young, Shields and Wheeler 1977; Frank 1978; McMillian, Lightman and Cohn 1981; Duncan and Shapiro 1983). A coherent picture has emerged whereby gas liberated during the collisions of residual main sequence stars and tidal disruptions by the hole is consumed by the hole, causing it to grow significantly. This gas accretion can efficiently generate AGN or quasar luminosities, provided the hole acquires a mass $\gtrsim 10^6 - 10^9 M_\odot$ (see Shapiro 1985 for a brief summary and references). Interestingly, at late times, the black hole accretion rate $\dot{M}$ and, hence, the luminosity L are found to decay like

$$L \sim \dot{M} \sim t^{-\alpha}, \qquad (14)$$

where α depends on the power-law stellar profile outside the core of the initial galactic nucleus. Typically, α is in the range $\alpha \sim 0.8 - 1$. The reason for the decay is that the black hole ultimately "eats its way" out into the halo of the galaxy where the star density profile falls off. If supermassive black holes are the central engines in quasars and/or AGNs, these sources should presumably exhibit a power-law decay with time as given by eqn. (14). It may therefore prove revealing to analyze observed quasar and AGN luminosity functions in light of this theoretical evolution curve.

4. APPROXIMATE SOLUTION

The evolutionary tracks of neutron star cores are plotted in Figure 6. [The plot for stellar-mass black hole cores is almost identical]. At any instant the cores are specified by two parameters: N_c and z_c (or v_c). The two distinct segments comprising each track correspond to the "point-mass" gravothermal and "finite-radius" collision-coalescence epochs, respectively. Two constraints confine the tracks to lie within the heavy lines in the figure: (1) $\tau = 330 t_r \leq H^{-1} = 2 \times 10^{10}$ yr; and (2) $N_c(z_{crit}) \geq 1$. Constraint (1) guarantees that secular core collapse to a relativistic state occurs in less than a Hubble time. Constraint (2) ensures that this collapse will not fizzle before reaching $z_c = z_{crit}$ because of the diffusion of all the core stars out into the halo.

The range of plausible initial core parameters as determined by the calculations of Colgate (1967) and Sanders (1970) is indicated by the box in the upper left-hand corner of the figure. These authors consider the evolution of dense but otherwise normal galactic nuclei initially composed of main sequence stars (see also Gold, Axford and Ray 1965; Spitzer and Saslaw 1966; Spitzer and Stone 1967; Colgate

THE BIRTH OF AGNs AND QUASARS

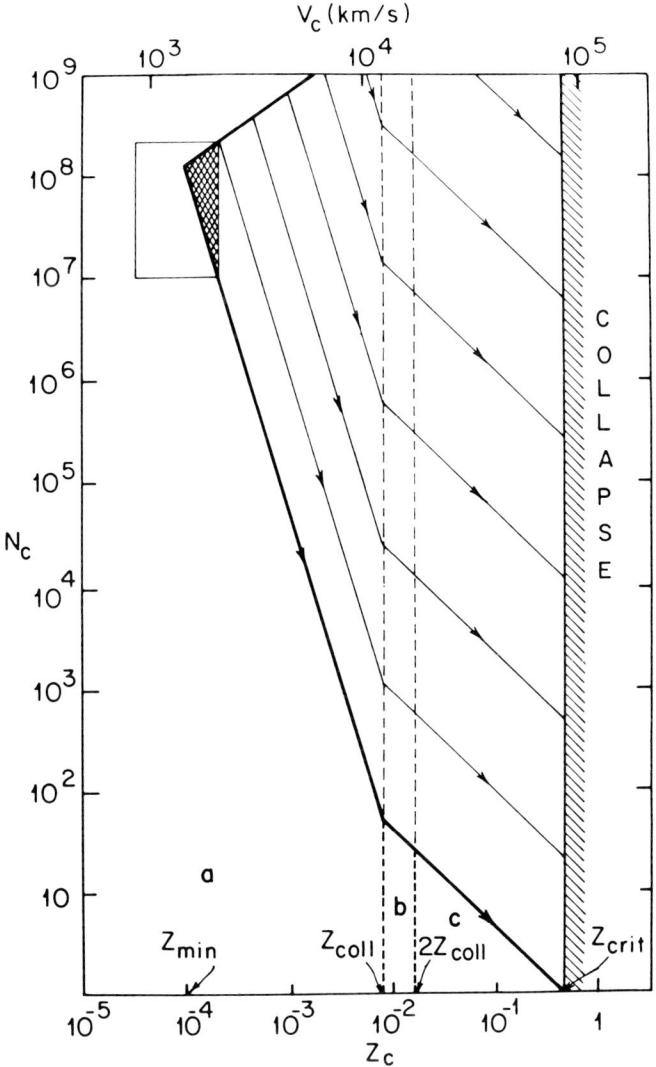

FIGURE 6. Dynamical evolution tracks in the N_c-z_c plane of dense cluster cores consisting of $1.4M_\odot$ neutron stars. During phase (a), the gravothermal catastrophe, the core evolves by two-body scattering to central redshift $z_{coll} \approx 8 \times 10^{-3}$. During phase (b) the neutron stars collide, coalesce and collapse to form $\sim 2.8M_\odot$ black holes. During phase (c) the remnant black holes collide and coalesce, driving the core to central redshift $z_{crit} \sim 0.5$. At this point the entire cluster becomes relativistically unstable and undergoes catastrophic collapse to a supermassive black hole in a few mean orbital periods. The range of plausible core parameters for dense galactic nuclei lies in the box shown in the upper left of the figure. The hatched region shows the fraction of these capable of evolving to a supermassive black hole in a Hubble time. (From Shapiro and Teukolsky 1985a)

1967; Sanders 1970; Spitzer 1971). They find that the evolution is invariably dominated by an epoch of stellar collisions and coalescence and the build-up of massive stars. These stars undergo supernova explosions that can leave behind compact stellar-mass remnants. Consequently, the endpoint of this "coalescence phase" is very likely a cluster of neutron stars or stellar-mass black holes (Begelman and Rees 1978). We infer from the calculations of Colgate (1967) and Sanders (1970) that such a cluster might originally consist of 10^7 -2 x 10^8 compact objects, each of $1-10 M_\odot$, confined to a region 0.01-0.1pc in radius and moving with a velocity dispersion 800-2000 kms^{-1}. These are the parameters that determine the location of the box in Figure 6.

5. THEORETICAL CONCLUSIONS AND OBSERVATIONAL STATUS

Several interesting consequences emerge from Figure 6. First, there exists a minimum velocity dispersion below which a core of compact stars cannot evolve to a relativistic state in a Hubble time,

$$v_{c,min} \approx 10^3 m_{c,*}^\delta \text{ km s}^{-1}, \quad \delta = (1-\alpha)/(7-3\alpha) = 0.061. \quad (15)$$

Second, a finite fraction of those compact star clusters likely to form in dense galactic nuclei are capable of evolving to a supermassive black hole in a Hubble time. Moreover, these clusters have roughly $10^7 \lesssim N_c \lesssim 10^9$ compact stars. Hence, following gravitational collapse, supermassive black holes with masses in the range $10^6 \lesssim M/M_\odot \lesssim 10^9$ can ultimately form. This is roughly the mass range of black holes that can generate AGN and quasar luminosities via gas accretion near the Eddington limit. (cf. eqn. 3). It is the range frequently cited in black hole models of such systems.

Direct observational evidence in support of the supermassive black hole formation scenario presented here does not yet exist. The principle difficulty is that the detection of dense stellar cores residing within 0.1 -0.01 pc of the centers of galactic nuclei requires extremely high resolution observations beyond the limits of most telescopes. Nevertheless, it is most intriguing that the high resolution observations of the nucleus of the Seyfert galaxy NGC 4151 and of M31 by the ballon-borne Stratoscope II telescope revealed stellar conditions close to those capable of generating dense clusters of compact stars that can undergo catastrophic collapse. Schwarzschild (1973) estimates that the stars in the central 3.5 pc of the core of NGC 4151 contain fully 4 x $10^9 M_\odot$ at a mean density of 2 x $10^9 M_\odot$ pc^{-3} and velocity dispersion of $\gtrsim 1.5$ x 10^3 kms^{-1}. Light, Danielson and Schwarzchild (1974) calculate that the inner 0.5 pc of the nucleus of M31 may contain $10^8 M_\odot$.

Planned observations of these and other nuclei with the high resolution optical detectors aboard the Space Telescope should, when used in conjunction with Figure 6, determine whether stellar conditions in these regions are suitable for triggering the black hole formation scenario we have described.

ACKNOWLEDGEMENTS

This work has been supported in part by National Science Foundation grant AST 81-16370 at Cornell University, Ithaca, NY.

REFERENCES

Ambartsumian, V.A. 1938, Ann. Leningrad State Univ. No. 22 (Astr. Series, Issue 4).
Begelman, M.C., Blandford, R.D., and Rees, M.J. 1984, Rev. Mod. Phys., 56, 255.
Begelman, M.C. and Rees, M.J. 1978, M.N.R.A.S., 185, 847.
Cohn, H. 1979, Ap. J., 234, 1036.
Cohn, H. 1980, Ap. J., 242, 765.
Colgate, S.A. 1967, Ap. J., 150, 163.
Duncan, M.J. and Shapiro, S.L. 1982, Ap. J., 253, 921.
Duncan, M.J. and Shapiro, S.L. 1983, Ap. J., 268, 565.
Fackerell, E.D. 1970, Ap. J., 160, 859.
Frank, J. 1978, M.N.R.A.S., 184, 87.
Gilden, D., and Shapiro, S.L. 1984, Ap. J., 287, 728.
Gold, T., Axford, W.I., and Ray, E.C. 1965, in Quasi-stellar Sources and Gravitational Collapse, ed. I. Robinson et. al. (Chicago: University of Chicago), p. 93.
Hills, J.G. 1975, Nature, 254, 295.
Ipser, J.R. 1969a, Ap. J., 156, 509.
Ipser, J.R. 1969b, Ap. J., 158, 17.
Ipser, J.R. and Thorne, K.S. 1968, Ap. J., 154, 251.
Light, E.S., Danielson, R.E., and Schwarzschild, M. 1974, Ap. J., 194, 257.
Lightman, A.P. and Shapiro, S.L. 1978, Rev. Mod. Phys., 50, 437.
Lynden-Bell, D., and Eggleton, P.P. 1980, M.N.R.A.S., 138, 495.
Marchant, A.B., and Shapiro, S.L. 1980, Ap. J., 239, 685.
McMillan, S.L.W., Lightman, A.P., and Cohn, H. 1981, Ap. J., 251, 436.
Rees, M.J. 1978, Phys. Scripta, 17, 193.
Rees, M.J. 1984, Ann. Rev. Astron. Astrophys., 22, 471.
Sanders, R. 1970, Ap. J., 162, 791.
Sargent, W.L.W., Young, P.J., Boksenberg, A., Shortridge, K., Lynds, C.R. and Hartwick, F.D.A. 1978, Ap. J., 221, 731.
Schwarzschild, M. 1973, Ap. J., 182, 357.
Seidl, F.G.P., and Cameron, A.G.W. 1972, Astrophys. Space Sci., 15, 44.
Shapiro, S.L. 1985, in I.A.U. Symposium No. 113, Dynamics of Star Clusters, eds. J. Goodman and P. Hut (Dordrecht: Reidel), p. 373.
Shapiro, S.L. and Teukolsky, S.A. 1985a, Ap. J. Letters, in press.
Shapiro, S.L. and Teukolsky, S.A. 1985b, Ap. J., in press.
Shapiro, S.L. and Teukolsky, S.A. 1985c, Ap. J., in press.
Spitzer, L. 1940, M.N.R.A.S., 100, 396.
Spitzer, L. 1971, in Galactic Nuclei, ed. D. O'Connell (Amsterdam: North-Holland), p. 443.

Spitzer, L., and Hart, M.H. 1971, Ap. J., 164, 399.
Spitzer, L., and Saslaw, W.C. 1966, Ap. J., 143, 400.
Spitzer, L. and Stone, M.E. 1967, Ap. J., 147, 519.
Strittmatter, P.A. and Williams, R.E. 1976 in Ann. Rev. Astron. Astrophys., 14, 373.
Ulrich, M.H., Boksenberg, A., Bromage, G.E., Clavel, J., Elvius, A., Penston, M.V., Perola, G.C., Petteni, M., Snijders, M.A.J., Tanzi, E.G. and Tarenghi, M. 1984, M.N.R.A.S., 206, 221.
Young, P.J., Westphal, J.A., Kristian, J., Wilson, C.P. and Landauer, F.P. 1978, Ap. J., 221, 721.
Young, P.J., Shields, G.A., and Wheeler, J.C. 1977, Ap. J., 212, 367.
Zel'dovich, Ya. B., and Podurets, M.A. 1965, Astron. Zh., 42, 963 [English translation in Sov. Astron.-A.J., 9, 742] (ZP).

DISCUSSION

PHINNEY: At the end of core collapse, a plausible initial cluster only has 10-100 $M_\odot$ of stars in the collision dominated core. Can you describe how the resulting 10-100 $M_\odot$ hole manages to drag another 10^6-10^9 $M_\odot$ in with it?

SHAPIRO: The fact that a nonneglible fraction of the total cluster mass can collapse to a black hole in a few crossing timescales is due to a collective dynamical instability predicted by ZP and termed "avalanche-type" contraction by them. Essentially, stars outside the relativistic core, which readily collapses, feel an ever increasing interior mass with time. This increase is due to inward moving stars on radial trajectories from the halo which are captured by the nascent hole and thus never return to the halo. Hence, stars initially remaining outside the core on only midly eccentric orbits have an increasing probability of being captured as they are pulled ever closer to the growing hole at the center. The growth runs away and a cascade ensues. An appreciable mass, whose actual size depends on the precise stellar distribution function outside the core at the onset of collapse, is thus able to collapse to a black hole on a dynamical timescale. This mass may be many times the mass of the central core.

REES: I'm still worried about the typical stars in the outer part of the cluster (whose specific orbital angular momentum may be a hundred times larger than can be accepted by a 10^8 $M_\odot$ hole) finally get swallowed. Could it be that they are swallowed only after "loss cone diffusion", which occurs on a timescale of order the initial relaxation time for the cluster?

SHAPIRO: Undoubtedly, once the initial collapse is over and a quasi-static equilibrium is re-established between the nascent black hole and the remaining orbiting stars, the combined effects of stellar collisions and loss-cone diffusion will cause the hole to grow further with time (see eqn. 14 and the associated discussion). However, the "avalanche" instability, which occurs on a dynamical timescale and precedes this secular growth phase, should by itself produce a supermassive black hole of aprreciable mass in a few orbital periods.

Perhaps, the results of a numerical experiment we performed will be convincing. Consider the catastrophic collapse of an unstable, relativistic, n=4 polytrope with

$z_c = 0.73$. This isotropic cluster has only a tiny relativistic core containing but 0.6% of the total cluster mass and an extensive Newtonian halo reaching out to a radius $R/M = 0.75 \times 10^4$. Moreover, the ratio of the mean to central mass density is 1.6×10^{-13}. These extreme core-halo conditions mimick the conditions that are likely to exist in a realistic cluster (see Figure 6.)

Yet, by the time our integrations terminate, over 8% of the total cluster mass-almost 13 times the core mass-resides inside the (apparent) horizon of the black hole. This is the avalanche instability at work, not loss-cone diffusion, since our calculations only follow the collisionless phase of the implosion.

Of course, there is nothing really mysterious about the avalanche instability. The critical specific angular momentum for the capture of a Newtonian particle by a Schwarzschild black hole is 4M where M is the hole mass. The fraction of stars at, say, the half-mass radius R_h in an <u>isotropic</u> cluster which have smaller angular momentum and can thus be captured is

$$F \simeq 1/2 \times (4M/R_h V_h)^2 \simeq 8M/R_h$$

where V_h is the rms velocity at R_h. Assuming that these distant stars are typical, F gives the fraction of all cluster stars which can be captured and thus FM gives an upper limit to the mass of the hole which can form from the catastrophic collapse of an isotropic cluster. For the n=4 polytrope discussed above, F is roughly 0.09, ($R_h/M=89$), in good agreement with the calculated mass fraction of 0.08 found from the detailed Vlasov integrations. For a $10^9 M_\odot$ cluster with R_h inside 0.1 pc, $F \lesssim 10^{-3}$ and the black hole formed has a mass of about $\lesssim 10^6 M_\odot$.

Finally, we remark that realistic clusters which have undergone extensive gravothermal core collapse prior to becoming relativistic have very anisotropic distributions of halo stars. These halo stars move predominantly in radial orbits with low angular momenta. They are thus far more likely to be captured than halo stars in isotropic clusters. This results in a further increase in the fraction F of the cluster mass which collapses to a central black hole during the initial implosion.

WANDEL: Can you estimate to what extent would these results be modified if one allowed a deviation from spherical symmetry, (e.g. a triaxial potential), and what amount of deviation can be tolerated without chancing the result qualitatively (that is, preventing the collapse)?

SHAPIRO: A small deviation.

PISMIS: Your model does not include any angular momentum. I should expect that the inclusion of even a small amount of angular momentum might change the course of the evolution of your system appreciably.

SHAPIRO: That speculation is by no means clear. For example the gravothermal catastrophe tends to transport angular momentum outwards, perhaps leaving the central relativistic regions of a rotating cluster nearly spherical prior to collapse. Otherwise, of course, the implosion might very well proceed differently (including the generation of a burst of gravitational radiation!) and inhibit the formation of a black hole. The answer must await a detailed calculation of the 2+1 dimensional relativistic Vlasov equation that can handle rotation. We are now tooling up to do just that.

FILIPPENKO: In your calculations, you showed that a cluster whose initial velocity dispersion is less than about 1000 km/sec will not collapse to a black hole within a Hubble time. This is disappointing to a number of us (e.g. Weedman) who have postulated that starburst galaxies may the progenitors of at least some types of active galactic nuclei. Are there any effects you haven't yet considered in detail which might lead to a shorter collapse time?

SHAPIRO: While our relativistic Vlasov integrations are quite exact, our homological model for the earlier gravothermal collapse of a neutron star cluster in only approximate. Moreover, we have ignored completely the possible frictional influence of residual gas in the cluster, for example, which can hasten the collapse. So the value of 1000 km/sec is by no means the final word. On the other hand, this figure seems to be quite insensitive to the parameters in the _present_ model. Also, a cluster of compact stars is only born in a galactic nucleus following an epoch of collisions and coalescence of the progenitor main sequence stars. Such an epoch is triggered only when the central velocity dispersion becomes at least comparable to the escape velocity of a typical star (600 km s) which presumably sets a lower limit to the velocity dispersion.

THE POSSIBLE MECHANISM OF THE FORMATION OF THE HARD SPEC-
TRUM OF ACTIVE GALACTIC NUCLEI

I.D.Novikov
Space Research Institute Academy of Sciences of
the USSR, Moscow 117810, Profsoyuznaja 84/32
USSR

B.E.Stern
Institute for Nuclear Research of the Academy
of Sciences of the USSR, Moscow 117312,
60-the October Anniversary Prospect, 7a,
USSR

ABSTRACT. The mechanism of the formation of the hard spectrum of active galactic nuclei due to the conversion of the energy of charged particles accelerated up to ultrarelativistic velocity is considered. In the presence of magnetic field charged particles produce photons, which in their tern create high-energy $e^{\pm}$ pairs. Multiple interactions of pairs with magnetic field and photons, and photon-photon interactions lead to cascade increase of the number of $e^{\pm}$ particles and may result in a large ratio of thus produced rest mass of $e^{\pm}$ pairs to incoming energy and in the relaxation of photon spectra to some specific shape with a cutoff near 1 MeV and a power - law slope at energies less than 30 KeV. The confrontations of the simulation and observational data are given.

1. INTRODUCTION

The clue to the understanding of processes occurring in active galactic nuclei (AGN) and quasars may be in the properties of their γ-and x-radiation. The observations reveal a number of peculiar features at the hardest ends of their electromagnetic spectra. Since the most energetic phenomena of the Universe are involved, here one could expect that it should be just the hard radiation which bears the essential information on both the source of energy and the processes of its transformation in the vicinity of the primary source.
    In a number of theoretical models it has been assumed that the energy source in AGN is a massive black hole ro-

tating in an external magnetic field (see for example Blanford and Znaek, 1977). For this electric currents must be present in the vicinity of a rotating black hole. To sustain the currents, new charge bearers (most likely, in the form of electron-positron pairs $e^{\pm}$) must be continuously created there.

The most effective mechanism of $e^{\pm}$ - pair creation under such conditions is $\gamma + \gamma \rightarrow e^- + e^+$ reaction (see Blanford, 1982; Rees, 1982; Lingenfelter and Ramaty, 1982; Begelman, Blanford and Rees, 1984). So, an essential element of the mechanism under discussion is the presence and mutual interactions of γ -quanta and electron-positron pairs.

Continuum γ - and x-radiations have been directly observed in some cases (Bignami et al., 1979 ; Rothschild et al., 1983). Indirect evidence of the presence of $e^{\pm}$ pairs under similar circumstances comes from the discovery of the 511 KeV annihilation line from the center of our Galaxy (see a review article by MacCallum and Leventhal, 1983), where a less powerful version of an AGN energy generator may be working. Moreover, these latter observations enable us to evaluate the effectiveness of the $e^{\pm}$-pair generation mechanism in question. It turns out that the ratio of $e^{\pm}$- pair rest-mass-energy generation to the total energy release should be no less than $10^{-4} - 10^{-3}$.

The purpose of this paper is to calculate the processes of interaction between γ -quanta and $e^{\pm}$ pairs in external magnetic fields occurring at active galactic nuclei. More specifically, we deal with the following issues.

Assume that in the centers of AGN charged particles are being accelerated. A particular mechanism of such acceleration is not relevant here. Accelerated particles create high-energy γ -quanta (due to curvature radiation or inverse Compton scattering).

The energy in the form of γ -quanta is injected into a region where the processes of multi-particle creation and interaction occur within a comparatively small volume in the presence of external magnetic fields. The electric field is short-circuited due to a high density of current bearers. We emphasize that our model does not involve any other particles or radiation sources beside those primarily injected. The problem is solved in a completely self-consistent manner. A particular form of the spectrum of primarily injected particles is of little importance because its influence rapidly attenuates.

Below we present the results of Monte-Carlo numerical simulations (Stern, 1985) and compare them with the observational data.

In the model being discussed it is assumed that the emission from AGN detected at energies below 1 MeV originates most likely as a reprocession of energy of $e^{\pm}$ pairs accelerated up to relativistic velocities. We actually calculate how this reprocession proceeds in time.

It should be mentioned that alternative models resulting in hard-radiation spectra of AGN compatible with those observed have been proposed and discussed in literature as well. These include the models in which the observed hard radiation originates from the inverse Compton scattering of soft photons off the thermal electrons, or the models with a prescribed (required to fit the observations) power-law spectrum of electrons emitting synchrotron photons in the magnetic field, that are compton-scattered from the same electrons to form the observed high-frequency emission. We do not discuss such models here and refer only to the paper by Kazanaz (1984) and by Zdiarski (1985) in which the references to earlier publications can be found.

Finally note that beside $\gamma + \gamma$ reaction $e^{\pm}$- pairs can be created in the course of interaction of γ - photons with the matter (H or He). This latter mechanism has been explored in our previous publication (Kardashev et al., 1983) and we do not dwell on it here either (see also Burns, 1983; Lovelace and Ruchti, 1983; Ahdronian et al., 1984).

2. A GENERAL SCENARIO

Consider a sequence of events occurring in the region under discussion where the above mentioned interactions of particles take place and which we will call an electron-positron cauldron.

1. The γ-rays which were injected into the cauldron interact with soft photons producing high energy pairs. An external source of soft photons is not necessary since they can arise from synchrotron radiation of pairs produced before,(and being cooled due to Compton scattering and synhrotron radiation), that is, the conversion of photons can be self maintained. It is important, that a photon passing through a magnetic field aquires some pitch angle (if magnetic field lines are curved), and pairs are produced at nonzero pitch angles.

2. The $e^{\pm}$ pairs emit photons of a wide spectrum when decelerated by synchrotron and Compton losses.

3. Synchrotron and comptonized photons produce many more new pairs in interactions with each other.

4. Multiple interactions of $e^{\pm}$ with magnetic field, $e^{\pm}$ with photons, and photons with photons lead to cascade increase of the number of particles, the yield of the rest mass of electrons and positrons to the total energy of the electron-positron-photon soup becomes significant. The energy of hard particles diminishes until electrons and

positrons become semirelativistic and all photons with energy $\gtrsim m_e c^2$ are converted into pairs or scattered into soft region. The particle spectra should relax to rather soft quasiequilibrium shape, and further evolution of the system should be much slower than the energy degradation was.

5. Electrons and positrons annihilate emitting a wide annihilation line, but if the magnetic field lines are open the essential part of pairs can escape to a large distance where positrons will annihilate with electrons of interstellar gas producing a narrow 511 KeV line.

This scheme except stage 4 is traditional in general and it is similar to the model of Blanford (1982).

Stage 4 may result in a large ratio of thus produced rest mass of $e^{\pm}$ pairs to incoming energy and in the relaxation of photon spectra to some specific shape. In this case (the results of calculations confirm that it is really so) one may speak about a specific phenomenon called in the present work an "electron-positron cauldron" by analogy to the term "synchrotron cauldron" and in order to emphasize the important role of pair production. Of course stage 4 may develop only if a particle can undergo several collisions with others with the cross section

$$\sigma_0 = \pi r_0^2 = 2.3 \cdot 10^{-25} \text{ cm}^2$$ until it escapes the cauldron.

In other words a large optical depth of the cauldron is required. The exact definition of the optical depth will be given below.

3. A METHOD OF NUMERICAL MODELLING OF THE ELECTRON-POSITRON CAULDRON

Numerical simulation of a realistic steady-state cauldron would be rather difficult. One of the principal difficulties is due to the fact that the energy, being for example injected at a constant rate into the center of the cauldron, spreads to its outer parts. As a consequence, one must solve a spatially non-homogeneous problem. One should account also for a number of competing particle-interaction mechanisms, while the energy spectrum of particles created covers many orders of magnitude.

Consider however a cauldron permeated with a sufficiently ordered magnetic field whose field lines stretch from the injection region to the outer parts of the cauldron. Consider further a portion of injected particles which while interacting with one another and the magnetic field and creating new particles (we shall call them the descendants) - all move together almost exactly along the field lines from the primary source to the periphery of the cauldron . We shall call such a portion (with all the newly born particles) an element.

Assume that the particles from one element do not encounter the particles from other elements injected either before or after it. In this formulation the spatial structure of the cauldron is determined by the time evolution of injected elements arriving along the magnetic field lines at this or that location. And one should not allow for the interaction between particles from different elements, i.e. between particles of different age (as measured from the time of injection).

The radiation escaping the cauldron is determined by the particle spectra in its peripheral elements.

Such a cauldron is much simpler to calculate. One is to calculate only the time evolution of a spatially homogeneous system beginning from an initial stage of particle injection up to the time of reaching the boundary of the cauldron when the radiation escapes.

Introduce a reference frame, comoving with an injected element along magnetic field lines, in which the net energy flux is zero. To simplify the problem we assume that the distribution of injected particles over momenta in this comoving frame is isotropic[*]. To return to the laboratory frame one has to perform the inverse Lorenz transformation for the final results.

Under the above simplifying assumptions the problem can be solved numerically with Monte-Carlo methods.

So the numerical simulation problem reduces to the following:

Let at $t = 0$ the infinite space with a magnetic field of a given value H homogeneously filled by isotropic high energy electrons and positrons of a given energy distribution (maybe monochromatic). Let us simulate all interactions between particles and synchrotron radiation of electrons and positrons to see how the system stage varies with time. Let the system state at the age t be considered as the final result for the radiation escaping from the cauldron with the corresponding optical depth. We suppose $H = const$ (does not depend on time) for every version of simulation.

It is natural to define the age t of electron-po-

---

[*] This assumption is justified when magnetic field lines are noticeably curved and the primary γ -quanta are injected into the cauldron under noticeable pitch-angles.

sitron-photon soup and the optical depth τ of the whole cauldron as a number of collisions of a particle with others to a given moment or to the moment of escaping, respectively. This definition depends on the energy distribution but we may use the following simplification:

Let the total energy of all particles be distributed between monochromatic "conventional" particles, $m_e c^2$ for each. Let these particles move with the velocity of light and interact with the cross section $\sigma_0 = \pi r_0^2 = 2.3 \cdot 10^{-25}$ cm^2. Then the optical depth of the cauldron is:

$$\tau = \int_{x_0}^{\infty} \frac{\rho_E(X)}{m_e c^2 \gamma(X)} \cdot \sigma_0 \, dX, \qquad (1)$$

where $\rho_E(x)$ is the density of particle energy at a point x, $\gamma(x)$ is the Lorentz factor of a frame system in which the angular distribution becomes isotropic and integration is along the particle path through the cauldron. The definition of the system age t is the same if integration is made till a given point x, or corresponding moment.

This simple definition proves to be very usefull - the state of the system is well described by the age t, for example, the maximal yield of pairs is usually achieved at t = 10.

A question arises to what extent the above simplifications concerning the properties of the cauldron restrict the applicability of our results.

The basic simplifying assumption was that the particles from different elements - i.e. of different age - do not interact with one another. But, as we shall see below, numerical simulations reveal that at large enough times t (say at t > 20) the distribution spectra of all sorts of particles change very little. This is a reason to believe that, with the interaction of particles of different age being accounted for, the final results will change very little. In other words, our results most likely apply to the case of random magnetic fields and diffusion of particles just as well.

4. THE ELECTROMAGNETIC PROCESSES

The processes taken into account were: synchrotron radiation, Compton scattering, pair production by two photons, annihilation of pairs and synchrotron reabsorption.

All energies are given below in $m_e c^2$ units.

4.1. Synchrotron radiation

The ultrarelativistic classical approximation was used in which the intensity of radiation at freequency ν is

$$dI(\nu) = d\nu \sqrt{3} \frac{e^3 H_\perp}{m_e c^2} \frac{\varepsilon}{\varepsilon_c} \int_{\varepsilon/\varepsilon_c}^{\infty} K_{5/3}(x) dx$$

where ε is the energy of emitted photon ($\varepsilon = h\nu/m_e c^2$), $\varepsilon_c = 3/2\, H_\perp/H_0 \cdot \gamma^2$, $H_0 = 4.41\cdot 10^{13}$ Gs, γ is the Lorentz factor of electron.

When $E \gtrsim H_0/H$ the classical approximation is not valid, but we may restrict the calculations by cases when primary energy $E_0 < H_0/H$. If $E_0 > H_0/H$ the energy quickly decreases within a relatively narrow bond below $E = H_0/H$ because of the repeated processes of the pair creation. The old system cannot remember its evolution at $E > H_0/H$.

4.2. Pair production by two photons

This is the main process producing pairs. (The pair production by photons on electrons or positrons was also taken into account in preliminary culculations but its contribution proves to be negligible). The cross section was precisely represented as:

$$\sigma_{\gamma\gamma} = \frac{\sigma_0}{\omega^2} \left\{ (2 + 2/\omega^2 - 1/\omega^4) \ln(\omega + \sqrt{\omega^2-1}) - \sqrt{1 - 1/\omega^2}(1 + 1/\omega^2) \right\},$$

where ω is the center of mass energy of the photon. The probability of interaction depends on the angle between photon momenta: $p \propto \sigma_{\gamma\gamma} (1-\cos\Theta)$.

4.3. Compton scattering

The Klein-Nishina cross section which is valid for any energy of electrons, positrons and photons was used.

4.4. Annihilation of pairs

Annihilation was precisely represented as:

$$\sigma_a = \sigma_o \frac{41}{4\beta_c E_c} \left\{ \frac{3-\beta_c^2}{\beta_c} \ln \frac{1+\beta_c}{1-\beta_c} + 2(\beta_c^2 - 2) \right\},$$

where β_c and E_c are the center of mass velocity and the energy of an electron. The probability of annihilation depends on the relative velocity of particles $\beta_r : p \propto \sigma_a \beta_r$.

4.5. Synchrotron reabsorption

Numerical simulation of this process is very difficult and it was approximated in a rather rough way. The probability of reabsorption of a photon with frequency ν by ultrarelativistic electrons is

$$\mu_r = \frac{c^2}{8\pi \nu^2} \int_0^\infty E^2 \frac{d}{dE} \left(\frac{N(E)}{E^2} \right) F(\nu) \, dE ,$$

where

$$F(\nu) = 3 \frac{e^3}{m_e c^2} H_\perp \nu/\nu_c \cdot \int_{\nu/\nu_c}^\infty K_{5/3}(x) \, dx ,$$

$$\nu_c = \frac{3}{4\pi} \frac{e H_\perp}{m_e c^2} \gamma^2 .$$

But the accurate account of the term $d/dE(N(E)/E^2)$ which depends on the slope of electron spectrum requires a lot of computer time. To estimate approximately the effect of reabsorption this process was simulated so as if the slope of electron spectrum were constant everywhere. The slope of electron spectrum was taken equal to zero.

5. THE RESULTS OF SIMULATION

In the present simulation the range of magnetic field 10^3-10^8 Gs was scanned. For the cauldron in the active galactic nuclei one should expect $H \approx 10^4$ Gs. For the Galactic center it probably belongs to 10^5-10^8 Gs range. The upper limit of scanned primary energy was E_0 H_0/H for the reason mentioned above. It is the upper bound $\mathcal{E}_m = H/H_0 \, E_0^2$ of the synchrotron spectrum from primary pairs rather than the primary energy E_0 that is important in the problem. The results are different for the cases $\mathcal{E}_m \gg 1$, $\mathcal{E}_m \approx 1$, and $\mathcal{E}_m < 1$. The main attention was paid to the case of high initial energies. In all versions of simulations the primary energy of pairs was monochromatic E_0.

POSSIBLE MECHANISM OF THE FORMATION OF THE HARD SPECTRUM OF AGN

Another parameter of the problem is the ratio of the energy density of particles to the energy density of magnetic field Ω which defines the relative yield of Compton losses. The calculations have been made for negligible value of Ω for $\Omega = 5$ and for $\Omega = 500$. For $\Omega \gg 1$ the result depends on Ω only slightly.

The spectral slopes are given below for the scale $dN(\varepsilon)/d\varepsilon = \varepsilon^{-\alpha}$; note that photon spectra are presented in figures with the scale $\varepsilon \cdot dN(\varepsilon)/d\log(\varepsilon)$ and a slope spectrum ($\sim \varepsilon^{\beta}$) in figures is $\beta = 2 - \alpha$.

5.1. The case $\varepsilon_m \gg 1$

Fig. 1-3 show the variation of photon spectrum with the age for different initial conditions. One can see in Fig. 1 that at small t (t = 0.06 ≪ 1 in Fig. 1) there occurs a typical synchrotron spectrum of an electron decelerating due to its radiation, $\alpha = 1.5$ ($\beta = 0.5$). Then the soft part of the spectrum is yielded by radiation of the next generations of pairs. At t > 1 the spectrum becomes more flat. The hard part of the spectrum falls due to pair production, a high energy cutoff appeares that moves to the region $\varepsilon \approx 1$ as the age grows. Fig. 2 shows the evolution of the spectrum for the case $\Omega = 5$ which is smaller than Ω in the case in Fig. 1. In the case $\Omega = 5$ one can see the synchrotron radiation of the next generations of pairs even at t = 0.06. There is a cutoff of the spectrum at soft energy due to a synchrotron reabsorption.

Fig. 3 shows the evolution of the spectrum for the case $H = 10^6$ Gs, $E_0 = 10^7$, $\Omega = 5$ up to t = 16. The slope of the spectrum β is near zero at the great time. Near $\varepsilon \approx 1$ a detail appears. This is an annihilation line which is partly widened due to scattering. See also Fig. 4. The main conclusion from the simulation is the following. In the case $\varepsilon_m \gg 1$ at t in the interval $3 < t < 30$ the photon spectrum has the standard shape: the slope $\beta \approx 0.2$ ($\alpha \approx 1.8$) and the cutoff at $\varepsilon \approx 1$.

The evolution of the spectrum of $e^{\pm}$ pairs with age is shown in Fig. 5. The beginning energy of the primary pairs is monochromatic.

The number of pairs at small ages undergoes an approximately exponential growth, its rate and duration depend on magnetic field, the primary energy, the presence of primary soft photon background and so on. Then the growth becomes approximately linear, the behaviour of the system becomes the same for different initial conditions. At $t \approx 10$ the yield of $e^{\pm}$ rest mass reaches the maximal value which is ≈ 0.1 of total energy, then slowly decreases (asympthotically as $2/t$) due to e^+, e^- annihi-

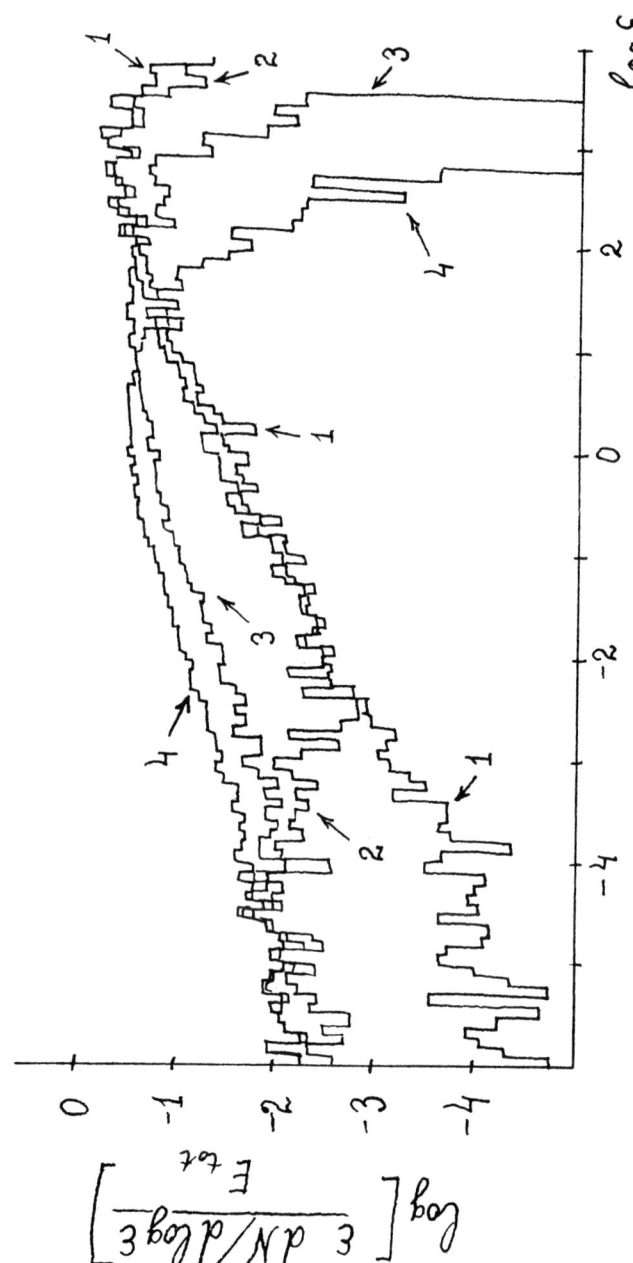

Figure 1. The evolution of the photon energy distribution with age. $H = 10^3$ Gs, $E_0 = 10^7$, $\Omega = 500$. 1 - t = 0.06, 2 - t = 0.3, 3 - t = 1.5, 4 - t = 3.

POSSIBLE MECHANISM OF THE FORMATION OF THE HARD SPECTRUM OF AGN

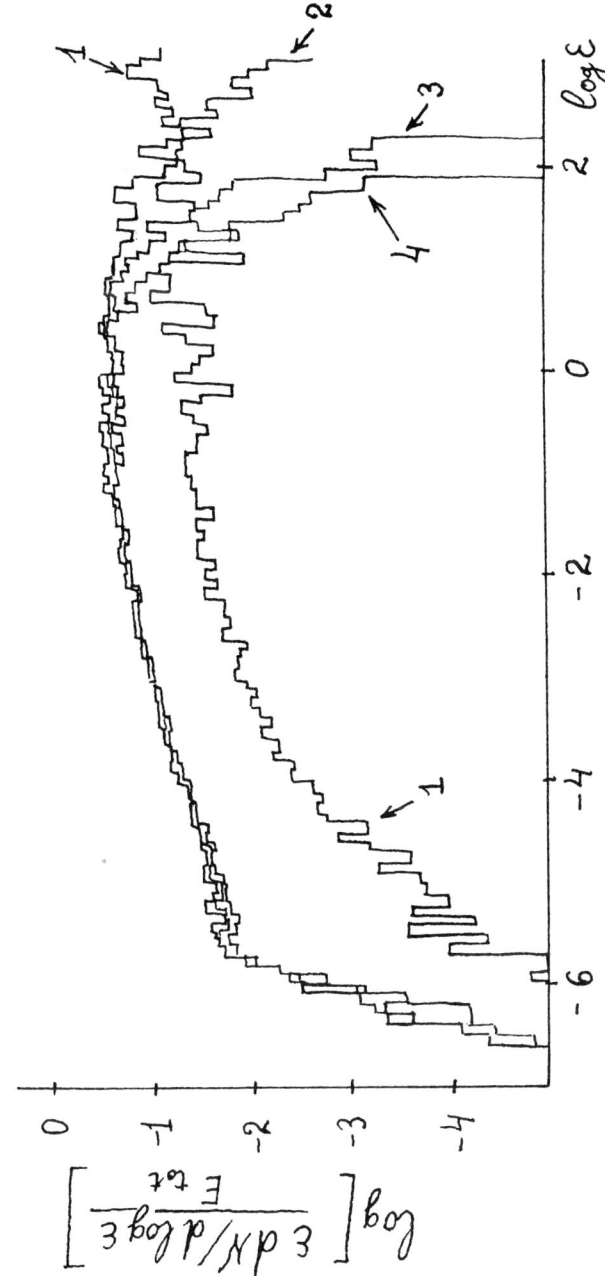

Figure 2. The evolution of the photon energy distribution with age.
$H = 10^5$, $E_g = 10^7$, $\Omega = 5$, $1 - t = 0.06$,
$2 - t = 0.3$, $3 - t = 1.5$, $4 - t = 3$.

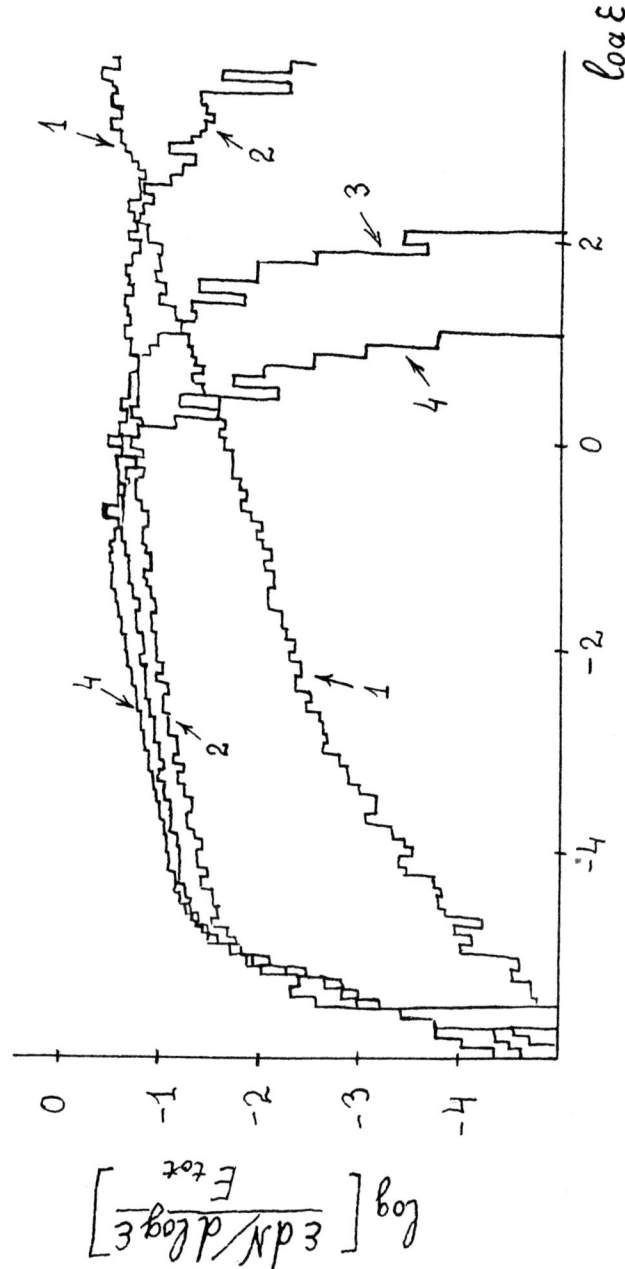

Figure 3. The evolution of the photon energy distribution with age. $H = 10^6$, $E_0 = 10^7$, $\Omega = 5$. 1 - $t = 0.06$, 2 - $t = 0.3$, 3 - $t = 3$, 4 - $t = 16$.

POSSIBLE MECHANISM OF THE FORMATION OF THE HARD SPECTRUM OF AGN

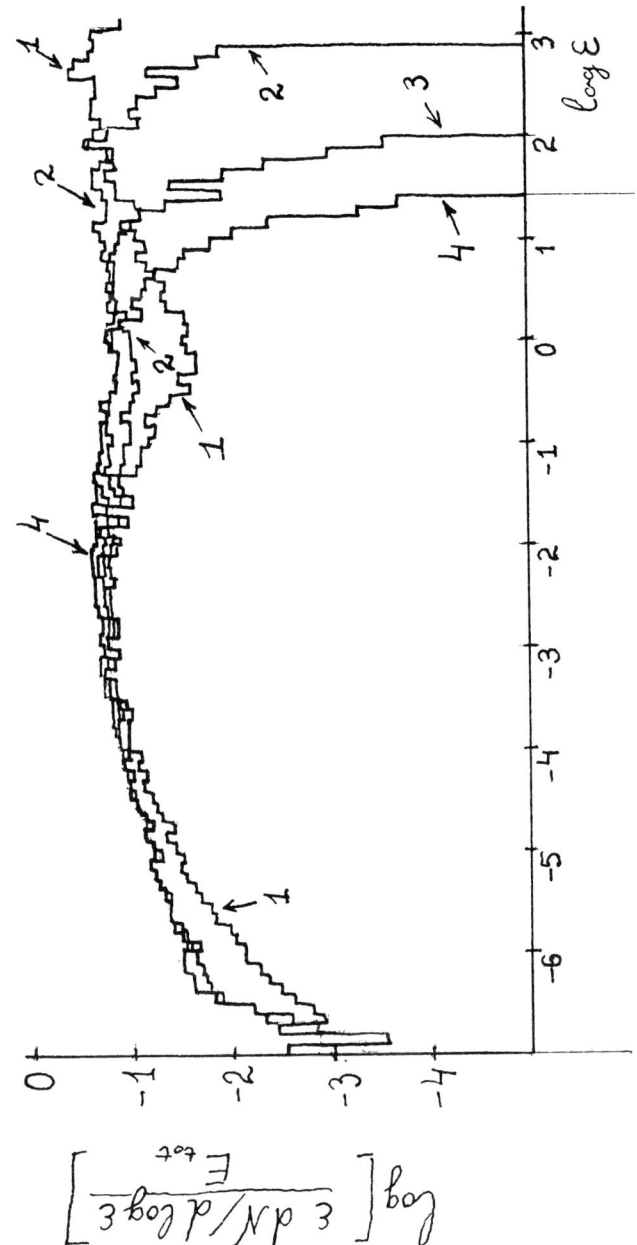

Figure 4. The evolution of the photon energy distribution with age. $H = 10^4$ Gs, $E_0 = 10^7$, $\Omega = 5$. $1 - t = 0.06$, $2 - t = 0.3$; $3 - t = 3$; $4 - t = 12$

lation (Fig. 6). The dependence of maximal rest mass yield on primary energy is stronger for the case of low Ω (Fig. 7). In this case the efficiency of pair production has a clear minimum at primary energy when the radiation of second generation electrons is a little softer than which is able to produce pairs. This minimum is smeared out at $\Omega \approx 1$ due to comptonization of synchrotron photons. The result for a wide flat spectrum of primary particles is close to that averaged over logarithm of primary energy.

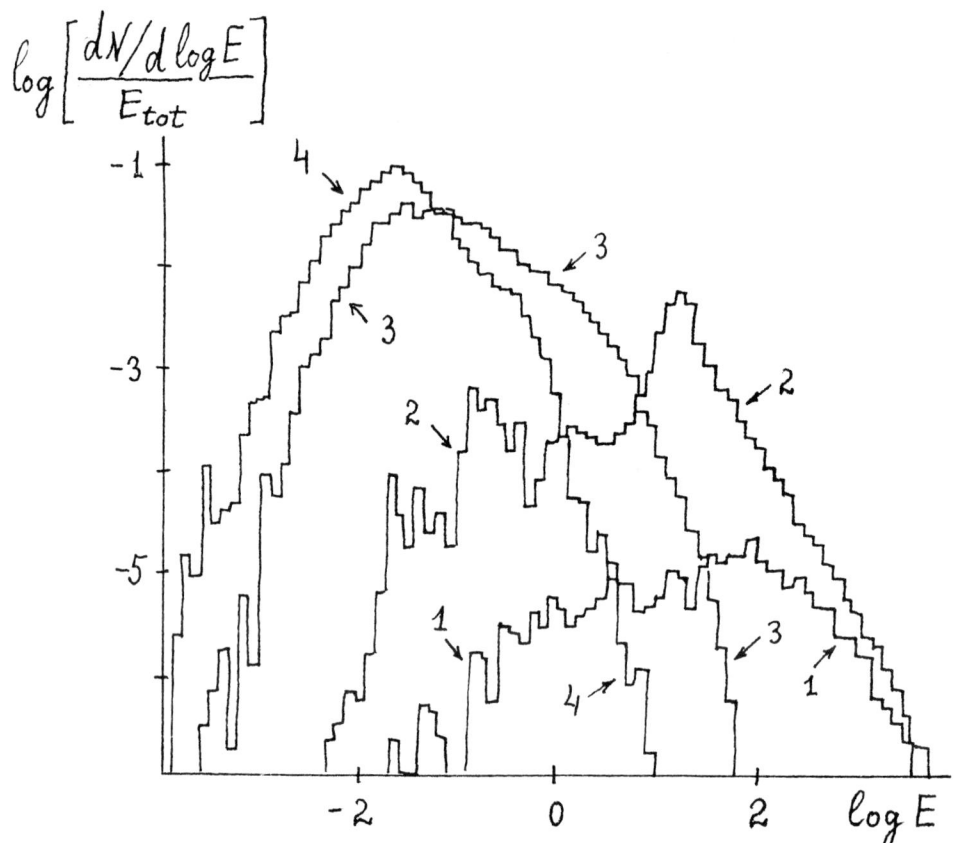

Figure 5. The variation of electron-positron energy distribution with age. The parameters are the same as in Fig. 3

5.2. The case

In this case the energy does not degrade in cascade way-
all pairs are produced by radiation of primary particles.
The pair production is the most effective (see Fig. 6,7),
the final photon spectrum is harder ($\alpha \approx 1.5$, see Fig.8).
This is the case considered in the paper Blandford (1982)

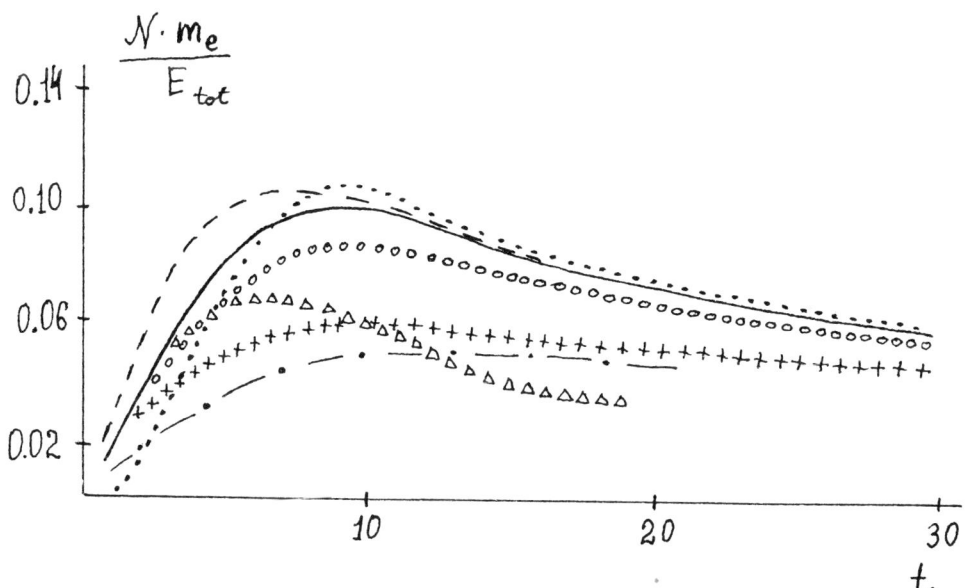

Figure 6. The yield of pairs versus age
───── $H = 10^6$ Gs, $E_0 = 10^7$, $\Omega = 5$,
+ + + $H = 10^6$ Gs, $E_0 = 10^7$, $\Omega = 10^{-2}$,
· · · $H = 10^8$ Gs, $E_0 = 3 \cdot 10^5$, $\Omega = 5$,
- - - $H = 10^5$ Gs, $E_0 = 10^4$, $\Omega = 5$, ($\mathcal{E}_m \approx 1$),
-·-·- $H = 10^6$ Gs, $E_0 = 10^3$, $\Omega = 5$ ($\mathcal{E}_{mm} < 1$),
o o o $H = 10^5$ Gs, $E_0 = 10^7$, $\Omega = 5$,
△ △ △ $H = 10^6$ Gs, $E_0 = 10^7$, $\Omega = 5$,

The last curve corresponds to the case when the way of electrons through the reactor is twice more durative than that
of photons. It leads to storage of electrons by factor 2
relatively to photons. This calculation roughly approximates
a reactor with field lines curved by $\theta \approx 1$ and gives a more
realistic result.

5.3. The case $\mathcal{E}_m < 1$

In this case synchrotron photons are too soft and can not produce pairs, but the pair production goes through comptonized photons if $\Omega \approx 1$, at the lower rate than the previous cases. The variety of resulting spectra is greater than that in the case of a large primary energy. An example is given in Fig. 8. In this case there are few maxima which are connected with the synchrotron radiation of the different generations of pairs.

Let us consider now the effect of synchrotron reabsorption. Reabsorption leads to a dropoff of the photon spectra in the soft region. A position of the dropoff depending on the magnetic field is satisfactory described as

$\mathcal{E}_{\gamma} \approx 10^{-9} H(Gs)^{0.7}$ for the case $\Omega = 5$, $\mathcal{E}_m \gg 1$. The effect of the reabsorption on electron-positron spectrum is not large - when the reabsorption is switched on the average energy of electrons and positrons increases by a factor 1.5 at small ages and the resulting efficiency of pair production increases by one percent for $H = 10^6$ Gs, $\Omega = 5$, $\mathcal{E}_m \gg 1$. The results in Figs. 5,6,7 are presented for the

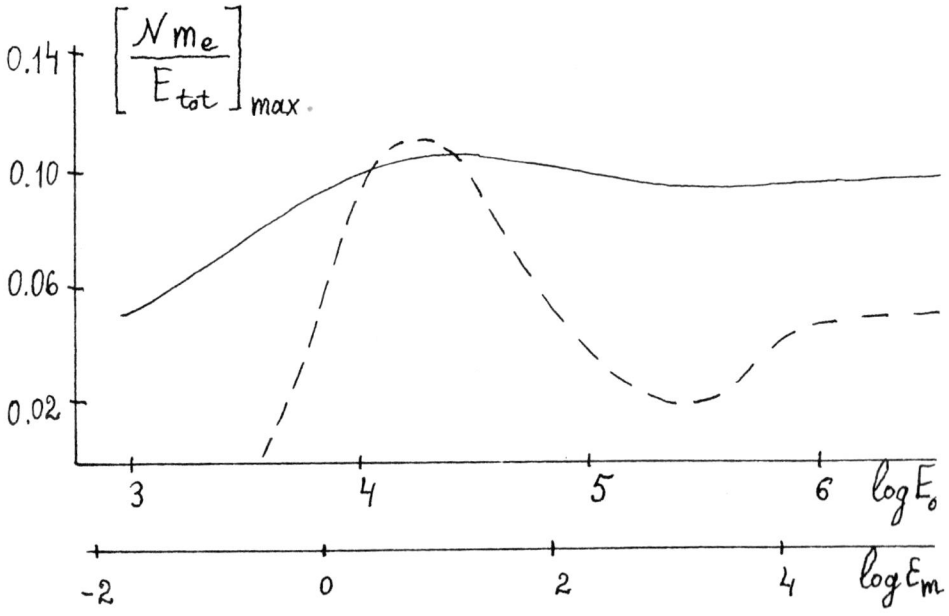

Figure 7. The maximal achived e^+e^- yield versus primary energy. The solid line $\Omega = 5$, dashed line - $\Omega = 10^{-2}$, $H = 10^6$ Gs

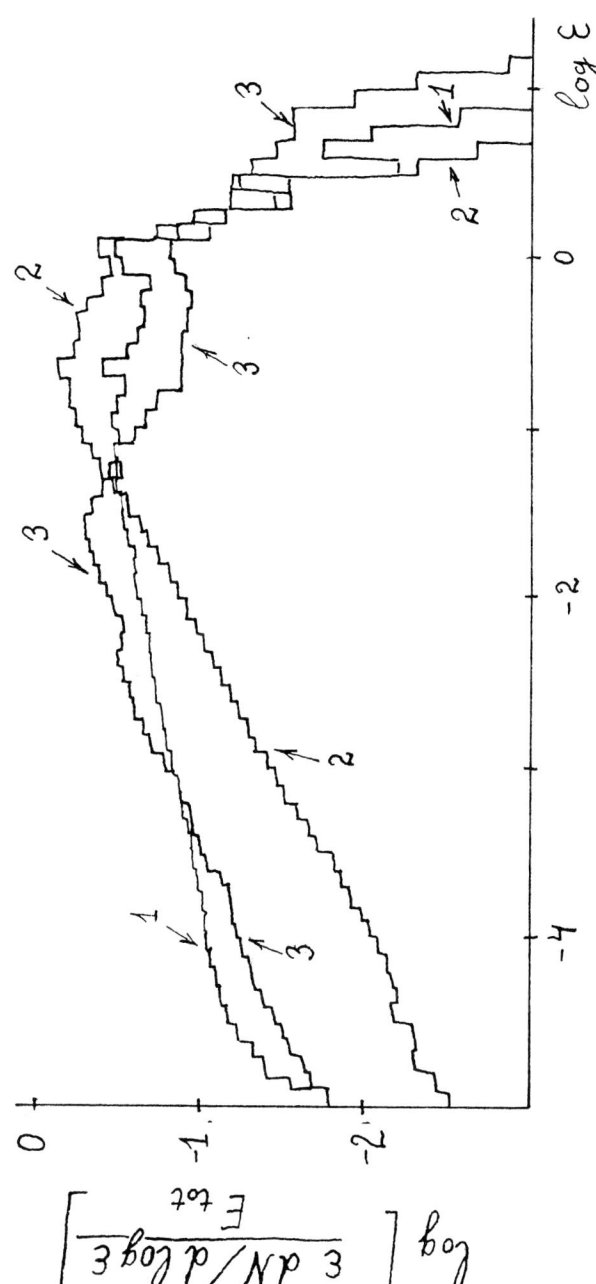

Figure 8. The old photons spectra (t=16) for various conditions:
1 - H = 10^6 Gs, E_0 = 10^6 Gs, Ω = 5, ε_m > 1; 2 - H = 10^6 Gs, E_0 = 10^4, Ω = 5, $\varepsilon_m \approx 1$; 3 - H = 10^6 Gs, E_0 = 10^3, Ω = 5, ε_m < 1.

case of switched off reabsorption. Remind that this process was described roughly but it can hardly lead to serious underestimation of the reabsorption effect.

As it can be seen from fig 1-4,6 the dependence of results on H is rather weak. As the magnetic field decreases, the efficiency of the pair production of the cauldron slightly falls since the energy range is expanded. For the same reason one should expect a stronger dependence of the efficiency on the primary energy for a weaker fields.

So the effect of e^+e^- cauldron gives similar results for various H, E_o (if $\varepsilon_m > 1$) and optical depths (for interval $3 < \tau < 30$).

6. THE CONDITIONS FOR REALISATION OF e^+e^- CAULDRON

Among astrophysical objects only pulsars and accreting black holes can produce a large acceleration power in a volume small enough to give rise for a e e cauldron. The pulsar can surely be an adequit accelerator, the question is only about its power required for cauldron formation. In the case of AGN only a black hole can have a power required for the explanation of their activities. As for a black hole we keep to a concept of particle acceleration by a rotating black hole surrounded by magnetic field which can be stored with accretion of a matter (Blandford, Znajek, 1977; Macdonald, Thorne, 1982).

Let us summarize the requirements for the magnetosphere of an object.

1. The magnetic field should be curved enough.
2. A large optical depth of an electron-positron-photon soup defined by (1). The optimal depth is $\tau \approx 10$, but good results are possible for rather wide range of depths.
3. A suffitient energy of radiation of accelerated electrons and positrons to produce pairs emitting synchrotron photons of energy above $m_e c^2$. Otherwise the cauldron is possible but of a less efficiency. See Fig. 7.
4. Open and not chaotic magnetic field lines. This is required specially for emission of narrow annihilation line. It is important for the explanation of the hard radiation of the Galactic center. But, of course, it is not necessary in the case of AGN. For example one can consider a case when magnetic field is pushed out in a form a loops loaded by pairs along the rotation axis of an accreting black hole.

Some constraints on the cauldron in the AGN follow from the requirement of a suffitient optical depth. To estimate the dimension of the reactor let us suppose that the energy flows into 4π solid angel from the radius R_0. Then the energy at $R > R_0$ is $\rho_E(r) = L_h/(4\pi R^2 \cdot c \cdot \beta)$, L_h is the injected high energy, c is a mean velocity of the energy flux in radial direction. From the expression (1) the optical depth is

$$\tau = \int_{R_0}^{\infty} \frac{L_h}{4\pi \beta\, c\, m_e c^2} \cdot \frac{1}{\gamma(R)} \cdot \sigma_0 \cdot \frac{dR}{R^2} \qquad (2)$$

If we suppose that $\sqrt{\beta} \cdot \gamma(R) \approx 1$, then

$$\tau \approx 10\, \frac{L_h}{10^{45}(\text{erg/s})} \left(\frac{R_0}{10^{14}\,\text{cm}}\right)^{-1} \qquad (3)$$

If a black hole is responsible for the cauldron, substituting $R_0 = R_g = 3\cdot 10^5$ cm $M/M_\odot$ we obtain a rough limit for a mass of a hole which can give rise for an effective cauldron with $\tau \gtrsim 10$:

$$M \lesssim \frac{L_h}{10^{45}(\text{erg/s})} \cdot 3\cdot 10^8\, M_\odot \qquad (4)$$

The objects promising for e^+e^- cauldron formation are active galactic nuclei with X-ray and γ luminosity not far from Eddington limit. Actually, substituting into (3) $L_h = L_{Edd} \approx 10^{38}$ erg/s$(M/M_\odot)$ and $R_0 = R_g = 3\cdot 10^5$ cm $M/M_\odot$ we obtain the optical depth $\tau \approx 300$. It is sufficient with a large excess for formation of a cauldron which can dump below the MeV the whole high energy luminosity produced by a black hole.

7. CONCLUSION

The photon spectra in Fig. 8 are the main results of our simulations for the confrontation with the observational data.

As it can be seen in Fig. 8 the main spectral features are the following.

There is a wide power law spectrum at $\log E < 1,5$ (<30 KeV). The slope of the spectrum depends on the initial ε_m. In the case $\varepsilon_m \gg 1$ the slope is $\alpha \approx 1,8-1,9$; if the primary energy is not very large ($\varepsilon_m \lesssim 1$) the spectrum may be more complicated with the slope down to 1,4-1,5. So the slope can be a test for the initial energy in the cauldron.

Another feature of the spectra is the cutoff in the MeV region. This cutoff of spectra from some AGN has been supposed in the papers of Rees (1978), Herterich (1979).

There is an indication that this cutoff is a general fearure of active nuclei - measured X-ray spectra being extrapolated into dozens MeV region give an excess on the diffuse γ-ray background, so a breake of spectra in MeV region is required (see Bignami et al., 1979). This cutoff is the direct consequence of the present model, the associated spectral features can be an annihilation line containing up to few percents of total luminosity. We want to emphasize that the real spectra may be more complicated because of the space variations of the magnetic field and the Dopler-shift due to the motion of the electron-positron-photon soup. The most probable candidate for e^+e^- cauldron is NGC 4151 (L 10^{45} erg/s), its measured γ-spectra (see Bignami et al., 1979) indicate a cutoff in MeV region. Otherwise the spectrum from 3C 273 (L $\approx 10^{47}$ erg/s) demonstrates the absence of e^+e^- cauldron as the intensive γ-flux in hundreds MeV region is observed.

The best limitation for 511 KeV line intensity from extragalactic objects is one for Cen-A (Gehrels, 1984) ($2.5 \cdot 10^{42}$ erg/s or 5% of its total luminosity), but this limit is too high to give constraints on this object. NGC 4151 seems to be more promising for observation of 511 KeV line.

It would be very interesting to find e^+e^- cauldrons among active galactic nuclei or quasars as it would give a good confirmation of widely discussed concept that these objects can be powerfull compact accelerators.

References

Aharonian, F.A., Vardanian, V.V., Kirillov-Ugryumov, V.G.: 1984, Astrofizika (USSR) 20(2), 223.
Begelman, M.C., Blandford, R.D., Rees, M.J.: 1984, Reviews of Modern Phys. 56, 255.
Blandford, R.D.: 1982, in Proceedings of AIP Conference "The Galactic Center", eds Reigler, G.R., Blandford R.D. N.Y., California Institute of Technology, p. 177.
Blandford, R.D., Znajek, R.L.: 1977, Monthly Notices Roy Astron. Soc., 179, 433.
Bignami, G.F., Fichtel, C.E., Hartman, R.C., Thompson, D.J.: 1979, Astrophys. J. 232, 649.
Burns, M.L.: 1983, in AIP Conference Proceedings, N 101, "Positron-Electron Pairs in Astrophysics", eds Burns, M.L., Harding, A.K., Ramaty, R., American Institute of Physics, p. 281.
Gehrels, N., et al.: 1984, Astrophys. J. 278, 112.
Herterich, K.: 1979, Nature 211, 472.
Kardashev, N.S., Novikov, I.D., Polnarev, A.G., Stern, B.E.: 1983, Astron. Z. 60, 209.
Kazanas, D.: 1984, Astrophys. J. 287, 112.
Lingenfelter, R.E., Ramaty, R.: 1982, in Proceedings of AIP Conference "The Galactic Center", eds. Reigler, G.R., Blandford, R.D., N.Y., California Institute of Technology, p. 148.
Lovelace, R.V.E., Ruchti, C.B.; 1983, in AIP Conference Proceedings, N 101 "Positron-Electron Pairs in AstroPhysics", eds. Burns, M.L., Harding, A.K., Ramaty, R., American Institute of Physics, p. 314.
Macdonald, D.A., Thorne, K.S.: 1982, Mon. Not. Roy. Astr. Soc. 198, 345.
MacCallum, C.J., Leventhal, M.: 1983, in AIP Conference Proceedings, N 101 "Positron-Electron Pairs in Astrophysics", eds. Burns, M.L., Harding, A.K., Ramaty, R. American Institute of Physics, p. 211.
Rees, M.J.: 1982, in Proceedings of AIP Conference "The Galactic Center", eds. Reigler G.R., Blandford, R.D., N.Y., California Institute of Technology, p. 166.
Rees, M.J.: 1978, Phys. Scr. 17, 193.
Rothschild, R.E., Mushotzky, R.F., Baity, W.A., Gruber, D.E., Matteson, J.L., Peterson, L.E.: 1983, Astrophys. J. 269, 423.
Stern, B.E.: 1985, Astron. Z, 62, 1000.
Zdiarski, A.A.: 1985, Astrophys. J. 289, 514.

DISCUSSION

PERRY: Can you change the parameters of your model to account for hard γ-ray spectra? For example, Bignami et al. 1981(Astron. Astrophys, 93, 71) report detection of 3C 273 at about 100 MeV.

NOVIKOV: The only way to get the γ-spectrum in a region of more than 1 MeV for an optically thick cauldron is to involve a global Lorentz factor $\gamma > 1$. The latter occurs if either (i) there is a bundle of magnetic field lines with the angular width $\Delta\theta \sim 1/\gamma$ or (ii) the loops of magnetic fields lines come out together with plasma having a great Lorentz factor. However, it seems in such a way one could get few MeV but not hundreds of MeV as in 3C 273. It is naturally to suggest that there is no well-developed cauldron in case of 3C 273. This means that either hard photons are emitted from the region far out of the Schwarzschild radius, or the field lines are straight and magnetic field can not serve as a particle collider.

LIVIO: Do you think your model could apply also to Cyg X-1? Because there is some evidence for the discovery of an annihilation line there.

NOVIKOV: The validity of our model depends on the value of ratio L/L_{Edd} (R/R_g), where L is a luminosity in a hard part of the spectrum, and R is a size of a cauldron. The model is applicable if the ratio is no less than $\sim 10^{-2}$ (the cauldron is optically thick in this case). The X-ray luminosity of Cyg X-1 is $0.1 \, L_{Edd}$, which is quite enough. One should expect a comparable luminosity in a range of hundreds of KeV as well, provided that there is no some unfortunate anisotropy. At first sight it seems that our model may well be applicable to Cyg X-1.

PETROSIAN: One of your assumptions is that particle pitch angle distribution is isotropic. If synchrotron radiation is the dominant low process by $e^{\pm}$ pairs, then pitch angle anisotropics may develop. What mechanism do isotropics the distribution?

NOVIKOV: We deal with the isotropization mechanism for photons only, that is, the curved magnetic field, moving in which a photon acquires a new pitch-angle. In fact decelerated electrons will locally be anisotropic, but an attempt to account for this anisotropy would drastically complicate calculations. In our view hardly would this drawback of the model affect the results. First, the electrons become anisotropic only after they have emitted most of their energy and are thus of less importance in the total energy balance; second, in the reference system where the total photon pulse is zero electrons are equiprobably moving in two opposite directions, this does not differ much from the isotropic case (it is important for us that particles be colliding); third, electrons still become partially isotropic via the Compton scattering.

REES: You need a source of soft photons, but is it necessary for your model that these come from synchrotron radiation? May be you would get the same final spectrum even if there were no magneitc field in the region?

NOVIKOV: Synchrotron radiation is enough for conversion of initial photons, but the existence of outer soft photons wouldn't make any harm, of course, and will only result in increasing of the conversion rate at first stage. These soft photons may well be noticeable in the final spectrum since the optical thickness for them are much less than for hard ones.

As for the case without magnetic field, it is crucially whether the initial photons (and their descendants) have enough angular dispersion, which will let them to scatter effectively. If dispersion is small, than cutoff in the spectrum will occur not at $m_e c^2$, but at $m_e c^2 (m_e c^2/\varepsilon)$, where ε is an energy of the soft photons. If there is enough angular dispersion, than, it seems, the hard part of the spectrum will be qualitatively the same as in our model. It is easy to calculate this case. But is it realistic? There seem to be no mechanism, which in such a natural way may play a role of collider, as a curved magnetic field could do it.

THE PROTOGALACTIC CONNECTION

Joseph Silk and Rosemary F.G. Wyse,

Department of Astronomy, University of California,
Berkeley, CA 94720.

Abstract: We argue that the physics underlying active galactic nuclei and the quasar phenomenon will not be understood without first having knowledge of the processes governing the gas to star transformation that occurs during galaxy formation. Galaxies actually in the process of forming are extremely difficult to observe, or perhaps even recognise, and we take a more indirect approach, utilising the observed scaling relations between various parameters of present day galaxies to infer information about the physical conditions at the epoch of galaxy formation. One example of such an approach is given, in which the characteristic parameters of galaxies are utilized as fossil records that help constrain the physical processes, involving both dissipation and dynamical relaxation, by which galaxies formed. Two contrasting models for the early evolution of a galaxy – a single massive burst of star formation, as opposed to many small bursts – may be distinguished by their different predictions for the power spectrum of the extragalactic background light in different bandpasses. A more definitive theory of protogalactic evolution may be attainable as the collection of data, its modelling and its interpretation proceed.

I. Introduction

To understand quasars and possibly also active galactic nuclei, one almost certainly needs to understand how galaxies form. Protogalaxies are likely to provide crucial clues to the origin of the massive black holes that most astronomers concur are embedded within galactic nuclei and whose fuelling is responsible for the QSO phenomenon.

First, in a general sense, the issue of how the massive black hole formed can be resolved more easily in a predominantly gaseous system, namely a protogalaxy, than in a predominantly stellar system, namely a galaxy. To initiate the catastrophic collapse of a stellar system that can trigger the formation of a black hole requires stellar collisions and disruptions at a density that exceeds hitherto measured values in galactic nuclei. Such a high density could not arise by dynamical relaxation alone, but requires very considerable dissipation. It seems far more natural to begin with a protogalaxy that inevitably undergoes very considerable gaseous dissipation during its initial collapse. Low angular momentum gas should accrete into a central core and be capable of forming a black hole. Note that the necessary gas density is only $\sim 100\, M_8^{-1}\, \mathrm{g\, cm^{-3}}$, where $M_8 = M_{\mathrm{hole}}/10^8\, M_\odot$, no higher than is attained in the course of the formation of ordinary stars.

Further arguments linking active nuclei with protogalaxies may be summarized as follows. The total number density of "dead" quasars and active nuclei is comparable to the number density of bright galaxies (Schmidt and Green 1983). The epoch of maximum activity of quasars, when the number density of luminous quasars greatly exceeded the local value, is at a redshift of between 2 or 3. This is soon after the presumed epoch of luminous galaxy formation. Some quasars appear to be embedded in galaxies with an exceptionally large ratio of far infrared to optical luminosity (Neugebauer et al. 1984). Such a ratio is interpreted, at least in nearby galaxies, to be indicative of a high rate of star formation. The fact that quasar emission line regions are metal rich tells us that extensive star formation has already occurred prior to black hole formation. Finally, a direct link is possible between radio jets and radio galaxies, and starbursts. It is likely that the process of fuelling the central engine is responsible for triggering a radio outburst; the enhancement of gaseous fuel can also trigger a starburst. Interacting galaxies have enhanced star formation rates, and radio jets may be capable of directly initiating a starburst, as apparently occurred in Minkowski's object (van Breugel et al. 1985). Evidently, to understand quasars and active galactic nuclei, one has to understand protogalaxies.

II. The Search for Protogalaxies

Protogalaxies have been notably elusive. The original motivation underlying search programs was that the solar metallicity level was produced over the duration of a freefall time for a typical massive galaxy. This yields the estimate

$$L_{pg} \simeq 3 \times 10^{46}(Z/Z_\odot)(10^9 \text{yr}/t_{burst}) \text{ erg s}^{-1}.$$

This estimate suggests that a protogalaxy would be as luminous as a quasar. Indeed, it has been proposed that some protogalaxies may be masquerading as quasars. A signature would be narrow emission lines, especially Lyman alpha, of width reflecting the underlying galaxy potential well of $\sigma \sim 300$ km s^{-1} and observed at an epoch corresponding to the freefall time, of about 10^9 years, for the outer regions of a massive galaxy.

Unfortunately, the observational situation is not clear. There are indeed examples of quasars with unusually narrow emission lines. If these objects were also found to be extended and strong infrared sources, they would be intriguing candidates for protogalaxies. Direct searches for protogalaxies have generally concentrated on the Lyman alpha emission line or Lyman break as a signature, and have hitherto been unsuccessful. The most stringent limits have been set by Koo and Kron (1980) who have ruled out the existence of sources with strong Ly α at redshifts z $\sim$ 5, where 'strong' means as bright as the line emission in a typical quasar at z $\simeq$ 2.5, and sources with a detectable Lyman continuum break at redshifts z $\sim$ 6.

Lyman alpha emission may well be quenched by dust, and broad band colors are an important diagnostic of actively starforming galaxies. Unfortunately, the initial burst may last for only 10^8 yr (the freefall time in the inner regions of a massive galaxy) and the effects of the burst on the spectrum of the protogalaxy will fade rapidly. Precisely how rapidly is a somewhat contentious issue. Wyse (1985) has included core helium burning and asymptotic giant branch stars for the first time, in a simple spectral synthesis model and concludes that the blue phase, in B−V color, will only last for 10^8 yr (Figure 1). This means that it may be very difficult to detect the fading of the initial burst of star formation in elliptical galaxies. The epoch of galaxy formation could conceivably have occurred very recently (z $\sim$ 1 or 2).

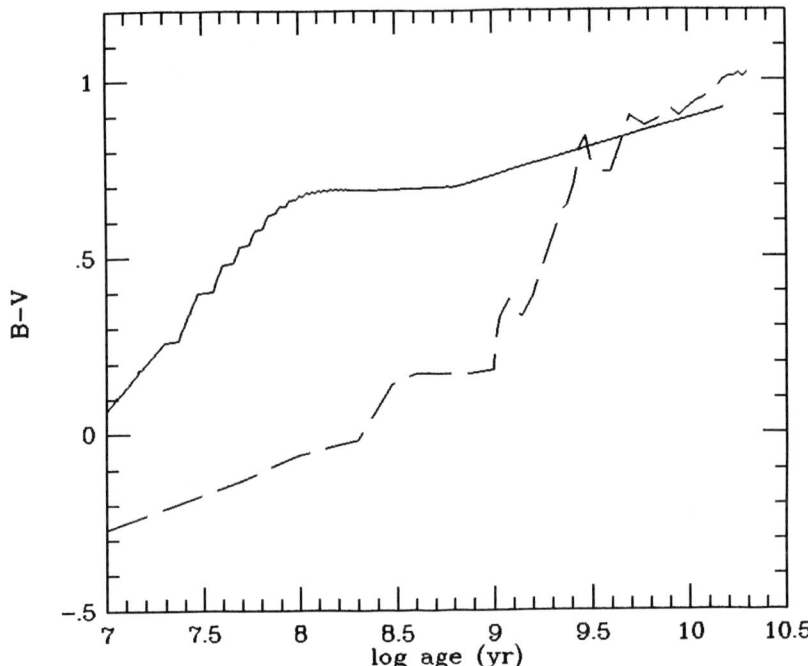

Figure 1. B−V color as a function of age of a model elliptical galaxy, formed in a burst of star formation of duration 10^9yr, modelled by superposing mini-bursts each of duration 10^7yr. Full line includes core helium burning and AGB phases, dashed line (Bruzual, 1983) neglects them.

Presently available data on distant galaxy colors do suggest that an increasing dispersion of blue colors appears towards fainter magnitudes and higher redshifts, to $z \lesssim 0.6$ (see Figure 2 for B−V). These blue galaxies are not well fit by models which formed their stars in a single burst at redshifts greater than unity, especially if one accounts for the fact the giant ellipticals have generally greater than solar abundance, while the models are of solar metallicity. Whether this is due to a spread in formation ages or to the effects of continuing star formation is not clear. The dispersion may be a direct consequence of the poorly understood variety in present day ultraviolet fluxes of giant ellipticals, which can produce a one-magnitude spread in broad band colors at redshifts which reflect the rest-frame UV (Bertola, Cappacioli and Oke 1982). The blueward excursion of the data may be a consequence of the combination of the observational technique, which involved looking through a fixed aperture, and color gradients — for higher redshifts more of the bluer, outer regions are sampled. Couch, Shanks and Pence (1985) found a blueing of $0^m.4$ in observed B–V over a factor two in radius for a brightest cluster member at a redshift of 0.57. An extension of Wyse's models to infrared colors is underway, which will allow testing of the models in the redshift : color plane with higher redshift data. Bruzual's (1983) more sophisticated spectral synthesis models (which unfortunately lack stars in post-helium ignition phases) are compatible with a burst at an assumed redshift of ~ 5, but he did not explicitly address the question of formation epoch. The uncertainties in the relative normalisations and slopes of number counts of galaxies to faint magnitudes found by different groups unfortunately detract from the usefulness of these at the present time (see Koo 1984).

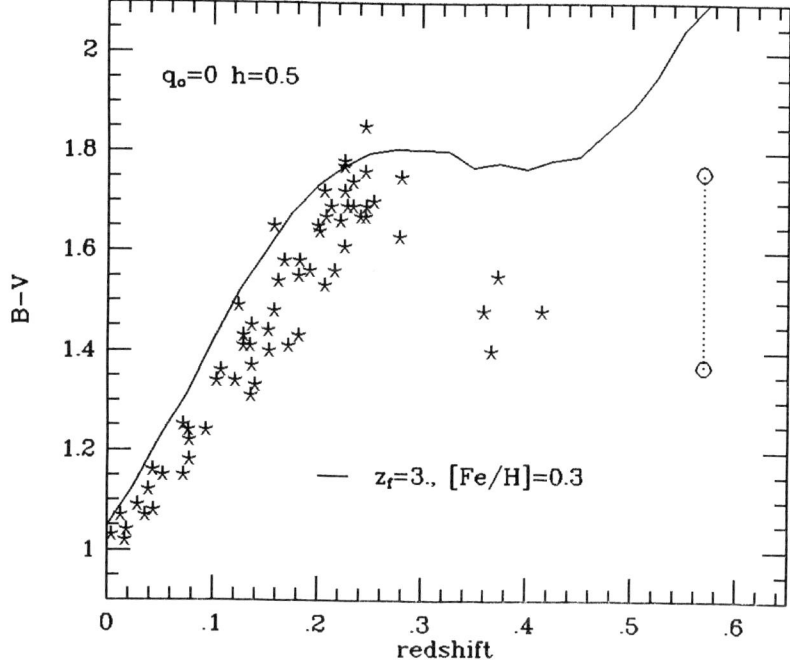

Figure 2. Observed B–V as a function of redshift for a model galaxy of twice solar metallicity and which formed at redshift 3. The observational data are from Kristian et al. ($\star$) plus the two aperture observations of one galaxy (at radii of 40 kpc and 80 kpc) of Couch et al. (o).

Of course, at high redshift only the most luminous galaxies can be seen. One might well expect these to have formed earliest, so that they are not representative of the average L_* galaxy, where $L_* = 3 \times 10^{10} L_\odot$ is the characteristic scale in the Schechter luminosity function. By isolating galaxy clusters at high redshift, and obtaining spectra of galaxies that are well down the luminosity function from the first brightest members that are ordinarily studied as cosmological distance indicators, it should be possible to obtain an indication of post-starburst evolution of elliptical galaxies. Such studies are presently in progress.

The discussion above has concerned models where massive galaxies form in a short-lived, luminous burst and evolve as isolated systems. There are equally plausible scenarios in which a protogalaxy forms in a series of small minibursts either in individual low mass clumps which subsequently merge together, or in a large but relatively quiescent system. Such models would by-pass any initially very luminous phase dominated by radiation from massive stars. Not all is lost, however, for radiation is still emitted and contributes to the diffuse background light. Analysis of the power spectrum of the extragalactic background light in different bandpasses should contain a wealth of information about primeval galaxy formation and evolution although, of course, any diffuse light from distant galaxies has yet to be detected. A simple merging scheme, where structure is built up successively by increasing the mass of bound systems by a constant factor at each stage, can yield insight into possible observable differences between single burst models and hierarchical schemes.

The adoption of a power law form of given slope for the underlying spectrum of primordial density fluctuations ($\delta\rho/\rho \propto M^{-\alpha}$), taken to be normalised by the galaxy-galaxy two-point correlation function, defines the epoch at which a given scale begins non-linear growth, and collapses to form a bound object. The luminosity evolution may be estimated by invoking synthesis of metals at each stage of the hierarchy, normalising to the observed mass–metallicity relationship of elliptical galaxies, $Z \propto M^{0.4}$ (Mould 1984), a typical giant elliptical having twice solar metallicity. Thus from an adopted basemass — that scale that first goes non-linear with the adopted power spectrum and is greater than the post-recombination Jeans mass $M_J \sim 10^6 M_\odot$ — and corresponding metallicity, sufficient energy is assumed to be radiated during each merger to increase the metal content to that associated with the next level of the hierarchy. The bolometric luminosity evolution is modelled by assuming that star formation occurs in a burst of duration about equal to a freefall time on the relevant scale (always $\gtrsim$ the lifetime of the most massive stars which are the sites of nucleosynthesis), the luminosity decaying after the burst by the power law behavior found by Wyse (1985). It is assumed that the next level of the hierarchy does not begin its collapse and star formation until after the cessation of star formation on the current level. This constraint on the relative freefall timescales of neighboring levels sets the value of the increase in mass at each stage, maximum efficiency minimising the time between successive bursts of star formation. The luminosity then increases up the hierarchy, and the resulting photometric evolution of *bound objects* as a function of the age of the universe, may be compared with a massive single burst model which reaches peak luminosity at the same time.

Figure 3(a) shows the restframe V luminosity evolution of the merging hierarchy. The decay after each merging, the steepness of which increases as the proportion of stars from previous levels increases, is due to the fact that the fading of old stars dominates the increase of luminosity as stars form to synthesise metals. Here the hierarchy has been truncated at $3 \times 10^{11} M_\odot$, typical of an L_* galaxy.

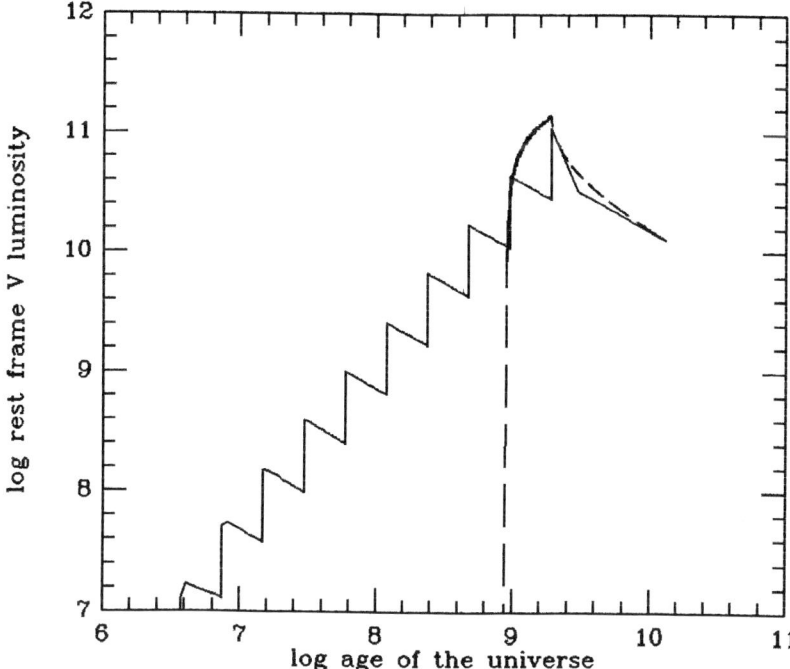

Figure 3 (a). Restframe V luminosity, in solar units, of a typical bound stellar system as a function of the age of the universe, for a hierarchy with primordial power spectrum of slope $\alpha = 1/3$, with cosmological parameters $\Omega = 1$ and h=0.5. The dashed line shows the behavior of a solar metallicity single burst model, which began star formation at redshift 5, and formed stars over a period of 10^9 yr.

The single burst model, which initiated star formation at a redshift of 5, is for a *solar* metallicity system, and so the total energy output is below that of the hierarchical model (both models have been normalised to give the same final luminosity) — the relevant feature is the different output at early times. It should be noted that the V-luminosity in both models is dominated by old stars, and that in the hierarchical model the increase at merging is dominated by the simple effect of adding together stars from several systems of the previous level. The role of young stars is discussed below. The restframe B−V color of bound stellar systems may also be derived, taking explicit account of the variation of the metallicity and its effect on colors using calibrations from stellar evolution calculations, and the result is given in Figure 3(b). The photometric evolution is evidently very different in the two cases, the slower build up of structure seen in the merging model resulting in less spectacular evolution, as may have been expected.

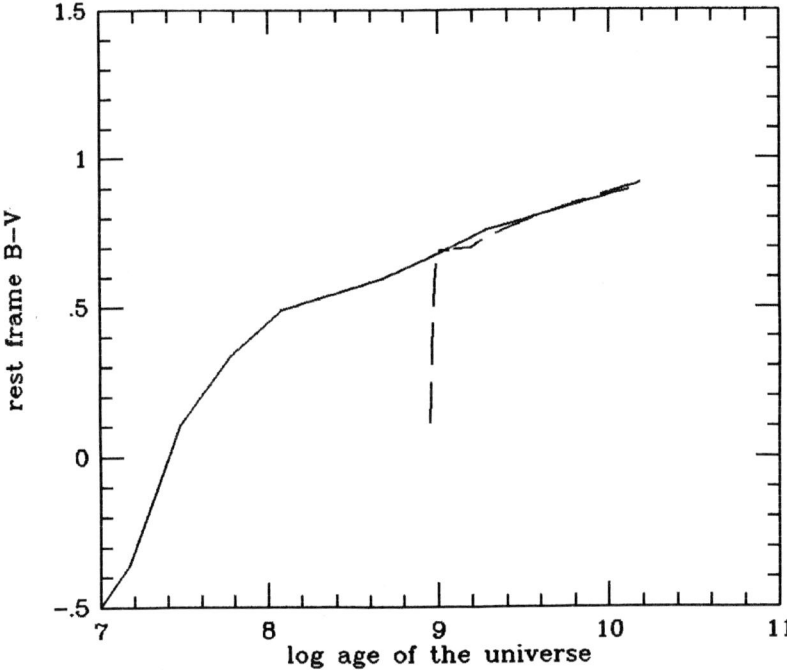

Figure 3 (b). Rest frame B-V color evolution. Here the colors of the single burst model have been modified to include the effects of twice solar metallicity.

The merging models here are normalised by the metallicity requirements of giant ellipticals, hence the *total* energy output per giant galaxy is fixed. However, since the timescale, and redshift interval, over which the energy output occurs is much larger in the hierarchical model, the extragalactic background light (EBL) in various bandpasses will be very different from the single burst case. The initial blue phase of each level occurs over a range of redshifts from ~ 30 to ~ 4, for the chosen cosmological parameters. Thus the light emitted in the optical regions of the spectrum will be presently observed over the near to far infrared. The contribution of any level to the intensity of the EBL at given wavelength can be easily calculated, as can that of the entire hierarchy. The single burst model does not contravene current observational limits to the EBL at any wavelength (Bruzual 1983; Wyse 1985) and the simple maximally efficient hierarchy will obviously be below that since the starlight is emitted over a much larger redshift range, and thus spread over a larger observed wavelength range. There is, however, no reason to believe that galaxy formation was efficient; indeed there are strong reasons for presuming it was an inefficient process. Such inefficiency is necessary if the dark material in galaxy halos, and for that matter, in the solar neighborhood, were to consist of stellar remnants — in such a situation, most of the debris produced by evolving stars must have been lost early in the evolution of the galaxy. The intergalactic gas in rich clusters is thought to contain a greater mass of heavy elements than could have been produced by present stellar evolution in the cluster galaxies; presumably intensive star formation at an early stage of galactic evolution accounted for this enhanced yield.

With this in mind, the interest here is in investigating the consequences of *inefficiency* in building up the hierarchy. For instance, only two out of every ten smaller scale masses may actually make it to the next level. Those which failed to merge would of course still contribute to the EBL intensity, causing an enhancement. This would also lead to a dwarf galaxy luminosity function which continues to increase towards decreasing luminosities, unlike that for giant galaxies, as has been observed, at least in the Virgo cluster (Sandage, Bingelli and Tammann 1985). An inefficient merging process is relevant to discussions of 'biased galaxy formation', where only extreme fluctuations on the scale of massive galaxies actually succeed in making observed (luminous) systems. Such inefficiency seems natural in any hierarchical scheme : small gas-rich galaxies are easily disrupted by a burst of massive star formation. This point will be developed further in section III.

One may place limits on the inefficiency by focussing on the V-band luminosity of the brightest phase of a particular level, which forms at given redshift (for all reasonable models the spectrum of the bright phase will be fairly flat). This will contribute to the EBL intensity at the redshifted wavelength $5500(1+z)$Å and one may use the (high) limits on the general infrared background from IRAS to constrain the number density of that level, and hence the inefficiency below the simple hierarchy. Using the luminosity evolution of Figure 3(a), combined with the number density evolution characterizing the hierarchy results in a limit on how inefficient a process the merging can be.

The intensity observed at $5500(1+z)$Å, due to stellar systems at that level of the hierarchy which collapsed and underwent their bright phase, of luminosity $\mathcal{L}_V(z)$ per unit wavelength at the V-band, during the redshift interval dz around redshift z, is

$$i_{V(1+z)} = \int \frac{\mathcal{L}_V(z)\,n(z)\,c\,dz}{4\pi H_o\,(1+z)^7\,(1+\Omega z)^{\frac{1}{2}}},$$

where n(z) is the number density of these sources at redshift z. This parameter is fixed relative to that of present day giant ellipticals by the evolution in number density back down the hierarchy, modified by the adopted inefficiency at each step. The luminosity of a given level is determined by the merging process, and hence the righthand side of the above expression is completely defined, once a power spectrum of primordial fluctuations and cosmological model are specified.

As a crude but illustrative example, consider the luminosity evolution of a given level as approximated by a delta-function in redshift, and take $\Omega = 1$. Further, write the number density of sources as $n(z) = n(0)\,(1+z)^3 \phi(z)$, where

ϕ represents the proper density evolution. The intensity per unit wavelength is then given by:
$$i_{V(1+z)} = \frac{c}{4\pi H_o} \mathcal{L}_V(z)\, n(0)\, \phi(z)\, (1+z)^{-\frac{9}{2}}.$$

The luminosity density may then be expressed in terms of the known present day bright galaxy value using the luminosity evolution of Figure 3(a) as a guide. Specifically, consider six levels back in the hierarchy, systems which were brightest at redshifts ~ 30 (when the age of the universe was $\sim 8 \times 10^7$ yr) and whose V-luminosity is now received around 20μ. The maximally efficient hierarchy, which merged four units at each stage, requires $4^6 \sim 4 \times 10^3$ of these systems to make one large galaxy, and each smaller system had $\sim$ one-tenth of the present V-luminosity of the final system (mass-to-light ratio increases up the hierarchy). Hence

$$i_{20\mu} \simeq 0.1 \frac{c\, \mathcal{L}_V(0)\, n(0)}{4\pi H_o} (4 \times 10^3)(1.9 \times 10^{-7})\epsilon^6,$$

where an inefficiency factor, ϵ, has been introduced. The simplest model sets ϵ to a constant for each level, as here. The second factor in this expression, the zero redshift contribution, is known to be $\simeq 2 \times 10^{-9}\mathrm{erg}^{-1}\mathrm{\AA}^{-1}\mathrm{cm}^{-2}\mathrm{s}^{-1}\mathrm{sr}^{-1}$. This gives

$$i_{20\mu} \simeq 1.6 \times 10^{-12}\epsilon^6\ \mathrm{erg}^{-1}\mathrm{\AA}^{-1}\mathrm{cm}^{-2}\mathrm{s}^{-1}\mathrm{sr}^{-1}$$
$$= 1.6 \times 10^{-9}\epsilon^6\ \mathrm{erg}^{-1}\mu^{-1}\mathrm{cm}^{-2}\mathrm{s}^{-1}\mathrm{sr}^{-1}.$$

The IRAS limit on the zodiacal light, which may be treated as an extreme upper limit on the EBL, at 25μ, is $\sim 2.5 \times 10^{-4}\mathrm{erg}^{-1}\mu^{-1}\mathrm{cm}^{-2}\mathrm{s}^{-1}\mathrm{sr}^{-1}$, (Hauser et al. 1984) leading to

$$\epsilon \lesssim 8.$$

Thus galaxy formation must be at least 10% efficient at each stage of the hierarchy.

A direct search for protogalaxies will be much more difficult if galaxies form by hierarchical merging in the way described above, rather than by a single burst. This is due to the much reduced luminosity from young, massive stars — and hence Lyman alpha emission from HII regions — at any level of the hierarchy, compared to the single burst model. Figure 4 shows the bolometric luminosity, in solar units, due to young stars (*i.e.* less than 10^7 yr old, typical of the main sequence lifetime of early O and B stars) for the models of Figure 3. The bursts of constant star formation (at merging in the hierarchy, and at formation for the single burst model) have been modelled by superposition of mini-bursts, each of duration 10^7 yr. Only galaxies which formed in a massive burst of star formation are likely candidates to be found by search techniques involving the identification of the signature of young stars.

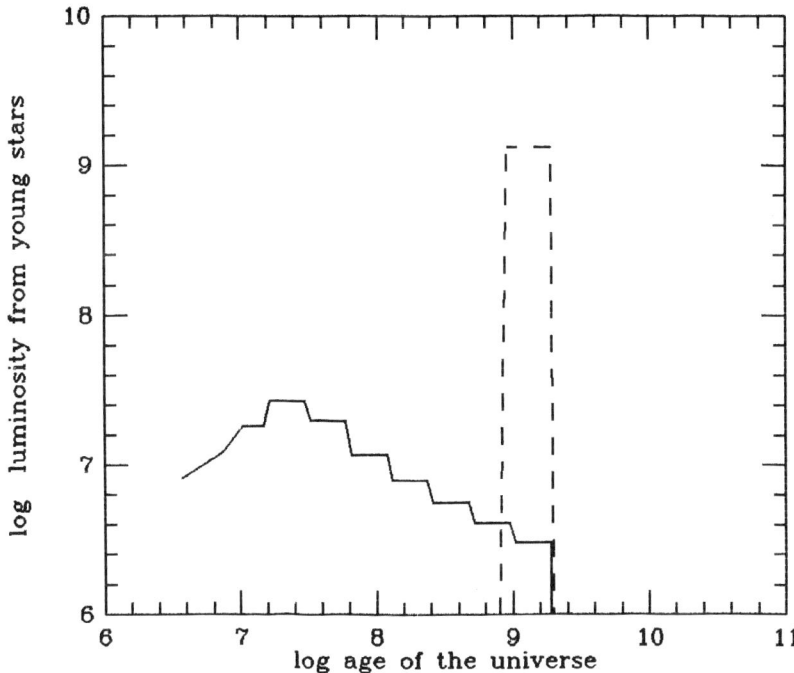

Figure 4. Bolometric luminosity from young stars in the merging hierarchy (full line) and the single burst model (dashed line).

As mentioned above, the power spectrum of the EBL will also differ between the two scenarios, since the fluctuations in the extragalactic component of the sky brightness depend on the spatial distribution of luminous sources. In principal, analysis of the power spectra at different wavelengths can yield the evolution in number density of stellar systems. The auto-correlation function (the Fourier transform of the power spectrum) contains the same information, and several groups have been investigating the evolution of clustering by analysing the angular correlation function of *galaxies* at faint magnitudes (Koo and Szalay 1984; Stevenson et al. 1985). The most straightforward parameterisation of the evolution as a function of redshift, or apparent magnitude, is in terms of the *slope* of the correlation function, when expressed as a power law in angular scale. The different groups agree that there is apparently little or no clustering evolution in samples of galaxies to 23rd B magnitude, which would agree with the merging models presented above since the hierarchy is truncated at redshifts corresponding to fainter magnitudes. However, there exists a large scatter between the clustering behaviors found in different parts of the sky. An analysis of the EBL provides a check that galaxies are the major contributors to the luminosity density of the universe, since the mean space luminosity density obtained from the galaxies and from the EBL should then agree. Cosmology enters through the angular size–redshift relation, and through the time–redshift relation once galaxy evolution is allowed. Schemes where the galaxy luminosity function is strongly redshift dependent at high redshifts, as for the hierarchy described here, are most easily discussed in terms of the EBL correlation properties, since the effects of the earlier phases are stronger. Improved data, especially in the far infrared, on the EBL may enable this approach to be pursued further.

III. Galaxies as Fossils

There is an alternative approach to direct observations in probing protogalactic evolution. Galaxies in many respects are fossils. Their characteristic properties were determined at, or very soon after, birth. These properties include their surface brightness, size, velocity dispersion, luminous mass, and metallicity. Consider first the dynamical characteristics of the old populations of galaxies. If we define the surface brightness within the half-light radius, we can convert this to mass density of the underlying old population by using an M/L ratio appropriate to the Hubble type. Use of either the core-averaged velocity dispersion, σ, or the maximum rotational velocity, v_r, multiplied by $\sqrt{2}$ yields an estimate of the velocity dispersion of the old spheroidal stellar component, presumed to be isotropic.

Armed with these definitions, we can make use of the observed correlations between luminosity, size and σ or v_r to obtain the representations of the data shown in Figures 5 and 6. The plot of surface brightness *versus* velocity dispersion discriminates Hubble type, while the plot of surface brightness *versus* luminous mass discriminates dwarf galaxies from giants.

To understand the significance of these plots, one must discuss the role of dissipation in protogalactic evolution. Dissipation of energy in protogalactic collapse is crucial to understanding the enhanced surface density of galaxies relative to galaxy groups and clusters that clearly stands out in Figure 5. Moreover, continuing star formation which necessarily involves gas and therefore gaseous dissipation, is indicated by the observed metallicity gradients and correlations between metallicity and luminosity. At the same time, it is evident that dynamical relaxation provides the most elegant explanation of the universal nature of the light profiles of giant spheroids. The low rotation of most luminous ellipticals also is suggestive of a dynamical, as opposed to hydrodynamical, origin.

A satisfactory model would include multiple gas clouds interacting dynamically and dissipatively. While the detailed construction of such a model would be a difficult undertaking, it is possible to extract a key result from fairly straightforward considerations. A collapsing gaseous protogalaxy will cool at the virial temperature. Gas shocks will heat the gas to the post-shock temperature, whether the gas is in very small clouds, or relatively uniform. In order for the gas to effectively form stars, it must be able to dissipate bulk kinetic energy and cool. There is a critical column density for cooling to occur that is independent of the local volume density ρ, and depends only on the virial temperature. The

Figure 5. Constraints on protogalactic evolution. In this plot of baryon density *versus* virial temperature (or equivalently, velocity dispersion), the cooling curve for a collapsing protogalactic cloud of primordial abundance demarcates the region where cooling occurs rapidly or slowly with respect to a dynamical collapse timescale. This curve reduces to the line marked 'photoionized IGM' if the intergalactic medium is highly ionized, as is the case after quasars turn on at $z=2$ or 3. The hatched area indicates the density range over which supernova-driven energy input can support a gaseous protogalaxy against collapse for several values of initial metallicity at onset of the starburst. Plotted data for galaxies of different morphological types, and groups and clusters, are based on statistical correlations (see Silk 1985 for references) between luminosity L and σ (for ellipticals and brightest cluster members), and L and maximum rotation velocity (for spirals, S0's and irregulars), and L and half-light radius (dwarf ellipticals), and number counts within the central cores for groups and clusters. The velocity dispersion is the line-of-sight central velocity dispersion for the ellipticals, and the one-dimensional velocity dispersion of a presumed isotropic velocity dispersion for the other systems. All correlations are converted to a (Σ,σ) relation by assuming that the systems consist predominantly of stars out to the half-light radii, with appropriate stellar mass-to-light ratios (Faber and Gallagher 1979). The region occupied by the Lyman α clouds seen in absorption against high redshift quasars is also shown.

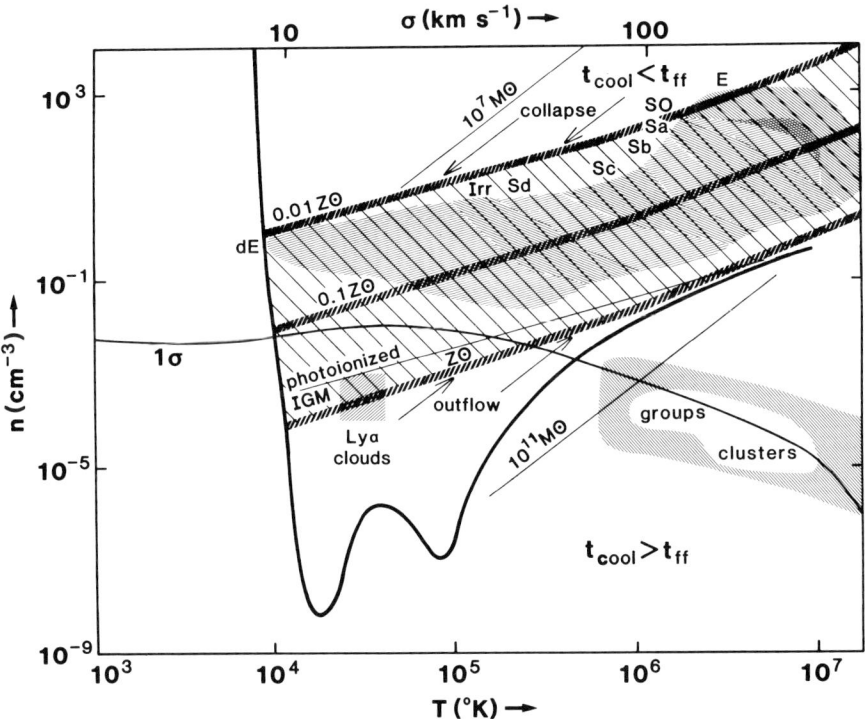

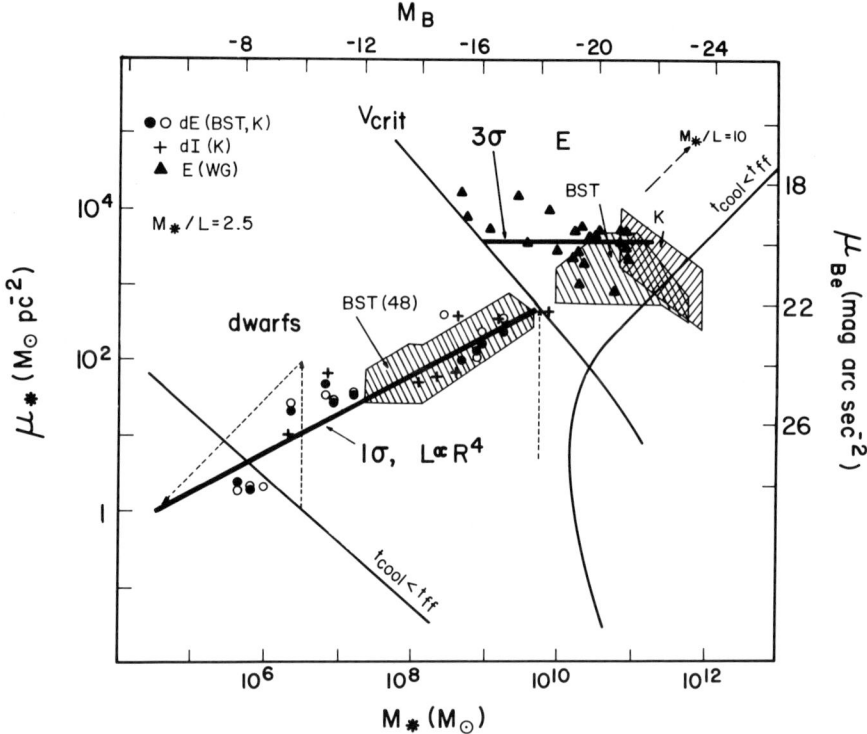

Figure 6. Surface brightness within the effective radius versus luminosity (assuming B−V=0.65), for a compiled sample of dwarf ellipticals (in the local group and in Virgo), dwarf irregulars, and ellipticals. A comparison of the supernova-driven wind-stripping theory with the observations is shown. (From Dekel and Silk 1985, where references for the data points are given.)

column density of a cloud must exceed

$$\Sigma_{cool} = \rho \sigma t_{cool},$$

where the cooling timescale is

$$t_{cool} = \frac{3}{2}kT[\Lambda(T)\rho]^{-1},$$

and $\Lambda(T)$ is the cooling rate per gram. Now, the post-shock temperature is given by

$$kT = \frac{3}{16}\mu\sigma^2,$$

where μ is the mean molecular weight per gram. The quantity $\Sigma_{cool}(T)$ is shown in Figure 5. It is immediately apparent that galaxies have $\Sigma > \Sigma_{cool}$, and underwent dissipation during collapse to the observed Σ, while groups and clusters could not have experienced much dissipation of bulk kinetic energy. Once stars formed, dissipation would effectively have stopped for the stellar component, whose surface density thereby yields a measure of the value of Σ when the dissipation phase ended.

To understand the detailed distribution of surface density, namely the distinction between low surface brightness dwarfs and high surface brightness giants, and the actual locus of the surface density, it is necessary to examine the role of energy input from newly forming stars. Supernovae occur, and the remnants overlap if the rate is sufficiently high. One knows the rate, however, if stars form with a solar neighborhood initial mass function, one supernova occurring for every 100 or so solar masses of gas consumed in forming stars. There is little evidence on the initial mass function in a protogalaxy, but abundance ratios in the oldest stars are near-solar and suggestive of either a solar neighborhood initial mass function, or perhaps one slightly favoring more massive stars. This supernova rate suffices to drive a wind, the supernova remnants overlapping before they have decelerated below the velocity dispersion characterizing the protogalaxy potential well, if

$$\sigma \lesssim 100\, n^{1/22} \text{kms}^{-1}.$$

Hence low σ galaxies are stripped of gas before they can efficiently form stars. The net result will be the formation of low surface brightness dwarfs as indicated in Figure 6.

While a wind will not be driven in a more massive galaxy, the stellar energy input can play an important role in regulating the rate of collapse of the interstellar gas. One appeals to protostellar energy input to prevent interstellar clouds in our own galaxy from free-fall collapse and an otherwise catastrophic star formation rate: a similar phenomenon should be occurring in protogalaxies.

During the initial collapse of an inhomogeneous, gaseous protogalaxy, stars may form in individual clouds, but the mean cooling time will be longer than the collapse time. However, as the density increases, the cooling time for the overall system will soon equal and become shorter than a dynamical time-scale. This means that strong dissipative interactions of colliding gas clouds will be initiated, and a starburst should develop. The associated energy input is capable of regulating the mean gas density of the protogalaxy: if the density increases, the rate of star formation rises, enhancing the energy input and lowering the density, while if the density decreases, the star formation rate declines, the energy input is lowered, and the gas can recollapse. The critical gas density self-regulated in this manner is very sensitive to the gas metallicity: low metallicity means inefficient cooling and hence allows a higher gas density to develop for a given star formation rate. Figure 5 shows that the resulting range in critical density attained for a plausible initial metallicity range; this yields the final stellar density once the gas is converted into stars, and corresponds well to the observed densities (within the effective radius) for all Hubble types.

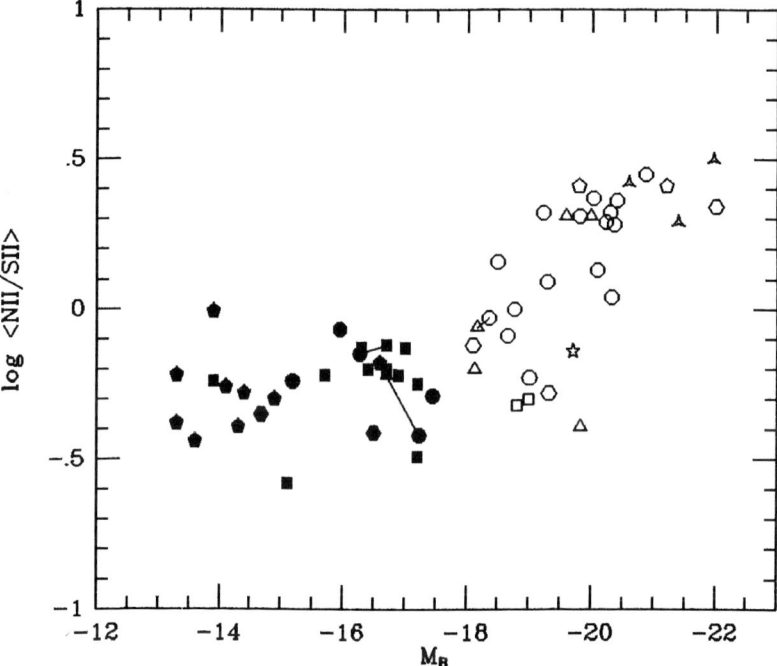

Figure 7. Mean ratio of nitrogen to sulphur for HII regions in disk galaxies, plotted as a function of the parent galaxy luminosity. This ratio has a constant value, which equals the prediction of stellar nucleosynthesis calculations if nitrogen is *primary*, for gas in galaxies below a critical luminosity ~ -18. The relative abundance of nitrogen increases with luminosity above this threshold, the expected behavior if there is a *secondary* component to nitrogen.

One consequence of this theory of protogalaxies is that the star formation burst occurred for much longer than a single dynamical time–scale t_d: in fact, the characteristic star–forming time–scale is of order $(\lambda/\sigma^2)t_d$, where $\lambda \approx (1000$ km s$^{-1})^2$ is the specific energy input associated with supernova heating. Such systems form massive galaxies: the mean metallicity approaches the solar neighborhood yield y $\approx 2.5Z_\odot$ if no mass loss occurred.

In the dwarfs, however, stripping is initiated after a time t_d, when a wind can first develop, and the metallicity at this stage inferred by the mass fraction that has formed stars prior to the initiation of the wind, is Z $\approx$ yσ^2/λ. Hence we see that chemical and dynamical evolution are intimately related for dwarf galaxies. The predicted dependence of Z on σ is in reasonable accord with the data. Wyse and Silk (1985) have noted moreover that the observed N/S abundance ratios in irregular and disk galaxies can be interpreted in terms of a primary enrichment phase in low σ ($\lesssim$ 100 km s^{-1}) dwarfs, followed by secondary nitrogen production in more massive galaxies (Figure 7).

Ideally, one would like to commence with appropriate initial conditions on Σ and σ at the onset of protogalactic collapse, and derive evolutionary tracks in the (Σ,σ) plane. Constant mass evolution implies that $\Sigma \propto \sigma^2$, while cloud collapse in the fixed potential well of a dark halo requires $\sigma \approx$ constant. One might hope that a satisfactory theory would enable one to understand the Hubble type distribution in the (Σ,σ) plane. Clearly, the disk/bulge ratios of spiral galaxies are a secondary phenomenon that is associated with the evolutionary tracks: one could well imagine that mergers, tidal interactions, gas infall and stripping in the protogalactic gas–rich phase all have some relevance in determining whether a large disk grows and survives. In all of this, the protogalactic environment also plays an important role: isolated systems will develop disks as gas accumulates while in strongly interacting systems, disk formation should be inhibited.

The initial conditions indicated in Figures 5 and 6 are appropriate to a theory which postulates primordial scale–invariant density fluctuations in a universe dominated by cold dark matter (Blumenthal et al. 1984). Normalization of the predicted fluctuation spectrum is accomplished by matching to the clustering scale of the large–scale galaxy distribution. It appears as though luminous galaxies are formed by collapse of the rare, 2σ or 3σ, fluctuations in such a theory, since only rare fluctuations on galactic scales attain sufficiently high density in the initial adiabatic collapse to enter the cooling regime where efficient star formation occurs. Such a result allows a dense universe ($\Omega=1$) to be reconciled with the observation that $\Omega_{lum} \approx 0.1$, since most of the matter in the universe does not participate in luminous galaxy formation.

However, cold dark matter only offers one viable scenario. In fact, the role of protogalactic starbursts is largely to mask the direct influence of initial conditions. It is apparent that if stellar energy input determines the strip in the (Σ,σ) plane occupied by luminous galaxies, a large variety of initial conditions become possible.

IV. Conclusions

We have argued that protogalaxies most likely reveal themselves as metal-poor starbursters. If galaxies form hierarchically, most of the luminosity associated with metal production may be produced by dwarf galaxies rather than by giants. The search for protogalaxies at high redshift may therefore be doomed to failure. Objects such as the extreme metal poor extragalactic HII region IZw18 could be closer to the prototypical protogalaxy, and would not be easily detected at high redshift.

The formation of a massive central black hole is greatly aided by the continuous resupply of gas as protogalactic clouds aggregate together. Central binary black holes should be common in such a scheme, as would isolated massive black holes, ejected in 3-body enounters. While most galaxies would be expected to have a modest ($\sim 10^6$ $M_\odot$) central black hole, only occasional rare, massive galaxies should have succeeded in growing the very massive ($\sim 10^8 - 10^9$ $M_\odot$) holes needed to account for the luminous quasars.

As for the study of protogalaxies, indirect means seem preferable. The post-starburst phase of distant ellipticals should become evident at high redshift, while the far infrared background light should provide a probe of the integrated emission from all protogalactic activity. The association of starbursts with active nuclei, quasars and radio jets needs to be clarified. It is likely that galaxy interactions simultaneously fuel all of these phenomena. It has been argued that some protogalaxies may be masquerading as quasars, but the number of possible narrow emission line very luminous objects at high redshift is far too small to be of much significance.

We have argued that the key to understanding protogalactic evolution lies in the fossilized characteristic parameters of galaxies. Interpretation of the locations of different galaxy types in the (Σ, σ) plane may eventually be as fruitful for galaxy formation theory as the Hertzsprung-Russell diagram has been for stellar evolution. Already, it seems clear that nature exercised a more profound influence over galaxies than nurture, that is to say, heredity and conditions at the epoch of formation exercised a dominant role over the origin of the characteristic properties of galaxies as compared to environmental effects subsequent to the gas-rich formation epoch. We are still far from a satisfactory theory of protogalaxies, but the boundary conditions of such a theory are reasonably well defined.

References

Bertola, F., Capaccioli, M. and Oke, J. B., 1982, *Ap. J.*, **254**, 494.
Blumenthal, G., Faber, S., Primack, J. and Rees, M. J., 1984, *Nature*, **311**, 517.
Bruzual, G., 1983, *Ap. J.*, **273**, 105.
Couch, W. J., Shanks, T. and Pence, W., 1985, *M.N.R.A.S.*, **213**, 215.
Dekel, A. and Silk, J., 1985, *Ap. J.*, in press.
Faber, S. and Gallagher, J., 1979, *Ann. Rev. Astr. Ap.*, **17**, 135.

Hauser, M. G. et al., 1984, *Ap. J. (Letters)*, **278**, L15.
Koo, D. C., 1984, in proceedings of RGO workshop *Astronomy with Automatic Plate Measuring Machines.*
Koo, D. C. and Kron, R. G., 1980, *P.A.S.P.*, **92**, 537.
Koo, D. C. and Szalay, A., 1984, *Ap. J.*, **282**, 390.
Kristian, J., Sandage, A. and Westphal, J. A., 1978, *Ap. J.*, **221**, 383.
Mould, J., 1984, *P.A.S.P.*, **96**, 773.
Neugebauer, G., et al., 1984, *Ap. J. (Letters)*, **278**, L83.
Schmidt, M. and Green, R. F., 1983, *Ap. J.*, **269**, 352.
Silk, J., 1985, *Ap. J.*, in press.
Stevenson, P. R. F., Shanks, T., Fong, R. and MacGillivary, H. T., 1985, *M.N.R.A.S.*, **213**, 953.
van Breugel, W., Filippenko, A. V., Heckman, T. and Miley, G., 1985, *Ap. J.*, **293**, 83.
Wyse, R. F. G., 1985, *Ap. J.*, in press.
Wyse, R. F. G. and Silk, J., 1985, *Ap. J. (Letters)*, in press.

DISCUSSION

PISMIS: During the early part of your talk you showed us a list of oveall initial parameters of a galaxy. I was surprised not to find the total mass in that list. It is evident that the total mass with which a galaxy gets started as such is an important parameter and decisive in the dynamical evolution of it. Although it is not a directly observable datum - as the total luminosity - what we estimate at present is the initial total mass presumably constan throughout the lifetime of the galaxy except for environmental effects which I don't believe are important enough as generally thought.

SILK: I am arguing that a critical observable parameter is the luminous mass of a galaxy, that is to say, the mass within an effective radius. Use of this and one other dynamical parameter (velocity, dispersion, for example) appears to control the observed characteristics of differing Hubble types.

RAKOS: In your first two diagrams B-V versus z, I suppose the color is calculated in the rest frame of galaxies.

SILK: Yes.

SEYFERT GALAXIES IN THE UNIVERSE

Donald E. Osterbrock
Lick Observatory, Board of Studies in Astronomy and Astrophysics
University of California, Santa Cruz, CA 95064

ABSTRACT. Recent optical spectroscopy of Seyfert galaxies, particularly at Lick Observatory, is discussed. The frame of reference is the luminosity function of Seyfert galaxies, the incompleteness of available surveys, and how complete samples can be found by combining several different types of surveys. The importance of spectral classification, or differentiating between various classes of active galactic nuclei, is stressed. All Seyfert galaxies and AGNs cannot be fitted into a single one-parameter sequence. Starburst galaxies and LINERs are briefly discussed. At the end some recent results on the frequencies of Seyfert galaxies with companions and in interacting systems are briefly summarized.

1. INTRODUCTION

Seyfert galaxies are important constituents of the universe. They make up a few percent of the population of giant field galaxies. At higher luminosities they form an increasingly large fraction of the population, and at the highest luminosities all the objects are QSOs, which seem to form a smooth continuum with Seyfert galaxies.

Seyfert galaxies, or at least many of them, can be fairly easily recognized up to large distances by their strong ultraviolet continuum and/or strong emission lines (on objective prism surveys), their high X-ray luminosities, or (most recently, in the IRAS survey) as "warm," extended objects in the infrared spectral region. Thus, the combination of high luminosity and straightforward recognizability makes them good objects for probing the universe to large distances.

Seyfert galaxies release or emit large amounts of energy, which evidently do not come from thermonuclear reactions in stellar interiors. Rather, the energy is almost certainly made available in gravitational energy "production" processes. How the mass is collected, and how the gravitational energy is built up and released into heat, radiation, relativistic kinetic energy, magnetic energy, etc. are among the most interesting problems in astrophysics.

All the experience of stellar astronomy teaches that we will not learn much about the objects, or about the universe, from simple galaxy counts or Seyfert-galaxy counts, or QSO counts, any more than we did from star counts. There is too much of a range in properties, including absolute magnitude, for counts to give more than the simplest information. Physical understanding, and spectral, as well as morphological, classification will be necessary to make real progress.

2. LUMINOSITY FUNCTION

Probably the best luminosity function for Seyfert galaxies published until very recently is that of Meurs (1982, see also Meurs and Wilson 1984). It is based on the first nine lists of the Markarian survey, covering about 1.1×10^4 square degrees of the sky. The sample is believed to be reasonably complete to apparent magnitude $m_p \approx 14.2$, and to be corrected acceptably for incompleteness to $m_p \approx 15.5$. According to it, at $M_p \approx -21$ (all absolute magnitudes are based on an assumed Hubble constant $H_o = 50$ km/sec/Mpc), about 10 % of all field galaxies are Markarian galaxies, and of these about 10 % (or about 1 % of all field galaxies) are Seyfert galaxies. However, by $M_p \approx -23$ essentially all galaxies are Markarian galaxies, essentially all of which are Seyfert galaxies. Both these results are in good agreement with earlier work (Huchra and Sargent 1973, Huchra 1977). The Meurs luminosity function converges smoothly with that of the "optically selected QSOs." This, together with their high degree of spectral similarity, strongly suggests that "Seyfert galaxies" and "QSOs" are basically similar objects, called by different names at different levels of luminosity. In their Palomar Bright Quasar Survey (which covered essentially the same area as the Markarian surveys) Schmidt and Green (1985) found both QSOs and Seyfert 1s; they adopted the terminology that objects brighter than $M_B = -23$ are *quasars*, and objects of lower luminosity are *Seyfert 1 nuclei* or *low-luminosity quasars*.

Meurs also derived separate luminosity functions for Seyfert 1 and Seyfert 2 galaxies. He found the Seyfert 1s to have a weak maximum at $M_p \approx -21$, and the Seyfert 2s at $M_p \approx -20$. The numbers of each type per unit volume are about the same, but to a given limiting apparent magnitude the number of Seyfert 1s is about double the number of Seyfert 2s. Meurs (1982) took his classifications from the unpublished list of Huchra, which uses only types 1 and 2. A better breakdown would be into types 1, 1.5, 1.8 or 1.9 and 2, a sequence of decreasing importance of the broad components of H I Balmer-line emission with respect to the narrow components. That this classification scheme is physically meaningful is indicated, for instance, by the fact that the average (weak) radio-frequency luminosities of Seyfert 1.5 galaxies are, in fact, intermediate between those of Seyfert 1s and 2s (Osterbrock 1984a, Meurs and Wilson 1984).

Actually, the sample of Seyfert galaxies used by Meurs, though the best available at that time, is probably still seriously incomplete. Many "new" Seyfert galaxies have been turned up in recent slit-spectral surveys, some made specifically to seek more Seyferts (Phillips, Charles and Baldwin 1983, Osterbrock and Dahari 1983), others radial-velocity surveys of complete samples of field galaxies (Huchra, Wyatt and Davis 1982). Many of the newly found Seyfert galaxies are Seyfert 2s, which are not so readily identified on objective-prism searches for galaxies with strong ultraviolet continua as Seyfert 1s are (e.g. Markarian, Lipovetskii and Stepanyan 1981, 1983). This more serious incompleteness in the statistics of Seyfert 2 galaxies is shown by comparing the early list published by Weedman (1977), in which only about 20% of the approximately 90 Seyferts are type 2, with the catalogue of Meurs (1982), in which about 30% of the 109 Seyferts are Seyfert 2s and with the still more recent unpublished list of Huchra (1983), in which about 45% of the roughly 400 Seyferts listed are type 2. No doubt the true fraction of Seyfert 2s is even larger. All of these are samples roughly to a given apparent magnitude; the fraction per unit volume that are Seyfert 2s will be still greater. If the one-magnitude average difference in luminosity holds up, this fraction would be about 75% Seyfert 2s. Perhaps we have now found all of them, but this seems unlikely. We shall return in several later sections of this review to the problem of using different methods in an attempt to find additional still undiscovered Seyfert galaxies.

In a very recent paper Cheng *et al.* (1985) have rediscussed the optical luminosity function of Seyfert 1 *nuclei*. Their sample is based on the first nine Markarian lists, and they tried to correct for the starlight contribution of the galaxy itself, and thus be left with the magnitudes of the Seyfert nuclei alone. They found good agreement of their derived luminosity function with that of the optically selected QSOs at $M_B \approx -23$, again suggesting the continuity of these objects. They did not find any turndown of the luminosity function toward fainter absolute magnitudes, as Meurs did (for the entire-galaxy magnitudes).

The sample used by Cheng *et al.* is based on the Seyfert 1 galaxies identified in the first nine Markarian lists, plus brighter NGC galaxies and Zwicky galaxies in the same area. To estimate its completeness, I used the largest list of

Seyferts available to me, the unpublished Huchra (1983) tabulation. Considering only galaxies with B magnitude 15.0 or brighter, and north of $-10°$, the approximate limits of the Markarian survey, there are 44 Markarian galaxies, 4 Zwicky galaxies, 15 NGC or IC galaxies, and 9 other galaxies (the latter three counting only objects that are not in the first fourteen Markarian lists) listed as Seyfert 1s. Of the NGC objects, seven or about half the total are recently recognized Seyfert 1 galaxies, as are seven or about 80% of the "other" galaxies. Many of them were only recently identified as Seyfert 1 galaxies from spectra taken in radial-velocity surveys. Thus, even this recent and very carefully worked out analysis probably still suffers from significant incompleteness. In addition, of course, "Seyfert 1" is a very broad category of objects (like Meurs and other authors, Cheng et al. include Seyfert 1 and 1.5 galaxies, but not Seyfert 1.8 or 1.9). Ultimately not only a complete survey but a detailed spectral classification will be necessary to understand fully active galactic nuclei.

3. STAR-BURST AND NARROW EMISSION-LINE GALAXIES

Not all galaxies with emission lines are Seyfert galaxies, nor are all spirals with emission lines. In fact, most are not. Only about 10% of the Markarian galaxies are Seyferts; nearly all the rest have emission lines similar to those in H II regions (Huchra 1977). Evidently they are galaxies with gas and significant numbers of OB stars in their nuclei, which photoionize the gas. The most luminous galaxies of this type with the strongest emission lines are called starburst galaxies (Weedman et al. 1981), while we generally call all such objects narrow emission-line galaxies (Shuder and Osterbrock 1981). An example of the spectrum of a fairly typical one, UM 283, is shown in Figure 1.

Measurements of the stronger emission lines in many star-burst galaxies have been published by Balzano (1983). Although most have [O III] $\lambda 5007/H\beta \leq 3$, a significant fraction of them are sufficiently highly ionized so that this limit is surpassed (Balzaro 1983). Evidently they are cases in which the ionizing photons come chiefly from hot O stars, as in the central parts of the Orion nebula. Thus, the statement that [O III] $\lambda 5007/H\beta \geq 3$ is a good basis for defining an emission-line galaxy as a Seyfert 2 (Shuder and Osterbrock 1981), while giving a good first approximation, is not true in all cases. It was based on a limited sample of narrow emission-line galaxies.

The spectrum of a counter example, Haro 4 = Mrk 36 is displayed in Figure 2. It has [O III] $\lambda 5007/H\beta \approx 5$. However, like all high-ionization narrow emission-line galaxies, it can be distinguished from a Seyfert 2 (with similar $\lambda 5007/H\beta$ intensity ratio) by the weakness of [N II] $\lambda 6583$ and especially by

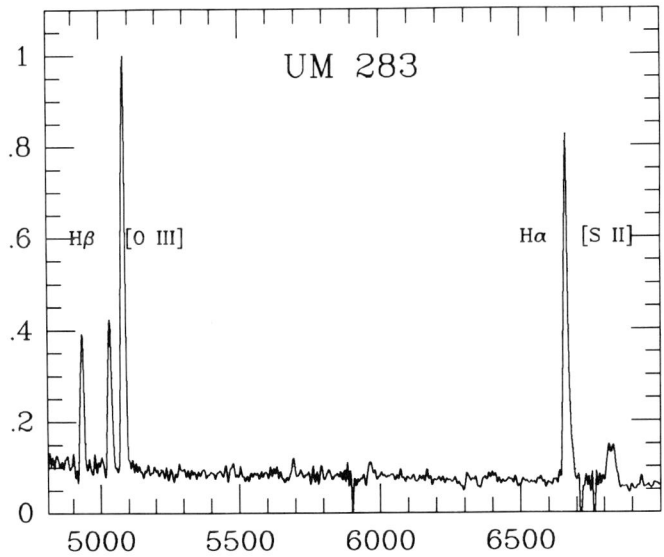

Figure 1. Lick Observatory CCD spectral scan of narrow-emission-line galaxy UM 283. Relative flux per unit wavelength interval versus wavelength, in the rest system of the Sun.

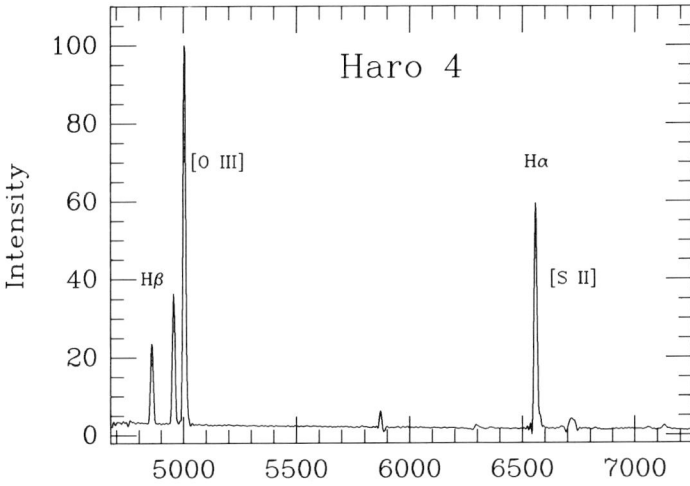

Figure 2. Lick Observatory CCD spectral scan of narrow-emission-line galaxy Haro 4 = Mrk 36. Relative flux per unit wavelength interval versus wavelength, in the rest system of Haro 4.

the extreme weakness of [S II] $\lambda\lambda 6717$, 6731 and practically complete absence of [O I] $\lambda 6300$. The [S II] and [O I] emission lines come preferentially from the large zones of partly ionized H that are characteristic results of the power-law photoionizing spectra of active galactic nuclei, but that are practically absent in classical H II region structures photoionized by radiation from hot stars. This is the physical difference that distinguishes a Seyfert galaxy from a starburst or narrow emission-line galaxy. It is the basis of the classification scheme of Baldwin, Phillips and Terlevich (1981). At Lick we are now measuring [O I] $\lambda 6300$ and [S II] $\lambda\lambda 6716$, 6731 ratios with respect to Hα for a fairly large number of narrow-emission line galaxies and Seyferts, in an effort to make this separation as quantitative as possible. Though these lines provide a very good way to distinguish Seyfert 2s from high-ionization narrow emission-line galaxies, they are weak and it is necessary to obtain good signal-to-noise ratio spectra.

In the blue spectral region He II $\lambda 4686$ is another very clear signal of photoionization by a power-law rather than a hot-star spectrum (Halpern and Steiner 1983, Ferland and Netzer 1983, Osterbrock and Dahari 1983). However, like [O I] $\lambda 6300$ it is a weak line, particularly in relatively low-ionization Seyfert 2s, and good signal-to-noise ratio spectra are required to detect it, or to set meaningful upper limits to its strength.

4. SEYFERTS IN CLUSTERS OF GALAXIES

It has long been known that the fraction of emission-line galaxies is much smaller in rich clusters than in the general field (Gisler 1978). This general result includes Seyfert galaxies as well as narrow emission-line galaxies. Dressler and Gunn (1983), however, found in the 3C 295 cluster (with $z = 0.46$) three Seyfert galaxies (one certain Seyfert 1, two probable Seyfert 2s) in addition to 3C 295 itself, which has a narrow-line radio-galaxy optical spectrum (indistinguishable from a Seyfert 2). As they pointed out, these galaxies make up a relatively high fraction of the most luminous galaxies in the cluster. They warned that statistics of small numbers are dangerous, but speculated that perhaps the conditions back at $z \approx 0.5$ were more favorable for the formation of active galactic nuclei than in the present epoch.

I questioned this conclusion (Osterbrock 1984b), pointing out that because of the large redshift of this cluster, only the most luminous galaxies have been observed. Among these objects, in the field, we know that the fraction of active galactic nuclei is quite large, as discussed in Section 2. Thus, comparisons should be made in comparable absolute-magnitude intervals. Also, in addition to look-back times, other physical characteristics such as density, internal velocity dispersion, interaction collision rates, and other still unrecognized parameters or initial conditions may be involved as well.

In a paper in press, Dressler, Thompson and Schectman (1985) have discussed a much wider sample of data from this point of view. The conclusion that the fractions of emission-like galaxies and of Seyferts in clusters are smaller than in the field remains unchanged and is now much more firmly based. They examined reasonably good spectra of 1095 galaxies in rich clusters, and of 173 field galaxies for comparison, and found that only 7% of the cluster members showed strong emission, compared with 32% of the field galaxies (to the same level of detection). Likewise, only about 1% of the cluster members, but 5% of the field galaxies, showed Seyfert characteristics.

However, Dressler, Thompson and Schectman (1985) also found another cluster of galaxies: DC 0428-53 (at $z = 0.05$), with a fraction of active galactic nuclei comparable to the 3C 295 cluster among its members. They therefore consider it possible that some nearby clusters are also, like the 3C 295 cluster, unusually active, and that physical conditions may be more important than redshift in determining this property.

5. LINERS AND CONTINUITY

In an interesting paper Heckman (1980) defined the class of low-ionization nuclear emission-line galaxies, or LINERs. Those of them that are spiral galaxies (at least) appear to be objects with low-luminosity sources of power-law ionizing continua (Ferland and Netzer 1983, Halpern and Steiner 1983, Osterbrock and Dahari 1983). In many ways they represent an extension of Seyfert 2s to lower ionization, lower luminosity, and possibly in some cases, larger distances of the ionized gas from the photoionizing source. The study of the continuity of active galactic nuclear features, as the "activity" gives to zero, will undoubtedly add to our understanding of this phenomenon.

Stauffer (1982a) and Keel (1983a) have surveyed well-defined samples of field spiral galaxies, and have found that nearly all of them show emission lines, particularly Hα and [N II] $\lambda\lambda$6548, 6583, on sufficiently good signal-to-noise ratio spectra. A significant fraction of them have [O I] λ6300 and [S II] $\lambda\lambda$6716, 6731 sufficiently strong to indicate the presence of a non-stellar photoionizing continuum (Keel 1983b). Though Stauffer (1982a) believed that "shock-excitation" (conversion of kinetic energy of mass motion to heat, with thermal collisional ionization) is significant in these objects, so far as I can see the observational data agree much better with photoionization (Keel 1983b).

A few of these galaxies also show very weak broad Hα components in their emission-line profiles. A few examples are NGC 2639, with Hα full width at zero intensity (FW0I) ≈ 3900 km s^{-1}, NGC 3998 with FW0I ≈ 4500 km s^{-1}, and NGC 4579 with FW0I ≈ 3500 km s^{-1} (Keel 1983b), as well as NGC 52732, with FW0I ≈ 6400 km s^{-1} (Stauffer 1982b). Other previously published examples of such emission-line galaxies with very weak broad Hα emission are NGC 5033, with FW0I ≈ 5700 km s^{-1} and NGC 2992, with FW0I ≈ 4600 km s^{-1} (Shuder 1980), M 81, with FW0I ≈ 3600 km s^{-1} (Peimbert and Torres-Peimbert 1981, Shuder and Osterbrock 1981), and Mrk 883, with FW0I ≈ 6400 km s^{-1} (Osterbrock and Dahari 1983). More examples are listed by Binette (1985). All these objects have relatively strong, low-ionization, narrow emission-line spectra. In nearly all of them the narrow lines are narrower than in typical Seyfert 2s. They might be called "Seyfert 1.95," but most of them, if the broad Hα were not detected, would more probably be classified as LINERs or narrow-emission-line galaxies. NGC 7213 has been called a LINER by Filippenko and Halpern (1984), but in fact its broad Hα component is quite strong and I would classify it as a Seyfert 1 galaxy. Evidently all facets of "activity" are not always correlated in the same way. Other specific examples are the very high-ionization Seyfert 2 galaxy Mrk 1388 (Osterbrock 1985a), and a group of narrow-line Seyfert 1s and 1.5s (Osterbrock and Pogge 1985). The Seyfert galaxies do not form a simple one-parameter sequence, and this must always be kept in mind in discussions of their origin and evolution.

6. WASILEWSKI SURVEY

A high-dispersion search for emission-line galaxies was carried out by Wasilewski (1983), who surveyed an area of 825 square degrees centered on the north galactic pole. He used a high-dispersion objective prism (∼ 400 Å mm^{-1} at Hβ) mounted on the Case Burrell Schmidt telescope at Kitt Peak. At this spectral resolution Wasilewski was able to see individual emission lines in galaxy spectra. The limiting magnitude for reliably recording the continuum was $B ≈ 16$, but objects with moderate to strong emission lines could be detected as faint as $B ≈ 17$. From his analysis Wasilewski derived the result that the survey was complete only to $B ≈ 15.7$, a significant improvement on the completeness limit $m_p ≈ 14.2$ found for the Markarian Seyfert galaxies by Meurs (1982).

In this field, Wasilewski (1983) found 132 emission-line galaxies; of these 23 were previously published Markarian galaxies, 13 were previously known emission-line galaxies from other sources (chiefly Haro galaxies), and 96 were newly discovered emission-line galaxies. Among these Wasilewski listed eight as previously unknown Seyfert galaxies, one of them a Seyfert 1 and the other seven Seyfert 2s. His classification was based on a few slit spectra, but mostly on relative intensities [O III]λ5007/Hβ estimated from the objective-prism plates. As he stated, these Wasilewski galaxies make up a signficant addition to the number of previously known Seyfert galaxies in this area.

At Lick we have obtained spectra of all of Wasilewski's Seyfert candidates, plus a number of his emission-line galaxies that he did not list as Seyferts, but that I selected as the most likely candidates for Seyferts possibly missed by him (Osterbrock 1984c, 1985b). Our spectra confirm that of the candidates, one is a Seyfert 1, another is a Seyfert 1.9 (with a weak, broad component of Hα emission), and two are Seyfert 2s. The spectrum of one of the latter, Was 31, is shown in Figure 3. The other four Seyfert 2 candidates in fact are NELGs, three of them with relatively strong [O III] λ5007/Hβ. The previously known Seyfert galaxies in the area, from our slit spectra plus published data, are four Seyfert 1s and four Seyfert 2s. Thus the Wasilewski survey has added significantly to our knowledge of Seyfert galaxies in this area. It illustrates that a high-dispersion objective-prism survey, based primarily on emission lines rather than ultraviolet continuum can bring in previously unknown Seyfert galaxies. This high-dispersion survey should be extended over a much larger area of the sky.

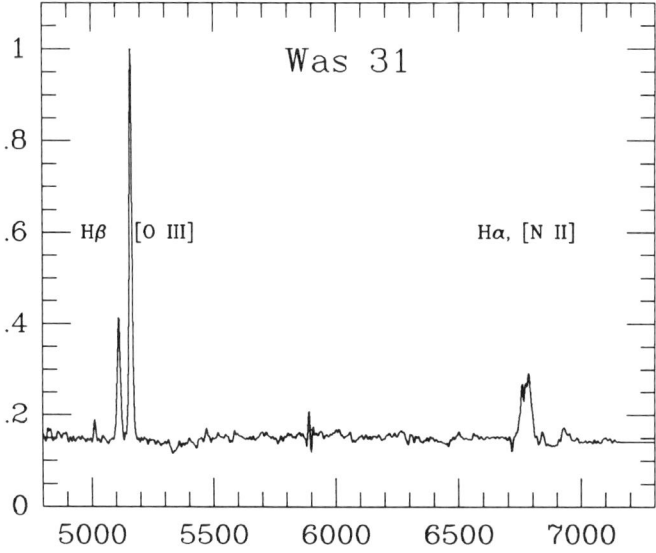

Figure 3. Lick Observatory spectral scan of Seyfert 2 galaxy Was 31. Relative flux per unit wavelength interval versus wavelength, in the rest system of the sun.

7. OTHER SURVEYS

With large telescopes and modern spectral detectors, the main problem in studying Seyfert galaxies is not obtaining the spectra, but rather identifying which galaxies to observe. As stated above, most of the Seyferts included in luminosity-function studies have come from the Markarian lists. All the Markarian galaxies were classified as such on the basis of their strong ultraviolet continuum, on objective prism plates with a dispersion of 1800 A mm^{-1} at Hγ (Markarian 1967). About 10% of the objects turn out to be Seyfert galaxies (Huchra 1977). Markarian and his collaborators have, in their later papers, listed the probable and possible Seyfert candidates, and their correctness and completeness have been impressive, but by no means perfect (see e.g. Osterbrock and Dahari 1983).

More recently Markarian and his collaborators have begun a second objective-prism survey for emission-line galaxies and QSOs (Markarian, Lipovetsky, and Stepanian 1983; Markarian and Stepanian 1984). Using fine-grain plates, three different objective prisms with dispersions 1800 Å mm^{-1}, 900 Å mm^{-1} and 280 Å mm^{-1}, multiple limiting exposures with different orientations, and their vast experience and accurate eyes, they have been able to find much fainter Seyfert galaxies than previously recognized. There are many examples down to $m_B \approx 17$ or 18. In many cases slit spectra are obtained with the 6-m telescope before the identification is published. Nearly every object from these lists that I have checked to date with independent Lick spectra has been correctly classified by Markarian. The level of completeness of the fields studied to date is very difficult to evaluate, and is not at all well known. It will be very important to continue this second Markarian survey over as large an area of the sky as possible, to push down the magnitude limit of our knowledge of Seyfert galaxies, and to increase the volume surveyed for luminosity-factor purposes.

An objective prism survey for emission-line galaxies and QSOs has been carried out with the Michigan Curtis Schmidt telescope at Cerro Tololo (MacAlpine, Smith, and Lewis 1977). To date five lists have been published, covering an area of about 670 square degrees, at $-2.5° < \delta < 7°$ toward the south galactic pole (MacAlpine and Williams 1981). The aim of the galaxy part of this survey was to find more Seyferts to a fainter magnitude level. The objective prism spectra were taken at a dispersion approximately 1740 Å mm^{-1} at Hβ, and were classified by inspection of both the continuum and the emission lines.

At Lick we have to date obtained slit spectra of 20 of these UM galaxies, some of them selected as the most likely Seyfert galaxy candidates from reported line strengths or reported presence of [Ne V] $\lambda3426$, others chosen from the list more haphazardly. Of these twenty galaxies, one is a Seyfert 1, one a Seyfert 1.5, one a Seyfert 2, and the other seventeen are NELG. The fraction of Seyferts, 15%, is based only on a small sample but is comparable with or greater than the fraction in the Markarian survey, though certainly significantly smaller than in the Markarian Seyfert candidates.

A more systematic study of the Michigan emission-line galaxies was carried out by Lewis (1983). He obtained slit spectra of about 100 galaxies from the first four lists. Of these about 3 - 5% are probably Seyfert 1 galaxies, while about 12% are Seyfert 2s. From the objective prism spectra, MacAlpine and Lewis had been able to isolate a set of Seyfert 1 candidates (essentially on the basis of a strong blue continuum and apparently relatively weak emission lines). Only about one fourth of these candidates turned out to be actual Seyfert 1s; on the other hand they did not miss any known Markarian or Zwicky Seyfert 1 galaxies in their fields, and did miss only one other known Seyfert 1 in the area. Of the Seyfert 2 candidates (essentially selected as objects with very strong [O III] ($\lambda 4959+ \lambda 5007$) nearly all turned out to be actual Seyfert 2s.

This objective-prism survey is complete to $m_p \approx 17$, but individual galaxies with strong emission lines have been found as faint as $m_p \approx 20$. The relatively larger number of Seyfert 2s is partly due to the fact that they could be recognized to a fainter magnitude level. A paper summarizing the results of this thesis, and giving the conclusions that can be drawn from them concerning the Seyfert-galaxy luminosity functions is in preparation, and no doubt will add greatly to our knowledge of these objects.

More recently the Case Western Reserve University astronomers have begun a low-dispersion objective-prism survey aimed at finding blue and/or emission-line galaxies, as well as extragalactic H II regions, QSOs and blue stars (Pesch and Sanduleak 1983, Sanduleak and Pesch 1984). The area covered in the two papers published to date is 210 square degrees, centered at $\delta=+31°$. The combination of fairly low dispersion ($\sim$1500 Å min^{-1} at $\lambda 4500$) fine optics, good seeing, fine-grain plates and careful guiding have enabled them to obtain unwidened spectra to a limiting magnitude $B \approx 18$. This figure refers to continuum detection for near-stellar objects; emission lines of even fainter galaxies can be seen if they are strong. In many cases they have been able to recognize and estimate the strength of [O III] $\lambda 3727$ and [O III] ($\lambda 4959 + \lambda 5007$) as well as the color of the continuum.

To date we have taken spectra of only a very few of these CG galaxies, but have not yet found a single Seyfert galaxy. I understand the same result has been found by D. Kunth for a somewhat larger sample of these objects (Sanduleak 1984). Nevertheless the faintness of the candidates in this survey makes it worthwhile to continue obtaining slit spectra of more of them. If we eventually learn to recognize Seyfert galaxies to $B \approx 18$, it will add greatly to our knowledge of the luminosity function. The Case low-dispersion survey area includes the Wasilewski region; this overlap region should be studied with the highest priority in any future surveys for Seyfert galaxies carried out by other methods.

8. IRAS GALAXIES

Very recently results of the IRAS survey have become available. Since Seyfert galaxies are known to be strong infrared sources, the IRAS lists were expected to provide new information on these objects. This has been spectacularly confirmed by Carter (1984) and de Grijp, Miley, Lub and de Jong (1985). They investigated the IRAS "warm galaxies," defined by colors between 12 μ, 25 μ, 60 μ, and 100 μ. The criteria were selected from comparable IRAS measurements of known Seyfert galaxies (Miley, Neugebauer, and Soifer 1985).

In his first sample Carter (1984) obtained optical slit spectra of 13 "warm" IRAS galaxies, and found 10 of them to be Seyfert galaxies and three NELGs. From his published scans and descriptions, I would agree with nearly all these classifications. Of the Seyferts he classifies four of Seyfert 1s, one as a Seyfert 1.5, one as a Seyfert 1.9, three as Seyfert 2s, and one as a Seyfert 2 - LINER transition case. I would agree with most of these classifications, but to me the "Seyfert 1.5," IRAS 1319 - 164 and the "Seyfert 1.9," IRAS 1833 - 654 both appear to be Seyfert 2s. It is very difficult to distinguish however, given the resolution and signal-to-noise ratio of his published scans. Our own spectrum of IRAS 1249 - 131 is shown in Figure 4; it is a Seyfert 1 galaxy with relatively narrow H I lines and strong Fe II, as Carter stated.

De Grijp *et al.* (1985) obtained spectra of 40 IRAS warm galaxies and found 18 (42%) of them to be Seyferts. Of these 5 (28%) are Seyfert 1s, and the remainder Seyfert 2s. Of a sub-sample of 24 of these IRAS warm galaxies that have "reliably determined flat infrared spectra," 17 (71%) are Seyferts, according to de Grijp *et al.* (1985). Most, but not quite all of the non-Seyferts have strong emission lines. The number of newly discovered IRAS Seyferts is comparable (per unit area) with the number of Seyferts already known, prompting de Grijp *et al.* to label them possibly "a new population of active galaxies."

To check these classifications, M. M. De Robertis and I have been obtaining optical spectra of those IRAS warm galaxies that we can reach at Lick Observatory. We have not been able to find all of them (only the coordinates are given in IRAS Circular 11, 1984), and two objects at the listed positions appear

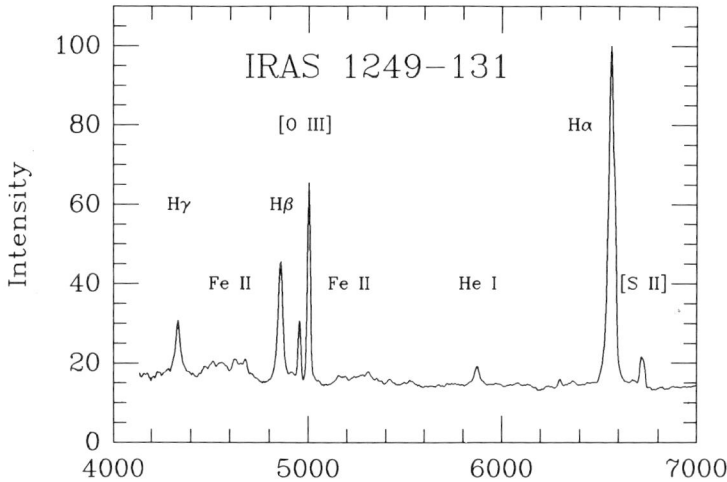

Figure 4. Lick Observatory spectral scan of narrow-line Seyfert 1 galaxy IRAS 1249 - 131. Relative flux per unit wavelength interval versus wavelength in the rest system of IRAS 1249 - 131.

to be stars with $z \approx 0.000$. Of the 16 IRAS warm galaxies we have observed to date, two are Seyferts 1s (one of them the "narrow-line Seyfert 1" IRAS 1249 - 131 mentioned above), five are Seyfert 2s, three are still uncertain to us, either Seyfert 2 or NELG, and 6 are NELGs. The spectrum of one of the Seyfert 2 galaxies, IRAS 0450 - 032, is shown in Figure 5.

The emission-line spectra of the Seyfert galaxies in this sample do not appear to be heavily reddened; the dust associated with them is evidently closer to the central source and does not greatly affect the spectrum emitted by the ionized gas. On the other hand, several of the NELG spectra among the IRAS warm galaxies do have large Hα/Hβ ratios, apparently the result of reddening. An example is the spectrum of IRAS 0428 - 097, shown in Figure 6. Note its strong Hα emission, and weak Hβ, with the broad, strong Hβ absorption line of the integrated-stellar galaxy continuum. We plan to follow this up quantitatively to see if the class of warm galaxies contain two somewhat different types of objects, normal Seyfert galaxies and heavily reddened NELGs. Several of the Seyfert 2s among the IRAS warm galaxies appear to be relatively low-luminosity objects, suggesting they are not so much a "new population of active galaxies" as an extension to fainter absolute magnitudes of a known population. Quantitative measurements will be necessary to settle this point.

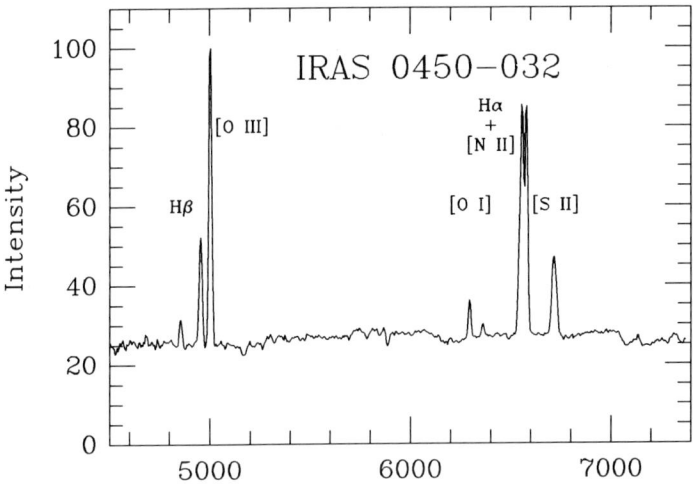

Figure 5. Lick Observatory spectral scan of Seyfert 2 galaxy IRAS 0450 - 032. Relative flux per unit wavelength interval versus wavelength in the rest system of IRAS 0450 - 032.

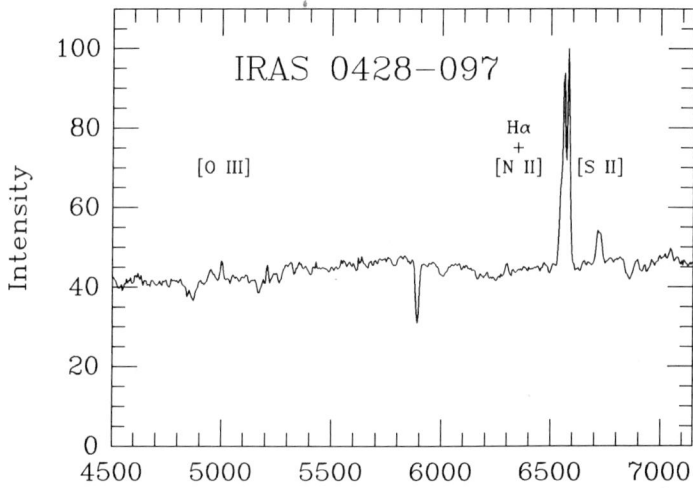

Figure 6. Lick Observatory spectral scan of narrow-emission-line galaxy IRAS 0428 - 097. Relative per unit wavelength interval versus wavelength in the rest system of IRAS 0428 - 097.

9. INTERACTIONS AND COMPANIONS

It has long been suspected that Seyfert galaxies have a tendency to occur in pairs or in interacting systems (e.g. Adams 1977; Vorontsov-Velyaminov 1977). Two quite recent studies have investigated this question quantitatively. Dahari (1984) found a definite excess of close companion galaxies in a well-defined redshift-limited sample of Seyfert galaxies. Using a quantitative definition of a companion (in terms of projected distance and redshift), he found the fraction of the Seyferts with physical companions to be about 15%, while the upper limit to the number of companions (defined by the same criteria) in a comparison sample of field galaxies in the same magnitude and redshift ranges is only 3%. The overabundance of *very close* companions of *comparable* size to the Seyfert galaxies was even greater. On the other hand Fuentes-Williams and Stocke (1984) reported finding no statistical evidence that Seyfert galaxies have an overabundance of companions compared with morphological similar spirals, but have not yet published their full paper.

Approaching the problem from the other side, Dahari (1985a) also obtained spectra of a sample of interacting and asymmetric galaxies. Here he found an excess (at the 90% confidence level) of Seyfert galaxies among interacting spirals, and a highly significant excess (at the 98% confidence level) among the *strongly* interacting spirals. However, in the subgroup of galaxies with extreme tidal distortions, he found no Seyfert galaxy but many NELGs.

Likewise Kennicutt and Keel (1984) investigated the effect of close galaxy-galaxy interactions in spiral galaxies by comparing spectroscopically well-defined complete samples of multiple systems and of single systems. They found galaxies with close companions have significantly higher emission-line luminosities, and significantly higher levels of ionization, on the average. They also found a considerably higher fraction of Seyfert galaxies among the galaxies with close companions.

Keel, Kennicutt, Hummel and van der Hulst (1985) further studied both a complete sample of spirals with bright companions, unbiased as to the morphological appearance, as well as another sample selected for visual evidence of tidal distortion. Both samples showed significant excesses of nuclear emission, star-burst as well as Seyfert. On the other hand they found LINER nuclei to be less common in both samples than in single-galaxy sample. They also found, as Dahari (1985a) did, that *very* disrupted systems possess *fewer* Seyfert nuclei.

Evidently galaxies with companions, particularly close companions, are in many cases interacting in such a way that mass, presumably in the form of interstellar matter, perhaps in the form of dwarf companions (Gaskell 1985) is delivered to the nucleus and becomes available as fuel for the active galactic nucleus. But in the cases of extreme disruption, it appears the galaxy is so strongly perturbed that the flow is largely diverted from the nucleus and is used up in star formation. Numerical simulations of simplified models support these ideas (Toomre and Toomre 1972). Stockton (1982) has related QSO activity and the observed close companions of QSOs to such a picture, and De Robertis (1985) has shown how density evolution of the population of QSOs can be understood in these terms.

I am greatly indebted to Drs. M.M. De Robertis, O. Dahari, J.S. Miller, and W. G. Mathews for many helpful discussions on the questions treated in this paper, and to Dr. G.M. MacAlpine for a summary of the Michigan Curtis Schmidt results on Seyfert galaxies. I am grateful to the National Science Foundation for continued partial support of this research, most recently under Grant AST 83-11585.

References

Adas, T.F. 1977, *Ap. J. Suppl.*, **33**, 19.

Baldwin, J., Phillips, M. M., and Terlevich, R. 1981, *P.A.S.P.*, **93**, 5.

Balzano, V.A. 1983, *Ap. J.*, **268**, 602.

Binette, L. 1985, *Astron. Ap.*, **143**, 334.

Carter, D. 1984, *Astronomy Express*, **1**, 61.

Cheng, F.-Z., Danese, L., De Zoth, G., and Franceschini, A. 1985, *M.N.R.A.S.*, **212**, 857.

Dahari, O. 1984, *A.J.*, **89**, 966.

_____. 1985a, *Ap. J. Supp.*, in press.

de Grijp, M.H.K., Miley, G.K., Lub, J., and de Jong, T. 1985, *Nature*, in press.

De Robertis, M. M. 1985, *A.J.*, in press.

Dressler, A., and Gunn, J.E. 1983, *Ap. J.*, **270**, 7.

Dressler, A., Thompson, I.B., and Schectman, S.A. 1985, *Ap. J.*, in press.

Ferland, G.J., and Netzer, H. 1983, *Ap. J.*, **264**, 105.

Filippenko, A.V., and Halpern, J.P. 1984, *Ap. J.*, **285**, 458.

Fuentes-Williams, T., and Stocke, J. *Bull. A.A.S.*, **16**, 989.

Gaskell, C.M. 1985, *Nature*, in press.

Gisler, G.R. 1978, *M.N.R.A.S.*, **183**, 633.

Halpern, J.P., and Steiner, J.E. 1983, *Ap. J.*, **269**, L37.

Heckman, T.M. 1980, *Astron. Ap.*, **87**, 152.

Huchra, J.P. 1977, *Ap. J. Suppl.*, **35**, 171.

Huchra, J.P. 1983, Unpublished draft list of Seyfert galaxies.

Huchra, J.P., and Sargent, W.L.W. 1973, *Ap. J.*, **186**, 433.

Huchra, J.P., Wyatt, N.F., and Davis, M. 1982, *A.J.*, **87**, 1628.

IRAS Circular No. 11, *Astron. Ap.*, **134**, C5.

Keel, W. C. 1983a, *Ap. J. Suppl.*, **52**, 229.

_____ . 1983b, *Ap. J.*, **269**, 466.

Keel, W.C., Kennicutt, R.C., Hummel, E., and van der Hulst, J. 1985, *Ap. J.*, in press.

Kennicutt, R.C., and Keel, W.C. 1984, *Ap. J.*, L5.

Lewis, D.W. 1983, University of Michigan Ph.D. Thesis.

MacAlpine, G.M., Smith, S.B., anhd Lewsi, D.W. 1977, *Ap. J. Supp.*, **34**, 95.

MacAlpine, G.M., and Williams, G.A. 1981, *Ap. J. Supp*, **45**, 113.

Markarian, B.E. 1967, *Astrofizika*, **3**, 55: *Astrophysics*, **3**, 24.

Markarian, B.E., Lipovetsky, V.A. and Stepanian, J.A., 1981, *Astrofizika*, **17**, 619: *Astrophysics*, **17**, 321.

_____ . 1983, *Astrofizika*, **19**, 29: *Astrophysics*, **19**, 14.

Markarian, B.E., and Stepanian, J.A. 1984, *Astrofizika*, **20**, 21: *Astrophysics*, **20**, 10.

Meurs, E.J.A. 1982, Leiden University Ph.D. Thesis.

Meurs, E.J.A. and Wilson, A.S. 1984, *Astron. Ap.*, **136**, 206.

Miley, G. K., Neugebauer, G., and Soifer, B.T. 1985, *Ap. J.*, in press.

Osterbrock, D.E. 1984a, *Q.J.R.A.S.*, **25**, 1.

_____ . 1984b, *Ap. J.*, **280**, L43.

_____ . 1984c, *P.A.S.P.*, **96**, 792.

_____ . 1985a, *P.A.S.P.*, **97**, 25.

———. 1985b, *Ap. J.*, in preparation.

Osterbrock, D.E., and Dahari, O. 1983, *Ap. J.*, **273**, 478.

Osterbrock, D.E., and Pogge, R.W. 1985, *Ap. J.*, in press.

Peimbert, M. and Torres-Peimbert, S. 1981, *Ap. J.*, **245**, 845.

Pesch, P., and Sanduleak, N. 1983, *Ap. J. Supp.*, **51**, 171.

Phillips, M.M., Charles, P.A., Baldwin, J.A. 1983, *Ap. J.*, **266**, 485.

Sanduleak, N. 1984, private communication to author.

Sanduleak, N., and Pesch, P. 1984, *Ap. J. Supp.*, **55**, 517.

Schmidt, M., and Green, R. F. 1985, *Ap. J.*, **269**, 352.

Shuder, J. M. 1980, *Ap. J.*,, **240**, 32.

Shuder, J.M., and Osterbrock, D.E. 1981, *Ap. J.*, **250**, 55.

Stauffer, J. R. 1982a, *Ap. J. Suppl.*, **50**, 517.

———. 1982b, *Ap. J.*, **240**, 32.

Toomre, A., and Toomre, J. 1972, *Ap. J.*, **178**, 623.

Vorontsov-Velyaminov, B. 1977, *Astron. Ap. Supp.*, **28**, 1.

Wasilewski, A. J. 1983, *Ap. J.*, **272**, 68.

Weedman, D.W., Feldman, F.R., Balzano, V.A., Ramsey, L.W., Sramek, R.A. and Wu, C.-C. 1981 *Ap. J.*, **248**, 105.

DISCUSSION

DE ZOTTI: You mentioned that the space densities of Seyferts may be underestimated because of the incmpleteness of the major surveys (and primarily of Markarian's). I would like to remind, however, that it is possible to reliably correct for incompleteness if the sample is "homogeneous" (Neyman and Scott, 1961). We (Cheng et al., 1985) have explicitly verified that our sample is "homogeneous" and have derived the appropriate correction. As a check on our results we have recently repeated our analysis adding a number of newly discovered Seyferts. We have found that our earlier results are not significantly effected.

OSTERBROCK: Yes - that is good and I agree. But of course correcting for incompleteness basically means assuming that the objects you do not see have the same distribution of properties of the objects you do see - it does not add anything to the sample. It is important to push the magnitude limits of completeness down as faint as possible, to increase the number of objects and thus improve the statistics even of rarer classes of objects, like Seyfert 1.8 and 1.9 galaxies, Seyferts with [FeX]λ6375, etc.

ULRICH: Regarding the narrow lines, what would you say are the differences between Seyferts 1 and Seyferts 2?

OSTERBROCK: On the average, the Seyfert 1 (and 1.5) galaxies narrow-line spectra show higher ionization - often the [FeVII]λ6087 and sometimes the [Fe X]λ6375 - than Seyfert 2 galaxies, though both have broad ranges with a considerable overlap. The narrow-line widths in some Seyfert 1s show a good correlation with ionization potential, but in others with critical density; in Seyfert 2s research that M.M. De Roberts and I are just in the process of completing seems to show there are less correlations with ionization potential, and more correlations with critical density.

KIANG: Do you have a <u>mean</u> spectrum of Seyferts? And if so, do you find the same difficulty with the emission lines being too strong relative to the Lyman continuum which Dr. Phynney pointed out in the case of the quasars?

OSTERBROCK: No, I am sorry, I do not. My work has largely been concerned with the emission lines and we do have mean emission-line spectra, but that is all.

KIANG: My reason for asking the question is this: Dr. J.H. You and I would like to point out that there is a little known mechanism of producing the emisssion lines without the Lyman continuum - what we call the Cerenkov Line Radiation. Provided there is a sufficient high density of relativistic electrons and of the atom concerned, Cerenkov radiation may be generated in the vicinity of the atomic lines, mimicking broad emission lines. This theory predicts a rigorous relation between the intensity-ratios $H\alpha/H\beta$ and $H\alpha/H\beta$. The observed points for quasars seem to fall close to the predicted curve; but those for the Seyferts do not.

SCHMIDT: Our search at Palomar (with Gunn and Schneider) for quasars near the redshift limit, by a CCD grism technique, has so far yielded dozens of emission-line galaxies as a byproduct. This extends the statistics of such objects to a redshift of 0.3 and may, in combination with the brighter surveys you discussed, allow study of their evolution.

OSTERBROCK: Excellent! I am sure that such "serendipitons" Seyfert galaxies at fairly low Z, turned up on spectral surveys designed primarily to find QSOs at high Z, will add significantly to our knowledge of their luminosity function. I would urge all observers to publish their results on discoveries of such Seyfert galaxies - and indeed to publish upper limits if none are found.

PISMIS: It is known that the spectra of Seyfert galaxies show variations. Are these variations sufficently large such that a Seyfert 1 may qualify for a Seyfert 2 or viceversa?

OSTERBROCK: To date we have only observed Seyfert galaxies for a few years - the most extrene variation I know has been in NGC 7603 which did appear to change from a Seyfert 1 "nearly to 2" (perhaps today I would say from 1 to 1.9) and back again. NGC 4151 has also shown large variations though (in my opinion) it would always have been classified as a Seyfert 1.5 (on spectra I have seen).

SCHULZ: Can some Seyfert 2 - spectra be explained by a mixture of photoionization by hot stars and shock heating?

OSTERBROCK: The average properties of Seyfert 2 galaxies, and the properties of many individual Seyfert 2 galaxies can be better explained by photoionization by a band spectrum then by shock heating. The main evidence is the combination of relatively low temperature (expecially from [OIII] $\lambda 5007/\lambda 4363$) with a wide range of ionization extending to [O III] and sometimes [NeV] and [FeVII]. That some object that has been called a Seyfert 2 can be better explained by a combination by photoionization by hot stars plus shock heating I cannot exclude, but I do not know of any such cases.

EVOLUTIONARY CONNECTION OF SEYFERT GALAXIES AND QUASARS

Daniel W. Weedman
Department of Astronomy
The Pennsylvania State University
525 Davey Laboratory
University Park, PA 16802
USA

ABSTRACT. Various recent data are summarized to yield a local luminosity function for Seyfert 1 galaxies. This luminosity function is evolved in different ways to find fits to other observations of faint, high redshift quasars. It is found that a simple form of pure luminosity evolution, exponential in look-back time, is consistent with all existing data. Extrapolations to redshift 4 and magnitude 26 are given for future tests. It is pointed out that such tests from ground or space must continue to rely on the unresolved quasar. An angular diameter-redshift diagram implies that there is no realistic possibility of detecting galactic envelopes of quasars at high redshift.

1. INTRODUCTION

In years past, we have wondered whether Seyfert galaxies are local examples of quasars, primarily in hopes that the physical processes in quasars could be understood by studying the closer, brighter Seyfert galaxies. Many studies have shown the basic similarities between the luminosity sources in these objects (e.g. Yee 1980, Kriss and Canizares 1982, Wu et al. 1983). This does not, however, prove that quasars are distant Seyfert galaxies. The interest in answering this fundamental question goes far beyond morphology or semantics.
   It is now reasonably certain that Seyfert nuclei and quasars are ultimately powered by gravitational accretion. But there are various mechanisms of producing compact accretors, so a basic question remains whether in both categories we observe the consequences from a massive accretor of primordial origin. Have such accretors existed in galactic nuclei ever since their origin, producing the power of quasars and active galactic nuclei? If so, physical models can be constructed to account for the results (e.g. Cavaliere et al. 1983). Or, did the high luminosity, high redshift quasars represent a population of accretion-powered sources long since become invisible? In that case,

local quasars and Seyfert galaxies would arise from unrelated, more recent mechanisms for producing accretors, such as the leftovers of nuclear starbursts (e.g. Balzano 1983).

It is in the search for answers to these astrophysical questions that we seek the correct parameterization of quasar evolution. Can we smoothly connect, with uniform evolution, the local Seyfert galaxy population and the most distant quasars? Is there a single, simple expression for evolution that can account for these objects everywhere in the universe? If not, is there evidence for more than one population in the redshift dependence of quasar events?

To summarize the current status of such questions, I will present a best-estimate summary of the local Seyfert galaxy luminosity function. This will then be treated with evolution parameters to see how well the properties of high redshift quasars can be connected. Finally, the most helpful new observations that could be made are suggested. First, however, I begin with a somewhat discouraging result: that we have little expectation in the foreseeable future of observing the galaxies presumed to surround quasars at high redshift.

2. THE GALACTIC COMPONENT OF QUASARS

A great deal of observational effort has gone into showing that some quasars are surrounded by galactic envelopes that, in the main, are similar to those of Seyfert galaxies. The demonstration of this in terms of the host galaxies extends to $z \sim 0.5$ (e.g. Malkan et al. 1984). If the envelopes of quasars are fitted by exponential disks, as for spiral galaxies, the median form from these authors is $I_b(r) = 2.85 \times 10^{-17} \exp[-r/6.4 \text{ kpc}]$ ergs cm^{-2} s^{-1} Å^{-1} arcsec^{-2}, in the b band. This was deduced by subtracting the flux of the unresolved nuclear component.

Within conventional cosmologies, such a galactic envelope will not be detectable beyond $z \sim 1$, even with the Hubble Space Telescope or an advanced generation of ground based telescopes. The reason for this is the rapidly diminishing surface brightness of galaxies at cosmological distances, even though metric angular sizes do not diminish beyond $z \sim 1$. As pointed out by Hubble and Tolman (1935), the surface brightness in I_λ units drops with $(1+z)^{-4}$, which becomes $(1+z)^{-5}$ if the bandpass term normally contained within the K correction is also included. The detectable diameter of a galaxy is the isophotal diameter, or the portion of the galaxy which is visible above whatever background surface brightness is present. For an exponential disk, assuming $q_0 = 0.1$, the isophotal diameter is given by Weedman and Huenemoerder (1985) as

$$\theta'' = 2(1+z)^2 \ln[(1+z)^{-5} I_\lambda(o,z)/S_{6500}]$$
$$\times 6.38 \times 10^{-4} H_o R\{10z - 90[(1+0.2z)^{1/2} - 1]\}^{-1}.$$

Here, R is the metric scale length, taken as 6.4 kpc for the galaxy considered. A red wavelength of 6500 Å is chosen for the calculation because this is the most efficient for space or ground based observations with CCD detectors. Transforming the profile of Malkan et al. to this wavelength gives $I_{6500}(o,0) = 2.4 \times 10^{-17}$ ergs cm^{-2} s^{-1} Å^{-1} arcsec^{-2}, the central surface brightness of the disk for $z \sim 0$.

Table I

Central Surface Brightness $I_\lambda(o,z)$ for Sbc Galaxy

z	$I_\lambda(o,z)$	z	$I_\lambda(o,z)$	z	$I_\lambda(o,z)$
0.05	2.4	0.8	1.6	2.0	1.0
0.1	2.5	1.0	1.2	2.3	1.1
0.2	2.6	1.3	0.98	2.6	1.1
0.4	2.8	1.6	0.97	3.0	0.96
0.6	2.6	1.8	1.0		

$I_\lambda(o,z)$ is the central surface brightness of an Sbc galactic exponential disk, at rest frame wavelength $\lambda = 6500$ Å $(1+z)^{-1}$ in units of 10^{-17} ergs cm^{-2} s^{-1} Å^{-1} arcsec^{-2}.

In Table I is given $I_\lambda(o,z)$, which is what the central surface brightness would be at other wavelengths λ observed for higher redshifts, such that $(1+z)\lambda = 6500$ Å. This adopts the intrinsic spectrum of an Sbc galaxy from Coleman et al. (1980). Note that presenting the intrinsic spectrum in this way removes the $(1+z)$ term normally contained in the K-correction. The only other parameter required is the limiting surface brightness which is detectable, S_{6500}. It is normally generously assumed that a limit on the order of 1% of the background is detectable, in which case $S_{6500} \sim 2 \times 10^{-19}$ for the best observations from either ground or space. (Because of detector noise, surface brightness detection limits do not improve significantly in space.)

Incorporating all of these parameters yields the results in Figure 1. Quasar envelopes can be resolved from the ground ($\theta \gtrsim 2''$) to $z \sim 0.8$, but the redshift limit extends only to $z \sim 1.1$ for space-quality resolution ($\theta \gtrsim 0\rlap{.}''2$). Note that doubling the scale size of the envelope would not increase either of these redshift limits by more than ~ 0.1 in z. Drastically increasing the central surface brightness is the only hope for significant increases in the redshift limit. For comparison, another curve is given in Figure 1 which shows the results if the central surface brightness is arbitrarily increased by a factor of 10. Doing this pushes the ground-based limit to $z \sim 1.8$ and the space limit to $z \sim 2.4$. If envelopes are ever seen for quasars at $z \sim 2$, therefore, either these are very different from the galaxies associated with local quasars, or our cosmological assumptions are very wrong.

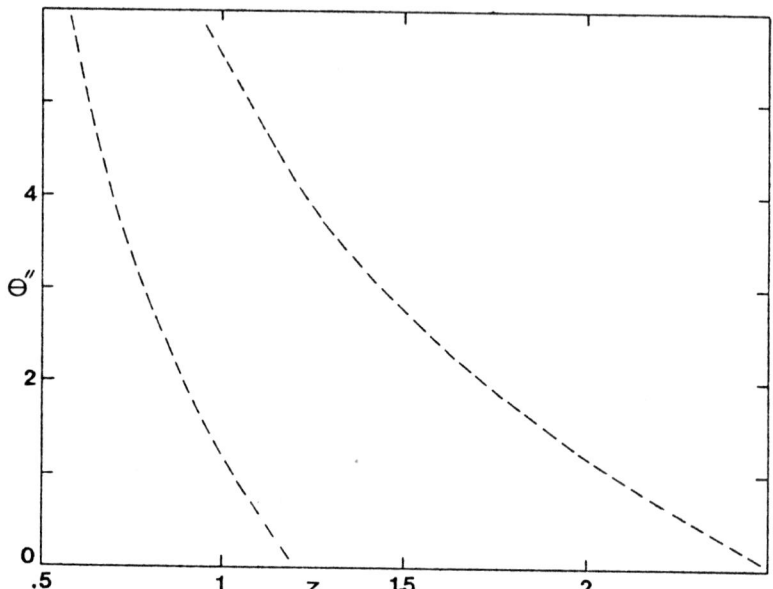

Figure 1: Angular diameter-redshift relation for galactic envelopes of quasars. θ" is the isophotal angular diameter in arcsec observable at redshift z to the faintest surface brightness limit which is realistic from ground or space. Lower curve is result for envelope like that observed around local quasars; upper curve is hypothetical galaxy for which surface brightness is increased by factor of 10.

3. THE LOCAL SEYFERT 1 LUMINOSITY FUNCTION

If it will not prove possible to demonstrate quasar similarities using the galactic component, we can only continue as before--using the unresolved nucleus. To begin with Seyfert galaxies, the first effort is to assemble the best local luminosity function. As we intend to compare with a distant quasar sample, most of which have strong and broad permitted emission lines, the local sample should omit Seyfert 2 galaxies. These have, by definition, no observed evidence of a broad line region. If such a region is there but obscured, it would also presumably be obscured in quasars at high redshift. I do not omit the local Seyfert nuclei with weak broad line regions; these are included as Seyfert 1 in the present accounting. The effect of this is to include a few nuclei that were not initially recognized by optical observes as Sy 1 until the presence of X-ray luminosity encouraged a closer look.

A variety of samples of Sy 1 galaxies are shown in Figure 2, to produce an overall luminosity function. Sources of the samples are described in the caption. Particularly useful are the limits from Keel's survey of low luminosity nuclei, which make it possible to extend the luminosity function to very faint values. Keep in mind that

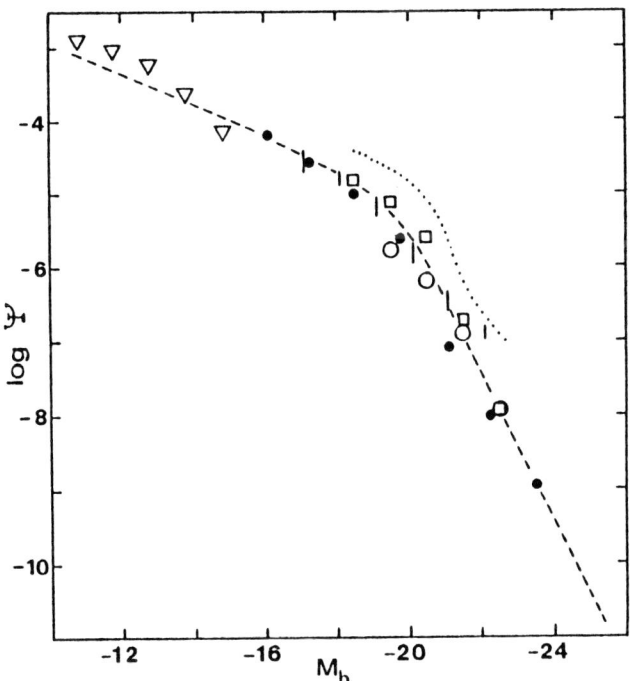

Figure 2: The integral luminosity function for nuclei of local Seyfert 1 galaxies. Ψ is the number of galaxies Mpc^{-3} brighter than absolute magnitude M_b, normalized to $H_o = 75$ km s^{-1} Mpc^{-1}. Triangles are upper limits using galactic nuclei observed by Keel (1983), squares are Markarian Seyfert 1 to 1.9 galaxies brighter than 15.5 mag., and open circles are objects with $z < 0.1$ from the Palomar Bright Quasar Survey, all treated as described in Weedman (1985a). Vertical bars are for nuclear magnitudes of Markarian galaxies from Fu-Zhen et al. (1985). Dotted curve is Seyfert 1 from Meurs and Wilson (1984). Filled circles are 2 keV X-ray luminosity function of Maccacaro et al. (1984), related to optical using $\alpha_{ox} = 1.35$. Dashed curve is the result finally adopted and given in Table II.

the summary luminosity function which is adopted--also given in Table II--is meant to refer only to the nuclei. For high luminosities, the nuclear luminosity dominates that of the entire galaxy, but at low luminosity values it is important but difficult to isolate the nucleus alone. There remains much room for improvement in this regard by the observers.

Table II

Local Luminosity Function for Seyfert 1 Galactic Nuclei

M_B	$\log \psi$	M_B	$\log \psi$	M_B	$\log \psi$
-25	-10.4	-20	-5.5	-15	-4.0
-24	- 9.4	-19	-5.0	-14	-3.75
-23	- 8.4	-18	-4.7	-13	-3.55
-22	- 7.4	-17	-4.45	-12	-3.35
-21	- 6.45	-16	-4.15	-11	-3.15

ψ has units of number Mpc^{-3}, brighter than absolute magnitude M_B, normalized to $H_o = 75$ km s^{-1} Mpc^{-1}.

4. EVOLUTION PARAMETERIZATION

There is disagreement within analyses of existing quasar samples as to whether all quasars can be explained with pure luminosity evolution, or whether luminosity-dependent evolution in luminosity or density is required (e.g. Schmidt and Green 1983, Marshall et al. 1983). More is at stake here than the tastes of the parameterizer. Luminosity-dependent evolution would appear to occur if there were more than one quasar population, such as a fainter, unevolving population and a brighter, evolving population. Consequently, I will here analyze the results of evolution in the context of two simple models.

The first model assumes pure luminosity evolution of the local Sy 1 luminosity function, parameterized as found by Marshall et al. (1983, 1984) or Weedman (1985a) to be approximately $L_{opt} \propto L_{opt}(z=0) \exp(6z/1+z)$, because $z/1+z$ is a measure of look-back time. From optical samples, the index seems to be from 5 to 7, depending on spectral shapes assumed. Analyses of X-ray derived samples (e.g. Maccacaro et al. 1984) yield a similar result, but with evolution of a lesser amount, $L_x \propto L_x(z=0) \exp(4.5z/1+z)$. This lower index can be explained if L_x/L_{opt} is luminosity dependent, but could also arise by erroneously assuming a quasar X-ray spectrum that is too flat. It can be shown, for example, that decreasing the assumed spectral index of the X-ray continuum by 0.5 raises the X-ray evolution index to that of the optical. There are reports that the previously assumed X-ray continua are too flat (Lawrence and Elvis 1985).

My second parameterization of evolution is designed to show the case of two quasar populations, arbitrarily divided at $M_B = -21$. The local Sy 1 luminosity function for $M_B > -21$ is assumed not to evolve; the brighter Seyferts, those with $M_B \leq -21$, are taken to evolve as above, with $L \propto L(z=0) \exp(6z/1+z)$. This is in the spirit of Schmidt and Green, who felt that low luminosity quasars evolved slower than high luminosity quasars. Obviously, a variety of evolution forms could be adopted for the multi-population case. My object in this exercise is to illustrate where the differences would be most conspicuous between the two extremes, because what we need just now is an indication of whether we can discriminate at all between such widely different concepts of quasar evolution.

Any existing parameterization depends on the cosmology adopted. These have generally been open Friedmann universes. The universe certainly is not empty, but the difference between evolution results for $q_o = 0$ and $q_o = 0.1$, say, is negligible compared to other uncertainties. Because the volume and look-back time equations are much simpler for $q_o = 0$, this value has generally been used, so I calculate the expectations of my models using that cosmology. (The astute reader notices that the angular diameter equation used $q_o = 0.1$. It is relatively simple for that situation, so one may as well be as realistic as possible when it isn't too much trouble.) All results are for $H_o = 75$ km s^{-1} Mpc^{-1}.

5. EXPECTATIONS FROM EVOLUTION

Results of evolving the local Seyfert 1 luminosity function with the two alternatives are shown in Figures 3-5. This is done as in Weedman 1985a, although the luminosity function and magnitude limits differ. All results use a continuum of form $f_\nu \propto \nu^{-0.5}$. What are displayed are

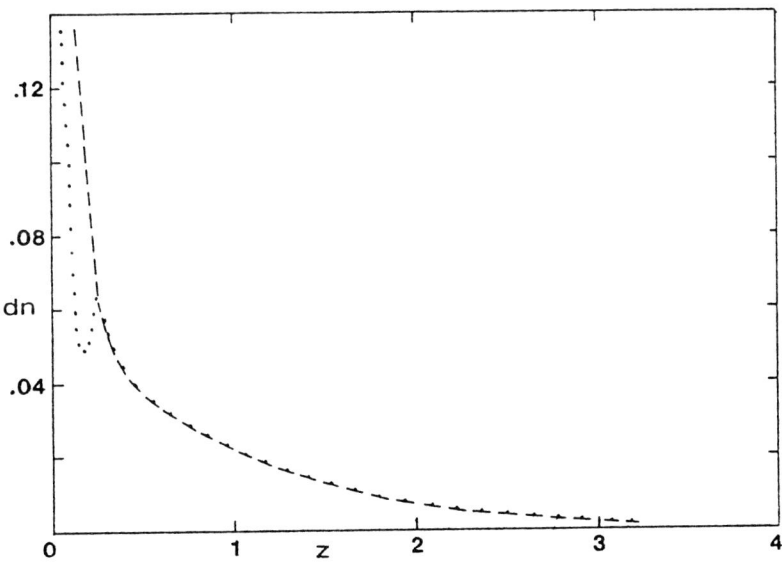

Figure 3: Differential quasar counts to blue magnitude 18. dn is the number of quasars deg^{-2} per redshift interval of 0.2, as a function of redshift z. Dashed curve is expectation for pure luminosity evolution model. Dotted curve is expectation for two-population evolution model.

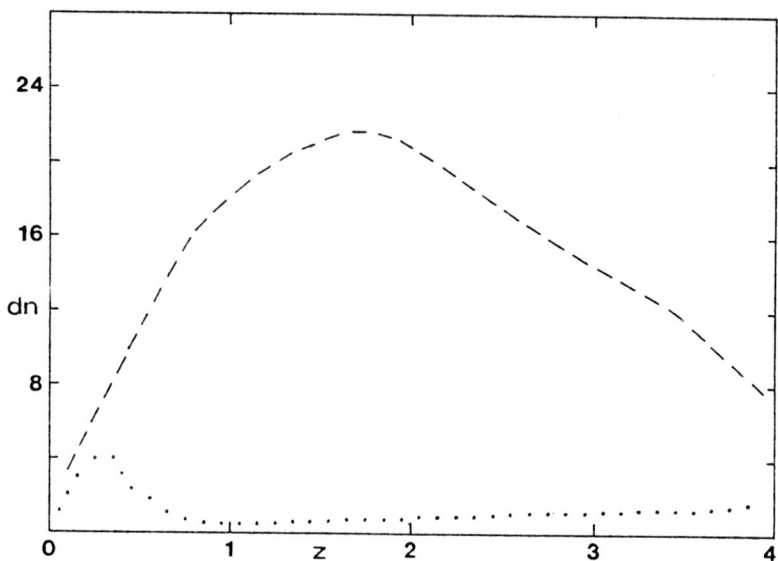

Figure 4: Differential quasar counts to blue magnitude 22. dn is the number of quasars $\deg^{-2}$ per redshift interval of 0.2, as a function of redshift z. Dashed curve is expectation for pure luminosity evolution model. Dotted curve is expectation for two-population evolution model.

results expected for limiting survey magnitudes of 18, 22 and 26. The first choice is that of the best existing samples, such as the Braccesi survey or Michigan-Tololo objective prism survey, based on wide-field Schmidt surveys. By magnitude 22, the limits of existing surveys of any kind are pushed. The calculations are made for magnitude 26 because there is a chance that the Hubble Space Telescope will survey to this limit.

Differences among the various limits are conspicuous. From Figure 3, it is seen why it is so hard to distinguish with existing samples pure luminosity evolution from any form of luminosity dependent evolution. My two models differ very little in what the observer sees. It is at faint magnitudes that differences are easiest to notice. When working at such magnitudes, one should make the task as easy as possible. I have a strong prejudice in favor of that redshift regime where quasars are easiest to observe spectroscopically, redshifts such that the two strongest lines (L α and C IV λ1550) are visible. Because my own emphasis has been on quasars in the interval $2 < z < 2.5$, I made the calculations shown in Figure 6. There, the expected numbers of quasars just within the interval $2 < z < 2.5$ are shown as a function of magnitude limit, compared to existing data (Weedman 1985b).

A lot of observational improvement is a realistic possibility, because quasars in this redshift interval are so conspicuous spectro-

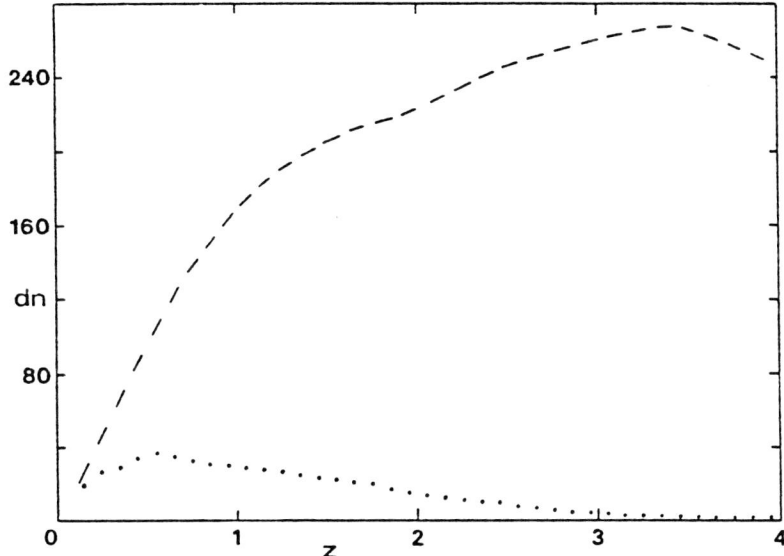

Figure 5: Differential quasar counts to blue magnitude 26. dn is the number of quasars deg^{-2} per redshift interval of 0.2, as a function of redshift z. Dashed curve is expectation for pure luminosity evolution model. Dotted curve is expectation for two-population evolution model.

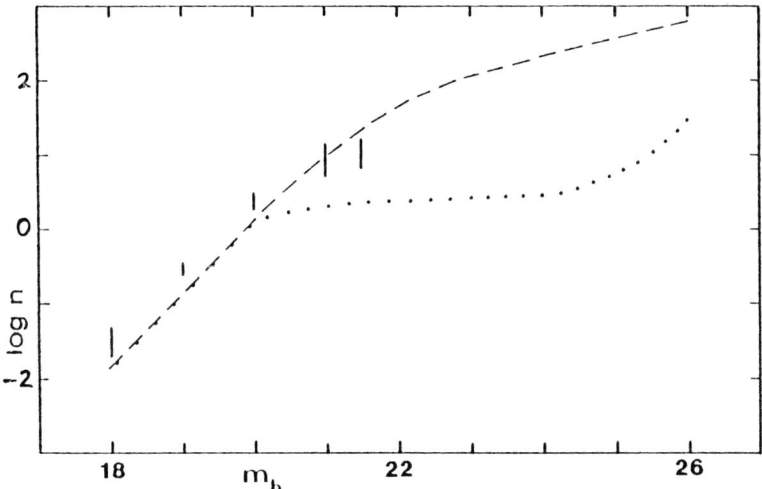

Figure 6: Integral quasar number counts for quasars only with $2.0 < z < 2.5$. n is the number of quasars deg^{-2} within this redshift interval which are brighter than blue magnitude m_b. Dashed curve is expectation for pure luminosity evolution model. Dotted curve is expectation for two-population evolution model. Vertical bars are observations from Weedman (1985b).

scopically. Hopefully, similar survey techniques will be available for Hubble Space Telescope, and we know improvements are possible from the ground (e.g. Schneider et al. 1983). So I wish to concentrate on this survey interval as the one most likely to determine the real evolution that is required to transform the local luminosity function to that of high redshift quasars. It appears that we are tantalizingly close to an answer. Because the difference between models approaches a factor of 10 at magnitude 22, even typical observers should be able to do this job. Note that there will be no great improvement in discrimination ability by going to 26 mag. By that limit, even an unevolved Seyfert population would begin to show at $z \gtrsim 2$. A definitive conclusion is not possible from the data in Figure 6, but the results are more consistent with the model for pure luminosity evolution. The crucial issue is the amount of flattening in quasar counts fainter than 22 mag. It is important to emphasize that some flattening is predicted even for pure luminosity evolution, because the local Seyfert luminosity function flattens at faint absolute magnitudes.

A final test is to compare expectations with total counts of faint blue starlike objects. Various surveys have attempted to make such counts precisely (e.g. Koo and Kron 1982). Assuming these counts extend to $z = 2.2$ (at which redshift the strong Lα emission can prevent a quasar's ultraviolet excess in broad band observations), I show in Figure 7 the total counts for the alternative models. Superposing data from the summary of Braccesi (1983) yields a result clearly favoring the model for pure luminosity evolution. There are problems with knowing the fraction of contaminating stars and dwarf blue galaxies, especially in counts at faint magnitudes. Nevertheless, the agreement between observations and expectations in Figure 7 for the one model is quite satisfying, given that the observed counts are virtually independent of the data which were used to construct the quasar luminosity function and parameterize its evolution. Of course, this solution is not unique; other parameterizations could be found to fit the data. This is why it remains necessary to consider various data sets, such as redshift distributions, counts within a redshift interval, and total counts in search of the best overall fit.

6. CONCLUSION

From this analysis comes the pleasing result that it is possible to connect the local Seyfert 1 luminosity function to all data for intermediate and high redshift quasars using a simple form of pure luminosity evolution. But the quest is far from over. Refining the form of evolution is not simply a minor detail; the very basic question of whether all quasars are a product of a well defined epoch and event in the universe remains unanswered. It is not likely to be answered by a few clever observations, even with the very best technology. Tedious searches for quasars, with existing telescopes pushed to their limits, must continue in order to refine our description of how the quasar phenomenon has changed with the aging of the universe.

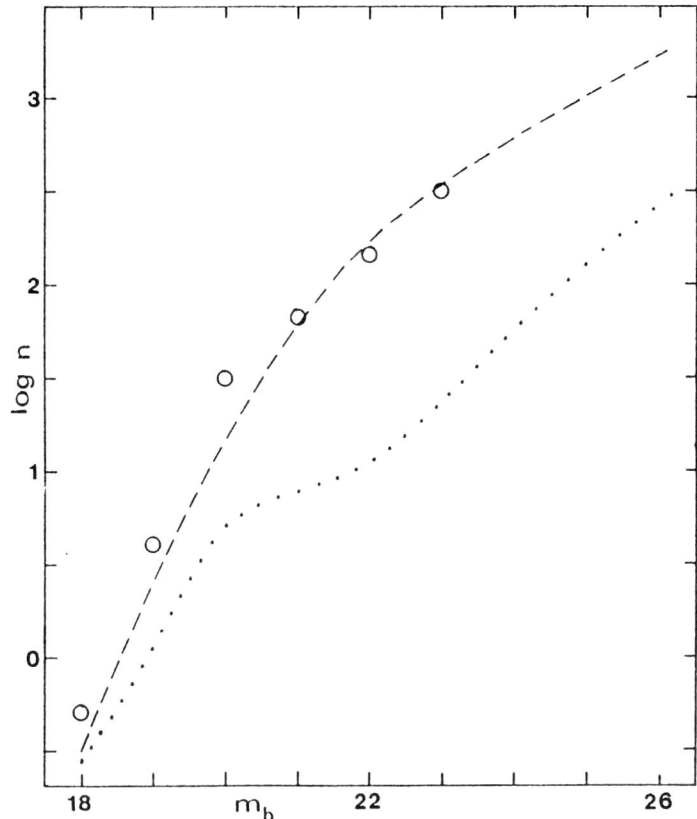

Figure 7: Integral counts for all quasars with $z < 2.2$. n is the number of quasars $\deg^{-2}$ brighter than blue magnitude m_b. Dashed curve is expectation for pure luminosity evolution model. Dotted curve is expectation for two-population evolution model. Open circles are observations from Braccesi (1983).

REFERENCES

Balzano, V.A. 1983, Ap.J., 268, 602.
Braccesi, A. 1983, IAU Symp. 104, Early Evolution of the Universe and its Present Structure, eds: G.O. Abell and G. Chincarini, (Dordrecht: Reidel) p. 23.
Cavaliere, A., Giallongo, E., Messina, A. and Vagnetti, F. 1983, Ap.J., 269, 57.
Coleman, G.D., Wu, C.-C., and Weedman, D.W. 1980, Ap.J. Suppl., 43, 393.

Fu-zhen, C., Danese, L., De Zotti, G., and Franceschini, A. 1985, M.N.R.A.S., 212, in press.
Hubble, E., and Tolman, R. 1935, Ap.J., 82, 302.
Keel, W. 1983, Ap.J. Suppl., 52, 229.
Koo, D. C. and Kron, R. G. 1982, Astr. Ap., 105, 107.
Kriss, G. A., and Canizares, C.R. 1982, Ap.J., 261, 51.
Lawrence A., and Elvis, M. 1985, in Astrophysics of Active Galaxies and Quasi-Stellar Objects, proceedings 1984 Santa Cruz conference, in press.
Maccacaro, T., Gioia, T.M. and Stocke, J. 1984, Ap.J., 283, 486.
Malkan, M.A., Margon, B., and Chanan, G.A. 1984, Ap.J., 280, 66.
Marshall, H.L., Tananbaum, H., Zamorani, G., Huchra, J.P., Braccesi, A., and Zitelli, V. 1983, Ap.J., 269, 42.
Marshall, H.L., Avni, Y., Braccesi, A., Huchra, J.P., Tananbaum, H., Zamorani, G., and Zitelli, V. 1984, Ap.J., 283, 50.
Meurs, E.J.A., and Wilson, A.S. 1984, Astr. Ap., in press.
Schmidt, M., and Green, R.F. 1983, Ap.J., 269, 352.
Schneider, D., Schmidt, M., and Gunn, J. 1983, B.A.A.S., 15, 957.
Weedman, D.W. 1985a, in Astrophysics of Active Galaxies and Quasi-Stellar Objects, proceedings 1984 Santa Cruz conference, in press.
Weedman, D.W. 1985b, Ap.J. Suppl., 57, in press.
Weedman, D.W., and Huenemoerder, D.H. 1985, Ap.J., 291, in press.
Wu, C.-C., Boggess, A., and Gull, T.R. 1983, Ap.J., 266, 28.
Yee, H.K.C. 1980, Ap.J., 241, 894.

DISCUSSION

FILIPPENKO: Cyril Hazard has perhaps the largest private collection of QSOs in the world. The last time I talked to Cyril, he told me that he has a whole bunch of QSOs with $z \simeq 3$ or greater, and that something strange happened to the number density near $z=2$ rather than $z \gtrsim 2.5$. Are you familiar with his work, and could you comment on the different results?

WEEDMAN: There is no doubt that UK Schmidt plates such as Hazard has have the greatest achievable concentration of quasars per plate. I have not seen enough quantitative results of his to fit them into the plots I have. It is certainly possible that the alteration of evolution is setting in by $z \simeq 2$. Learning this precisely is just why we need a lot of quantitative survey results.

MEURS: I should comment on the difference you quote between the Seyfert LF of Andrew Wilson and me and the various nuclear LF's you are showing. Our LF is based on integrated magnitudes, and is therefore explicitely a non-nuclear LF. Because of the different magnitudes involved there has to be a separated position for our LF in your figure, rather than to describe it to our incompleteness correction. Moreover, nuclear magnitudes which I now am getting for our sample combined with the same incompleteness correction lead to a nuclear LF that agrees well with the estimates of Fu-Chen et al. that also are included in your plot.

WEEDMAN: I would agree except that your function is also too high at bright absolute magnitudes where the nuclear and total magnitudes should be about the same. In any case, we are after a nuclear magnitude function so the fact that everyone is consistent there is the most important result.

MARSHALL: I have also determined the band of the luminosity function implied by Huchra's sample of Seyfert galaxies from the CfA redshift surveys (these proceedings). This sample does not suffer from the large, uncertain incompleteness present in the Markarian surveys. I find a similar bend which matches yours and Cheng et al. data well which evolves via pure luminosity evolution to fit the number counts very well.

WEEDMAN: I am pleased that our results are consistent and I encourage the audience to look at your poster paper.

SCHMIDT: I expect taht the combination of pure luminosity evolution with your local Seyfert luminosity function will lead, together with other extragalactic sources, to a total 2 keV X-ray background that is larger than the observed background.

WEEDMAN: I defer to Bill Keel on this, because I know he and some others have considered this. My impression is that there was no problem about the background.

KEEL: Elvis, Soltan, and I have derived the contribution of low-level AGN to the X-ray background. Under fairly general assumptions, the contribution of these objects peaks in the Seyfert luminosity regime, and convergers (below the measured flux) even if the luminosity function extends to very low luminosities. Thus, as far as indirect arguments based on narrow emission-line fluxes can tell, the AGN LF presented by Weedman (and a pure luminosity-evolution model) do not conflict with the X-ray background measurements.

PETROSIAN: Your model B will give better agreement with the quasar data if the evolution of the high luminosity objects are made stronger and if the Seyfert galaxy luminosity function is extended to higher luminosities. Otherwise there will not be any quasars (up to $z=2.5$) of $M_B > -29$. This is already in agreement with observation for the assumed model, $q_0 = 0.1$.

WEEDMAN: That may well be so. I emphasize that these parameterizations are illustrative examples and that no attempt was made to maximize fits to all quasars data.

WANDEL: I should comment on the question about the X-ray background, that the work done by E. Boldt (to be presented in this meeting) shows that the residual X-ray background (after substracting identified present contributions) is very flat, which, if AGNs at high redshifts contribute dominantly to the background, indicates that those hypothetical AGNs should have a flatter X-ray spectrum than the observed spectral index of -0.7.

WEEDMAN: My feeling is that it is now hopeless to try and fit quasars into the problem of the background. The reason is that nothing is known about the X-ray spectrum of high luminosity or high redshift quasars. I understand that new reductions at CfA imply steeper quasar X-ray spectra than for the lower luminosity Seyferts. This diminishes the quasar contribution to the background. The background is certainly the most exciting problem in X-ray astronomy.

EVOLUTIONARY CONTINUITY OF THE OPTICAL AGNs

A. Cavaliere
Astrofisica, Dip. Fisica, II Universita' di Roma
E. Giallongo
Ist. Astronomia, Universita' di Padova
F. Vagnetti
Ist. Astronomico, I Universita' di Roma

ABSTRACT. With the optical data base now rapidly expanding, it is time to begin bridging the gap between astrophysical issues and phenomenological descriptions concerning the evolutionary properties of the AGNs. We review problems, achievements and perspectives of such 'intermediate' models that are parameterized in terms of quantities of direct astrophysical meaning. Specific issues here discussed are: the continuity between Seyfert 1 nuclei and QSOs; a first parameterization of the interactions as triggers of activity; the rise of the QSOs; the evolutionary peculiarity of BL Lac Objects.

1. PROBLEMS

The current data base concerning the luminosity function (LF) of the optically selected AGNs and its evolutionary properties, is reported and discussed in many contributions to this Volume. The coverage of the L-z plane - bounded severely by statistics, and by flux limits more open to forthcoming instrumental progress - is still far from uniform; but integral observables like the number counts and the binned z-distributions are converging to a definite shape. In addition, within the privileged strips at $z \simeq 0$ to 0.1 and at $z \simeq 2$ to 2.5 the LF begins to emerge almost directly in some differential detail. Meaningful constraints to astrophysical issues may be contemplated, once some real ambiguities in the observations (cf Sect. 7) will be cleared up.

The present content of the data may be examined at three levels of increasing information - and of decreasing certainty.

At a basic level, there is wide consensus that the steep rise of the bright optical counts demonstrates a large increase - at given L - of the apparent number of the QSOs as they are traced back into the past light cone.

At the next level, two pieces of related evidence stand out: the flattening of the counts at magnitudes B > 20 (strengthened by Boyle et al. 1986, by Marano, Zamorani, Zitelli 1986 and by Koo, 1986); and the constraint set by the XRB (X-ray background) to the numbers of medium and low-L sources, in view of the tight correlation of the optical and the X-ray emissions (Zamorani et al. 1981, Setti and Woltjer 1982). The simplest interpretation is in terms of a 'luminosity evolution' (LE): in astrophysical terms, the QSOs statistically conserve their number since $z \simeq 2$, but they fade or dim off on average (i.e., averaging over individual variance and any recurrent activity) with a time scale $\tau \simeq 3$ Gyr, reducing by now and here to much weaker luminosities $L(0) \sim 10^{-2} L(z=2)$ such as characterize the Seyfert 1 nuclei. This view brings to the forefront the issue of continuity of QSOs with the Seyfert 1, 1.5 stressed and discussed by Cavaliere, Giallongo and Vagnetti (CGV) 1984a,b and also considered by Danese, De Zotti, Franceschini 1985 and by Weedman 1986. The next questions concern: down to what luminosity continuity holds; whether any recurrency duty cycle D could vary so much with z as to simulate some effective 'density evolution' (# potential AGNs/# normal galaxies $\sim$ few % D^{-1}). HST observations at intermediate $z \simeq 0.5$ will be instrumental in clarifying the continuity issue.

At a third level, a complex and intriguing network of constraints is emerging, but has not yet consolidated.

For example, the effectiveness of a luminosity-type evolution depends on the shape of the LF: the effective number of objects that governs the counts is given for a LF $N(L)dL \propto L^{-\gamma} dL$ by $[L(z)/L(0)]^{\gamma-1}$ ('evolution function' in the range where it is defined, cf Fig 0). At high L the LF is now being observed almost directly at high redshifts (Koo 1985, Weedman 1985): a check is feasible that its slope is so steep ($\gamma \simeq 3.5$) to allow steep counts with a moderate increase of $L(z)/L(0)$. But by a similar argument the flatter shape of the local LF at $M_B > -21$ would be related (in a given cosmology) with the flatter section of the counts, if the continuity were so straightforward as to imply an evolutionary scale τ constant at all z and at all luminosities. On the other hand, in these conditions the luminosity increase in look back time $L(z)/L(0) = f[T(z)/\tau]$, however moderate, would still imply large masses in the local Seyfert 1 nuclei, the primary power being produced (cf Rees, 1985) by the release of gravitational energy upon accretion onto onto a central black hole: $M \propto D \tau L(z_{max})$ $> 3 \cdot 10^9 M_\odot$ (with efficiency $\varepsilon \lesssim 0.1$, $D \gtrsim \tau/t_o$, $z_{max} \simeq 2$) for the common local nuclei of a modest $M_B \simeq -18.5$ like NGC 4151.

Thus the true evolution - the average scale $\tau(L,z)$ plus some of the detailed behaviour of $L(z)/L(0)$ at $z < 2$ - is increasingly constrained by a framework of local and distant observations currently in progress or planned for the near future, in particular with HST.

Observational constraints already bound the behaviour of early QSOs at $z > 2$. Even the present data could not easily accomodate an evolution still proceeding at a pace $\tau <$ few Gyr to $z \simeq 2.5$ to 3 with numbers conserved, because of the resulting excess in many observables: faint counts and XRB, z-distributions at faint magnitudes, masses of local AGNs. But results from recent surveys reported by Koo 1985 and by Hazard and McMahon 1985 concur with previous surveys (cf Osmer 1983) to directly indicate a sudden dearth of objects, even of high power, beyond $z \simeq 2 - 2.5$.

Thus the remarkable pattern emerging from the phenomenological end is one where at all epochs the bright AGNs as a population succeed in evolving on a time scale $\tau(z) < H(z)^{-1}$, considerably _faster_ than the universe as a whole.

Important questions arise from the astrophysical end: (i) how the QSO birth is related with galaxy formation; (ii) what equilibrium reaches a maturing AGN between exhaustions of its mass stockpile and refuelings by interactions of its host galaxy with companions; (iii) what role may play in the late stages a steady accretion of gas from the surrounding galactic body. The general heading is constituted by the interactions of a compact region - black hole plus surrounding material - which is self-gravitating and tendentially autonomous, with its wider environment changing over the cosmic time.

Is there any prospect for convergence to a meeting point from the two ends ?

2. METHODOLOGY

Proceeding from the theoretical end, one does expect complexity. As an example related to issue (ii) above, $L(z)/L(0)$ with its scale $\tau(L,z)$ represents in fact a statistical convolution of individual behaviours of yet unknown _variance_: obviously, if the individual τ_i were $<<$ average τ the latter scale would rather reflect a coordination exerted by the changing cosmic environment, presumably through the conditions prevailing in groups and associations of galaxies with their large intrinsic variance. Variance in the host galaxies obviously affects issue (iii). As for (i), spreads in QSO turn-on epochs may be expected from spreads in the collapse epochs of the parent protogalaxies.

So, if one were to stress complexity, one would

represent the epoch-dependent luminosity function as a multiple convolution

$$N(L,z) = \int dp\, f(p)\, \frac{1}{\dot{L}(L,t;p)} \int_{\Delta t} dt'\, \dot{L}(L',t';p) S(L',t';p) \quad (1)$$

folding in the QSO birth rate S, their secular light change $\dot{L}$ once born, and the distributions f(p) of several parameters like the mass of the hole and of its fuel supply, type and mass of the host galaxy.

But from a balanced standpoint one would take advantage of some sharp features already present in the data to guide keen unfolding or constraining of the relevance of the possible astrophysical ingredients. For example, the related facts that N(S) and N(L) are steep for large values of S and L, the fact that $\tau \ll H_0^{-1}$ and likely also $< H(2)^{-1}$, the indications that dN/dt is steep beyond $z \simeq 2$ at high L: all suggest some bound to the relevant variances.

We feel that the bookkeeping - what and where over the L-z plane is important - is eased by representing the change of N(L,z;p) in the formally equivalent, differential terms

$$\frac{\partial N}{\partial t} + \frac{\partial}{\partial L}(\dot{L}\, N) = S \quad (2)$$

* independent convolutions when important. Note the linearity.

In the same vein, one should strike a delicate balance between phenomenology and astrophysics, and concentrate first not on one very specific theoretical model for L(t) and S(L,t), but rather on representations that embody features shared by <u>classes</u> of models, parameterized by quantities of direct astrophysical meaning. All in the prospect of singling out key trends first, and with sober expectations as to deconvolving very specific features soon and/or uniquely.

Figs. 1 to 5 and their captions illustrate an <u>example</u> of such 'intermediate models' (hereafter model MM) that complies with the body of existing data starting from the following astrophysical assumptions: Most AGNs are born before $z \simeq 2$. The objects, once born, fade with a time scale $\tau \simeq 3$ Gyr (Ho = 50 km/s Mpc, low density universe) for $L \gg L_s \simeq 10^{43}$ erg/s by steady exhaustion of an initial stockpile of fueling mass. But at lower L a steady supply intervenes (10^{-3} $M_\odot$/yr leaking from the galactic body) to stabilize the output: $\tau(L) = \tau/(1-L_s/L)$. At $z > 2$ the objects are born at a fast rate $\tau_s \simeq 1$ Gyr $> H(2)^{-1}$ and soon

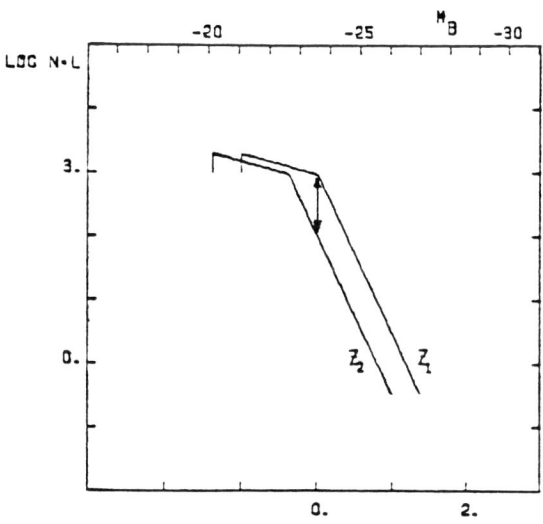

Fig. 0
The effective numbers are amplified in a LE by a steep LF, by a factor $[L(z)/L(0)]^{\gamma-1}$. The resulting slope of the integral counts is given approximately by min $[(\gamma-1), (\gamma-1) \, T(z=0.15, \Omega_o)/\tau]$ in the case of an exponential $L(z)/L(0)$.

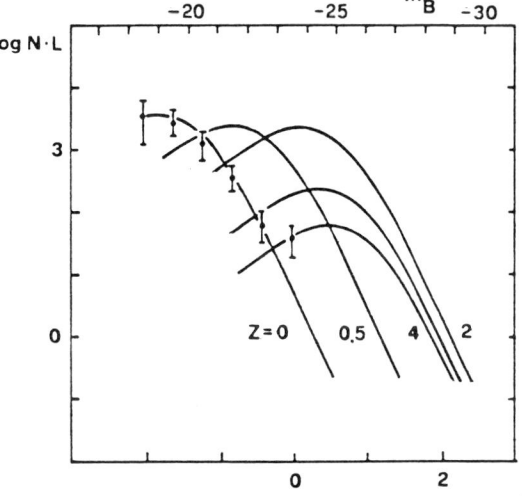

Fig. 1
Model MM: LFs at various z. Up to z = 2 the LFs are given by Eq. 3.5 in CGV 1984b. Parameters: $\tau = 3$ Gyr, $\gamma_2 = 3.7$, $L_s = 0.008\, L^*$. For z = 2 to 4, the objects are born with a birth function $S \propto N(L, z=2) \exp\{[t-t(2)]/\tau_s\}$, $\tau_s = 0.5$ Gyr. Local data from Cheng et al. 1985: the normalization of the LF is a byproduct of the maximum likelihood test using the L-z values of the objects in the samples BQS+AB+BF to provide the above values of τ and γ_2. $H_o = 50$ km/s/Mpc and $\Omega_o = 0.1$. $L^* = 10^{30}$ erg/s Hz at 2500 Å rest frame.

Fig. 2
Model MM: integral number counts. Data points from Marshall et al. 1984, and from Setti 1984.

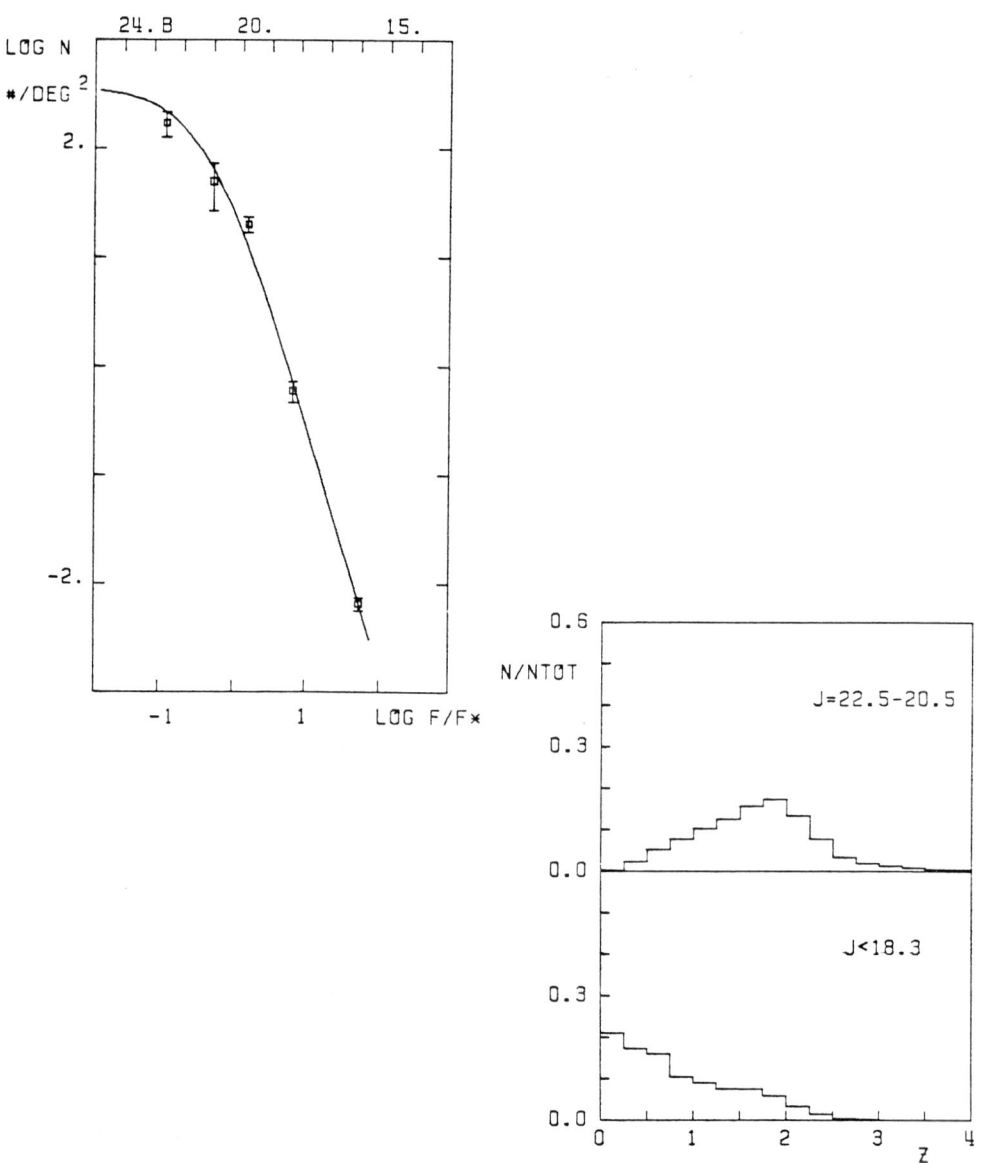

Fig. 3
Model MM: z-distributions in two bins of apparent J magnitude. No corrections for Lyα absorption.

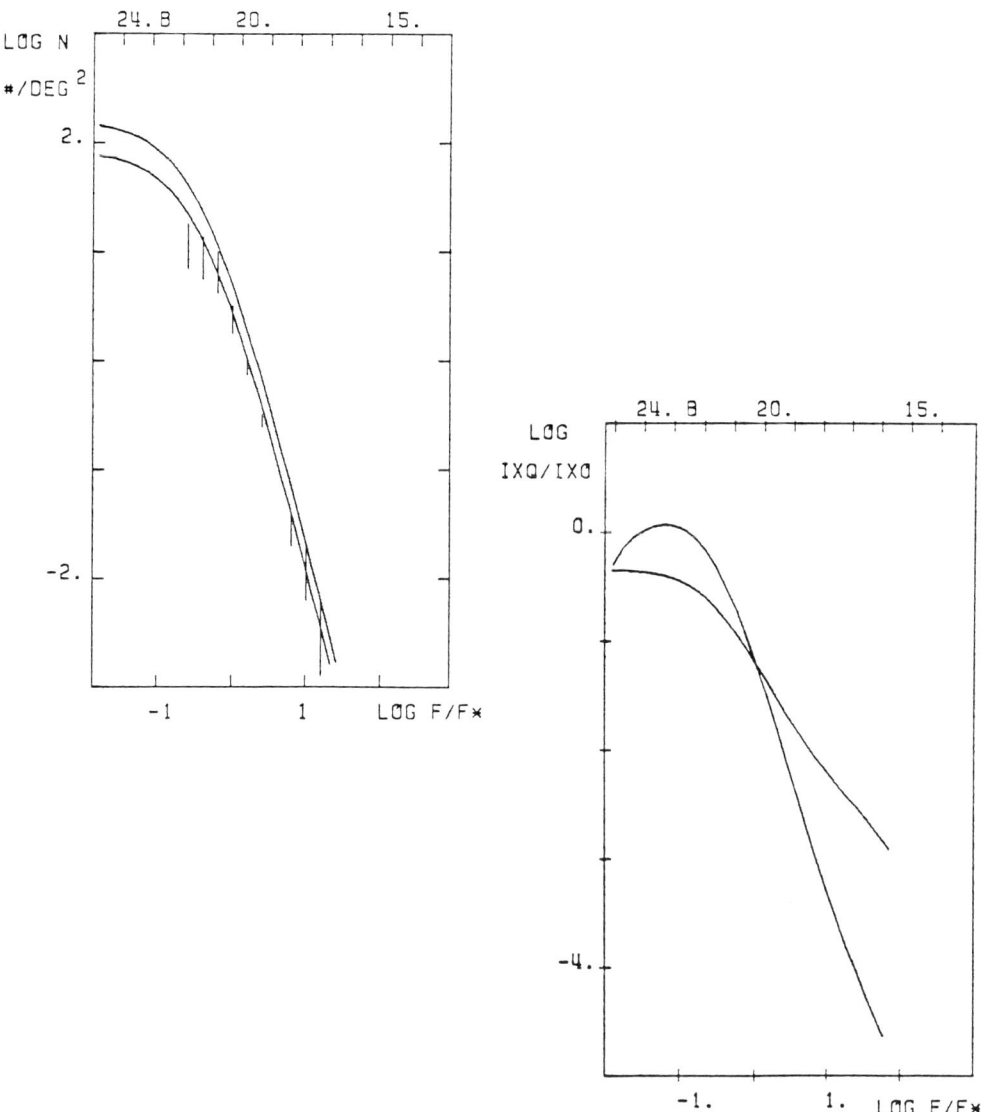

Fig. 4
Model MM (lower curve), and its variant with number conserved up to $z = 2.5$ (upper curve), compared with the $z = 2-2.5$ data by Weedman 1985.

Fig. 5
Model MM: XRB production from evolving AGNs shown as an integral (total=41%) and a differential function of the corresponding optical magnitude B.
$\alpha_{opt} = \alpha_x = 0.5$, L_x/L_{opt} after Avni and Tananbaum 1982.

begin to fade.

Our comparison with the data will assume the observed optical luminosity to be proportional to the bolometric output; it must be cautioned that the present knowledge of the optical spectra in the relevant region is ironically poor, especially because of the possible widespread presence of UV bumps (cf Elvis 1985).

The above points will be discussed, modified and tested in the next Sections.

3. BRIGHT AGNs

The findings by Osmer, by Koo and by Hazard and McMahon indicate that many bright QSOs turn-on (at least as objects with a conventional spectrum) at $z \gtrsim 2$. The steep and persistent slope of N(L) at high and intermediate L (cf Figs. 1 and 4) confirms that not many bright objects are likely to have added since then. All this suggests S = 0 over the corresponding region of the L-z plane as a relevant working hypothesis.

This is a __strong__ astrophysical assumption: of the kind to be tested first within the present approach. Because it eliminates the hard core of the integral structure of Eq. 1 reducing it to finite terms

$$N(L,z;p) = N(L(0),0;p) \, dL(0)/dL, \tag{3}$$

it links directly the local with the distant luminosity functions without necessarily requiring constant shape (cf Fig. 1), nor equal turn-on times for all objects (it holds since the last relevant birth events; but the linearity of Eq. 2 allows easy superpositions, if necessary). Because of the very number conservation, it easily moderates the numer of weak sources contributing to the faint counts and to the XRB. This is, of course, a generalized LE. The underlying continuity assumption is being and will be tested from ground and with HST even at intermediate $z \sim 0.5$ (cf Osterbrock 1984, Dressler 1985).

In this framework, the objects obviously fade in time, $\dot{L} < 0$, and the next question concerns the time scale $\tau = L/|\dot{L}|$. Figs. 1 to 5 show that $\tau = \text{const} \simeq 3$ Gyr is a fair starting point to satisfy the integral observables. To the next approximation, do the data already require $\tau(z)$?

One may reconsider from this viewpoint the two time-honored and remarkably enduring phenomenological forms $L = L(0) \exp(T/\tau)$ and $L = L(0) (1+z)^k$. They differ less than they seem, as it is easily realized under the form

$$\dot{L} = -L/\tau \qquad \tau = \frac{\tau_o}{\tau_o t/t_o} \qquad [k=(\tau_o H_o)^{-1} \atop \text{for both } \Omega_o = 0,1] \qquad (4)$$

which suggest to test against the data the interpolation

$$\tau(t) = \tau_o(t/t_o)^n. \qquad (5)$$

$0 < n < 1$, i.e. τ shorter into the past.

This may represent the simplest transition from exhaustion of a given mass stockpile (constituted by circumnuclear stars or by dwarf satellite galaxies orbiting the host) on a fixed <u>internal</u> dynamical time scale, to an exhaustion forced by the <u>environmental</u> effects, presumably interactions of the host galaxy with companions causing orbit perturbations(cf Roos 1981, Norman and Silk 1983, Gaskell 1985); impulsive interactions occur preferentially in groups and subclusters with high density and low velocity dispersion, but both conditions change over cosmic time if hierarchical clustering reshuffles the members into larger associations with lower density and larger specific binding (White and Rees 1978).

The value of n may be constrained by means of a maximum likelihood test over the L-z plane. We use first the objects in the classical samples BQS + AB + BF (Schmidt and Green 1983, Braccesi, Formiggini and Gandolfi 1970, Marshall et al. 1984), with the results given in Table I.

Table I
Results of a triparametric maximum likelihood test of the evolutionary law in Eq. 5, against different samples of QSOs (H_o = 50 km/s Mpc, Ω_o = 0, Galactic absorption after Sandage 1973, L_s/L^* = 8 10^{-3}). The parameters tested are: γ_2 = slope of the bright LF, τ_o = local time scale for fading, n: cf Eq. 5. Formal best fit values are given, the statistical uncertainty for n is ~ 0.1.

samples	L_{cut}/L^*	#	γ_2	τ_o/t_o	n =>	$N(L^*,0) \cdot Gpc^3$
BQS+AB+BF	.14	164	3.5	.16	.07	4.6
	.37	150	3.5	.20	.48	9.7
	.6	141	3.6	.24	.75	16.2
US+AB+BF	.14	89	3.3	.16	.11	7.4
	.37	85	3.4	.16	.09	7.3
	.6	80	3.4	.16	.14	7.0

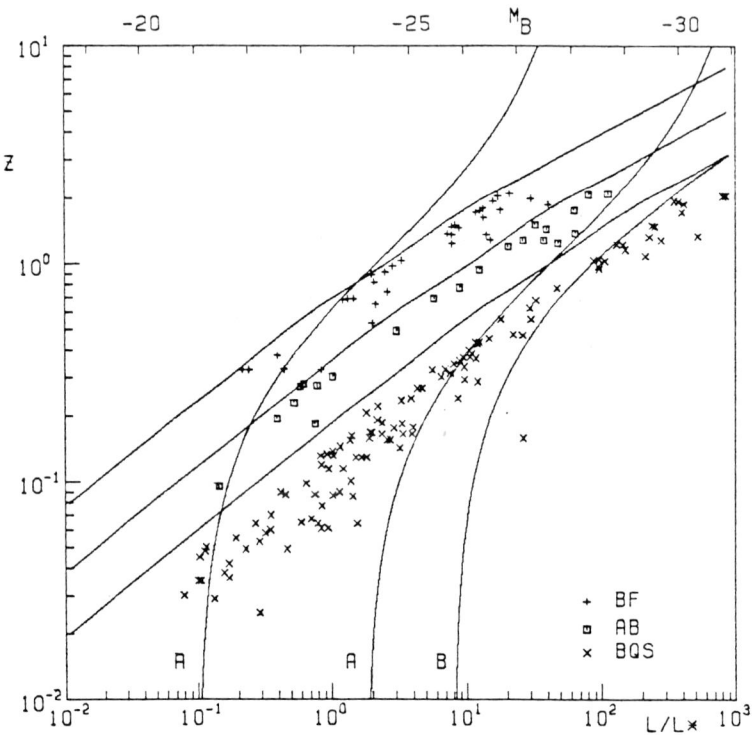

Fig. 6
Distribution of the samples BQS, AB, BF, over the L-z plane. Galactic absorption after Sandage 1973. Curves A and B show the evolutionary paths of single objects for the cases $n=0$ (A) and $n=1$ (B): the maximum likelihood test tends to optimize their correspondence with the bright envelope of the data. Also shown are the isoflux lines at B=16.6, 18.25, 19.8.

It is seen that the best fit value of n increases significantly from $\simeq$ 0.1 to $\simeq$ 0.8 upon increasing the lower cut off L_{cut} which is designed to avoid the morphological incompleteness of the BQS at the Seyfert level discussed by the Authors; for $\Omega_0 = 1$, n increases from $\simeq$ 0.5 only for $L_{cut}/L\ast > 0.6$, i.e. for $M_{Bcut} < -23$. But an examination of Fig.6 clarifies that this behaviour is to be traced back to the apparent gap at $z \simeq 0.8 - 0.9$ in the BQS sample also pointed out by the Authors, that becomes relatively more prominent when the low L subsample is progressively cut off. In fact, when the BQS sample is replaced in the analysis by the US (Mitchell, Warnock III, Usher 1984), n remains stable at values < 0.5 at the 90% confidence level. The result is preliminary, as the latter sample is rather small.

Here one may have reached a limit of the present data: improving the coverage of the corresponding region in the L-z plane will be hard (the BQS covers $\sim$ 1/4 sky) if highly desirable.

In fact, there is more at stake than a specific fitting. Values of n are to be viewed in the light of the copious if still ambiguous evidence for correlation of activity of the AGNs with interactions undergone by their hosts (see e.g. Fried, 1986; Dahari 1984, De Robertis 1985, Heckman et al. 1985, Keel et al. 1985). To set a perspective: Values of $n \ll 1$, if confirmed, would indicate either highly indirect effects of the interactions; or their near invariance [$\tau(t) H(t) \propto t^{n-1}$] to a degree which is surprising, even though some statistical smoothing of the changes in group conditions is expected from the large known variance in the clustering time scales (cf Forman and Jones 1982, Cavaliere et al. 1984).

Note one further invariance property of $N(L,z)$ at high L: τ will be statistically distributed, e.g. because of different hole and supply masses; $f(\tau)$ is expected to decrease for decreasing τ, that is fast objects are exceptional. Then $N(L,z)$ flattens as $z \rightarrow 0$ at the faint end, conserving its shape elsewhere (Padovani 1985).

4. SEYFERT TYPE OBJECTS

More complex is the situation in the range $M_B \gtrsim -21$, where a number of processes may superpose to continuity. However, a stringent constraint is set in this range by the XRB.

In fact, the integrated contribution from AGNs to the full intensity in the few keV range is not to exceed some 75%, considering other classes of sources like Seyferts 2, clusters and normal galaxies. But a more stringent limit (pointed out first by Cavaliere et al. 1981) is set by the flat, smooth spectral shape of the XRB in the range: a few keV to $\simeq$ 10 keV. This would be spoiled past conceivable repair (with a dip at lower energies in the residual spectrum and no presently known process to fill it) even by a 30 - 40% contribution from the Seyferts with their definitely steeper spectral slope (Mushotzky 1984). The partial contribution to be expected from AGNs with QSO-type luminosities depends on spectral and statistical uncertainties, but it hardly reaches 25% of the AGN total (CGV 1984b; also Schmidt 1985). If the spectral slope of these objects is steeper and more dispersed compared to Seyferts (as indicated by the small, low z sample of Elvis, Wilkes and Tananbaum 1985) the QSO subset contributes less to the total intensity but relatively more to the spectral dip.

A uniform luminosity evolution may be carefully adjusted to fit tolerably the local LF and the distant data. But adopting for L_x/L_{opt} values and variance as given by Avni and Tananbaum 1982 (and confirmed by Kriss and Canizares 1985), a contribution of 75% to the integrated XRB is produced from z < 2.2. Filippenko 1986 argues that the optical LF of the AGNs is likely to extend down to $M_B \simeq -17$: the corresponding XRB production (cf Elvis, Soltan and Keel 1984) then may rise up to ~ 90% (z < 2.2). Such values are unacceptably high, and remain so unless the average L_x/L_{opt} or its variance were found to be significantly smaller for the relevant low luminosity objects.

On the other hand, on astrophysical grounds the uniformity of the evolution, that is the invariance of τ, is very likely to break down at $L \sim 10^{43}$ erg/s: such powers require an accretion of only $\dot{M} \sim 10^{-3} M_\odot$/yr ($\varepsilon \sim 10^{-1}$), an amount of gas likely to spill down toward the nucleus in the form of gas from the galactic body to sustain such levels of activity for times much longer than a few Gyr. Such considerations underly the low L behaviour of model MM.

This provides a physical interpolation between the two extremes discussed by Weedman at this Meeting: a single, uniform population or two disjoint sets. Stretching without tearing the LFs with increasing lookback times, it conserves the unity of the AGN population while giving a minimum evolution to the weak nuclei that minimizes their XRB production, still within the known limits to the faint counts. The total XRB produced is 41% (and appreciably less with $\Omega_0 = 1$) out to z = 4, with 30% from objects of $M_B > -23.8$ and 11% from QSOs proper. The mass accreted in objects of $M_B = -18.5$ is $M = 3\ 10^8 M_\odot$ (with a bolometric correction by a factor ~ 6 after Soltan 1982), an order of magnitude less than with a uniform evolution.

Model MM shall be completed relating its lower LF to the LF of the host galaxies: assuming $L_s \propto L_{galactic\ body}$, and folding with a distribution of L_s, $N(L_s) \propto L_s^{-1.25}$, gives the preliminary results in Figs. 7 - 8. The XRB increases < 5% for the planned extension to $M_B = -17$.

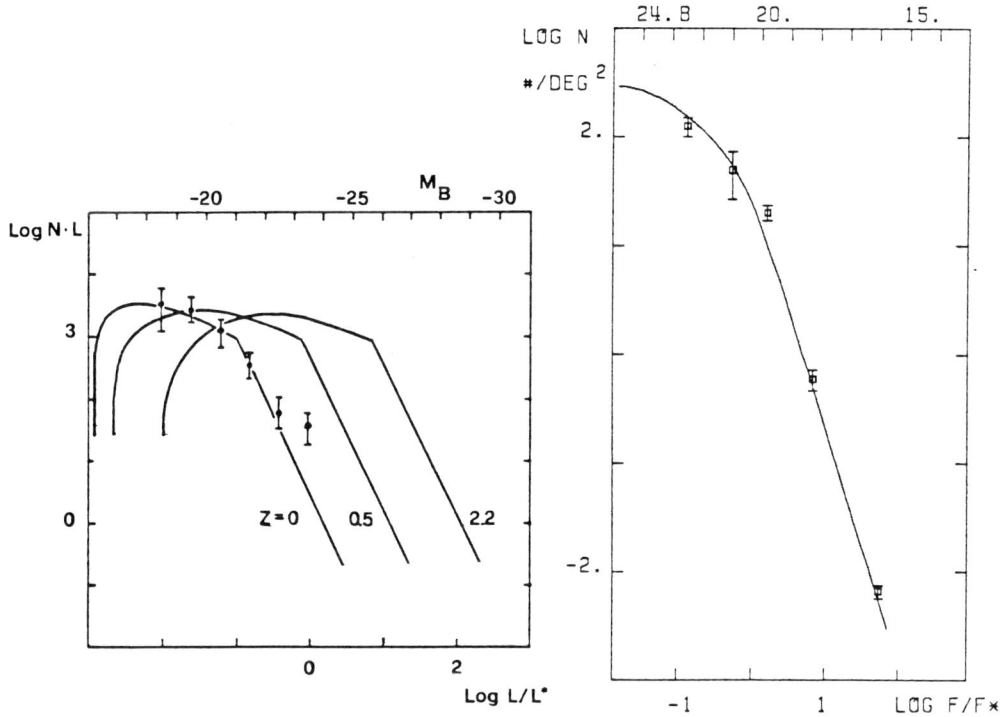

Fig. 7
LFs for a model similar to MM, convolved with a distribution of L_s ($\propto L_s^{-1.25}$)

Fig. 8
Integral number counts for the case in Fig. 7.

The considerations to follow rise the possibility that some fraction of low L objects may emerge at epochs close to the present, also a way to reduce XRB production.

5. EARLY QSOs

While continuity (i.e. number conservation) apparently dominates the behaviour for $z < 2$ at least at high L, it is obviously broken at some higher z when the objects turn on. And indeed at $z \gtrsim 2$ there are signs of a new turn of events. The evidence reported by Osmer, by Koo and by Hazard and McMahon indicates changes occurring in the earlier population of AGNs (at least for those meeting canonical selection criteria): The apparent object number certainly does not increase further; the proportion of intrinsically weak objects decreases; even the bright, relatively more persistent QSOs of $M_B < -27$ may decrease in number by $\sim 10^{-1}$ from $z \simeq 2.5$ to 3.5.

The latter finding needs statistical confirmation. But such a dramatic drop occurring in a range of cosmic time $\Delta t \simeq 1$ Gyr $< H(2.5)^{-1}$ is provocative enough to warrant some comments concerning epochs and time scale associated with the birth function $S(L,z)$.

Figs. 9a and 9b illustrate the phenomenological effects expected from (a) a birth function superposed on LE, i.e. the objects are born down to $z \simeq 2$ and soon begin to fade away; (b) a luminosity 'anti-evolution' (i.e. the objects brighten up to $z \simeq 2$ before undertaking the main fading course), with an apparent sharp drop due once again to the steep slope of the LF.

Physically, the latter behaviour might be enforced by dust obscuration in intervening galaxies as discussed by Ostriker and Heisler 1984. The difficulty with overproduction of XRB pointed out by the Authors and by Setti and Zamorani 1984 still persists if the true evolution beyond $z \simeq 2$ is of a LE type, see Sect. 4; unless one is prepared to assume additionally a real cut off at higher z (meaning by itself a birth function in cosmic time) for the weak objects, with an observable result of the type in Fig. 9c.

But then a more appealing scheme to that same effect has been discussed by CGV 1984a,b: with the inevitable source $S(L,z)$ active for $z \gtrsim 2$, the objects once born brighten up with a differential time scale $\tau_b \propto L^{-p}$ as one physically expects in runaway processes of black hole collapse, and growth in mass and output (cf Phinney, Shapiro, 1985). Then one can understand the formation of a steep, self-similar $N(L) \propto L^{-(p+1)}$ at high luminosities from a flatter distribution that forms at substantially lower L, and would essentially persist, with minor alterations, at

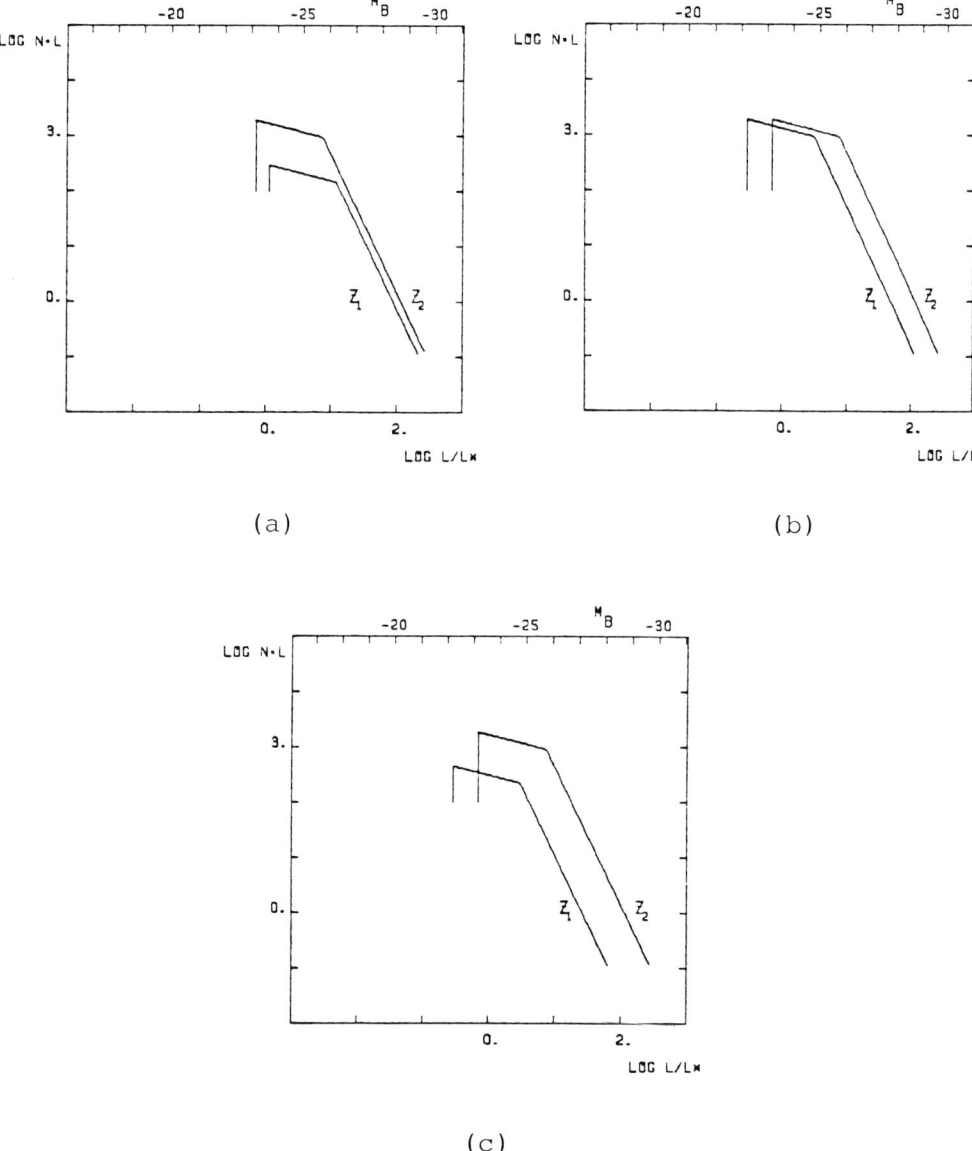

Fig. 9
Three schematic modes of evolution at high z's
($z_1 > z_2 \simeq 2$):
(a) object birth superposed to LE;
(b) luminosity 'anti-evolution': the objects brighten up toward us;
(c) object birth superposed to luminosity 'anti-evolution'.

the faint end of the local LF.

The latency time implied by $\tau_b \propto L_{in}^{-p}$ means an effective lag from the start of the process of hole collapse and brightening. This would begin at $z \gtrsim 4-8$ ($\Omega_0 = 1$ to 0), in connection with protogalaxy settlings; it would proceed on a time scale $< H(2.5)^{-1}$ if the stochastic spread intrinsic to hole formation $<$ 1 Gyr. Such short scales would be echoed, as it were, in the sharp QSO rise at $z \simeq 2.5$ (Cavaliere and Szalay, 1985); weaker objects may emerge much later into observable nuclear activity.

6. BL LACS: AN EVOLVING POPULATION ?

The case of the BL Lac objects is interesting both in itself and as an opposite extreme to the AGN case stressing the degree of order implied by the steep counts of the latter.

In fact, these objects stand apart from other AGNs for the remarkable flatness of their optical and X-ray counts, apparent from the limited data and upper limits now available (Setti and Woltjer 1982, Maccacaro et al. 1984).

On the other hand, BL Lacs are also peculiar for their steep spectra in the IR - soft X-rays range, for their sharp time variability, for their strong and variable optical polarization (Angel and Stockman 1980). Difficulties raised in single objects by the last two features for the application of conventional mechanisms of non-thermal radiation can be solved or alleviated by the beaming hypothesis (Blandford and Rees 1978).

It is then interesting to see how beaming as a caracterizing feature affects also the population behaviour. Urry and Shafer 1984 have computed the ensuing luminosity functions: even when the LF of the parent objects is a steep single-power-law LF, that of the boosted objects acquires at its lower end a long, very flat ($\gamma_1 \lesssim 1.3$) plateau: the many faint objects near the faint end of the parent LF are redistributed over a range of boosted apparent L corresponding - for a given value of the bulk Lorentz factor Γ - to the range of viewing angles that make the beamed component dominant over the isotropic one.

The increased disorder corresponding to $NL \Rightarrow$ const for most of the objects must correspond to flatter counts even when the parent population evolves strongly. Specifically, for a LE with constant τ the evolution function for the boosted objects: $[L(z)/L(0)]^{\gamma_1-1} = [f(T/\tau)]^{\gamma_1-1} \propto \exp[(\gamma_1-1)T/\tau]$ contains an effective time scale $\tau/(\gamma_1-1)$ much longer than that $[\tau/(\gamma-1)]$ for the parent population at comparable apparent luminosities.

Indeed, CGV 1985 compute in detail the counts resulting if the parent population is constituted by the radio-loud

Fig. 10
Optical LF's for evolutionary (eb) and for non-evolutionary (ub) BL Lacs, with the parent population (P) for the former. Bulk Lorentz Γ = 4 and 7 respectively. Ratio of beamed to isotropic power = $5\ 10^{-3}$ and $2\ 10^{-2}$ respectively.

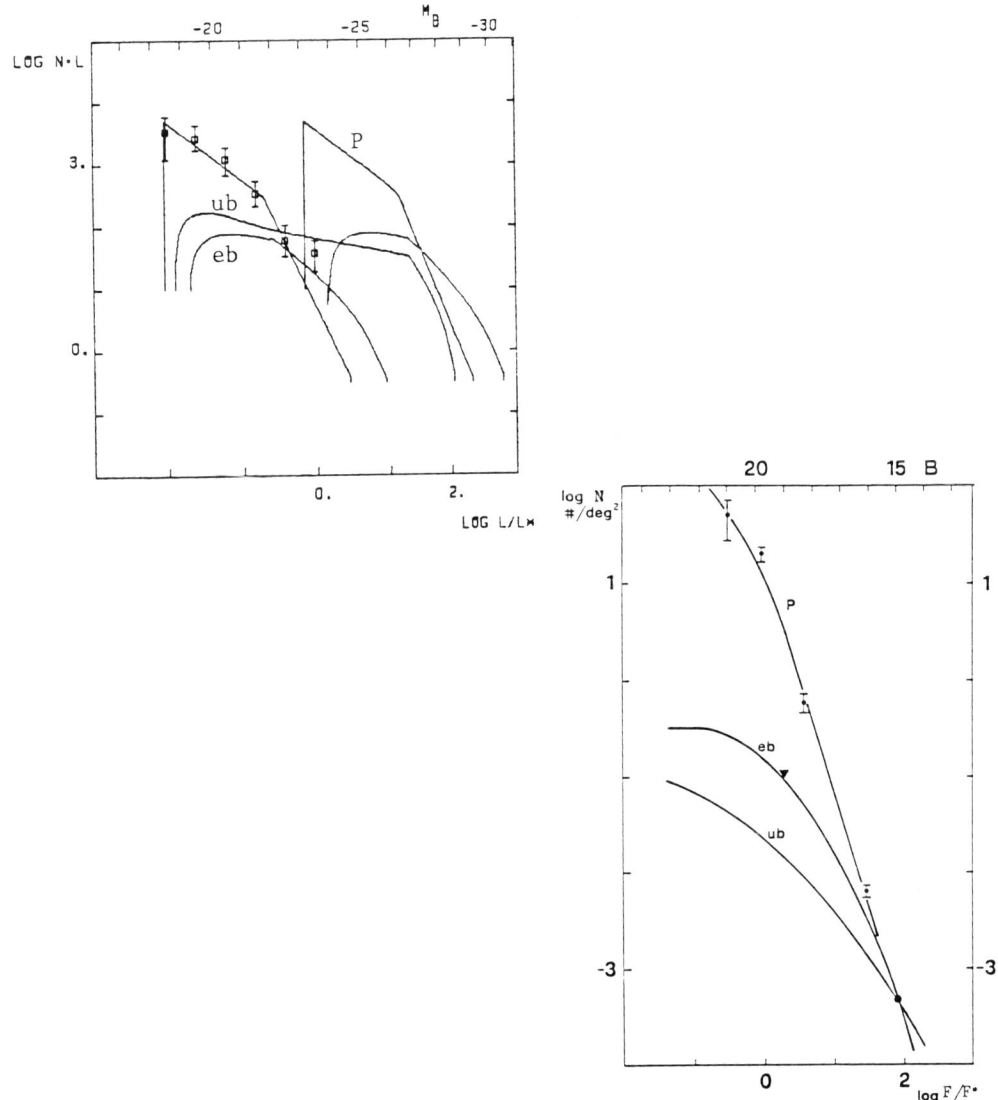

Fig. 11
Integral number counts corresponding to the LF's in Fig. 10. Data point and upper limit from Setti and Woltjer 1982. For the counts of the parent population see Fig. 2 and its caption.

subset of the AGNs evolving as in Sect. 3 (the problem admits only one free parameter, corresponding to the fraction of the whole population that actually supports a beam). The results are shown in Figs. 10 and 11.

The other astrophysically significant parent population may be constituted by the bright ellipticals proposed by Perez-Fournon and Biermann 1984. Low upper limits to their isotropic non-thermal powers and lack of intrinsic evolution require a stronger boosting to reach apparent luminosities relevant for BL Lacs and hence imply exceedingly flat counts. These are also represented in Fig. 11, see also Fig. 10.

Whether the optical (radio-selected) BL Lacs can be unified to the rest of the AGNs as for their intrinsic evolution will be tested by counts at $B \sim 16 - 17$ over $< 1/2$ sky. Similar considerations, with a weaker boosting, may apply to the X-ray selected BL Lacs of Maccacaro et al. 1984.

7. CONCLUDING REMARKS

From the above discussion the following general points emerge.

Continuity (i.e. LE) is consistent with the data at $M_B \lesssim -21$ and up to $z \simeq 2$, both in a low and in a high density ($\Omega_o = 1$) universe to within a factor < 2, the current uncertainty in most data points. The time scale over which these objects fade is ~ 3 Gyr, quite shorter than H_o^{-1} and also shorter than $H(2)^{-1}$.

Interactions are morphologically documented in many cases. The statistical data, within their current uncertainties, do not enforce yet the progressive change of the fading time scale that is expected in the simplest interaction scenario. If interactions are important, recurrency is likely: a short duty cycle can be elicited by comparison of the nuclear masses of currently active and of (observationally more difficult) inactive galaxies.

The emergence of the bright objects at $z \simeq 2.5$, if so sharp as the present scanty data suggest, may be rooted into earlier epochs of protogalaxy collapses and nuclei formation. During the ensuing time lag the objects are likely to brighten up to visibility at a rate increasing with luminosity, which produces the break in the LF. Alternatively, sharp disguising spectral changes or initial intrinsic obscuration must take place at $z \simeq 2 - 2.5$; but unambiguous evidence of transitional objects is still lacking.

The behaviour of the weaker objects may be more complex: within the limits set by the XRB, the unity of the AGN population may be maintained if these objects are similar, though generally smaller, gravitational machines for which a minimal, steady accretion mode prevails. In an $\Omega_o=1$ universe the contributions to XRB are somewhat reduced, but the spectral shape is still a stringent constraint if the ratio L_x/L_{opt} does not decrease sharply toward low L. While waiting for the very faint counts to take some definite shape, a more recent emergence may be considered for a consistent fraction of the weaker AGNs: relevant data at intermediate z are likely to come sooner from HST.

Radio selected BL Lacs, compatibly with the scanty optical counts, may well be intrinsically evolutionary objects similar in kind to the rest of the optical AGNs, but singled out also as for their statistical behaviour by a small beamed component, dominant when directed toward us.

Given that the most conspicuous part of the optical nuclear activity is powered by accretion onto massive black holes, a simple scenario envisages the unity of the basic underlying machine, and three fueling modes in succession: (i) runaway sucking in of material left over from the initial collapse; (ii) a maturity phase where interactions of the host galaxy feed recurrently the engine; (iii) a terminal stage fueled by steady leakage from the host. With some optimism, evolutionary signs may be discerned for phases (i) and (iii), while phase (ii) remains statistically more elusive, save for one general point: merging between equal galaxies would be consistent rather with a behaviour of the 'density evolution' kind, while continuity by itself is more consistent with perturbative effects feeding a basically invariant number of activity centers.

Sobering considerations follow. The bright end of the luminosity function might be altered by minilensing (cf Vietri 1985) if star-sized objects contribute $\Omega_o \simeq 0.2$, and the source sizes < 1 light-week. Uncertainties of 0.2 mag may alter the counts by a factor $\simeq$ 1.5 where their slope is steep. Limits to more specific modeling are apparently set by statistical flukes at bright magnitudes, where the progress is necessarily limited. Faint samples cover small areas and may be vulnerable to real angular fluctuations (the case of the BF sample has been widely discussed at this Meeting); here fibre optics may help powerfully, provided that these fluctuations do not exceed 100% over ~ few deg^2.

On the other hand, real ambiguities derive from the little known shape and extension of the UV 'bumps' that may affect most the spectra seen in the optical window, and the estimates of the bolometric powers (cf Elvis 1985). Moreover, the XRB constraints depend on extrapolations of the ratio L_x/L_{opt} toward high z and low L_{opt} (cf Elvis, Soltan, Keel 1984, Kriss and Canizares 1985). In these respects, the wide spectral range accessible to HST, and the X-ray surveys planned with ROSAT, may assess realistic expectations of more specific advances.

REFERENCES

Angel,J.R.P., and Stockman,H.S. 1980, Ann.Rev.Astron. Astrophys. $\underline{8}$,321.
Avni,Y., and Tananbaum,H. 1982, Ap.J.(Letters),$\underline{262}$,L17.
Blandford,R.D., and Rees,M. 1978, in 'Pittsburgh Conference on BL Lac Objects', A.M. Wolfe Ed. (Pittsburgh: University of Pittsburgh),p.328.
Boyle,B.J., Shanks,T., Fong,R., and Peterson,B.A. 1986, this Volume.
Braccesi,A.,Formiggini,L., and Gandolfi,E. 1970 Astr.Ap., $\underline{5}$,264.
Cavaliere,A., Danese,L., De Zotti,G., and Franceschini,A. 1981, Astr.Ap.,$\underline{97}$,269.
Cavaliere,A., Giallongo,E., and Vagnetti,F. (CGV) 1984a in 'X-ray Astronomy 84' M. Oda and R. Giacconi Eds. (Institute of Space and Astronautical Science, Tokyo), p.435.
--------- 1984b, preprint to appear in Ap.J.
--------- 1985, in preparation
Cavaliere,A.,Santangelo,P.,Tarquini,G.,Vittorio,N. 1984, in 'Cluster and Groups of Galaxies', F. Mardirossian, G. Giuricin, and M. Mezzetti Eds., Astrophysics and Space Science Library, D. Reidel Company, $\underline{111}$,499.
Cavaliere,A. and Szalay,A.S. 1985, in preparation.
Cheng,F.Z.,Danese,L.,De Zotti,G., and Franceschini,A. 1985, M.N.R.A.S., $\underline{212}$,857.
Dahari,O. 1984,Astron.J. $\underline{89}$,966.
Danese,L., De Zotti,G., and Franceschini,A. 1986, this Volume.
De Robertis,M. 1985,Astron.J. $\underline{90}$,998.
Dressler,A. 1985, preprint.
Elvis,M. 1985, preprint.
Elvis,M., Soltan,A., and Keel,W.C. 1984,Ap.J.,$\underline{283}$,479.
Elvis,M., Wilkes,B., and Tananbaum,H. 1985,Ap.J.,$\underline{292}$,357.
Filippenko,A.V. 1986, this Volume.
Forman,W., and Jones,C. 1982, Ann.Rev.Astron.Astrophys., $\underline{20}$,547.
Fried,J. 1986, this Volume.
Gaskell,C.M. 1985,Nature,$\underline{315}$,386.
Hazard,C., and McMahon,R. 1985,Nature,$\underline{314}$,21.
Heckman,T.M., Bothun,G.D., Balick,B., and Smith,E.P. 1985, Astron.J.,$\underline{89}$,958.
Keel,W.C., Kennicutt jr.,R.C., Hummel,E., and van der Hulst,J.M. 1985, Astron.J. $\underline{90}$,708.
Koo,D. 1986 this Volume.
Kriss,G.A. and Canizares,R.G. 1985, preprint.
Maccacaro,T., Gioia,I.M., Maccagni,D., and Stocke,J.T. 1984, Ap.J.(Letters) $\underline{284}$,L23.
Marano,B., Zamorani,G., and Zitelli,V. 1986, this Volume.
Marshall,H.L., Avni,Y., Braccesi,A., Huchra,J.P., Tananbaum,

H., Zamorani,G., and Zitelli,V. 1984 Ap.J., 283,50.
Mitchell,K.J., Warnock III,A., Usher,P.D. 1984, Ap.J. (Letters), 287,L3.
Mushotzky,R.F. 1984, Adv.Space Res., 3,Nos.10-12,157.
Norman,C., and Silk,J. 1983 Ap.J.,266,502.
Osmer,P.S. 1983, in 'Quasars and Gravitational Lenses', J.P. Swings Ed. (Liege: Institut d'Astrophysique), p.51.
Osterbrock,D.E. 1984,Ap.J.(Letters), 280,L43.
Ostriker,J.P. and Heisler,J. 1984 Ap.J.,278,1.
Padovani,P. 1985, Thesis, Univ. Padova.
Perez-Fournon,I. and Biermann,P. 1984,Astr.Ap.,130,L13.
Phinney,E. 1985, this Meeting.
Rees,M., 1986, this Volume.
Roos,N. 1981, Astr.Ap.,104,218.
Sandage,A. 1973, Ap.J.,183,711.
Schmidt,M. 1986, this Volume.
Schmidt,M., and Green,R.F. 1983, Ap.J.,269,352.
Setti,G. 1984, in 'X-ray and UV Emission from Active Galactic Nuclei', W. Brinkmann and J. Trümper Eds., MPE report 184, Garching, p.243.
Setti,G., and Woltjer,L. 1982, in 'Astrophysical Cosmology', H.A. Bruck, G.V. Coyne, and M.S. Longair Eds. (Pont. Acad. Scientiarum Scripta Varia, 48,315).
Setti,G., and Zamorani,G. 1984, COSPAR/IAU Symposium 'Advances in High Energy Astrophysics and Cosmology',(Rojen),in press.
Shapiro,S.L. 1986, this Volume.
Soltan,A. 1982, M.N.R.A.S.,200,115.
Urry,C.M., and Shafer,R.A. 1984, Ap.J., 280,569.
Vietri,M. 1985, Ap.J., in press.
Weedman,D.W. 1985 , Ap.J. Suppl., 57,423.
Weedman,D.W. 1986, this Volume.
White,S.D., and Rees,M. 1978, M.N.R.A.S.,183,361.
Zamorani,G., et al. 1981, Ap.J.,245,357.

DISCUSSION

MARSHALL: First, Boyle et al. (these proceedings) find a surface density at B=20 that is consistent with our BF result (within 2 sigmas) but from a much larger area for many positions on the sky. Second, the LF that I derive for a sample of low-luminosity active nuclei (these proceedings) predicts the observed number counts quite well and does predict a fairly sharp break near B=20, implied by the BF point.

CAVALIERE: From contributions and discussions at this Meeting, it appears now that the BF counts are somewhat high by some 2 sigmas. All sensible models, including those I discussed here and Danese's contribution, predict counts flattening at B~19 and sistematically 1-2 sigmas below the BF point.

PETROSIAN: Is your statement, that for pure luminosity evolution models one predicts fewer Seyfert galaxies, in disagreement with Weedman's conclusion?

CAVALIERE: My statement is as follows. Pure (i.e. uniform) LE tends to overproduce X-ray background from AGNs, even with a very flat faint end of the local LF (N$\propto$L**-1.2 : 75% of the XRB at a few keV, these proceedings), and more so if the local LF is extended below MB ~ 18.5. If the local LF is (arbitrarily) cut off at such MB, then few Seyferts are predicted at MB > -20 at z ~ 0.5. Non uniform LE, decreasing toward low L ~ 10**43 erg/sec on the plusible ground that the small mass inflow needed could be mantained for very long times, constitute a physical interpolation between the two extreme cases discussed by Weedman. This produces considerably less XRB, the local LF can be extended down to MB ~ -17, and a substantial number of Seyferts is predicted at z ~ 0.5 with MB ~ -19.

SURVEYS OF LOCAL AGN's[*]

P. Véron
Observatoire de Haute Provence
04870 Saint Michel l'Observatoire, France

ABSTRACT. We review the optical, X-ray and infrared surveys of local AGN's, discussing the limitations and biases of each of them. We conclude that no single survey may lead to a complete sample but that all of them must be used simultaneously to build a meaningful luminosity function.

I. INTRODUCTION

We define an AGN (active galactic nucleus) as an emission line nucleus the ionization of which cannot be due to hot stars. (This definition is not quite satisfactory as it excludes the BL Lac objects).
Emission line nuclei belong to four main types : the Seyfert 1 and Seyfert 2 nuclei, the Liners and the H II regions (or star bursts). The H II regions are hydrogen clouds ionized by hot stars. Galaxies of type 1 have Balmer lines which are broader than the forbidden lines ; their width is typically a few thousands Km s^{-1}. In Seyfert galaxies of type 2, Balmer and forbidden lines have the same profile ; [O III] λ 5007 is much stronger than Hβ , while Hα and [N II] λ 6583 are of comparable strength. All Seyfert 1 galaxies have a star-like, often variable nucleus, while class 2 nuclei are resolvable (>1 arc sec) and are clearly not as bright compared to their surrounding galaxy as are the class 1 nuclei (Khachikian and Weedman 1971).
Seyfert 1 nuclei share all the characteristic properties of quasars and can be considered as low-luminosity quasars. Seyfert 2 galaxies have a high excitation emission-line spectrum which is probably due to photoionization by a non-thermal continuum extending far into the ultraviolet (Osterbrock and Miller 1975). This emission-line spectrum is similar to the narrow-line components of the Seyfert 1 nuclei.
The liners (low ionization nuclear emission region) have been defined by Heckman (1980) ; they are now believed to be also ionized by a non-thermal continuum.
Seyfert (1943) has published a list of six galaxies mainly characterized by

[*] Paper presented at the conference on "Structure and evolution of active galactic nuclei". Trieste, April 10-13, 1985.

the presence of strong emission lines of high excitation in their nucleus. One of them (NGC 1275) is now believed to be a BL Lac object (Véron 1978), one is a Seyfert 2 galaxy (NGC 1068), and four are Seyfert 1 (NGC 3516, 4051, 4151, and 7496).

In addition to the six objects studied by himself, Seyfert listed six additional galaxies known at that time to have emission lines and believed by him to belong to the same class; three of those (NGC 3227, 5548, and 6814) are now called Seyfert 1; the three others (NGC 2782, 3077, and 4258) probably have an nuclear H II region rather than an active nucleus.

Thus it is not surprising that Seyfert could not give an unambiguous definition describing this heterogeneous set of objects.

A number of surveys (optical, X-ray and infrared) have been carried out to find AGN's; they are discussed below.

II. OPTICAL SURVEYS

II. 1. UV excess objects. The Byurakan objective-prism surveys.

- The first Markarian survey.

In the middle of the sixties, the first Byurakan survey was started, aimed at the discovery of galaxies with strong ultraviolet radiation from their nuclear region. This survey was made using the 40-52" Schmidt telescope of the Byurakan Observatory and a 1°.5 objective-prism (dispersion ~ 2500 Å mm^{-1} at H β) with a IIaF emulsion. The limiting magnitude varies from plate to plate in the range 16.5-17.5. The observations which extend from the north pole to declination -15°, excluding the region with $|b| < 20° - 30°$, cover a region of approximately 15000 square degrees. They were completed in 1978.

The survey discovered 1500 galaxies with a strong ultraviolet continuum. The data were published in a serie of 15 lists, the first one in 1967 by Markarian, the last one in 1981 (Markarian et al.).

A number of spectrographic investigations have shown that an important fraction of these objects are Seyfert galaxies. A recent catalogue of active galaxies (Véron-Cetty and Véron, 1984) lists 91 Seyfert 1 and 32 Seyfert 2 Markarian galaxies; these numbers are lower limits, on one hand because some Markarian galaxies lay hidden in the catalogue under another name, and on the other hand because the 1500 Markarian objects have not yet been all observed spectroscopically. Huchra and Sargent (1973) have estimated that 10 % of the Markarian galaxies are Seyfert galaxies.

The dispersion used dit not usually permit the detection of emission lines. Seyfert 1 galaxies usually have a strong UV continuum and it is these systems that one would expect Markarian's technique to be most sensitive to. Seyfert 2, on the other hand, which do not have excessively bright UV continua, could easily elude the UV excess search method. Thus the deficiency of Seyfert 2 galaxies is most probably a result of survey selection effects (Wasilewski 1983). Markarian appears to be finding all galaxies with U-B < -0.3, plus a decreasing fraction of the redder galaxies (Huchra 1977). These redder galaxies can be either unreddened Seyfert 1 nuclei of very low absolute luminosity ($M_B > -19$) which are faint compared with the bulge of the parent galaxies, or reddened Seyfert 1 nuclei which have no UV excess.

- The Kazarian survey.

Using the same telescope and technique, Kazarian has started a similar survey, with a similar limiting magnitude, varying from 16 to 18 magnitude from plate to plate. Five lists totaling 580 objects have already been published (Kazarian 1979 a, b; Kazarian and Kazarian 1980, 1983 a, b). Very little follow-up spectroscopy of these objects has yet been published.

- The second Markarian survey.

Starting in 1978, a second spectral survey has been undertaken by Markarian at the Byurakan Observatory, using the same telescope and objective prism, but the observations are made with fine grain Kodak IIIaJ and IIIaF plates; the limiting magnitude is 19-20 mag, extending 2 or 3 magnitudes fainter than the first survey. This region will cover the region of the sky between 8^h and 17^h in right ascension, and 49° and 61° in declination. To achieve an effective and uniform survey, each area, measuring 4 x 4° is photographed several times on baked IIIaJ and IIIaF plates; these observations are complemented with photographs through a 3° and a 4° prisms (giving dispersions of 900 and 280 Å mm^{-1} at H_γ respectively), in conjonction with different filters to find weak and low contrast emission lines (Markarian et al. 1983). The first field, containing 107 objects, has been published (Markarian and Stepanian 1983); five of them are confirmed Seyfert galaxies, 3 Seyfert 1 and 2 Seyfert 2 (Markarian et al. 1983). The second field contains 110 objects of which 20 are possible quasars and four possible Seyfert galaxies (Markarian and Stepanian 1984). The third field contains 94 objects; spectroscopic observations of 53 of them have revealed no new Seyfert (Markarian et al. 1984).

II. 2. Emission-line objects.

- The Tololo survey.

A survey of southern emission-line galaxies and quasars has been carried out at Cerro Tololo using the thin (1°.8) prism (Blanco 1974) on the 61/91 cm Curtis Schmidt Telescope. The motivation for this low dispersion, objective-prism survey arose directly from the success achieved by Markarian using the Byurakan Schmidt telescope.

The dispersion was 1740 Å mm^{-1}, the plate scale 96.7"mm^{-1}. The survey plates were all exposed for between 75 and 90 minutes; the limiting magnitude is 17.5-18.0 (Smith 1975). Line identification was the primary criterium for object selection (MacAlpine et al. 1977 a). The result of the survey has been published by Smith et al. (1976) and MacAlpine et al. (1977 a, b, c; 1978; 1981); these last five lists contain altogether 655 emission-line objects. At galactic latitude $|b| > 50°$, about three emission-line galaxies have been discovered in each 10 square degrees of the sky.

The technique used has proven very effective in finding high-redshift QSOs, but there is indication that no more (and perhaps many less) emission-line galaxies can be discovered in this way than in a Markarian-like survey. Indeed the emission lines from more than 80 % of Tololo galaxies seem to be produced by hot stars; about 2 % are Seyfert 1 and less than 10 % are class 2 Seyferts. It appears that virtually all of the Tololo galaxies would also be discovered in a Markarian-type survey, but only about one-third of the Markarian galaxies have sufficiently strong emission lines to be included in the Tololo listings (Bohuski et al. 1978).

- The Pesch and Sanduleak survey.

A large scale objective-prism survey for blue and emission-line galaxies has been started by Pesch and Sanduleak (1983) with the 61/91 cm Burrell Schmidt telescope of Case Western University, located at the Kitt Peak Station of the Warner and Swasey Observatory, with a 1°.8 prism ($\sim$ 1500 Å mm^{-1} at H$_\beta$), on IIIaJ plates. The limiting magnitude is B $\sim$ 18.0. A first list of 71 blue and/or emission-line galaxies has been published. The dispersion used being similar to that of the Tololo survey, this survey probably suffers from the same limitations.

- The Wasilewski survey.

The 61/91 Burrell Schmidt telescope, equipped with a 4° objective-prism and the IIIaJ Kodak emulsion has been used to survey a 825 square degrees region near the north galactic pole. The dispersion and resolution (with 1 arcsec seeing) at H$_\beta$ were 400 Å mm^{-1} and 4 Å respectively (Wasilewski 1983). This method allows the detection of the [OIII] λ 5007 line in low redshift (z < 0.07) objects. The limiting magnitude is B $\sim$ 16, the survey is complete to B $\sim$ 15.7.

Down to m$_{pg}$ = 15.7, Wasilewski identified 132 emission-line galaxies (96 news and 36 previously known objects), of which 16 are Seyfert galaxies, five of them Seyfert 1 galaxies, and 11 Seyfert 2 galaxies. Eight of these Seyfert galaxies were previously known, but of the 8 new ones Wasilewski discovered, one is a Seyfert 1 (Was 26), while seven are Seyfert 2 galaxies (Osterbrock 1984).

About 2 % of the galaxies in the absolute magnitude range M$_B$ = -18 to -21 are Seyfert galaxies according to these data; about half of them are Seyfert 2, while as we have seen above, 75 % of the Markarian galaxies are Seyfert 1 (see also Meurs and Wilson 1984). Weak narrow emission lines are more readily detected with higher dispersion because the continuum is diluted while the lines are not which explains the success of the Wasilewski's survey in finding Seyfert 2 galaxies. This suggests that the Markarian's surface density of Seyfert 2 galaxies may be low by as much as a factor 3.

II. 3. Compact or high surface brightness galaxies.

- The Zwicky survey.

During the 1960's, Zwicky made an extensive examination of the Palomar Sky survey plates that resulted in his "Catalogue of galaxies and of clusters of galaxies". In the course of this study, he picked up numerous examples of what he labelled as "compact galaxies or compact part of galaxies" (Zwicky 1971). This second catalogue contains more than 3700 objects; thirty one of them turned out to be galaxies with an active nucleus (22 Seyfert 1, 7 Seyfert 2, 1 BL Lac object and one possible Seyfert); they are listed in table 1. Although this is a lower limit as a spectroscopic survey of all these galaxies has not been

Table 1
AGNs in the Zwicky's list of compact galaxies

		z			$\bar{M}_B$
IV Zw 1	0000+21	0.023	S1	Mark 334	-21.1
III Zw 2	0007+10	0.089	S1	PGC 0007+10	-23.3
Zw 0033+45	0033+45	0.048	S1		-22.0
Zw 0039+40	0039+40	0.073	S2	Mark 957	-22.7
IV Zw 29	0039+40	0.102	S1		-22.8
I Zw 1	0050+12	0.061	S1		-23.8
II Zw 1	0119-01	0.054	S1		-22.4
V Zw 85	0137+31	0.065	S2		-22.2
V Zw 317	0301+31	0.058	S1		-21.2
III Zw 55	0338-01	0.025	S2	NGC 1410	-20.6
II Zw 14	0430+05	0.033	S1	3C 120	-21.4
VII Zw 118	0702+64	0.079	S1		-23.3
I Zw 26	1122+54	0.020	S1	Mark 40	-20.3
I Zw 27	1129+53	0.027	S2	Mark 176	-21.4
VII Zw 490	1254+57	0.041	S1	Mark 231	-23.1
I Zw 171	1342+56	0.038	S?	Mark 273	-21.9
I Zw 81	1406+49	0.051	S1		-20.9
I Zw 92	1439+53	0.038	S2	Mark 477	-21.8
Zw 1518+59	1518+59	0.079	S1	SBS 1518+593	-22.4
I Zw 112	1524+41	0.008	S2	NGC 5929	-19.4
I Zw 121	1535+54	0.039	S1	Mark 486	-22.0
III Zw 77	1622+41	0.034	S1	Mark 699	-21.3
VII Zw 653	1636+85	0.063	S1		-21.6
I Zw 187	1727+50	0.055	BL		-21.6
VII Zw 742	1747+68	0.063	S1	Kaz 163	-21.4
VII Zw 838	1845+79	0.057	S1	3C 390.3	-22.3
II Zw 101	2105+03	0.026	S2		-21.2
II Zw 136	2130+09	0.061	S1	PG 2130+099	-23.2
II Zw 171	2209+18	0.070	S1	PG 2209+184	-22.3
II Zw 187	2301+22	0.040	S1	Mark 315	-22.1
III Zw 127	2359+03	0.026	S1	Mark 543	-21.2

undertaken, the yield (< 1%) is rather small. The large fraction of Seyfert 1 galaxies (and BL Lac objects) is not unexpected as these objects have a starlike nucleus.

Sargent (1970) has obtained spectrograms of 141 Zwicky compact galaxies; they were selected for their blue appearance on the Palomar Sky survey prints and small angular size; nine of them are Seyfert galaxies ($\sim$ 6 %); this rather high success rate is due to the bias toward blue, very compact objects. The Sargent's sample was selected from the 5 first Zwicky's lists which contain 1146 objects.

Since the time of Sargent's spectroscopic investigation, 13 additional Seyfert galaxies have been recognized in these 5 first lists, which shows that probably 2 % (22/1146) of all Zwicky galaxies are Seyfert.

- The Arakelian survey.

Arakelian (1975) has compiled a list of 621 high surface brightness galaxies (Thirty of them were already known to be Markarian galaxies and have not been assigned a new number). A very small number of Seyfert galaxies have been recognized in this sample : two Seyfert 2 (Akn 347 and 539) and three Seyfert 1 (Akn 120 and 564, and Mark 359).

- The Fairall survey.

1065 galaxies have been selected from the British SRC IIIaJ Sky survey on the basis of compact appearance or bright nucleus. All declination zones from the south celestial pole to $\delta = -20°$ have been covered, excluding fields very close to the plane of the Milky Way ($\sim$ 200 fields). The chief objective of this program was to discover Seyfert galaxies (Fairall 1984a and references therein). Five Seyfert 1 galaxies, 9 Seyfert 2 and 23 possible Seyfert have been found (Fairall 1984 b). Here, again, the fraction of Seyfert galaxies among the sample is very small ($\sim$ 1.5 % excluding the possible Seyfert). The small fraction of Seyfert 1 is rather surprising.

II. 4. Spectroscopic surveys of nearby galaxies.

Three major spectroscopic surveys of the nucleus of bright galaxies have been recently published :

- Heckman et al. (1980) have obtained spectrophotometric observations with an Image Dissector Scanner (IDS) mounted on the 2.1m telescope of the Kitt Peak National Observatory. The sample is an optically complete one consisting of all galaxies listed in the "Second Reference Catalog of Bright Galaxies" (de Vaucouleurs et al. 1976). Normal, metal rich HII regions, associated with young stars are found to occur in a large fraction of the nuclei of galaxies of late Hubble type (Sbc or later) and rarely in early-type galaxies (Heckman 1970 a). A class is defined of low ionization nuclear emission-line regions (Liners) which have optical spectra dominated by emission-lines from low ionization species. These Liners occur frequently in normal galaxies, particularly those of early Hubble-type. They are characterized by lines whose widths are similar to those in the narrow-line region of Seyfert galaxies (a few hundreds Km s^{-1}), but whose luminosity is generally much less. It was suggested that a continuity

exists between Seyfert galaxies and Liners (Heckman 1980 b). Liners are now
believed to be ionized by a power law continuum, like the Seyfert 2 galaxies,
although of a steeper index, rather than shock heated (see for instance
Binette 1985).

- Spectra of 139 field spiral and 50 galaxies with an IDS (Stauffer 1982 a). The
main conclusions of the preceding survey have been confirmed (Stauffer 1982b).

- Spectrophotometry has been obtained of the nuclei of an optically complete
sample of 93 spiral galaxies. The survey sample consists of all galaxies in the
Second Reference Catalog with integrated magnitude B <12.0, declination in
the range $-15° < \delta < +40°$, and type from S0/a to Scd. The observations were
carried out with an IDS. All the nuclei exhibit emission lines, either HII
regions or Liners; a sharp transition between the two occurs between Sbc and
Sc, with Liners predominent at earlier Hubble type. The Liner seems to be
present in every nucleus not containing an HII region of much higher luminosity,
which would render a Liner undetectable if present.

- Véron-Cetty and Véron (1986) have made a spectroscopic survey of all the
320 galaxies in the Revised Shapley-Ames catalog (Sandage and Tammann, 1981)
south of $\delta = +20°$, brighter than $M_B = -21.0$ and with a galactocentric radial
velocity $V_o < 3000$ Km s^{-1}. The observations have been carried out with an
IDS and a Boller and Chivens spectrograph attach to the Cassegrain focus of
the 1.5 m ESO telescope at La Silla. The dispersion was 171 Å mm^{-1} and the
resolution ~ 10 Å (FWHM). The [NII] $\lambda 6583$ emission line was detected when its
equivalent width was larger than ~ 1.5 Å. The emission-line spectra were
divided into 5 classes : the Seyfert 1 and Seyfert 2; the N Type in which
[NII] $\lambda 6583 > H\alpha$ and which are weak Seyfert-like galaxies (in most cases the
only visible lines are [NII], Hα and [SII] $\lambda\lambda$ 6715, 6731, therefore we cannot tell
between a Liner or a Seyfert 2 galaxy) ; the composite spectra for which both
a HII region and a Seyfert-like nebulosity are simultaneously present in the
aperture (NGC 1808 (Véron-Cetty and Véron 1985) is an example); and the HII
regions.
   We have reobserved a number of these galaxies around Hα with an higher
dispersion (60 Å mm^{-1}, with a better signal-to-noise ratio, using the ESO 3.6 m
telescope. These spectra allow the measurement of the line-width. We have
plotted on fig. 1 the intensity ratio Hα /[NII] λ 6583 versus the line-width
(uncorrected for the instrumental broadening). This figure clearly shows that
there exist two main types of spectra; the first one has narrow lines
(<300 Km s^{-1} FWHM) and strong Hα (H α/6583 >1.7); they correspond to the
H II regions ionized by hot stars. The second has broad lines (> 300 Km s^{-1})
and weak Hα (Hα/6583 <1.2). The few points in the intermediate region of the
diagram correspond to the composite spectra as shown by the different
profiles of Hα and [NII] lines in these cases.
   The table gives the distributions of the various spectral types for the
Hubble type of the galaxies from E to Sc. Altogether the sample contains 8
Seyfert 1 and 12 Seyfert 2 galaxies (6 % of all galaxies), which corresponds to
a volume density for Seyfert galaxies of 3×10^{-5} Mpc^{-3}), while that of
Seyfert 1 galaxies alone is $\sim 1.2 \times 10^{-5}$ Mpc^{-3}. But 103 (or more than 30 % of
all sample galaxies) have a Seyfert-like spectrum. The fraction of Seyfert-like
is largest for Hubble type Sa-Sab; it drops drastically for Sc galaxies. This is

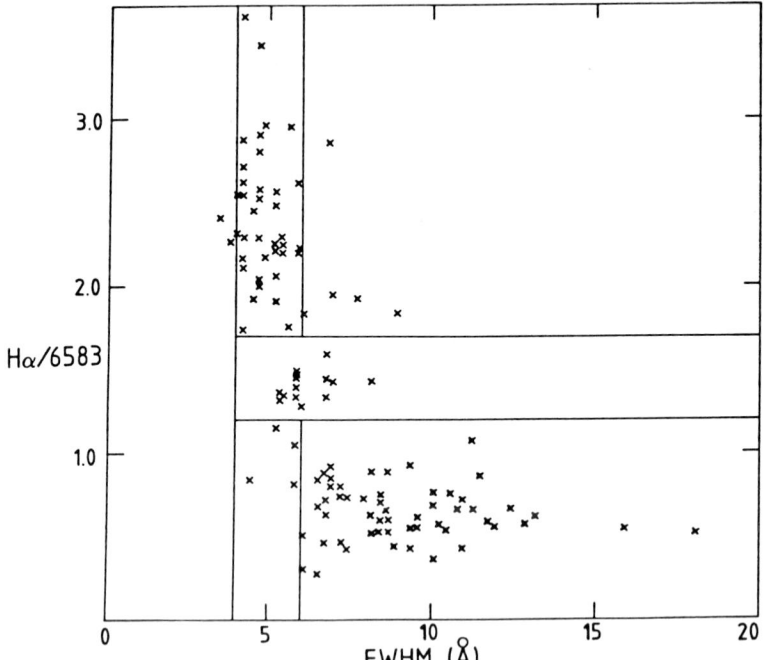

Fig. 1. Plot of the intensity ratio Hα /[NII] λ 6583 versus the width of the lines in the nucleus of a representative sample of Shapley-Ames galaxies. The points are not ramdomly distributed but are grouped into two main regions, corresponding to the HII regions ionized by hot stars, with Hα /6583 > 1.7 and linewidth < 6 Å (300 Km s^{-1}) and to the Seyfert-like nuclei with Hα /6583 < 1.2 and linewidth >6 Å. Most of objects in the intermediate regions have been shown to be composite with both an HII region and a Seyfert-like nebulosity simultaneously present.

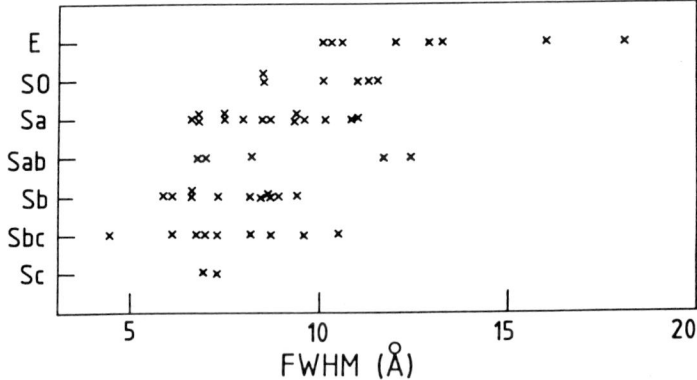

Fig. 2. Plot of the line width of the emission line in the nucleus of Seyfert-like galaxies versus their Hubble type. There is a trend for earlier type galaxies to have broader emission lines.

in good agreement with previous studies.

		S1	S2	N	C	HII	no em.
E + S0	76	0	0	21	0	0	55
Sa + Sab	56	3	5	26	4	1	17
Sb	53	1	3	11	3	17	18
Sbc	53	3	3	10	5	22	10
Sc	80	1	1	2	1	58	17

A new result which may turn out to be of some significance is that the line width of Seyfert-like galaxies tends to be larger for earlier Hubble types as shown by fig. 2 : the line-width for E galaxies is invariably larger than 500 Km s^{-1} (10 Å, FWHM); it is in general smaller than this value for all other types.

Sectroscopic studies of nearby, bright galaxies have shown that a large fraction of them display weak, Seyfert-like activity which would probably escape any other method.

III. X-RAY SURVEYS

The presence of dust in the nuclei of active galaxies can alter the way in which we perceive these objects. Dust can absorb the optical and ultraviolet continuum (reradiating this in the infrared), and block the soft X-rays. Objects with substantial amounts of dust absorption will not display blue colors; in extreme cases where massive amounts of dust lie between the object and the observer, the object may not appear to be "active" at optical wavelengths. An important effect of internal reddening is to diminish the apparent magnitude of the source beyond the survey limits. Current optical method of discovering active galaxies are biased against such objects. Hard X-ray surveys should mitigate this effect (Rudy 1984).

- Uhuru.

A survey of the entire sky has been made by the Uhuru (SAS A) X-ray satellite observatory. The fourth and final Uhuru catalogue lists positions and intensities (2-6 keV) for 339 sources brighter than $\sim 3.4 \times 10^{-11}$ ergs cm^{-2}s^{-1} (Forman et al. 1978).

- Ariel V.

The Ariel V (3A) survey covers 90% of the high galactic latitude sky ($|b| > 10°$) and contains 142 sources brighter than 3.0×10^{-11} ergs cm^{-2}s^{-1} (2-10 keV); twenty three are still unidentified (McHardy et al. 1981).

- The HEAO 1 A-2 hard X-ray survey.

The HEAO 1 experiment A-2 has performed a complete X-ray survey of the 8.2 steradians of the sky at $|b| > 20°$ down to a limiting sensitivity of 3.1×10^{-11} ergs cm^{-2} s^{-1} in the 2-10 keV band. Of the 85 detected sources (excluding the LMC and SMC sources), 17 have been identified with galactic objects, 61 with extragalactic objects, and 7 remain unidentified. The extragalactic objects subdivide almost equally between clusters of galaxies (31) and single galaxies (30). All sources with flux $\geq 4.0 \times 10^{-11}$ ergs cm^{-2} s^{-1} are identified (Piccinotti et al. 1982).

- The HEAO 1 A-1 survey.

This is an all-sky catalogue of the brightest X-ray sources in the energy range from 0.25 to 25 keV. Positions and intensities for 842 sources have been catalogued. The limiting flux corresponds to 3.3×10^{-12} ergs cm^{-2} s^{-1} in 2-6 keV, or 4.8×10^{-12} ergs cm^{-2} s^{-1} in 2-10 keV, both for a Crab-like spectrum; the catalogue is more than 90 % complete at a flux level of 2.9×10^{-11} ergs cm^{-2} s^{-1} (2-10 keV). 428 sources in the catalogue are still unidentified. (Wood et al. 1984).

- The HEAO 1 A-2 soft X-ray survey.

The HEAO A-2 low energy X-ray detectors have surveyed over 95 % of the sky in the spectral bands 0.18-0.44 and 0.44-2.8 keV, down to a typical sensitivity of 1.0×10^{-11} and 3.0×10^{-11} ergs cm^{-2} s^{-1} in each band respectively. 114 sources are catalogued; 32 remain unidentified. None of the previous surveys were sensitive to photon energies ≤ 1.5 keV, rendering them relatively insensitive to sources whose spectra could be characterized by steep power laws. Sources detected at 1 keV resemble the distribution of previous X-ray catalogs; sources detected at 0.25 keV include a heavier proportion of stellar identifications (Nugent et al. 1983).

A number of the X-ray sources listed in these catalogues have been identified with "active galaxies". They are listed in table 2, excluding, however, QSOs, BL Lac objects, radiogalaxies (NGC 5128) and one starburst galaxy (M 82). This table thus contains 47 non elliptical, Seyfert galaxies. A number of these objects have been detected through their X-ray emission, and some of them turn out to have narrow emission lines and look like Seyfert 2 galaxies (for instance NGC 2992, 5506 and 7582) (Ward et al. 1978). Lawrence et Elvis (1982) have suggested that these objects could be examples of Seyfert 1 nuclei in which the broad emission-line region is heavily obscured by dust. The observations of weak broad wings on the permitted lines of some of them (Shuder, 1980; Véron et al. 1980) show that this is indeed the case.

A few such galaxies have been studied in detail : 3A 0557-383 (Fairall et al. 1982), MCG-6.30.15 (Pineda et al. 1980), IC 4329 A (Disney, 1973), NGC 5506 (Griffiths et al. 1979), UGC 10683 B (Wilson et al. 1981), ESO 103-G55 (Philipps et al. 1979), NGC 7172 (Sharples et al. 1984) and NGC 7582 (Ward et al. 1978).

Kriss et al. (1980) have shown that Seyfert 2 galaxies are much fainter X-ray emitters than Seyfert 1. In table 2, out of the 47 objects, 38 are Seyfert 1 galaxies, and nine only are Seyfert 2. Of those, NGC 1068 is probably the only genuine Seyfert 2; it is also the galaxy with the smallest X-ray luminosity in the table. The others (NGC 2110, Mark 78, MCG-5-23-16,

Table 2

X-ray Seyfert galaxies

			z	V	B-V	U-B	Uhuru 4	3A	HEAO A1	HEAO A2H	HEAO A2S	A1 count
Mark 335	0003+19	S1	0.025	13.85	0.34	-0.75	x		x			0.0028
III Zw2	0007+10	S1	0.089	15.40	0.56	-0.83			x	x		0.0062
NGC 526a	0121-35	S1	0.018	14.60	0.97	0.09	x	x	x	x		0.0072
F9	0121-59	S1	0.045	13.25	0.20	-0.95	x	x	x	x		0.0071
GP X 2	0206+52	S1	0.049						x			0.0019
Mark 590	0212-00	S1	0.027	13.81	0.67	-0.55			x	x		0.0027
NGC 985	0232-09	S1	0.043	14.28	0.36	-0.95			x?			0.0036
NGC 1068	0240-00	S2	0.003	10.83	0.87	0.13			x			0.0032
4U 0241+61	0241+62	S1	0.044	16.4			x	x	x		x	0.0104
NGC 1566	0418-55	S1	0.004	13.17	0.76	-0.04	x					-
3C 120	0430+05	S1	0.033	15.05	0.67	-0.77	x	x	x	x		0.0058
NGC 2110	0549-07	S2	0.007	15.2				x	x			0.0072
MCG 8-11-11	0551+46	S1	0.020	14.62	0.75	-0.65	x	x	x			0.0078
3A 0557-383	0556-38	S1	0.034	14.5			x	x	x	x		0.0070
Mark 376	0710+45	S1	0.056	14.62	0.55	-0.58	x?	x?				0.0071
Mark 78	0737+65	S2	0.038	14.58	0.99	-0.31			x			0.0040
Mark 79	0738+49	S1	0.022	14.27	0.47	-0.78		x	x			0.0038
NGC 2992	0943-14	S1	0.007	13.78	1.06	0.40	x	x	x	x		0.0098
MCG-5-23-16	0945-30	S2	0.008	13.70	1.05	0.45	x	x	x			0.0084
NGC 3227	1020+20	S1	0.003	11.79	0.82	0.29			x	x		0.0087
NGC 3281	1029-34	S2	0.011	14.02	1.15	0.58			x			-
NGC 3783	1136-37	S1	0.009	13.43	0.56	-0.70	x	x	x	x		0.0064
NGC 4051	1200+44	S1	0.002	12.92	0.67	-0.12			x			0.0029
NGC 4151	1208+39	S1	0.003	11.85	0.71	-0.26	x	x	x	x		0.0079
Mark 205	1219+75	S1	0.070	15.24	0.40	-0.94			x			0.0032
Ton 1542	1229+20	S1	0.064	15.30	0.27	-1.02			x?			0.0048
NGC 4593	1237-05	S1	0.009	13.15	0.80	-0.19	x	x	x	x		0.0084
NGC 5033	1311+36	S1	0.003	12.03	0.93	0.37		x	x			0.0067
MCG-6.30.15	1333-34	S1	0.006	13.58	0.95	0.34			x	x		0.0085
IC 4329 A	1346-30	S1	0.014	14.44	0.96	0.51	x	x	x	x		0.0126
Mark 279	1351+69	S1	0.031	14.58	0.69	-0.45		x	x			0.0061
NGC 5506	1410-02	S2	0.007	14.38	0.87	0.14	x	x	x	x		0.0037
NGC 5548	1415+25	S1	0.017	13.73	0.62	-0.46	x	x	x	x		0.0127
Mark 290	1534+58	S1	0.029	14.96	0.60	-0.62			x?			0.0042
H 1613+06	1615+06	S1	0.038	15.66	0.89	-0.30					x	-
UGC 10683 B	1702-01	S1	0.031	15.55	1.29	0.23	x		x			0.0068
Mark 507	1748+68	S2	0.053	15.45	0.88	0.19			x?			0.0030
ESO 103-G35	1833-65	S1	0.013	14.53	0.98	0.44			x	x		0.0058
ESO 141-G55	1916-58	S1	0.037	13.64	0.19	-0.94	x	x	x	x		0.0053
NGC 6814	1939-10	S1	0.005	14.21	1.12	0.37		x	x			0.0077
Mark 509	2041-10	S1	0.035	13.	0.23	-0.93		x	x	x		0.0105
NGC 7172	2159-32	S2	0.008	13.61	1.07	0.59		x		x		-
NGC 7213	2206-47	S1	0.006	12.08	1.01	0.45		x	x	x		0.0065
NGC 7314	2233-26	S1	0.006	13.11	0.84	0.36		x	x	x		0.0031
NGC 7469	2300+08	S1	0.017	13.04	0.38	-0.72	x	x	x	x		0.0098
Mark 926	2302-08	S1	0.047	14.18	0.58	-0.75	x	x	x	x		0.0061
NGC 7582	2315-42	S2	0.005	13.57	0.87	0.17		x	x	x		0.0094

An intensity of 10^{-3} counts cm^{-2} s^{-1} in the last column, which is the limiting flux in the HEAO A1 catalog, corresponds to 3.3×10^{-12} ergs cm^{-2} s^{-1} in 2-6 keV, or 4.78×10^{-12} ergs cm^{-2} s^{-1} in 2-10 keV, both for a Crab-like spectrum.

NGC 3281, NGC 5506, Mark 507, NGC 7172 and NGC 7582) could be obscured Seyfert 1; some of them show obvious signs of obscuration (see for instance NGC 7172, Sharples et al. 1984).

Keel (1980) demonstrated a remarkable paucity of optically discovered edge-on Seyfert galaxies, i.e. those with axial ratio less than 0.5. Moreover, Cheng et al. (1983) have shown that the observed colours of Seyfert 1 nuclei are tightly correlated with the inclination of the surrounding galaxies, qualitatively accounting for the Keel's effect.

Lawrence and Elvis (1982) have defined a subset of the 2 A catalogue (Cooke et al. 1978), an earlier version of the Ariel V survey, to be a complete sample of extragalactic X-ray sources brighter than 3×10^{-11} ergs cm^{-2} s^{-1} (2-10 keV) at $|b| > 10°$. The identifications of this sample are complete for galaxies down to $\sim 16^{th}$ magnitude (Ward et al. 1978); the sample contains 16 active spiral galaxies. Seven of them have a positive U-B index, excluding their discovery in a Markarian-type survey; they could be considered as X-ray discovered active galaxies, although IC 4329 A had been discovered previously by chance to be a Seyfert 1 galaxy (Sandage 1978; Disney 1973); six of them have a b/a aspect ratio $\lesssim 0.5$, while none of the UV excess objects have such a small b/a. This confirms that the deficiency of optically discovered, edge-on Seyfert galaxies reported by Keel (1980) is indeed due to a lack of UV excess in these galaxies due to dust extinction in the plane of the galaxies. Two additional galaxies in this sample (MCG 8.11.11 and Mark 926) have been discovered thanks to their X-ray emission, although they have a strong UV excess; this is due to the fact that they are located in regions of the sky which had not yet been optically surveyed for active galaxies; one of them (Mark 926) has been since independently rediscovered as an UV excess galaxy.

The X-ray spectra of Seyfert 1 galaxies are generally consistent with a single power law of energy index 0.62 ± 0.15 over a wide range of energies (0.75-165 keV). In the spectrum of the most luminous Seyfert 1 galaxies ($L_x > 5 \times 10^{43}$ ergs s^{-1}), there is no apparent absorption from cold gas in the line of sight in excess of that expected from our own galaxy. The low-luminosity Seyfert 1 galaxies, especially those which show signs of optical obscuration (faint UV excess, high Balmer decrement, weak broad Balmer components, high inclination on the line of sight), as for instance NGC 2992, 5506, 7582 and MCG-5.23.16 (Lawrence and Elvis, 1982), possess X-ray spectral slopes indistinguishable from those of luminous Seyfert 1 galaxies, but display evidence of cold gas with column densities greater than 10^{22} cm^{-2} in the line of sight (Mushotzky et al. 1980; Mushotzky 1982; Rothschild et al. 1983; Petre et al. 1984; Maccacaro et al. 1982). Observations of galactic X-ray sources have shown that optical extinction due to dust and hydrogen column densities are proportional with $A_V = 4.5 \times 10^{-22} N_H$ mag., where N_H is the column density of hydrogen in atoms cm^{-2} (Reina and Tarenghi, 1973; Jenkins and Savage, 1974; Gorenstein, 1975).

However, two soft X-ray sources have been identified with AGNs :
- H 1613 + 06, with a Seyfert 1 galaxy (V = 15.7 , z = 0.038); its X-ray spectrum is steep, with an energy index $\alpha = 2.5 \pm 0.8$ (Pravdo et al. 1981).
- H 1814 + 63, with a quasar (V= 14.1, z = 0.297); its X-ray spectrum is a power-law with an energy index $\alpha = 1.31 \pm 0.30$ (Pravdo et al. 1984).

These two sources are similar to the Seyfert 1 galaxy NGC 4051 which has been observed using the imaging proportional counter (IPC) on the Einstein

Observatory (Giacconi et al. 1979); the X-ray spectrum of NGC 4051 has been found to be unusually soft compared with other Seyfert 1 galaxies; the best fit energy spectral index is 1.5 (Marshall et al. 1983).

On a time scale of years or less, the fluxes and spectra of the soft X-ray AGNs change; they may represent the high state of an AGN accretion disk (Pravdo et al. 1984).

Conclusions :

- Seyfert 2 galaxies are weak X-ray sources and cannot be efficiently discovered by X-ray surveys.
- There are probably two classes of Seyfert 2 galaxies : the genuine Seyfert 2 which are very faint X-ray emitters and the X-ray Seyfert 2 which are most probably Seyfert 1 galaxies with a much absorbed nucleus.
- Seyfert 1 galaxies are strong X-ray sources; their X-ray spectrum is a power-law; the dispersion of the spectral indices is small. Moreover, there is a rough correlation between the X-ray flux and the non-thermal optical flux. Therefore, any hard (> 5keV) X-ray survey is very efficient in discovering Seyfert 1 galaxies, including those which have either an intrinsically faint or an heavily reddened nucleus.
- As the result of a high column density (> 10^{22} cm-2) on the line of sight is to flatten the X-ray spectrum at energies below $\sim$ 5keV, and as high column density objects are also those which have a smaller UV excess, soft X-ray surveys are inefficient in finding those galaxies which are discriminated against by the current optical surveys based on UV excess.
- The available hard X-ray surveys show that may be as much as half the Seyfert 1 galaxies escape optical detection.

- The Einstein Observatory Medium Sensitivity Survey.

The Einstein Observatory Medium Sensitivity Survey is an X-ray survey of the high galactic latitude sky ($|b|$ > 20°) at sensitivities in the range 7×10^{-14} to 5×10^{-12} ergs cm-2 s-1 (0.3 - 3.5 keV), carried out with the Imaging Proportional Counter (IPC) aboard the Einstein Observatory. 338 IPC fields have been analyzed; only the central part (32 arcmin2) of each field of view was used. The total area covered is $\sim$ 90 square degrees.

The total number of serendipitous sources found is 112; of them 56 are quasars or Seyfert galaxies, the others being stars or clusters of galaxies. Of the 56, 17 are quasars (M_B <-24.0) and 39 Seyfert galaxies (M_B > -24.0). The redshifts of the Seyfert galaxies span the range 0.02 - 0.87, the medium redshift being z = 0.24. The faintest galaxies have an absolute magnitude M_B = -20.0 (Maccacaro et al. 1982; Stocke et al. 1983; Maccacaro et al. 1984; Gioia et al. 1984).

The X-ray detected AGN's are, for the most part, similar to optically or radio selected AGNs. However, about 10 % of them have reddish colors (B-V $\geq$ 0.8) and would not have been easily discovered by standard colour techniques. Most of these reddish objects are of low optical absolute luminosity (M_V > -23.3) which suggests that the measured colours (through a 7".2 diameter aperture) are dominated by the stellar continuum. Not all of them have detected broad Balmer lines, therefore some of them are probably Seyfert 2 or obscured Seyfert 1 (Gioia et al. 1984). However, because of the low energy band used (0.3-3.5 keV), this survey probably discriminates against such objects which may have a high column density on the line of sight.

A few other X-ray surveys of AGNs have been carried out with the Einstein Observatory. References can be found in Maccacaro (1984).

- The X-ray satellite ROSAT.

The X-ray satellite ROSAT (Trümper 1984), to be launched in 1987, will provide sensitivity similar to the Einstein Medium Sensitivity Survey, but over the whole sky, rather than 90 square degrees. The bandwidth will be 0.1-2 keV, therefore, again, the obscured Seyfert 1 will be discriminated against.

IV. THE INFRARED IRAS SURVEY

The Infrared Astronomical Satellite (IRAS) (Neugebauer et al. 1984) has detected many thousands of galaxies, principally at wavelengths of 60 and 100 microns. Most galaxies detected in flux-limited samples at these wavelengths have relatively steep infrared spectra with indices ~ -2.5 ($S_\nu \alpha \nu^\alpha$). This cold infrared emission ($\sim 30°K$) is believed to be produced in the disks of spiral and interacting galaxies.

"Warm " objects, selected to have spectral indices between 25 and 60 microns of $-1.25 < \alpha < -0.5$ and identified with a galaxy on the Palomar Sky Survey and ESO/SRC plates, not already known to be a Seyfert, have been published in IRAS Circular 11 (1984). Spectra were taken of 40 of the 54 objects in this list (de Grijp et al. 1985; see also Carter, 1984). Of those, 24 have reliably determined flat infrared spectra between either 60 and 25 µm or between 60 and 100 µm. Of these flat spectrum objects 17 (71 %) have Seyfert spectra (13 Seyfert 2 and 4 Seyfert 1) (the redshift range is 0.01-0.16). This should be compared with a Seyfert rate of ~ 5 % for Markarian objects. Hence, a flat infrared spectrum is a remarkably efficient predictor of the Seyfert phenomenon. Extrapolation of this result to the all-sky survey indicates that the IRAS data base contains several thousand Seyferts. Assuming a Hubble constant of 50 Km s^{-1} Mpc^{-1}, the space density of the infrared Seyfert is $\sim 5 \times 10^{-5}$ Mpc^{-3}.

There is no significant difference in the redshift distribution of the infrared and Markarian sample; both have median redshifts of 0.03. Most (75 %) of the infrared Seyfert are Seyfert 2 which have a relatively small UV excess and are therefore discriminated against by the Markarian technique.

The IRAS point source catalogue as presently available on magnetic tape has been cross-referenced with a number of catalogue of celestial objects, including the "catalogue of quasars and active nuclei" (Véron-Cetty and Véron, 1984) which contains 2251 quasars and 554 Seyfert and related galaxies. Dennefeld and Véron-Cetty (1985) have extracted all 280 IRAS sources identified with an object in the quasar catalogue. The histogram of the V magnitude of these objects is strongly peaked at about V = 14.5, with a very broad tail extending up to V = 20. (Figure 3a). Of the 41 objects with V $\geqslant$ 16.0, 27 are easily shown to be wrongly identified, the correct identification being the bright nearby galaxy (most of them are quasars found near bright NGC galaxies by Arp). Four objects fainter than V = 17.0 are left. With a mean

SURVEYS OF LOCAL AGNs

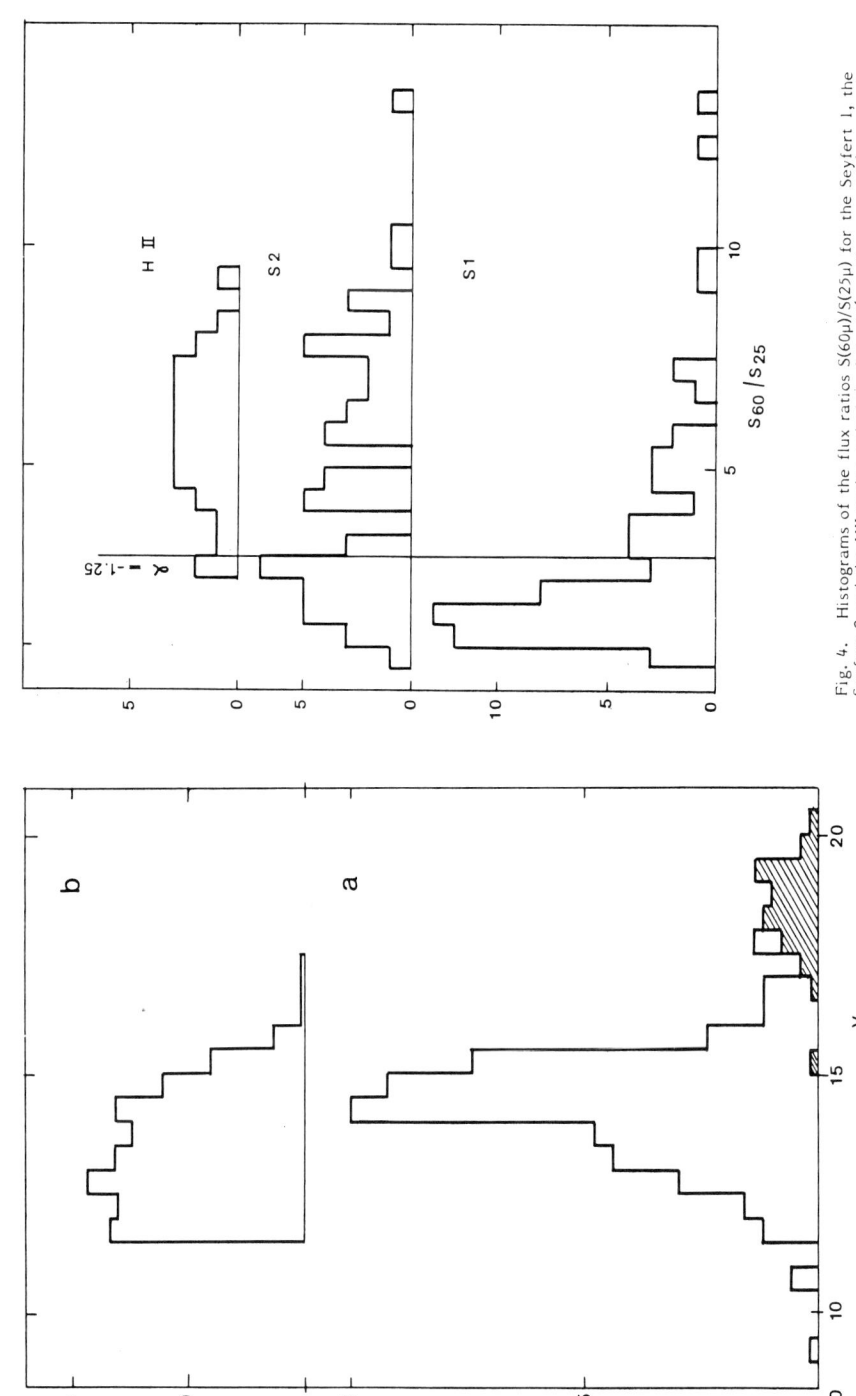

Fig. 4. Histograms of the flux ratios $S(60\mu)/S(25\mu)$ for the Seyfert 1, the Seyfert 2 and the HII region galaxies of the Véron-Cetty and Véron catalogue measured by IRAS. Excluding the objects with $\alpha > -1.25$ ($S(60)/S(25) > 3.0$) (as de Grijp et al. did) eliminates most of the HII regions but also a large fraction of the Seyfert galaxies.

Fig. 3. a) Histogram of the V magnitude of the galaxies in the Véron-Cetty and Véron catalogue identified with an IRAS source. The hatched area corresponds to objects which are most likely incorrect identifications.
b) Fraction of the galaxies in the same catalogue identified with an IRAS sources versus the apparent V magnitude of these galaxies.

surface density of IRAS point sources equal to $\sim$ 1.0 (Rowan-Robinson et al. 1984) and a search area of 1'.5 in radius, the expected number of chance coincidence in the quasar catalogue is $\sim$5. It is therefore safe to say that there is no object fainter than V = 17.0 in the quasar catalogue detected in the infrared by the IRAS satellite. Figure 3b gives the percentage of quasars or Seyfert-like galaxies detected in the infrared versus the V magnitude; it shows that more than 80 % of all objects brighter than V = 13.5 are detected while less than 5 % of the objects fainter than V = 15.0 are. (However, two of the brightest Seyfert 1 nuclei known, namely NGC 4151 (V = 11.85) and F9 (V = 13.25), are not detected). This suggests a rather good correlation between the optical and infrared luminosities for active galactic nuclei and means that very few faint quasars will be found among the IRAS sources.

The comparison between the histograms of the spectral indices between 60 and 25 μm for the Seyfert 1, Seyfert 2 and starburst (HII regions) galaxies (fig. 4) shows that the limit chosen by de Grijp et al. (1985), $\alpha(60, 25) = 1.25$, is very effective in eliminating the starburst galaxies, but at the price of losing more than one third of the Seyfert 1 and about two third of the Seyfert 2. These number should certainly, however, be treated with caution as the Véron-Cetty and Véron catalogue is a compilation of known objects and is not a complete sample in any sense. Moreover, it appears that there is a correlation, for Seyfert 1 galaxies, between the U-B color index and the $\alpha(60,25)$ spectral index in the sense that the object with a flat infrared spectra have a large UV excess; therefore the Seyfert 1 galaxies missed by the de Grijp et al. method are precisely those which have no UV excess, being reddened by dust and which escape detection by the usual optical methods. The Seyfert 1 galaxies detected in the infrared are the same as those detected by the Markarian method.

The de Grijp et al. method will allow the detection of Seyfert galaxies in large number, but by itself it will not produce a complete sample. The use of two-colour diagrams (such as $\alpha(60,100\mu)$, $\alpha(25-60\mu)$ for instance) may help separating starburst from Seyfert galaxies.

Thermal dust emission is the principal source of the infrared flux in Seyfert 2 galaxies (Rieke 1978; Rieke and Lebofsky, 1979); the IRAS survey may contain a large number of them as shown by de Grijp et al. who have found 13 Seyfert 2, for 4 Seyfert 1, a proportion similar to that of Wasilewski (1983) (see paragraph II.2).

V. CONCLUSION

Numerous attempts have been made to built either the luminosity function of galaxies which have a Seyfert nucleus, using the integrated magnitude of the galaxies (see for instance Meurs and Wilson, 1984) or the luminosity function of the nuclei of Seyfert 1 galaxies (or miniquasars) (the most recent is by Cheng et al. 1985). This can be done only if complete samples are available.

This review shows that :
- UV excess surveys are efficient in finding Seyfert 1 galaxies, but they are biased against low luminosity nuclei and nuclei reddened by dust.
- hard (E > 5keV) X-ray surveys allow the detection of low luminosity and reddened nuclei.
- soft X-ray surveys may contain a population of soft X-ray Seyfert 1

galaxies.
- the IRAS survey most probably detect the reddened Seyfert 1, but it may not be easy to isolate them from the much more numerous starburst galaxies.
- spectroscopic survey of complete sample of galaxies may be the best way to find the Seyfert 1 nuclei of very low luminosity.

- Seyfert 2 galaxies are best detected either by moderate dispersion (~ 400 Å mm^{-1}) objective prism survey such as the Wasilewski (1983) one or by the IRAS survey.
- AGNs of very low activity can be found only through systematic spectroscopic observations of complete sample of galaxies.

All methods have their biases; they should all be combined together if one want to build meaningful luminosity functions needed to understand the nature and evolution of these objects.

REFERENCES

Arakelian, M.A.: 1975, Soobshsh. Byurakansk. Obs. 47, 3.
Arakelian, M.A., Dibai, E.A., and Esipov, V.F. : 1976, Astrophysics 12, 456.
Binette, L.: 1985, Astron.Astrophys. 143, 334.
Blanco, V.M.: 1974, Publ. astr. soc. Pacific 86, 841.
Bohuski, T.J., Fairall, A.P., and Weedman, D.W.:1978, Astrophys.J. 221, 776.
Carter, D.: 1984, Astronomy Express 1, 61.
Cheng, F.Z., Danese, L. and de Zotti, G.: 1983, Mon. Not.roy.astr.Soc. 204, 13P.
Cheng, F.Z., Danese, L., de Zotti, G., and Franceschini, A.: 1985, Mon.Not.roy. astr. Soc. 212, 857.
Cooke, B.A., Ricketts, M.J., Maccacaro, T., Pye, J.P., Elvis, M., Watson, M.G., Griffiths, R.E., Pounds, K.A., McHardy, I., Maccagni, D., Seward, F.D., Page, C.G., and Turner, M.J.L.: 1978, Mon.Not.roy.astr. Soc. 182, 489.
Dennefeld, M., and Véron-Cetty, M.P.,: 1985, in preparation.
Dibai, E.A., Doroshenko, V.T., and Terebizh, V.Y.: 1976, Astrophysics 12, 459.
Disney, M.J.: 1973, Astrophys.J. Letters 181, L55.
Fairall, A.P.: 1984a , Mon.Not.roy.astr.Soc. 210, 69.
Fairall, A.P.: 1984b , in "Astronomy with Schmidt telescopes", IAU Coll. 78, Cappaccioli, M. ed. p. 397.
Fairall, A.P.: 1982, Mon.Not.roy.astr. soc. 198, 13 P.
Forman, W., Jones, C., Cominsky, L., Julien, P., Murray, S., Peters, G., Tananbaum, H., and Giacconi, R.: 1978, Astrophys.J. Suppl. 38, 357.
Giacconi, R., Branduardi, G., Briel, U., Epstein, A., Fabricant, D., Feigelson, E., Forman, W., Gorenstein, P., Grindlay, J., Gursky, H., Harnden, F.R., Henry, J.P., Jones, C., Kellogg, C., Koch, D., Murray, S., Schreier, E., Seward, F., Tananbaum, H., Topka, K., van Speybroeck, L., Holt, S.S., Becker, R.H., Boldt, E.A., Serlemitsos, P.J., Clark, G., Canizares, C., Markert, T., Novick, R., Helfand, D., and Long, K.: 1979, Astrophys. 230, 540.
Gioia, I.M., Maccacaro, T., Schild, R.E., Stocke, J.T., Liebert, J.W., Danziger, I.J., Kunth, D., and Lub, J.: 1984, Astrophys.J. 283, 495.
Gorenstein, P.: 1975, Astrophys. J. 198, 95.

Griffiths, R.E., Doxsey, E.R., Johnston, M.D., Schwartz, D.A., Schwarz, J., and Blades, J.C. : 1979, Astrophys.J. Letters 230, L 21.
de Grijp, M.H.K., Miley, G.K., Lub, J., and de Jong, T. : 1985, Nature 314, 240.
Heckman, T.M. : 1980a, Astron. Astrophys. 87, 142.
Heckman, T.M. : 1980b, Astron. Astrophys. 87, 152.
Heckman, T.M., Balick, B., and Crane, P.C. : 1980, Astron. Astrophys. Suppl. 40, 295.
Huchra, J.P.: 1977, Astrophys. J. Suppl. 35, 171.
Huchra, J.P., and Sargent, W.L.W. : 1973, Astrophys. J. 186, 433.
IRAS Circ. N° 11 : 1984, Astron. Astrophys. 134, C5.
Jenkins, E.B., and Savage, B.D. : 1974, Astrophys. J. 187, 243.
Kazarian, M.A. : 1979a, Astrophysics 15, 1.
Kazarian, M.A. : 1979b, Astrophysics 15, 117.
Kazarian, M.A., and Kazarian, E.S. : 1980, Astrophysics 16, 7.
_____ 1983a, Astrophysics 18, 285.
_____ 1983b, Astrophysics 19, 119.
Keel, W.C. : 1980, Astron. J. 85, 198.
Keel, W.C. : 1983a, Astrophys. J. 269, 466.
Keel, W.C. : 1983b, Astrophys.J. Suppl. 52, 229.
Kriss, G.A., Canizares, C.R., and Ricker, G.R.: 1980, Astrophys. J. 242, 492.
Lawrence, A., and Elvis, M.: 1982, Astrophys.J. 256, 410.
MacAlpine, G.M., Smith, S.B., and Lewis, D.W.: 1977a, Astrophys. J. Suppl. 34, 95.
MacAlpine, G.M., Smith, S.B., and Lewis, D.W. 1977b, Astrophys.J.Suppl. 35, 197.
MacAlpine, G.M., Lewis, D.W., and Smith, S.B. : 1977c, Astrophys.J. Suppl. 35, 203.
MacAlpine, G.M., and Lewis, D.W.: 1978, Astrophys.J.Suppl. 36, 587.
MacAlpine, G.M., and Williams, G.A.: 1981, Astrophys.J.Suppl. 45, 113.
Maccacaro, T.: 1984, MPE Report 184, Brinkman, W., and Trümper, J. eds.p.63.
Maccacaro, T. Feigelson, E.D., Fener, M., Giacconi, R., Gioia, I.M., Griffiths, R.E., Murray, S.S., Zamorani, G., Stocke, J., and Liebert, J.: 1982, Astrophys. J. 253, 504.
Maccacaro, T., Perola, G.C., and Elvis, M.: 1982, Astrophys. J. 257, 47.
Maccacaro, T., Gioia , I.M., and Stocke, J.T.: 1984, Astrophys. J. 283, 486.
Markarian, B.E.: 1967, Astrophysics 3, 55.
Markarian, B.E., and Stepanian, D.A.: 1983, Astrophysics 19, 354.
Markarian, B.E., and Stepanian, D.A.: 1984, Astrophysics 20, 10.
Markarian, B.E., Lipovetski, V.A., and Stepanian, D.A.:1981, Astrophysics 17,321.
_____ 1983, Astrophysics 19,14.
_____ 1984, Astrofizika 21, 35.
Marshall, F.E., Holt, S.S., Mushotzky, R.F., and Becker, R.H.: 1983, Astrophys. J. Letters 269, L31.
McHardy, I.M., Lawrence, A. , Pye, J.P., and Pounds, K.A.: 1981, Mon.Not. roy. astr. Soc. 197, 893.
Meurs, E.J.A., and Wilson, A.S.: 1984, Astron. Astrophys. 136, 206.
Mushotzky, R.F.: 1982, Astrophys. J. 256, 92.
Mushotzky, R.F., Marshall, F.E., Boldt, E.A., Holt, S.S., and Serlemitsos, P.J.: 1980, Astrophys. J. 235, 377.
Neugebauer, G., Habing, H.J., van Duinen, R., Aumann, H.H., Baud, B., Beichman, C.A., Beintema, D.A., Boggess, N., Clegg, P.E., de Jong, T.,

Emerson, J.P., Gautier, T.N., Gillett, F.C., Harris, S., Hauser, M.G., Houck, J.R., Jennings, R.E., Low, F.J., Marsden, P.L., Miley, G., Olnon, F.M., Pottasch, S.R., Raimond, E., Rowan-Robinson, M., Soifer, B.T., Walker, R.G., Wesselius, P.R., and Young, E.: 1984, Astrophys.J. Letters 278, L1.
Nugent, J.J., Jensen, K.A., Nousek, J.A., Garmire, G.P., Mason, K.D., Walter, F.M., Bowyer, C.S., Stern, R.A., and Riegler, G.R.: 1983, Astrophys. J. Suppl. 51, 1.
Osterbrock, D.E.: 1984, Astrophys. J. Letters 280, L 43.
Osterbrock, D.E., and Miller, J.S.: 1975, Astrophys. J. 197, 535.
Pesch, P., and Sanduleak, N.: 1983, Astrophys. J. Suppl. 51, 171.
Petre, R., Mushotzky, R.F., Krolik, J.H., and Holt, S.S.: 1984, Astrophys.J. 280, 499.
Philipps, M.M., Feldman, F.R., Marshall, F.E., and Wamsteker, W.: 1979, Astron. Astrophys. 76, L 14.
Piccinotti, G.,, Mushotzky, R.F., Boldt, E.A., Holt, S.S., Marshall, F.E., Serlemitsos, P.J., and Shafer, R.A.: 1982, Astrophys.J. 253, 485.
Pineda, F.J., Delvaille, J.P., Grindlay, J.E., and Schnopper, H.W.: 1980, Astrophys. J. 237, 414.
Pravdo, S.H., Nugent, J.J., Nousek, J.A., Jensen, K., Wilson, A.S., and Becker, R.H.: 1981, Astrophys.J. 251, 501.
Pravdo, S.H., and Marshall, F.E.: 1984, Astrophys. J. 281, 570.
Reina, C. and Tarenghi, M.:1973, Astron. Astrophys. 26, 257.
Rieke, G.H.: 1978, Astrophys. J. 226, 550.
Rieke, G.H., and Lebofsky, M.J.: 1979, Ann.Rev. Astron. Astrophys. 17, 477.
Rothschild, R.E., Mushotzky, R.F., Baity, W.A., Gruber, D.E., Matteson, J.L., and Peterson, L.E.:1983, Astrophys.J. 269, 423.
Rowan-Robinson, M., Clegg, P.E., Beichman, C.A., Neugebauer, G., Soifer, B.T., Aumann, H.H., Beintema, D.A., Boggess, N., Emerson, J.P., Gautier, T.N., Gillett, F.C., Hauser, M.G., Houck, J.R., Low, F.J., and Walker, R.G.: 1984, Astrophys. J. Letters 278, L 7.
Rudy, R.J.: 1984, Astrophys. J. 284, 33.
Sandage, A.: 1978, Astron. J. 83, 904.
Sandage, A., and Tammann, G.A.: 1981, "A Revised Shapley-Ames Catalog of Bright Galaxies", Carnegie Institution of Washington Publication 635, Washington D.C.
Sargent, W.L.W.: 1970, Astrophys.J. 160, 405.
Seyfert, C.K.: 1943, Astrophys. J. 97, 28.
Sharples, R.M., Longmore, A.J., Hawarden, T.G., and Carter, D.: 1984, Mon. Not.roy.astr.soc. 208, 15.
Shuder, J.M.: 1980, Astrophys. J. 240, 32.
Smith, M.G.: 1975, Astrophys.J. 202, 591.
Smith, M.G., Aguirre, C., and Zemelman, M.: 1976, Astrophys. J.Suppl. 32, 217.
Stauffer, J.R.: 1982a, Astrophys.J. Suppl. 50, 517.
Stauffer, J.R.: 1982b, Astrophys. J. 262, 66.
Stocke, J.T., Liebert, J., Gioia, I.M., Griffiths, R.E., Maccacaro, T., Danziger, I.J., Kunth, D., and Lub, J.: 1983, Astrophys.J. 273, 458.
Trümper, J.: 1984, MPE Report 184, Brinkman, W., and Trümper, J.eds.p.254.
de Vaucouleurs, G., de Vaucouleurs, A., and Corwin, H.G.: 1976, "The Second Reference Catalog of Bright Galaxies". Austin : University of Texas.
Véron, P.: 1978, Nature, 272, 430.

Véron, P., Lindblad, P.O., Zuiderwijk, E.J., Véron, M-P., and Adam, G.: 1980, Astron. Astrophys. 87, 245.
Véron-Cetty, M-P., and Véron, P.: 1984, ESO Scientific Report N° 1.
    ―――――――――――――――――――――― 1985, Astron. Astrophys. 145, 425.
    ―――――――――――――――――――――― 1986, in preparation.
Ward, M.J., Wilson, A.S., Penston, M.V., Elvis, M., Maccacaro, T. and Tritton, K.P.: 1978, Astrophys.J. 223, 788.
Wasilewski, A.J.: 1983, Astrophys. J. 272, 68.
Wilson, A.S., Wood, K., Ward, M.J., Griffiths, R.E., and Mushotzky, R.E.: 1981, Astron. J. 86, 1289.
Wood, K.S., Meekins, J.F., Yentis, D.J., Smathers, H.W., McNutt, D.P., Bleach, R.D., Byram, E.T., Chubb, T.A., Friedman, H., and Meidav, M.: 1984, Astrophys.J. Suppl. 56, 507.
Zwicky, F. : 1971, "Catalogue of selected galaxies and of post-eruptive galaxies", Publ. by F. Zwicky, CH-3073 Guemligen, Switzerland.

DISCUSSION

FOSBURY: How much of the line-width in the nuclei of "composite" spectrum galaxies could be due to the normal rotation of the ISM?

VERON: One well studied case of composite spectrum is NGC 1808. Slit spectra obtained by good seeing show the Seyfert-like component of the emission lines to be unresolved (<1".5) although a few hundred kilometers in widths. The HII region is quite extended and rotate as expected with the galaxy.

PISMIS: In your table listing the frequency of activity in spirals you have made no distinction between barred and non-barred spirals. How is the situation with regard to barred and non-barred ones?

VERON: We have not yet made a detailed analysis of the difference of barred and non- barred spirals with regards to the occurence of Seyfert-like activity, but, as far as we can tell, there are no gross differeces.

FILIPPENKO: Regarding Fosbury's question, I have found from any long-slit spectra that rotation does indeed often contribute to the line widths measured in aperture spectra. This is especially true in early-type galaxies; there are total velocity shifts of up to 300-400 km/s over the central 2"... A Particularly striking example is NGC 5005 at a position angle of 70°.

VERON: We have not obtained many slit spectra of these emission line galaxies; however in one case, namely NGC 5728 (SBb), the Seyfert 2 nebulosity is about 10 arcsec in extend; the velocities of the different parts of this nebulosity cover a range of over 500 km/s, but this cannot be interpreted as a rotation; in fact the largest velocities are observed right on the nucleus. This does not enclude that rotation may be the main widening agent of the lines in many of these objects, but it is certainly not always the case.

DISCOVERY OF NARROW AND VARIABLE LINES IN THE ULTRAVIOLET SPECTRUM OF THE SEYFERT GALAXY NGC 4151, AND AN OUTLINE OF OUR PREVIOUS RESULTS

M.H. Ulrich[1], A. Altamore[2], A. Boksenberg[3], G.E. Bromage[4], J. Clavel[5], A. Elvius[6], M.V. Penston[3], G.C. Perola[2] and M.A.J. Snijders[3].

[1] European Southern Observatory, 8046 Garching bei München, Federal Republic of Germany
[2] Istituto Astronomico dell'Università, Via Lancisi 29, 00161 Roma, Italy
[3] Royal Greenwich Observatory, Herstmonceux Castle, Hailsham, East Sussex BN27 1RP, U.K.
[4] Astrophysics Group, Rutherford and Appleton Laboratory, Chilton, Didcot, Oxfordshire OX11 OQX, U.K.
[5] Observatoire de Meudon, 92190 Meudon, France, and ESA Villafranca Tracking Station, Apdo 54065, Madrid, Spain
[6] Stockholm Observatory, 13300 Saltsjöbaden, Sweden

ABSTRACT. Extensive observations of NGC 4151 when the nucleus is at a minimum have led to the discovery of two emission lines which have a full width at half maximum of less than 7 and 16 Å respectively (after correction for the instrumental profile) and which vary in intensity by a factor of 3 in less than 10 days. These lines appear on either side of the CIVλ1550 line at λ_{rest} = 1518.5, 1594.4 Å.

Regardless of their identifications these lines are too narrow to be emitted by the whole broad line region. They must come instead from two localized regions which have a special excitation mechanism.

No identifications for these lines, for example with transitions excited by fluorescence, have been found. An attractive - but speculative - possibility is that the regions emitting them are associated with a two-sided jet reminiscent of SS 433.

1. PROLOGUE

I would like to present the discovery of narrow, variable lines in the ultraviolet spectrum of the nearby Seyfert galaxy NGC 4151, and set this discovery in the context of our previous findings on the broad emission line region in this active nucleus.

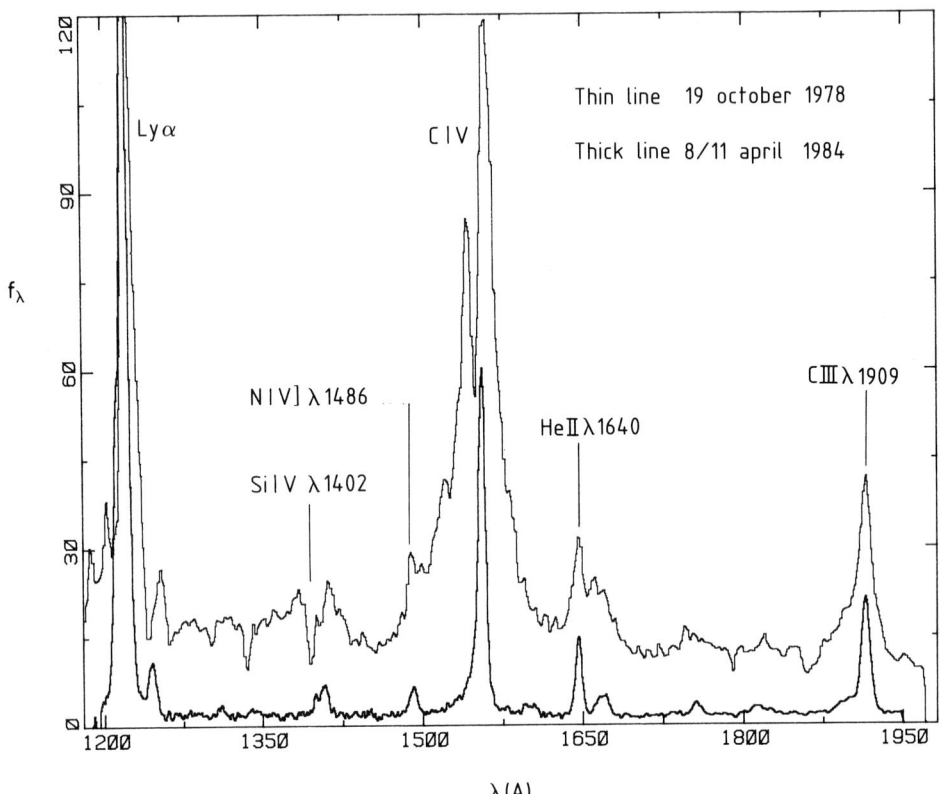

Figure 1: Extreme of the variations observed so far in the ultraviolet spectrum of NGC 4151. Thin line: maximum of 19 October 1978. Thick line: minimum of 8/11 April 1984. The absorption lines so prominent when the nucleus is brigt are blue-shifted by 800 km s^{-1} in the rest frame of NGC 4151. Ordinates in 10^{-14} erg cm^{-2} s^{-1} Å^{-1}.

The results presented here are part of the work done in a collaboration of several European astronomers, which I organized in 1978.

We have observed NGC 4151 on ~80 different days with the satellite IUE. Each observation usually consists of taking 2 well exposed spectra at low resolution (~ 6 Å) in each of the wavelength ranges of IUE: 1200-1950 Å and 2200-3100 Å. The reproducibility between well exposed spectra is about 5%.

One of our aims is to determine the distribution of the broad line gas and its velocity field (inflow, outflow, rotation or parabolic motions), since the gas kinematics provides a clue to the mass of the central object. In active nuclei matter is ejected under the form of radio sources, radio jets, and ionized matter causing broad absorption lines. And in the "classical black hole scenario" matter is also being accreted. Thus, there must be some sort of circulation of matter and replenishment which we wish to investigate.

Our study can easily be divided into two periods, 1978-1980 and 1981-1984, which correspond to two regimes of the nucleus of NGC 4151: Bright and active in the first period, while in the second period the nucleus was found in a low and nearly constant state for several consecutive months. In the second period we succeeded in improving our observing procedure by organizing long (up to 2.5 months) series of observations with the appropriate time resolution (5-8 days).

2. PERIOD 1978-1980: UNRAVELLING THE STRUCTURE OF THE BROAD LINE REGION

2.1. Results

a) The variations of the intensity and of the profiles of the emission lines observed in 1978-1980 indicate that the emissivity of the dense gas varies in response to the intensity variations of the ultraviolet continuum (see Figure A1 in the Appendix). The difference between the light curves of I(CIV) and of the continuum between May 3 and June 1 1979 provides evidence for a time delay of 10 to 20 days caused by the light travel time between the photoionizing continuum source and the line emitting gas, and between the different points in the dense nebula and the observer (Ulrich et al., 1984). It is important to confirm the time delay as it gives a lower limit to the size of the broad line region (in contrast to the time scales of the variations which give upper limits).

b) We have found no evidence that outflowing motions dominate the velocity field of the broad line gas and consider parabolic or chaotic motions likely. Some gas, however, is outflowing as evidenced by the relatively strong blue wing of the CIV emission line and by the blueshifted absorption lines.

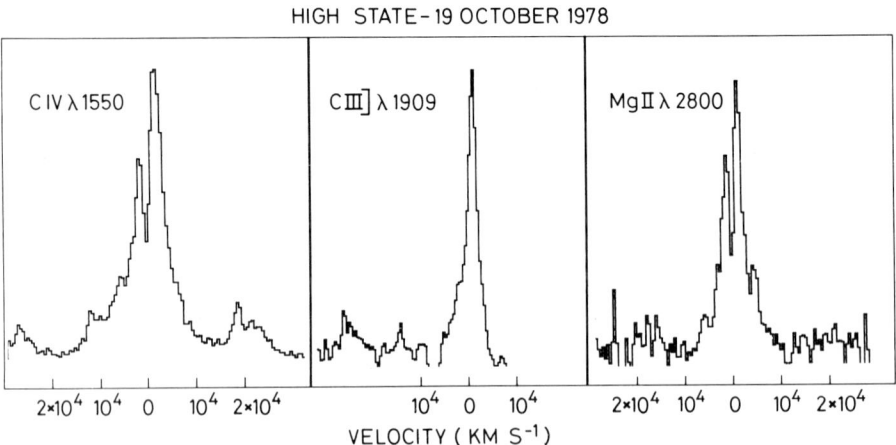

Figure 2: Differences between line profiles when the nucleus is bright (Ulrich et al., 1984). Regarding the difference in the wings of the CIVλ1550 and MgIIλ2800 lines it must be pointed out that in the presence of a radial gradient, clouds of identical characteristics but different radial distances can produce lines with different profiles.

c) The time scale of the variations of the broad component of the CIV line together with its width gives a value of the mass of the central object of $\sim 5 \times 10^8$ $M_\odot$. If the assumptions underlying this calculation are correct, the hypothetical central black hole emits much below the Eddington limit.

d) The emission lines emitted under different physical conditions have different profiles, in particular CIVλ1550 is broader than MgIIλ2800. These differences appear to be larger than the uncertainties caused by blends of weak FeII lines near MgIIλ2800 and by the differences in signal to noise ratio between the lines. In particular at high states the CIV line shows a prominent blue wing which has no counterpart in MgIIλ2800 (Figure 2).

e) The time variations of the CIV line intensity and profiles show that the gas with the highest velocity (the one emitting the wings of the line) responds first to the variations of the continuum intensity, consistent with and suggesting a radial gradient in velocity with the highest velocities being closer to the center and decreasing outward.

2.2. Significance of the Differences Between Line Profiles

Excellent theoretical calculations of line intensities in quasars and Seyferts have been published by several authors. The intensities and the line ratios are usually expressed as a function of the ionizing parameter Γ_i defined by $\Gamma_i = L(\Pi R^2 N_0)^{-1}$ with L the ionizing flux of the central source, R the distance of the cloud from the continuum source and N_0 the nuclear density. It is interesting to consider these calculations in conjunction with the differences between the profiles of the CIVλ1550 and MgIIλ2800 lines. Figure 3 from Kwan (1984) gives the ratio I(CIV)/I(MgII) for a range of N_0 and Γ_i which are relevant to the case of NGC 4151. For N_0 and Γ_i constant the ratio of the two lines is constant and thus the two lines have the same profiles. An increase of Γ_i however produces a larger increase in CIV than in MgII. From this it can easily be deduced that, for N_0 constant, the wings of the CIVλ1550 line are emitted by high velocity clouds which have a relatively large value of Γ_i. They are therefore located closer to the photoionization source than are the clouds contributing to the more central parts of the CIVλ1550 and MgIIλ2800 line profiles.

Consequently, in the presence of a radial gradient of velocity, lines with different profiles can be produced by clouds of identical characteristics but located at different radii.

Figure 3 also shows that a change in the central continuum source intensity L causes variations of smaller amplitude in I(MgII) than in I(CIV). There is a second effect which produces smaller amplitude variations in I(MgII) than in I(CIV): Since the ratio I(CIV)/I(MgII) decreases from the center outward then, very roughly, the MgII line is on average emitted further out than is the CIV line, and therefore the

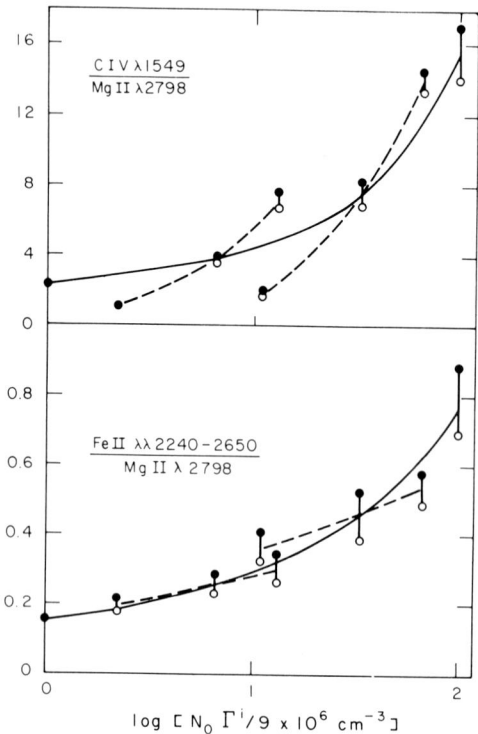

Figure 3: Correlations of line ratios with $N_o \Gamma_i$ from Figure 6 in Kwan, 1984. Note in particular that the line ratio CIV/MgII increases when $N_o \Gamma_i$ increases. For explanation of symbols see Kwan, 1984.

light travel time effects between the central continuum source and the different parts of the broad line region cause the time scale of the variations of I(MgII) to be larger than that of I(CIV).

2.3.

After our first set of observations, 1978-1980, it became clear that the best way to proceed in order to confirm and extend our results was to observe NGC 4151 for long periods of time with a time resolution of 5 to 10 days.

The most interesting episodes to observe are the times immediately following a minimum. At a minimum the broad components of the lines are very weak, probably because of an insufficient ionizing flux. During the rise which follows a minimum one can observe the evolution of the profile and of the intensity of the lines - in particular CIVλ1550 - as the burst of ionizing continuum spreads throughout the nebula. A blue or red wing in the CIV line just after a minimum would reveal an outflow or inflow of matter and a particularly broad component would indicate that the gas with the largest velocities is the first illuminated and is therefore in the innermost parts of the broad line region.

It is especially important to determine the characteristic time scales of the components of different widths as this gives the relationship velocity vs radial distance R and consequently the mass of the central object M (unless there is a significant outflow component in the observed velocities):

$M \sim R V^2/G$ if motions are circular,

$M \sim 2 R V^2/G$ for free-fall motions.

In my opinion the best results on the gas kinematics will be obtained from the CIV line for three reasons: (i) it is the strongest emission line, (ii) it is a strong function of Γ_i and (iii) it is relatively free of blends with weak lines.

3. 1981-1984: DISCOVERY OF THE NARROW VARIABLE EMISSION LINES L_1 AND L_2

The observations performed between March 1981 and June 1984 were aimed at improving our determination of the velocity field and distribution of the gas. We have not made the expected progress on these points because, unexpectedly, during that period the nucleus of NGC 4151 was most often found in a low state - a state which we had previously encountered only once in the years 1978-1980, on 1980 April 21. The 1981-1984 observations provide excellent data on the behaviour of the emission lines and of the continuum at low states and have led to the discovery of a new phenomenon:

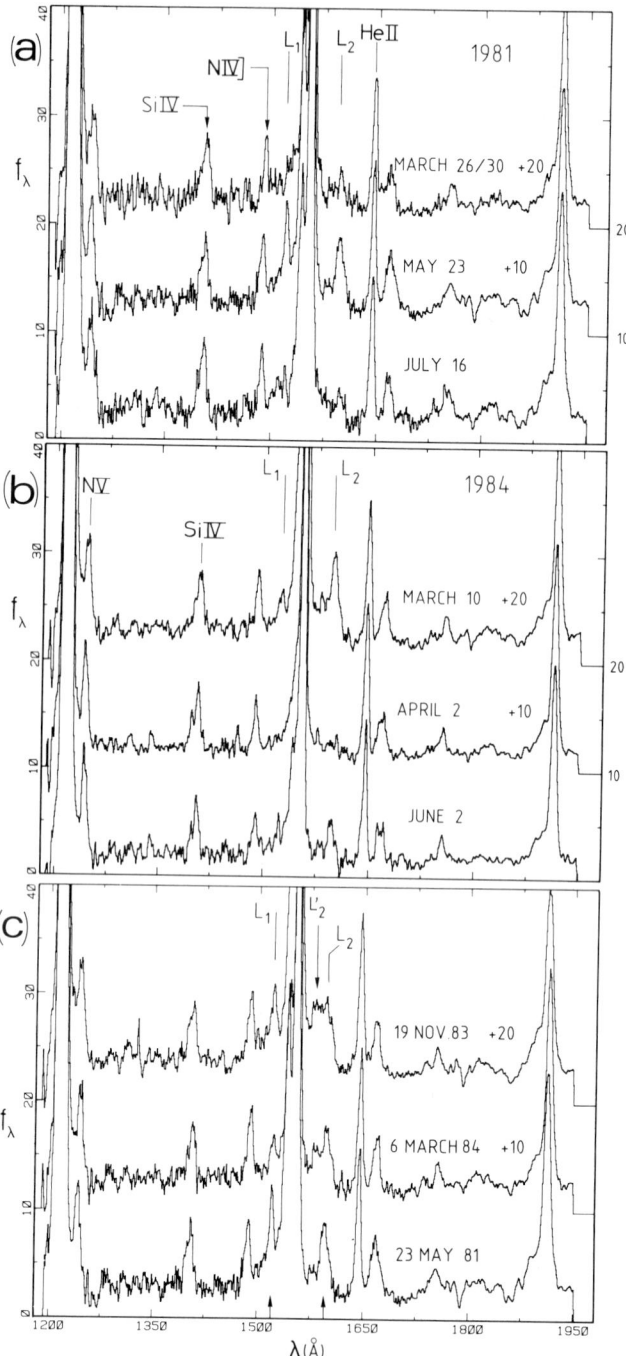

Fig. 4

At low states the spectra of NGC 4151 show 2 unidentified lines L_1 and L_2 on either side of CIVλ1550 which are both narrow and rapidly variable. The behaviour of these lines does not follow the model developed from the observations of the broad components of the emission lines, in particular CIVλ1550. These lines do not suggest to make a change in our model representing the variations of the bulk of the broad line region gas but rather to add a new component to our model possibly in the form of a jet (Ulrich et al., 1985).

The lines L_1 and L_2 are best observed when the nucleus of NGC 4151 is at low states. Such a low state was encountered first on 21 April 1980 and, remarkably, low states became more frequent in 1981-1984. The low states have two characteristics which, up to the present time, have always appeared together: (i) weakness of the wings of the CIV line and of the other permitted lines; (ii) low level of the ultraviolet continuum with $f_\lambda(1440-1470 \text{ Å}) \lesssim 4 \times 10^{-14}$ erg cm^{-2} s^{-1} Å^{-1}. The weakness of the CIV wings is probably due to an insufficient photon flux.

The examples of spectra in Figure 4 illustrate the behaviour of the lines L_1 and L_2.

The properties of the lines are outlined in the next section.

3.1. Variability and Time Delay

The largest fractional change in the line intensities was observed between May 12 and May 17, 1981. In this 5-day interval $I(L_1)$ varied by a factor ~3 and $I(L_2)$ by a factor ~2.

The intensities of L_1 and L_2 are somewhat correlated in the sense that L_1 and L_2 increase and decrease together and appear and disappear together. Their intensity ratio is not constant and $I(L_2)/I(L_1)$ can be found between 1.8 and 4.

Figure 4: Examples of NGC 4151 spectra at 8 epochs, all low states. **a,b:** Spectra arranged in chronological order showing the variations of the lines L_1 and L_2. **a:** 1981 March 26/30 shifted by +20, May 23 shifted by +10 and July 16. **b:** 1984 March 10 shifted by +20, April 2 shifted by +10 and June 2. Ordinates in 10^{-14} erg cm^{-2} s^{-1} Å^{-1}. **c:** Three low-state spectra arranged by increasing contamination of L_2 by the feature L_2'. When L_2' is nearly as large as, or larger than L_2 (for example, 1983 November 19) no attempt has been made to measure $I(L_2)$. The spectrum of 19 November 1983 has been shifted by +20 and that of 6 March 1984 by +10. Ordinates in 10^{-14} erg cm^{-2} s^{-1} Å^{-1}. (From Ulrich et. al., 1985.)

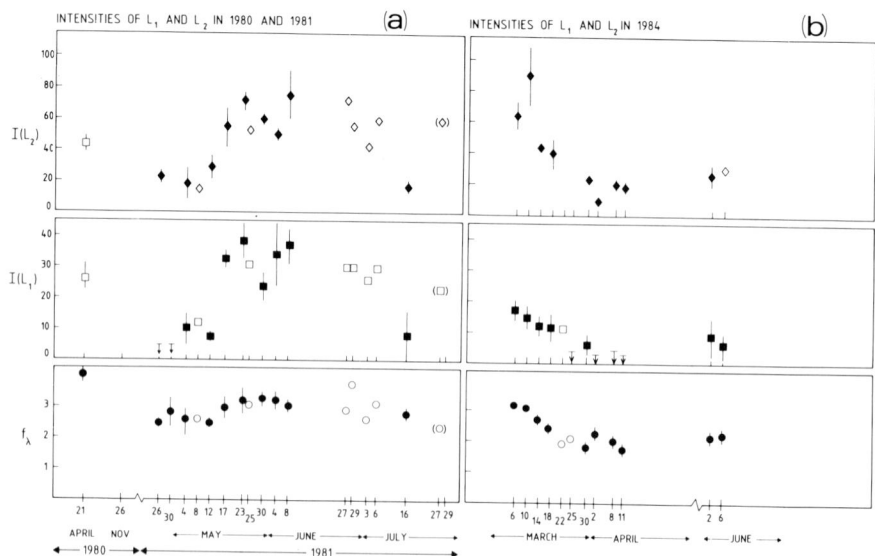

Figure 5: The continuum in the window 1440-1470 Å and the intensities of L_1 and L_2. **a,b:** Observations in 1981 and 1984. The nucleus was in a low state at all epochs of observations. The low state 1980 April 21 is added to **a**. Closed circles, average of the values obtained from two different images taken during same shift; extremities of the error bars represent the individual measurements; open circles, value measured on one spectrum only; ordinates, f_λ is in 10^{-14} erg cm^{-2} s^{-1} Å^{-1}; $I(L_1)$ and $I(L_2)$ are in 10^{-14} erg cm^{-2} s^{-1}. (From Ulrich et al., 1985.)

DISCOVERY OF NARROW AND VARIABLE LINES IN THE UV SPECTRUM OF NGC 4151

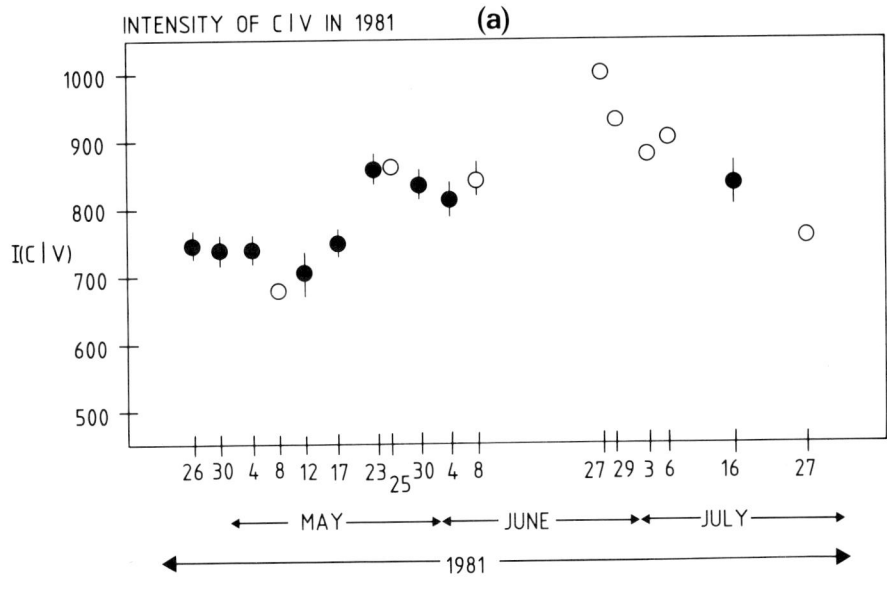

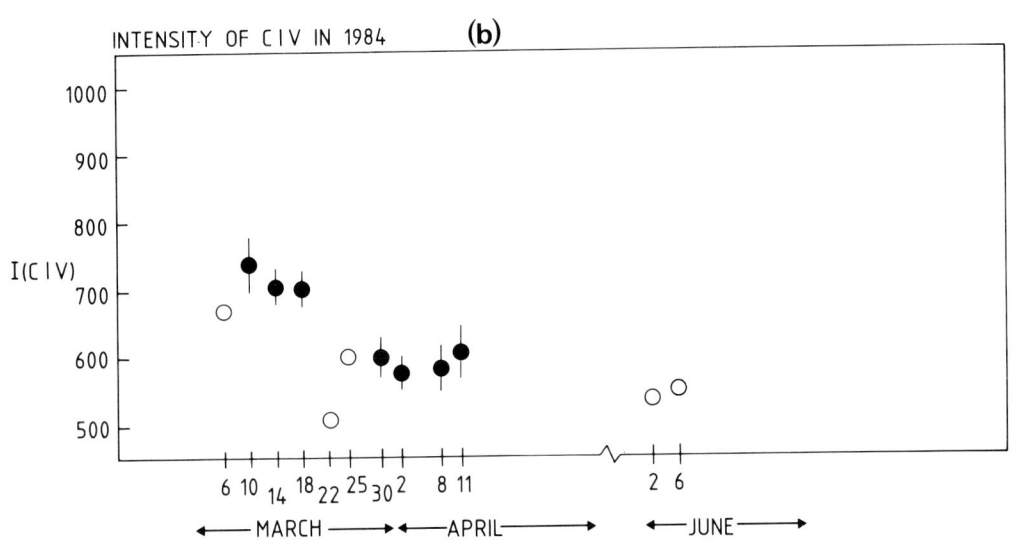

Figure 6: Same as Figure 5 for the intensity of CIVλ1550 measured in the interval 1470-1620 Å.

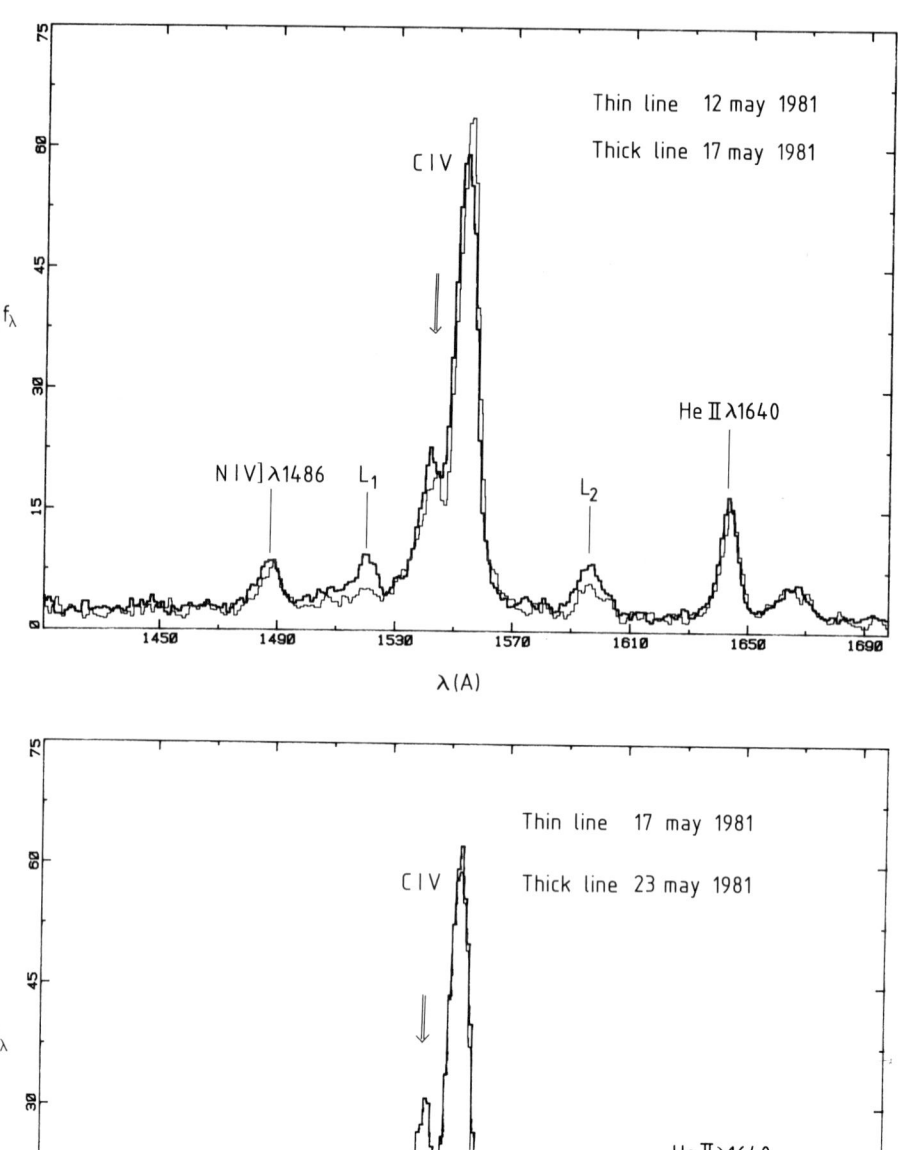

Fig. 7

a) The time delay between the intensity variations of the two lines is
the amount by which the light curve of L_1 can be shifted in time
(backward or forward) and still coincides in shape (not in amplitude) with the light curve of L_2. This time delay is <10 days from
the observations of May 1981. A better determination of this time
delay is an important motivation for further observations of NGC
4151 as it can strongly constrain the models for producing L_1 and
L_2 (Figure 5).

The intensity variations show some degree of correlation with the
small fractional variations of the continuum at 1450 Å. The continuum may have two components: a relatively large constant component
possibly emitted by a blue stellar population, and a small component undergoing large fractional changes (seen diluted by the intense constant component) and which may be related to the variations of L_1 and L_2.

b) The behaviour of L_1 and L_2 when the nucleus goes into medium or
high states is not known, except that their intensity does not
increase like that of the CIVλ1550 wings at their wavelengths.

It is also interesting to note that L_1 and L_2 are present on the
1984 spectra with the same wavelengths and in the same range of
intensity as they had in 1981, in spite of the bright active phases
of the nucleus in 1982 and 1983.

3.2. The Maximum Intensity of L_1 and L_2

The maximum intensity of L_1 and L_2 is of the order of 1/50 of the
total maximum intensity of the CIV lines. This would correspond to
about 10^{-4} $M_\odot$ of gas if L_1 and L_2 were components of CIVλ1550 and
taking $T = 3 \times 10^{4}°K$, $N_o = 10^{+9}$ cm^{-3} and all C in C^{+3}. No lines
similar to L_1 and L_2 have been found in the vicinity of Lyα,
CIII]λ1909 and MgIIλ2800.

3.3. Accompanying Phenomena

The intensity variations of L_1 and L_2 are accompanied by small
variations of the CIV line: in the weak blue wing and on the blue side
of the narrow component (Figure 7 and Figure A2 in Appendix).

Figure 7: Brightening of the unidentified lines L_1 and L_2, and of the
blue wing of CIVλ1550 between three consecutive dates of observations
in May 1981. The thick arrow points to the part of the CIV line which
varies most. The CIV absorption line becomes more prominent as the
blue flank of the emission line becomes stronger. Top panel: thin
line, May 12; thick line, May 17. Bottom panel: thin line, May 17,
thick line, May 23. (An ion event on one of the two spectra of May 23
causes a spurious shoulder to the red of OIII]λ1663). Ordinates in
10^{-14} erg cms^{-2} s^{-1} Å^{-1}.

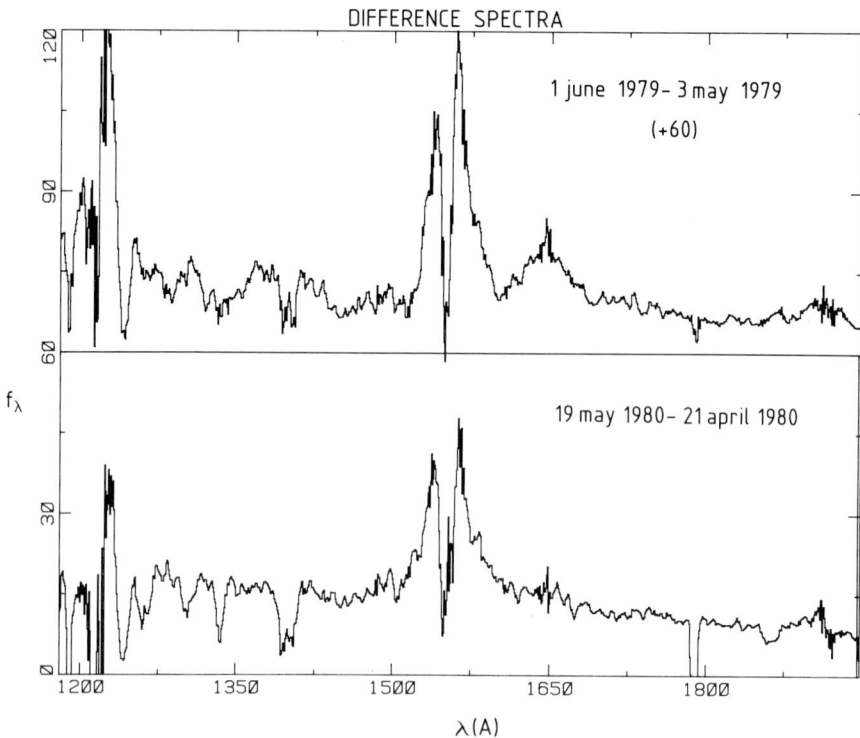

Figure 8: Differences between spectra give the variable component of CIVλ1550 and other emission lines. Top panel: Difference between the spectra of 1 June 1979 and 3 May 1979 (shifted by +60). Bottom panel: Difference between the spectra of 19 May 1980 and 21 April 1980. The full width at half maximum of the variable component of CIV is ~35 Å (top panel) and ~40 Å (bottom panel). Ordinates in 10^{-14} erg cm^{-2} s^{-1} Å^{-1}.

In addition, a feature L'_2 at $\lambda_{observed}$ = 1575 Å is present and blended with L_2 in October-November 1983. It is also weakly present in March 1984. So far this feature has never been seen without being blended with L_2.

3.4. Wavelengths

The wavelengths of L_1 and L_2 have remained constant at 1519.5 Å $\pm$ 1.5 Å and 1599.5 $\pm$ 1.5 Å since May 1981 in the rest frame of NGC 4151 (z = 0.00326) (Figures A3 and A4 in Appendix).

3.5. Widths

The FWHM, after correction for the instrumental profile, is $\leq$ 7 Å and $\leq$ 16 Å for L_1 and L_2 respectively. In contrast, the broad variable component of the CIV line emitted by the whole broad line region is $\sim$40 Å (Figure 8).

An important consequence of the narrowness of L_1 and L_2 is that these lines do not originate from the entire region that emits the broad variable component of CIVλ1550, but must come instead from a region or regions with a localized ionization and excitation mechanism.

Table 1: Characteristics of the narrow and variable emission lines L_1 and L_2

	Unidentified Lines in NGC 4151	
	L_1	L_2
$\lambda_{observed}$ (Å)	1523.5	1599.6
λ_{rest} (Å)	1518.5	1594.4
FWHM (Å)	$\lesssim$ 7	$\lesssim$ 16
Dimension (light days)	5	10

If L_1 and L_2 are components of CIVλ1550:

V = shift in km s^{-1}	-6100	+8500

2 main constraints on models:

- Time delay between the variations of $I(L_1)$ and $I(L_2)$ < 10 days
- Variations of the shift: $\Delta V/V$ < 0.10 in 3 years
- Mass $\sim 10^{-4}$ $M_\odot$

We have as yet not found any plausible identifications for L_1 and L_2. An unidentified line at λ_{rest} = 1595 Å is present in the spectrum of the symbiotic star T Corona Borealis (J. Clavel and P.L. Selvelli, private communication) and could be the same line as L_2, or it could be a coincidence.

It must be emphasized that even <u>if L_1 and L_2 are emitted by ions at the redshift of NGC 4151, the problem of explaining the remarkable temporal behaviour and narrow width of the lines will remain</u>.

L_1 and L_2 as Components of CIVλ1550?

In the absence of plausible identifications it is interesting to speculate that L_1 and L_2 are blue-shifted and red-shifted components of CIVλ1550. The shifts represent velocities of -6100 and +8500 km s^{-1}.

The kinematic interpretations of L_1 and L_2 in terms of free flying clouds in free-fall or being ejected are unsatisfactory because the change in radial distance in 3 years implied by the observed velocities ($\sim$ 7000 km s^{-1}) would result in a gradual increase (in the case of free fall) or decrease (in the case of ejection) of L_1 and L_2 intensities, which is not observed. The interpretation of L_1 and L_2 as 2 clouds in rotation around the central object is not acceptable either because the upper limit on the change of the line of sight velocity in three years moves the clouds to a radial distance incompatible with the upper limit of ten days between the correlated variations of L_1 and L_2.

There is the exciting but speculative possibility that L_1 and L_2 are excited by a two-sided jet like the components of the hydrogen and helium lines in SS 433.

If the asymmetry of L_1 and L_2 in velocity shift is due to the transverse Doppler effect, and if the regions emitting L_1 and L_2 are moving at the same velocity v in opposite directions, then one obtains v = 0.095c and the jet direction is at 75° of the line of sight to the observer. This implies a large misalignment between the angular momentum of the central object and the rotation axis of the galaxy.

CONCLUSIONS

If indeed L_1 and L_2 are excited by a jet, then they constitute the first observations of the interactions of the relativistic particles with the gas in the immediate vicinity of the central engine (broad line gas or intercloud gas). It is very important to continue observing the CIVλ1550 region of the spectrum in NGC 4151. The best observations are those which set strong constraints on the models:

a) a stricter upper limit of the time delay between the intensity variations of L_1 and L_2,

b) a better determination of the change of the wavelengths of L_1 and L_2 in time (if any),

c) a study of the profile and structure of L_1 and L_2 with a higher wavelength resolution than possible with IUE. Observations with Space Telescope will of course be of tremendous importance.

At the same time the search for narrow variable lines in <u>other Seyfert galaxies</u> should be pursued actively. Finding lines with the same behaviour as L_1 and L_2 and at exactly the same rest wavelengths in other galaxies will prove that L_1 and L_2 are not components of CIVλ1550 but are emitted by as yet unidentified ions. Finding lines with the same behaviour but at different wavelengths (still in the vicinity of CIVλ1550) will prove that the lines are shifted components of the CIV line. It is this latter situation which seems to be the most likely at the present time.

Finally, another important observation is to search for variations of the <u>VLBI source</u> in intensity or in structure occurring simultaneously with the variations of L_1 and L_2.

REFERENCES

Bromage, G.E., et al. 1985, Mon. Not. Roy. astr. Soc. (in press).
Kwan, J. 1984, Ap.J. <u>283</u>, 70
Ulrich, M.H. et al. 1984, Mon. Not. Roy. astr. Soc. <u>286</u>, 221
Ulrich, M.H. et al. 1985, Nature <u>313</u>, 745.

APPENDIX

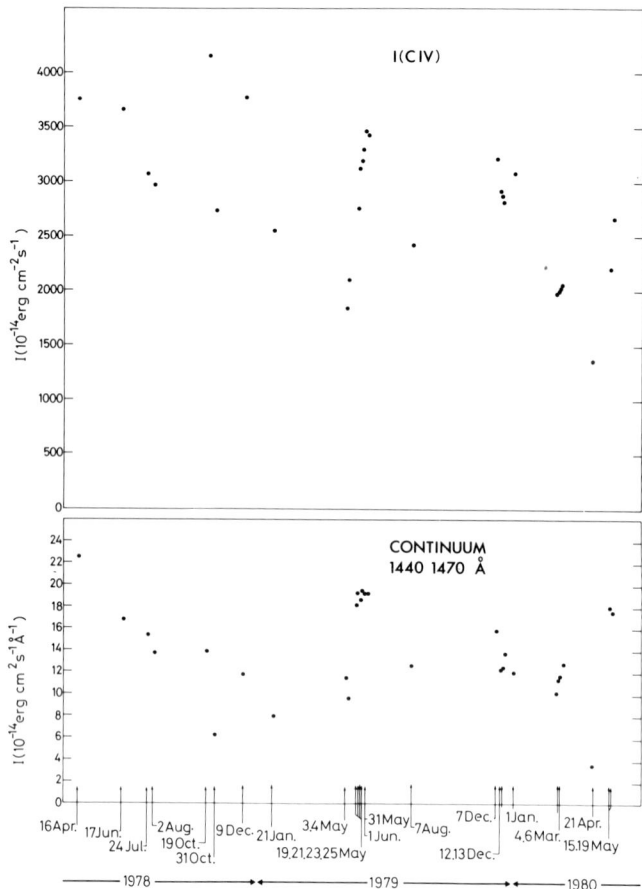

Figure A1: The light curves of CIVλ1550 and of the continuum in the range of 1440-1470 Å in NGC 4151 during the period 1978-1980. The nucleus was never observed at a low state during that period except on 21 April 1980. In May 1979 the continuum leveled off after May 4 while I(CIV) continued to increase suggesting a time delay of ~15 days between the variations of the ionizing continuum and of the CIV line intensity. (From Ulrich, M.H., XI Texas Symposium of Relativistic Astrophysics, ed. Evans, D.S., 291. Academy of Sciences, New York, 1984.)

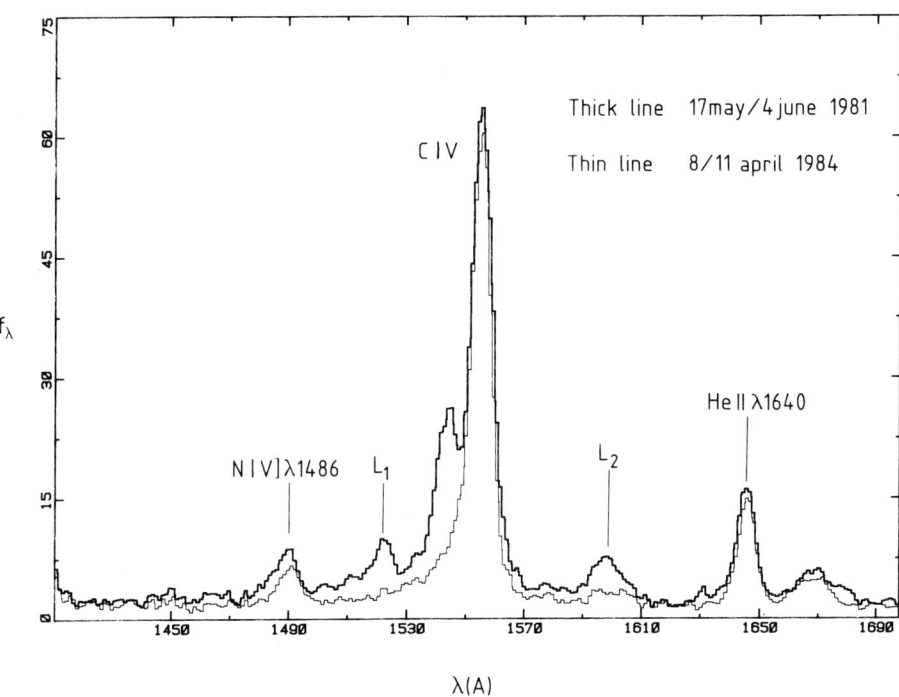

Figure A2: Examples of spectra at low states. Thin line: average of the 4 spectra of 1984 April 8 and April 11 (also shown in Figure 1). Thick line: average of the 9 spectra of May 17, 23, 25, 30 and June 4. Ordinates in 10^{-14} erg cm^{-2} s^{-1} Å^{-1}.

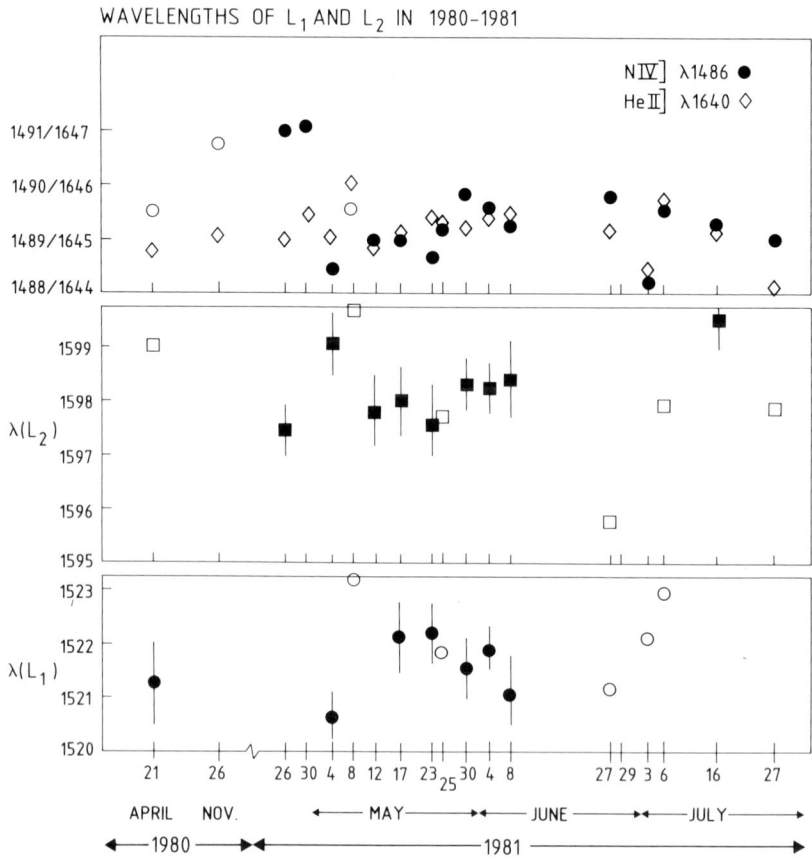

Figure A3: Observed wavelengths of L_1 and L_2 measured during the low states of 1980-1981. The observed wavelengths of NIV]λ1486.5 and HeIIλ1640.5 are given for comparison. Closed circles, average of the values obtained from two different images taken during same shift; extremities of the error bars represent the individual measurements. Open circles, value measured on one spectrum only. Ordinates in Angstroms.

DISCOVERY OF NARROW AND VARIABLE LINES IN THE UV SPECTRUM OF NGC 4151

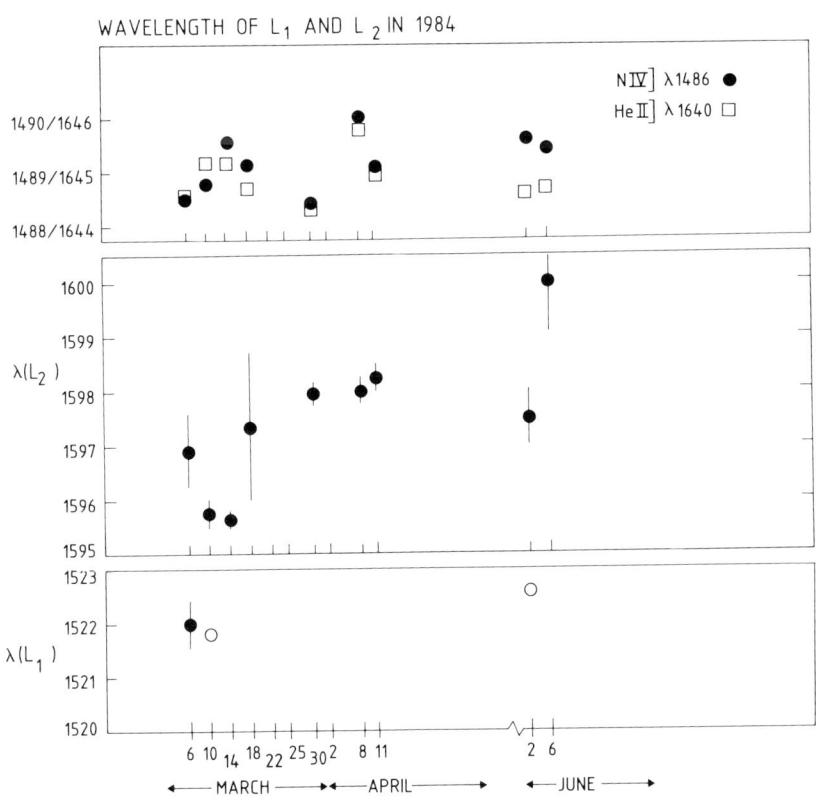

Figure A4: Wavelengths of L_1 and L_2. Same as Figure A3 for 1984.

DISCUSSION

OSTERBROCK: 1) The seqnal-to noise ratio of the spectra showing the two unidentified lines L1 and L2 is very good. Thus the large difference in their measured FWHM's is no doubt significant; and makes it seem unlikely that they are two different lines of the same ion. 2) If L1 and L2 were components of CIV 1549, coming from small condensations with peculiar velocities, other strong lines might have similar components at the same velocity shifts. If, say MgII λ 2800 (or any other line) had such components, with the same relative intensities with respect to the main line, would they have been detectable in your data?

ULRICH: The components of Lyα and MgII corresponding to L1 and L2 for CIV have not been detected. I am now in the process of measuring the upper limits.

WILKES: There have been many reports of variations in the optical spectra of NGC 4151. Do you have any optical spectra at about the same time as your IUE observations which might show the L1 and L2 components in other emission lines?

ULRICH: In our group we do not have optical observations exactly simultaneous with our IUE observations. But there may be optical observations performed by other astronomers which are not yet available in print or preprint form.

BECKMAN: Are the MgII and CIV absorption lines due to our own galactic corona? Are there SiIV or other absorption features of a similar nature?

ULRICH: The MgII and CIV abosorption lines are intrinsic to NGC 4151, and blueshifted by approximately 800 Km/s with respect to NGC 4151. They blend with the absorption lines in our Galaxy. The absorption line spectrum of NGC 4151 is presented and discussed in Bromage, G.E. 1985 (Mon. Not. Roy. Astron. Soc., in press).

LARGE SCALE IONIZED GAS IN RADIO GALAXIES AND QUASARS

R.A.E. Fosbury*
Space Telescope European Coordinating Facility
European Southern Observatory
Karl-Schwarzschild-Str. 2
D-8046 Garching bei München
FRG

ABSTRACT. Many early-type active galaxies and quasars are surrounded by extensive regions of low density, highly ionized gas which can be considered the analogy of the much smaller Narrow Line Region in the spiral Seyferts. Observations of this gas carry information about the galaxy and its active nucleus and can now be interpreted in the context of X-ray observations of hot gas and 21 cm observations of cold gas in the same systems.

1. INTRODUCTION

Interstellar gas in galaxies and quasar environments is known to exist in three thermally stable phases (Lepp et al., 1985). For the conditions of pressure and exciting mechanisms appropriate to the active ellipticals I wish to consider, these phases can be called: coronal with $T > 10^5$ K, warm with $T \sim 10^4$ K and cold with $T < 100$ K. In elliptical galaxies, both "normal" and active, all three phases have now been observed using a variety of observational techniques.

Using soft X-ray observations from the Einstein Observatory satellite, several authors have reported hot ($T \sim 10^6$ K), extended haloes around elliptical galaxies (see in particular Forman, Jones & Tucker, 1985) previously thought to be more-or-less devoid of gas. These coronae, with masses of order 10^{9-10} $M_\odot$, have cooling times much less than the Hubble time and so will recycle gas, in cooler phases, back to the galaxy.

The warm gas is accessible to observation through its emission in optical and UV lines and also by its absorption of the light from background quasars (Stocke et al., 1984). It is the nature of this warm gas, particularly in active ellipticals and low redshift quasars, which I wish to address in this talk.

---

* Affiliated to the Astrophysics Division, Space Science Dept., European Space Agency

Cool gas, observed in the 21 cm line, has been known in some ellipticals for about a decade now and the statistics of its distribution as a function of the galaxy luminosity are quite different from that in later type galaxies, suggesting an external rather than an internal origin for at least this gaseous phase (Knapp, Turner & Cunniffe, 1985).

What are the problems associated with the detection of large scale ionized gas in early-type galaxies and what might we expect to learn from observations of it? The particular problems I shall address are: which types of object exhibit the phenomenon, how is the gas ionized, how does it interact with other phases of the ISM, what is its origin and chemical composition, how is it evolving and what effect does it have on the galaxy and finally, can it be used as a probe to tell us something about the parent galaxy and its nuclear activity?

2. WHICH GALAXIES EXHIBIT THE PHENOMENON?

Figure 1 shows [OIII] line minus continuum images of two Southern radio galaxies, PKS0349-27 and PKS0634-20, where highly ionized gas can be seen out to distances of tens of kiloparsecs. The physical state and the velocity fields of this gas have been studied by Danziger et al. (1984) and by Fosbury et al. (1984), but these images were obtained recently by S. di Serego Alighieri and collaborators using the European Space Agency Photon Counting Detector (PCD) (di Serego Alighieri, Perryman & Macchetto, 1985) on the ESO 2.2 m telescope on La Silla. Both of these galaxies are powerful classical double radio sources and classified as ellipticals and it is perhaps worth drawing the analogy here with the usually spiral Seyfert galaxies where the spatial extent of the Narrow Line Region, and the region of influence of the active nucleus in general, is typically a few hundred parsecs or so. If the nucleus can be held directly responsible for ionizing this vast region of tenuous gas then the analogy is a good one: there are, however, problems with such a simple picture and we should be careful not to push it too far.

Before we ask how general this phenomenon is in early-type galaxies, it would be noted that most of the discoveries of these extended regions of ionized gas were made with long slit spectrographs and not using narrow band imaging techniques. With an historical perspective, it is clear that this led to the belief that the phenomenon was a rare curiosity. Once the imaging detectors had become sufficiently sensitive and it was realized that the poor spatial sampling given by a single long slit spectrograph was missing a significant fraction of objects, a clearer idea of the statistics began to emerge.

For nearby elliptical galaxies of relatively normal appearance, the existence of ionized gas near the nucleus is common. Phillips et al. (1985) - see also the paper by Elaine Sadler in this conference - have shown that at least half of all nearby ellipticals contain $10^3 - 10^5$ $M_\odot$ of ionized gas, often associated with dust, in a region of size ~1 kpc around their nuclei. This gas has a spectrum characteristic of LINERS and, where spatially resolved, appears to be rotating with velocities of a few hundred km/s. Demoulin-Ulrich, Butcher & Boksenberg

(1984), found six out of a sample of twelve known emission line ellipticals to have rotating gaseous disks of sizes ranging up to 3 kpc. Bertola et al. (1984) have discussed the relative velocity fields of gas and stars in several nearby ellipticals. Realistic estimates of the mass of ionized gas are difficult to make since, in the extended emission, the electron densities are so low that the usual plasma diagnostic forbidden line ratios are of little use. The masses quoted may be best interpreted, therefore, as lower limits.

For active and radio ellipticals there is a large and observationally somewhat inhomogeneous literature from which it appears (Tadhunter 1985) that more than half the galaxies with detected nuclear emission lines show distributions of ionized gas on scales of kiloparsecs, ranging in some cases up to more than 100 kpc. A wide range of ionization is seen but the state is often very high with strong HeII and [Ne V] lines and therefore not characteristic of normal, stellar photoionized HII regions. Extensive studies of the gas associated with radio structures have been made by Miley, Heckman, van Breugel and co workers (see, for example, van Breugel et al. 1985) but it is clear that the ionized gas is not exclusively associated with the radio jets and can exist also in the form of distorted disks, rings and filaments (Fosbury et al. 1982, Danziger et al. 1984, Fosbury et al. 1984).

The massive early-type galaxies residing in the centres of clusters have been shown to contain ionized gas which is most likely to be the result of a cooling flow of, initially hot, intracluster gas (Fabian, Nulsen & Canizares 1984). In the most distant radio galaxies whose optical spectra have been examined, the [OII] emission often has a very large equivalent width and shows signs of having a considerable spatial extent (Spinrad & Djorgovski 1984). Although the strength of the [OII] lines has resulted in low ionization classifications, these objects appear also to show signatures, e.g. [Ne V], of a high ionization component and may be rather similar to the nearer objects which have been studied in more detail.

Since the early work of Kristian (1973) which established that the lower redshift quasars showed resolved images if the core was of sufficiently low luminosity to avoid saturation of the detector, there have been a number of studies of samples of these objects looking for underlying galaxies or any form of spatial structure or "fuzz". The most recent survey (Stockton & MacKenty (1985); reported by Stockton (1985)) was of a sample of 60 quasars with $z \leqslant 0.45$ examined with a narrow filter centered on the redshifted [OIII] line. This found 15 objects with extended emission, all of which came from the subsample of 40 showing at least moderately strong narrow line emission. Stockton reports that the line images are highly structured, suggesting that the gas is mostly density rather than ionization bounded. The line-free continuum images are, however, much smoother. Boroson & Oke (1984) reported fairly detailed studies of a smaller number of low redshift quasars and were able to separate them into two groups on the basis of the continuum colour and emission lines shown by the extended structure. This separation correlates with other properties of the quasars, including radio morphology, the presence of FeII lines and the Balmer line width. Such a separation suggests to me that we are seeing an extension to higher redshift of the differing properties of the spiral Seyferts and the elliptical radio galaxies but an extrapolation of this kind is not yet conclusively established.

The conclusions to be drawn from these and other studies of early-type galaxies and quasars appear to be: the presence of moderately strong narrow (nuclear) emission lines is a necessary but not sufficient condition for seeing extended emission; the ionized gas is usually highly structured (blobs and filaments) and shows relative line intensities which are not characteristic of normal HII regions; the early-type galaxies show extended emission regions which are typically an order of magnitude larger than the active spirals. It is also interesting, and probably important, that relative velocities of the gas within a given object are rarely, if ever, larger than a few hundred km/s: this suggests either that the gas motions are governed by gravity, rotation or infall, and/or that there is a pervasive medium which rapidly limits the speed of the clouds of warm gas whatever their origin might be.

3. HOW IS THE GAS IONIZED?

The location and nature of the ionization source responsible for the extended, and indeed also the nuclear emission, in these active galaxies is an intriguing question and only in a few cases has it satisfactorily been answered. In a number of cases, the wide range of ionized species exhibited by a single spectrum can be used to argue in favour of photoionization by a spectrum extending well down into the soft X-ray region: power-law models with a spectral index of about -1 give a rather satisfactory fit to the relative line intensities. Indeed, in the case of the narrow line radio galaxy PKS2158-380, Fosbury et al. (1982) have compared the optical and UV line spectrum with the IUE observations of the UV continuum to show that the nucleus is capable of ionizing the whole of the extended emission region provided that the gas is distributed over a significant area of the galaxy's "sky".

Such a simple picture of nuclear photoionization, while appealing and probably satisfactory for the more luminous quasars (Bergeron et al. 1983, di Serego Alighieri, Perryman & Macchetto 1984), does run into problems for some of the radio galaxies which do not presently show strong nuclear continua. In particular, PKS0349-27 (Danziger et al. (1984) shows highly ionized gas at more than 50 kpc from the nucleus (H_o = 50 km/s/Mpc used throughout) and there is a conflict in the space of ionization parameter, nuclear luminosity and gas filling factor for the observed line luminosities. From the objects I have looked at, there appears to be a positive correlation between the state of ionization and the surface brightness of the emission; an observation which would be in conflict with the simple central source of ionization picture. An anisotropy in the ionizing radiation field, produced either by the radiation mechanism itself or by absorption, could remove some of the difficulties. There are also the cases where there is a detailed correlation between radio and optical structures which argue in favour of local sources of ionization. The problem is to elucidate the nature of such sources; shocks are appealing but the line spectra are generally not at all like that expected from a collisionally heated gas. Deep ultraviolet imaging of a carefully chosen sample of objects will be of great help in resolving this dilemma.

Some information about the physical condition of the gas may be obtained without a detailed understanding of the ionization mechanism. Classical optical/UV plasma diagnostic techniques can be used together with auxiliary information from high spatial resolution radio polarization measurements and soft X-ray observations. Temperature measurements are usually more firmly established than density estimates which is why the ionized gas masses are only poorly known. Where the density can be obtained using some auxiliary constraint, the gas mass is typically of order 10^8 $M_\odot$. Such a figure may serve as a useful working number to be used with caution until better techniques for density measurement are established. A programme of detailed spectrophotometry of regions whose ionization mechanism can be understood will allow us to start examining the chemical composition of the gas and perhaps answer some of the questions about its origin and past history. Power law photoionization models with solar abundances are quite successful in fitting the line ratios but since the input parameters are strongly coupled, all that can be said is that the element abundances with respect to hydrogen are neither spectacularly high nor low.

4. THE ORIGIN OF THE GAS

Clues about the origin of the gas may be obtained from its present spatial distribution, its kinematic state and, when it is possible to determine it with any confidence, its chemical composition. Whatever the answer, it is presumably linked intimately with the origin of the hot gas in the X-ray haloes and the cold gas seen in the neutral hydrogen line. Although in some cases the evidence for tidal interactions, collisions and subsequent merging is fairly conclusive (for quasars, see the review by Hutchins 1983), the widespread occurrence of the phenomenon argues that we look also for a mechanism which involves material with an internal origin. It may even prove that the interactions that we do see act primarily as mechanisms to transfer pre-existing gas into the warm phase and thus render it visible. A uniformity of chemical composition, if it could be established, would support this idea rather than one in which gas was always donated by merging companions with widely differing masses and star formation histories.

What is the evidence from the observations of morphology and kinematics? Line images of radio galaxies and quasars with extended emission generally show a complex and highly non-uniform structure. On examination of a sufficient number of objects, however, some order may be discerned. For the sake of further description, I try to divide the gas into "disk" and "polar" components. It may be distorted but, when velocity measurements are available, it is possible to interpret some part of the structure as a rotating disk. In addition, many objects show a distribution of excited gas along the axis of the double radio source, although not necessarily restricted to the immediate vicinity of the radio jets, which appears to be distinct from the disk and I shall therefore refer to as polar. A well-known object which exhibits such a distinction rather nicely is NGC5128 (Centaurus A) with the famous warped gas/dust disk and also the broad polar cones of [OIII] emission (Phillips et al. 1984) and the filaments along the radio axis (Graham & Price 1981). The disk in Cen A is not highly excited however, so we should be careful about comparing it with the very highly ionized systems. Detailed analyses of the gaseous and stellar kinematics of this galaxy are now concluding that the material, in the disk at least, must have come from a relatively recent collision (Bland 1985, Wilkinson et al. 1985, but see also Bertola, Galletta & Zeilinger 1985). In addition, there are a few objects which show gas in the form of rings or shells: see, for example, PKS0634-20 in Fig. 1. The relationship between these gaseous rings and the presumably stellar shells reported by Malin & Carter (1980) is fascinating although we have yet to find a galaxy which exhibits both.

We have already noted that the relative gas velocities are generally restricted to a few hundred km/s. The observations often show a velocity field which could be interpreted as rotation with the characteristic steep rotation curve near the nucleus which would be expected in these early-type galaxies. Part of the velocity structure is chaotic, however, and several objects show line widths of a few hundred km/s at locations well away from the nucleus. PKS0349-27 (Danziger et al. 1984) does show somewhat higher velocities but here there is good evidence for a current tidal interaction with two companion galaxies having similar systemic velocities to the radio galaxy. One must always be careful, of course, about interpreting the observed velocities as pure rotation since a radial component is so hard to distinguish. If material were being driven outwards by the nucleus, the radio jets for instance, a bigger range of observed velocities might be expected. It may be, however, that the velocity of the warm gas is being limited by the presence of a surrounding medium: there are good physical reasons why the sound speed in the hot haloes is of the same order as the rotation or free-fall speeds.

Although much more careful comparisons need to be made in individual objects, it appears that the gas pressures in the hot, warm and cool phases are very similar. Indeed, making the assumption that they are so may be a good way to determine other physical properties. Fosbury et al. (1978) argued that the warm and cold gas around the active elliptical galaxy NGC1052 may be in pressure equilibrium. Also, where the central source photoionization model can be shown to work, the pressure of warm gas required to give an appropriate ionization parameter is similar to that for the hot gas derived from the X-ray observations.

In summary, I believe that there is compelling evidence in some cases that the extended emission phenomenon is related to an interaction or collision between two or more galaxies. Such an association may not, however, be universal and the X-ray observations argue that the presence of significant quantities of gas in early-type systems is the norm. It still appears that the presence of an active nucleus and extended optical/UV emission lines may be more probable in colliding systems.

5. THE EVOLUTION OF THE GAS AND ITS EFFECT ON THE GALAXY

Since gas is a dissipative medium, the structures we see must evolve. If they have an external origin, the disks will, through a stage of differential precession, finally settle into a stable configuration in a principle plane of the galaxy ellipsoid. This may not be a simple planar disk if the galaxy potential is tumbling. It has been argued (Fosbury et al. 1982) that PKS2158-380 has a disk which is in a severely warped phase. There is also the likelihood that the physical state of the gas will evolve as well through processes of star formation and changes in the sources of ionization. Some of the problems with the nuclear photoionization model may arise because there is no guarantee that the nuclear UV flux we see now represents the temporal average seen by the surrounding gas. Indeed, in some cases, this gas could be considered a fossil remnant of past nuclear activity.

One of the important observational projects we need to undertake is the careful examination of the stellar population associated with the gas. When does it form stars and how does the IMF compare with that in spiral galaxies? This requires UV spectrophotometry with a sensitivity greater than can be achieved with the IUE satellite.

Although the mass of gas (and dust) associated with the early-type galaxies is probably a very small fraction of the total mass, its effects can be dramatic, both on the observed colours of the stellar population and, perhaps, on the nucleus by acting as a fuel source.

6. CONCLUSIONS

We have seen that spatially extended regions of ionized gas, having total masses up to about 10^8 $M_\odot$ are a common feature of early-type galaxies and a significant fraction of quasars. It can occur over regions greater than 100 kpc in size but to be well developed, appears to require an active nucleus which emits a moderately strong narrow emission line spectrum. The spectrum is often not like that of normal HII regions and there is a tendency for the ionization state to correlate positively with the surface brightness of the line emission. The source of ionization is not always known but is presumably associated, at least indirectly, with the active nucleus; indeed, in the quasars, the nucleus provides ample ionizing radiation. The morphology and kinematic state of the gas suggests in some cases the existence of a collision or merger in the recent past.

The recent discovery of massive hot coronae around many normal ellipticals (Biermann & Kronberg 1983, Nulsen, Stewart & Fabian 1984, Forman, Jones & Tucker 1985) together with the known presence of cold HI disks around some, suggests that the warm gas, visible at optical and UV wavelengths may simply be an intermediate phase in a cyclic process which is occurring in all such galaxies. When a powerful source of ionization is present, the gas which is cooling through temperatures of around 10^4 K becomes easily detectable by its line radiation.

Like the temperature of the hot corona, the macroscopic velocity field of the warm gas may be used to investigate the mass distribution within the galaxy although the way in which the two phases interact needs to be considered carefully. What evidence there is suggests that rotation curves are flat and supports the notion, derived from the X-ray data, that mass-to-light ratios in early-type galaxies are high.

Acknowledgements

I should like to thank Alan Stockton for providing much information about extended emission around quasars. Clive Tadhunter helped greatly in the preparation of this talk, in particular by examining the frequency of occurrence of the phenomenon in radio galaxies. Sperello di Serego Alighieri kindly made available the data for Fig. 1 prior to publication.

References

Bergeron, J., Boksenberg, A., Dennefeld, M. & Tarenghi, M., 1983. Mon. Not. R. astr. Soc., **202**, 125.
Bertola, F., Bettoni, D., Rusconi, L. & Sedmak, G., 1984. Astron. J., **89**, 356.
Bertola, F., Galletta, G. & Zeilinger, W.W., 1985. Astrophys. J., in press.
Biermann, P. & Kronberg, P.P., 1983. Astrophys. J., **268**, L69.
Bland, J., 1985. D. Phil. Thesis, University of Sussex.
Boroson, T.A. & Oke, J.B., 1984. Astrophys. J., **281**, 535.
van Breugel, W., Miley, G., Heckman, T., Butcher, H. & Bridle, A., 1985. Astrophys. J., **290**, 496.
Danziger, I.J., Fosbury, R.A.E., Goss, W.M., Bland, J. & Boksenberg, A., 1984. Mon. Not. R. astr. Soc., **208**, 589.
Demoulin-Ulrich, M.-H., Butcher, H.R. & Boksenberg, A., 1984. Astrophys. J., **285**, 527.
Fabian, A.C., Nulsen, P.E.J. & Canizares, C.R., 1984. Nature, **310**, 733.
Forman, W., Jones, C. & Tucker, W., 1985. Astrophys. J., in press.
Fosbury, R.A.E., Mebold, U., Goss, W.M. & Dopita, M.A., 1978. Mon. Not. R. astr. Soc., **183**, 129.
Fosbury, R.A.E., Boksenberg, A., Snijders, M.A.J., Danziger, I.J., Disney, M.J., Goss, W.M., Penston, M.V., Wamsteker, W., Wellington, K.J. & Wilson, A.S., 1982. Mon. Not. R. astr. Soc., **201**, 991.
Fosbury, R.A.E., Tadhunter, C.N., Bland, J. & Danziger, I.J., 1984. Mon. Not. R. astr. Soc., **208**, 955.
Graham, J.A. & Price, R.M., 1981. Astrophys. J., **247**, 813.
Hutchins, J.B., 1983. P.A.S.P., **95**, 799.
Knapp, G.R., Turner, E.L. & Cunniffe, P.E., 1985. Astron. J., **90**, 454.
Kristian, J., 1973. Astrophys. J., **179**, L61.
Lepp, S., McCray, R., Shull, J.M., Woods, D.T. & Kallman, T., 1985. Astrophys. J., **288**, 58.
Malin, D.F. & Carter, D., 1980. Nature, **285**, 643.
Nulsen, P.E.J., Stewart, G.C. & Fabian, A.C., 1984. Mon. Not. R. astr. Soc., **208**, 185.
Phillips, M.M., Taylor, K., Axon, D.J., Atherton, P.D. & Hook, R.N.,

1984. Nature, **310**, 554.
Phillips, M.M., Jenkins, C.R., Dopita, M.A., Sadler, E.M. & Binette, L., 1985, in preparation.
di Serego Alighieri, S., Perryman, M.A.C. & Macchetto, F., 1984. Astrophys. J., **285**, 567.
di Serego Alighieri, S., Perryman, M.A.C. & Macchetto, F., 1985. Astron. Astrophys., in press.
Spinrad, H. & Djorgoviski, S., 1984. Astrophys. J., **285**, L49.
Stocke, J.T., Liebert, J., Schild, R., Gioia, I.M. & Maccacaro, T., 1984. Astrophys. J., **277**, 43.
Stockton, A., 1985, in Proceedings of the Third Asian-Pacific Regional Meeting of the IAU, Kyoto, Japan.
Stockton, A. & MacKenty, J.W., 1985, in preparation.
Tadhunter, C.N., 1985. D. Phil. Thesis, University of Sussex, in preparation.
Wilkinson, A., Sharples, R.M., Fosbury, R.A.E. & Wallace, P.T., 1985. Mon. Not. R. astr. Soc., in press.

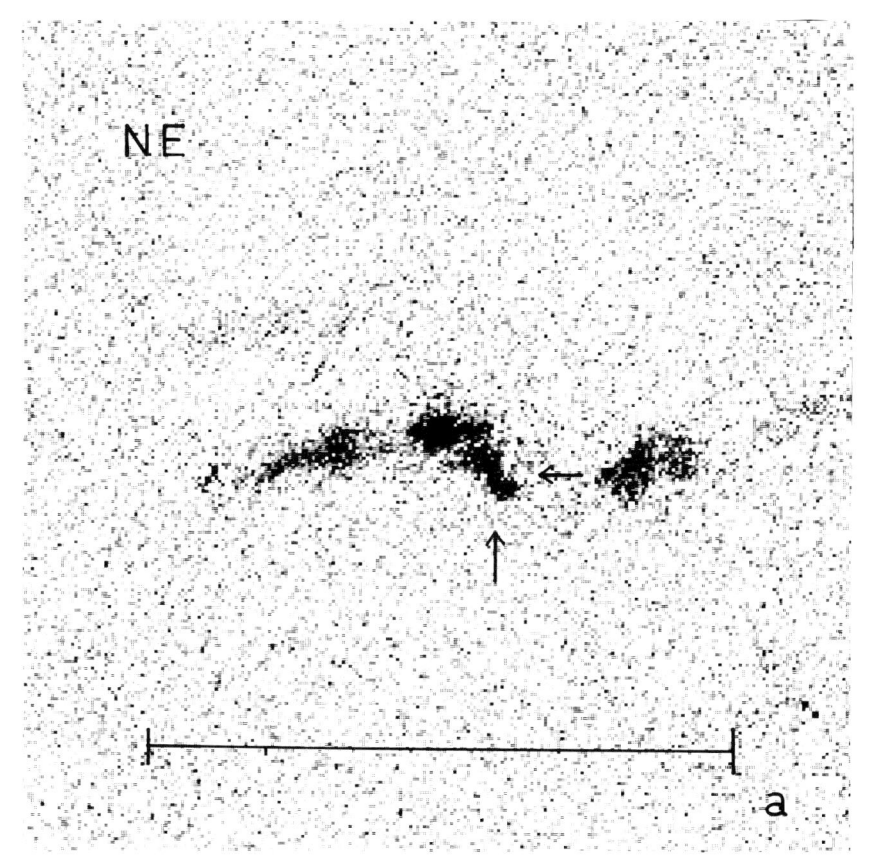

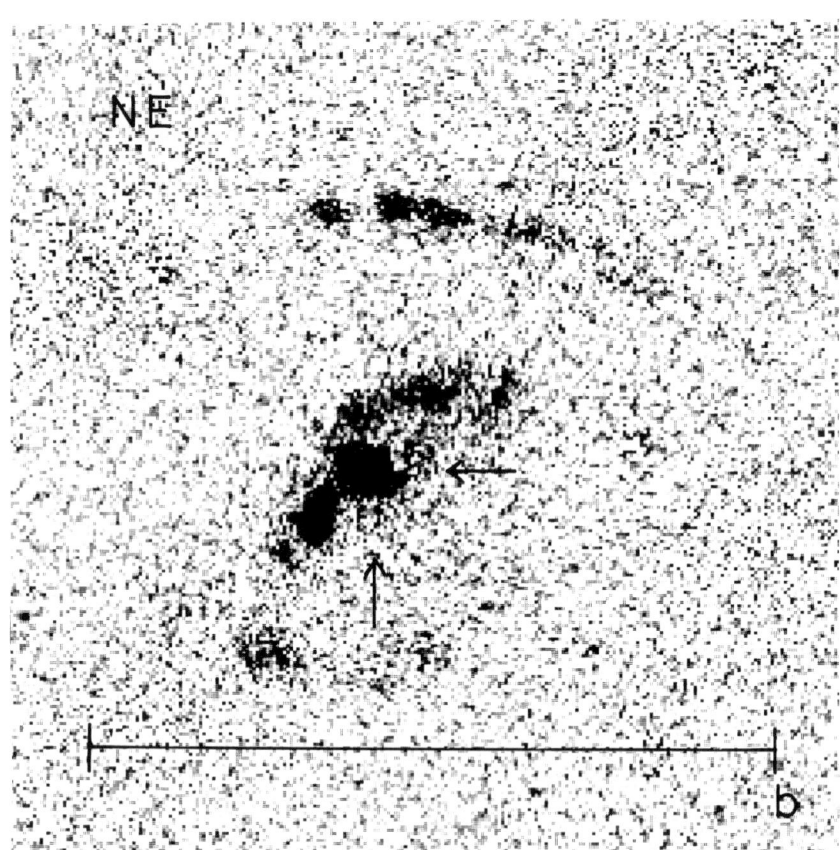

Figure caption.

Figure 1. [OIII] λ 5007 line minus continuum images of the two radio galaxies (a) PKS0349-27 and (b) PKS0634-20. These were obtained in April 1985 by S. di Serego Alighieri using the European Space Agency Photon Counting Detector on the European Southern Observatory 2.2 m telescope on La Silla, Chile. The filters used for the line and neighbouring continuum had FWHM of approximately 35 Å and 180 Å respectively. The scale bar on each picture represents a distance of 100 kpc at the galaxy (H_o = 50 km/s/Mpc) and the positions of the nuclei are marked.

DISCUSSION

SILK: What is the fate of the gas seen around active galaxies? Is it consistent with normal star formation?

FOSBURY: We know that stars form in the disk of Cen A. Our observations occasionally reveal a weak blue continuum associated with the extended emission which may be associated with young stars. The total mass of warm gas is, however, only $10^7 - 10^8$ $M_\odot$ in a typical case.

OSTERBROCK: Does the extended emission in the active ellipticals show, in any of the objects, ionization as high as [NeV], [FeVII], or even [FeX]?

FOSBURY: We often find HeII λ 4686 and sometimes find [NeV] 3424: their ratio is, of course, a strong function of the ionization parameter.

BURBIDGE: Is it your belief that all of the extended gas in radio galaxies arises outside the galaxies?

FOSBURY: In the "normal" ellipticals, the recent X-ray data have demonstrated a cyclic mechanism for feeding stellar mass loss into a halo of hot gas and subsequently cooling this to form a neutral disk. In the active objects there is more evidence for tidal interactions and mergers. It must be said, though, that we don't really know if the gas is rotating, inflowing or outflowing. We do not, however, see any high gas velocities which might indicate ejection.

BURBIDGE: Do all your active galaxies have near neighbours?

ROOS: It seems possible that all AGNs are triggered by mergers. Neverthe;ess, this does not imply that all RG's and QSO's should have a visible neighbour. The activation of a nucleus by a merger is most likely to occur during the final stages of the merging process, when the nuclei have a separation much smaller that 1 kpc. Also the galaxy responsible for tirggering a nucleus into activity may be much smaller than the active galaxy.

SCHULZ: I would like to comment further on the similarity between the large scale gas in RGs and the narrow like region in Seyferts. In my spatially resolved spectra of the NLR in NGC 4151 there appear components which are most easily explained by gas which is spiraling inwards. Furthermore, the ionization level is high even outside the NLR where very-narrow line gas (FWHM 20-50 km s^{-1}) of much higher excitation than normal HII regions can be seen. It should also be noted that Seyferts are preferentially early type spirals which have bulges resembling "small elliptical galaxies". Thus in many respects the NLR in Seyferts looks like a scaled-down version of your extended gas in Seyferts. It is suggestive that in both cases infall of some gaseous debris plays a major role to explain the observations.

HOST GALAXIES OF QUASARS

J.W. Fried
Max-Planck-Institut für Astronomie
D-6900 Heidelberg 1
West Germany

SUMMARY. Properties of host galaxies of low redshift quasars are reviewed. There is good evidence that radio-loud quasars sit in bright elliptical galaxies, whereas optical and X-ray quasars reside in spiral galaxies. The host galaxies of radio-loud quasars are brighter by ~ 2 mag than those of radio-quiet ones; the brightest ones have luminosities comparable to those of brightest cluster galaxies. The neighborhood of quasars has higher than average galaxy density; tidal interactions with the host galaxies are frequently observed.

A preliminary analysis of a survey of 15 radio-loud quasars with redshifts from 0.3 to 3.4 yields the result expected for cosmological redshifts that host galaxies and neighboring galaxies are detected only for the low-redshift objects.

Very deep surfacephotometry of 2 radio-loud low-redshift quasars shows that in one case (1635+119, z = 0.146) the host galaxy can very well be fitted by an elliptical galaxy over 3 orders in intensity; the host galaxy of 4C14.83 (z = 0.237) is strongly asymmetric and can not be well fitted by either elliptical or spiral galaxy.

1. INTRODUCTION

Though quasars had originally been defined to be unresolved point sources - at least on POSS plates -, direct imaging with high spatial resolution has undoubtedly shown that all low-redshift (z $\lesssim$ .5) quasars are resolved into a point source and surrounding host galaxy. Clues to the still poorly understood quasarphenomenon can be derived from studies of the immediate neighborhood of the quasar, including not only the host galaxy but also nearby objects. During the last years a large amount of observational material - both surface photometry and spectroscopy - has been collected on low-redshift quasars; the main conclusions drawn from these studies are summarized in section 2. New results on a sample of radio-loud quasars in the redshiftrange from 0.3 to 3.4 are presented in section 3, and surfacephotometry of the host galaxies of 2 radio-loud quasars with unprecedented S/N is presented in section 4.

2. LOW-REDSHIFT QUASARS

2.1 Direct imaging

Since the early work of Kristian (1973), who showed that some quasars are surrounded by a faint fuzz, several high resolution imaging studies (Hutchings et al. 1981, 1982, 1984, Wyckoff et al. 1981, 1983, Gehren et al. 1984, Malkan 1984, Malkan et al. 1984) have shown that all low-redshift quasars are resolved if observed carefully enough. Decomposition into point source and underlying galaxy is done either by subtracting a properly scaled stellar image or by fitting a model to the observed azimuthally averaged radial intensity profile. There is general consensus that this fuzz has the photometric properties of a galaxy. However there is much confusion concerning the morphological types of these galaxies; since much of this confusion seems to be due to mixing of data of host galaxies of different quasar types (optical, radio, X-ray), the published data are reviewed according to these types. Except for the large sample of 78 quasars of Hutchings et al. (1984) I concentrate on data samples obtained with linear digital detectors.

a) optical (radio-quiet) quasars

Hutchings et al. (1984) have collected data for 23 optical quasars; they find evidence for spiral structure for 67% of these. In the sample of Gehren et al. there is only 1 well resolved quasar; radial intensity profile and structure of the host galaxy are characteristic of spirals. Both groups agree that the host galaxies of optical quasars are fainter by $\sim$ 2 mag than the host galaxies of radio-loud quasars. Gehren et al. derive $\langle M_r \rangle = -20.8$ and as metric diameter $\langle D \rangle = 37$ kpc. Colors for 3 optical quasars measured by Malkan (1984) are the same as early-type spiral galaxies.

b) radio-loud quasars

The sample of Hutchings et al. (1984) contains 26 radio-loud quasars; spiral structure is indicated for 12% of these objects. For 9 radio-loud objects in the sample Gehren et al. (1984), the azimuthally averaged radial intensity profiles of the host galaxy can be well fitted by a Hubble law. They find that radio-loud quasars are located in giant elliptical galaxies with $\langle M_r \rangle = -22.8$ and metric diameters $\langle D \rangle = 128$ kpc. Thus the radio quasars reside in larger and brighter galaxies than the optical quasars. The same conclusions have been reached by Malkan (1984) for a sample of 23 quasars.

c) X-ray quasars

Hutchings et al. (1984) find evidence for spiral structure for 59% out of 29 X-ray emitting quasars. Malkan et al. (1984) conclude from radial profiles, linear extent, ellipticity and color of 24 quasars that X-ray quasars typically reside in normal spiral galaxies.

So there is good evidence that optical and X-ray quasars sit in normal spiral galaxies, whereas radio-loud quasars sit in giant ellipticals. (Bl-lac objects have up to now been studied too sparsely to draw statistically reliable conclusions). It should be stressed that due to the frequently found tidal interactions (see section 2.3) such morphological classifications may grossly oversimplify the real situations. Thus there appears a morphological continuity from Seyfert galaxies to optical quasars and radiogalaxies to radio-loud quasars. This continuity is also reflected in the nucleus to galaxy luminosity ratios. Gehren et al. 1984 and Hutchings et al. 1984 find $\langle L_N/L_G \rangle \simeq 2$, though with large scatter; radio-loud quasars appear to have higher $\langle L_N/L_G \rangle$ than radio-quiet ones. Yee (1983) finds $\langle L_N/L_G \rangle \lesssim 1$ for Seyfert 1 and $\langle L_N/L_G \rangle \lesssim 0.1$ for Seyfert 2 galaxies.

Correlations between the luminosities of the nuclei and host galaxies have been found by Hutchings et al. (1984) and Gehren et al. (1984) in the sense that the more luminous nuclei reside in more luminous galaxies. The mass of the galaxy and the nuclear engine therefore appear to be correlated.

The Hubble Diagram of quasar host galaxies (Wehinger et al. 1983) shows a large scatter which is probably due to an intrinsic scatter of absolute magnitudes. The brightest host galaxies are comparable in luminosity to brightest cluster galaxies.

2.2 Spectroscopy

Spectra of the fuzz surrounding the quasars (c.f. Boroson and Oke (1984)) have definitely shown that this fuzz consists of starlight sharing the redshift of the quasar. For a sample of optically selected quasars Boroson et al. (1982) find continuum emission from starlight arising presumably in spiral galaxies. Fuzz spectra for 8 radio-loud high luminosity quasars obtained by Boroson and Oke (1984) show either a blue continuum with strong emission lines (group I) or a red continuum with weak or absent emission lines (group II). This separation is well correlated with properties of the nucleus such as no FeII emission and steep spectral radio index for group I, strong FeII emission and flat radio spectral index for group II. For group II objects the fuzz emission can be attributed to a bright elliptical galaxy population, whereas for group I objects the blue colors require significant contribution from a young population which could result from a burst of starformation. The luminosities derived from the spectra are consistent with the fuzz being a luminous galaxy.

2.3 Tidal interactions and environment

The deep imaging surveys have shown that the local extragalactic environment of quasars has higher than average density. For example, Gehren et al. (1984) find an excess of galaxies (m = 21...23) relative to the predicted background density by factors ranging from 2 to 16 within 150 kpc projected distance from the quasar. The physical association of quasar and companion galaxies has been shown spectroscopically by Heckman et al. (1983) (since the distances are undisputed for galaxies, these observations present very strong evidence for quasars being at their cosmlogical distances too). French and Gunn (1983) and Yee and Green (1984) have provided statistically relevant evidence that galaxies cluster around quasars. Yee and Green find indications that radio-loud quasars have a higher than average clustering amplitude than radio-quiet quasars and that the spatial covariance function steepens at small distances from the quasars. These results and the velocity differences of $\approx$ 1000 km/sec between quasar and companion galaxies found by Heckman et al. (1984) indicate that quasars preferentially reside in high-central-density groups or small clusters. However, since these studies are severely limited either in magnitude or in magnitude and spatial coverage of quasar environment these conclusions are preliminary. Clearly, better data are needed.

Given the high local galaxy density, interactions with the host galaxy of the quasars have to be expected and are actually observed. Impressive examples are given by Bothun et al. (1982), Hutchings and Campbell (1983), Stockton and MacKenty (1983) and MacKenty and Stockton (1984). Gehren et al. (1984) find clear evidence for tidal interactions for 3 quasars out of their sample of 17 or 18%. Hutchings et al. (1984) find evidence for interaction for 8% of their sample, and probably interacting are 28%; more interactions are indicated for radio-loud (35%) and X-ray (31%) than for optical quasars (17%). This supports the speculation that nuclear activity may be associated with interactions. For other active galaxies there is also evidence that a large fraction are interacting or disturbed, but not statistically substantiated (c.f. Balick and Heckman 1982).

3. HIGH REDSHIFT QUASARS

Up to now studies such as the ones referred to had been concentrated on low-redshift objects because of the simple reason that iosphotal diameters decrease rapidly with redshift. For example, a galaxy which has an isophotal diameter of 40" at z = 0.2 will be easily detectable, but the same galaxy will have only 4" diameter at z = 1.5 and be buried under the seeing disk. The K-correction will further decrease the isophotal diameter. We therefore expect not to resolve the host galaxies of quasars for redshifts in excess of 0.8 by ground-based observations if the quasars are at cosmological distances.

Though a negative result had to be expected, I observed a sample of 15 radio-loud quasars in the redshift range 0.3 to 3.4. Except for S50014+81 (z = 3.4, Kühr et al. 1984) all objects were selected from the list of Hewitt and Burbidge for observability during the allocated telescope time. Only the two low-redshift objects (z = 0.3, 0.5) in the sample could be resolved.

Since the low-redshift quasars are located in groups or clusters of galaxies, these images were searched for galaxies. No galaxies in excess of the predicted background density could be found in a preliminary analysis. Again this negative result was anticipated, since cluster galaxies with M = -23 would appear at $r \lesssim 25$ for $z \gtrsim 1.5$, well beyond the detection limit of m = 24.

These new data on high-redshift objects are consistent with cosmological distances of the quasars. A full analysis is in preparation (Fried and Gehren 1985).

4. TWO WELL-STUDIED LOW-REDSHIFT QUASARS

In order to establish defintely the morphological difference between radio-quiet and radio-loud quasars, I took several deep CCD exposures for each object of a sample of 6 quasars. Unfortunately all frames obtained for radio-quiet quasars were lost due to flexure problems, but excellent data down to about μ_r = 28 mag/☐ " do now exist for 2 radio-loud quasars. Fully 2-dimensional fits of model galaxies show:

<u>1635 + 119</u>: The host galaxy is well fitted by a giant elliptical over 3 orders of intensity; a spiral galaxy clearly does not fit the data.

<u>4C14.82</u>: The host galaxy is strongly asymmetric at low light levels ($\mu_r \lesssim 24$). Therefore neither elliptical or spiral galaxy are a good fit. Since 4C14.82 has been found to be located in a group or cluster (Gehren et al. 1984), this asymmetry could be the result of a tidal interaction.

A full analysis will be published elsewhere

LITERATURE

Balick, B., Heckman, T. 1982, Ann. Rev. Astron. Astrophys. **20**, 431
Boroson, T., Oke, J. 1982, Nature **296**, 397
Boroson, T., Oke, J. 1984, Ap.J. **281**, 535
Boroson, T., Oke, J. Green, R. 1982, Ap.J. **263**, 32
Bothun, G., Mould, J., Heckman, T., Balick, B., Schommer, R., Kristian, J. 1982, A.J. **87**, 1621
French, H., Gunn, J. 1983, Ap.J. **269**, 29
Fried, J., Gehren, T. 1985 in preparation
Gehren, T., Fried, J., Wehinger P., Wyckoff, S. 1984, Ap.J. **278**, 11
Heckman, T., Bothun, G., Balick, B., Smith, E. 1984, A.J. **89**, 957
Hutchings, J., Campbell, B. 1983, Nature **303**, 584
Hutchings, J., Crampton, D., Campbell, B., Gower, A., Morris, S. 1982 Ap.J. **262**, 48
Hutchings, J., Crampton, D., Campbell, B., Pritchet, C. 1981, Ap.J. **247**, 743
Hutchings, J., Crampton, D., Campbell, B. 1984, ap.J. **280**, 41
Kühr, H., Liebert, J. Strittmatter, P., Schmidt, G., MacKay, C. 1983, Ap.J. Letters **275**, L33
Kristian, J. 1973, Ap.J. Letters **179**, L1
MacKenty, J., Stockton, A. 1984, Ap.J. **283**, 64
Malkan, M., Margon, B., Chanan, G. 1984, Ap.J. **280**, 66
Malkan, M. 1984, Ap.J. **287**, 555

Stockton, A., MacKenty, J. 1983, Nature **305**, 678
Wyckoff, S., Wehinger, P., Gehren, T. 1981, Ap.J. **247**, 750
Wyckoff, s., Wehinger, P., Gehren, T., Fried, J., Spinrad, H., Tapia, S., 1983 "Quasars and Gravitational Lenses", 24th Liege Astrophysical Colloquium
Yee, H. 1983, Ap.J. **272**, 473
Yee, H., Green, R. 1984, Ap.J. **280**, 79

DISCUSSION

VIGNATO: Does the pointspread function subtraction give unique results?

FRIED: The scaling of the psf is determined interactively by trial and error; variation of the scaling factor within reasonable limits leads to magnitude differences of the host galaxy of about 0.5 mag. The radial intensity profile, however, is essentially unaffected outside the seeing disk.

PISMIS: I shall take the liberty to comment on a matter concerning semantics, namely on the use of the expression "host galaxies" for the faint outer regions detected around some quasars. An imaginative person has coined the word and I admit that it is a "cute" one but is it necessary to introduce a new terminology? I think not. Astronomy has a good many words of that sort: misnomers that have persisted all along. In that same vein should we also refer to, say, Seyfert galaxies as hosts to the active nucleus?

OBSERVATIONS OF FAINT QUASARS: SHAPE AND EVOLUTION OF THE LUMINOSITY FUNCTION

David C. Koo
Space Telescope Science Institute
3700 San Martin Drive
Baltimore, Maryland 21218
U.S.A.

ABSTRACT. We highlight results from a study of faint quasars in SA 57 using 4-m photographic multicolor photometry, astrometry, variability, and CCD spectroscopy. Quasar candidates with B < 22.6 were chosen as stellar-like objects with colors unlike normal stars. Together, these techniques allow additional corrections for white dwarfs, hot subdwarfs, and narrow-emission-line galaxies in the sample and yield a quasar density of about $150/deg^2$ to B = 22.6. A major surprise was that no quasars were found with redshifts above 2.54 among the 8 bona-fide quasars fainter than B = 21.5, a result strongly supported by the relative paucity of non-ultraviolet excess candidates. We discuss the importance of our faint counts and redshifts in defining the shape and evolution of the luminosity function (LF) of quasars. In general, the simplest fit to our faint data is made by assuming mainly luminosity evolution of the local LF of Seyfert 1 galaxies, with luminosity increasing by $(1 + z)^{4.0}$ (low q_o, spectral index $\alpha=1$) up to redshifts about 2 to 3, beyond which we find a decline in the volume density.

1. INTRODUCTION

Quasars fainter than 20th magnitude provide important clues to the shape and evolution of the LF of quasars. In our first survey of such quasars using deep 4-m plates for multicolor photometry of SA 68 at 0015+15 (Koo and Kron 1982), we discovered a flattening of the slope of the number counts, a result which strongly excludes a pure density evolution model for quasars. Although we showed that the counts were almost perfectly fit by a pure luminosity evolution model proposed by Braccesi et al. (1980), more complex scenarios could not be excluded without redshifts. Recently, we have completed a similar survey in SA 57 at 1305+29 near the North Galactic Pole (Koo, Kron, and Cudworth 1985). The availability of plates taken at six epochs spanning 10 years allowed us to explore the astrometric and variability property of our candidates, as was done by Kron and Chiu (1981). Of more importance, we have undertaken a spectroscopic survey of a subsample of these candidates (Koo and Kron 1985, preliminary results were

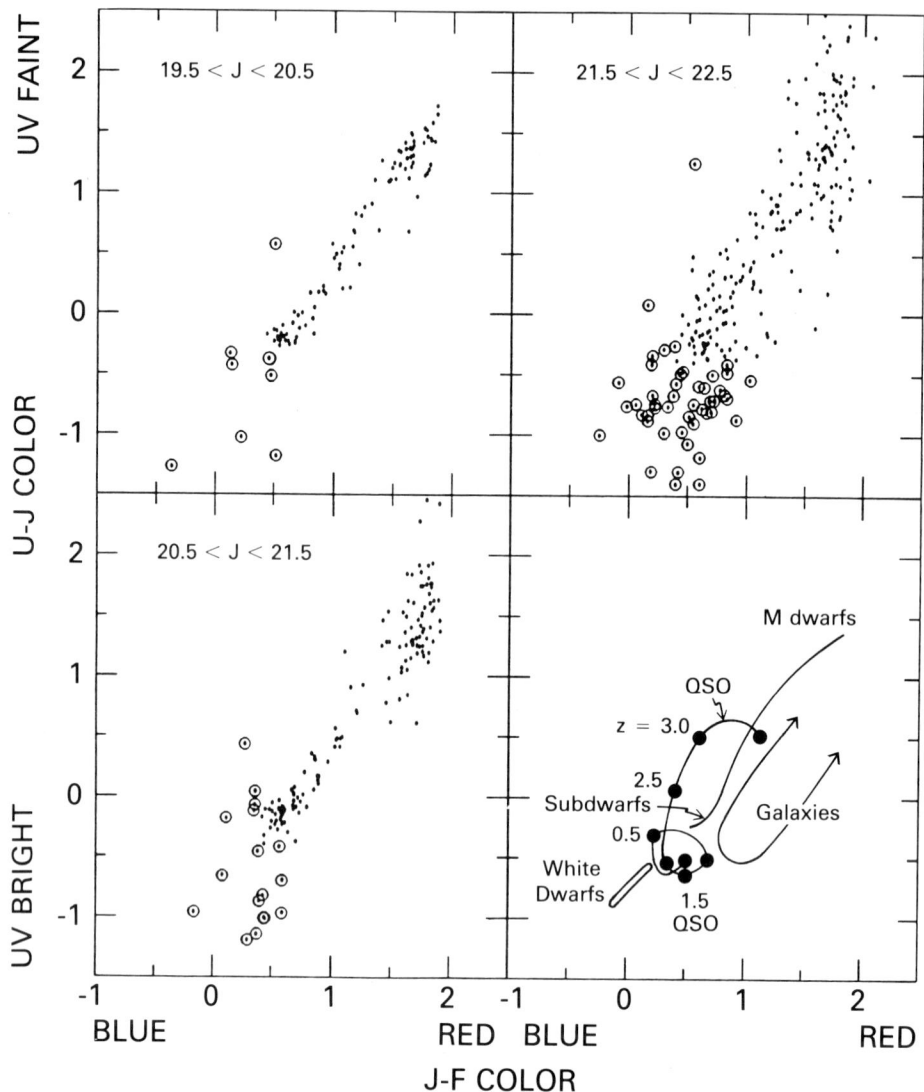

Figure 1: a) Two-color plot of all stellar images in SA 57 over an area of 0.30 deg^2 in the blue (J) magnitude interval 19.5 to 20.5 (B ~ J + 0.1). Circled dots are those chosen as quasar candidates. U-J is close to the standard U-B; J-F is close to B-V. Note that the vertical axis is reversed from the usual UBV diagram. b) Same as for a) except for J between 20.5 and 21.5. c) Same as for a) except for J between 21.5 and 22.5. d) Two-color plot showing positions of blue subdwarf halo stars, red main sequence dwarfs, hot white dwarfs, faint galaxies, and quasars at a variety of redshifts marked at every 0.5 interval along the dashed path.

reported by Koo 1983a and 1983b). The aim of this paper is to highlight some of the results of these recent surveys, especially with regard to the shape and evolution of quasars.

2. PHOTOMETRY AND SELECTION OF QUASARS

The quasar candidates in both SA 68 and SA 57 were extracted from a much larger photometric catalog of stars and galaxies covering 0.3 deg^2 in each field. These catalogs contained photographic UJFN photometry (our UJFN bands are close to the standard UBVI bands) derived from prime focus plates taken at the Mayall 4-m telescope at Kitt Peak National Observatory and reached a limit of about B = 24. Photoelectric standard stars in UBV were kindly provided by Sandage and used for calibrating our fields. Image sizes discriminated stellar-like objects, such as Galactic stars and quasar candidates, from fuzzy galaxies to a brighter limit of 23 mag. Details of the completeness, photometric errors, and techniques used in the scanning, photometry, calibration, and classification of images are discussed by Kron (1980), Koo (1981), and Koo and Kron (1982).

In each field about 2000 stellar-like objects were found. Quasar candidates were selected from these by using our equivalent of the UBV two-color diagram to pick out all objects which were separated from the loci expected for faint but normal Galactic stars (see also UBV work by Usher 1981 for a brighter survey in SA 57 using Schmidt plates). We emphasize that the technique is not limited to UV-excess objects. The near-infrared band (N), though available and in principle an added discriminant (Braccesi et al. 1980), was not found to be useful in finding additional candidates without increasing the number of contaminations by Galactic stars. In SA 57, 76 candidates were found to J < 22.5 which can be compared with 65 in SA 68.

Our selection technique is best visualized in Fig. 1a to 1c, which show the two-color position of all stellar-like objects (dots) in the designated blue (J) magnitude range and those objects selected as quasar candidates (open circles). The selection was a subjective one based upon the two-color dispersion of probable subdwarfs at each magnitude interval. We may easily have missed a few objects that lie close to the subdwarfs, but these would increase our sample by about only 10% to 20%. Fig. 1d displays the expected positions for a variety of objects as well as the approximate two-color path taken by an "average" quasar spectrum at different redshifts. The quasar colors are for illustrative purposes and should not be taken too seriously since observed quasars exhibit a wide variety in their spectral energy distributions and hence colors, especially beyond a redshift of about 3 when the exact colors are sensitive to a "forest" of Lyman alpha absorption. We expect our multicolor approach of finding quasars to be useable up to redshifts beyond 3.

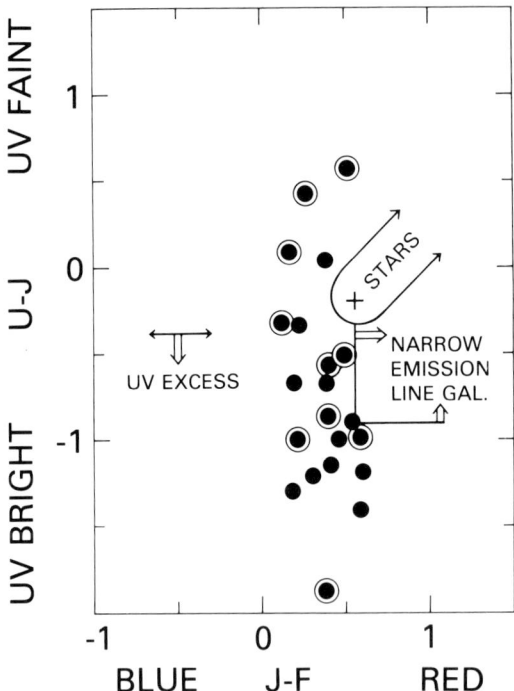

Figure 2: Two-color plot, U-J vs J-F, showing positions of quasar candidates showing significant variability at two or more epochs among the six available. Circled objects are those which have been confirmed spectroscopically to be bona-fide quasars.

3. VARIABILITY AND PROPER MOTIONS

Our photographic data in SA 57 includes eight 4-m blue plates taken at 6 epochs spanning a total of 10 years. We plan to use these plates to search among all stellar objects for variables and objects of zero proper motion in order to obtain an independent sample of very faint quasars free from the selection effects of broadband color or low resolution slitless techniques. Such an investigation should place excellent constraints upon the possibility that very high redshift or unusual quasars are hiding among the Galactic stars. At present we have only studied the 76 quasar candidates in SA 57. A similar study was first made for a smaller overlapping central region of 0.1 deg^2 by Kron and Chiu (1981).

For objects as faint as J = 22, we can detect variability larger than 0.05 mag rms per plate and proper-motions as small as 0.2 to 0.3 arc sec rms per century. This latter value is of high enough precision to detect motions at the 2 to 3 sigma level of typical halo subdwarfs to distances about 10 kpc. Details of this aspect of the program can be found in Koo, Kron, and Cudworth (1985).

Fig. 2 shows the result of our variability study. Among the 76

OBSERVATIONS OF FAINT QUASARS

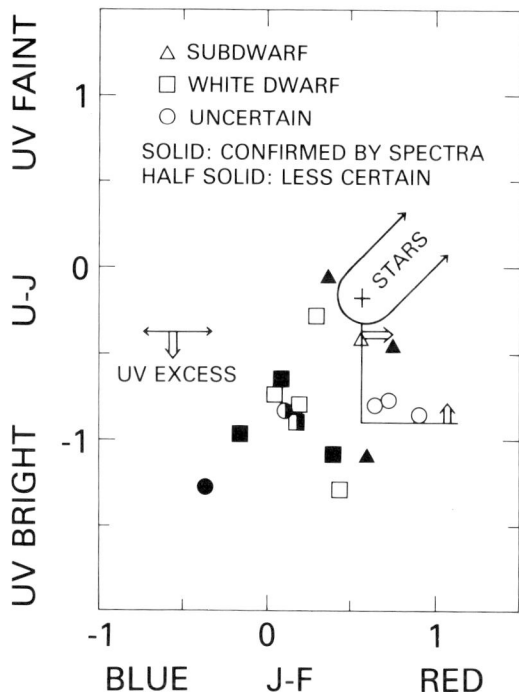

Figure 3: Two-color plot, U-J vs J-F, for all quasar candidates showing significant proper motions. Reduced proper motions are used to further divide the proper-motion sample into likely subdwarfs and likely white dwarfs. Objects which have been confirmed to be Galactic stars are circled.

quasar candidates, almost one-third (21) exhibited significant variability, where significant was defined to be at least two occurrences, among the 6 available epochs, of magnitude deviations from the mean at the two to three sigma level. Among these 21 variables, about half (10) have already been confirmed with spectroscopy to be broad-emission-line quasars (see discussion below). No such variable objects were found among the spectroscopic sample of 15 non-quasars: white dwarfs, subdwarfs, and narrow-emission-line galaxies. Thus among the final subsample of 45 quasar candidates (see section 6) which are believed to be bona-fide quasars, the 21 significant variables represent nearly 50% completeness. This completeness can certainly be raised by lowering the variability threshold, by securing more epochs, or by improving the photometry or statistical tests of variability.

The result of our astrometry measurements can also be displayed in the two-color plot, Fig. 3, which shows those quasar candidates possessing significant (>2.5 sigma) proper motions. The reduced proper motions further classify these Galactic stars into white dwarfs or subdwarfs as marked in the Fig. Again, spectroscopic followups have generally confirmed our classifications, although some of these

stars do show unusual colors, possibly associated with binary systems
or photometric errors. The main point is that white dwarfs and a few
subdwarfs are well discriminated when proper motions are used; this
allows a further refinement of our classification of the quasar
candidates.

4. SPECTROSCOPY

Moderate resolution spectroscopy of our quasar candidates has been the
most difficult set of observations but vital to check the nature and
redshift distribution of the true quasars. Almost all spectra were
measured with the "Cryogenic Camera System" on the 4-m at Kitt Peak
National Observatory. The main advantage of this system is that
multiple spectra can be acquired simultaneously with the use of a
custom-made aperture mask covering about a 5 arc min diameter field
and having a dozen miniature slits or several dozen round holes. A
transmission-grating prism combination (grism) is used for the
dispersing element; a low-readout-noise charge coupled device (CCD) of
800 x 800 pixels is the detector. In general the spectral coverage is
between 4500A to 7500A and the resolution is about 15A FWHM spanned by
four pixels. Exposures lasted one to several hours for each aperture
mask.

Table I. Summary of Spectroscopic Survey in SA 57

Blue Mag	No. Obs.	QSO	Star	Narrow	Unknown
< 20.5	11	7	4	0	0
20.5 - 21.5	6	3	3	0	0
21.5 - 22.5	19	9	2	6	2

Table I summarizes the current status of our spectroscopic survey
in SA 57. Spectra of quasar candidates in two other fields (SA 68 and
1720+50) have also been measured and used in deriving the LF of faint
quasars (see below) but are not included in Table I. The QSO class
includes only those objects which show one or more broad emission
lines. The star class includes only those objects which show
significant absorption features at Hα and/or Hβ; broad absorption
lines were assumed to belong to white dwarfs. The narrow class refer
to objects whose spectra showed only unresolved emission lines,
usually Hβ and [O III] 4959, 5007 and/or the doublet [O II] 3727.
Based upon their redshifts between 0.3 to 0.6, their very blue
intrinsic colors, their blue absolute magnitudes being close to that
of field galaxies, the ratio of the strengths of the emission lines,
and their near-stellar image size (as a group, these objects are
slightly more extended than the stars or genuine quasars), we suspect
that the narrow class objects are compact Markarian or Haro galaxies

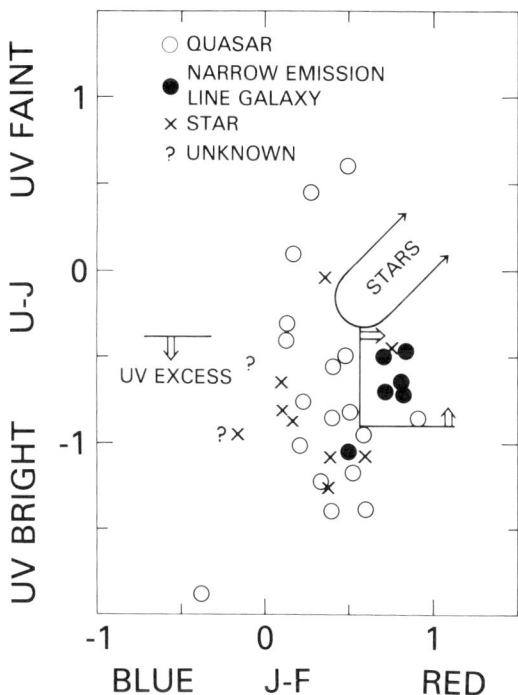

Figure 4: Two-color plot, U-J vs J-F, for all quasar candidates in SA 57 with spectra.

seen at high redshift. Seyfert 2 galaxies and very bright extragalactic HII regions of low metallicity are also possibilities. Finally the unknown class includes spectra with signal-to-noise ratios too low for any identification or with no strong absorption or emission features.

For the complete sample of 13 candidates brighter than J = 20.53, 8 are confirmed quasars and 5 are Galactic stars, 3 of which are brighter than 19.3. Of the 4 remaining candidates to 21.5, 2 are quasars and 2 are Galactic stars. The 19 candidates fainter than J of 21.5 are of special interest, for they represent the deepest spectroscopic sample of quasars with excellent photometry and reasonable completeness. Among these, only 9 appeared to be bona-fide quasars, 6 are narrow emission line galaxies, 2 are probably stars, and two have no recognizable features. Although the redshifts remain uncertain for several of the faint quasars with only a single broad emission feature, in most cases our best guess is that the redshifts are low, typically around 1.3 to 2.4. We judged the broad line to be Mg II 2798 or [CIII] 1909, since both Lyman α and CIV 1550 are both strong and ideally placed within our spectral range. The highest redshift among this faint sample is only 2.54; the one quasar with higher redshift of 3.07 was in fact the second brightest quasar at B = 19.6 with a red U-J of 0.57 and J-F of 0.51.

Fig. 4 displays the two-color positions of all quasar candidates in SA 57 which have received spectroscopic attention. The narrow class generally occupies a region redward of J-F about 0.55 and U-J near -0.9; 3 other such candidates found in SA 68 also lie in the same portion of the plot. This two-color segregation between the narrow class and quasar class enables us to statistically correct our candidate list in both fields and thus further refine our sample and counts of faint quasars. Further details of our spectroscopic survey are discussed by Koo and Kron (1985).

5. NUMBER COUNTS OF FAINT QUASARS

Exploiting the combination of photometry, astrometry, variability, and spectroscopy, we can provide a more accurate assessment of the nature of each of our quasar candidates originally chosen on the basis of colors. For an individual object, the classification may occasionally be in error, but statistically the reclassifications should be far superior to the use of only colors. Until we extend our variability and proper-motion survey to all stellar objects in our field or until other techniques, such as faint CCD slitless surveys (Koo and Kron 1980, Schneider, Schmidt, and Gunn 1983), are undertaken, we cannot exclude the possibility that quasars may in fact be hiding in large numbers among the Galactic stars in the two-color plot. Based upon the presentation of Marano (this conference), however, the numbers are low, probably about 20%. Among photographic slitless surveys covering SA 57 by Hoag (private communication) and Weedman (1985), none have yielded a quasar candidate not already found by our color technique. The results of our SA 57 surveys are tabulated in Table II.

Table II. Classification of Quasar Candidates in SA 57 (0.30 deg^2)

Blue (J) Mag	All	White Dwarfs	Sub-Dwarfs	Narrow	QSO	No./deg^2
<20.5	11	1	3	0	7	23 ± 9
20.5-21.5	17	2	2	1	12	40 ± 12
21.5-22.5	48	5	1	16	26	87 ± 17

Our total count of 65 quasar candidates in SA 68 (Koo and Kron 1982) is close to that of 76 in SA 57. Unfortunately, we do not have sufficient data in SA 68 to apply the spectroscopic, astrometric, and variability corrections to make a precise comparison. For purposes of obtaining counts, however, we note that the major correction at faint magnitudes are the narrow-emission-line galaxies, which can be identified by colors alone. After applying corrections for Galactic stars in proportion to that found in SA 57, we find 7, 42, and 87 quasars/deg^2 in SA 68 for the three bins of Table II.

These new values of faint quasar counts are approximately 30% lower than our original estimates made in SA 68 and thus strengthen our conclusion that a flattening of the number counts and convergence of the light of quasars have been detected (the effect of 20% incompleteness is negligible and would make our uncorrected counts closer to the true values). This conclusion can be made independent of the possibility that substantial reddening exists in our field (Braccesi 1983); moreover, since the two fields agree so well, and since the North Galactic Pole field in SA 57 is unlikely to possess much reddening, we can feel more confident that our counts in both fields are representative rather than statistical flukes. At magnitudes brighter than 20.5 or so, our counts appear to be low with respect to that in the Braccesi field studied by Marshall et al. (1984), but in support of our low counts, the recent work by Marano, Zamorani, and Zitelli (1984) and Marano (this conference), who used similar two-color techniques with equally deep plates in another field, yielded counts fully consistent with our own. Our counts also agree with the results of the excellent survey of Boyle et al. (this conference), which indicate a quasar density of $50/deg^2$ brighter than J of 21; we obtain 40 ± 12. (Kron and Chiu 1981 claimed a total of 80 quasars/deg^2 to B of 21 based upon 8 likely quasars found in an area of 0.1 deg^2; our new work confirms their quasar identifications but suggests that their counts were high by a factor of two relative to our new average over an area three times larger.) We are presently studying 6 other 4-m fields to evaluate the extent to which field-to-field fluctuations are a calibration problem, a result of extinction, or a genuine feature giving clues about large scale clustering of quasars at high redshift (Oort 1983). Differences are within factors of two, an insignificant amount when discussing the reality of the flattening of the quasar counts. In fact, the predictions of pure density evolution models (e.g. Marshall et al. 1984) would exceed the <u>total number of stellar objects</u> (about 5 to 10 times the number of quasar candidates) by J of 22.5! Thus our conclusion that some flattening has been detected is a secure one.

6. COLORS AND REDSHIFTS OF FAINT QUASARS

A major surprise even before we had measured redshifts was the noticeable paucity of non-ultraviolet excess quasar candidates, which presumably would have redshifts generally greater than 2.3 or so (see Fig. 1d). As a check, we located in the sample of Véron-Cetty and Véron (1984) all quasars which had a redshift of 2.5 or greater and UBV photometry. The results of the search are shown in Fig. 5. Indeed most, but not all, of these high redshift quasars did not have an ultraviolet excess. The numbers are too small for accurate estimates of our incompleteness, but it is reassuring to find relatively few such quasars (presumably found by radio or slitless techniques) in the two-color positions of stars. At most, one half of the high-redshift quasars are missed with our multicolor method. Among quasars with redshifts 3 or greater, at least half were located in the upper left portion of the UBV plot, well separated from positions of most faint Galactic stars as well as ultraviolet excess

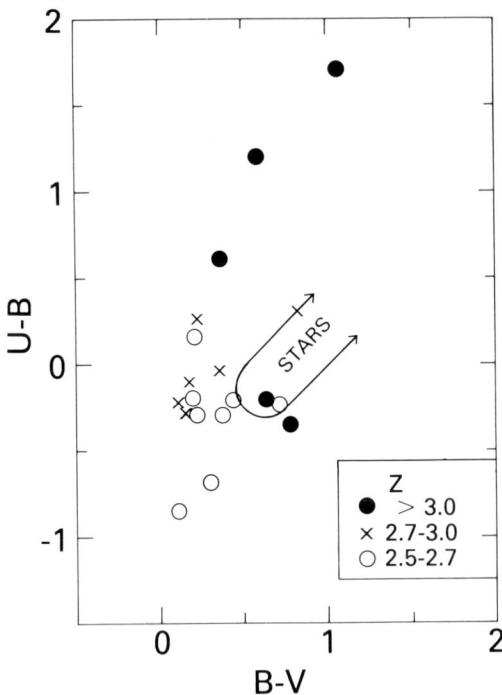

Figure 5: Two-color plot, U-B vs B-V, of all quasars which have redshifts greater than 2.5 and UBV photometry in the catalog of Véron-Cetty and Véron (1984). Note that the vertical axis has been flipped with respect to the usual UBV plot so that the plot approximately matches the previous UJF plots.

quasars. The corresponding area of our plots show few such objects (see Fig. 1a to 1c). The implication is that the vast majority of faint quasars are actually at relatively low redshifts. This conclusion has been confirmed by spectroscopy and provides an important hint regarding both the shape and evolution of faint quasars.

With redshifts and an estimate of our incompleteness, we can of course directly plot the LF. Including redshifts for quasar candidates in fields other than SA 57, we have obtained a total of 23 redshifts for genuine quasars, including one from Usher, Warnock, and Green (1983) and a few from Kron and Chiu (1981). Although our redshift survey is far from being totally complete, we have presumed that our present sample of redshifts is representative of the entire population. Dividing our 23 redshifts into four redshift bins, each with three magnitude bins, we plot the resultant very noisy LF in Fig. 6. Also included for comparison is the recent determination of the local LF of the nuclei of Seyfert 1 and 1.5 galaxies by Cheng et al. (1985) as well as that of galaxies. The bright parts of the quasar LF for $z < 2.2$ was generously provided by H. Marshall and is

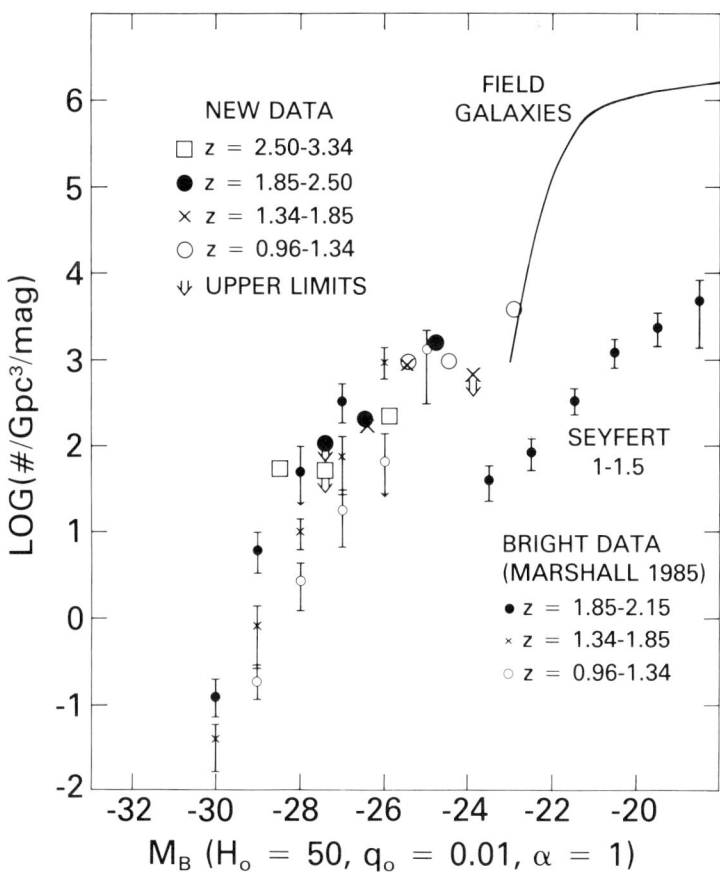

Figure 6: Luminosity function of quasars, Seyfert 1 and 1.5, and galaxies assuming a Hubble constant of 50 km-sec^{-1}-Mpc^{-1}, a deceleration parameter q_o of 0.01, and a spectral index for the continuum of 1 (i.e. no K correction). The bright portions of the quasar LF were kindly provided by Marshall (private communication); the Seyfert portion is from Cheng et al. (1985); the galaxies are from the Schechter LF parameters derived from a recent redshift survey (Ellis 1982). Our new faint quasar data are shown as large symbols with upper limits plotted as downward-pointing arrows at positions corresponding to a single detection; error bars have not been included but a factor of 2 or 3, i.e. ±0.3 to ±0.5 in the ordinate, should be assumed to account for our small number of objects in each bin and incompleteness (mainly for $z > 2.5$).

based upon the ultraviolet-excess quasar data used by Marshall (1985). Before we discuss the constraints posed by our data on the shape and evolution of quasars, a brief overview of the major models is presented.

7. OVERVIEW OF QUASAR EVOLUTION MODELS

In the last few years a number of models for quasar evolution have been proposed, mainly in the form of simple analytical expressions to describe the number-magnitude-redshift distribution of quasars. As emphasized by Weedman in his contribution to this conference, the parameterization of the LF and evolution of quasars is not merely an exercise of statistics but may actually have profound implications for our understanding of the physical mechanisms that produce quasars as well as for our interpretation of the relationship between Seyferts and quasars. The following summarizes the models used for quasars and the importance of our faint data in the form of counts, colors, and redshifts to discriminate between them.

The major contending models have been the pure density evolution model (PDE), the "pure" luminosity evolution model (PLE), the luminosity-dependent density evolution model (LDDE), and possibly some hybrid or more complex form of the evolution involving both luminosity and density (LDE) and maybe even the shape of the LF. There is also the question of a redshift "cutoff" for quasars, i.e. a lack of high redshift quasars; this may be viewed as a relatively abrupt change or a special epoch in what would otherwise be a smooth extrapolation of the behavior of the evolution at smaller redshifts.

Until recently, PDE as originally proposed by Schmidt (1968) was probably the most widely accepted, not only because all available data could be fit within this framework but also because of its conceptual simplicity. In this picture, the past LF was the same in shape as the local LF but shifted uniformly, i.e. by an amount independent of luminosity, to higher densities. One could thus legitimately talk about a much higher density of quasars in the past, independent of qualifiers regarding luminosity or the relationship between Seyferts and quasars.

Within a few years, Mathez (1976,1978) suggested that PLE models provide equally valid descriptions of quasar data. In this picture, the shape is again preserved and the evolution involves a uniform shift to brighter luminosities in the past of long-lived individual objects or a statistical ensemble of short-lived events.

The LDDE has been recently suggested by Schmidt and Green (1983) for optical quasars and is a generalization of the two-population model of Petrosian (1973) or Weedman (this conference). This form of the evolution is also popular among radio astronomers (e.g. Peacock and Gull 1981) and basically suggests that the amount of density evolution is dependent upon luminosity (or radio power), usually with the most luminous sources evolving most rapidly. Although this model does involve a change in the shape of the LF, only two parameters are usually used to describe the evolution rather than some more complex

OBSERVATIONS OF FAINT QUASARS

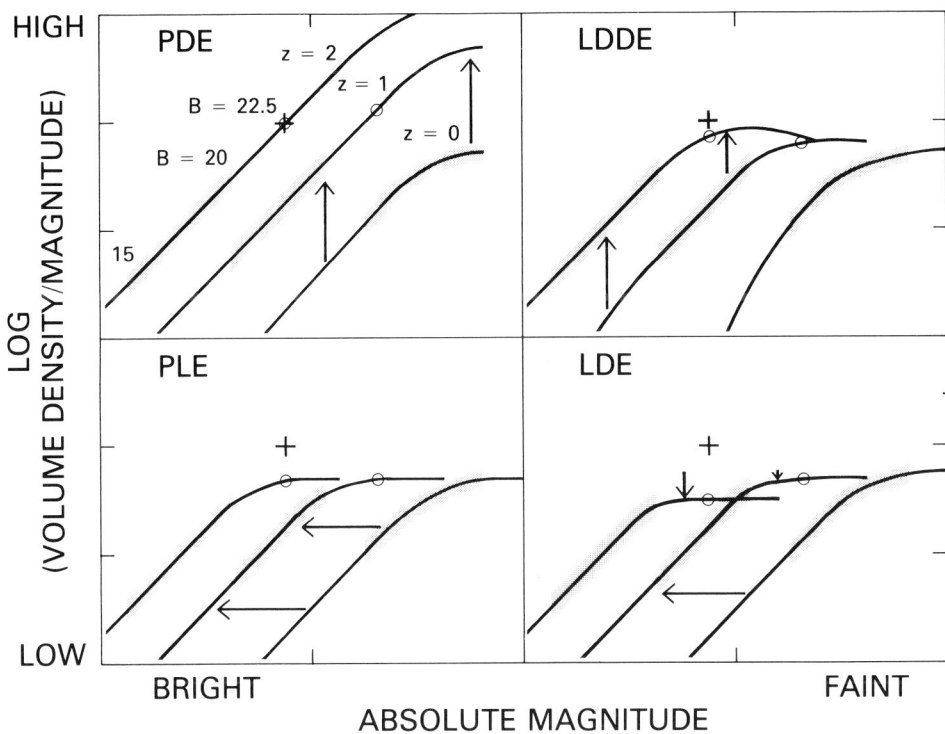

Figure 7: Panel showing schematically the differential LF at three different redshifts for four popular models of quasar evolution. The shaded regions represent the available observational data from B = 15 to 20. Circles indicate predictions for B ~ 22.5. Cross is used as a fiducial mark.

function. Note that in contrast PDE and PLE can in principle require only one evolution parameter.

The LDE models, on the other hand, can be as simple as a two-parameter change, one for luminosity and one for density, but with the shape of the LF preserved or as complex as one desires. The simple form has been suggested by Jaffe (1982) and Condon (1984) for radio sources and more recently for optical quasars by Koo (1983b).

Fig. 7 shows explicitly why all these models are difficult to discriminate when only the quasar data brighter than B = 20 is available. Such quasars possess a LF shape very close to a power law with a slope that is almost identical at all redshifts. The similarity is best seen in the recent work by Marshall (1985). Relying mainly upon UV-excess broadband color surveys of quasars with redshift information, Marshall (1985) shows how well all the data for redshifts z < 2.2 and B = 20 can be fit by various models, although the best fit is attained for PLE of a power-law LF with a low luminosity cutoff that changes with redshift and a power-law evolution function. He also demonstrates how the use of an exponential evolution function exp(Kτ), where τ is the lookback time, leads to a conclusion that LDDE, i.e. K depends on luminosity, is the better fit while the use of a power-law evolution function $(1 + z)^K$ leads to no dependence of K on luminosity.

Since a substantial part of the sky has already been surveyed to B = 16 by Schmidt and Green (1983), one cannot hope to improve substantially our knowledge of the extensions of the bright end of the observed LF at different redshifts and thereby discriminate between the models. At magnitudes fainter than 20, however, the differences between the models rapidly diverge and become relatively easy to detect. In particular, PDE predicts a continuing steep rise whereas the other three models would show a flattening of the counts, qualitatively in agreement with our faint counts. Surveys between B = 16 and 20 would also be very useful to define the LF shape with higher precision (e.g. Mitchell, Warnock, and Usher 1984).

8. SHAPE OF THE LUMINOSITY FUNCTION

The flattening of the quasar counts fainter than B = 20 is direct evidence that the steep power-law LF of distant quasars must turn over and that the PDE model must be abandoned. The existence of 22nd mag quasars with redshifts below 2 means that the constraint of a minimum luminosity cutoff adopted in the PLE models of Mathez and Nottale (1982) or Marshall et al. (1984) must also be dropped or modified when applied to faint surveys. As easily seen in Fig. 6, our faint data shows a turnover which qualitatively resembles that seen in the local Seyfert LF of Cheng et al. (1985) (see also contribution by Weedman and Marshall at this conference). At no point do we observe a density in excess of 10^4 Gpc^{-3} mag^{-1}, implying that quasars even in the distant past were always a small fraction of the number of galaxies. If all galaxies did undergo a quasar stage, it must have occurred at much higher redshifts or the quasar stage must have a very short lifetime. We also note that the suggestion that primeval galaxies may be masquerading as quasars (Meier 1976) is not borne out by our quasar survey--the narrow-emission-line objects, which would normally be excellent candidates for primeval galaxies, had small redshifts.

9. EVOLUTION OF THE LUMINOSITY FUNCTION

Given that the shape of the LF of high-redshift quasars is so similar to that of local Seyfert 1 galaxies, the most straightforward interpretation of the change of density and luminosity of the "characteristic bend" is that quasars have undergone mainly luminosity evolution, at least up to redshifts of about 2.5. The suggestion that a density decrease may also be occurring (Koo 1983b) is no longer as compelling (see Fig. 8) since the new Seyfert 1 LF is significantly lower in density than that given by Véron (1979). To check if this decrease has occurred, we will compare our observed counts against the predictions of a simple PLE model without any density decrease. This model is constructed by merely evolving the local Seyfert 1 LF by an amount which Marshall (1985) found to fit bright quasar data. More specifically, the Cheng et al. (1985) values were adopted for the local LF and then extended to the bright end by averaging the local LF derived by Schmidt and Green (1983) for their model HH1 and HL1. To be conservative, the faint end extension was maintained at $5 \cdot 10^3$ Gpc^{-3} mag^{-1}. For a q_o of 0.1, luminosity was evolved at $(1 + z)^{4.0}$; similarly, for q_o of 0.5, the evolution function was $(1 + z)^{3.7}$. These exponents are 0.5 larger than those of Marshall (1985) since we have assumed a spectral index $\alpha=1$ rather than $\alpha=0.5$. Table III gives the predicted counts for redshifts between 0.5 and 2.5 and 0.5 to 3.0 for each model:

Table III. Faint Counts for Pure Luminosity Evolution of Local AGN Objects (No./deg^2)

Blue Mag	Obs. SA 57	$q_o = 0.1$; z =		$q_o = 0.50$; z =	
		0.5-2.5	0.5-3.0	0.5-2.5	0.5-3.0
19.5 - 20.5	20±8	18	24	11	14
20.5 - 21.5	40±12	57	77	32	40
21.5 - 22.5	87±17	130	175	66	83

If q_o is 0.1 or smaller, the high predicted counts to redshift of 2.5, below which the observed counts should be reasonably complete, suggest that indeed some density decrease has probably occurred by that redshift and that relatively few if any redshifts above 2.5 can be added. If q_o is as high as 0.5, the predicted counts for z between 0.5 and 2.5 are lower than observed. Adding quasars between z of 2.5 and 3.0 to the model gives an excellent fit to the observed counts; thus a density decrease is not needed to fit our data. This conclusion depends upon our counts being complete up to z of 3.0; with incompleteness, the model must then be extended to redshifts beyond 3.0. At this stage, however, the lack of high redshifts among the observed faint quasars becomes a constraint. In all these models, approximately equal numbers of quasars are found in each 0.5 redshift interval, i.e. equal numbers of quasars between redshifts of 1.5 to 2.5 and redshifts of 2.5 to 3.5. The lack of redshifts higher than 2.54 in our spectroscopic survey of faint quasars and the very small fraction of non-ultraviolet excess quasars in the complete sample both support claims that a decrease of quasars may be occurring well before a redshift of 3.5 (Osmer 1982). This conclusion is also supported by the brighter slitless work of Hazard (1984) and the preliminary CCD results reported by Schneider, Schmidt, and Gunn (1984).

Like the PLE models, the LDDE models also predict substantial

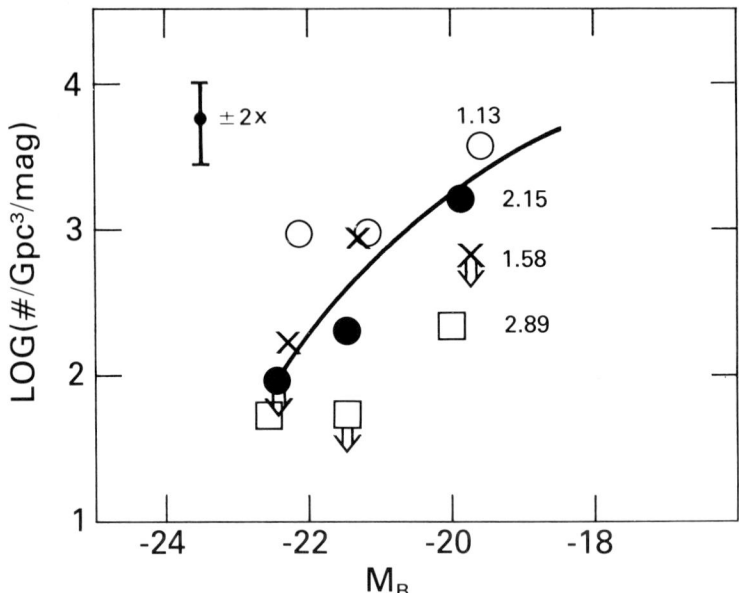

Figure 8: Plot showing the LF of our faint quasar data at the indicated redshifts, shifted in luminosity to local values from those plotted in Fig. 6. The shift was $(1+z)^{4.0}$ in luminosity where $q_o \sim 0$ and spectral index = 1 were assumed. The line follows the local Seyfert 1 LF derived by Cheng et al. (1985). As in Fig. 6, errors in log-density of ± 0.3 to ± 0.5 should be assumed.

numbers of high-redshift quasars, which are not seen in our sample, and thus also suffers from being a good predictor of faint quasar evolution for redshifts greater than 2.5.

10. CONCLUSION

In the review of quasar surveys and cosmic evolution by Véron (1983), he explicitly stated that "the two main questions yet to be answered are:
   1) At which redshift is the evolution changing sign?, or in other words, at which cosmic time did quasars form?
   2) What is the maximum volume density of quasars of all luminosities? Is it significantly smaller than the volume density of galaxies?, i.e. did all galaxy nuclei contain, at one time, a quasar (or Seyfert 1 nucleus)?"
   Though the final word should await even fainter surveys to check on the Seyfert 1 question and more checks of the completeness and redshift distribution of our survey to B = 22.5, our faint quasar data suggests some answers. Based upon using a variety of optical techniques to find and determine the nature of faint quasars, including multicolor photometry, astrometry, variability, and

spectroscopy, we are confident that the counts of quasars rapidly flattens beyond B = 20 to 21. Confirmed with some redshifts, we conclude that the maximum volume density of quasars up to redshifts of 2.5 and to relatively faint luminosities has always been comparable to (or even less than) that of local Seyfert 1 galaxies. Quasars were never more numerous in the past, contrary to popular belief. The average brightness, however, was brighter by a factor of $(1 + z)^{4.0}$. Unless the lifetimes of quasars is less than a couple percent of the age of the universe at redshifts 2, the number of quasars have never been a large fraction of the number of galaxies found today. This answers the second question.

Beyond a redshift of 2.5 or so (and possibly at smaller redshifts if $q_o < 0.1$), the quasar density appears to be dropping, at least under the assumption of conventional cosmologies (see Petrosian's contribution to the conference). Our redshift survey yielded no redshifts greater than 2.54 among the faint quasars with spectra. Moreover, the colors of faint quasar candidates are consistent with few having redshifts of 3. Yet most models predict that the faint quasars should be predominantly of such redshifts. We conclude that quasars were already less numerous, but on average much brighter, by redshifts of 3.

Thus in answer to Véron's first question, quasars appear to have formed not at one epoch but rather over an extended period, lasting from the birth of the earliest quasars ($z > 3.8$) to redshifts about 2.5 (if $q_o = 0.5$) or to more recent times (if q_o is small). Evolution may never have changed sign: quasars may have always been brighter but less numerous in the past. The reader is referred to the models of Roos and Cavaliere (this conference) for plausible physical interpretations of our observations.

11. ACKNOWLEDGEMENTS

I am grateful to H. Marshall for providing the bright quasar data shown in Figure 6. He and R. Kron are thanked for helpful discussions. Kitt Peak National Observatory has been generous in allocating telescope time as well as giving financial help over many years. Much of this work was done while the author was a Fellow at the Department of Terrestrial Magnetism of the Carnegie Institution of Washington.

Added Note: After this conference, six more candidates were observed with the spectroscopic system described in Section 4. Of most interest, we purposely chose to observe the reddest (U-J) candidate in Fig. 1c. This J ~ 22.0 object was a quasar with redshift ~3.1, which is consistent with its red colors and is now the highest redshift among the candidates fainter than J = 21.5 and observed spectroscopically. Since we have now demonstrated that our technique can indeed locate faint quasars of high redshift and since few such faint candidates lie in the vicinity of this quasar, our conclusions regarding the paucity of high redshift quasars and the shape and evolution of the quasar LF are strengthened.

REFERENCES

Braccesi, A. 1983 in IAU Symp. 104 on 'Early Evolution of the Universe and Its Present Structure' ed. G. O. Abell and G. Chincarini (Reidel, Dordrecht) p. 23.
Braccesi, A., Zitelli, V., Bonoli, F., and Formiggini, L. 1980 A. A., 85, 80.
Cheng, F. Z., Danese, L., De Zotti, G., Franceschini, A. 1985 M.N.R.A.S., 212, 857.
Condon, J. J. 1984 Ap. J., 287, 461.
Ellis, R. G. 1982 in Proceedings of Erice School on 'The Origin and Evolution of Galaxies' ed. Jones, B. J. T. and Jones, J. E. (Reidel, Dordrecht).
Hazard, C. 1984 Proceedings of Manchester Conference
Jaffe, W. 1982 Ap. J., 262, 15.
Koo, D. C. 1981 Ph.D. Dissertation, Univ. of Calif., Berkeley.
   1983a in IAU Symp. 104 on 'Early Evolution of the Universe and Its Present Structure' ed. G. O. Abell and G. Chincarini (Reidel, Dordrecht) p. 105.
   1983b in Proceedings of 24th Liege Astrophysical Coloquium, 'Quasars and Gravitational Lenses' p. 240.
Koo, D. C. and Kron, R. G. 1980 P.A.S.P., 92, 537.
   1982 A. A., 105, 107.
   1985 in preparation.
Koo, D. C., Kron, R. G., and Cudworth, K. M. 1985 P.A.S.P., submitted.
Kron, R. G. 1980 Ap. J. Suppl., 43, 305.
Kron, R. G. and Chiu, L.-T. 1981 P.A.S.P., 93, 397.
Marano, B., Zamorani, G., Zitelli, V. 1984 The Messenger, 38, 6.
Marshall, H. L., Avni, Y., Braccesi, A., Huchra, J. P., Tananbaum, H., Zamorani, G., and Zitelli, V. 1984 Ap. J., 283, 50.
Marshall, H. L. 1985 Ap. J., submitted.
Mathez, G. 1976 A. A., 53, 15.
   1978 A. A., 68, 71.
Mathez, G. and Nottale, L. 1982 A. A., 113, 336.
Meier, D. L. 1976 Ap. J., 207, 343.
Mitchell, K. J., Warnock, A. III, and Usher, P. D. 1984 Ap. J. (Letters), 287, L3.
Oort, J. 1983 in Proceeding of 24th Liege Astrophysical Coloquium, 'Quasars and Gravitational Lenses', p. 301.
Osmer, P. 1982 Ap. J., 253, 28.
Peacock, J. A. and Gull, S. F. 1981 M.N.R.A.S., 196, 611.
Petrosian, V. 1973 Ap. J., 183, 359.
Schmidt, M. 1968 Ap. J., 151, 393.
Schmidt, M. and Green, R. F. 1983 Ap. J., 269, 357.
Schneider, D., Schmidt, M., and Gunn, J. 1983 B.A.A.S., 15, 957.
   1984 B.A.A.S., 16, 488.
Usher, P. D. 1981 Ap. J. Suppl., 46, 117.
Usher, P. D., Warnock, A. III, and Green, R. F. 1983 Ap. J., 269, 73.

Véron, P. 1979 A. A., **18**, 46.
   1983 in Proceedings of 24th Liege Astrophysical Coloquium, 'Quasars and Gravitational Lenses', p. 210.
Véron-Cetty, M.-P. and Véron, P. 1984 ESO Scientific Report, **1**, 1.
Weedman, D. W. 1985 Ap. J. Suppl., **57**, 523.

DISCUSSION

REES: How do you know that most of the objects aren't "young galaxies" (as discussed by, for instance, Meier) rather than actual nuclei? A young galaxy with a high supernova rate could give broad lines (and even a small amount of variability).

KOO: I don't know for sure. Meier suggested narrow-strong emission lines and no variability as likely characteristics of primorial galaxies masquerading as quasars. Most of the high Z quasars show neither set of features. Excluding objects by reclassification will only enhance our conclusion that few high Z quasars are found z>2.5, if our candidates are actually young galaxies.

DE ZOTTI; You mentioned that low luminosity objects may be missed because their images may appear fuzzy. What is your estimate of the absolute luminosity above which your counts should be reasonably complete?

KOO: We estimate that fuzz may be detected to $z \sim 0.5$ in an AGN with $M_B \sim -23$ at $B \sim 22$. Stellar-like AGN brighter than this will probably overwhelm any underlying galaxy (but this depends on assumption of mild galaxy evolution of course).

OSTERBROCK: Emission line widths are often a good diagnositc criterion. What is the instrumental profile FWHM for your spectra?

KOO: 15 Å FWHM with 4 Å per pixel. This is unfortunately not good enough to resolve differences between Seyfert 2 and some narrow emission line galaxies.

WANDEL: The results of the subsample where the candidates are selected by the color-method could be reflecting evolution in color (for example, if quasars with a higher z had a color signature which is closer to that of stars, there would be more quasars which are not selected out of the stars among high z quasars than among lower z ones, which would have the effect of reducing the appearant luminosity evolution) a possible test for this selection effect could be comparing the results with different selection methods. Such an effect could be suggested by the work of B. Marano (reported in this meeting) who finds 20% more quasars with the GRISM than by the color method).

KOO: Yes. I agree fully that independent checks be made for quasars hiding among the stars.

SCHMIDT: As you know, we also find evidence for a maximum redshift of, at most, three for quasars fainter than B 20. This does not invalidate the luminosity-dependent density evolution models Green and I published in 1983, which were partly based on the content of grism surveys in the redshift range 1.8-3.1. Admittedly, it would be better to only use the quasars in the grism surveys in the redshift range 1.8-2.5. This will lead to a somewhat steeper luminosity dependence of the density evolution.

KOO: Both the luminosity-dependent-density evolution models that you published with Green and the published predictions of pure-luminosity evolution models of e.g. Mathez cannot be extrapolated much beyond 2 or 2.5 without some modification.

PETROSIAN: 1) For the determination of the luminosity function did you use only the 45 objects you mentioned? If so how large are the error bars? 2) Since the color criterion is your first step of selection how sure are you about non existence of $z>3.0$ quasars with non-power law spectra?

KOO: 1) No. The luminosity function was based upon a spectroscopic sample that includes quasars found in other areas, namely SA68 (Koo and Kron 1982) and Hercules.
2) Although some high-redshift ($z>2.5$) quasars are in fact expected to be lost among Galactic stars, many (≥ 50%) should still be found in the color-color plot. Few are seen by color-colr plots or by spectroscopic follow up of candidates that span a wide range in U-B color. We hope our future variability, astrometry, and slitless surveys will uncover the hidden quasars, if they are there.

A NEW COMPLETE SAMPLE OF FAINT, OPTICALLY SELECTED QUASARS:
COMPARISON WITH PREVIOUS RESULTS*

B. Marano[1], G. Zamorani[2] and V. Zitelli[1]

1-Dipartimento di Astronomia
  via Zamboni 33, 40126 Bologna (Italy)
2-Istituto di Radioastronomia
  via Irnerio 46, 40126 Bologna (Italy)

ABSTRACT. We present the results on a new sample of faint, optically selected quasar candidates. Two different search techniques (multicolour and grism) have been applied to the same field of 0.69 square degrees. The cross-check of the two lists of candidates allows us to estimate the level of completeness of each technique. We find about 50 quasars per square degree with $J<21.0$. In the magnitude range $J=21-22$ our results are in very good agreement with those of other existing surveys.

1. INTRODUCTION

The basic evolutionary properties of the luminosity function of quasars are best constrained by combining samples covering different regions of the redshift - magnitude plane (see, f.i., Schmidt 1985). This requires that the samples to be combined are not affected by selection effects, which could introduce systematic differences between the observed and the real number - density of quasars as a function of redshift and/or luminosity. Therefore, the final goal of all the searches of quasars directed to the study of the evolutionary properties is to produce samples covering, without significant losses, large areas of the redshift - luminosity plane.
　　Among the methods proposed to select quasars in the optical range, two have been particularly successful:
1) The ultraviolet excess method (UVX), based on the observational evidence that almost all the known quasars with $z<2.2$ have $U-B<-0.4$. Since high galactic latitude stars with this colour are relatively rare objects, especially at magnitudes fainter than $B=18$, this method produces quite complete and reasonably uncontaminated samples of quasar candidates with $z<2.2$ (Braccesi et al. 1970, 1980).
2) The search for emission line objects and/or objects with blue continuum on objective prism or "grism" plates. This method is highly

*Based on observations collected at the European Southern Observatory, La Silla, Chile.

efficient in finding high redshift quasars and most of our present knowledge on this population of objects is based on slitless spectroscopy. However, the level of completeness of this technique, both in terms of limiting magnitude and emission lines equivalent width, is very difficult to assess (Clowes 1981).

Koo and Kron (1982) applied to the search of quasars a multicolour selection criterion, which represents an extension of the UVX method and takes advantage of the large measuring capabilities of computer controlled microdensitometers. This method requires the measure of the colours of <u>all</u> the objects in the field brighter than the limiting magnitude of the survey. Since high galactic latitude stars occupy a well defined and relatively narrow locus on the U-B vs B-V (or, equivalently, U-J vs J-F) diagram, each stellar object lying outside this locus is considered a quasar candidate. This method is less biased, but not fully unbiased (see Section 2), against high redshift objects: non-UVX quasars are selected only if they show anomalous colours with respect to the bulk of stars.

Multicolour search technique and slitless spectroscopy are, in a sense, complementary, the latter being most efficient when the former can suffer from important losses. The two methods are therefore most effective when both applied to the same field, in order to increase the combined level of completeness.

In Section 2 we briefly discuss the completeness of the multicolour technique, indicating which redshift and luminosity ranges may, at least partially, escape this selection. The basic data reduction procedure that we have performed on plates taken at the ESO 3.6 m telescope is described in Section 3. In the next two Sections we finally give our results on colour selected and grism selected quasar candidates.

2. COMPLETENESS OF THE MULTICOLOUR SELECTION

Even assuming that losses due to failures of the procedure (as, for example, photometric errors) are absent or negligible, the completeness of multicolour surveys is not fully established over the entire redshift - luminosity plane. We stress here that it is likely that the number of "lost" objects can only marginally change the integral number counts of quasars. However, if well defined ranges in redshift and luminosity are mostly affected by these losses, important biases in the determination of the luminosity function at different epochs are introduced.

Two are the main categories of objects which may not be found by the multicolour selection method:
a) High redshift objects. For $z>2.2$ the Ly α line enters the J (or B) band and the ultraviolet excess disappears. Some high redshift objects have been found well outside the locus of normal stars and have then been selected by colour criteria (Kron and Chiu 1981). However, the present knowledge of the colour properties of high redshift quasars is still quite poor. For example, the Asiago Catalogue of Quasars (Barbieri et al. 1982) lists only about ten objects with $z>2.5$ having

reliable colour information. Therefore, it is possible that a number of high redshift quasars could be lost simply because they show just the same colours as stars. The high efficiency of slitless spectroscopy in revealing objects with Lyα within the adopted passband can succesfully be applied to recover lost objects in this region of the Hubble plane.

b) Low absolute luminosity objects (M_B>-23.0). First, we have to face a problem of definition: are all the phenomenologies grouped under the term "low luminosity AGNs" to be connected with the quasar phenomenon? To which extent can we operate a sharp distinction between AGNs and normal galactic nuclei?

Second, there is now increasing evidence that many low luminosity AGNs, discovered because of the presence of broad emission lines in their optical spectra or because of strong X-ray luminosity, would have not been found by colour methods:

i) Marshall (1985) has shown that most of the broad emission line galaxies discovered in the CfA survey (Huchra et al. 1983) are absent from the BQS sample (Schmidt and Green 1983), despite their being brighter than the magnitude limit of the BQS survey.

ii) The X-ray selected AGNs in the Einstein Medium Survey have a large variety of colours and spectral properties (Stocke et al. 1983). Many show narrow emission lines or "reddish" J-F colour (no U magnitude is available). The region of the Hubble plane covered by these X-ray selected objects is significantly more extended toward the low absolute luminosities than the colour selected quasar samples covering the same range in apparent magnitude (see, f.i., Braccesi et al. 1970, 1980).

This scenario becomes even more complex if we also consider the evidence for "mini-Seyfert" nuclei in normal galaxies (Filippenko 1986) and the changes from Seyfert I to Seyfert II condition observed in the spectra of some active galaxies (Alloin et al. 1985). The completeness of the selection of low luminosity objects is therefore still an open problem and will probably be achieved only through multifrequency surveys (optical, X-ray and, possibly, infrared).

It is worth noticing, at this point, that, at an apparent magnitude of J=22.5, only low luminosity quasars have z<2.2. Therefore, for the reasons discussed above, it is likely that multicolour criteria become somewhat inefficient in selecting AGNs at these faint magnitudes, so that samples fainter than this limit would suffer from both contamination (see Koo 1986) and incompleteness.

3. OBSERVATIONS AND DATA REDUCTION

Our observations were done in a high galactic latitude field, centered at R.A.=$3^h13.8^m$ and DEC.=-55°25′ (b^{II}=-50°). Harris and van den Bergh (1974) give a reddening of E(B-V)=0.00 for the nearby globular cluster NGC1261 ($3^h10.3^m$, -55°25′). In this field we have obtained various sets of deep exposures in the U (IIIaJ+UG1), J (IIIaJ+GG385) and F (IIIaF+GG495) bands at the prime focus of the ESO 3.6 m equipped with the triplet corrector. This telescope configuration gives a corrected,

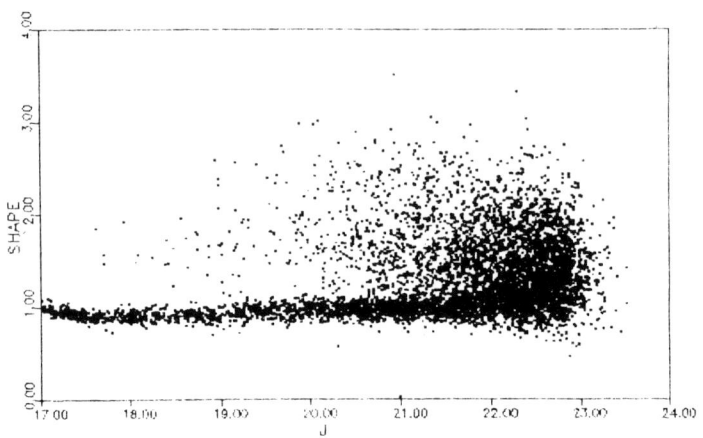

Figure 1. Shape parameter versus J magnitude for about 6000 objects. The shape parameter, which is the average of the normalized moments obtained in six plates, is such that the stellar objects cluster around the value of 1.0. The separation between stellar and extended objects becomes somewhat uncertain at magnitudes fainter than J=22.0.

unvignetted field of 60 arcminutes in diameter with a linear scale of 18.2 arcsec/mm. The whole useful field on six plates, two for each passband, was scanned with the ESO PDS microphotometer. We adopted a square aperture of 50 microns (0.9 arcsec), representing a compromise between spatial resolution and number of pixels to be stored and analyzed. An automatic procedure, following the schemes proposed by Herzog and Illingworth (1977) and by Kron (1980), was applied to obtain positions and fluxes of all the objects brighter than a given flux cut-off. The plates were linearized using sixteen calibration spots obtained during the exposure on the sky.

A comparison between different plates in the same band showed that this procedure left some residual non-linearity in the flux scale, giving plate to plate systematic differences of 0.1-0.2 magnitudes on a three magnitude range. The six considered plates were then calibrated using the following procedure:
a) A U,B,V photoelectric sequence, reaching B=17, was obtained in the same field with the ESO 1 m telescope.
b) B and V sequences were extended to B=22 through some CCD exposures.
c) J and F sequences were then obtained from B and V magnitudes applying the transformation formulae given by Kron (1980).
d) A deep U sequence was obtained from a 3.6 m plate exposed with a Pickering-Racine photometric wedge giving a difference of 4.0 magnitudes between primary and secondary images. The reliability of the wedge derived magnitudes was tested (and confirmed) by comparing with each other the sequences derived in the J band with the CCD and the wedge exposures.

As a result, we expect a systematic error in the magnitude scales not larger than 0.1 over the range J=18.5-22.0. A comparison between plates in the same band shows that the typical random error at J=21 is

A NEW COMPLETE SAMPLE OF FAINT, OPTICALLY SELECTED QUASARS

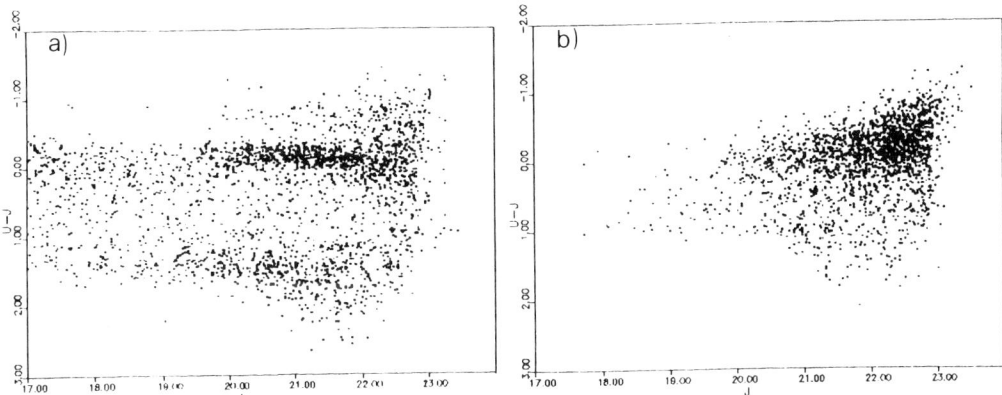

Figure 2. U-J versus J magnitude for "stellar" (Fig.2a) and "extended" (Fig.2b) objects. The stars are splitted in two main families, the blue subdwarfs and the M dwarfs. The objects with ultraviolet excess are clearly visible above the locus of the blue subdwarf stars. Note the large number of ultraviolet galaxies appearing at faint magnitudes.

0.12 magnitudes for each single measurement. The average of two plates therefore gives an r.m.s. error of 0.08 magnitudes.

For the present purposes, the working list was restricted to all the objects brighter than J=22.5 in an area of 0.69 square degrees. The first moment of all the images in this list was then used to classify the objects. Figure 1 shows the mean of the normalized moments obtained from the six considered plates versus the J magnitude. Stellar objects populate a well defined domain in the lower part of the diagram, while galaxies are spread over a large area, corresponding to higher values of the moment. According to their position on this diagram, objects were then separated in three classes, "stellar", "extended" and "intermediate" objects. The meaning of the "intermediate" class changes somewhat with the magnitude. The few bright (J<21) "intermediate" objects would be commonly defined as "fuzzy". At fainter magnitudes the poorer signal to noise ratio produces a larger spread in the moments and, as a consequence, an increasing number of objects for which the morphological classification is actually uncertain.

Figure 2 shows the U-J versus J diagram for "stellar" and "extended" objects. As typical of high galactic latitude fields, stars are clearly splitted in two main families, blue subdwarfs and M dwarfs. The colours of galaxies undergo a systematic change with magnitude and a large number of ultraviolet galaxies appears at magnitudes fainter than J=22. This last fact, together with the increasing difficulty in separating faint compact galaxies from true "stellar" objects, sets an important limit to the possibility of selecting quasars as ultraviolet excess objects at extremely faint magnitudes (see also Section 2). For these reasons, we limit our present search to objects brighter than J=22.0.

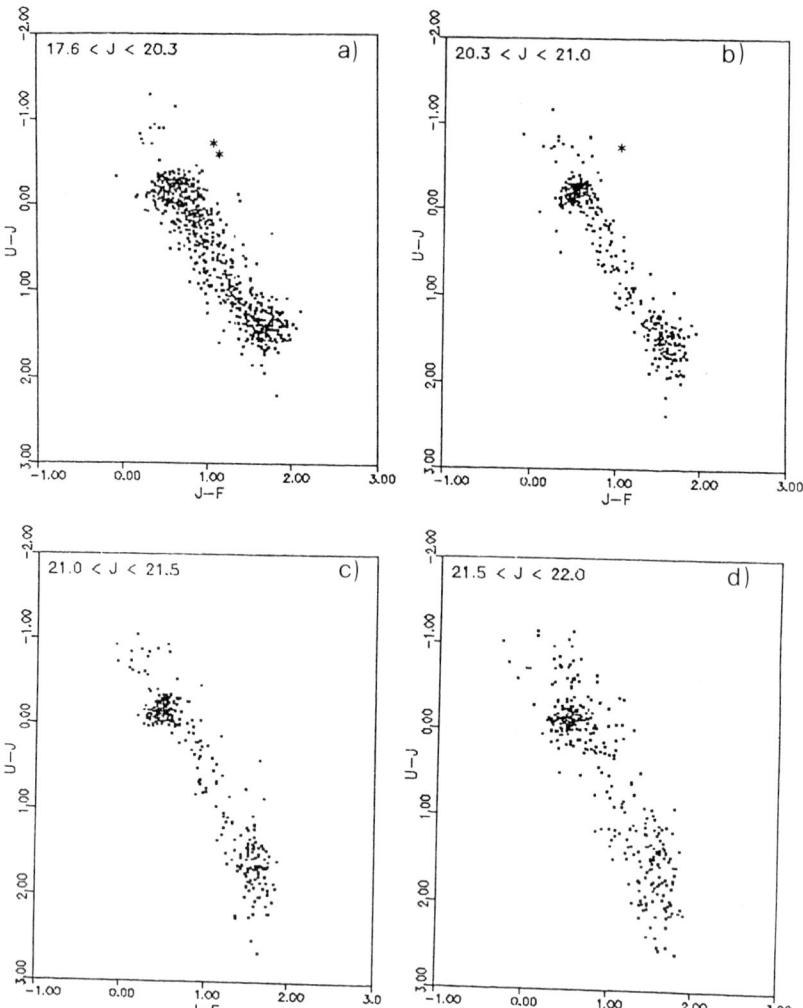

Figure 3. U-J versus J-F for the "stellar" objects in four different magnitude ranges. The colours of the objects indicated by an asterisk are uncertain due to a possible photometric error in one of the six plates (see text). The relatively large dispersion of the data points in Fig.3a) is due to bright objects (J<18.5), for which the magnitude errors tend to increase.

4. COLOUR SELECTED QUASAR CANDIDATES

Figure 3 shows the two colour diagram (U-J vs J-F) for the stellar objects in four magnitude ranges. Adopting the extended colour criterion suggested by Koo and Kron (1982), we will call quasar candidates all the objects whose colours are different from those of normal stars. The values plotted in Fig. 3 are the average of the colours obtained in two different years. Variability, unless strongly

A NEW COMPLETE SAMPLE OF FAINT, OPTICALLY SELECTED QUASARS

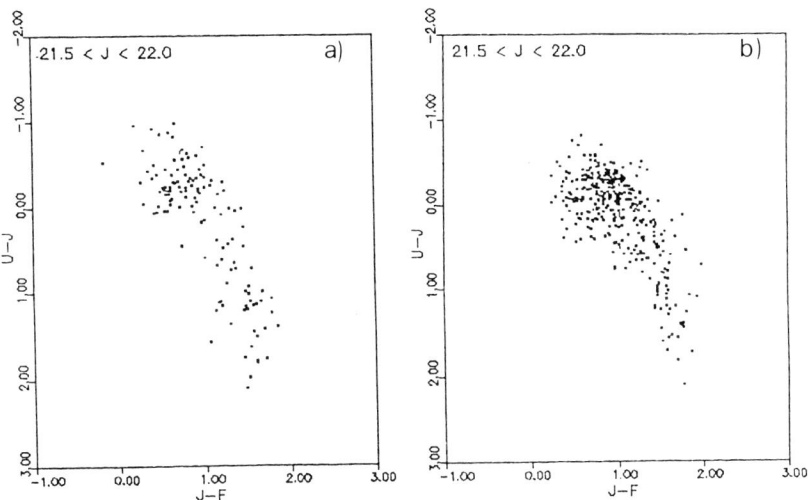

Figure 4. U-J versus J-F for "intermediate" (Fig.4a) and "extended" (Fig.4b) objects in the magnitude range 21.5<J<22.0. The similarity of the regions of the plane occupied by the two classes of objects suggests that many of the objects in the intermediate class are actually galaxies.

wavelength dependent, should not cause significant displacements in colours. However, a photometric error in one plate can determine such a displacement. We have then checked if any of our candidates shows a large magnitude difference in one band only. If the colours of these objects, as measured in one of the two years, are within the region occupied by the stars, we will assume that most likely they are not quasar candidates. These objects are indicated by an asterisk in the figure. The selection of the candidates in the four panels of Fig.3 is relatively easy and straightforward. The same is true, down to J=21.5, for the so-called "intermediate" objects. The situation is much less clear for these objects in the last magnitude bin. As already said in Section 3, because of the large number of blue faint galaxies and the increasing difficulty in classifying the images at faint magnitudes, both really stellar and extended objects coexist in the class of intermediate objects for J>21.5. This is confirmed by Figs.4a) and 4b), which show the colour-colour diagram for intermediate and extended objects in the magnitude range 21.5-22.0. The regions of the plane occupied by the two classes of objects are similar, suggesting that many of the objects in the intermediate class are actually galaxies. Very ultraviolet objects (U-J<-0.70), however, are much more numerous in the intermediate class than in the extended class, and we have considered them as good quasar candidates (7 objects). On the other hand, if we adopt, in Fig.4a), a criterion more similar to that used for stellar objects in the same magnitude range, we would have a total of 15 candidates. The "real" number of candidates should then be intermediate between these two estimates.

A summary of the counts of quasar candidates obtained using the above criteria is given in Table I. In a first spectroscopic run we obtained spectra of the 13 candidates brighter than J=20.3. Eleven are real quasars, two stars. The two objects marked with an asterisk in Fig.3a) turned out to be stars.

A comparison of our results with other similar surveys is shown in Table II. From this table a very good agreement is apparent between our results, those obtained by Koo and Kron (1982) in the Selected Area 68, and those presented by Koo (1986) for the Selected Area 57. Note that, in order to make a meaningful comparison, for both S.A. 68 and 57 we have used the number of quasar "candidates" rather than those corrected after spectroscopy (Koo 1983 and 1986). Our data are in excellent agreement also with the results presented at this Conference by Boyle et al. (1986). By performing multi-aperture fiber optics spectroscopy on a sample of about 400 objects selected on the basis of ultraviolet excess over three square degrees, they find a surface density of 50 confirmed quasars per square degree at $J<21.1$. In conclusion, three completely independent surveys lead to very similar and consistent results for the number density of quasars in the magnitude range 21-22.

Given the relatively large area covered by our survey (0.69 square degrees), we can tentatively compare our results with those obtained from other samples at brighter magnitudes. A very well studied sample is the so-called BF sample (Braccesi et al. 1980, Marshall et al. 1984). In this sample, covering 1.7 square degrees, there are (20.3 ± 3.4) spectroscopically confirmed quasars per square degree with $B<19.8$. At the same magnitude limit we should, therefore, find 14 quasars in our area, while we actually find 5. The reason for this difference, significant at about 2.7 sigma, is not understood. A shift in the magnitude scale of the order of 0.5 magnitudes would be

Table I. Number of Colour Selected Quasar Candidates.

Magnitude Range	Stellar Objects		Intermediate Objects	
	UVX	non-UVX	UVX	non-UVX
$J<20.3$	11	1	1	0
$20.3<J<21.0$	17	3	2	1
$21.0<J<21.5$	21	0	7	0
$21.5<J<22.0$	34	2	7-15	1

Table II. Integral Density of Colour Selected Quasar Candidates.

Magnitude	This Survey	S.A.68	S.A.57
$J<21.0$	51	41	==
$J<21.5$	91	82	93
$J<22.0$	154-166	136	==

required in either ours or in Braccesi's data in order to let the observed difference disappear completely. Otherwise, it is possible that at least part of the effect is due to some real large fluctuation in quasar distribution. It is interesting to note that a suggestion of a possible overdensity of quasars in the Braccesi field comes also from an analysis of the X-ray data (see Schmidt 1986).

5. GRISM SELECTED QUASAR CANDIDATES

In the same area of sky on which we performed the multicolour selection of quasar candidates, we obtained also three grism plates at the ESO 3.6 m telescope. The exposure times were 60, 120 and 165 minutes. The wavelength range (3400-5300 A) is such that Ly α would not be detected at redshifts larger than about 3.3. The two best plates were then visually inspected by us searching for quasar candidates. The criteria adopted for this first broad selection were either the presence of possible emission lines or the shape of the continuum. The preliminary list constructed in this way consisted of 146 objects. All the objects for which we have spectroscopic observations were selected at this stage.

All the 146 objects have been scanned with the PDS machine at the ESO on the best of our grism plates. The tracings of the scans have then been looked at and a tighter selection of candidates has been performed, accepting as candidates only objects showing "bona fide" emission lines. The results for different magnitude ranges are summarized below.

a) J<20.3

More than half of the objects in the original list of candidates were brighter than J=20.3. Most of them are stars, lying within the blue subdwarf region in the colour-colour plane (U-J vs J-F). All the known spectroscopic quasars, with only one exception, would have been selected as good candidates, on the basis of the PDS tracing. The only quasar that would have been missed at this stage has z=0.99. This redshift is such that the MgII line is just outside the available wavelength range, while the CIII] line (at about 3800 A) is not clearly visible on the spectrum. Table III gives the relevant data for additional quasar candidates from the grism plates which were not selected on the basis of the multicolour analysis. For these objects no spectroscopic confirmation is available yet.

b) 20.3<J<21.0

In this magnitude range 34 objects were selected in the preliminary list. Fifteen of these were then extracted as likely candidates. A comparison between colour and grism candidates gives the following results:
a) ten of 20 colour candidates in the stellar class are also grism candidates;
b) none of the three colour candidates in the intermediate class are grism candidates;
c) five grism candidates were not selected on the basis of the colour-colour diagram. Two of these are almost definitively quasars,

Table III. Grism Candidates not Colour Selected (J<20.3)

J	U-J	J-F	Notes
20.28	-0.10	0.29	1)
18.87	-0.90	0.20	2)
19.71	0.08	0.30	3)
18.09	0.11	0.84	4)
18.72	-0.24	0.77	4)
19.54	-0.10	0.19	4)

1) Quasar with redshift of the order of 2.75 on the basis of two emission lines identified with Lyα and Lyβ.
2) Quasar with strong emission lines. This object was lost in the original selection on the direct plates, because it lies at about three arcseconds from a bright (17 mag.) star. Its magnitudes and colours have been recovered later on.
3) Probable quasar. This object is just outside the limit of the grism plate scanned with the PDS at ESO. An emission line is clearly visible in a poorer plate.
4) Possible quasar.

showing strong emission lines; for the other three the classification is more uncertain. These five objects are within the blue subdwarf region in the colour-colour plane, even if at least two of them are at the border of such region.

c) $21.0 < J < 21.5$

In this magnitude range 20 objects were selected in the preliminary list. Twelve of these were then extracted as likely candidates. A comparison between the colour and the grism candidates gives the following results:

a) eight of 21 colour candidates in the stellar class are also grism candidates;

b) none of the seven colour candidates in the intermediate class are grism selected;

c) three grism candidates were not selected on the basis of the colour-colour diagram. Two of them are very likely to be quasars, showing strong emission lines; for the other object the classification is more uncertain. One of the two likely quasars is classified as "extended" on the basis of the shape parameter.

In addition to these quasar candidates, we also found a small number of galaxies with prominent Hβ+[OIII] complex. No conclusion on the density of these emission line galaxies is possible on the basis of these data, because of the very limited redshift range available for detecting Hβ on our plates (z<0.1). Moreover, the original visual search of candidates was not intended to be complete at all for this class of objects. It may be interesting to note, however, that, if selected through other means (for example X-ray), these objects might be called low luminosity Active Galactic Nuclei.

In summary, the results of our grism search for quasars are as follows:
a) In the bright magnitude range (J<20.3) our search of quasar candidates on the grism plates appears to be about 90% complete: only one out of eleven spectroscopically confirmed quasars would have been lost.
b) For fainter magnitudes (down to J=21.5) the success of the grism selection (when compared with "stellar" colour selected candidates) is of the order of 45% (18 out of 41 colour candidates were also selected on the basis of the grism plates).
c) In the same magnitude range it is interesting to note that none of the ten "intermediate" colour candidates was selected from the grism. This suggests the existence of a difference between the two classes of objects. A possibility is that the redshifts of the objects in the two classes are, on the average, different: relatively nearby quasars may appear not completely stellar and may not have strong emission lines in the wavelength range covered by our grism plates.
d) In all magnitude ranges (down to J=21.5) the number of likely grism candidates not colour selected is about 30% of the grism candidates which were selected also on the basis of the colour diagram. This fraction, however, can not be acritically used as a correction factor to the counts of quasars based on colour selection. In view of the higher efficiency of grism searches for high redshift (and then possibly non-UVX) quasars, this 30% can be reasonably assumed as an upper limit for the fraction of quasars missed by colour techniques.

6. CONCLUSIONS

We have presented results on a sample of faint, optically selected quasar candidates. Two different search techniques (multicolour and grism) have been applied to the same field of 0.69 square degrees.

Our counts of quasar candidates are in excellent agreement, in the magnitude range J=21-22, with those of other surveys. Three independent surveys agree on a number of about 50 colour selected quasars per square degree with J<21.0.

At the bright magnitude limit of our survey, we find a smaller number of quasars than in the Braccesi's sample. The difference, larger than a factor of two, is significant at about 2.7 sigma level.

By comparing our grism results with the multicolour selection, we find that the grism selection is about 90% complete for J<20.3. This level of completeness decreases to about 35% at J=21.5. Some of the likely grism candidates, however, were not selected by the multicolour technique. A few of them are in the outskirts of the locus of the blue subdwarf stars, but others have colours typical of normal stars.

REFERENCES

Alloin,D., Pelat,D., Phillips,M., and Whittle,M. 1985, Ap.J., 288, 205.

Barbieri,C., Capaccioli,M., Cristiani,S., Nardon,G., and Omizzolo,A. 1982, "The Asiago Catalogue of Quasi Stellar Objects", Mem.Soc.Astron.Ital., 53, 511.
Boyle,B.J., Fong,R., Shanks,T., and Peterson,B.A. 1986, This Volume.
Braccesi,A., Formiggini,L., and Gandolfi,E. 1970, Astron. and Astrophys., 5, 264.
Braccesi,A., Zitelli,V., Bonoli,F., and Formiggini,L. 1980, Astron. and Astrophys., 85, 80.
Clowes,R.G. 1981, M.N.R.A.S., 197, 731.
Filippenko,A. 1986, This Volume.
Harris,W.E., and van den Bergh,S. 1974, Astron.J., 79, 31.
Herzog,A.D., and Illingworth,G. 1977, Ap.J.Suppl., 33, 55.
Huchra,J.P., Davis,M., Latham,D.W. and Tonry,L.J. 1983, Ap.J.Suppl., 52, 89.
Koo,D.C. 1983, Proceedings of the 24th Liege International Astrophysical Colloquium on "Quasars and Gravitational Lenses", p. 240.
Koo,D.C. 1986, This Volume.
Koo,D.C., and Kron,R.G. 1982, Astron. and Astrophys., 105, 107.
Kron,R.G. 1980, Ap.J.Suppl., 43, 305.
Kron,R.G., and Chiu,L.T.G. 1981, P.A.S.P., 93, 397.
Marshall,H.L. 1985, Ap.J., in press.
Marshall,H.L., Avni,Y., Braccesi,A., Huchra,J.P., Tananbaum,H., Zamorani,G., and Zitelli,V. 1984, Ap.J., 283, 50.
Schmidt,M. 1986, This Volume.
Schmidt,M., and Green,R.F. 1983, Ap.J., 269, 352.
Stocke,J.T., Liebert,J., Gioia,I.M., Griffiths,R.E., Maccacaro,T., Danziger,I.J., Kunt,D., and Lub,J. 1983, Ap.J., 273, 458.

DISCUSSION

DE ZOTTI: A check of the reality of surface density fluctuations of ∼20th magnitude QSOs suggested by optical counts can be provided by an analysis of the counting rates of HEAO-1 A2 detectors with $1°.5 \times 3°$ field of view: a factor of 2 overdensity should result in a detectable excess flux unless QSO have a very soft X-ray spectrum.

MACCACARO: Is the discrepancy between the quasars counts from your survey and those from faint Braccesi survey limited to the total number of objects or the discrepancy extends to the redshift distribution as well?

MARANO: No, within the limit of the statistics (our sample is made by 10 objects), we don't find any evidence of discrepancy in the redshift distributions.

MARSHALL: I have a few points: 1) It seems that the horse hasn't been flogged enough. I agree that there may be an overdensity of quasars in the BF region; it now remains to determine how much it is. Schmidt (these proceedings) claims a factor of five, you estimate a factor of two and Boyle et al. (these proceedings) find a factor of 1.5 or about a 2σ result. I tend to believe the latter because their sample is actually larger than the BF sample (Marshall et al. 1984). 2) We reached a bit fainter into the subdwarf region than either you or Koo and Kron (see Koo, these proceedings) but not so far as Boyle et al. I find more quasars that way, at the expenses of getting more stars (although the stellar contamination rate is somewhat low overall). 3) It is important to compare the redshifts distributions, as I have done with the recent sample of Mitchell, Warnock and Usher (1984). My latest pure luminosity evolution model (Marshall 1985) predicts 2.7σ fewer objects but the correct redshift and luminosity distributions.

MARANO: Concerning your second point, I like to stress that both Braccesi's sample (BF) and ours (considering also the grism selection) contain only a few quasars lying at the upper border of the blue subdwarf region. These objects can not introduce major changes in the counts. Moreover, should we have lost many objects with the colour selection, we would have recovered them with the grism, as shown by the high rate of success of the grism selection in this magnitude range.

SCHMIDT: Could you say what fraction of the spectroscopically confirmed quasars are fainter than $M_B=-23$ (for $H_0=50$)?

br;MARANO: Only one of the 11 spectroscopically confirmed quasars brighter than J=20.3 has an abosolute magnitude fainter then $M_B=-23$. This is also the only candidate, in this magnitude range, having an "intermediate" (or "fuzzy") shape (see Table I).

EVOLUTION OF THE LUMINOSITY FUNCTION OF EXTRAGALACTIC OBJECTS

Vahé Petrosian
Center for Space Science and Astrophysics
Stanford University
Stanford, CA 94305

ABSTRACT. A non-parametric procedure for determination of the evolution of the luminosity function of extragalactic objects and use of this for prediction of expected redshift and luminosity distribution of objects is described. The relation between this statistical evolution of the population and their physical evolution, such as the variation with cosmological epoch of their luminosity and formation rate is presented. This procedure when applied to a sample of optically selected quasars with redshifts less than two shows that the luminosity function evolves more strongly for higher luminosities, indicating a larger quasar activity at earlier epochs and a more rapid evolution of the objects during their higher luminosity phases. It is also shown that absence of many quasars at redshifts greater than three implies slowing down of this evolution in the conventional cosmological models, perhaps indicating that this is near the epoch of the birth of the quasar (and galaxies). However, it has been shown that the same is not true in all cosmological models, in some of which the epoch of birth could be at much higher redshifts.

I. INTRODUCTION

I will describe the steps and procedure required for determination of the evolution of the luminosity function of extragalactic objects and apply the results to quasars and, in particular, address the question of the evolution at redshifts higher than three. The purpose of such a study, of course, is to learn about the formation and evolution of the population. There are four distinct steps involved in this procedure. These are:

Selection of complete samples with known observational selection effects.
Choice of a cosmological model.

Statistical analysis of the sample and determination of the evolution of the luminosity function.
Use of the luminosity function for determination of the formation rate and the physical evolution of the objects.

These steps are described in the next section and are applied to a sample of quasars in Section III. A short summary is presented in Section IV.

II. THE PROCEDURE

A. The Selection of the Sample

Ideally, one requires a complete sample with known selection effects. The simplest case is obtained when the sample is limited by magnitude at one wavelength band. Then the data will consist of objects with known redshifts z and flux densities $f(\nu)$ greater than some limiting flux density f_0 and can be represented by the following distribution function

$$n(f,z) = \sum_{i=1}^{N} \delta(f - f_i)\delta(z - z_i) , \qquad (1)$$

where N is the total number of objects in the sample. Unfortunately, more than one criterion is needed for identification of the objects. If f above stands for some optical flux density, then the radio or X-ray flux densities of the object will determine their membership in a radio or X-ray sample, so that the distribution in general is multivariate rather than bivariate as in Equation (1). This aspect of the problem is not a source of difficulty but adds to the complexity of the calculations. We shall limit our discussion to optically selected samples. Even in this case, however, there are additional selection criteria like color in selections based on UV excess or line strength in selections based on slitless spectra. These put additional known limits on the sample such as z < 2.2 for UV excess samples or z > 1.8 for samples based on slitless spectra and perhaps some other unknown limit. We shall ignore the latter, which remains controversial, even though there has been considerable discussion about it in the past.

B. The Cosmological Models

The observed distribution [Eq. (1)] is not only a reflection of the luminosity function but also of the cosmological model. In general, the effects of the cosmological model cannot be separated from the evolution of the luminosity function. One needs to specify one of these two unknowns to determine the other. It is customary to assume a cosmological model and derive the luminosity function in that model. And it turns out that, because of the wide dispersion of the luminosity function and the small differences between the conventional cosmlogical models at moderately low redshifts, z < 2, changing the cosmological model does not affect the outcome significantly. However, when dealing with the luminosity function at higher redshifts, in particular, for addressing the questions in regard to the turn-on redshift of the objects (quasars or galaxies), the differences in the cosmological models, especially if one is not limited to the conventional models with zero cosmological constant, becomes significant. Consequently, and because the new cosmological scenarios like the inflationary models suggest wider possibilities for the cosmological model, I shall consider three widely different cosmological models.

The first two models will be based on an inflation scenario that requires a negligible space curvature now and in the past since the epoch of inflation. This means that if we neglect the contribution of zero rest mass (or relativistic) particles, the cosmological constant Λ can be expressed in terms of the density parameter Ω (of non-relativistic matter) as $\lambda = \Lambda/3\ H_0^2 = 1 - \Omega$ (cf., e.g., Peebles 1984). Since $\Omega > 1$ or $\lambda < 0$ models can be ruled out because of their short ages (note that for $H_0 = 100$ km s^{-1} Mpc^{-1}, the $\Omega = 1$ model is already in difficulty), I shall consider two extreme models: $\Omega = 1.0$, $\lambda = 0$, and $\Omega = 0$, $\lambda = 1.0$. I shall also consider a third model that is not based on the inflationary scenario but is a closed-world model with negligible curvature now but in which the curvature was important in the near past giving rise to a quasistatic period of the expansion. For the parameters of this model I use $\lambda = 1.2$ and $\Omega = 0.1$. These three models are called the Einstein-deSitter, the deSitter, and the Lemaître models, respectively.

Given the cosmological model, we can then calculate for each object its intrinsic flux (or luminosity) at a specified rest frame frequency

$$F_i(\nu) = 4\pi D_L^2(z,\alpha,\Omega,\lambda)\ f_i(\nu)\ , \tag{2}$$

and a new distribution (dropping the frequency dependence)

$$n(F,z) = \sum_{i=1}^{N} \delta(F - F_i)\delta(z - z_i)\ , \tag{3}$$

where D_L is the luminosity distance and depends on the redshift, the cosmological parameters, and the spectral index $\alpha = -d\ln f(\nu)/d\ln \nu$. The variation of D_L with z (for $\alpha = 0.5$) is shown in Figure 1 for the above three models. Note that for the Lemaître model at a redshift $z \simeq 2$ one has reached the so-called antipode of the closed universe where $D_L \to 0$ and for a given luminosity F the flux density $f \to \infty$. However, such a drastic brightening of the sources near the antipode will be diminished by the presence of inhomogeneities in the distribution of matter (such as galaxies and clusters). The exact form of the D_L vs z curve then depends on the characteristics of these inhomogeneities. For a detailed discussion of this the reader is referred to Petrosian and Salpeter (1968). The dashed line in Figure 1 is an example with some assumed size and distribution of the inhomogeneities. However, for simplification of the calculations, I will assume the solid line, which on the average will give a result similar to that of the more realistic dashed line. This will satisfy my purpose here, which is to show the extent of the differences between these widely different cosmological models.

C. The Luminosity Function

The bivariate luminosity function $\psi(F,z)$ can formally be related to the observed distribution by

$$\psi(F,z) = w(F_i, z_i) n(F,z) \quad , \tag{4}$$

where w is the weight of each object. These weights would be unity if there were no selection bias. However, because of the selection biases, objects of given F and z, which are less likely to be present in the sample, carry higher weights. The problem then is reduced to determination of the weights. The usual procedure, however, has been to parameterize the luminosity function and then find the value of the parameters bypassing the difficult but more accurate procedure of determining the weights. The use of the parameters determined in this manner for prediction of the expected number of objects outside the range of the parent sample (e.g., extension to deeper samples or higher redshifts) may give misleading results. The optimum procedure is to use a non-parametric procedure in the determination of the weights and as far as possible use these weights for further predictions.

The simplest non-parametric approach is to divide the area of the F-z plane accessible to the particular sample into various bins (as in Figure 2a) and from this find the ratio of the luminosity function at different bins. If the sample is large, the bins could be numerous and from the ratios of the numbers $n_{i,j}$ in different bins one can construct the differential luminosity function. In general, this is not the case, and the number of sources in the sample is small. It is more convenient to define the cumulative luminosity functions

$$\Phi(F,z) = \int_F^\infty \psi(F',z) dF' \quad , \quad \sigma(z,F) = \int_0^z \psi(F,z') dz' \quad , \tag{5}$$

which increase stepwise at values F and z of each source, respectively. The size of the steps are equal to the weights w. Graphic illustration of the cumulative functions is more convenient and more illuminating than the illustration of the delta fuction representation (Eq. (4)) of the differential luminosity function.

Referring to Figure 2a, we can then evaluate the ratios of the cumulative functions at different values of F and z. For example, consider the objects in the vertical strip between z_{i-1} and z_i and with $F > F_{min}(z_i)$, where $F_{min}(z)$ is the minimum value of F an object with redshift z, must have in order to be included in the sample (i.e., to be below the heavy diagonal line). For these objects then

$$\frac{\Phi(F_j,z_i)}{\Phi(F_{j-1},z_i)} = 1 + \frac{n_{i,j}}{N_{i,j-1}} \quad , \quad N_{i,j} = \sum_{r=0}^{j} n_{i,r} \quad . \quad (6)$$

It is clear that the repeated application of this equation at different F_j (starting from F_0) yields

$$\Phi(F_j,z_i) = \Phi(F_0,z_i) \prod_{r=1}^{j} \left(1 + \frac{n_{i,r}}{N_{i,r-1}}\right) \quad , \quad F_1 > F_j > F_{min}(z_i) \quad . \quad (7)$$

This procedure can then be repeated for all z_i as well as for horizontal strips at all F_j to obtain the $\sigma(z_i, F_j)$.

This method, however, ignores the few objects that may lie in the triangular regions bounded by the heavy lines of Figure 2a. As it will be shown more clearly below, in order to utilize all the information in the data fully, we can go to the limit of small bins so that each bin contains one object ($n_{i,j} \rightarrow 1$), in which case Equation (7) gives a result identical to that of Lynden-Bell's (1971) c^- method. Then we can extend this equation to objects in the triangular region if we define for them a new $N_{i,j}$. For example, for the object shown by the open circle, $N_{i,j}$ is equal to the number of objects in the shaded area. Clearly, an object in the triangular region carries more weight since the limit of the sample has excluded the object in the complementary triangle shown by the dashed lines. The fact that $N_{i,j}$ for this object may be smaller reflects this higher weight.

With this procedure we can then obtain two series of histograms $\Phi(F,z)$ and $\sigma(z,F)$, which can then be converted into the delta function form of the differential luminosity function or can be smoothed out and differentiated to yield $\psi(F,z)$. In general, because of the absence of low-luminosity objects at high redshifts and high-luminosity objects at low redshifts, the histograms $\Phi(F,z)$ at different z's [or $\sigma(z,F)$ at different F's] will have small overlapping regions. This makes it difficult to produce a complete description of the luminosity function throughout the accessible region of the F-z plane. Combining large sky area surveys with deeper but limited area surveys can alleviate this problem. In any case, without further assumptions about the luminosity function, we cannot extend it to the region outside the observed parts of the F-z plane for prediction of the expected distribution of sources in samples with different selection criteria. The above-mentioned histogram, however, may be helpful in choosing among various forms of the function $\psi(F,z)$.

The non-parametric procedure explained above becomes extremely useful if one can assume that the luminosity function is separable

$$\psi(F,z) = \psi(F)\rho(z) \quad , \quad (8)$$

which means that the two variables F and z are stochastically independent. In principle, a sizable sample can be used to determine the stochastic independence of the variables. In practice, however, this is difficult, and most procedures require some kind of binning of the objects. The alternate possibility is comparison of these data with numerically simulated data sets. The discussion of this is beyond the scope of this presentation, and I shall not dwell on it here. I will assume that, even if F an z are known to be stochastically dependent, there exist two other parameters, F_s and z_s, which are functions of F and z and are stochastically independent. In this case the data set can be transformed into these new variables and the analysis carried out in terms of them. For convenience I shall drop the subscript s from the discussion below and assume Equation (8), keeping in mind that F and z may no longer refer to the observed luminosity and redshift but may be arbitrary functions of them.

Given Equation (8), we can define new cumulative luminosity functions Φ and σ (see Eq. (5)) which are now functions of only one parameter. Note taht as defined here $\rho(z)$ is not the density per unit co-moving wolume V but is the marginal distribution in z. The density is equal to $\Phi(0)\rho(z)/(dV/dz)$.

$$\Phi(F) = \int_F^\infty \psi(F')dF' \quad , \quad \sigma(z) = \int_0^z \rho(z')dz' \quad . \tag{9}$$

This also means that in Equation (6) both the numerator and the denominator can be integrated over the redshift from the minimum redshift of the sample to a maximum redshift $z_{max}(F_j)$, which yields

$$\frac{\Phi(F_j)}{\Phi(F_{j-1})} = 1 + \frac{n_j}{N_{j-1}} \quad , \quad n_j = \sum_{k=o}^{i} n_{k,j} \quad , \quad N_j = \sum_{k=o}^{i} N_{k,j} \quad , \tag{10}$$

where the bin i is determined by the luminosity F_j such that z_i is the maximum redshift that an object with luminosity F can have and still be in the sample. Now, in the limits of small bins, $F_j = F_{j-1} - dF$ and $\Phi(F_j) = \Phi(F_{j-1}) + \psi(F_j)dF$, Equation (10) can be written as

$$\frac{\psi(F)}{\Phi(F)} = -\frac{d\ln\Phi(F)}{dF} = \frac{n(F)}{N(F)} \quad , \tag{11}$$

where now

$$n(F') = \int_0^{z_{max}(F)} n(F',z')dz' \quad , \quad N(F) = \int_F^\infty n(F')dF' . \tag{12}$$

Similarly, we can write

$$\frac{\rho(z)}{\sigma(z)} = \frac{d\ln\sigma(z)}{dz} = \frac{m(z)}{M(z)} , \qquad (13)$$

with

$$m(z') = \int_{F_{min}(z)}^{\infty} n(F', z') dF' , \qquad M(z) = \int_{0}^{z} m(z') dz' . \qquad (14)$$

The various quantities entering Equations (11) to (14), are illustrated in Figure 2b.

The data n(F,z) given by Equation (3) implies that n(F) or m(z) are just a series of delta functions at the luminosity or redshift of the objects in the sample and that N(F) and M(z) are histograms that increase by one every time an object is crossed (decreasing F or increasing z). Thus, Equations (11) and (13) can be written symbolically as

$$\frac{d\ln X}{dx} = \frac{\delta(x - x_i)}{N_i + \theta(x - x_i)} , \qquad (15)$$

where $\theta(x) = \int \delta(x) dx$ is the step function. Integration of (15) then yields

$$\delta \ln X_i = \ln\left(1 + \frac{1}{N_i}\right) . \qquad (16)$$

By this procedure we obtain two monotonically increasing histograms for the cumulative functions $\sigma(z)$ and $\Phi(F)$. Note that the quantity on the right hand side of equation (16) is well defined only if $N \to N_i$. As we start with the first object at lowest redshift (or highest luminosity), M (or N) will be zero. For sufficiently densely packed samples already, for the second object M could be equal to 1, and $\ln\sigma$ and $\delta\ln\sigma$ would be well defined. If not, we proceed to the first object with M or N ≠ 0. If Φ_0 and σ_0 are the values of Φ and σ just below this object, then

$$\sigma(z_i) = \sigma_0 \prod_{j=1}^{i}\left(1 + \frac{1}{M_j}\right) , \qquad \Phi(F_i) = \Phi_0 \prod_{j=1}^{i}\left(1 + \frac{1}{N_j}\right) . \qquad (17)$$

Once the cumulative functions are known, one may wish to smooth them out and differentiate to obtain the differential functions ψ and ρ. Or one may wish to keep the integrity of the data and express ψ and ρ in the delta function form of Equation (4). In Equations (11) and (13) if we replace the quantities n and m by this delta function representation, we obtain

$$\rho(z) = \sum_i \frac{\sigma(z)}{M(z)} \delta(z - z_i) \,, \quad \psi(F) = \sum_i \frac{\Phi(F)}{N(F)} \delta(F - F_i) \,, \quad (18)$$

which, when compared with Equation (4), means that the weights are

$$w(F_i, z_i) = \frac{\Phi(F_i)\sigma(z_i)}{N(F_i)M(z_i)} \,. \quad (19)$$

Note that F and z may not represent the real luminosity or redshift but some other parameters that are stochastically independent, in which case one can transform Equations (18) or (19) into the real luminosity and redshift domain.

Given these functions from the parent sample, we can make limited prediction about other samples. For example, for a sample with a different magnitude limit than that of the parent sample, the redshift or luminosity distributions, $m'(z)$ and $n'(F)$, can be obtained as follows:

$$m'(z)dz = \rho(z)dz\Phi\left(F'_{min}(z)\right) = \sigma(z)d\ln\sigma(z)\Phi\left(F'_{min}(z)\right) \,,$$

$$(20)$$

$$n'(F)dF = \psi(F)dF\sigma\left(z'_{max}(F)\right) = -\Phi(F)d\ln\Phi(F)\sigma\left(z'_{min}(F)\right) \,,$$

where $F'_{min}(z)$ [or $z'_{max}(F)$] is the minimum (maximum) value of luminosity (redshift) that an object, with redshift z (luminosity F) must have in order to be in the new sample.

For z, F, $F_{min}(z)$ and $z_{max}(F)$ within the range of the parent sample histograms representing some integrals of m' and n' can be calculated without further assumptions or extrapolations. However, for extending such predictions to regions outside the observed domain of the parent sample, one needs further assumptions. For example, to determine the number of expected objects at redshifts larger than the highest so far observed in any complete sample, we need to extrapolate $\sigma(z)$ to higher z values. In the next section we shall need to carry out such extrapolations.

D. The Source Function and Physical Evolution

The aim of the investigation of the luminosity function and its statistical evaluation is to determine the physical evolution, F(t), of the objects with cosmic time and the source function S(F,t), which describes the rate of their formation as a function of cosmic time. The luminosity function can be expressed in terms of cosmic time $\psi(F,t) = \psi(F,z)dz/dt$ through the redshift-time relation of the specific cosmological model. As mentioned above $\psi(F,t)$ stands for the total number of object (of luminosity F) within a specified co-moving volume and is related to $\dot F(t)$ and $S(F,t)$ through the equation of continuity:

$$\partial\psi(F,t)/\partial t + \partial[\dot F\psi(F,t)]/\partial F = S(F,t) \quad, \quad \dot F \equiv dF/dt . \quad (21)$$

Clearly this single equation is not sufficient to determine both $S(F,t)$ and $F(t)$ (or $\dot F$). We need further information or assumptions. Very little attention has been paid to this equation except recently by Cavaliere (see this proceedings) and his colleagues. They have assumed various forms for $F(t)$ and $S(F,t)$ and compared the derived $\psi(F,t)$ from the solution of Equation (21) with the observed luminosity function such as that derived by Schmidt and Green (1983). I think it will be more profitable to reverse this procedure in the sense of solving Equations (21) for $S(F,t)$ (or $\dot F$) for a given $\psi(F,t)$ and an assumed $\dot F$(or $S(F,t)$). As our discussion in the previous part shows, the analysis of the data yields directly the cumulative luminosity function rather than the differential function $\psi(F,t)$. Therefore, if we integrate Equation (21) over F and t, we can express the cumulative source function in terms of the cumulative luminosity function. For the purpose of the illustration, let us consider the simple but plausible case where the luminosity function is separable, as in Equation (8), and $\dot F$ is independent of the cosmological epoch. Then, integrating Equation (21) over time and luminosity and noting that $\rho(t=0) = 0$ and $\psi(F=\infty) = 0$, we obtain

$$\mathcal{S}(F,t) = \int_F^\infty dF' \int_0^t dt' S(F',t') = \rho(t)\Phi(F) + \dot F\sigma(t)\psi(F) . \quad (22)$$

Now, with the help of Equation (18), we can relate the cumulative source function $\mathcal{S}$ to the data directly as

$$\mathcal{S}(F,t) = \sum_i \sigma(z)\Phi(F)\left[\frac{\delta(z-z_i)}{M(z)}\frac{dz}{dt} + \frac{\delta(F-F_i)}{N(F)}\frac{dF}{dt}\right] . \quad (23)$$

The use of such equations is beyond the scope of the present work. I am presenting these equations to indicate the complexity of the problem and to show how far we are from a direct determination of the source function $S(F,t)$.

Luminosity or Number Evolution: There has been considerable discussion in the past and in this symposium in regard to whether the evolution of quasars and active galactic nuclei can be represented by a luminosity function that undergoes a pure density or a pure luminosity evolution. To begin with, I would like to point out that when referring to density one is talking about the number of objects in a unit co-moving volume, which is proportional not to the density of objects at different epochs but to number of objects within a specified co-moving volume. (For a closed universe this could be the total number of objects). Consequently, number evolution is a more appropriate term than density evolution.

A thorough discussion of the possible evolutionary forms of the luminosity function was given in a paper by Lynds and Petrosian (1972). Most of the evolutionary forms discussed in this symposium, notably those by Schmidt, Weedman, and Koo, were fully covered in that earlier paper. Most importantly, however, what Roger Lynds and I stressed was the difference between what we called the statistical and the physical evolutions. The confusion between these remains the main source of controversy. What one normally calls the evolution of the luminosity function, which deals with the mathematical representation of the data by the function $\psi(F,z)$, is a statistical evolution of the population. The physical evolution of sources are described by the function $\dot{F}$ and the source function S in Equation (21).

I suggest that it is more appropriate to tailor the nomenclature to the physical processes rather than to the mathematical representation. To illustrate how this can be done, let me compare the time scales associated with the three terms in Equation (21). The first term has a time scale of the order of Hubble time τ_H ; $\partial\psi/\partial t \simeq \psi/\tau_H$. The second term is of the order of ψ/τ_F, where $\tau_F = F/\dot{F}$, and the time scale of the third term is determined by the time scale τ_S of the formation rate of the objects: $S(F,t) \simeq \psi/\tau_S$.

If $\tau_F \simeq \tau_H \gg \tau_S$, which could be the case for $t > t_c$ if all the sources are created prior to an epoch t_c, then any evolution of ψ is a reflection of the physical luminosity evolution. The total number of objects (integrated over all luminosities) is a constant. Therefore, the term luminosity evolution is an appropriate term here. However, it should be stressed that this does not guarantee a pure luminosity evolution. The pure luminosity evolution requires that all sources, irrespective of their initial state or environment, evolve the same way, $F(t) = F_0 g(t)$.

In the other limiting case, where $\tau_S \simeq \tau_H \gg \tau_F$, each source evolves very rapidly so that the luminosity function is determined primarily by the rate of formation of the sources. This may be called a number (or density) evolution. However, this again does not mean a pure density evolution model, which requires $S(F,t)$ to be a separable function of F and t.

III. EVOLUTION OF QUASARS

I will now use the procedure developed in the preceding section to determine the statistical evolution of the optical luminosity function of the quasars and, comment on the physical evolution of the sources. As discussed above, this procedure requires a knowledge of the appropriate stochastically independent parameters before one can determine the global evolution of the luminosity function over the wide ranges of the observed luminosity and redshift. Lacking this knowledge

I consider evolution over limited redshift and luminosity ranges, in which case I can assume that redshift and luminosity are stochastically independent. This assumption is a good approximation (and more readily testable) over small ranges of F and z.

The main new result I would like to concentrate on here is the question of cutoff of the luminosity function at high redshifts ($z > 3$). However, before doing this I will briefly describe the evolution of the luminosity function at low redshifts ($z < 2$), in which case the difference between the various cosmological models is insignificant as compared to the dispersion of the luminosity function.

A. Evolution at "low" redshifts ($z < 2$)

In one of the first analyses of a complete optically selected sample (Petrosian 1973), I had concluded that the distribution of the low luminosity, low redshift quasars can be described by a non-evolving luminosity function while a strong evolution was required for higher luminosity (higher redshift) objects. Now the new more extensive PG sample essentially confirms this earlier result (Schmidt and Green 1984), except that Schmidt and Green describe the distribution by a luminosity function whereby the evolution of the number of sources becomes monotonically stronger at higher and higher luminosities. In a short paper presented at the 1982 Liege conference (Petrosian and Jankevics 1982), it was shown that when the PG sample is divided into two (high and low luminosity) parts one finds that the low luminosity part shows no evolution while the high luminosity part shows strong evolution of number of quasars (cf. Figure 3). In the same paper it was also shown that the pure luminosity evolution with the parameters derived by Marshall et al (1983) does not agree with the PG sample. (A different set of pure luminosity evolution parameters can be found for a reasonable agreement; H. Marshall, private communication.) But there is no escaping of the fact that whatever the evolution of the luminosity function it is small or non-existant at low F but become significant at high values of F. Figure 3 clearly demonstrates this where, assuming a pure density evolution, it is shown that for low luminosities $\sigma(V) \propto V$ [i.e. $\rho(V)$ and $\rho(z)$ are constants] while for high luminosities $\sigma(V) \propto V^4$, implying a strong evolution.

This kind of behavior can be described by the following simple physical conditions. Suppose the rate of physical evolution of quasars is independent of the cosmological epoch t but depends on the luminosity F: e.g., $\dot{F} \propto F^{-\alpha}$. This is a reasonable assumption as it demands that quasars spend a shorter time in a higher luminosity state than in a lower one. For example, an equal energy consumption at different luminosity states implies $\alpha = 2$.

Let us first consider the luminosity function for high luminosities. At sufficiently high values of F (say $F > F_{cr}$), the lifetime $\tau_F = F/\dot{F}$ could be much shorter than the Hubble time τ_H so that on the right hand side of Equation (21) the first term is much smaller than the second and can be ignored. The rest of the equation can then be integrated to give

$$\psi(F,t) \propto F^{-\alpha} \int_F^\infty S(F',t)dF' \quad , \quad F \gg F_{cr} \quad , \quad (24)$$

which shows that a strong evolution (variation with t) is possible and that such an evolution is a reflection of the evolution of the creation rate of quasars. The observation that $\sigma(V) \propto V^4$ for high luminosity sources implies that S(F,t) decreases rapidly as the universe expands. For values of $F \ll F_{cr}$, however, τ_F could become larger than the age of the universe so that now the second term in Equation (21) is negligible leading to the solution

$$\psi(F,t) = \int_0^t S(F,t')dt' \quad , \quad F \ll F_{cr} \quad . \quad (25)$$

Because S(F,t) was larger in the past (low t) and has decreased rapidly since then, it is obvious that for large values of t (low redshifts) the integral of S will be nearly independent of t explaining the observed absence of strong evolutions for weak sources.

Note that for $F \ll F_{cr}$ the luminosity function $\psi(F,t)$ is a reflection of the luminosity dependence of the source function (Eq. 25). But at high luminosities $\psi(F,t)$ is steeper than S(F,t) by a power of $1 - \alpha$, indicating steepening of the luminosity function at high values of F because α is expected to be greater than unity. Such a steepening of the luminosity function is observed and can be used to determine the physical evolution of the quasars such as the luminosity dependence of the source function and the power law index α.

B. High redshifts and the redshift cutoff

One of the important cosmological questions has to do with the epoch of formation of galaxies and other structures. Theoretical arguments indicate that this epoch could correspond to a time anywhere from redshift 2 to 1,000. Because quasars are the only objects bright enough to be observed at such high redshifts, it is then expected that the evolution of their luminosity function at high redshift could shed light on this question.

There have been various attempts to extrapolate the evolution of the luminosity function to high redshifts and compare its consequences with observation, with the obvious conclusion that the strong evolution obtained from the data in the redshift ranges 1 to 2 cannot continue to very high redshifts. One of the most convincing results comes from Osmer's (1982) slitless spectroscopic search for quasars in the redshift range 3.7 to 4.7. This search yielded no quasars within this redshift range, prompting the conclusion that this may be the epoch of galaxy formation.

I would like to reconsider this problem, using the more rigorous method described in section II, for the three cosmological models mentioned there. As is shown in figure 1, the cosmological models begin to diverge from each other significantly (relative to the

EVOLUTION OF THE LUMINOSITY FUNCTION OF EXTRAGALACTIC OBJECTS

dispersion in the data) only at large redshifts and are expected to give different results.

1) **The Sample**: The sample of sources I use are selected from the complete samples compiled by Schmidt and Green (1983) which, in addition to the PG sample, includes the Braccesi AB sample (Braccesi, Formiggini and Gandolfi 1970) four slitless spectroscopic samples (Hoag and Smith 1978; Lewis, McAlpine and Weedman 1979; Osmer and Smith 1980; Sramek and Weedman 1978) and the Kron and Chiu (1981) sample. To this I have added the few objects from the deeper Braccesi BF sample (Marshall et al 1983). In order to avoid extensive extrapolation over large cosmological distances and luminosities, I have limited the sample to the bright end of the luminosity which necessarily means limiting to high redshifts. The PG sample is sparse around redshift of 0.5 to 1.0 This turns out to be a convenient point of separating the sample into two sets, a high and a low luminosity set. For each cosmological model the lower cutoff of luminosity is chosen to correspond roughly to that of the same object in the PG sample which has a redshift of 0.944. This same luminosity cutoff is then used for all the other samples. Consequently, the selected sub-samples are somewhat different in the three cosmological models. There are about sixty objects in each case within the redshift range of 1 to 3. This implies a co-moving volume change of less than one order of magnitude and a luminosity range of slightly larger than one order of magnitude for the whole sample.

2) **The procedure**: I will assume stochastic independence betweeen luminosity F and redshift z, or co-moving volume V(z), which, in view of small ranges of the parameters just mentioned (and as we will show by simple binning) is a reasonable assumption. Then the application of the procedure described in sectin II-C is straightforward except that combining of samples with complicated selection criteria requires some modification of that procedure. All the samples have the same luminosity limit but different magnitude limits. This does not affect the procedure. But the fact that the slitless spectroscopy has a lower redshift cutoff at z = 1.8 complicates the procedure. It is not clear how rapidly the efficiency of discovering sources increases from a small value for z < 1.8 to a constant and significant value above this redshift and whether this efficiency is constant over the whole redshift range when Ly-α falls in the proper bandwidth of the plates. I will assume a zero efficiency outside this range and a constant efficiency throughout the range. The results presented below will, of course, be altered if this is not the case.

The lower redshift cutoff does not change the equations for the evaluation of the cumulative luminosity function $\Phi(F)$. However, Equation (15) for the number (or the co-moving density) evolution at z > 1.8 is altered as follows:

$$\frac{d\sigma}{\sigma - \sigma(z=1.8)} = \frac{\delta(z-z_i)}{M(z>1.8)} dz \quad , \qquad (26)$$

so that

$$\sigma(z_i > 1.8) = \sigma(z = 1.8) + \prod_{j+1}^{i} \left(1 + \frac{1}{\tilde{M}_j}\right), \qquad (27)$$

where $\tilde{M}$ now includes only objects with $z > 1.8$ and $F > F_{min}(z)$. This modification requires some interpolation of $\sigma(z)$ around $z = 1.8$ which I will not go into here.

3) The results: I will first present some of the histograms $\Phi(F)$ and $\sigma(z)$ and then extrapolate them to evaluate the expected numbers at high redshifts.

a) The cumulative luminosity function $\Phi(F)$ is shown in Figure (4) for all three models. The shapes of these luminosity functions are approximately the same (showing steepening at high luminosity) for all three models except that the luminosity scale is different. To somewhat justify the assumption of the stochastic independence of F and z the total sample was divided into three redshift bins with equal numbers in each bin and the luminosity function $\Phi(F)$ was evaluated for each bin. Figure (5) shows this result for the deSitter model ($k = 0$, $\Omega = 0.0$, $\lambda = 1.0$). The small number of objects in each sub-sample (about 20) makes a detailed comparison difficult, but rough similarities between the three histograms show that the assumption of the stochastic independence (over the small redshift and luminosity range) will not lead to large errors.

b) The cumulative number evolution functions $\sigma(z)$'s are shown in figure (6). Clearly the predictions of the three cosmological models are quite different. However, as evident, the three models show similar strong evolution in the range $1 < z < 2$ mentioned above but begin to diverge at higher redshifts. In particular, the Lemaître model ($\lambda = 1.21$, $k = +1$, $\Omega = 0.1$) shows a very slow increase of $\sigma(z)$ with redshift for $z > 2$ while the other two models show that the strong evolution continues to redshift of up to 3 and may even continue further on up to the highest redshift in the samples. The value of $\sigma(z)$ for $z > 3$ is uncertain (dashed histogram) because there are very few (about 4) objects in this range. Consequently, it is difficult to extrapolate this curve beyond $z = 3.5$. For the deSitter and the Einstein-deSitter models I show two possible extrapolations. One of these extrapolates by fitting a straight line (dashed) to the upper portion of the $\ln\sigma - \ln V$ curve obtaining $\sigma(V) \propto V^\beta$ with $\beta = 2.8$ and 3.7 for the two models, respectively. (Note that this is a slower evolution than $\sigma(V) \propto V^4$ found at lower redshift, perhaps indicating the slowing down of the evolution.) The second extrapolation assumes no evolution which means $\sigma(V) \propto V$ (dotted lines). For the Lemaître model the extrapolation seems to be fairly obvious ($\ln\sigma(z) \propto 0.37z$) as the curve in the $2 < z < 3$ range is well defined.

c) Predictions: As an example I have calculated the predicted cumulative redshift distribution for a sample limited to 19.5 blue magnitude (for definition of the magnitude see Schmidt and Green

1983). This is done using Equation (29) with $F_{min}(z)$ evaluated for 19.5 magnitude. The result is shown by the inset in figure (6a) where the predicted values (dashed and dotted lines) for the two extrapolations of $\sigma(z)$ beyond redshift $z = 3$ are compared with the observations (solid lines). The agreement at low redshifts is not surprising since the observed sample in this comparison was part of the parent sample used in deriving $\sigma(z)$ and $\Phi(F)$. The important feature here, however, is the deviation of the strong evolution case (dashed line) from observations at redshift $z > 3$. This shows that, in this model, the evolution is already slowing down at $z > 3$. Note that no other extrapolation, except that for $\sigma(z)$, was needed for this calculation.

Further comparison between the prediction in the three cosmological models and observations are shown in Table I. For the first two models in this table there are two predictions based on the two extrapolations shown in figures 6a and 6b. The prediction at 19.5 limiting blue magnitude is compared with the data tabulated by Schmidt and Green (1983) (their table III, HS and SW List) taken from Hoag and Smith (1977) and Sramek and Weedman (1978). The first line on Table I shows that when the redshift range is divided into two parts each containing about the same number of objects (9 in one and 10 in the other in this case) the predictions agree quite well with the observation. However, if we compare numbers expected above and below redshift 3, we find that the first two models agree with observation if the sources do not evolve beyond redshift $3(\sigma(V) \propto V)$. However, the third model gives acceptable results for the natural extrapolation shown in figure 6c.

The obvious conclusion from this is that in Lemaître type models the source evolution is found to have slowed down beyond $z > 2$ and could continue at this rate to a redshift of 3.5 while in the other two (inflationary) models the evolution must slow down and stop beyond redshift 3. This latter, however, is not a firm result because selection effects can account for absence of observed objects beyond redshift 3. An indication of this is shown on the third line of Table I where I compare the predicted and observed ratios of the number of high and low luminosity objects in the same sample. Because the luminosities of objects are different in the three cosmological models, the luminosity dividing the two bins will also be different. The latter is selected such that the observed ratio of the number of the objects in the two bins is about unity. As is evident too few low luminosity objects are observed as compared to the predictions of the two models. This could be caused by a bias against discovery of high luminosity objects by slitless spectroscopy. The so-called Baldwin effect (cf. e.g. Wampler et al 1984) can cause such a bias as the line equivalent width is smaller at higher luminosities making the discovery of such objects more difficult.

The strongest evidence for the decrease of the numbers at high redshifts is provided by Osmer's (1982) observations where the claim is that it goes deeper by 1.2 magnitude than the earlier Hoag and Smith (1971) observation (with a limiting magnitude of 19.5) and was designed to discover quasars in the redshift range $3.7 > z > 4.7$. None was discovered in a search limited to a 5.1 sq. deg. area. Using the derived luminosity function and the extrapolations of $\sigma(z)$ mentioned above, I have estimated the expected numbers in this redshift range for a limiting blue magnitude of 20.7 for the three cosmological models. The predicted numbers are based on the observed numbers (at 19.5 magnitude) given in the parenthesis.

This, in agreement with Osmer's (1982) conclusion, clearly rules out the conventional Einstein-deSitter model with a strong evolution and, to a lesser degree, no evolution in this model and the strong evolution in the deSitter model. The deSitter model with no evolution beyond z = 3 and the Lemaitre model are acceptable considering the many uncertainties in the limiting magnitude, the completeness and possible selection bias of the samples.

IV. SUMMARY AND DISCUSSIONS

The study of statistical evolution of the luminosity function of extragalactic objects is important for understanding of the physical evolution of the objects and provides important clues about the variation with time of the birth rate and the luminosity of the objects and, in general, about the formation of structures in the universe. I have explained the necessary steps for a complete description of this evolution.

The first and the most difficult step in this study is an accurate statistical analysis of the distribution of objects in the redshift and luminosity domain once a sample and a cosmological model are selected. Parametric approaches suffer from the fact that they are not unique and that could lead to misleading results if extrapolated beyond the observed range of the variables. The non-parametric procedure described here works for a luminosity function which is separable into functions of some variables which need not be the basic variable, namely, the luminosity and redshift. Discovery of such stochastically independent variables is the difficult step which I have avoided here by considering the data for small ranges of the basic variables, in which case the assumption that these variables are stochastically independent is a good approximation.

Using results from earlier works, I have argued that at lower redshifts (mean redshift of about one) and independent of the cosmological model the luminosity function shows stronger evolution at higher luminosities than at lower luminosities. One interpretation of these results is that the activity in galaxies which leads to the quasar phenomenon was much more prevalent in the past and that the rate of the physical evolution of the luminosity of the quasars is a strong function of the luminosity in the sense that much shorter time is spent at a high luminosity phase than a low one.

The next question I have addressed here is how far into the past this ever increasing activity can be extrapolated and whether the existing data tells us about the cosmological epoch when this activity started. Here I have shown that the conclusion depends strongly on the cosmological model and that in conventinal cosmological models at redshifts greater than three the activity must have been occurring at a much slower rate than the simple extrapolation would indicate, but that there are cosmological models where the activity could have been present at larger redshifts. Therefore, the epoch of the formation of quasars (and galaxies) remains an unknown.

ACKNOWLEDGMENT

I would like to thank Professor Paul Switzer of the Department of Statistics at Stanford University for helpful discussions on the statistical matters and acknowledge partial support from the National Aeronautics and Space Administration under Grant NGR 05-020-668.

REFERENCES

Braccesi, A., Formiggini, L. and Gandolfi, E. 1970, Astron. and Astrophys. 5, 264.
Hoag, A.A. and Smith, M.G. 1971, Ap. J. 217, 362.
Kron, R.G. and Chiu, L.-T.G. 1981, Publ. A.S.P. 93, 397.
Lewis, D.W., MacAlpine, G.M. and Weedman, D.W. 1971, Ap. J. 233, 787.
Lynden-Bell, D. 1971, Mon. Not. R. Astr. Soc. 155, 95.
Lynds, R. and Petrosian, V. 1972, Ap. J. 175, 591.
Marshall, H.L., Tananbaum, H., Zamorini, G., Huchra, J.P., Braccesi, A. and Zitelli. V. 1983, Ap. J. 269, 42.
Osmer, P.S. 1982, Ap. J. 253, 28.
Osmer, P.S. and Smith, M.G. 1980, Ap. J. Suppl. 42. 333.
Peebles, P.J.E. 1984, Ap. J. 284, 439.
Petrosian, V. 1973, Ap. J. 183, 359.
Petrosian, V. and Jankevics, A. 1983, Quasars and Gravitational Lenses, Proc. of the 24th Liege Int'l. Astrophys. Colloq., Ed. J.P. Swings, pp. 250-257.
Petrosian, V. and Salpeter, E.E. 1968, Ap. J. 151, 411.
Schmidt, M. and Green, R.F. 1983, Ap. J. 269, 352.
Sramek, R.A. and Weedman, D.W. 1978, Ap. J. 221, 468.
Wampler, E.J., Gaskell, C.M., Burke, W.L. and Baldwin, J.A. 1984, Ap. J. 276, 403.

TABLE I

Comparison of the Observed and Predicted Number of Objects in the Three Cosmological Models

Model	k = 0* $\Omega = 1.0, \lambda = 0.0$	k = 0* $\Omega = 0.0, \lambda = 1.0$	k = +1 $\Omega = 0.1, \lambda = 1.21$	Observed
a) 19.5 mag.				
$\dfrac{N(2.2<z<3.5)}{N(1.8<z<2.2)}$	0.71 1.5	1.3 2.0	1.2	9/10
$\dfrac{N(3<z<3.5)}{N(1.8<z<3)}$	1.8/18 7.4/18	1.1/18 6.7/18	2.2/18	1/18
$\dfrac{N(\text{High F})}{N(\text{Low F})}$	2.1	3.0	1.2	10/9
b) 20.7 mag. **				
$N(3.7<z<4.7)$	5.4 (10) 46	0.53 (2) 4.7	1.8 (9)	0

*First ratios from extrapolation assuming no evolution ($\sigma \propto V$); the second ratios from the evolutionary extrapolation (dashed lines Figure 6a, 6b).

**The calculated numbers are based on the number (shown in the parenthesi of objects in the 19.5 magnitude limited sample.

EVOLUTION OF THE LUMINOSITY FUNCTION OF EXTRAGALACTIC OBJECTS

FIGURE CAPTIONS

Figure 1. Luminosity distance $D_L(z)$ versus redshift for three cosmological models. For the $k = +1$, $\Omega = 0.1$, $\lambda = 1.21$ model the dashed line is a more realistic relation, but for simplification of calculations I have assumed the solid line. Spectral index α is assumed to be 0.5.

Figure 2. Schematic representation of the distribution of object in the luminosity (F) - redshift (z) plane. The heavy solid line is the F-z relation at the limiting apparent flux value f_0. For each F (or z) this line defines the maximum z (or minimum F) that an object with this F (or z) can have and be in the sample.

(a) Defines the parameters when the analysis is carried out by binning the sample. $n_{i,j}$ is the number of objects in the bin with $F_j < F < F_{j-1}$ and $z_{i-1} < z < z_i$.

(b) Defines the parameter $z_{max}(F)$, $F_{min}(z)$ and $M(z)$ and $N(F)$ used in the text.

Figure 3. The cumulative density evolution function σ versus volume V (up to redshift z) for "low" redshift quasars (z < 2, PG sample, Schmidt and Green 1983) at high and low luminosities separately. Note for constant number (or co-moving density) $\sigma(V) \propto V$. The cosmological model assumed here is $\Omega = 0$, $\lambda = 0$, $k = -1$ but this result is not sensitive to the model. V_0 and V_0' are arbitrary volumes.

Figure 4. The cumulative luminosity function versus the blue absolute magnitude as defined by Schmidt and Green (1983) for the three cosmological models depicted on Figure 1. For clarity, the histograms are shifted to the right by the specified amounts. The relative values of $\Phi(M)$ are arbitrary.

Figure 5. The luminosity function for the $k = 0$, $\Omega = 0.0$, $\lambda = 1.0$ cosmological model for objects in three non-overlapping redshift ranges. For clarity two of the histograms are shifted up or down by the quantity X indicated. The rough similarity of these histograms shows that the assumption of stochastic independence of the luminosity (or absolute magnitude M_B) and redshift over the small ranges of these parameters will not lead to erroneous results.

Figure 6. The cumulative number (or co-moving density) evolution function $\sigma(z)$ versus redshift z, or volume V(z) up to redshift z, for the three cosmological models. The dashed portions of the histogram have larger uncertainties.

(a) $k = 0$, $\Omega = 1.0$, $\lambda = 0.0$.
(b) $k = 0$, $\Omega = 0.0$, $\lambda = 1.0$.
(c) $k = 1$, $\Omega = 0.1$, $\lambda = 1.21$.

In Figures (a) and (b) two possible extrapolations for $z > 3$ have been shown with one of them assuming no number or density evolution, $\sigma(V) \propto V$. The inset in (b) shows the predicted cumulative number of sources between $z_{min} = 1.8$ and z expected at blue limiting magnitude of 19.5 and for the two extrapolations shown in the main part. In (c) the steps of the histogram at $z > 2$ are smaller than the width of the line.

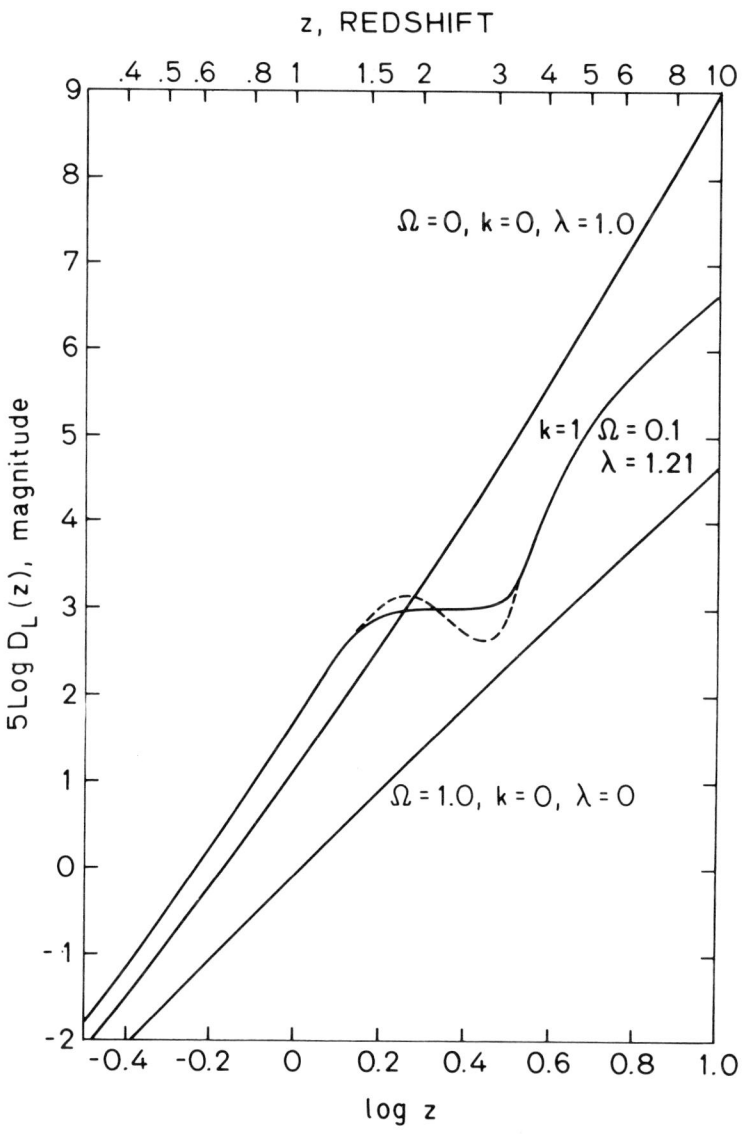

Figure 1

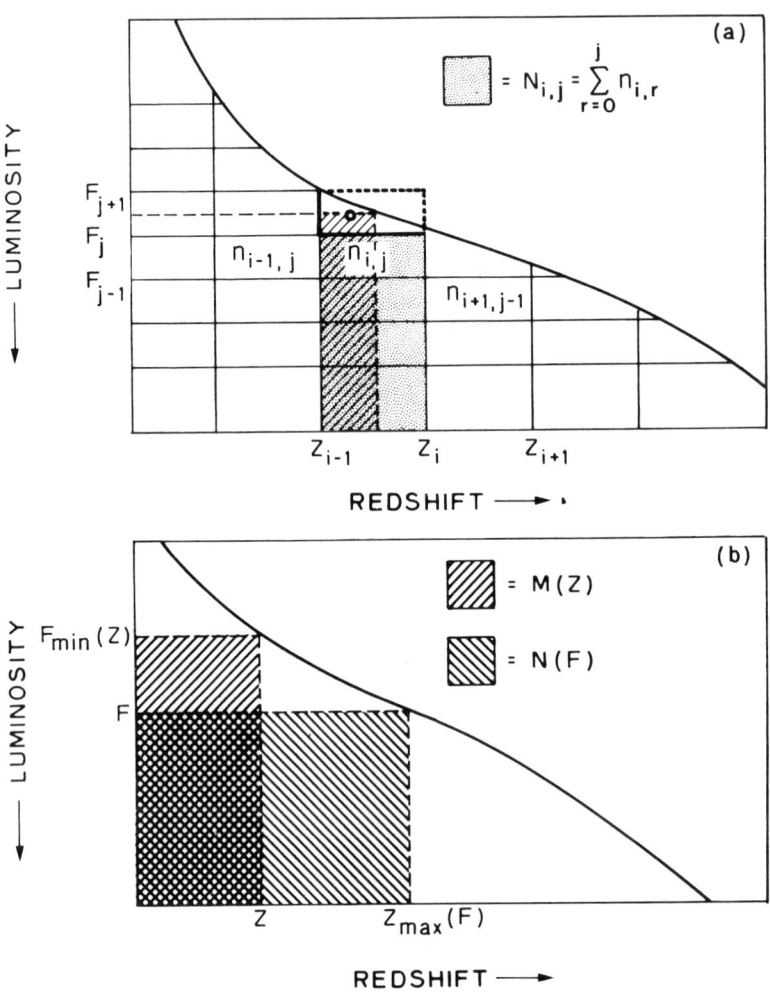

Figure 2

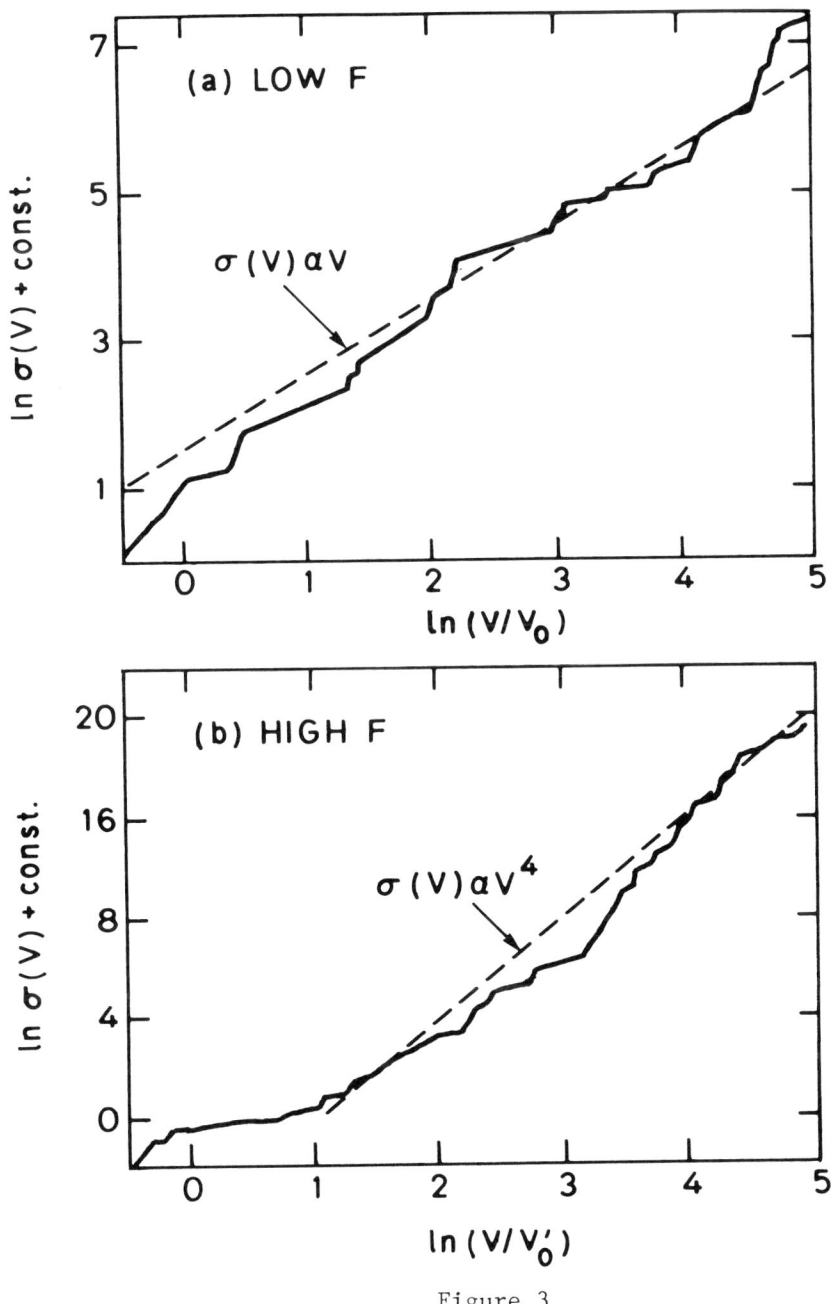

Figure 3

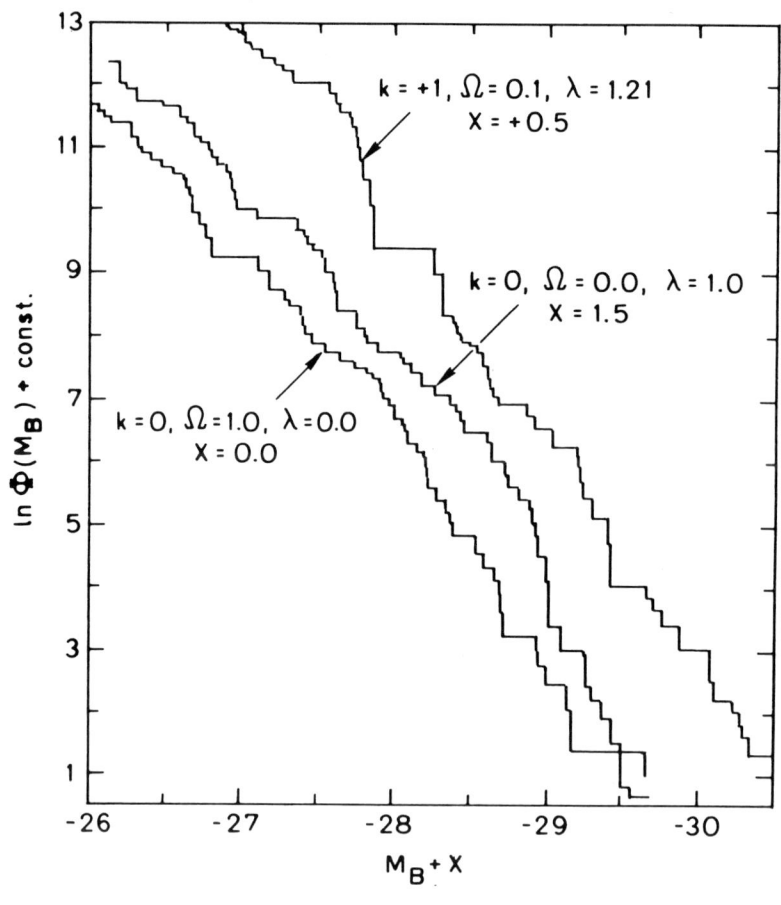

Figure 4

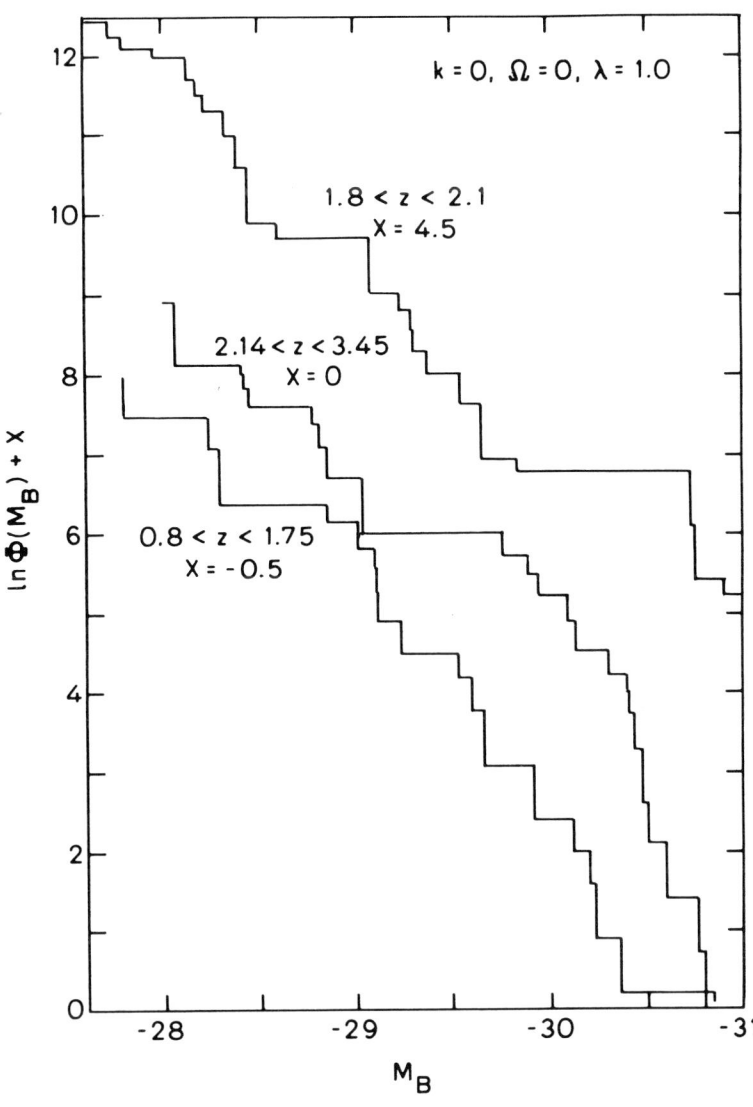

Figure 5

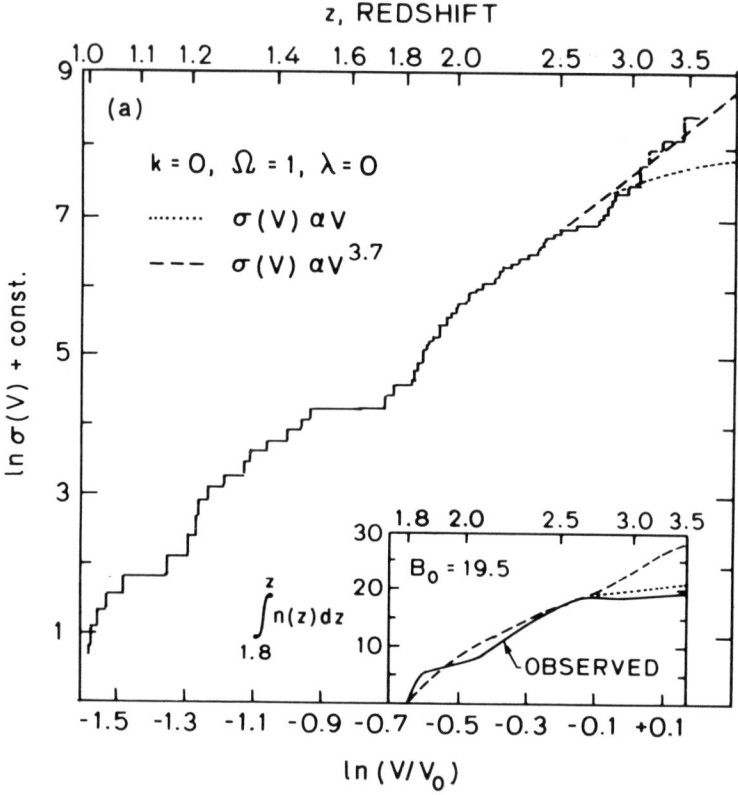

Figure 6a

EVOLUTION OF THE LUMINOSITY FUNCTION OF EXTRAGALACTIC OBJECTS

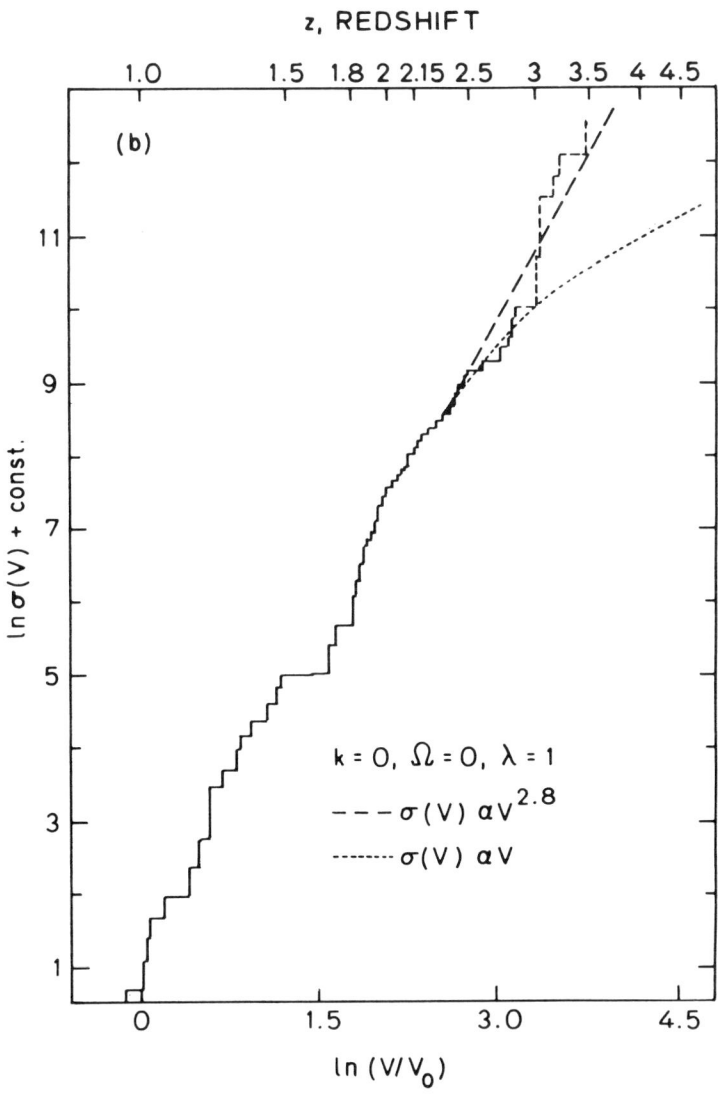

Figure 6b

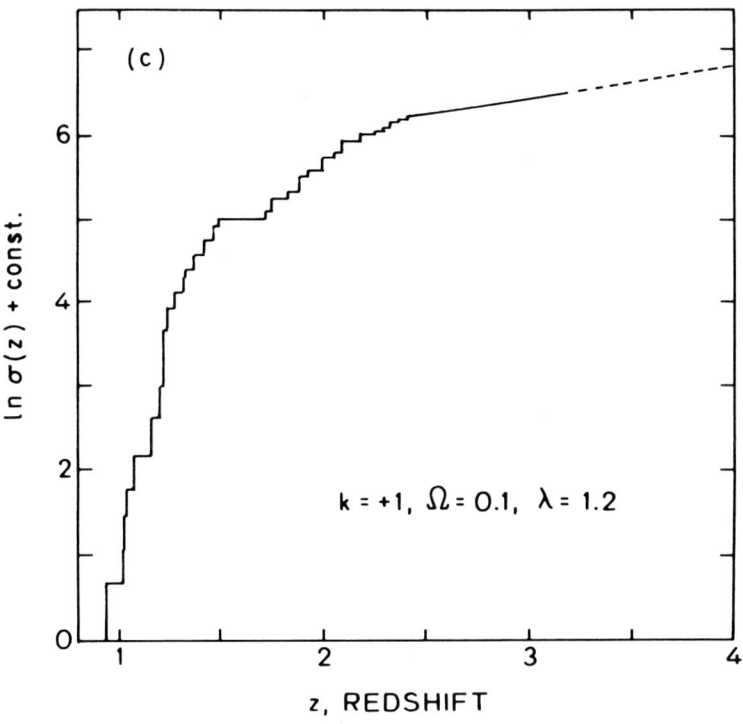

Figure 6c

DISCUSSION

MARSHALL: 1) I understand that your method is similar to that used by Schmidt and Green (1973) except that you use a non parametric model of the evolution function whereas they parametrized it, allowing them to use the V_m^{-1} method to get the luminosity function and you are forced to use the generalization of the C method of Lynden-Bell. 2) Did you try to include the data from the AB and/or BF quasars samples?

PETROSIAN: 1) Yes, the method I use is fully non-parametric and makes no assumption about the form of the luminosity function or its evolution, while some parametrization is necessary in the V_m^{-1} method. 2) Yes, I did include the AB and BF quasars which were in the relevant redshift and luminosity range.

SPECTRAL EVOLUTION IN YOUNG ACTIVE GALACTIC NUCLEI

Elihu Boldt and Darryl Leiter[*]
Laboratory for High Energy Astrophysics
NASA/Goddard Space Flight Center
Greenbelt, Maryland 20771 USA

ABSTRACT. The spectral evolution of AGNs is discussed within the context of a scenario where the cosmic X-ray background (CXB) is dominated by these sources. Attention is drawn to the fact that individually observed AGN X-ray spectra are significantly steeper than that of the CXB. The remarkably flat spectrum thereby required for the "as-yet" unresolved sources of the residual CXB is interpreted as an observational constraint on an earlier stage of AGN evolution. Assuming black hole disk accretion, a picture emerges where young AGNs are compact Eddington limited thermal X-ray sources and where canonical AGNs represent later stages in which they have become appreciably less compact, exhibiting the importance of non-thermal disk-dynamo processes.

1. INTRODUCTION

We examine the constraints on spectral behavior in young AGNs from a point of view based on two assumptions, one theoretical and the other semi-empirical. The theoretical assumption is that the AGN phenomenon is powered by disk accretion into a supermassive black hole (see review by Rees 1984). The other assumption is that the cosmic X-ray background (CXB) is dominated by the integrated contributions of AGNs in all their various stages of evolution. In particular, for this discussion, we ignore the possibility that any substantial portion of the CXB is intrinsically diffuse (e.g. as could arise from a hot intergalactic medium (see review by Fabian 1981) or some non-standard cosmology).

2. OBSERVATIONAL SETTING

The spectrum of the CXB has been very well determined with HEAO 1 and may be characterized over the band 3-100 keV by an optically thin thermal model with kT = 40 keV (Marshall et al. 1980; Rothschild et al. 1983). However, bright AGNs observed over this same band exhibit

[*]Now at Sciences Division, FSTC, Charlottesville, Virginia 22901 USA

spectra quite different from that of the CXB; most have power-law spectra with an energy spectral index compatible with 0.7 (Mushotzky 1982; Rothschild et al. 1983). In general, these bright AGN are within the present epoch (i.e. low redshift). Although many high-redshift AGNs (quasars) have been detected in X-rays with the HEAO 2 Einstein Observatory telescope the data are restricted to energies below 3 keV, and spectral determinations are relatively infrequent and uncertain (Elvis et al. 1985). Furthermore, most sources of the CXB are too faint to have been detected with HEAO 2. Hence, at this stage the spectrum characteristic of the complete ensemble of young AGNs (including possible precursor quasars) can best be inferred from the spectrum of the CXB. More precisely, we must account for that portion of the CXB arising from sources of known spectral properties (within the present epoch) in order to isolate the residual CXB to be identified with young sources at high redshifts. This has been done under a variety of approximations (DeZotti et al. 1982; Leiter and Boldt 1982; Worrall and Marshall 1984), all consistent with the same general picture emerging. Specifically, the sources of the residual CXB are in this way found to be characterized by spectra that are extremely flat at the lowest energies and then fall off with an e-folding energy of about $23(1+z)$ keV, in their proper frame. For $z > 3.5$, where there is an apparent paucity of canonical quasars (Osmer 1982; Schmidt and Green 1983), this characteristic energy would be greater than 100 keV.

3. FUNDAMENTAL PROPERTIES

What is the irreducible parameterization to be used in defining a young AGN? The physical size is one basic consideration. Assuming that an AGN is powered by black hole disk accretion the size (R) of the associated radiating system during its earliest stage of development is measured on a scale given by the gravitational radius ($GM/c^2 = 1.5 \times 10^5$ $M/M_\odot$ cm); the possible growth of much more extended structure, such as jets, would require special conditions and the passage of a suitably long interval of time. Depending on the spin of the black hole, the inner radius of the accretion disk is only a few times the gravitational radius (see Figure 1). Clearly the spin itself is a fundamental parameter. If the black hole core of a young AGN were of pre-galactic origin its spin could be zero at the start of the accretion disk formation process (Leiter and Boldt 1982). Completing our characterization of a young AGN involves making an assumption about the immediate environment of the black hole. Specifically, we assume galactic conditions appropriate for the formation of an ample accretion disk.

Having defined the basic parameters of a young AGN we investigate what constraints they imply as regards radiation. When accretion disks are first formed, possibly at the onset of galaxy formation, the accretion rate into the central black hole is here taken to be limited only by the Eddington luminosity $L = L_{Edd}$ ($L_{Edd} = 1.3 \times 10^{38}$ $M/M_\odot$ ergs/s). As long as the size is less than a thousand gravitational radii the resulting compactness parameter of the source (L/R) is then

SPECTRAL EVOLUTION IN YOUNG ACTIVE GALACTIC NUCLEI

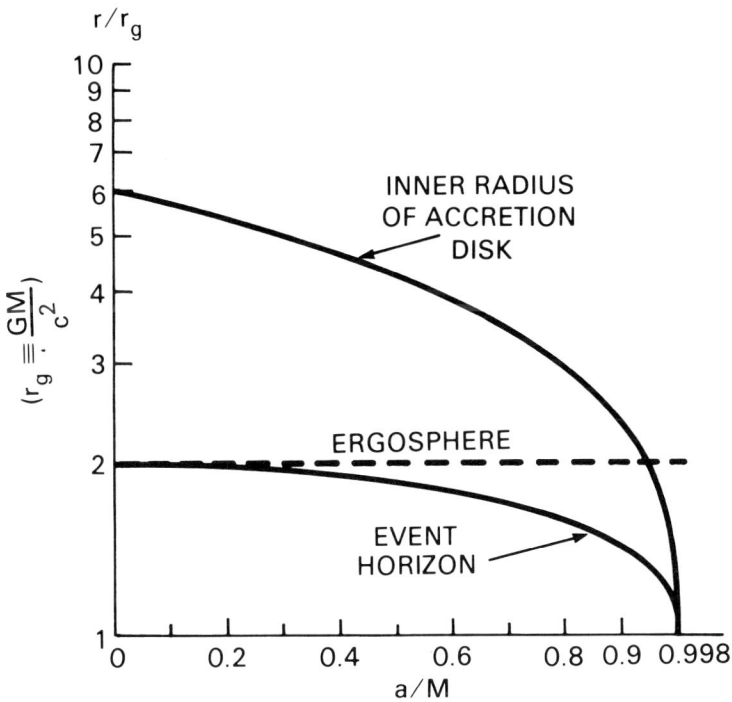

Figure 1. The inner radius of a black hole accretion disk and the event horizon (units of $r_g \equiv GM/c^2$) versus the black hole specific angular momentum per unit mass a/M (see Bardeen 1972).

greater than 10^{30} ergs/(s cm). However, as pointed out by Cavaliere and Morrison (1980), when the compactness parameter is this large energy losses limit the efficient disk-dynamo acceleration of relativistic electrons; non-thermal electron emission mechanisms are suppressed. The thermal disk emission that then dominates the radiation from a young AGN is itself severely constrained as regards temperature and radiative transfer. In particular, for hard X-rays we note from Herterich (1974) that the optical depth for electron-positron pair production in photon-photon collisions is essentially equal to the compactness parameter divided by the critical value of 10^{30} erg/(s cm); a high value of (L/R) thereby implies an effective phase transition limiting the temperature of the radiating plasma of the accretion disk to $kT < mc^2$ (Cavallo and Rees 1978; Lightman 1982; Svensson 1982; Leiter and Boldt 1982; Rees 1984). Furthermore, for a highly compact source with luminosity close to the Eddington limit, the optical depth for photon-electron Thomson scattering is larger than (1 + 2p), where p is the ratio of positrons to protons in the radiating plasma. From this we infer that the thermal spectrum of the X-radiation emerging from a young AGN would be significantly flattened by comptonization. The amount of comptonization involved is to be determined by identifying this spectrum with that ascribed to the apparently flattened thermal type spectrum of the residual cosmic X-ray background. For compact AGN sources with $(L/R) = 10^{30}$ erg/(s cm) Zdziarski (1984) finds that the Thomson optical depth is about 4, yielding a comptonized thermal spectrum remarkably compatible with that of the residual CXB.

4. EVOLUTION

The early evolution of a young AGN may be traced via the mass growth of the underlying black hole. A zero-spin Schwarzschild black hole evolves to a canonical Kerr spinning black hole after increasing its mass via disk accretion to a level equal to 2.5 times the initial value (Thorne 1974). The rate of mass increase is governed by the associated accretion driven luminosity as

$$\Theta(\dot{X}/X) = (L/L_{Edd}) [1 - \varepsilon(X)]/\varepsilon(X)$$

where X is the ratio of the growing black hole mass to that at the beginning of accretion, $\varepsilon(X)$ is the efficiency for converting accreted mass (M_a) to radiation given by

$$\varepsilon \equiv L/(c^2 \dot{M}_a)$$

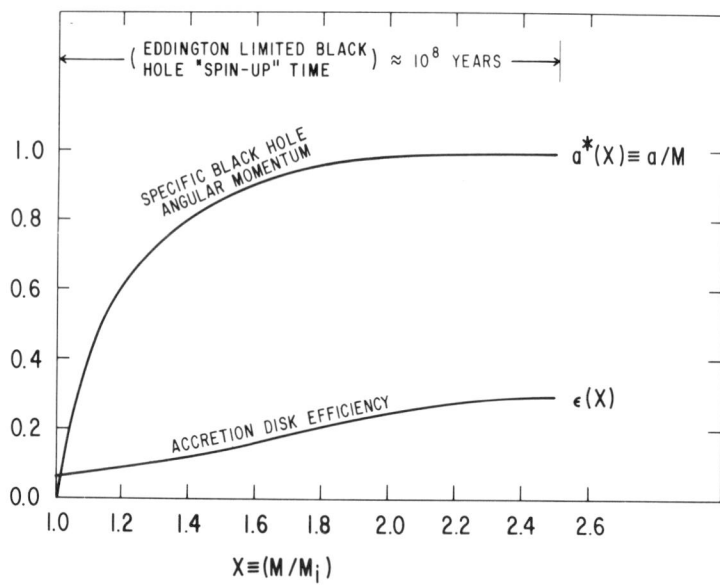

Figure 2. Early evolution of AGN. A plot of black hole specific angular momentum per unit mass $a^*(x)$ and accreted mass-to-radiation conversion efficiency $\varepsilon(x)$ versus the parameter $x \equiv (M/M_i)$, where M_i is the initial mass of a Schwarzschild black hole at the beginning of accretion disk formation (see Thorne 1974).

and $\Theta \equiv 4.5 \times 10^8$ years. The evolution of spin and radiation efficiency with respect to mass growth is shown in Figure 2. For a situation corresponding to an Eddington limited luminosity, it takes about 10^8 years to reach the asymptotic state of spin where 1) the radiation efficiency reaches a maximum value of 0.32 (at zero-spin the efficiency was only 0.057), and 2) the maximal penetration of the inner region of the accretion disk causes energetic Penrose processes within the ergosphere (see Figure 1) to be "switched on", possibly serving to trigger and/or surge disk-dynamo action (Leiter and Boldt 1982). The evolutionary behavior for the case of constant luminosity (hence L/L_{Edd} decreasing with time) is exhibited in Figure 3. For this situation we take the value of L/L_{Edd} at zero time to be unity. Under this condition we note that the lifetime of the "young" state (e.g. identified by $L/L_{Edd} > 0.4$) is approximately 2×10^8 years and that after about a billion years, comparable to the evolutionary time scale for quasars, L/L_{Edd} falls somewhat below 0.2. After 2×10^{10} years L/L_{Edd} drops to 0.01.

Two extreme stages of AGN evolution may be identified. Early on there are the highly compact thermal sources characteristic of the residual CXB, those having $L/R > 10^{30}$ ergs/(s cm) and $L = L_{Edd}$. Eventually, a stage is reached where the compactness parameter becomes less than the critical value; the efficient disk-dynamo acceleration of relativistic electrons can then occur. This transition can be initiated by the development of extended structure (i.e. much larger than the

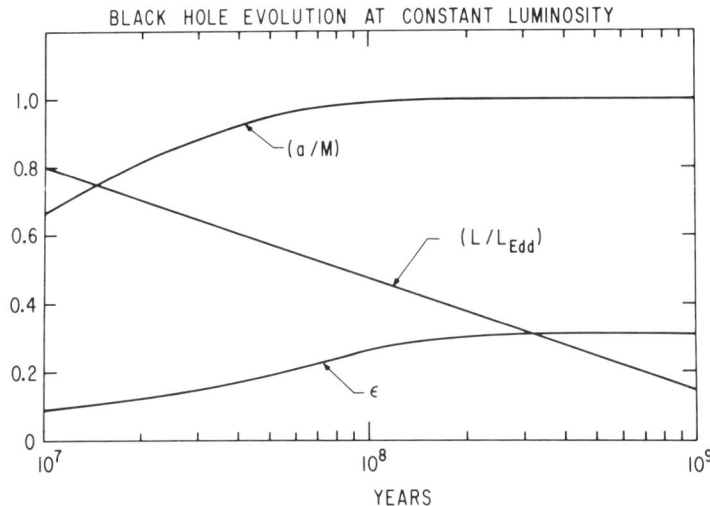

Figure 3. The temporal evolution of an accreting black hole, assuming constant luminosity, is plotted for the specific angular momentum per unit mass (a/M), the radiation efficiency (ϵ) and the ratio (L/L_{Edd}) of invariant bolometric luminosity (L) to the increasing value of the Eddington limit (L_{Edd}). (L/L_{Edd}) = 1 at the beginning of accretion (t = 0) when the object is a Schwarzschild black hole.

gravitational radius) and/or a decrease in L/L_{Edd}. For a compact source size on the order of 10 gravitational radii, the transition occurs at $L = 10^{-2} L_{Edd}$. As emphasized by White et al. (1984), during the time when $L > 10^{-2} E_{Edd}$ in hard photons for such a source, a significant pair photosphere would be present. Using optical line data, Wandel (this meeting) estimates that $L = 10^{-2} L_{Edd}$ for a large sample of Seyfert-1 nuclei and nearby quasars. In this connection we note that, apart from BL Lac type objects, X-ray bright AGNs observed within the present epoch have a universal power-law spectral form (Mushotzky 1982), independent of luminosity (see Figure 4). The

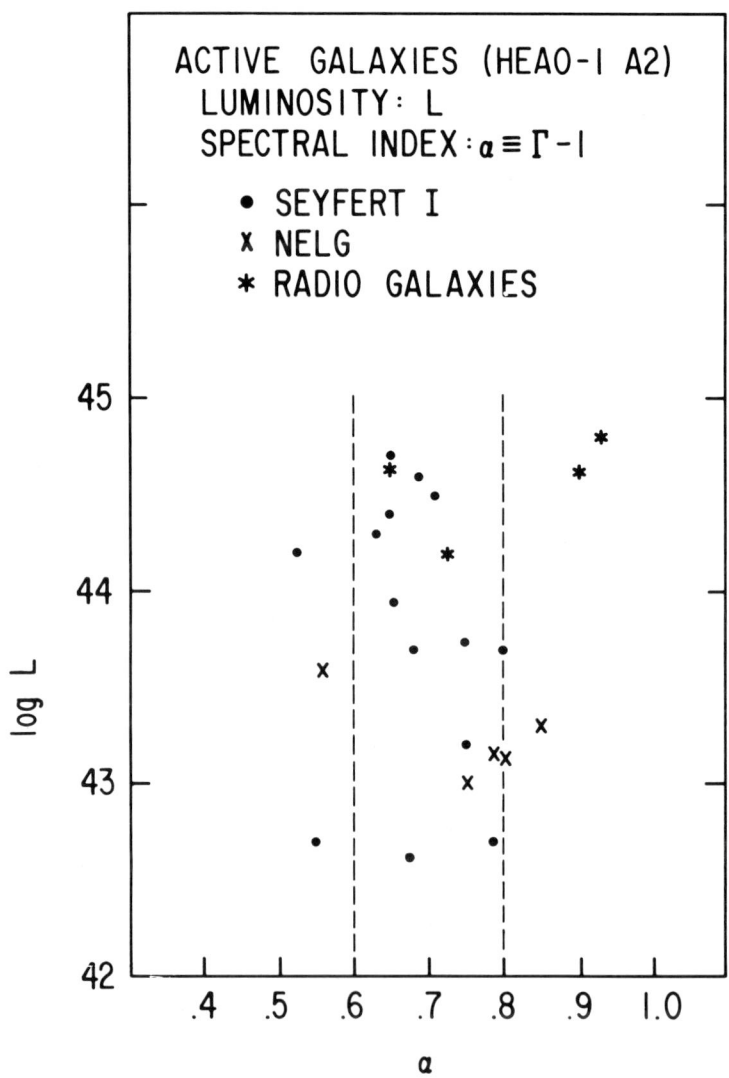

Figure 4. Active galaxies measured with the HEAO-1 (A2) experiment (Mushotzky 1982) plotted as regards X-ray luminosity (ergs s^{-1}) and energy spectral index.

particular energy spectral index $\alpha = 0.7$ characteristic of the broadband X-rays from such objects could be due to a radiative process involving a non-thermal electron spectrum that is buffeted into a unique distribution function by the feedback into the plasma of electron-positron pairs created in the photon-photon absorption of its own inverse-Compton radiation (Kazanas 1984). The associated spectral index is thereby essentially independent of source luminosity provided only that 10^{30} ergs/(s cm) > L/R > 10^{27} ergs/(s cm); see Zdziarski and Lightman (1984).

5. X-RAY SPECTRUM

The 3-100 keV spectrum of the overall extragalactic sky (i.e., before removing the contributions from sources in the present epoch) may be approximated by the following expression:

$$dI/dE = A \ (E/3 \ keV)^{-\alpha} \ exp(-E/B)$$

where E is photon energy (in keV), A = 5.6 keV/(keV cm^2 s sr), B = 40 keV and $\alpha = 0.29$. This is to be contrasted with the spectra for known sources in the present epoch. We here characterize these discrete source spectra in terms of the same spectral form used above for the overall CXB. Accordingly, the ensemble average spectrum for clusters of galaxies may be described by an optically thin thermal model fixed by kT = B = 7 keV (Stottlemyer and Boldt 1984) and $\alpha = 0.5$; however, the contribution of such clusters to the CXB is only a few percent. Present-epoch AGNs (at z < 1), such as Seyfert galaxies, make a much more substantial contribution to the CXB and, as already noted, exhibit power-law spectra characterized by an energy spectral index value distributed in a narrow interval about $\alpha = 0.7$; for the brightest among these we already know that B > 100 keV (Rothschild et al. 1983). Although an ensemble of quasars (mostly radio-loud) has been found to exhibit a composite X-ray spectrum (fit by $\alpha = 0.83$ (+0.32, -0.22)) compatible with that for Seyfert galaxies (Worrall and Marshall 1984) there is now some evidence that the low energy spectra (E < 3 keV) for quasars in a true optically selected sample are steeper (Elvis 1985), with $\alpha = 1.2$ as more appropriate. Taking $\alpha = 0.7$ (+0.1, -0.1) for present-epoch AGNs, kT = 7 keV for clusters of galaxies and $\alpha = 0.9$ as appropriate for the complete ensemble of canonical quasars, the foreground spectrum arising from the composite of all these sources has been estimated (Leiter and Boldt 1982, Appendix D). Subtracting this foreground from the spectrum of the overall CXB yields that of the residual CXB to be identified with young AGNs. As shown in Figure 5, for E < 10 keV this residual background is clearly flatter than that of the total CXB. For E >> 40 keV the residual CXB is expected to be vanishingly small relative to the total (Boldt 1981; Rothschild et al. 1983). In terms of the spectral form used here the residual CXB itself must have $\alpha < 0.2$. As shown in Figure 6 a good fit is obtained with $\alpha = 0$ and B = 23 keV. Considering energies of emission well below (1+z)23 keV the sources of this residual CXB do indeed appear to have

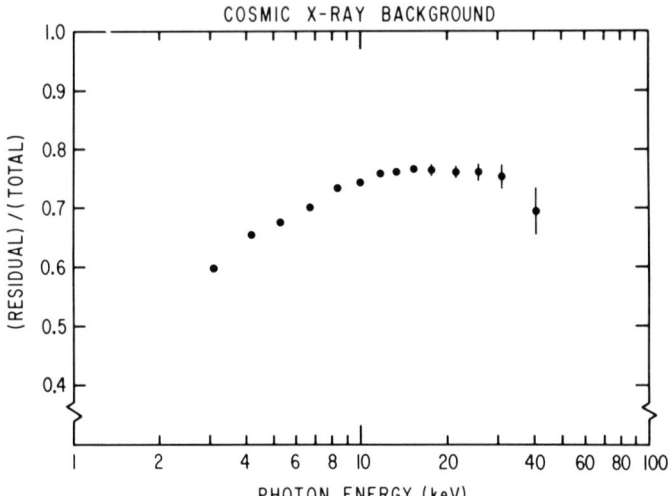

Figure 5. The ratio of the spectral density of the estimated residual CXB (Leiter and Boldt 1982) to that of the total CXB observed with HEAO-1, plotted as a function of photon energy, in keV. Statistical errors are indicated for those data points where they exceed the size of the dots.

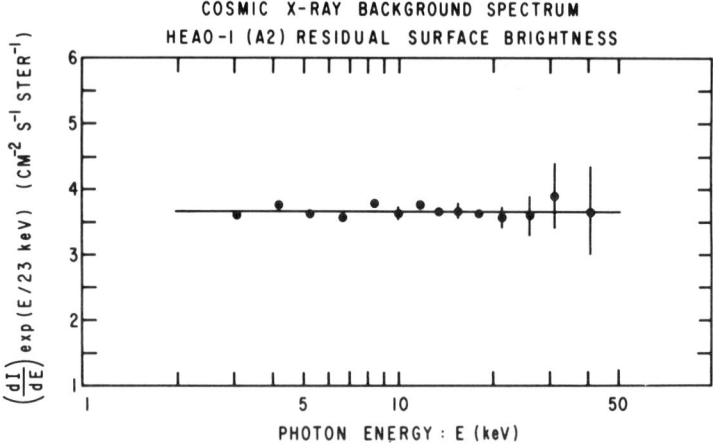

Figure 6. The residual energy spectrum for the CXB multiplied by $\exp(E/23\,\text{keV})$ as a function of photon energy (Leiter and Boldt 1982). In general, the statistical errors associated with the data points increase with energy and are indicated only for representative energies ($\sim$ 5 keV, 10 keV and > 20 keV).

the flat spectrum we have identified with young AGNs characterized by a compactness parameter comparable with the critical value (or somewhat larger). If the main sources of the residual CXB are resolved by AXAF, they should exhibit just such a spectrum.

6. OPTICAL CHARACTERISTICS

Although the bolometric luminosity of a young precursor quasar could be comparable to that of a canonical quasar they are likely to differ greatly in optical characteristics. The spectral steepness of the continuum emission from the central source is extremely important in this regard. This is illustrated in Figure 7, where we exhibit the total emission rate $Q(E)$ of photons ($E < h\nu < 100$ keV) for two different power-law spectra as a function of the energy lower limit (E) for the integration. Both of these spectra correspond to the same luminosity over the band 13.6 eV (1 rydberg) to 100 keV. For a lower energy limit comparable with 1 rydberg, the Q for the steep ($\alpha = 1$) power law exceeds that for the flat ($\alpha = 0$) spectrum by two orders of magnitude. This has two very pronounced observational consequences. First, the UV continuum photon number flux from the quasar associated with $\alpha = 1$ would be much larger than that from a young AGN characterized by $\alpha = 0$. Second, since the photoexcitation of the surrounding gas clouds by the central continuum source depends most critically on the total number flux of ionizing UV photons (Boldt and Leiter 1984), the amount of atomic line emission from many young AGNs could be quite small relative to that for quasars. In fact, not until well into the X-ray band ($E > 3$ keV) does the Q for the flat spectrum exceed that for the steep power law. As indicated in Figure 7, unit optical depth for a 3 keV X-ray occurs at a thickness of about 0.1 g/cm^2, comparable with that inferred for the X-ray absorbing clouds associated with NGC 4151 (Holt et al. 1980). For the steep spectrum, however, most photons with energies exceeding a rydberg reside at $h\nu < 40$ eV, a regime where photoelectric absorption by hydrogen is assured even when the gas cloud is much thinner.

In comparing the two ionizing continua displayed in Figure 7, the main effect of interest here is that of the pronounced increase in the output of UV photons with rising spectral steepness ($\alpha = 0$ to $\alpha = 1$). This leads to an increase in the production of ions (such as H^+) from species with low ionization potentials and results in an enhancement in associated line emissions (e.g., Lyα). Although the primary ionization of such species is generally dominated by UV photons, X-ray interactions can make significant contributions to secondary ionization and heating. For thick clouds (> 0.01 g/cm^2), the input of X-rays can lead to appreciable line emission from ions of high ionization potential, such as the lithium-like ions O_{VI} and Ne_{VIII} (see Boldt and Leiter 1984). With the possible exception of lines from such ions, however, the usual atomic lines and optical continuum radiation characteristic of canonical quasars would tend to be suppressed in objects where the central source of ionizing radiation has the flat spectrum which we infer is a fundamental attribute of the youngest AGNs. For this reason

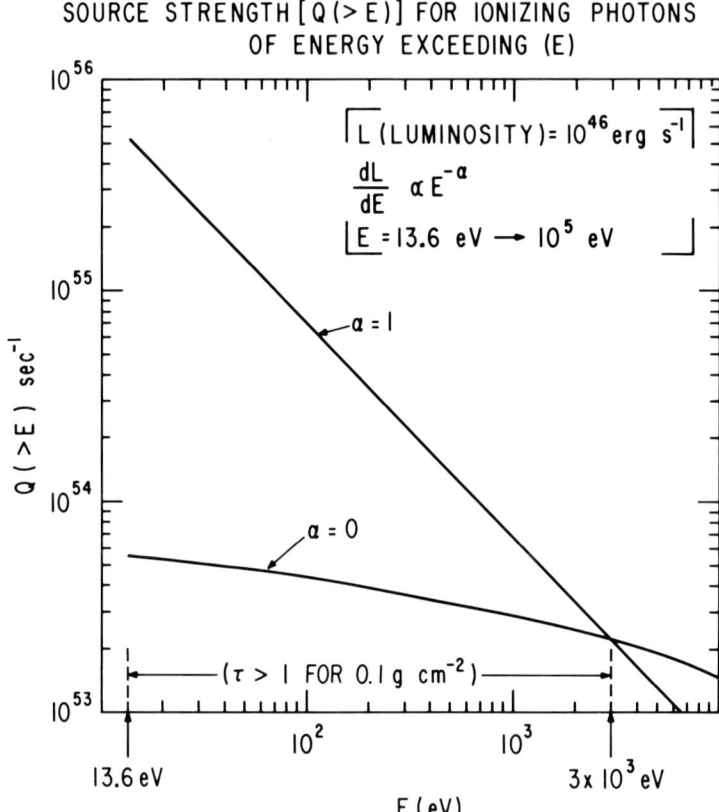

Figure 7. The total number of photons (up to 100 keV) emitted per second Q(>E) by the central source of continuum radiation is plotted as a function of the energy lower-limit (E) for the integration. The curves shown correspond to underlying power-law energy spectra of index $\alpha = 0$ and $\alpha = 1$, as indicated. Both curves correspond to a luminosity $L = 10^{46}$ erg s^{-1} over the band 13.6 eV-100 keV. The band of energies corresponding to an optical depth greater than unity ($\tau > 1$) is shown for a shell thickness of 0.1 g cm^{-2}.

we speculate that the young AGN sources dominating the residual CXB at high redshifts (e.g. at redshifts beyond that where quasars have been observed) may not readily lend themselves to detection in the optical band, even if they are resolved out in the X-ray band with AXAF.

Based on recent work by Amri Wandel (reported at this meeting) concerning broad emission lines, there now appears to be some direct evidence to support the trend called for in this paper whereby young AGNs radiate at a luminosity level closer to the Eddington limit than do those AGNs at the present epoch. In a large sample of Seyfert-1 galaxies and low redshift quasars (z < 0.6) Wandel finds that the empirical mass of the central black hole implies L(optical) ≈ 10^{-2} E_{Edd}, with a low dispersion. The Hβ line which is used to determine the mass cannot be observed from the ground for 3.2 > z > 0.8. However, applying the same method to the sufficiently high redshift object S5 0014+81 (at z = 3.4, and the most luminous quasar yet detected) the measured Hβ line (Kuhr et al. 1984) yields L(optical) ≈ 0.3 L_{Edd}. Considering bolometric luminosity, this object might very well be radiating close to its Eddington limit. In any event, as pointed out by Kuhr et al. (1984), the equivalent widths for all emission lines normally associated with the broad-line region are significantly smaller than observed for lower luminosity quasars at smaller redshifts.

7. CONCLUSIONS

A comparison of the non-thermal spectrum characteristic of present-epoch AGNs with the spectrum of the CXB indicates a process of spectral evolution whereby young AGNs accounting for the residual background would involve compact thermal objects with kT $\gtrsim$ 100 keV (if z $\gtrsim$ 3.5). Young AGNs such as this could be relatively weak in the optical line emissions typical of canonical AGNs. This sort of spectral evolution is a natural consequence of the expected decrease (with age) in the AGN compactness parameter (L/R), provided that

1) the non-thermal radiation from AGNs arises from electrons accelerated by an ordered electromagnetic field configuration (e.g. as generated by a disk-dynamo), and
2) young AGN sources of the residual CXB have L/R > 10^{30} ergs/(s cm).

In this sense, the AGN spectral evolution implied by the CXB provides us with evidence that black hole disk accretion (with eventual disk-dynamo action) is the power generator for the AGN phenomenon.

ACKNOWLEDGMENTS

One of us (E.B.) is particularly grateful to Amri Wandel for presenting this paper at the meeting in his absence, on short notice, and for stimulating discussions.

REFERENCES

Bardeen, J. 1972, in Black Holes, ed. C. DeWitt and B.S. DeWitt (New York: Gordon & Breach), p. 225.
Boldt, E. 1981, Comments Ap., **9**, 97.
Boldt, E., and Leiter, D. 1984, Ap. J., **276**, 427.
Cavaliere, A., and Morrison, P. 1980, Ap. J., **238**, L63.
Cavallo, G., and Rees, M. 1978, M.N.R.A.S., **183**, 43.

DeZotti, G., et al. 1982, Ap. J., **253**, 47.
Elvis, M. 1985, Proc. U.S./Japan Symp. (Jan. 1985); in press. CFA Preprint #2126.
Elvis, M., Wilkes, B., and Tananbaum, H. 1985, Ap. J., **292**, in press.
Fabian A., 1981, Proc. Tenth Texas Symp. Rel. Astr., Ann. N.Y. Acad. Sci., **375**, 235.
Herterich, J. 1974, Nature, **250**, 311.
Holt, S., et al. 1980, Ap. J., **241**, L13.
Kazanos, D. 1984, Ap. J., **287**, 112.
Kuhr, H., et al. 1984, Ap. J., **284**, L5.
Leiter, D., and Boldt, E. 1982, Ap. J., **260**, 1.
Lightman, A. 1982, Ap. J., **253**, 842.
Marshall, F., et al. 1980, Ap. J., **235**, 4.
Mushotzky, R. 1982, Ap. J., **253**, 28.
Osmer, P. 1982, Ap. J., **253**, 28.
Rees, M. 1984, Ann. Rev. Astron. Astrophys., **22**, 471.
Rothschild, R., et al. 1983, Ap. J., **269**, 423.
Schmidt, M., and Green, R. 1983, Ap. J., **269**, 352.
Stottlemyer, A., and Boldt, E. 1984, Ap. J., **279**, 511.
Thorne, K. 1974, Ap. J., **191**, 507.
White, N., Fabian, A., and Mushotzky, R. 1984, Astron. Astrophys., **133**, L9.
Worrall, D., and Marshall, F. 1984, Ap. J., **276**, 434.
Zdziaraski, A. 1984, Ap. J., **283**, 842.
Zdziarski, A., and Lightman, A. 1984, Ap. J. (Letters), submitted.

DISCUSSION

MACCACARO: From what we presently know on optically and X-ray selected quasars it seems that the major contributors to the XRB are the quasars at moderate redshifts (1.5-2.5), not those at very high redshifts ($z>3.5$). Therefore I do not see how you can constrain the spectrum of these very high redshift quasars with XRB arguments.

BOLDT: Although canonical quasars might account for much of the CXB, at least for $E<3keV$, present X-ray data on these sources indicate that their spectra are steeper than that of the background. Furhtermore, after subtracting any substantial estimated contribution from such sources, it becomes evident that the spectrum of the resulting residual background is necessarily even flatter (i.e., agrivating the discrepancy). It is in this sense that the residual background sets a constraint on those sources of the CXB (possibly at relatively high redshifts) that have not as yet been identified.

SILK: There may well be a substantial contribution to the diffuse X-ray background from hot gas in superclusters. This would naturally lead to thermal emission with a flat spectrum rolling off exponentially at 20 or 30keV, much as you find for the residual XRB component.

BOLDT: Pravdo et al (1979, Ap.J. __234__, 1) have used HEAO-1 data to set upper limits on the emission from hot gas in superclusters (i.e. all the X-ray emission was consistent with originating in the individual constituent clusters). However, the possibility that the residual CXB arises from a more pervasive hot intergalactic gas (heated at $z=5-10$) can still not be ruled out with presently available data (see Fabian 1981).

WILKES: Can you reconcile the predictions of your model that the X-ray slope of AGNs be between 0.6 and 0.8 with the Einstein IPC soft X-ray slopes observed for QSOs which cover a wide range up to 2.3?

BOLDT: The principal conclusion of this study is that the sources of the residual CXB have the relatively flat spectrum expected for the comptonized radiation emerging from compact thermal objects corresponding to an early stage of AGN development. Beyond a certain evolutionary phase (where L/R decreases below the critical value) the AGN spectra are expected to have an increasingly important non-thermal component. We cited the observational result that AGN within the present epoch, such as Seyfert galaxies, exhibit power-law spectra characterized by an energy spectral index 0.7 (+.1,-.1) and noted that this could happen for a large range in L/R below the critical value. As emphasized by Elvis (1985), the much steeper soft X-ray spectrum observed for the quasar PG 1211+143 might represent the upper end of a "thermal bump" that peaks in the XUV.

SCHMIDT: It would be of interest if rough source counts could be predicted for these young AGNs. This would tell us at what X-ray flux level we should expect these objects to turn up. In this connection it may be relevant that until now the X-ray sources in the Medium Sensitivity Surveys of Maccacaro et al. are all optically identified.

BOLDT: We estimate that the number of such objects would be over 200 per square degree (Leiter and Boldt 1982), more than an order of magnitude larger than the sky surface density of sources detected in the deep surveys carried out with the HEAO-2 Einstein Observatory. Normalizing to the residual CXB, as extrapolated below 3keV, we find $S(1-3keV) \times 10^{-14}$ ergs/(s cm^2) or somewhat less for a representative young AGN. This intensity is below that for the faintest reliably detected sources in the HEAO-2 deep surveys but is well within the capability anticipated for AXAF.

QUASAR EVOLUTION AND LUMINOSITY FUNCTION: AN X-RAY PERSPECTIVE

Yoram Avni
Harvard-Smithsonian Center for Astrophysics
Cambridge, Massachusetts   02138, U. S. A.
and
Weizmann Institute of Science
Rehovot 76100, Israel

ABSTRACT. X-ray properties of optically-selected QSOs are reviewed, and their implications for the QSO evolution and luminosity function are summarized. Analyses of the explicit dependence of $\alpha_{o,x}$, the x-ray to optical luminosity ratio parameter, on redshift z and optical luminosity L_o, are described. In particular, results from a recent analysis of complete, magnitude-limited samples of optically-selected QSOs are presented. Previous results obtained from the study of heterogeneous samples are confirmed. It is shown that the large majority, probably all, of optically-selected QSOs are x-ray loud: no more than a few precent can be x-ray quiet. Thus, x-ray emission seems to be a universal property of QSOs. It is found that comparisons of the optical evolution and luminosity function, combined with the $\alpha_{o,x}(z,L_o)$ dependence, with x-ray-selected samples, are numerically sensitive to the details of the input ingredients. There is a residual discrepancy of about a factor of two between the QSO x-ray number counts calculated from available simple pure luminosity evolution models for the optical evolution and luminosity function combined with the best estimate $\alpha_{o,x}(z,L_o)$ dependence, and observed counts from x-ray selected samples. Directions for further research, which will contribute to our understanding of the full bivariate optical-x-ray evolution and luminosity function, are discussed.

I. INTRODUCTION

Most recent studies of the cosmological evolution and luminosity function of QSOs concern the optical "evolution and luminosity function" (ELF), that describes the differential number-density distribution of QSOs in redshift and optical luminosity, and which is defined by:

$$dN = dV(z) \; \psi_o(z,L_o) \; dL_o. \qquad (1)$$

dN is the differential number of QSOs, z is redshift, dV(z) is an element of co-moving volume, L_o is the optical luminosity, which in order to be sepcific we take to be the monochromatic optical luminosity at 2500Å at the source rest frame (erg sec^{-1} Hz^{-1}), and

$\psi_o(z, L_o)$ is the optical ELF. We consider only Friedman cosmologies and assume $H_o = 50$ Km sec^{-1} Mpc^{-1} and $q_o = 0$ unless noted otherwise.

The emphasis on the optical ELF is, in large part, due to practical considerations. At present, QSOs can be positively identified as such only through optical spectroscopy. Studies of complete samples of optically selected QSOs yield the optical ELF from observations in just one spectral region, namely the optical region. On the other hand, a derivation of the radio or x-ray ELF requires complete optical identifications of complete samples of radio or x-ray selected QSOs, and, therefore, involves observations in two spectral regions. Clearly, a derivation of any bivariate ELF (such as optical-radio or optical-x-ray ELF) requires observations in more than one spectral region; hence, the relative ease of deriving the optical ELF.

Following the launch of the Einstein Observatory (Giacconi et al., 1979), much attention has been focused on the properties of QSOs in the x-ray region. The flux sensitivity and angular resolution of the Einstein Observatory have made it possible to positively detect x-ray fluxes from many previously known QSOs, or to derive meaningful upper limits for the x-ray flux from such QSOs, and to discover many previously unknown QSOs. Thus, the population properties of QSOs at x-ray energies could be investigated. We address in this review the x-ray aspects of the QSO ELF.

We define the conditional x-ray luminosity function as follows. The differential number-density distribution of QSOs in redshift, optical luminosity, and x-ray luminosity can be written in the form

$$dN = dV(z) \, \psi_o(z, L_o) dL_o \, \Phi_x(L_x | z, L_o) dL_x \qquad (2)$$

where L_x is the x-ray luminosity, which in order to be specific we take to be the monochromatic x-ray luminosity at 2 keV at the source rest frame (erg sec^{-1} Hz^{-1}), and $\Phi_x(L_x|z,L_o)$ is the conditional x-ray luminosity function given z and L_o. The function Φ_x describes the distribution of L_x and the way it depends on z and L_o, and connects the optical ELF to the bivariate optical-x-ray ELF and to the x-ray ELF. Φ_x contains information on such questions as the correlations between x-ray and optical luminosities and the relations between the x-ray and optical evolution rates. These are fundamental ingredients for physical models of QSOs and for cosmological implications of QSOs.

X-ray luminosities of QSOs are commonly described by the parameter $\alpha_{o,x}$ (Tananbaum et al., 1979), defined by

$$\frac{L_x}{L_o} = \left[\frac{\nu_x}{\nu_o}\right]^{-\alpha_{o,x}} = 10^{-2.605 \alpha_{o,x}} \qquad (3)$$

where ν_x and ν_o are, respectively, the frequencies corresponding to 2 keV and 2500Å. The dynamic range of the presently detected values of $\alpha_{o,x}$ is 1.0-2.0, which corresponds to ~ 3 orders of magnitude in L_x/L_o. The choice of the parameter $\alpha_{o,x}$ was motivated by the finding that L_x is positively correlated with L_o.

For a given, fixed L_o, $\alpha_{o,x}$ can replace L_x as an independent variable. Therefore, dN of Equation (2) can also be written as

$$dN = dV(z)\, \psi_o(z,L_o)\, dL_o\, \phi_x(\alpha_{o,x}|z,L_o)\, d\alpha_{o,x} \qquad (4)$$

where $\phi_x(\alpha_{o,x}|z,L_o)$ is the conditional distribution function of $\alpha_{o,x}$ given z and L_o. The function ϕ_x describes how $\alpha_{o,x}$ depends on z and L_o on the average, and how $\alpha_{o,x}$ is scattered around this average dependence; these can be derived by analyzing x-ray observations of previously known optically-selected QSOs. Hence, the interest in and importance of studying the correlations of $\alpha_{o,x}$ on z and L_o and the distribution function of $\alpha_{o,x}$: they are equivalent to studying the conditional x-ray luminosity function.

The first study of a sample of previously known QSOs with the Einstein Observatory (Tananbaum et al, 1979) yielded two fundamentally important conclusions. (1) X-ray emission from QSOs is a common phenomenon: an appreciable fraction of QSOs are x-ray emitters. (2) X-ray emission from QSOs is energetically strong: the integrated x-ray luminosities for the detected QSOs are comparable, by order of magnitude, to the integrated optical luminosities. These findings implied that much can be learned about QSOs from x-ray observations, and that QSOs can be efficiently discovered through their x-ray emission. It has also become evident that accounting for the x-ray emission and its systematic properties must be included in any model for the QSO phenomena.

A larger sample of previously known QSOs has been further studied with the Einstein Observatory by Zamorani et al. (1981) (see also Ku, Helfand and Lucy, 1980). The correlations of $\alpha_{o,x}$ with z, L_o, and radio luminosity L_R (monochromatic at 5000 MHz) have been specifically addressed. Two main conclusions have been derived. (1) $\alpha_{o,x}$ is different for "radio-loud" QSOs and "radio-quiet" QSOs, such that QSOs which are relatively brighter in the radio are relatively brighter in x-rays. For this reason, and since most QSOs are presently not detected in the radio, we consider only optically-selected QSOs, and thereby only the dependence of $\alpha_{o,x}$ on z and L_o. (2) $\alpha_{o,x}$ depends on redshift and/or optical luminosity, such that for higher z and/or higher L_o, L_x/L_o is lower ($\alpha_{o,x}$ higher). This finding has induced the study of the conditional x-ray luminosity function.

We summarize in this review recent results obtained at the Harvard-Smithsonian Center for Astrophysics on the QSO conditional x-ray luminosity function as reflected in the $\alpha_{o,x}$ systematics, list some open problems, and indicate directions for further study.

II. PREVIOUS RESULTS

The dependence of $\alpha_{o,x}$ on z and L_o for optically-selected QSOs has been studied by Avni and Tananbaum (1982) using a heterogeneous sample of previously known QSOs. This sample consists of a "mixed-bag" of QSOs from the CFA survey, Seyfert 1 nuclei with available nuclear magnitudes from Kriss, Canizares, and Ricker (1980), and several QSOs that comprise a complete sample (B_{lim} = 19.2) from Marshall et al. (1983a). As a whole, the heterogeneous sample does not have any well-defined completeness properties in the optical. Formally, since we are considering the conditional distribution of $\alpha_{o,x}$ given z and L_o, such a completeness is not required for the type of analysis carried out. However, it should be kept in mind that complete samples have the advantage of being manifestly free from any biases which may be introduced by "human" selection.

This heterogeneous sample has a total of 73 QSOs, with 41 of them positively detected in x-rays by the Einstein Observatory, and the other 32 have yielded flux upper bounds. The distribution of these QSOs in z and L_o has a rather strong correlation between z and L_o, which implies that it is difficult to separate a dependence of $\alpha_{o,x}$ on L_o from a dependence on z. We note that in this respect a heterogeneous sample has an advantage over a complete optically-selected sample with a single magnitude limit, since such a complete sample would have an even stronger correlation between z and L_o: both geometrical volume effects and cosmological evolution would concentrate the QSOs of such a sample at the magnitude limit.

A simple, linear dependence of $\alpha_{o,x}$ on cosmological look-back time $\tau(z)$ (in units of the present age of the Universe) and on $\log L_o$ was assumed:

$$\alpha_{o,x} = \{A_z[\tau(z)-0.5] + A_o[\log L_o-30.5] + A\} + \{\text{residual}\}. \qquad (5)$$

A_z and A_o are the coefficients of the linear dependence. The subtraction terms 0.5 and 30.5 are chosen to make A the typical value of $\alpha_{o,x}$ for the central values of $\tau(z)$ and $\log L_o$ in the sample. The terms in the first curly brackets represent the average dependence of $\alpha_{o,x}$ as a function of z and L_o, $\bar{\alpha}_{o,x}(z,L_o) = <\alpha_{o,x}|z,L_o>$. The term in the second curly brackets represents the distribution of the residuals of $\alpha_{o,x}$ relative to the average dependence, namely the distribution of $\alpha_{o,x} - \bar{\alpha}_{o,x}(z,L_o)$. This distribution must have an average of 0 by construction, and was assumed to be Gaussian (with mean 0) characterized by a dispersion σ. Since only about half of the QSOs in the sample were positively detected, the regression analysis was carried out using a parametric version of the Detections and Bounds (DB) method (Avni et al., 1980). The choice of the Gaussian form for the distribution of the residuals was motivated, in part, by the fact that it defines the most straightforward generalization of the standard regression analysis to the domain of the Detections and Bounds problem. The validity of the Gaussian distribution for this particular sample was confirmed a posteriori.

The results of the DB regression analysis for the two coefficients A_z and A_o are described in Figure 1 (adapted from Avni and Tananbaum, 1982), where the best estimate parameters and the $\Delta S = 4$ error-contour are displayed. The best estimate corresponds to $A_z \simeq 0$, which indicates that $\alpha_{o,x}$ depends predominantly on L_o rather than on z; however, a substantial dependence on z cannot be ruled out. The narrow and elongated shape of the error-region is due to the intrinsic z-L_o correlation in the sample and means that, in fact, it is difficult to separate a dependence of $\alpha_{o,x}$ on L_o from a dependence on z.

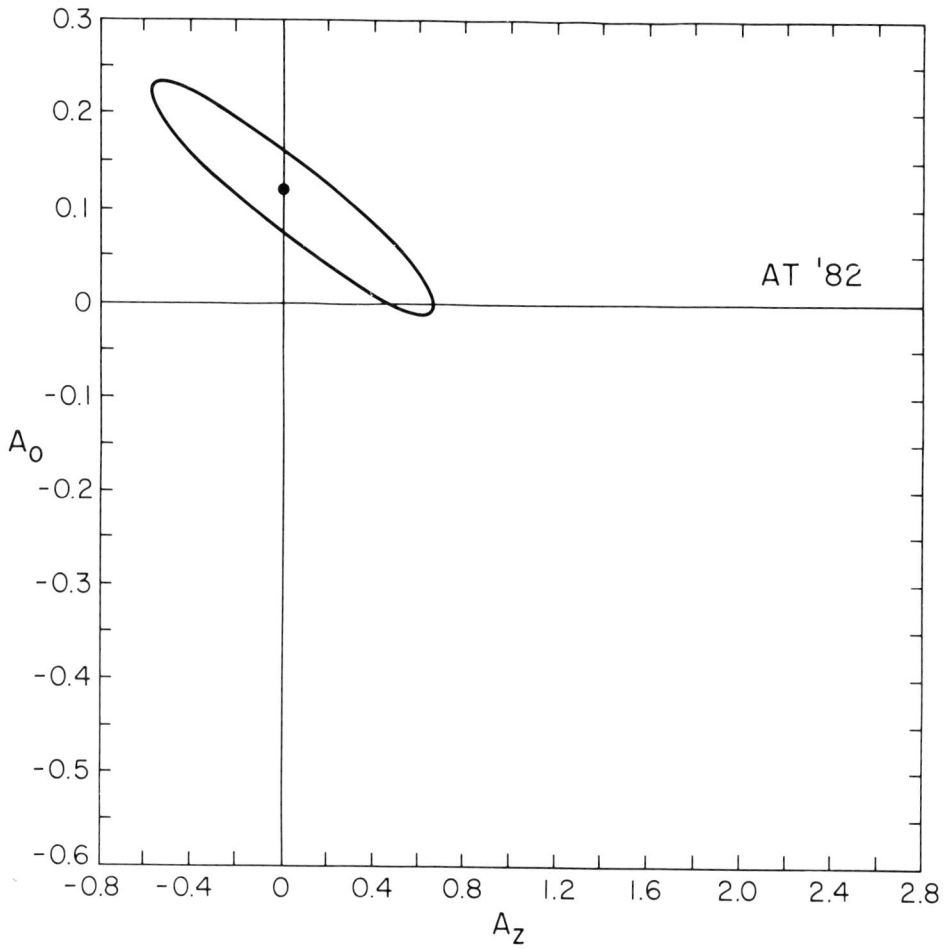

Figure 1: Best estimate and $\Delta S=4$ error contour for the correlation parameters A_z and A_o that describe the average dependence of $\alpha_{o,x}$ on $\tau(z)$ and log L_o, for the old heterogeneous sample. (Adapted from Avni and Tananbaum, 1982).

If the explicit dependence of $\alpha_{o,x}$ is only on L_o, not on z, as given by the best-estimate parameters, then L_x depends on L_o as

$$L_x \propto L_o^{1-2.605A_o} \propto L_o^{0.7}. \tag{6}$$

Such a dependence was shown by Tucker (1983) and by Schlosman, Shaham, and Shaviv (1984) to be consistent with models for QSOs which involve accretion onto massive black holes. We note, however, that the functional dependence which we have derived refers specifically to the conditional properties of L_x at given L_o, a point which was not explicitly addressed by those authors.

The range of allowed parameter values (A_z, A_o) implies that, for a wide class of models for the cosmological evolution of QSOs, the x-ray evolution rate is expected to be slower than the optical evolution rate. For example, in the case of pure luminosity evolution with an exponential dependence on look-back time

$$L_o \propto e^{\gamma_o \tau(z)}, \quad L_x \propto e^{\gamma_x \tau(z)}, \tag{7}$$

the evolution parameters γ_o, γ_x are related through

$$\gamma_x = \gamma_o - 2.605 \, (\gamma_o A_o + 2.3 \, A_z). \tag{8}$$

Within the error-region given in Figure 1 for (A_z, A_o), one finds $\gamma_x < \gamma_o$ for typical observed values of γ_o. This has been supported by a comparison of the optical evolution rate (Marshall et al., 1983b) with the x-ray evolution rate (Maccacaro et al., 1983) in the framework of pure luminosity evolution. Such a difference in evolution rates would have far reaching implications. The bolometric luminosity could evolve still differently, and care must be exerted in comparing theoretical evolution models with observations, and in evaluating cosmological implications of QSOs.

Several issues which required further study were raised by the analysis of Avni and Tananbaum (1982).

(1) The heterogeneity of the sample used, which was "humanly" selected. Although it was formally argued that the sample was adequate for the type of analysis carried out, was it still possible that a dependence of $\alpha_{o,x}$ on some yet unidentified QSO property introduced a "hidden" bias into the sample?

(2) The assumed Gaussian distribution of the $\alpha_{o,x}$ residuals. The actual distribution of the residuals relative to the best-estimate average dependence was compared with the expected best-estimate Gaussian distribution, and shown to be acceptable. However, this comparison indicated (but did not require) a skew distribution of residuals. Would a larger data set require a non-Gaussian distribution? Would such a different distribution affect the derived average dependence? As a special case, would the nondetected QSOs indicate the existence of a separate population of "x-ray quiet" QSOs, not properly described by a localized distribution?

(3) The numerical values used for some of the parameters. The analysis assumed traditional values for the optical and the x-ray spectral indices, $\alpha_o = 0.5$ and $\alpha_x = 0.5$, and also $q_o = 0$. To what extend do the results depend substantially on the choice of those values?

In addition, when the $\alpha_{o,x}(z, L_o)$ correlation was combined with a specific model for the optical ELF (assuming, in particular, a pure power-law luminosity function), the calculated slope for the x-ray luminosity function did not agree with the slope derived from the Einstein Observatory Medium Sensitivity Survey (Maccacaro et al., 1983). What is the source of this discrepancy, and how can it be resolved?

The following sections describe a more recent study (Avni and Tananbaum, 1985), which is based on new observational data, and which addresses those questions.

III. NEW SAMPLES

The recent study of Avni and Tananbaum (1985) is largely based on two samples with well defined completeness properties in the optical. The first sample is the Bright Quasar Sample (BQS) of Schmidt and Green (1983), which is a complete optically-selected sample to an average B_{lim} of 16.2 magnitudes. Sixty-six of the 82 QSOs in this sample outside of the declination zone $30° < \delta < 60°$ have been observed by the Einstein Observatory in a collaborative effort of the CFA group and Schmidt and Green (Tananbaum et al., 1985). The choice of the observed QSOs has been dictated by considerations of scheduling of the Einstein Observatory. Thus the 66 observed QSOs can be considered as a fair, unbiased, representative group of the full BQS sample viz-a-viz their redshift, optical and x-ray properties, and can therefore be viewed as a complete optically-selected sample. Of the 66 QSOs, 57 have been positively detected in x-rays, and the other 9 have yielded flux upper bounds.

The second sample is the Braccesi faint sample (BF) of Marshall et al. (1984). This is a complete optically-selected sample to a B_{lim} of 19.8 magnitudes, observed with the Einstein Observatory. This sample contains 35 QSOs, with 13 x-ray detections and 22 flux upper bounds.

In addition, a third sample, without any completeness properties in the optical, has been used. This sample is an updated version of the original heterogeneous CFA survey, excluding those QSOs which are included in the BQS or BF samples, excluding the Kriss, Canizares, and Ricker (1980) Seyferts, and including several additional QSOs from the CFA survey that have since been processed. This sample will be called henceforth the "new heterogeneous" sample (HET) (Avni and Tananbaum, 1985). It consists of 53 QSOs, with 24 x-ray detections and 29 flux upper bounds.

The distribution of QSOs from the BQS sample in the (z,L_o) plane exhibits a very strong correlation between z and L_o, as expected for a complete sample with a single magnitude limit. (Actually, the BQS sample contains many separate fields with different limiting magnitudes, but the magnitude limits are very close to each other.) Similarly, there is a very strong correlation between z and L_o for the QSOs of the BF sample. When the two samples are taken together, however, they span a rather wide dynamic range in the two dimensional (z,L_o) plane, and the intrinsic correlation is much weaker. The new heterogeneous sample partially fills in the gap between the BQS and BF samples in the (z,L_o) plane, and weakens the intrinsic z-L_o correlation even further. Since the BQS and BF samples are based upon UV excess selection, they are confined to $z \leq 2.2$. The HET sample also provides QSOs at $z > 2.2$. In total, the three samples contain 154 QSOs, with 94 x-ray detections and 60 flux upper bounds.

The explicit dependence of $\alpha_{o,x}$ on z and L_o, using the functional form given by Equation (5) and assuming a Gaussian distribution of residuals, has been studied separately for each of the three samples, BQS, BF, and HET. The best estimate coefficients A_z, A_o and the $\Delta S=4$ error contours in the (A_z, A_o) plane, are described in Figure 2 which is adapted from Avni and Tananbaum (1985). The results for the three samples are consistent with each other. In particular, since the results for the two complete samples, BQS and BF, are consistent with each other, these samples can be joined together. The best estimate parameters and error region for the joint BQS+BF sample are described in Figure 3, adapted from Avni and Tananbaum (1985).

By comparing Figures 1, 2, and 3, we find that the recent results for each of the two complete samples, BQS and BF, as well as for their joint sample, BQS+BF, are consistent with the previous results of Avni and Tananbaum (1982) for the "old" heterogeneous sample, and with the results for the new heterogeneous sample. (The two heterogeneous samples are not independent of each other since they share many QSOs in common.) In particular, the best estimate results for the joint BQS+BF sample are very close to the best estimate results for each of the heterogeneous samples. This establishes that, in fact, a heterogeneous sample is adequate for the type of analysis carried out, and that there are no strong dependencies of $\alpha_{o,x}$ on yet unidentified QSO properties which would have introduced a hidden bias into the heterogeneous samples. We, therefore, form the joint BQS+BF+HET sample, which we call henceforth the "total" sample. The results for (A_z, A_o) from the total sample are also described in Figure 3. These new results are now fully consistent with and very similar to the results obtained previously from the old heterogeneous sample by Avni and Tananbaum (1982), described in Figure 1. The new results, however, are largely based on complete optically-selected samples, and the error region is now smaller by better than a factor of two. Again, the best estimate indicates that $\alpha_{o,x}$ depends predominantly on L_o rather than on z, but, again, a substantial dependence on z cannot be ruled out.

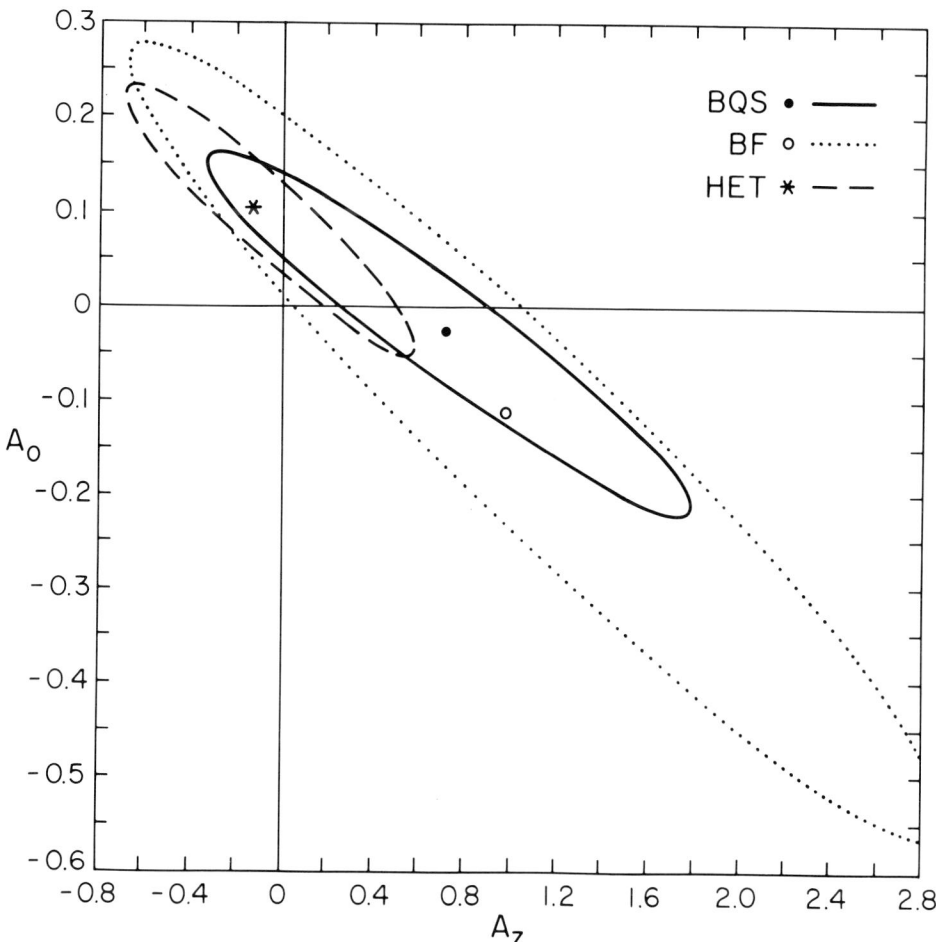

Figure 2: Best estimate and $\Delta S=4$ error contour for the correlation parameters A_z and A_o that describe the average dependence of $\alpha_{o,x}$ on $\tau(z)$ and $\log L_o$. Filled circle and solid line: Bright Quasar Sample. Open circle and dotted line: Braccesi Faint Sample. Star and broken line: New Heterogeneous Sample. (Adapted from Avni and Tananbaum, 1985).

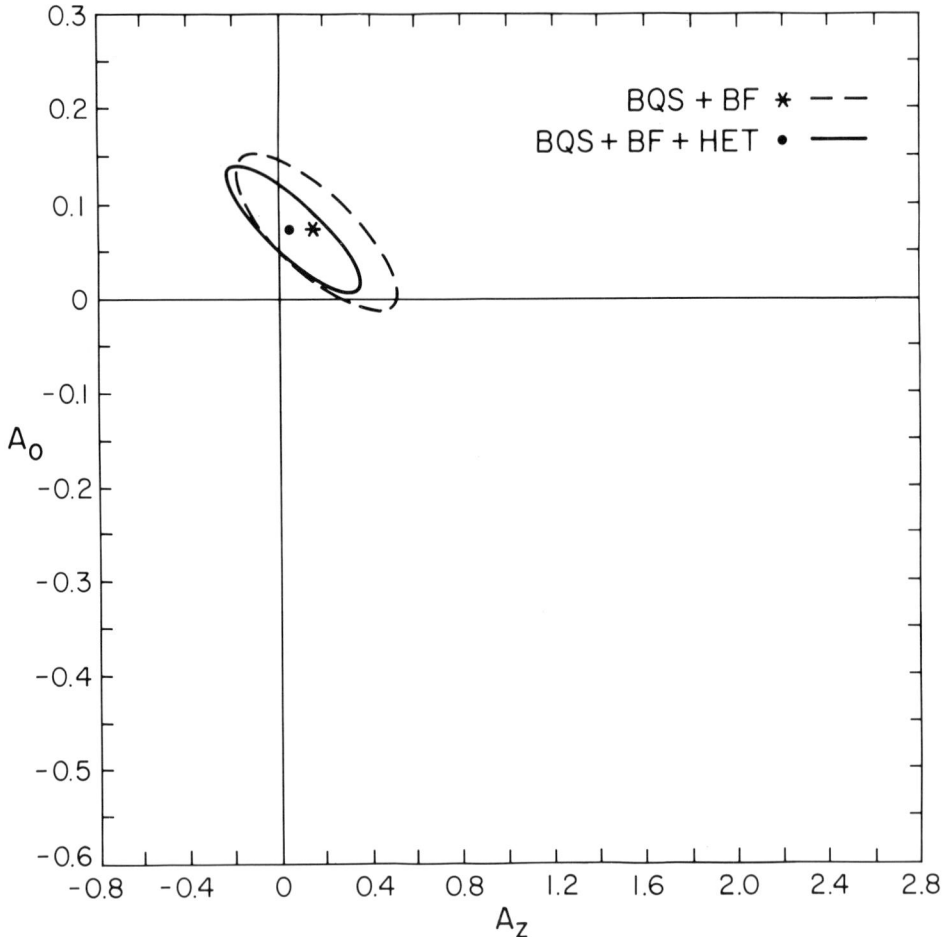

Figure 3: Best estimate and $\Delta S=4$ error contour for the correlation parameters A_z and A_o that describe the average dependence of $\alpha_{o,x}$ on $\tau(z)$ and $\log L_o$. Star and broken line: joint BQS+BF sample. Filled circle and solid line: total BQS+BF+HET sample. (Adapted from Avni and Tananbaum, 1985).

Several additional comments can be made on the basis of a number of comparisons of our separate results. The error region for (A_z, A_o) derived from the BQS sample is substantially longer and wider than the region derived from the old heterogeneous sample, in spite of the facts that the two samples are comparable in size and that the number of detections in the BQS sample is larger than the number of detections in the old heterogeneous sample. This results, in part, from the tighter intrinsic z-L_o correlation in the complete BQS sample relative to the old heterogeneous sample. It highlights the specific disadvantage of analyzing any single complete sample by itself. If we consider, in addition, the enormous reduction of the size of the error region when BQS and BF are analyzed jointly, it follows that it would be better to increase the size of the sample by studying a larger number of intermediate size complete samples with different magnitude limits — so as to obtain a fuller coverage of a wide 2-dimensional dynamic range in the (z, L_o) plane — rather than by studying a smaller number of larger complete samples.

We also note the apparently puzzling fact that the best estimate for (A_z, A_o) from the BQS+BF sample does not lie between the corresponding best estimates from BQS and BF separately. This follows from the existence of two additional parameters in the regression analysis, A and σ. In particular, an offset in the best estimate for A from BQS relative to that from BF causes the "sideways" shift in the best estimate for (A_z, A_o). We have explicitly checked that all the results from the separate samples are consistent with each other in the full 4-dimensional parameter space.

It is also interesting to note that the best estimate for (A_z, A_o) from the BQS sample (as well as that from the BF sample) lies outside the error region derived from the joint BQS+BF sample, and also outside the error region derived from the total BQS+BF+HET sample. This highlights the importance of using all available information and the advantage of combining all available samples, and cautions against using best-estimate results from any single sample.

We have thus found that the previous results of Avni and Tananbaum (1982) are confirmed by new, independent data from complete samples. We have furthermore considered how the results are affected when q_o is changed from 0.0 to 0.5, and found that there is no appreciable qualitative effect. We have also studied the effects on the results of changing the assumed optical spectral index from its traditional value of $\alpha_o = 0.5$. We found that varying α_o in the range from 0.0 to 1.0 does not have an appreciable systematic effect for the present sample; however, for samples containing ~4 times the number of QSOs a more precise knowledge of the optical spectrum will be required. At this point, we have not studied the effect of changing the x-ray spectral index from its traditional value of $\alpha_x = 0.5$; the observed diversity of QSO x-ray spectra (Elvis, Wilkes, and Tananbaum, 1985) and the lack of systematic results for the spectral indices do not enable a meaningful quantitative study at present. The results for (A_z, A_o) from the present total sample imply the same general conclusions for the x-ray properties of optically-selected QSOs as those of Avni and Tananbaum (1982), namely, that if $\alpha_{o,x}$ depends only on L_o as indicated by the best estimate, then

$$L_x \propto L_o^{1-2.605 A_o} \propto L_o^{0.8} \tag{9}$$

(a slight change in the value of the power-law index), and that from the full range of permissible values for (A_z, A_o), the x-ray evolution rate is expected to be slower than the optical evolution rate within the framework of a wide class of QSO evolution models, including pure luminosity evolution. We wish to emphasize that the relationships which we have derived apply to the observed optical and x-ray luminosities. As will be discussed in Section VI, there are several effects which require rigorous and detailed study, and which can, in principle, have a quantitative influence on those relationships, by modifying the relations between observed and intrinsic properties. These effects include time variability (the x-ray and optical observations have not been simultaneous), and anisotropies in the optical and/or x-ray emission. Further study should fully evaluate such effects.

IV. NO X-RAY-QUIET QSOs

The availability of the new total sample, BQS+BF+HET, which spans a wider 2-dimensional dynamic range in the (z,L_o) plane and contains a larger number of QSOs than the old heterogeneous sample, makes it possible to address more detailed questions regarding the $\alpha_{o,x}(z,L_o)$ dependence and the distribution of residuals. We consider here specifically two related questions. The first question is whether the distribution of $\alpha_{o,x}$ residuals departs from a simple Gaussian in the sense of requiring significant skewness. The second question is whether a population of optically-selected, x-ray-quiet QSOs exists.

It may be recalled that the study of Avni and Tananbaum (1982) has indicated (but did not require) a skew, non-Gaussian distribution of residuals, in the sense of a shorter tail at low $\alpha_{o,x}$ (high L_x) and a longer tail at high $\alpha_{o,x}$ (low L_x). This could mean either of two extreme possibilities, or some combination of them. The first possibility is that the dynamic range of the $\alpha_{o,x}$ values for the QSO population is rather narrow, roughly coinciding with the dynamic range of 1.0 to 2.0 actually found for those QSOs positively detected in x-rays, and the distribution of the $\alpha_{o,x}$ residuals is correspondingly "localized" (approximately -0.5 to +0.5) and skew. This possibility is of a "technical" nature, which nevertheless does have important numerical implications for the QSO bivariate optical-x-ray ELF and for the QSO x-ray ELF. The second possibility is that a certain fraction of the population of optically-selected QSOs has much lower x-ray luminosities, and correspondingly much higher values of $\alpha_{o,x}$, than implied by the above dynamic range. We refer to such a possible part of the QSO population as "x-ray-quiet." This possibility is of a deeper significance, since it is related to the universality of x-ray emission from QSOs. It is not easy to distinguish a priori between these two possibilities given the prevalence of x-ray non-detections, which amount to $\sim 40\%$ of the QSOs in the sample, since the actual values of $\alpha_{o,x}$ for the non-detections could either correspond to a localized distribution of $\alpha_{o,x}$ or be well above the values of 1.0 to 2.0.

To answer those questions, we have carried out a regression analysis $\alpha_{o,x}(z,L_o)$ assuming the same average dependence as before but allowing for a more general distribution of residuals. Formally we write — as in Equation 5 —

$$\alpha_{o,x} = \{A_z[\tau(z)-0.5] + A_o[\log L_o-30.5] + A\} + \{residual\}. \qquad (10)$$

The distribution of the residuals is now assumed to include two components, and is described schematically by Figure 4, adapted from

Avni and Tananbaum (1985). A fraction 1-P of the population of optically-selected QSOs is "x-ray-loud", and corresponds to a localized distribution of residuals with mean zero by construction. A fraction P of the population is "x-ray-quiet", and is described formally by a δ-function contribution to the distribution of residuals at very high values of $\alpha_{o,x}$. The precise formal description of the x-ray-quiet population is not important, as long as the corresponding values of the $\alpha_{o,x}$ residuals are well above the dynamic range of the localized distribution describing the x-ray-loud population. With this representation, the first curly brackets of Equation (10) define the average $\bar{\alpha}_{o,x}(z,L_o|\text{x-ray-loud})$ for the "loud" component of the population.

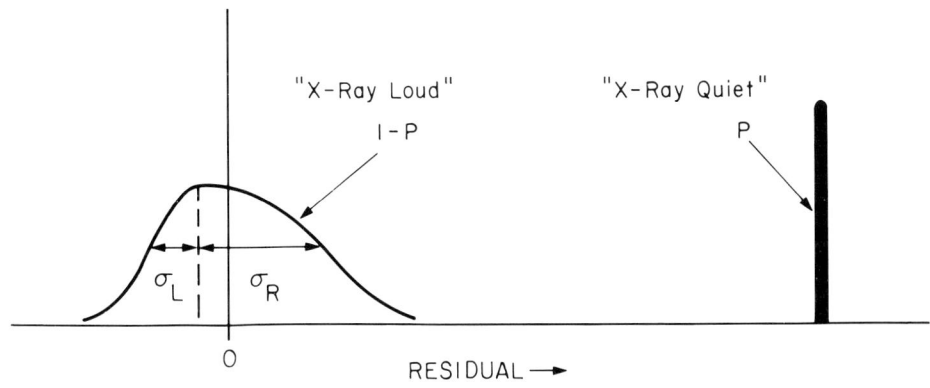

Figure 4: A schematic representation of the distribution of $\alpha_{o,x}$ residuals that allows for the existence of x-ray-quiet QSOs and for a skew distribution for x-ray-loud QSOs. (Adapted from Avni and Tananbaum 1985).

We describe the distribution of residuals for the x-ray-loud component by a functional form which is a simple modification of the Gaussian form, which admits a transparent measure of skewness, and which is convenient for analytic and numerical manipulations. This functional form is made out of two "half-Gaussians" with a common height which are "glued together" at their maxima (see Figure 4). Since this distribution must have a mean of zero by definition, it is characterized by two independent parameters, which can be chosen to be σ_L and σ_R, the usual σs for the left and right half-Gaussians respectively. It is also possible to characterize the distribution by another set of two independent parameters: σ, defined as the formal r.m.s. width of the distribution, and a skewness parameter $R = \sigma_R/\sigma_L$. Thus, the full representation of $\alpha_{o,x}$ is defined by six parameters: A_z, A_o, A, P, R, and σ.

An analysis of the total sample BQS+BF+HET, using the DB method (Avni et al., 1980), assuming the traditional values of $q_o = 0.0$, $\alpha_o = 0.5$, $\alpha_x = 0.5$, and considering all the six regression parameters as free parameters, has yielded the following best-estimate values: P=0, R=3.3, σ=0.21, A_z=0.02, A_o=0.09, and A=1.54. The best-estimate value of P=0 means that the data indicate that there are no optically-selected x-ray-quiet QSOs. Furthermore, treating P as a single interesting parameter (Avni 1976), the DB regression analysis yields an upper limit of $P \leq 8\%$ at the 95% confidence level. Thus, no more than a few percent of optically-selected QSOs can be x-ray-quiet. While these specific quantitative results depend on the functional representation chosen and on the numerical values of the assumed parameters, the qualitative conclusion is not very sensitive to these assumptions. Thus, the large majority, probably all, of optically-selected QSOs are x-ray-loud, in the sense of having $\alpha_{o,x}$ values in the approximate range of 1.0 to 2.0. In this sense x-ray emission seems to be a universal property of QSOs. This affirms the importance of x-ray observations for studying QSOs, and confirms the x-ray emission as an effective tool for discovering or selecting QSOs.

The best-estimate value of R=3.3 means that the data indicate a large amount of skewness in the distribution of $\alpha_{o,x}$ residuals, with a longer tail at high $\alpha_{o,x}$ (low L_x) and a shorter tail at low $\alpha_{o,x}$ (high L_x). Treating R as a single interesting parameter, the DB regression analysis yields a very significant skewness: $R > 1$ at 99.999% confidence level. This result is also supported by an a posteriori comparison of the actual nonparametric best-estimate distribution of residuals relative to the parametric best estimate average dependence (determined by A_z, A_o, A), with the expected parametric best-estimate distribution of residuals (determined by σ, R). From this comparison, presented in Figure 5 (adapted from Avni and Tananbaum, 1985), a KS test marginally rejects the pure Gaussian (R=1) best fit at $\gtrsim 94\%$ confidence, while the skew (R=3.3) distribution is perfectly acceptable (KS probability ~50%).

We note that while the new samples require a non-Gaussian distribution of residuals, such skewness has an insignificant effect on the average dependence $\bar{\alpha}_{o,x}(z, L_o)$. When the values P=0 and R=1 are enforced in the regression analysis, the best-estimate parameters which result are $A_z=0.05$, $A_o=0.07$, $A=1.54$ and $\sigma=0.20$. These are not significantly different from the values derived in the full, 6-dimensional analysis. Thus, the previous conclusions regarding the correlation of x-ray and optical luminosities, and regarding the relative x-ray and optical cosmological evolution rates, remain unaffected by the finding of the significant skewness. This skew, non-Gaussian behavior is a new qualitative property of the conditional x-ray luminosity function $\Phi_x(L_x|z, L_o)$.

We also wish to emphasize that the results presented in this section, namely that there are "no" optically-selected x-ray-quiet QSOs and the skewness of the distribution of $\alpha_{o,x}$ residuals, as well as the results presented in the previous section, are entirely independent of the QSO optical ELF $\psi_o(z, L_o)$. Thus, these results are not affected by many of the present uncertainties in $\psi_o(z, L_o)$ discussed in the next sections.

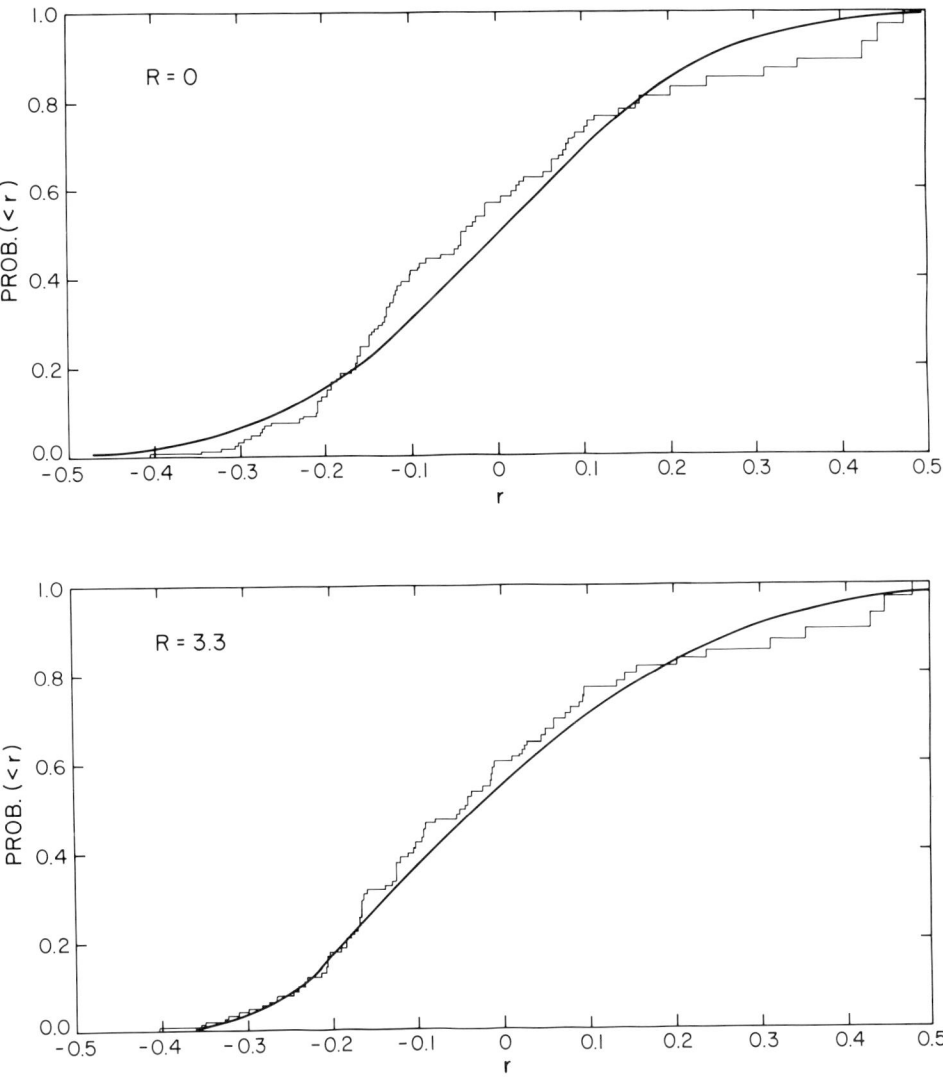

Figure 5: A comparison of the actual nonparametric best-estimate distribution of $\alpha_{o,x}$ residuals (light-line histogram) relative to the parametric best-estimate average $\alpha_{o,x}(z,L_o)$ dependence, with the parametric best-estimate distribution of residuals (heavy-line curve). Upper panel: pure Gaussian (R=1) enforced. Lower panel: best estimate derived from fit (R=3.3). (Adapted from Avni and Tananbaum 1985).

V. COMPARISON WITH X-RAY SELECTED SAMPLES

The conditional x-ray luminosity function $\Phi_x(L_x|z,L_o)$, or, equivalently, the average dependence $\bar{\alpha}_{o,x}(z,L_o)$ together with the distribution of $\alpha_{o,x}$ residuals, connect the optical ELF with the x-ray ELF. Thus, if one assumes a specific model for $\psi_o(z,L_o)$, one can calculate $\psi_x(z,L_x)$, which can then be compared with observations of samples of x-ray selected QSOs, in particular with x-ray number counts.

There is at present a considerable uncertainty regarding the choice of an appropriate model for $\psi_o(z,L_o)$. There are two main reasons for this uncertainty. The first reason is that a number of samples of optically selected QSOs do not agree with each other on the optical number counts at B≈20 magnitudes, with a discrepancy that ranges up to a factor of ~4 (Marshall et al., 1984; Boyle et al., 1986; Koo 1986; Marano 1986). It has been argued that the BF field presents a fluctuation (statistical or physical) in the number density of QSOs; however, this has not been established, in particular in view of the apparent agreement — within statistics — between the samples of Marshall et al. (1984) and Boyle et al. (1986). Thus, if the point of view is taken that the BF field is anomalous, the validity of other samples can also be questioned. The second reason is that a wide range of functional forms for representing $\psi_o(z,L_o)$ have been suggested, with different approaches to describe the cosmological evolution of QSOs. It is commonly accepted that pure density evolution (DE) does not provide a good description of the QSO population, since several DE models are in sharp conflict with optical number counts at faint magnitudes and violate constraints imposed by the x-ray background. However, a variety of other types of evolution are being employed: luminosity dependent density evolution (Schmidt and Green 1983; Schmidt 1986), pure luminosity evolution (LE) (Marshall 1985, 1986; Weedman 1986; Danese et al., 1986), luminosity dependent luminosity evolution (Cavaliere 1986), and combined luminosity and density evolution (Koo 1986). The different models correspond to very different global behavior of $\psi_o(z,L_o)$, and diverge in important details.

In view of this situation, it is not entirely surprising that none of the suggested representations for $\psi_o(z,L_o)$ is at present apparently consistent with all available samples. The specific luminosity dependent density evolution models suggested by Schmidt and Green (1983) are not consistent with the Braccesi faint (BF) sample (Marshall 1983, Landman 1984). Simple pure luminosity evolution models with an exponential dependence on look-back time are not fully consistent with the detailed properties of the available complete samples (Marshall 1985). The detailed pure luminosity evolution models of Danese et al. (1986), are not consistent with the differential optical number counts at B=19.8 magnitudes from the BF sample. Thus, a great deal of care must be exerted when drawing conclusions that are based on any specific choice of representation for the optical ELF. Of course, calculations of the x-ray ELF and comparisons with x-ray selected samples are still very valuable, since they are important for obtaining a better understanding of all the issues involved, and for eventually leading to a consistent representation of the full bivariate optical-x-ray ELF.

As mentioned earlier, the slope of the x-ray luminosity function calculated from a pure luminosity evolution model for $\psi_o(z,L_o)$ with an exponential dependence on look-back time and with a pure power-law luminosity function, combined with the $\alpha_{o,x}(z,L_o)$ dependence derived with Gaussian residuals, does not agree with the slope determined from the Einstein Observatory Medium Sensitivity Survey (Maccacaro et al., 1983). Furthermore, it has been known for some time to a number of workers in the field that the QSO x-ray number counts calculated from the above representations do not agree with the counts determined from the Medium Survey; this inconsistency is mentioned and briefly discussed by Maccacaro (1984). However, no specific quantitative details have previously been published.

We have calculated the expected x-ray number counts for a few combinations of representations for $\psi_o(z,L_o)$ with representations for $\alpha_{o,x}(z,L_o)$. Our purpose here is to explore the sensitivity of the calculated counts to some of the input ingredients. We have concentrated on two pure luminosity evolution models for $\psi_o(z,L_o)$ from Marshall (1985), and assumed $q_o=0$. The first model invokes an exponential dependence of the optical luminosity on look-back time $e^{5.9\tau(z)}$, a pure power-law present epoch luminosity function $\tilde{L}_o^{-3.6}$ ($\tilde{L}_o$ is the present epoch optical luminosity), and a normalization of 8.7 Gpc^{-3}. This model will be henceforth denoted by "$e^{5.9\tau(z)}$". The range of validity of this model in the (z,L_o) plane is characterized by a luminosity cutoff $L_o \geq 0.6 \; 10^{30}$ erg sec^{-1} Hz^{-1}, by a present epoch luminosity cutoff $\tilde{L}_o \geq 10^{29}$ erg sec^{-1} Hz^{-1}, and by a redshift range $0 \leq z \leq 2.2$. The second model invokes a power law dependence of the optical luminosity on redshift $(1+z)^{3.5}$, a pure power-law present epoch luminosity function $\tilde{L}_o^{-3.6}$, and a normalization of 22 Gpc^{-3}. This model will be henceforth denoted by "$(1+z)^{3.5}$." The range of validity of this model is characterized by $L_o \geq 0.6 \; 10^{30}$ erg sec^{-1} Hz^{-1}, $\tilde{L}_o \geq 2 \; 10^{29}$ erg sec^{-1} Hz^{-1}, and $0 \leq z \leq 2.2$. We emphasize that the derivation of these two models (Marshall 1985) is strongly based upon the BF sample, and they are therefore subject to the same uncertainties as the BF sample.

The range of validity in the (z,L_o) plane of any given model for $\psi_o(z,L_o)$ is determined by the completeness properties of the optical samples on which the derivation of $\psi_o(z,L_o)$ is based, and by the functional form of $\psi_o(z,L_o)$. This range must be taken into account when a comparison is made with the observed QSO x-ray number counts. Only those x-ray selected QSOs whose optical luminosities and redshifts are within the range of validity should be included in the x-ray number count for such a comparison. This is an important effect, since about half of the QSOs of the Medium Sensitivity Survey from Maccacaro et al. (1984) are excluded by the requirement $L_o \geq 0.6 \; 10^{30}$ erg sec^{-1} Hz^{-1}, as shown in Figure 6, which is adapted from Avni and Tananbaum (1985). For the comparisons to be made here, we have therefore re-derived the x-ray number counts from the Medium Survey data taking into account the constraint imposed by the optical evolution models.

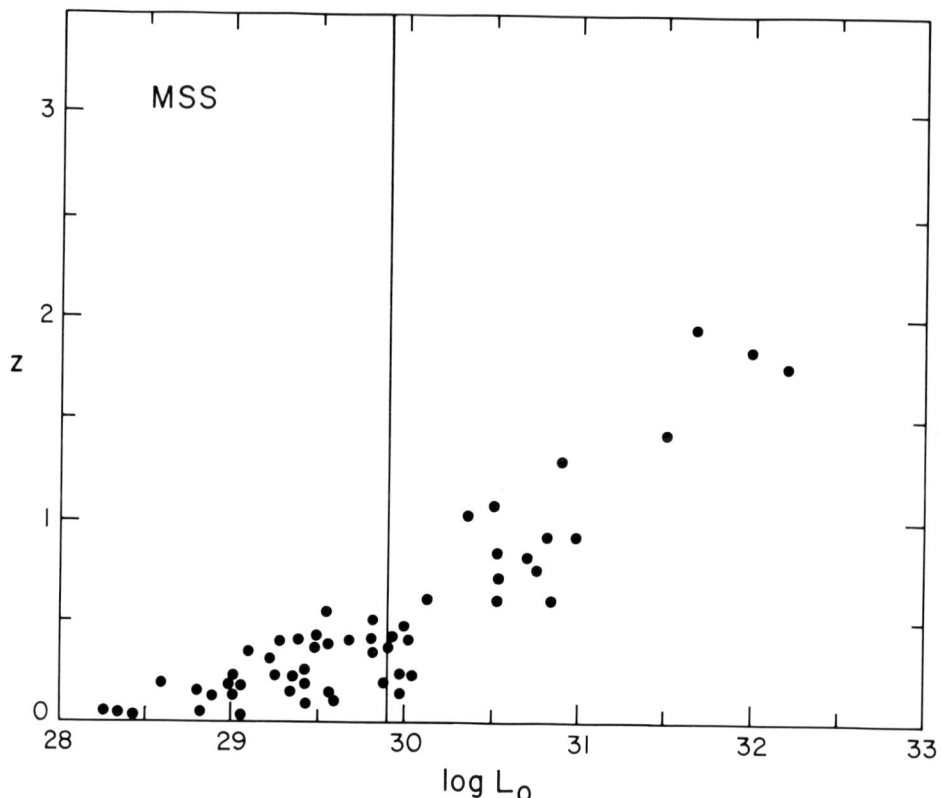

Figure 6: Distribution of x-ray selected QSOs from the Einstein Observatory Medium Sensitivity Survey in the (z, L_o) plane. Vertical line: $L_o = 0.6 \; 10^{30}$ erg sec^{-1} Hz^{-1}. (Adapted from Avni and Tananbaum 1985).

Table 1, adapted from Avni and Tananbaum (1985), presents a comparison between calculated and observed QSO x-ray number counts $N(>S_x)$ (per steradian), for two values of S_x, where S_x is the integrated 0.3-3.5 keV flux in erg sec^{-1} cm^{-2}. The notation "R=1" denotes the best estimate $\alpha_{o,x}(z,L_o)$ average dependence and distribution of residuals, obtained when P=0 and R=1 are enforced, i.e., with a Gaussian distribution of residuals. "R=3.3" denotes the overall best estimate $\alpha_{o,x}(z,L_o)$ average dependence and distribution of residuals, allowing for the more general skew distribution of residuals. "MSS" denotes the observed number counts which we have derived from the Medium Survey data.

Table 1
Comparison of Calculated and Observed X-ray Number Counts

Model or Data	$S_x(0.3-3.5$ keV) erg sec^{-1} cm^{-2}	
	$1.56 \ 10^{-13}$	$1.0 \ 10^{-12}$
$e^{5.9\tau(z)}$, R=1	$1.5 \ 10^4$ str^{-1}	$7.0 \ 10^2$ str^{-1}
$e^{5.9\tau(z)}$, R=3.3	$1.2 \ 10^4$	$2.3 \ 10^2$
$(1+z)^{3.5}$, R=3.3	$0.78 \ 10^4$	$1.7 \ 10^2$
MSS	$0.38 \ 10^4$	$1.3 \ 10^2$
$(1+z)^{3.5}$, R=3.3, $\sigma=0.18$	$0.55 \ 10^4$	$1.1 \ 10^2$

The calculated number counts for "$e^{5.9\tau(z)}$, R=1" are the values corresponding to the "traditional" exponential model for $\psi_o(z,L_o)$ and to Gaussian residuals. Those values are higher than the observed number counts by factors of 4-5. The calculated number counts for "$e^{5.9\tau(z)}$, R=3.3" correspond to the same optical ELF, combined with the skew distribution of residuals. The effect of replacing the "R=1" Gaussian distribution by the "R=3.3" skew distribution is very large — reduction by a factor of 3 — for $S_x=10^{-12}$ erg sec^{-1} cm^{-2}, but is less pronounced — reduction by 20% — for $S_x=1.56 \ 10^{-13}$ erg sec^{-1} cm^{-2}. We therefore find a strong numerical sensitivity to the shape of the distribution of residuals for a large part of the dynamic range of S_x covered by the Medium Survey. The calculated number counts for "$(1+z)^{3.5}$, R=3.3" correspond to the same skew distribution of residuals, combined now with the power law model for $\psi_o(z,L_o)$. The effect of replacing the exponential evolution model by the power law evolution model is substantial — the calculated x-ray number counts are reduced by 30%.

An immediate conclusion which emerges from the few cases we have studied is that the calculated number counts are sensitive both to the shape of the distribution of $\alpha_{o,x}$ residuals and to the functional form of the optical ELF. As a further demonstration of this conclusion, we have also calculated the x-ray number counts for the representation "$(1+z)^{3.5}$, R=3.3", changing — in an ad hoc fashion — the value of σ from 0.21 to 0.18. (When the QSOs with the lowest values of L_o are excluded from the $\alpha_{o,x}(z,L_o)$ analysis, lower values of σ are indicated.) The results are presented in the last line of Table 1. This reduction of the value of σ leads to a reduction of 35% in the calculated number counts, and the counts calculated for $S_x=10^{-12}$ erg sec^{-1} cm^{-2} are below (by 20%) the observed counts.

A second conclusion which follows is that the particular combination of optical ELF and $\alpha_{o,x}(z,L_o)$ representation studied is not consistent with the observed x-ray number counts, by 25% at 10^{-12} erg sec^{-1} cm^{-2} and by a factor of 2 at $1.56 \ 10^{-13}$ erg sec^{-1} cm^{-2}, in spite of the fact that the new results go a significant way towards resolving the number count discrepancy. In view of the numerical sensitivities mentioned above, detailed work has to be done on all aspects of this problem to fully understand the QSO bivariate optical-x-ray ELF.

VI. DISCUSSION

The study of Avni and Tananbaum (1985) has yielded a number of consequences regarding the x-ray properties of optically selected QSOs. Heterogeneous samples of optically selected QSOs were established as adequate for the study of the explicit dependence of $\alpha_{o,x}$ on z and L_o. It was shown that for such studies, it is better to increase the size of the data base by observing a larger number of intermediate size complete samples with different magnitude limits rather than a smaller number of larger complete samples. The previous results of Avni and Tananbaum (1982) regarding the $\alpha_{o,x}(z,L_o)$ dependence have been confirmed, leading to the same implications for the correlations between observed x-ray and optical luminosities, and for the observed relative rates of x-ray and optical cosmological evolution. It was shown that changing the value of q_o from 0.0 to 0.5 does not have an appreciable qualitative effect on the results. It was found that varying the assumed slope of the optical spectrum, α_o, between 0.0 and 1.0, does not have an important effect for the present size sample. Most importantly, it was shown that the large majority, probably all, of optically selected QSOs are x-ray loud: no more than a few percent of optically selected QSOs can be x-ray quiet. The distribution of the $\alpha_{o,x}$ residuals relative to the $\bar{\alpha}_{o,x}(z,L_o)$ average dependence was shown to be significantly skew, a property which is important for constructing or constraining the QSO bivariate optical - x-ray ELF. All these results are independent of any choice of the QSO optical ELF.

We have also found that when an optical ELF is combined with an $\alpha_{o,x}(z,L_o)$ dependence and compared with x-ray selected samples, such a comparison is numerically sensitive to the details of the functional form of $\psi_o(z,L_o)$ and to the shape and width of the distribution of $\alpha_{o,x}$ residuals. At present, there is a residual discrepancy of about a factor of 2 between the calculated x-ray number counts, corresponding to available simple pure luminosity evolution models for $\psi_o(z,L_o)$ combined with the best estimate $\alpha_{o,x}(z,L_o)$ average dependence and distribution of residuals, and the observed counts from the Medium Survey. This lack of consistency needs to be resolved.

We now discuss briefly some directions for further research, related to the derivation of the optical ELF, to the study of the $\alpha_{o,x}(z,L_o)$ dependence, and to the derivation of the x-ray ELF, which are important for fully understanding the properties of the QSO population in the optical and x-ray regions.

Regarding the optical ELF $\psi_o(z,L_o)$, perhaps the most pressing issue is understanding the inconsistency in the optical number counts at $B \approx 20$ magnitudes between the various optical samples, and in particular deciding whether the BF sample is a fluctuation (physical or statistical) in number density. It is an important issue since the optical number count curve seems to flatten near B≈20 magnitudes, and many derived quantities depend sensitively on the detailed behavior of N(<B) at this point. This calls for observations of randomly chosen fields of about the same area as the BF field, to about the same limiting magnitude, with the same selection criteria. One needs first to understand the data base before detailed representations for $\psi_o(z,L_o)$ are constructed. Another problem which needs to be addressed is the shape of the luminosity function at low L_o, and the connection of the "QSO" evolution rate to the "Seyfert" (and related low luminosity AGNs) evolution rate. There is a large numerical sensitivity to the details of the shape of the luminosity function and to the rate of evolution at low L_o. In addition, effects of measurement errors of the observed magnitudes have not yet been fully and conclusively evaluated.

With respect to the study of the explicit dependence $\alpha_{o,x}(z,L_o)$, there are three factors which require detailed and rigorous study, since they affect directly the shape and width of the distribution of $\alpha_{o,x}$ residuals, and affect indirectly the average $\bar{\alpha}_{o,x}(z,L_o)$ dependence: measurement errors, time variability, and anisotropic emission. Time variability has an effect because the optical and x-ray luminosities have not been observed at the same time. When the dependence of L_x on L_o is studied, time variability of L_o needs to be explicitly dealt with. Anisotropic emission has an effect if the angular distribution of L_x is different from the angular distribution of L_o, since this will cause the observed luminosities, "directed" towards us, to be different from the intrinsic, direction-averaged luminosities. The potential importance of those effects stems mostly from the large numerical sensitivity to the shape and width of the distribution of $\alpha_{o,x}$ residuals which we have shown above. In fact, a paper contributed to this symposium (Franceschini et al., 1985) shows that by using substantially lower values of σ, the agreement with x-ray selected samples can be greatly improved.

Further progress in the study of $\alpha_{o,x}(z,L_o)$ can also be obtained by using x-ray observations of previously known QSOs from the Einstein Observatory Data Bank, to increase substantially the size of the sample, and to improve the coverage of the two dimensional dynamic range in the (z,L_o) plane of the x-ray observed QSOs. Observations by other x-ray satellites will also be beneficial for that purpose. Progress can also be made by improving the treatment of the non-detections, by using a detailed probability distribution for the observed counts in the detector rather than the discrete 3σ upper limits. The effects of the departure of the x-ray spectral index α_x from the traditional value of 0.5 need also be studied, and the systematics of QSO x-ray spectra are required for that. Different functional forms for the average $\bar{\alpha}_{o,x}(z,L_o)$ dependence and for the distribution of residuals should also be explored.

Regarding the derivation of the x-ray ELF $\psi_x(z,L_x)$, much progress will be made by increasing the size of the Einstein Observatory Medium Sensitivity Survey, and by completing the optical identifications for the Einstein Observatory High Sensitivity Survey. For the MSS QSOs, observations of those color indices, that are commonly used to select

QSOs in UV excess samples of optically selected QSOs, are important. Comparisons of optically selected samples with x-ray selected samples should also take into account the range of validity of the assumed models in the color domain. Effects of measurement errors, time variability, and anisotropic emission, should be considered rigorously when studying the $\alpha_{o,x}$ properties of x-ray selected QSOs.

In summary, significant progress has been made in the study of the properties of the QSO population in the x-ray region, and of the implications of the x-ray observations for the cosmological evolution and luminosity function of QSOs. In particular, it was shown that most, probably all, optically selected QSOs are x-ray loud (no more than a few percent can be x-ray quiet). Further study along some well defined directions will undoubtedly teach us much more.

ACKNOWLEDGMENTS

It is a pleasure to thank H. Tananbaum for the collaboration which has yielded the results described here. I thank M. Schmidt and G. Zamorani for helpful discussions and comments. T. Maccacaro and I. Gioia kindly provided an unpublished, expanded version of the sky area coverage table for the Einstein Observatory Medium Sensitivity Survey. This research was supported by NASA Contract NAS8-30751 and by the MINERVA Foundation, Munich, West Germany.

REFERENCES

Avni, Y. 1976, Ap.J., **210**, 642.
Avni, Y., Soltan, A., Tananbaum, H., and Zamorani, G. 1980, Ap.J., **238**, 800.
Avni, Y., and Tananbaum, H. 1982 Ap.J. (Letters), **262**, L17.
Avni, Y., and Tananbaum, H. 1985, Ap.J., to be submitted.
Boyle, B.J., Shanks, T., Fong, R., and Peterson, B.A. 1986, these proceedings.
Cavaliere, A. 1986, these proceedings.
Danese, L., DeZotti, G., and Franceschini, A. 1986, these proceedings.
Elvis, M., Wilkes, B.J., and Tananbaum, H. 1985 Ap.J., in press.
Franceschini, A., Gioia, I., and Maccacaro, T. 1986, these proceedings.
Giacconi, R., et al. 1979, Ap.J., **230**, 540.
Koo, D.C. 1986, these proceedings.
Kriss, G.A., Canizares, C.R., and Ricker, G.R. 1980, Ap.J., **242**, 492.
Ku, W.H.M., Helfand, D., and Lucy, L.B. 1980, Nature, **288**, 323.
Landman, U. 1984, M.Sc. Thesis, Weizmann Institute of Science.
Maccacaro, T. 1984, in Proceedings of X-ray and UV Emission from Active Galactic Nuclei (Garching, West Germany, July 1984), eds. W. Brinkman and J. Trumper, p. 63.
Maccacaro, T., Avni, Y., Gioia, I.M., Giommi, P., Griffiths, R., Liebert, J., Stocke, J., and Danziger, J. 1983 Ap.J. (Letters), **166**, L73.
Maccacaro, T., Gioia, I.M., and Stocke, J.T. 1984, Ap.J., **283**, 486.

Marano, B. 1986, these proceedings.
Marshall, H. 1983, Ph.D. Thesis, Harvard University.
Marshall, H. 1985, Ap.J., submitted.
Marshall, H. 1986, these proceedings.
Marshall, H.L., Tananbaum, H., Zamorani, G., Huchra, J.P., Braccesi, A., and Zitelli, V. 1983a, Ap.J., **269**, 42.
Marshall, H.L., Avni, Y., Tananbaum, H., and Zamorani, G. 1983b, Ap.J., **269**, 35.
Marshall, H.L., Avni, Y., Braccesi, A., Huchra, J.P., Tananbaum, H., Zamorani, G., and Zitelli, V. 1984, Ap.J., **283**, 50.
Schlosman, I., Shaham, J., and Shaviv, G. 1984, Ap.J., **287**, 534.
Schmidt, M. 1986, these proceedings.
Schmidt, M., and Green, R.F. 1983, Ap.J., **269**, 357.
Tananbaum, H., Avni, Y., Branduardi, G., Elvis, M., Fabbiano, G., Feigelson, E., Giacconi, R., Henry, J.P., Pye, J.P., Soltan, A., and Zamorani, G. 1979, Ap.J. (Letters), **234**, L9.
Tananbaum, H., et al. 1985, Ap.J., to be submitted.
Tucker, W.H. 1983 Ap.J., **271**, 531.
Weedman, D.W. 1986, these proceedings.
Zamorani, G., Henry, J.P., Maccacaro, T., Tananbaum, H., Soltan, A., Avni, Y., Liebert, J., Stocke, J., Strittmatter, P.A., Weymann, R.J., Smith, M.G., and Condon, J.J. 1981, Ap.J., **245**, 357.

DISCUSSION

PETROSIAN: The sensitivity of the results to the values of the parameters you indicated is interesting. It suggests that, as far as possible, (at least in the initial steps of this kind of analysis) one should use a non-parametric method.

AVNI: I agree that non-parametric methods are generally prefeable, in particular for deriving best-estimate results, and - in fact - we have been using them where possible. It should also be noted, however, that parametric methods usually enable well defined estimates of error-regions for the parameters and for derived quantities.

Properties of the sharp metal-rich absorption lines observed in QSO spectra

Jacqueline Bergeron

Institut d'Astrophysique
98 bis, boulevard Arago
F-75014 Paris, France

I. Introduction

Many reviews have been devoted to the topic of the absorption lines in QSO (see e.g., Weymann et al., 1981), revealing the importance of these absorption line systems as a tool to probe the environment of QSO and galaxies, to study any intervening material not otherwise detectable and to investigate the cosmological evolution of these populations. We only consider here the sharp metal-rich absorption systems, now fairly generally believed to be of intervening origin. We outline their properties, with emphasis on their degree of ionization, and compare them to those of halos or extended disks around nearby galaxies. We present some evidence for cosmological evolution of the density per unit redshift z of these systems and try to find whether the differences between the high z absorbers and galaxy halos are a consequence of cosmological evolution in the ionization state or whether there are two populations or two phases among the metal-rich systems. In the assumption of ionization of the absorbers by the UV diffuse background due to QSO emission, we present results obtained from the consideration of models of photoionized clouds and outline the constraints on the ionization parameter when comparing the models to the observed absorption line equivalent widths and the derived column densities. Finally, we briefly discuss the possibility of a link between the metal-rich systems and the Lyα forest.

II. Available samples

The major published surveys have been done in the blue, roughly in the wavelength range λλ3500-5000, mainly to detect absorption lines as soon as they enter the visible range and also because QSO are brighter in the blue at least for z<2.5.

The number of detected absorption lines for a specific ion is a function of the abundance of the element and its degree of ionization. Aside from hydrogen lines, the larger samples are obtained for CIV. The wavelength interval referred above correspond to a redshift range zz1.3-2.5 for CIV and about one absorption system is then detected in average per (fraction of) line of sight sampled (see e.g. Weymann et al., 1979, Young et al., 1982a).

To study global properties of absorption systems, one must consider homogeneous unbiased data at sufficiently high resolution to get individual lines of the major doublets. Observations for about 50 lines of sight can be used to construct homogeneous unbiased samples with fairly small equivalent width lower limit, $w_{obs} \sim 0.7$ Å (Bergeron and Boissé, 1984, hereafter BB1). Available MgII samples at low redshift, $<z>\sim 0.5$, are smaller by about one order of magnitude with roughly 5 doublets detected in 50 lines of sight (Tytler et al., 1985). This strongly contrasts with the host of Lyα absorption lines found at high redshift, up to nearly 200 in one line of sight (Atwood et al., 198.), most of them belonging to the Lyα forest (metal-poor population). This clearly shows that different approaches are necessary to study samples based on different lines.

The number density per unit redshift of absorbers can be written (see e.g., Bahcall 1979)

$$\frac{dN}{dz} = \frac{c}{H_0} n_0 \sigma_0 \frac{(1+z)^{1+\alpha}}{(1+2q_0 z)^{0.5}}$$

where n and σ represent the density and projected cross-section of the absorber. The subscript o refer to the present epoch and a possible cosmological evolution is represented by a power law of index α. The number density dN/dz is an increasing function of z even when $\alpha=0$, and this contributes for about a factor of 2 to the order of magnitude difference observed for dN/dz between MgII and CIV samples.

For a given observing wavelength range, the redshift interval sampled increases with decreasing rest wavelength of the line studied, favouring again CIV samples over MgII ones, but still not explaining the factor 10 observed so that there are probably intrinsically fewer MgII than CIV systems independently of z.

III. Differences in the degree of ionization between the sharp absorption systems and the halos of galaxies

Absorption systems of large redshift, $z \approx 2$, are of much higher degree of ionization than high latitude Galactic gas as first clearly emphasized by Wolfe (1983). The main caracteristics of the high z absorbers and clouds in our Galaxy are schematically summarized below :

high z absorbers

- in most systems CII is absent or much weaker than CIV.
- the small fraction of systems with large $w_r (\geqslant 1$ Å) usually have CII at least as strong as CIV
- systems with strong CII and weak CIV are extremely rare.

gas in our Galaxy

- even in high latitude gas CII is always detected and is as strong as CIV
- most systems from the Galactic plane show strong CII and weak CIV.

QSO-galaxy pairs allow to probe the properties of the gas at large distances from the galaxy centre. About half a dozen such cases are reported in the literature, with most often detection of the CaII doublet. In none of these cases there is convincing evidence of a high ionization state. For the QSO-galaxy pair 3C232/NGC3067, we analyzed the UV observation of Snijders (1980) : a strong MgII is present, w(MgII) $\simeq$ 6 Å, but CIV is not detected, w(CIV)<1.5 Å, at projected radius of 16.5 kpc ($H_0=50$ km s^{-1} Mpc^{-1}) or a radius of 61 kpc in the plane of the disk. For the QSO-galaxy pair PKS1327-206/MCG03-34-084, Kunth and Bergeron (1984) found a strong NaI absorption at a projected radius of 20 kpc (or 42 kpc in the plane of the galaxy) but did not detect the CaII doublet. The weakness of the CaII absorption is confirmed by new observations and in the UV a strong FeIIUV1 absorption is detected (Kunth and Bergeron, in preparation).

These differences in degree of ionization between extended gaseous envelopes around galaxies and the high z absorbers are not straightforwardly understandable if the absorption systems are assumed to be associated to galaxies except if there is either a cosmological evolution in the degree of ionization of the absorption systems, or two populations (or two phases) among the metal-rich absorbers.

IV. Cosmological evolution

We briefly summarize results concerning a search for cosmological evolution. Since the number density per unit redshift of the absorption systems is strongly dependent on the absorption line choosen, statistical studies must be performed on homogeneous samples constructed for a specific line.

Aside from Lyα (possible confusion between the Lyα forest and Lyα metal systems), the only absorption line(s) for which large unbiased samples of metal-rich systems are available is the CIV doublet. It is then not yet possible to search directly for a cosmological evolution of the ionization degree of the absorbers (defined for example by the line ratio MgII/CIV).

From an unbiased sample of 38 CIV doublets detected in 39 lines of sight BB1 found a possible cosmological evolution:

$$\frac{dN}{dz} \propto (1+z)^\gamma \quad \text{with } \gamma \simeq 1.8.$$

both from linear regression fits and cumulative distribution studies. However the uncertainty on γ is large, $|\Delta\gamma| \simeq 1.5$, and models without cosmological evolution are also compatible with the observations whatever the value of q_0. The difference found between the density per unit redshift of the CIV ($<z>=1.8$), (Young et al., 1982a, BB1), and MgII ($<z>=0.5$) (Tytler et al., 1985) samples can be entirely due to cosmological evolution if γ is indeed close to 1.8.

To determine γ with a better accuracy, one primarily needs to increase the redshift interval sampled. This requires observations either in the red, z(CIV)>2.5, or in the UV, z(CIV)<1.2, these latter awaiting the launch of the Hubble Space Telescope.

If the value of γ is confirmed to be substantially larger than unity, this could imply a cosmological evolution of either the degree of ionization of the absorbers, or of their average cross-section, or the existence of two populations one of which at least evolving with time.

V. Bimodal state of ionization

The CIV and MgII doublets can be tracers of regions with different degrees of ionization and/or different opacities.

The dependence of dN/dz on the rest equivalent width w_r is not similar for CIV and MgII (BB1). At the low w_r end $dN/dzdw_r$ is steeply increasing for CIV systems while there is a flattening for MgII systems. This different behaviour for ions of low and high degree of ionization seems also to be present for SiII and SiIV although the samples are fairly small. At the large w_r end there is an excess of MgII systems. The CIV and MgII samples considered by BB1 for these studies did not have similar average redshift. To better understand whether the dependence of w_r distribution with degree of ionization is due to cosmological evolution or to the existence of two populations among the metal-rich absorbers, it appeared necessary to study CIV and MgII doublets from the same absorption systems.

To this aim Boissé and Bergeron (1985, hereafter BB2) observed in the red a small sample of 18 QSOs with 21 known CIV doublets (with absorption redshift smaller than the QSO emission redshift) and they detected 8 MgII or FeII systems. Very strong absorption from low excitation ions is also present in high z systems : they found FeII lines at $z \simeq 2$ with w_r as large as the largest values previously reported (z=1.15 in the QSO 0453-423 observed by Carswell et al., 1977). Therefore, although the sample of BB2 is small, we suspect that the excess in $dN(MgII)/dzdw_r$ at the large w_r end found at low z is also present at least up to z of about 2. In this sample, there is also an anti-correlation between the degree of excitation, $CIV/MgII \equiv w_r(CIV1548)/w_r(MgII2796)$, and the strength of the absorption lines. In the systems with $w_r(CIV1548) \geqslant 1$Å, the low excitation lines are always present and usually stronger than CIV1548 by a factor 1.5 to 2.5; in the systems with $w_r(CIV1548)<1$Å, these low excitation lines, when detected, are usually smaller than CIV1548 but most often they are not detected implying CIV/MgII>2 to 3.

The absence of weak MgII doublets can be interpreted as a turn-over in the w_r distribution of the MgII systems, but it may also be due to an observational bias. A real break in $dN(MgII)/dzdw_r$ would imply few systems with MgII/CIV in the range 0.1 to 0.5. To check upon this point, one should be able to detect weak MgII absorption lines, $w_r(MgII2796)$ ~ 0.1 Å, hence need to obtain QSO spectra at high resolution and signal to noise ratio.

To study this problem Bergeron and D'Odorico started a high resolution survey of MgII and FeII lines from known CIV systems. Results obtained for the bright BL Lac object 0215+015 are indeed surprising. Seven absorption systems have been previously detected in this object (Blades et al.,

1985). The very strong system at z=1.345 is of unusual low excitation with neutral elements present and CIV/MgII=0.23; two other systems are fairly strong with detected CII absorption. For all systems MgII and/or FeII lines were in the wavelength range observed λλ5650-6820 by Bergeron and D'Odorico (in preparation). The spectrum obtained at higher dispersion has a spectral resolution R=17000. The sensitivity can be inferred from the detection of a weak unresolved NaID absorption at $z \approx 0$ with observed equivalent of 80 and 33 mÅ for the doublet (see Fig. 1a). Very strong absorption lines of FeII, MgII and MgI from the z=1.345 are observed ; multiple structure is present with at least 4 components, two of which being highly saturated in the FeII2600 line, as shown in Fig. 1b, and in the MgII doublet. Not a single absorption line from any of the other systems was detected. Results are given in Table 1 for the stronger absorption systems, using 50 mÅ as a conservative upper limit for w_{obs} and taking the measurements of Blades et al., (1985) for the absorption lines in the blue of λ5650.

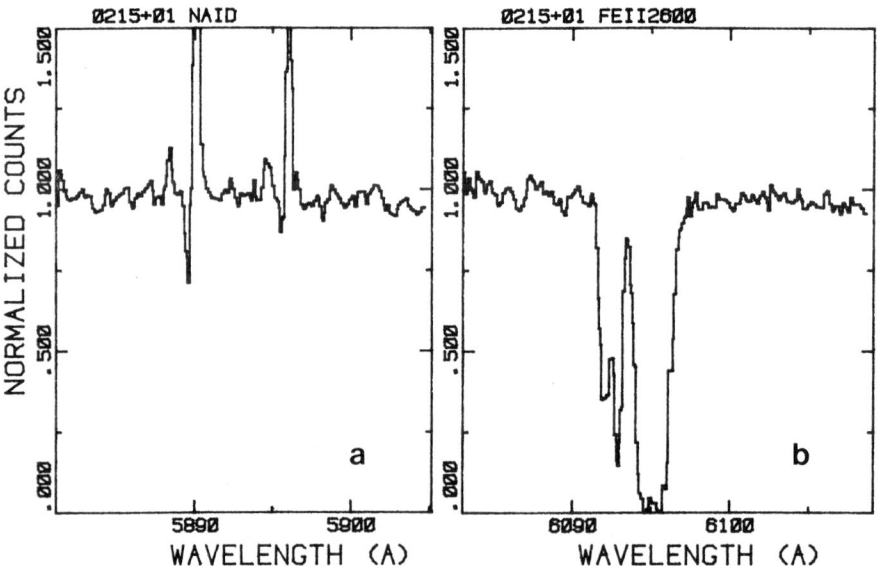

Fig.1 . High resolution absorption profiles of the NaID (z=0.0) and FeII2600 (z=1.345) lines in the BL Lac object 0215+015.

Table 1 . Absorption lines in 0215+015

z	w_r(CII1334)	w_r(CIV1548)	w_r(FeII2382)	w_r(MgII2796)
1.254		0.42		<0.022
1.345	1.27	0.44	1.92	1.89
1.549	0.10	0.89	<0.020	
1.649	0.28	0.90	<0.019	

An upper limit of 30 mÅ for w_r(MgII2796) leads to a column density $N(MgII) < 7.4 \; 10^{11}$ cm^{-2} assuming that the MgII doublet falls in the linear part of the curve of growth. The absence of MgII2796 or FeII2382 (when observed this lines has an equivalent width w_r(FeII2382)≈0.6 to 0.9 w_r(MgII2796)) in fairly strong absorption systems suggests indeed that there is a turn-over in the low w_r end of the MgII equivalent width distribution, with very few systems with an excitation parameter CIV/MgII in the range 2 to 20. These results also suggest the existence of a bimodal state of ionization among the metal-rich absorbers.

Another striking result concerns the absence of MgII or FeII absorption when CII is detected. This is probably due in part to a large abundance ratio C/Mg (the cosmic value equals 18) but also to a difference in the degree of ionization of C and Mg as will be discussed below.

From the strength of Lyα absorption, one can derive an estimate of HI column density thus of the opacity of the absorber in the Lyman continuum. The occurence of MgII/FeII absorption in 0215+015 appears to be linked with very strong Lyα absorption and large opacities in the UV continuum ($\tau(Lyc)>10$). This is also true for absorption systems in the sample studied by BB2 : two MgII/FeII systems have a large enough redshift for Lyα to be in the optical range and for both $w_r(Ly\alpha)$ exceeds 10 Å.

Among the metal-rich absorbers at high z, one can therefore distinguish :
- a high excitation population (CIV/MgII>10) which comprises the majority of the absorption systems ; their opacity in the Lyman continuum is fairly low,
- a population of lower excitation (CIV/MgII ≤1) with large HI column densities and large opacities, $\tau(Lyc)>10$. These absorbers have properties very similar to those observed in galactic extended disks or halos.

If the lower excitation absorbers are indeed associated to galaxies, one can derive the average halo radius (assuming spherical symmetry) using the galaxy number density in the luminosity function (Felten, 1977). Similar results are obtained by different authors (Young et al., 1982a, BB1, Tytler et al., 1985) for low redshift MgII samples

$$R = (28 \pm 8)h^{-1} \text{ kpc} \simeq 2.5 \, R_H$$

where h is the Hubble constant in unit of 100 km s^{-1} Mpc^{-1} and R_H is the Holmberg radius averaged over the luminosity function. For comparison the 21 cm average radius for HI disks around spiral galaxies is only of about 1.2 R_H (Krumm and Salpeter, 1979).

An attempt by Briggs and Wolfe (1983) to detect 21 cm absorption in the spectrum of radio QSOs with known MgII absorption led to few positive results. Although low excitation systems have "large" HI column densities as mentioned above, $N_{HI} > 10^{18}$ cm-2, those are most often still too small to give a measurable 21 cm absorption. More crucial the neutral hydrogen observed by Lyα absorption may be within a transition region at fairly high temperature (T⩾3000 K) and not in cold clouds, and very large HI column densities would then be necessary to detect a 21 cm absorption. In the QSO sample observed by Briggs and Wolfe (1983) the two low excitation systems at high z with detected 21 cm absorption have Lyα line with $w_r \geqslant 20$ Å, implying $N_{HI} \geqslant 8 \cdot 10^{20}$ cm^{-2} in the assumption of radiation damping. They represent extreme cases of the low excitation absorbers.

A low excitation population at high z and of large HI column density has also emerged from a low resolution Lyα survey made by Wolfe and Smith (private communication). They find a number of strong Lyα absorptions ($w_r(Ly\alpha) \geqslant 8$ Å) about 10 times larger than expected from an extrapolation of $dN(Ly\alpha)/dzdw_r$ at lower w_r. These damped Lyα absorbers have large HI column densities and low degree of excitation with CIV/CII⩽1. The class of absorbers we define as the low excitation population should include the systems discovered by Wolfe and Briggs but also extend to systems with smaller $w_r(Ly\alpha)$.

The existence of a low excitation population at high z with distinct properties from the bulk of the absorption systems shows that a search for global properties of the absorbers done with MgII and CIV samples at different z should be considered with great caution.

To get a deeper understanding of the comparative properties of galaxy halos and metal-rich absorption systems one is in need of low z CIV samples which will be availabe with the coming of the Hubble Space Telescope. A UV survey will also reveral whether the high excitation population observed at high z is also present at low z.

VI. Is Mg II tracer of optically thick systems compatible with photoionization by the diffuse UV background ?

The diffuse UV background is one important source of ionization at high z for clouds with moderate densities, $n \leq 1$ cm^{-3}. Even if the metal-rich absorbers are associated to galaxies, it could be the dominant source of ionization for material at large radial distances where the density of ionizing stars is very small.

The occurence of MgII absorption only in systems with large HI column densities could impose strong constraints on the model parameter. The ionization state is governed by the ionization parameter, $U \equiv N(\nu > \nu_{Lyc})/nc$, where $N(\nu > \nu_{Lyc})$ represents the total number of ionizing photons reaching the absorbing cloud per unit time. Models computed with the code PHOTO as described in Stasinska (1984) have been applied to the study of the absorption systems (Bergeron, Stasinska and Collin, in preparation). A slab geometry has been adopted, and for a diffuse ionizing flux of

$$J(\nu_{Lyc}) = 1\ 10^{21}\ \text{erg cm}^{-2}\ \text{s}^{-1}\ \text{st}^{-1}\ \text{Hz}^{-1},$$

one gets $U = 2\ 10^{-5}\ n^{-1}$.

The diffuse UV background has been determined at z=1.8, using estimates from Sargent et al. (1980) and Gondhalekar (1983) and assuming a spectral dependence $I_\nu \propto \nu^{-\alpha}$ ($U \propto \alpha^{-1}$).

Results independent of the absolute abundances must be first considered to avoid introducing too many unknown parameters. The relative strength of CIV and MgII absorptions is a function of the relative abundance C/Mg, assumed of cosmic value (18), the HI column density of the slab, the ionization parameter and the spectral shape of the diffuse UV flux. We report here preliminary results obtained with $\alpha = 1.5$ for values of U in the range $2\ 10^{-5}$ to $2\ 10^{-1}$.

The lack of MgII detection in optically thin absorption systems ($N_{HI} < 10^{17}$ cm^{-2}) set the constraint

$$\frac{CIV}{MgII} \equiv \frac{w_r(CIV1548)}{w_r(MgII2796)} > 5 \quad \text{or} \quad \frac{N(CIV)}{N(MgII)} > 50,$$

assuming that the CIV and MgII absorption lines are optically thin (both doublet ratios close to 2). Such a constraint is always satisfied for $U>10^{-3}$ ($n<2\ 10^{-2}$ cm^{-2} at $z = 1.8$). For smaller values of U the above ionization ratio becomes much too low, with CIV/MgII already as small as 0.2 for $U = 4\ 10^{-4}$ and $\alpha=1.5$, which does not correspond to the observed value from known optically thin absorption systems at high z.

For $U>10^{-3}$ the region of the absorbing clouds with $N_{HI}<10^{17}$ cm^{-2} has a high degree of ionization (MgII/Mg$\ll$1) and a sharp increase in MgII/Mg occurs when N_{HI} increases from 10^{17} to 10^{18} cm^{-2}. For $N_{HI}>10^{18}$ cm^{-2} and U in the range 10^{-3} to $2\ 10^{-1}$, the MgII column density is always larger than 10^{13} cm^{-2} for an abundance Mg/H of at least 2 % of the cosmic value. These detectable column densities of MgII are always associated to detectable CII1334 absorption. MgII and CII are even more dominant in the optically thick phase for $U<10^{-3}$, but for $U\leqslant 3\ 10^{-4}$ the weakness of CIV absorption relative to CII is incompatible with most known absorption systems, even extreme systems such as $z = 1.345$ in the BL Lac object 0215+015 mentioned above.

In the assumption of ionization of the absorbers by the diffuse UV background, MgII is indeed a tracer of high opacities ($\tau(Lyc)>10$) at high z, in agreement with the observations.

Is CII also a tracer of high opacity ? This is an important question since, if CII and MgII are both dominant ions in the same zones, observations of MgII would not be crucial to determine the degree of ionization and opacity in the high z absorbers. We already reported above, in the case of 0215+015, the lack of MgII and/or FeII detection from systems with known CII absorption.

First, considering optically thin regions ($N_{HI}<10^{17}$cm^{-3}), there is a narrow range of ionization parameter for which CII is detectable. The equivalent width ratio w_r(CII1334)/w_r(CIV1548) is larger than 0.2 (or N(CII)/N(CIV)>0.4) for $U\leqslant 3\ 10^{-3}$. Therefore CII but not MgII is detectable in optically thin absorbers for U roughly in the range 10^{-3} to $3\ 10^{-3}$. This is also true in the case of moderate opacities ($N_{HI}<10^{18}$ cm^{-2}), for about the same range of U, since there is no substantial variation of N(CII)/N(CIV) for $10^{17}\leqslant N_{HI}<10^{18}$ cm^{-2}. Therefore CII is not an unambiguous tracer of large opacities contrary to Mg II.

The observed absorption line ratios also allow to set an upper limit for the ionization parameter. The NV doublet is rarely detected, except for systems with a redshift very close to the QSO emission redshift. We will not discuss these special cases ($z_a \sim z_e$) for which the UV continuum emitted by the QSO may be an additional important source of ionization for the absorbing clouds. The absence of NV absorption set constraints on U for the higher excitation absorbers. An equivalent width ratio NV/CIV$\equiv w_r$(NV1238)/w_r(CIV1548)<0.2 (N(NV)/N(CIV)<0.4) implies U<4 10^{-2} in optically thin regions. For optically thick systems with $N_{HI}>10^{18}$, N(CIV) is always much larger than N(NV) for U$\leq$ 2 10^{-1}, but when U>4 10^{-2} the NV column density is always larger than 10^{15} cm^{-2} if the abundance N/H reaches at least 2% of the cosmic value (N/H=1 10^{-4}). For velocity dispersions smaller than 100 km s^{-1} both NV and CIV doublets would then be optically thick and if U>4 10^{-2} NV should be detected when CIV is observed.

The range of possible values for the ionization parameter for all metal-rich systems is therefore strongly constrained by the observed ionization states, with 4 10^{-4}<U<4 10^{-2} for $\alpha=1.5$.

The total column densities given by the photoionization models are roughly proportional to UN_{HI} for $N_{HI}\leq$ 10^{17} cm^{-2}. For U=4 10^{-3} (n=5 10^{-3} cm^{-3} at z=1.8) one gets $N_H \approx 1.2 \ 10^3 N_{HI}$ in the optically thin region ; the dimension of the absorber along the line of sight would then be around 8 kpc for $N_{HI}=10^{17}$ cm^{-2}, reaching 40 kpc for $N_{HI}=10^{19}$ cm^{-2}.

If the diffuse UV background can be the dominant source of ionization at large z, is this also the case at low z ? If one assumes that the QSOs which contribute most to the UV background are those at z>2, one obtains $J(\nu_{Lyc}) \propto (1+z)^{3+\alpha}$. Comparing absorbing clouds with average redshift of 1.8 and 0.5, one should consider a decrease by a factor 10 to 20 in the ionizing flux for 1<α<2. If the lower limit on the absorbing clouds density found at high z (n=5 10^{-4} cm^{-3} for $\alpha=1.5$), also applies at low z one gets the constraint U<3 10^{-3} at z=0.5. As seen above the equivalent width ratio CII/CIV>0.2 for U$\leq$3 10^{-3} even in optically thin clouds. This also applies to the equivalent width ratio MgII/CIV when U<1 10^{-3}. Therefore if there is not an important additional source of ionization one expects that low excitation ions will almost always be present in low z absorbers.

VII. Nature of the high excitation absorbers

The low excitation systems observed at large z have properties similar to those of extended disks or envelopes associated to galaxies, but, as discussed above, this is not true for high excitation systems. If these two classes of absorbers mainly differ by their absolute density, the systems of smaller density having higher ionization state, the high excitation absorbers could be associated to the outer parts of galaxy halos. Using the upper limit derived for the ionization parameter, $U_{max} \approx 4 \; 10^{-2}$, one obtains for $N_{HI} \sim 10^{17}$ cm^{-2} halo sizes of roughly 1 Mpc.

Not considering here further the idea of superhalos, we now briefly discuss the possible existence of a link between the high excitation systems and the Lyα forest. As outlined in section IV there may be a cosmological evolution for the CIV systems dN/dz $\propto (1+z)^{1.8}$, but further observations are needed to restrict the large range of possible values for the index γ and confirm the existence of a cosmological evolution. Several redshift evolution studies have been made for the Lyα forest (see e.g. Peterson 1983 and references therein, Young et al. 1982b, Atwood et al. 1985), all suggesting a cosmological evolution with γ in the range 1.5 to 2.4.

A link between the Lyα forest and metal-rich systems of high excitation is suggested by the similarity and smooth connection of the equivalent width distributions dN/dzdw$_r$ of these two populations as first pointed out by BB1. A recent study by Tytler (private communication) of a large sample of Lyα absorption lines with derived velocity dispersion shows the existence of a single HI column density distribution dN/dzdN$_{HI}$ for the Lyα forest and Lyα metal systems with a small overlap in N$_{HI}$. Another property common to the two classes of systems is the lack of correlation between w$_r$ and z (BB1). For the Lyα population alone Atwood et al. (1985) also found no evidence of redshift evolution in the distribution function of HI column density or velocity dispersion.

The Lyα forest and the metal-rich systems indeed differ by their heavy element abundances, even if the Lyα forest clouds may not be completely metal-free (Chaffee et al. 1985 ; Shaver, private communication). Few heavy element abundances have been derived for the metal rich absorbers

and the distribution of Z/H is not known. In particular it
would be important to find whether there is a smooth
distribution of Z/H toward low abundances ($[Z/H] \sim -2$) or
if there is really a discontinuity of Z/H between the two
populations. The difference in abundances together with the
uniqueness of the HI column density distribution could be
understood if the two populations were arising from the same
initial intergalactic perturbations but would relate to a
different range of size or mass, the metal-rich systems
being of larger mass. In the smaller or less dense systems
star formation could not or would not have yet occured.

One argument against the suggestion of a link between the
Lyα forest and the metal-rich absorbers relates to the
multiple velocity structure observed for the CIV systems
(peak in the two-point correlation function around 100 to
200 km s^{-1}) with no such effect for the Lyα forest (Sargent
et al. 1980). However blending of Lyα lines may be
responsible for this difference. The CIV doublets have a
smaller opacity than Lyα, and fine splitting of Lyα is
usually not observed for systems showing marked fine
splittings of CIV. Therefore a two-point correlation
analysis performed on Lyα-metal lines instead of CIV would
not have reveal a favored velocity scale. To check whether
line splittings preferentially occur for systems with fairly
large HI column density for both Lyα metal and Lyα forest
absorbers, one needs to study Lyman lines of higher order
than Lyα. Indeed high resolution data of Lyδ to LyK obtained
by Chaffee et al (1985) reveal double structure for a strong
Lyα forest system with a total HI column density of $7\ 10^{16}$
cm^{-2}.

References

Atwood, B., Baldwin, J.A., Carswell, R.F. : 1985,
  Astrophys. J., **292** , 58
Bahcall, J.N. : 1979, Scientific Research with the Space
  Telescope, IAU Colloquium, **54** , 215
Bergeron, J., Boissé, P. : 1984, Astron. Astrophys., **133** ,
  374
Blades, J.C., Hunstead, R.W., Murdoch, H.S., Pettini, M. :
  1985, Astrophys.J., **288** , 580
Boissé, P., Bergeron, J. : 1985, Astron. Astrophys., **145** ,
  59
Briggs, F.H., Wolfe, A.M. : 1985, Astrophys.J., **268** , 76
Carswell, R.F., Smith, M.G., Whelan, J.A.J. : 1977,
  Astrophys.J., **216** , 351
Chaffee, F.H., Jr., Foltz, C.B., Röser, H.-J., Weymann, R.J.
  Latham, D.W. : 1985, Astrophys.J., **292** , 362

Felten, J.E. : 1977, Astron.J., **82** , 861
Gondhalekar, P.M. : 1983, Mon.Not.R.Astron.Soc., **204** , 997
Krumm, N., Salpeter, E.E. : 1979, Astrophys.J., **228** , 64
Kunth, D., Bergeron, J. : 1984, Mon.Not.R.Astron.Soc., **210** , 873
Peterson, B.A. : 1983, Quasars and Gravitational Lenses, 24th Liege International Astrophysical Colloquium, p.563
Sargent, W.L.W., Young, P.J., Boksenberg, A., Tytler, D. : 1980, Astrophys.J.Suppl., **42** , 41
Snijders, M.A.J. : 1980, Proceedings Second European IUE Conference, Tübingen, Germany, ESA SP-157, p.IXXI
Stasinska, G. : 1984, Astron. Astrophys. Suppl., **55** , 15
Tytler, D., Boksenberg, A., Sargent, W.L.W., Young, P., Kunth, D. : 1985, preprint
Weymann, R.J., Williams, R.E., Peterson, B.M., Turnshek, D.A. : 1979, Astrophys.J., **234** , 33
Weymann, R.J., Carswell, R.F., Smith, M.G. : 1981, Ann. Rev. Astron. Astrophys., **19** , 41
Wolfe, A.M. : 1983, Astrophys.J., **268** , L1
Young, P., Sargent, W.L.W., Boksenberg, A. : 1982a, Astrophys.J.Suppl., **48** , 455
Young, P., Sargent, W.L.W., Boksenberg, A. : 1982b, Astrophys.J. **252** , 10

DISCUSSION

SILK: Could your discussion of the MgII/Mg ionization be used to indirectly measure the flux of ionizating photons at $z=2$?

BERGERON: The ionization state of the absorbers determines the range of possible values for the ionization parameter (ratio of the ionizing photon density to the particle density) and sets constraints on the spectral dependence of the ionizing radiative flux. However, an absolute measure of this flux cannot be derived from study of the ionization state except if the absolute density of the absorbers could be known otherwise.

PERRY: I am delighted to see that with your new and excellent data you are finding cosmological evolution in the heavy element systems. As you know, in 1981, Khare-Joshi and I (Mon. Not. Roy. Astron. Soc. 1982, 199, 785) showed that such an evolution was required, based on the data then available to us. We discussed the need for spectrally selected samples, and attempted some simple selections on the data. However, it was clear then that just such a study as you are now presenting would be necessary to establish if the effect we found was certain. I would also like to draw your attention to the rotation measure studies of QSOs which also support such evolution in the foreground absorber population.

BERGERON: As stressed in my talk the use of samples mixing absorption lines from different ions to derive statistical properties of the absorbers can be very misleading. Indeed the number density per unit redshift of MgII, CIV and Lyα lines is a function of the element abundances and the ionization state of the absorbers. Further, as we also discussed, ions of low degree of ionization, such as MgII, are not always present in high redshift absorption systems with detected CIV doublets. We therefore think that only large homogeneous samples based on a given absorption line will allow to find whether there is indeed a cosmological evolution of the metal-rich absorbers.

RADIO GALAXY POPULATIONS: A PROGRESS REPORT

H. van der Laan, P. Katgert, M.J.A. Oort
Leiden Observatory
University of Leiden, The Netherlands

1. INTRODUCTION

This is a progress report on the project on faint radio galaxy populations that is being carried out at Leiden Observatory. The aim of the project is to chart the active (radio) galaxy population in space and radio luminosity. The final goal is to use these data in a study of the evolutionary behaviour of individual active galaxies, population characteristics like birth rate functions etc. The status of the project was most recently reviewed by Van der Laan and Windhorst (1981), Van der Laan et al. (1983) and Windhorst (1984).

The primary samples used in the project derive from deep, complete flux-limited samples defined at 1.4 GHz. Optical identification on deep optical material is followed by colorimetry (optical and near-infrared), spectroscopy and high-resolution radio mapping.

In this paper we report on the current status of the radio and optical work; the present information on the nature of the radio- and optically- faint radio galaxies; and on the population evolution as described by the redshift dependence of the radio luminosity function (RLF).

2. RADIO SOURCE COUNT AND IDENTIFICATION STATISTICS

Since our 1983 report (Van der Laan et al. 1983) several new Westerbork 1.4 GHz surveys have become available. First the so-called Leiden-Berkeley Deep Survey (LBDS) of areas with multicolour Mayall 4m plates, covering an area of 5.5 sq. degr. down to 0.8 mJy (in the following all flux densities and luminosities (expressed in W/Hz) refer to 1.4 GHz) (Windhorst, van Heerde and Katgert, 1984). Secondly, a reobservation of one of the LBDS areas (0.7 sq. degr.) down to 0.3 mJy (Oort and Windhorst, 1985). Finally, a survey of an area of 8.5 sq. degr. down to a flux density limit of 1.3 mJy (of some of the Einstein x-ray observatory deep survey areas) by Katgert-Merkelijn et al. (1985).

These new data were combined with those from earlier Westerbork

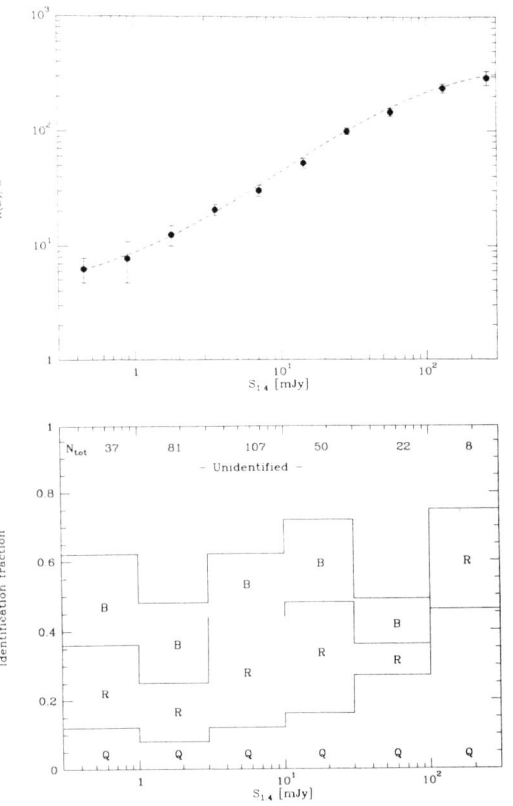

Fig. 1a. The combined 1.4 GHz Westerbork radio source count based on ~ 1200 sources

Fig. 1b. The identification statistics in the Leiden-Berkeley Deep Survey (Kron, Koo and Windhorst 1985)

surveys to yield the Westerbork 1.4 GHz source count shown in Fig. 1a, based on about 1200 sources. This count provides the best evidence for the onset of the "upturn" below about 10 mJy or, at least, the absence of continued convergence towards low flux densities. With its large statistical weight, the combined count shows convincingly that the upturn is very smooth, and not abrupt as suggested by some of the individual surveys. An obvious question is what causes the upturn, or rather, whether it can be attributed to a specific class of objects.

This last question can only be answered through optical information about the radio sources in the relevant flux density interval. The best optical data (as regards quality, quantity and diversity) are those for the LBDS and the Oort-Windhorst extension of one of the LBDS areas. In that case, optical identifications and photometry are based on deep Mayall 4m plates (U $\leq$ 23, J $\leq$ 23.5, F $\leq$ 22.5 and N $\leq$ 21) (Windhorst, Kron and Koo 1984, Kron, Koo and Windhorst 1985, and Windhorst et al. 1985).

After a morphological separation of stellar and non-stellar identifications, the latter are divided in two classes i.e. of red and blue galaxies, on the basis of colour (and, to some extent, optical morphology). The red galaxies are those with colours like the ones of classical ellipticals, while the class of the blue radio galaxies contains everything (non-stellar) with colours bluer than those of classical ellipticals (see Kron, Koo, Windhorst 1985).

The relative occurrence of the three identification classes (Q, R and B) down to F ~ 22.5, as a function of flux density, is shown in Fig. 1b. It is clear that the blue radio galaxies occur mostly, and most abundantly, at low flux densities (say, below 10 mJy), although some are found at higher fluxes. In view of this fact, it is tempting to speculate about a possible connection between the blue radio galaxies and the upturn of the radio source count. At present, without more direct supporting evidence, this has to remain a conjecture, however.

3. NATURE OF THE FAINT RADIO GALAXIES

a. Red radio galaxies

In the LBDS optical spectroscopy has been obtained for a representative subset of 30 red galaxies. From the spectroscopy it can be concluded that the red galaxies form a homogeneous class with absolute magnitude $M_F = -23.2 \pm 0.5$ ($H_o = 50$). This is to be compared with similar data for giant radio ellipticals at higher flux densities, e.g. in the S > 2 Jy so-called BDLF catalogue (Bridle et al 1972, Véron and Véron 1983) which gives $M_F = -23.9 \pm 1.0$; and in the 4C/B2 samples discussed by Katgert-Merkelijn et al.(1980), who find $M_F = -23.7 \pm 1.0$ (the larger dispersion in the latter two samples is due to inaccurate magnitude estimates).

On this basis we conclude that the red radio galaxies in the LBDS are (giant) ellipticals with absolute magnitudes as expected from the local bivariate luminosity function of elliptical radio galaxies (Auriemma et al. 1977). Therefore, their space distribution can be studied on the basis of photometric redshifts, if required (see section 4).

b. Blue radio galaxies

For these there is still only a limited amount of spectroscopy in the LBDS, and the available data may not be representative. Even with that proviso, some conclusions can already be drawn.

First, this class contains some <u>normal</u> spirals (generally nearby i.e. $z \lesssim 0.1$) with $M_F = -21.5 \pm 1.0$, and with radio luminosities log P $\lesssim 23$, as expected from the local bivariate luminosity function for spirals, derived by Hummel (1981).

In addition, there are - both at bright and faint magnitudes - blue radio galaxies that are <u>not</u> normal spirals, with $M_F = -22.5 \pm 1.5$ and log P > 23 (up to 26) which have no counterparts in the nearby spiral sample of Hummel. There is some indication that they have peculiar optical morphologies and/or occur in pairs.

The possibility of (not yet) representative data, the large spread in M_F and the possible inhomogenety of the class preclude the use of photometric distance estimates, and hence the determination of the blue galaxy RLF (let alone its redshift dependence) at the present moment.

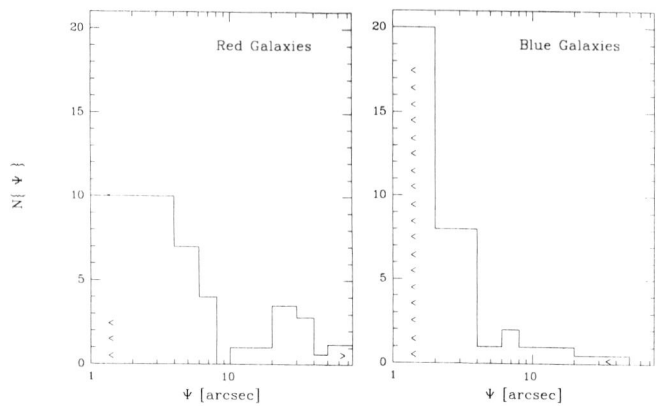

Fig. 2. The angular size distributions, as determined at 1.4 GHz with the WSRT and at 1.5 GHz with the VLA, of the blue and red radio galaxies in the Leiden-Berkeley Deep Survey

Recent high-resolution mapping with the VLA at 1.5 GHz shows a remarkable difference between the radio structures (sizes) of red and blue radio galaxies (see Fig. 2). Since the redshift distributions of the two classes are quite similar, the blue galaxies appear to have considerably smaller linear radio sizes than the red ones (note that, if anything, the blue ones are at smaller distances than the red ones, which would make the difference larger). The 1" upper limits do not imply (super-)compact structures, in agreement with the absence of (very) opaque radio spectra.

The true nature of these "radio-bright" blue galaxies is still unclear, but it is unlikely that they should be identified with Seyferts. Although radio luminosities of both classes are similar (Meurs and Wilson 1984), the space density of the blue radio galaxies (if only roughly known) would seem to be much larger than that of Seyferts (perhaps by as much as a factor ten or more).

4. RADIO LUMINOSITY FUNCTION AND EVOLUTION

As discussed above, we cannot derive a reliable RLF for the blue radio galaxies at the present time. However, using - where necessary - photometric redshift estimates, and combining the LBDS data with radio galaxy samples at higher flux densities, we have derived the RLF of the red (elliptical) radio galaxies (see Fig. 3). The total number of galaxies involved is about 400.

The present result differs considerably from that discussed by Van der Laan and Windhorst (1981), mainly because these authors combined red and blue radio galaxies. A decrease of the amount of evolution is the obvious result of leaving out the blue ones. There is also a small difference, in particular at higher luminosities, between the present result and that discussed by Van der Laan et al. (1983) and Windhorst (1984). The reason is that the local ($z < 0.29$) RLF is now very largely based on spectroscopic redshifts, especially for

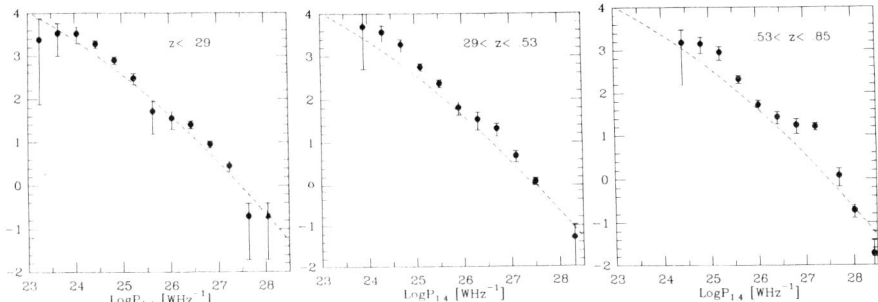

Fig. 3. The 1.4 GHz RLF of the red (elliptical) radio galaxies

log P > 26, and as a result has a slightly larger amplitude. This leads to a slightly smaller evolution.

The main conclusion about the RLF of the red galaxies is that the increase of the space density with redshift is negligible out to z ~ 0.3, and modest out to redshifts of about 0.8. This might seem to be in contrast with results by other groups on strong evolution of the luminous sources. However, all those studies included quasars as well, which we have left out deliberately.

We refer to the paper by Roos (this volume) for a model, based on galaxy mergers, to explain the z-dependence of the RLF.

Finally, we should mention that there is one source count model in the literature (Condon, 1984) in which the radio galaxy population has two components, viz. normal spirals and ellipticals. However, in that model the spirals only start to contribute significantly to the count around about 1 mJy, and therefore that model does not represent correctly the population in the 1 to 30 mJy flux range, since it ignores the "radio-bright" blue galaxies.

References

Auriemma, C., Perola, G.C., Ekers, R., Fanti, R., Lari, C., Jaffe, W., Ulrich, M.H., 1977, Astron. Astrophys. 57, 41
Bridle, A.H., Davis, M.M., Fomalont, E.B., Lequeux, J., 1972, Astron. J. 77, 405
Condon, J.J., 1984, Astrophys. J. 284, 44
Hummel, E., 1981, Astron. Astrophys. 93, 93
Katgert-Merkelijn, J.K., Lari, C., Padrielli, L., 1980, Astron. Astrophys. Suppl. 40, 91
Katgert-Merkelijn, J.K., Robertson, J.G., Windhorst, R.A., Katgert, P., 1985, Astron. Astrophys. Suppl. (in press)
Kron, R.G., Koo, D.C., Windhorst, R.A., 1985, Astron. Astrophys. 146, 38
Van der Laan, H., Windhorst, R.A., 1982, in "Astrophysical Cosmology", Eds. Brück, Coyne, Longair, p. 349

Van der Laan, H., Katgert, P., Windhorst, R.A., Oort, M.J.A., 1983, in "The Early Evolution of the Universe and its Present Structure", IAU Symposium 104, Eds. Abell, Chincharini, p. 73
Meurs, E.J.A., Wilson, A.S., 1984, Astron. Astrophys. 136, 206
Oort, M.J.A., Windhorst, R.A., 1985, Astron. Astrophys. 145, 405
Véron, P., Véron-Cetty, M., 1983, Astron. Astrophys. Suppl. 53, 219
Windhorst, R.A., 1984, Ph.D. Thesis Leiden University
Windhorst, R.A., Van Heerde, G.M., Katgert, P., 1984, Astron. Astrophys. Suppl. 58, 1
Windhorst, R.A., Kron, R.G., Koo, D.C., 1984, Astron. Astrophys. Suppl. 58, 39
Windhorst, R.A., Miley, G.K., Owen, F.N., Kron, R.G., Koo, D.C., 1985, Astrophys. J. 289, 494

DISCUSSION

WEEDMAN: Is there any evidence for evolution from the optical data alone?

VAN DER LAAN: The work by Kron, Koo and others suggests it is surprisingly weak.

KOO: The reddest radio galaxies to $z \sim 0.8$ are indistinguishable from the colors of a giant elliptical found today. In other words, we have yet to find any positive evidence for color evolution. The blue galaxies with spectra are not young ellipticals seen at high redshift with extensive star formation, as originally proposed a few years ago by radio astronomers.

VAN DEL LAAN: You are quite right. The blue galaxies are not the progenitors of the giant ellipticals. They are an apparently distinct active galaxy populatin of moderate power, in flux-limited samples mostly seen at modest values of z.

FOSBURY: How is the radio radiation generated in the field galaxy mergers? There are some examples, nearby, of relatively (radio) powerful mergers which should make good models to study e.g. NGC 6240.

VAN DER LAAN: Roos (1985 and this volume) has modelled the total power produced. The radio emission is a fraction of this, the determination of which requires a quite detailed astrophysical scenario. There are examples in the literature but there is no consensus concerning the dominance of any one process.

BURBIDGE: I'm glad to see that the evolution of radio galaxies is itself evolving in time towards less evolution. Dare we extrapolate? More seriously, are there differences in radio structure between what you call the red and the blue galaxies?

VAN DER LAAN: Yes there are. Westerbork surveys followed by VLA structure studies show blue RGs in the same flux density range to be more compact, with a less symmetric but more amorphous structure than the red RGs. The blue galaxies have typical sizes of $10^{1\pm 0.3}$ kpc, i.e. they are not ultracompact and generally not optically thick, but are much smaler than the red RGs.

MARSHALL: Were the models of Nico Roos also applied to optically selected quasars? And, if so, are the adjustable parameters similar?

ROOS: Yes, the model was also applied to the optical counts of QSO's. The parameters, were indeed similar, the main difference is that, in order to reproduce the counts at faint magnitudes we had to extend the population of rapidly evolving galaxies somewhat to the less luminous galaxies.

SCHULZ: Is there a similarity between the blue radiogalaxies (which are thought to be mergers) and those quasars which show tidal structures in [OIII], for instance in radiostructure?

VAN DER LAAN: In Roos model, quasars, AGNs and radio galaxies are diverse manifestations of the same processes, with the instantaneous circumstances determining the features which cause them to be classified as one or the other. The radio emission on the other hand is the cumulative effect of long term activity.

PFLEIDERER: How sure can you be that flattening of the source counts within the small observed area is not only due to clustering of fainnt sources?

VAN DER LAAN: Good question. We have mutually consistent evidence now for two regions and there is supporting, indipendent data from VLA surveys. But we are surveying additional areas, to be certain.

PETROSIAN: Your last slides seem to indicate that both the Blue and Red population are evolving and that they well can alone account for the source counts. How do you know the evolution of the blue population and what about the contribution of quasars to the counts.

VAN DER LAAN: That figure is from Roos'work and shows the

total source count contributions form "field galaxy" mergers and from rich cluster galaxy mergers. The radio emission is presumed a fixed fraction of the total power; quasars and radio galaxies are not distinguished in that sum (see Roos, this volume).

SOME THEORETICAL ASPECTS OF AGNs

Martin J. Rees
Institute of Astronomy
Madingley Road, Cambridge CB3 OHA
England

ABSTRACT. Some comments are made on the primary power source in QSOs and radio galaxies. The implications of high-z QSOs for theories of galaxy formation are briefly discussed.

1. PHYSICAL PROCESSES IN INDIVIDUAL AGNs

1.1. Comments on accretion

Processes of runaway evolution in galactic nuclei lead inexorably to the formation of massive black holes. Some categories of AGN may correspond to precursor stages; however, the most powerful phenomena can be most readily interpreted in terms of black hole processes. This topic has been addressed elsewhere (see, for instance Rees 1984a, b) and I shall not repeat this discussion in the present written text.

Other contributions to this meeting have considered the physics of accretion onto massive black holes. Even if the material accretes steadily, with constant angular momentum, the resultant primary radiation output is sensitive to uncertain details of the flow. The well-known "donut" configuration analysed by Abramowicz and collaborators, with a high value of $\dot{M}$, are supported "vertically" by radiation pressure, and radiate $L \simeq L_{Eddington}$. If the effective viscosity were very low (and the storage time in the donut correspondingly high) the material would be dense enough to thermalise the outgoing radiation. However, the thermalisation requirement is a very stringent one. If we define $\alpha = (v_{inflow}/v_{free\ fall})$, then one requires $\alpha < 10^{-3}(\dot{M}/\dot{M}_{crit})$ to thermalise radiation even at the densest part of the donut ($\dot{M}_{crit}$ being defined as $L_{Eddington}/c^2$). A still lower value of α would be needed in order to maintain thermalisation out to a photosphere at much larger radius (because the density must, for a stable entropy gradient, fall off at least as steeply as r^{-3}).

We do not know what α is likely to be. However, as Abramowicz has discussed at this meeting, non-axisymmetric instabilities may expel angular momentum so efficiently that they maintain an effective α far too high to permit thermalisation: so the inflowing material may behave in an irregular "cauldron-like" fashion: a long time exposure would still reveal a somewhat flattened axisymmetric configuration (though not necessarily with a completely empty "funnel" along the spin axis), but large irregular non-axial motions would make the density distribution and flow pattern variable on a dynamical timescale. The radiation from such a "cauldron" could be predominantly Comptonised bremsstrahlung; it would not, however, approximate a black body spectrum, and would instead resemble the spectra calculated for dissipative spherical accretion.

Despite these uncertainties, a massive black hole fuelled at a high (though not necessarily steady) rate still offers an acceptable basis for modelling the typical QSO. But something rather different is involved in the strong radio galaxies (e.g. Cygnus A). These have the distinctive property that the "kinetic" power required to energise the extended radio lobes (transmitted by the jets in the form of relativistic particles or Poynting flux) approaches the luminosity of a QSO, but the radiative luminosity of the nucleus itself is far lower. Is there a mechanism that could generate an intense plasma outflow, even if the accretion rate and nuclear luminosity were low?

1.2. Electromagnetic energy extraction from spinning holes

There is indeed another possible source of power over and above the gravitational energy released by infalling matter: this is the rotational energy of a spinning black hole, which can in principle be extracted, as was first recognised by Penrose (1969). Astrophysically plausible mechanisms for extracting this energy depend on exploiting the remarkably close analogy between a black hole and an ordinary electric conductor. This analogy is most simply illustrated, for a Schwarzschild hole, by calculating the electric field due to a point charge held at rest near the hole (Hanni and Ruffini 1972). As the charge, with radial coordinate r_c, is moved closer to the Schwarzschild horizon ($r_S = 2r_g = 2GM/c^2$), the field lines get progressively more distorted: they "wrap around" the hole so that as $r_c \to r_S$ they appear to emanate from $r = 0$, the field being essentially radial for $r - r_S \gg r_c - r_S$. It is as though the charge has spread itself over the hole's "surface". For a charge in free fall, the spreading happens in a time $\sim (r_S/c)$. Comparing this with the "classical" estimate of the time $r_S^2/4\pi\sigma$ taken for a charge to spread over a sphere of radius r_S and conductivity σ, we find that the effective resistance of a black hole is of order 100 ohms (cf. Znajek 1978).

A spinning (Kerr) black hole behaves like a spinning conductor (Blandford and Znajek 1977), in the sense that there are constraints on the orientations of any stationary electric and magnetic fields near the horizon. This analogy, spelt out in detail by Macdonald and Suen (1984), is sufficiently close that a "unipolar inductor" mechanism can indeed tap the spin energy of a hole.

Specific models for radio sources based on this general concept were developed by Rees et al. (1982) and Phinney (1983); this topic is reviewed in detail by Begelman, Blandford and Rees (1984).

Even a low-level and inefficient accretion flow can "anchor" a magnetic field that threads the hole, and thereby tap the hole's spin

energy, the current being provided by $e^+ - e^-$ pairs resulting from
γ-ray collisions near the hole. The extracted power naturally goes
predominantly into a relativistic bifurcated outflow. The power extracted is of order $B^2 r_c^2 c$: for a field $\sim 10^4$G, which can be confined
by plasma of density only 10^{-11}gm cm^{-3}, this can be $\sim 10^{45}$ erg s^{-1}.
This mechanism seems specially appropriate for strong radio galaxies
such as Cygnus A where the energy flowing along the jets dominates the
radiative output of the AGN itself. Electron-positron pairs moving
with Lorentz factors ~ 100 would transport some kinetic energy, but
most of the power outflow would initially be in the form of Poynting
flux associated with the magnetic field coiled round the jet axis, and
"frozen-in" to the pair plasma. This Poynting flux may be converted
into fast particles where the jet encounters ambient material (perhaps
on the scale of the VLBI radio components). The expected magnetic
field in the jet has the kind of configuration that could cause magnetic confinement and collimation. The plasma around the hole that
supplies the currents and anchors the field is just a catalyst: in
principle, the power output of a radio galaxy could be sustained with
zero accretion rate if some of the hole's spin energy were channelled
into the surrounding plasma to compensate for its (small) radiative
losses.

According to this idea, radio galaxies harbour massive black holes
formed long ago via catastrophic collapse (maybe during a quasar phase
of activity). The holes lurked quiescent, the galaxy being swept
clean of gas, for billions of years. Then some event, perhaps interaction with a companion, triggered renewed infall – maybe at a low rate,
but sufficient to reactivate the nucleus by applying a magnetic field.
This 'engaged the clutch', tapping the hole's latent spin energy, and
converting it into non-thermal directed outflow – Poynting flux and
$e^+ - e^-$ plasma – which ploughs its way out to scales $\sim 10^{10}$ times
larger. If this is indeed what happens in Cygnus A and M87, then these
very large-scale manifestations of AGN activity could offer the most
direct evidence for inherently relativistic effects.

1.3 What hope of a "unified model"?

There are therefore two quite distinct ways in which massive black
holes can generate a high luminosity: straightforwardly by accretion,
or via the electromagnetic process just described, where the energy
comes from the hole itself. The latter process tends to give purely
non-thermal phenomena, whereas accretion yields an uncertain mixture of
thermal and non-thermal power. The properties of an AGN must depend,
among other things, on the relative contributions of these two mechanisms, which depend primarily on $\dot{M}/\dot{M}_{crit}$ and the spin of the hole.
The properties of AGNs must depend on other parameters – the nuclear
mass M, the orientation and properties of the host galaxy etc. Ideally,
one would like a unified model which explains the multifarious types of
AGN in the same way that our theories for the Hertzsprung-Russell diagram
do this for stars.

Conditions around black holes are extreme, but the relevant physics
is essentially "well known" (in the sense that it can be learnt from,

for instance, Landau and Lifshitz). Moreover the key problem is at least well posed: axisymmetric plasma dynamics in a specified gravitational field, the aim being to calculate how much power is derived from accretion, and extracted from the hole's spin, and to find the form in which these respective contributions emerge. Such calculations play the same part in the modelling of AGNs that nuclear physics does in theories of stellar structure and evolution. The evidence that black holes have anything to do with AGNs is circumstantial, but the same is true for other cherished beliefs in astrophysics: the evidence that stars are powered by nuclear energy is also "merely" circumstantial. However the confrontation of models with observations — indirect even for stars — is admittedly much more ambiguous for AGNs: in stars the energy percolates to the observable surface in a relatively steady and well-understood way; in AGNs, on the other hand, it is reprocessed into all parts of the electromagnetic spectrum on scales spanning many powers of ten, in a fashion dependent on poorly-known environmental and geometrical effects within the host galaxy. The massive black hole hypothesis isn't infinitely "elastic", and could be disproved in several ways. It would, for instance, be in serious trouble if very regular periodicities were found in AGNs, or if HST studies of stellar velocity dispersions places upper limits $\ll 10^8$ $M_\odot$ on the central masses in any radio galaxies with large energy content.

2. IMPLICATIONS OF QSOs WITH THE HIGHEST REDSHIFTS

2.1 QSOs as a probe of the early universe

The most encouraging theme of this meeting has been the impressive progress in quantifying the shape and z-dependence of the AGN luminosity function. However, this is bound to remain for a long time a rather bewildering subject — we can't realistically expect all the data to fit any simple scheme with just a few parameters. One key issue is the lifetime of individual QSOs — does a typical QSO at $z \simeq 2$ shine for $\sim 2 \times 10^9$ yrs, or are there many successive generations of short-lived objects in the timescale over which the overall population declines? If the former alternative were correct, QSOs would involve such large masses that their characteristic luminosities were "sub-Eddington"; in the latter case, the remnant masses would be smaller but much more common. So this issue could be clarified either by developing detailed models for QSOs, or by seeking further evidence on massive black holes in nearby galaxies.

Ideally, one would like a theory that accounted both for the luminosity function of quasars and for the way their properties depend on z. The luminosity function depends on all kinds of environmental effects in the host galaxy. However, the fact that all classes of powerful AGNs display the same steep overall z-dependence suggests that the enhanced propensity of galactic nuclei to develop high-power engines is the primary phenomenon. There have been many schemes to derive an evolution law from the time-dependence of simple models (see, for instance, Cavaliere, Giallongo and Vagnetti (1985) and references cited therein).

Decay in the fuelling rate (arising from depletion of gas in the galaxy, or from depletion of stars in orbits that permit tidal capture) may be part of the story. But insofar as evolution proceeds autonomously within each galaxy, such models tell us only about the life-cycle of a single object, which may be much shorter than the timescale over which the population changes. There would be a spread in the latter owing to the spread in the formation epoch of galaxies. This spread is likely to be large enough to constrain any very steep z-dependence in AGN mean properties: even if each individual object experienced sharp discontinuities over its life cycle, this behaviour would be convolved with a function that could not plausibly change much more rapidly than the cosmological expansion timescale.

2.2 A high-z cut-off?

I'd like to conclude by venturing some comments on an aspect of AGN evolution that has specially important ramifications for the general process of galaxy formation — the nature of the objects with the highest redshifts, and the implications of the claimed "cut-off" beyond $z \cong 3.5$. The existence of high-z QSOs obviously sets a lower limit to redshift at which galaxies formed — or, more precisely, developed to the stage of having well-defined nuclei. There are no other firm arguments that pin down this redshift.

The cosmic microwave background photons are relics of the recombination epoch ($z = 1000$), when the primordial radiation shifted to wavelengths longward of the optical band. The universe thereafter stayed dark until there was light from the first bound systems — galaxies or their precursors. We do not know how long the "dark age" lasted: in some models, it could be several billion years before the first non-linear structures condensed (at $z \cong 3$); in other schemes, galactic and pregalactic activity could extend back to much earlier eras — much larger redshifts. The largest redshift known is $z = 3.78$ for PKS2000-330 This sets a lower limit to the redshift at which some bound systems formed, and allows us to probe the intervening medium at all smaller redshifts.

Studies of quasar evolution, particularly at high z, have recently been made by several authors, including Schmidt and Green (1983), Osmer (1982), Koo (1984), and Hazard and McMahon (1985); the subject has been discussed at this meeting by Weedman, Veron and others. There are not enough data to yield a unique evolutionary law; moreover, there are possibly important z-dependent selection effects. A common feature (first recognised for radio sources almost 20 years ago (Longair 1966)) is that it is the highest power sources that evolve most steeply. Green and Schmidt (1983) show that the data back to $z = 3.1$ are well fitted by 'density evolution' with an exponential decay whose time-constant is shorter for the most luminous sources. Other authors (e.g. Cheney and Rowan-Robinson 1981 and Koo 1984) favour 'luminosity evolution', according to which the luminosity function has a standard shape, steepening sharply for objects brighter than some magnitude M_{max}, but M_{max} is a function of z. The latter hypothesis would predict far fewer intrinsically faint quasars at high z than the former; however, this distinction, though it could have important astrophysical consequences, has no effect on the types of magnitude-limited surveys so far carried out.

The record redshift stuck at 3.53 for 12 years, even though the total number of known quasars multiplied several times, and there were several surveys geared specifically to the detection of high-z objects. Although uncertain selection biases preclude a precise statement, it seems that Schmidt and Green's exponential law cannot continue back to $z > 3.5$: indeed Osmer argues that his failure to find new objects at higher z implies an actual decrease in the comoving density for z beyond 3.5. Moreover, this decrease may start at a lower z for less luminous quasars. The few objects found at $z \gtrsim 3.5$ are all brighter than 19th magnitude, and searches down to 21st magnitude fail to reveal many more, whereas for $z = 2$ the number still increases by 4 per magnitude down to magnitude 21. In other words, those few objects that are known to exist at $z \gtrsim 3.5$ are all of ultrahigh luminosity. The QSO pheonmenon at the earliest epochs is apparently close to being an all-or-nothing affair: either the luminosities are $\gtrsim 10^{47}$ erg s^{-1}, or there is no QSO at all. This contrasts with the lower-z situation when the highest luminosity objects are merely the tail of a steep luminosity function. (We should parenthetically note the possible effect of gravitational lensing, which could be significant at high z if the intrinsic luminosity function were very steep.)

If QSOs are indeed hyperactive galactic nuclei, then they cannot light up until some galaxies have formed, at least to the extent of having developed a well-defined centre at which runaway gravitational collapse can occur. It would be of prime importance if we could infer something about the era of galaxy formation beyond merely setting a lower limit to its redshift. However, there is a possible timelag between the formation of the galaxy and the triggering of observable quasar-like activity.

2.3 Attenuation of QSO radiation by gas and dust in young host galaxies

Supposing that there were a cut-off in QSOs beyond some redshift, there are at least the following reasons why galaxies could still have formed much earlier:

(a) Massive black holes may take $\sim 10^9$ yrs to form in a galactic nucleus. Such might be the case if they formed from a star cluster, as Shapiro has proposed at this meeting. (It is often argued that radiation pressure restricts the rate at which black holes can grow by accretion, enforcing a growth timescale $\gtrsim 10^8$ yrs. This argument is, however, unconvincing. There is no necessity that any specific amount of energy be radiated per unit mass swallowed: "donuts" permit steady accretion with arbitrarily high $\dot{M}$ and low efficiency; spherical accretion with high $\dot{M}$ leads to a large "trapping radius", so that most of the radiation is advected into the hole rather than escaping; and an entire supermassive star which goes dynamically unstable can collapse to a black hole on a free-fall timescale, without needing to radiate any energy at all.)

(b) There may be intergalactic absorption at higher z. This could attenuate optical radiation, but would have no effect on radio emission nor on hard X-rays. Therefore, a sharp turnover in counts in either of these bands would still offer some constraint on the density at high z.

(c) Gas and dust in the host galaxy could prevent nuclear radiation from escaping. The total amount of obscuring material needed is less if it immediately surrounds the source than if this material pervades intergalactic space.

If there were indeed high redshift QSOs, which were undetected for reasons (b) or (c), then the absorbed or scattered radiation would eventually reach us in some waveband or other. In case (b), this would be an isotropic background, difficult to detect; in case (c), the host galaxies would be strong infrared sources.

The effects of powerful AGNs on their immediate surroundings have been discussed by Begelman (1985) and reviewed by Phinney at this meeting. In summary, there are three important processes:

(i) <u>Photoionization</u>. A typical QSO can ionize gas of density n within a volume V such that $\langle n^2 \rangle V < 10^{70} cm^{-3}$. This process could soak up much of the luminosity, but does not yield a gas temperature much above $T_{gas} \simeq 10^4$ °K.

(ii) <u>Compton heating</u>. This process can yield high T_{gas} (and thereby expel gas from the galaxy) provided that the QSO emits X-rays; the efficiency, however, is $\lesssim \tau_{Thompson} (h\nu_{max}/m_e c^2)$.

(iii) <u>Pressure of wind, relativistic plasma, etc</u>. This can be converted efficiently into energy of escaping gas. If the gas density is too high to be maintained ionized (cf. (i) above) then Lyman α radiation can be converted into kinetic energy with efficiency up to $\sim (\Delta\nu/\nu)$, which can be as large as a few percent.

In a young galaxy at $z \simeq 3$ (whether it is a proto-spiral or elliptical) we may expect gas densities $n = 1 - 10$ cm^{-3} throughout a spherical region of radius 10 kpc. Such an object would be highly luminous, because of the output from young stars and supernovae (Meier 1976). If a QSO lies in a young galaxy, its primary UV radiation may be "soaked up" by surrounding gas. If dust were present, only an IR (and hard X-ray) source would be detectable.

The above considerations suggest that typical QSOs at $z \gtrsim 3$ may not be readily detectable except in the infrared. One may conjecture further that only *ultra*-luminous QSOs would be able to blow away (or shine through) the gas in their host galaxy. This could explain why the luminosity function flattens at the highest z: only at smaller z does the gas and dust material in the host galaxy get tenuous enough to allow a more typical quasar to photoionize or expel it.

Quite apart from the issue of whether the turn-on of QSOs is coincident with galaxy formation rather than requiring a time delay, the observations confront us with the question of whether some objects interpreted as quasars on the basis of objective prism data could instead be young galaxies, in the sense that their luminous output is not

dominated by a non-stellar nucleus, but comes form a population of
young stars and gas spread through a region several kiloparsecs in size.
This can be settled in favour of the quasar interpretation in individual
instances if high polarization or rapid variability is discovered.
However, the typical quasar possesses neither of these attributes, and
we cannot be sure that its optical spectrum would be any different from
that of a primordial galaxy unless and until we knew what line widths,
etc., to expect for the latter. (Angular resolution of the kind possible
with the HST might just settle the question.) Some of the objects with
broad absorption features studied by Hazard *et al.* (1984), for instance
0330-380 and 1336+135, are completely bereft of the usual strong
emission lines. These could be AGNs whose luminosity is reprocessed
within the host galaxy by gas with a higher density than the surviving
gas in typical lower-z galaxies.

2.4   Implications for clustering and the intergalactic medium.

The mean spacing of 22nd magnitude QSOs is 10 - 20 Mpc (in comoving scale)
— i.e. a proper separation of 3 - 6 Mpc at $z \cong 2$. Clustering has been
looked for, particularly by Osmer (1981), and is interesting because
it offers a way to discriminate between different cosmogonic schemes
(cf. Rees (1983) for a review). Specifically, we can envisage three
modes of clustering:

(1) <u>Simple gravitational clustering of pre-existing galaxies</u>, with
$\delta\rho/\rho \propto (1 + z)^{-1}$. In this scheme, QSOs at high z would be *less*
clustered than present-day galaxies.

(2) <u>Galaxies form at "high σ peaks" of initial fluctuation spectrum</u>.
The probability that these high peaks occur is very sensitive to the
(smaller amplitude) fluctuations on larger mass scales (Rees 1983, Kaiser
1984). Galaxies could therefore be correlated *ab initio*, and the
clustering amplitude at $z \cong 2$ could be as large as it is locally

(3) <u>"Pancake" model, where gaseous superclusters collapse and fragment
into galaxies</u>. In this scheme, most galaxy formation would be at z < 2.
(Indeed, one of the main problems with the "pancake" model is to reconcile the existence of *any* high-z galaxies with the fact that clusters
and superclusters are not enormously denser than the background universe,
and cannot therefore have stopped expanding too early.) Only rare and
exceptional clusters corresponding to perturbations on the high-
amplitude tail of the (Gaussian?) distribution, could have formed at
$z \gtrsim 3$. However, each supercluster could contain several QSOs, so the
highest-z QSOs should be in a few "patches" with angular size $\sim 2° - 3°$
and a filling factor that decreases exponentially with increasing z.

The lack of a "Gunn-Peterson trough" in QSO spectra implies that
any diffuse intergalactic medium along the line of sight to QSOs must
be predominantly ionized. If, as is commonly argued, QSOs were the
responsible heating agents, then large regions of gas would remain cold

and neutral until the QSOs density had built up to some threshold. Thus, if we are really probing the era of 'first light', spectra of QSOs with $z \cong 3.5$, should display broad troughs due to Lyman α absorption in regions not yet ionized. There seems no evidence of this. Perhaps, therefore, we must already conclude that the intergalactic gas was heated earlier by some other agency — either a separate population of low-luminosity AGNs, or by pregalactic objects of the kind postulated in hierarchical models for galaxy formation.

REFERENCES

Begelman, M.C. 1985. *Astrophys.J.* (in press).
Begelman, M.C., Blandford, R.D. and Rees, M.J. 1984. *Rev.Mod.Phys.* 56, 255.
Blandford, R.D. and Znajek, R.L. 1977. *MNRAS*, 179, 433.
Cavaliere, A., Giallongo, E. and Vagnetti, F. 1985. *Astrophys.J.* (in press).
Cheney, J. and Rowan-Robinson, M. 1981. *MNRAS*, 195, 497.
Hanni, R.S. and Ruffini, R. 1972. in *"Black Holes"* ed. B. Dewitt and C. Dewitt, p R75 (Gordon & Breach).
Hazard, C. and McMahon, R. 1985. *Nature*, 314, 238.
Hazard, C., Morton, D.C., Terlevich, R.J. and McMahon, R. 1984. *Astrophys.J.*, 282, 33.
Kaiser, N. 1984. *Astrophys.J. (Lett.)*, 284, L9.
Koo, D.C. 1984. in *"Quasars and Gravitational Lenses"*, ed. J.P. Swings (Institut d'Astrophysique, Liège) p. 240.
Longair, M.S. 1966. *MNRAS*, 133, 421.
Macdonald, D.A. and Suen, W-M. 1984. *Phys.Rev.D.* (in press).
Meier, D.L. 1976. *Astrophys.J.*, 204, 869.
Osmer, P.S. 1981. *Astrophys.J.*, 247, 762.
Penrose, R. 1969. *Revista Nuovi Cim.*, 1, 252.
Phinney, E.S. 1983. Cambridge University Ph.D. Thesis.
Rees, M.J. 1983. in *"Clusters and Groups of Galaxies"*, ed. F. Mardirossian *et al.* (Reidel) p. 485.
Rees, M.J. 1984a. in *"X-ray and UV Emission from Quasars and AGNs"* ed. W. Brinkman (published by MPI, Munich).
Rees, M.J. 1984b. *Ann.Rev.Astr.Astrophys.*, 24, 471.
Rees, M.J., Begelman, M.C., Blandford, R.D. and Phinney, E.S. 1982. *Nature*, 295, 17.
Schmidt, M. and Green, R.F. 1983. *Astrophys.J.*, 269, 352.
Znajek, R.L. 1978. *MNRAS*, 185, 833.

DISCUSSION

WANDEL: To what extent could the obscuration of eraly quasars by their dust and gas envelopes lead to a color-evolution effect, affecting the observed luminosity function?

REES: It obviously could have such an effect. In so far as the observations set an upper limit to the z-dependence of the colours, one would have to postulate that the observation was an "all or nothing" effect: either the quasar is able to blow away or evaporate essentially all the dust, or it is completely obscured.

FILIPPENKO: In a radio galaxy whose extended lobes are a Mpc from the nucleus, the relativistic particles must somehow be reaccelerated along the jets (or they will never reach the lobes). What do you think is the most likeky mechanism for this?

REES: If the jets are initiated on scales $\ll 1$ pc, any initial random relativistic motions of electrons would have been quenched by radiative and/or adiabatic losses by the time the outflowing plasma reached the locations of observed radio components. (This holds for VLBI scales as well as the lobes of big double sources). The fact that radio emission is observed then indeed means, as you imply, that "in Situ" acceleration is occurring. This could be caused by encounters with obstacles, internal shocks, or reconnection of magnetic flux in shear layers. Whatever happens in supernova remmants, where velocities are 0.02 c could work with far higher efficiency when bulk velocities are relativistic.

NOVIKOV: You mentioned only a rotating black hole as an engine of AGN. Is there any other possibility?

REES: You are yourself much better qualified to answer this question than I am! I suspect that the electromagnetic extraction of a hole's spin energy is specially relevant to radio galaxies. Other AGNs could be powered by accretion, whether or not the holes are rotating. And some may not involve black holes at all, but merely one of the possible precursors.

SCHULZ: How much gas has to be present near a "starved" black hole to contain and collimate the jet produced by the extraction of the hole's rotational energy?

REES: The maximum power scales as B^2, and the gas around the hole merely has to be sufficient to confine and sustain this field. The gravitational binding energy of the surrounding gas must be at least as large as the electromagnetic energy.

POSTER PAPERS

MEGAMASERS IN NUCLEI OF GALAXIES

Willem A. Baan
Arecibo Observatory

Powerful maser emission has been detected in a small number of galaxies. Most of these have Seyfert or starburst nuclei and exhibit other signs of galactic activity. Broad OH emission has been seen in the prototype megamaser IC 4553 (= Arp 220) (Baan, Wood, and Haschick 1982; Baan and Haschick 1984; Norris et al. 1985), NGC 3690 and Mrk 231 (Baan 1985). Powerful H_2O emission has been observed in five galaxies (Baan 1985) of which NGC 3079 is the most luminous. Recently formaldehyde emission has been detected in IC 4553 and possibly in NGC 3079 (Baan, Güsten and Haschick 1985). The galaxies with maser emission are usually prominent infrared emitters. Some have perturbed disks with dust lanes and are showing evidence of bursts of star formation; others are part of a strongly interacting system.

No definite class of masing galaxies has yet emerged. However, most extragalactic maser emission can be interpreted in terms of an amplification model, where continuum radiation from a (centrallly located) background source is amplified by excited foreground molecular gas. Therefore the maser emission should be superposed exactly on the continuum structure; this has been verified for several galaxies. The saturation of the maser features will depend on the excitation state of the molecular material. Most of the masing galaxies show the characteristics which would allow such an amplification process to work (Baan 1985):
   a. a centrally located continuum source,
   b. a nearly edge-on molecular disk, thus increasing the probability of seeing inverted molecular clouds along the line-of-sight, and
   c. a powerful pump for exciting the molecular material within the disk.

There is ample infrared flux in the OH masing galaxies to excite the molecular masing regions (Baan 1985). The OH megamasers can run at a maser photon-to-infrared photon conversion efficiency of only a few percent. Such an efficiency is typical for unsaturated masers observed in our Galaxy. The velocity widths of the emission features (~ 300 km/s) indicate that a large fraction of the molecular structure is involved in the

masing process. Smaller pockets of molecular gas with a strong inversion
may produce narrow saturated features within the broad emission lines.

The infrared luminosity of the H_2O masing galaxies is sufficient
to pump the water molecules at a conversion efficiency of 50-100 percent
(Baan 1985). Such an efficiency is typical for saturated masers. However
it is plausible that collisional processes dominate in exciting the some H_2O
gas. The weaker and broader emission in the spectra of some H_2O sources is
likely due to more extended H_2O clouds. The narrow features superposed on
these broader structures are probably saturated and originate in highly
excited pockets of gas.

Formaldehyde emission at 6 cm has been seen in our Galaxy in SgrB2
and in the compact HII region NGC 7538. The emission observed in IC 4553 and
possibly in NGC 3079 is likely due to amplification of background continuum
by foreground inverted gas. Collisional processes cause the inversion of
the H_2CO molecules (Baan, Güsten, and Haschick 1985).

The spectral classification of the nuclei is slightly different
for the OH and H_2O masing galaxies. For the three OH masers the nuclear
classification is S and/or H. All three galaxies are edge-on having
strong HI absorption lines and in particular IC4553 has a very prominent
dust lane obscuring the nucleus (Schild 1985; Norris 1985). The existence
of dust and a low ionization or Seyfert spectrum can effectively conspire to
hide an HII type spectrum and the classification may be questionable.
However, for the OH amplification model it is not critical whether there is
a starburst or a Seyfert nucleus, as long as it provides a sufficient IR
flux to pump the molecular gas and both types of nuclei can do that.

The H_2O galaxies are not all exactly edge-on, especially NGC 1068
and the Circinus galaxy. The nuclei of these galaxies are classified as S
and/or L. However, for the H_2O galaxies the masing regions are not global
and concentric with the nucleus like in the OH galaxies. Pockets of excited
H_2O can be anywhere in the disk. Most of the prominent H_2O features are
redshifted or blueshifted relative to the systemic velocity of the galaxy.
It is plausible that H_2O masing occurs in shocked regions, which are
possibly forming stars and which are moving relative to the local disk gas.
Collisional processes can thus be the dominant pumping agents, since
collisional pumping is enhanced by the shocks.

The similarities and differences of the OH, H_2O, and H_2CO masers
allow some valuable insight into the physical conditions of molecular
material in active galaxies. The Seyfert or starburst nature of the host
galaxies appears to play an important role in the pumping of the molecular
gas. Furthermore the extragalactic masers appear to represent a new form of
astronomical "image processing". Unlike with gravitational lensing, in this
instance the processing is provided by diffuse clouds of interstellar
molecules.

ACKNOWLEDGEMENTS

I would like to thank Aubrey Haschick for his invaluable participation in obtaining the above described results. The Arecibo Observatory is part of the National Astronomy and Ionosphere Center, operated by Cornell University under contract with the National Science Foundation.

REFERENCES

Baan, W.A. 1985, Nature, 314, -.

Baan, W.A. and Haschick, A.D. 1985, Ap.J., 279, 541.

Baan, W.A., Güsten, R. and Haschick, A.D. 1985, preprint.

Baan, W.A., Wood, P.A.D. and Haschick, A.D. 1982, Ap.J. (Letters), 260, L49.

Norris, R.P. et al. 1984, Mon.Not.R.Astron.Soc., 213, 821.

Schild, R.E. 1985, Sky and Telescope, January, 24.

QUASAR CANDIDATES IN THE FIELD OF S.A. 94 (2h53m+0°20')

C. Barbieri (1), S. Cristiani (1, 2)
1) Istituto di Astronomia, Università di Padova, Vicolo dell' Osservatorio 5, I-35100, Italy.
2) European Southern Observatory, Casilla 19001, Santiago 19, Chile.

1. Introduction

Inspection by eye of objective-prism plates taken with Schmidt telescopes is a well established technique (Smith, 1978, 1981) to identify large numbers of quasar candidates. The combination of thin prism and IIIa-J emulsion is particularly effective in picking up QSOs in the redshift interval 1.8 to 3.5, namely when the strong Ly_α emission falls in the sensitivity range of the system. Quasars of lower redshift are also detected from the C IV 1549 Å, C III] 1909 Å, Mg II 2798 Å or by their ultraviolet continuum. Although selection effects, only partially alleviated by the adoption of automated searching techniques, strongly affect this kind of surveys, this method remains the most successful for discovering qusars. When complemented by different techniques, it can provide reasonable completeness (e.g. Cristiani, Véron-Cetty, Véron 1984).

Here we present the results obtained from the visual inspection of two UK-Schmidt plates covering a field centered on the SA 94. The search produced 208 candidates for 40 of which slit spectra were taken subsequently with the ESO telescopes at La Silla (4 were already known in the literature to be quasars). An effort was also made to put the magnitude scale on a firm basis.

The field was initially selected for several reasons: its high (-49°) Galactic latitude, its declination which allows observations both from northern and southern observatories, the presence in it of a good photometric calibration and, finally, the existence of various surveys, optical and radio, covering this area of the sky.

The centre of the field has been chosen to be coincident with the star N.° 7 (R.A. $2^h53^m22^s4$; dec. 0°18'59" (1950.0)) of Purgathofer's (1969) list.

2. The Search

The search for candidates has been carried out by both authors on the best objective-prism plate at our disposal, UJ 6752P, obtained with a seeing between 1 and 2 arcsec. The two preliminary lists of objects have then been merged. After a careful rediscussion of each entry and a control with direct and other objective-prism plates to discard all spurious cases due to overlaps or plate defects, we were finally left with a set of 208 candidates over the total area of 39 sq. deg. of the plate. Candidates that clearly appeared to be galaxies either on the prism or on the direct plate have been excluded.

In order to obtain magnitudes of the candidates on the direct plates, a photometric sequence has been obtained down to B and $V \simeq 21$, by means of photoelectric and CCD observations. To derive the B magnitude of the candidates in the central unvignetted 22.9 sq. deg. of the B7221 plate, scans of small areas containing the objects have been carried out with an Optronics S3000 machine. The digital output has then been transformed in an intensity scale using step wedges, and the final B magnitudes have been computed with the IHAP system of ESO Garching, using our photometric sequence as calibration. The resulting standard error has been estimated to be 0.15 magnitudes, down to $B = 19.7$.

3. Spectroscopy

The candidates have been classified in two ranks. Of the original sample of 208 objects, 74 fall in the upper, higher probability rank and 134 in the lower. Priority in the spectroscopic follow-up has been given to the candidates contained in the unvignetted area and, among those, to the first group, of which almost all the objects brighter than $B = 18.5$ and several fainter ones have been observed

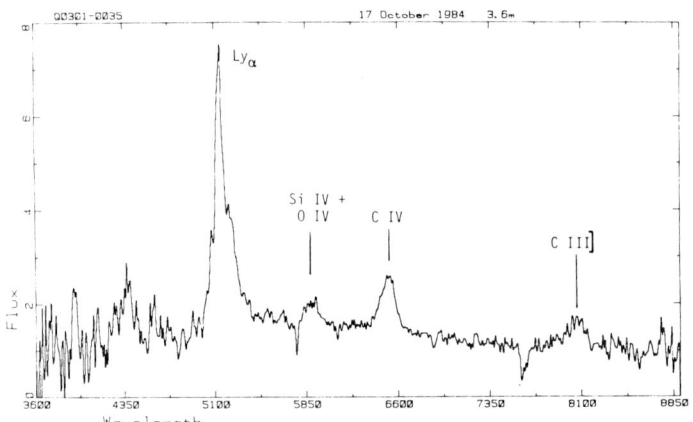

Figure 1: Spectrum of a confirmed candidate.

However, in order to have a better idea of the completeness of our sample, an effort has been spent in observing also part of the candidates of lower quality, at least down to $B = 18.4$. Some objects in the vignetted area also have been spectroscopically checked. Slit spectra have been obtained at ESO La Silla during 1983 and 1984 with the 3.6-m and the 2.2-m telescopes, using Boller and Chivens spectrographs with IDS, RPCS or CCD as detectors. The resolution of the spectra ranged between 8 and 20 Å.

Of the 44 candidates for which slit spectroscopy is available, 32 (24 quasaras or Seyfert 1 galaxies) pertain to class 1, and 12 (3 quasars or Seyfert 1) to class 2, corresponding to an overall $3/5$ success rate (75% for class 1, 25% for class 2). This confirmation rate seems lower than those quoted by other authors, for instance Osmer and Smith (1980) or Clowes and Savage (1983), which vary between 80 and 93% (however, Clowes and Savage sample of 30 objects checked with slit spectroscopy cannot be considered strictly representative of the entire sample of candidates). This discrepancy probably reflects our effort in selecting, for the sake of completeness, also objects with blue spectra rather than only emission-line candidates.

4. Comparison with other Observations

Some quasars were already known in the field of S.A. 94. Bolton and Wall (1970) first carried out optical identifications of the Parkes 2700-MHz survey in the ±4° declination zone and, later on, two of their objects were confirmed as quasars: PKS 0256-005 ($z = 1.998$, Peterson and Bolton 1972) and PKS 0300-004 ($z = 0.693$, Browne and Savage 1977). Savage and Wright (1981) added PKS 0245+013 ($z = 2.31$). Condon and Dressel (1978) reported a radio source, PKS 0241+011, in the spiral arms of the galaxy NGC 1073, which they identified with a blue stellar object (BSO), and tentatively concluded that this was a quasar. Arp and Sulentic (1979) confirmed that PKS 0241+011 is a quasar with a redshift of 1.400, and discovered two additional quasars, BSO 1 ($z = 1.941$) and BSO 2 ($z = 0.601$), again inside the spiral arms of the galaxy NGC 1073. In addition to this, Arp (1980) published another quasar close to NGC 1087, called U1 ($z = 2.147$). Finally, in a recent paper, Mitchell et al. (1984) three low-redshift, bright ($B < 17.2$) quasars: US 3150, US 3472, US 3605 and a Seyfert galaxy US 3498.

In our search, we have rediscovered PKS 0256-005, PKS 0300-004 and NGC 1087 U1. PKS 0245+013 has been missed: it looks very faint (cer $B > 19.5$) on the direct plates and almost invisible on the objective-prism one. Of the three quasars around NGC 1073, only one has been selected, BSO 2, the brightest ($B \simeq 18.9$), of course. The remaining two, in fact, are considerably fainter (BSO 1 $B \simeq 19.6$ and PKS 0241±011 $B \simeq 20$) and, in addition, the presence of the galaxy confuses the spectra of the nearby objects. The four bright objects reported by Mitchell et al. have been missed.

Luyten (1958) carried out a search for blue stars in the S.A. 94. Of the 83 objects found in our field, only one (Q0252+0118 = LB 2837, a Seyfert 1 with $z = 0.141$) has been selected by us. An "a posteriori" check of the 6 cases reported with $U - B < -0.3$ revealed 3 probable candidates, LB 180, LB 2772 and LB 2828, one marginally interesting, LB 2844, one unexceptional, LB 2850, and, finally, LB 2845, which appears like a normal star with no UV excess both on our prism plate and on the POSS prints. LB objects with $U - B > -0.3$ are, in general, unexceptional when checked on our objective-prism plate.

Part of our field overlaps also the PHL survey (Haro and Luyten

1962). Of the five objects, PHL 1447 (U - V = -0.5); PHL 4829, PHL 4291, PHL 4295 (-0.3 ≤ U - V ≤ -0.2) and PHL 8504 (U - V = 0.0), which are present on our plate, none has been selected.

A much more significant comparison can be established with the survey of Huang and Usher (1984) in the same field of S.A. 94. Huang and Usher found 677 candidates by means of a colour method, subdivided into three classes. Of those 677, 72 are contained in our list. Limiting-magnitude and selection criteria are different; therefore, in order to set a meaningful comparison, we can restrict the discussion to the candidates listed by Huang and Usher in their colour class 1 (where most quasars and white dwarfs fall), brighter than 19.0. In this subset, 40% of the objects selected by Huang and Usher are present also in our survey. On the other hand, of the 24 confirmed quasars or Seyfert 1 of our survey, which should be included in the Huang and Usher table, 16 have been picked up by them and 8 missed. Among those 8, however, 4 are QSOs with redshift larger than 2.5, undetectable by an UVX survey.

These results can be explained in terms of a bias towards emission-line objects still present in our survey in spite of our efforts and with the difficult evaluation of the properties of overexposed spectra on our plates. The great majority of the objects selected lie, in fact, in the range from $B = 17.5$ to $B = 19.5$. Fainter candidates can still be detected, but the lack of confidence in the classification prevents from any further work on them. However, for the bright ($B < 17.5$) objects, it well known that they are, in general, white dwarfs; a blue spectrum is, in this case, a very poor indicator of QSO nature. Further spectroscopical observations on December 1984 and January 1985 revealed, for instance, that LB 2772 (courtesy of B. Marano), LB 2811, LB 2813, LB 2847 and LB 2850 all have Balmer absorption lines at $z = 0$ (LB 2772 is clearly a WD).

5. Discussion

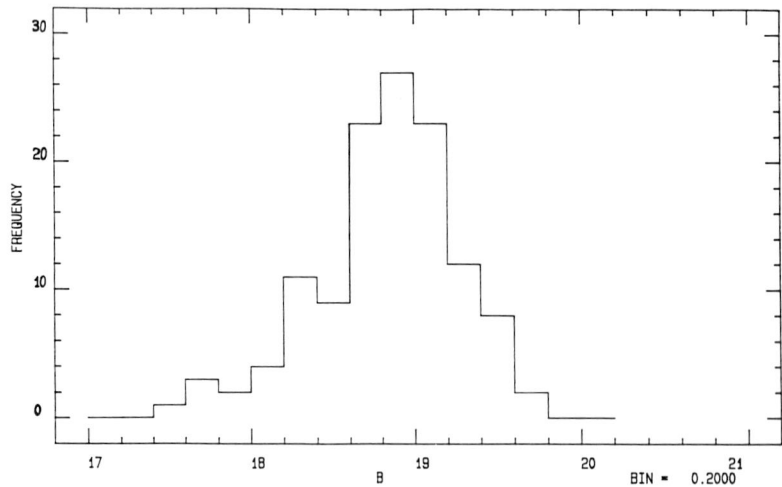

Fig. 2

Figure 2 shows the histogram number of candidates versus magnitude for the 126 objects contained in the 22.9 sq. deg. of the unvignetted area. The completeness limit with respect to magnitude seems to be placed, according to this histogram, around B = 18.8 to B = 19.0. It is also apparent that there is no sharp cut-off in the counts at fainter magnitudes. This is reasonable, since objects with strong emission lines can be still picked up considerably beyond the completeness limit.

Applying the confirmation rate observed in the sample of objects for which slit spectroscopy is available to the complete sample of candidates brighter than B = 19.0, we obtain an expected density of 1.6 quasars (Seyfert 1 excluded) per sq. deg. brighter than B = 19.0. A significative comparison can be carried out with the Clowes and Savage survey (1983): at a magnitude limit roughly corresponding to our B = 19.0 (0 < 2.5 log C < 1 in their notation, where C is a sort of continuum magnitude), they find a density of 1.1 candidates per sq. deg. Those figures should also be compared with an expected surface density of about 3 quasars per sq. deg. brighter than B = 19.0 and with z < 2.5 derived from UVX surveys.

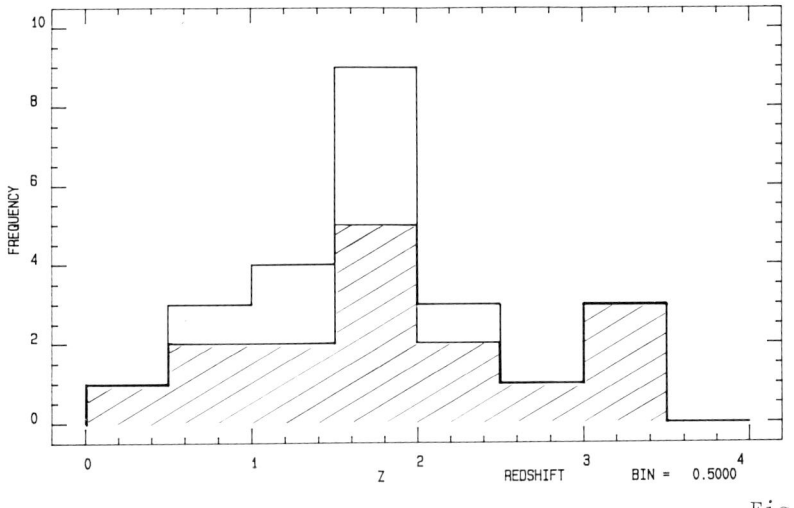

Fig. 3

Figure 3 shows the histogram number of objects versus redshift for all the confirmed quasars and restricted to those in the vignetted area and brighter than B = 19.0. This histogram can be compared again with the corresponding ones in Osmer and Smith (1980) and Clowes and Savage (1983). It is apparent that our distribution in redshift is considerably flatter, especially if we restrict the discussion to the quasars brighter than B = 19.0 located in the central unvignetted area.

Several parameters can determine these two results. In our opinion, the most important are: 1) UKSTU objective-prism surveys are more sensitive than the CTIO survey to smaller equivalent widths, as is already well known; 2) we have included in our survey also candidates with blue spectra in addition to the emission-line ones, while Osmer and Smith (1980) and Clowes and Savage (1983) selected only emission-line objects; 3) the histogram of figure 5 in its hatched part describes a sample truncated at the magnitude completeness limit; beyond this limit, the presence of Ly$_\alpha$ in the sensitivity range of the system becomes determinant for the visibility of the object on the objective-prisme plate and, therefore, a more preferential selection for redshifts between 1.8 and 3.5 takes place. In fact, in the overall sample of confirmed quasars of figure 3, which includes also objects fainter than 19.0, the peak in the bin $z = 1.5 \div 2.0$ is more prominent.

On the basis of these results and with the help of some observational material, which, in part, has been already accumulated and will be, hopefully, completed in the future, we are confident to refine our study of the S.A. 94 field. The use of complementary techniques (multicolour, variability) will take advantage of the considerable information resulting from this preliminary search to calibrate and control these alternative methods.

References

Arp, H., 1980: 9th Texas Symp., Munich (Ann. N. Y. Acad. Sci.) ed. J. Ehlers, J. J. Perry, M. Walker, p. 94.
Arp, H., Sulentic, J. W., 1979: Astrophys. J. 229, 496.
Bolton, J. B., Wall, J. V., 1970: Austr. J. Phys. 23, 789.
Browne, J. W. A., Savage, A., 1977: Mon. Not. R. Astron. Soc. 179, 65P.
Clowes, R. G., Savage, A., 1983: Mon. Not. R. Astron. Soc. 204, 365.
Condon, J. J., Dressel, L. L., 1978: Astrophys. J. 221, 456.
Cristiani, S., Véron-Cetty, M. P., Véron, P., 1984: Astron. Astrophys. 135, 122.
Haro, G., Luyten, W. J., 1962: Bol. Obs. Tonantzintla 3 (N° 22), 37.
Huang, K. L., Usher, P. D., 1984: Astrophys. J. Suppl. 56, 393.
Luyten, W. J., 1958: A Search for Faint Blue Stars - XVI Some Special Regions, Monnesota, Minneapolis.
Mitchell, K. J., Warnock, A. III, Usher, P. D., 1984: Astrophys. J. 287, L3.
Osmer, P. S., Smith, M. G., 1980: Astrophys. J. Suppl. 42, 333.
Peterson, B. A., Bolton, J. G., 1972: Astrophys. J. 173, L19.
Purgathofer, A. Th., 1969: Lowell Observatory Bull. 147, 98.
Savage, A., Wright, A. E., 1981: Mon. Not. R. Astron. Soc. 196, 927.
Smith, M. G., 1978: Vistas Astron. 22, 321.
Smith, M. G., 1981: in Investigating the Universe, ed. F. D. Kahn (Reidel. Dordrecht), p. 151.

IMPROVEMENT OF THE HUBBLE DIAGRAM FOR QSO'S LEADING TOWARDS q_o.

J.E. Beckman and M. Kidger,
Instituto de Astrofísica de Canarias
Universidad de La Laguna
Tenerife
Spain.

ABSTRACT. Using the predictions of the Lynden Bell/Hills model of the energy source of QSO's in which discrete packets of matter (e.g. stars) are converted into radiation near a massive black hole we show how the variability observations of QSO's over long periods may be used to upgrade the Hubble diagram for these objects. In a class of objects of low shot rate (see the article by Kidger and Beckman in these proceedings) extensive simulations using a shot noise model indicate that individual events may be distinguishable. This enables us to calibrate the relation between shot-rate and the variance of apparent luminosity and hence derive shot-rates for objects with measured light curves. From these shot rates absolute luminosities are in principle derivable, and here we indicate the observational difficulties still impeding progress in our path towards reliable values of q_o using this method.

PROCEDURE FOR SHOT RATE AND BOLOMETRIC LUMINOSITY.

With a purely stochastic model the light curve of a QSO should obey the relation

$$\lambda \, \sigma^2_L = k \tag{1}$$

where σ^2_L is the variance of the luminosity, λ is the average rate of light pulses (the "shot rate") and k is a constant   Fahlman and Ulrych (1975, 1976). Given measurements of both λ and σ^2_L we could derive k, and making the strong assumption that k is invariant for QSO's we can use (1) to infer λ for any QSO with measured σ^2_L. This assumption is equivalent to a statement that QSO luminosities differ principally due to the differences in the average density of the matter surrounding their central black holes. Rearranging equation (1) in logarithmic form:

$$\log k = 2 \, \text{Log} \, \sigma_L + \log \lambda \tag{2}$$

If the light curve of an object comprises a series of overlapping pulses of general form p(t) with time t, then

$$\sigma^2_L = \int_{t_1}^{t_2} p^2(t)\, dt \qquad (3)$$

where t_1 and t_2 are time limits to a given pulse.

To use (2) and (3) in the case of BL Lac (see article by Kidger and Beckman in these proceedings) we need as inputs z, the red-shift, m_B the mean apparent magnitude and measured values of λ and σ. However to obtain absolute luminosity we need to understand the conversion from measured blue magnitude to an implied bolometric magnitude, which for a QSO's or BL Lac object is a non-trivial exercise. We also need the "normal" k-correction and extinction corrections. For BL Lac itself, the extinction from within our galaxy is computed at 0.92 mag (see Tapia et al. 1976). The absolute luminosity of an object can be expressed as

$$L = 10^{0.4x} \qquad (4)$$

in units of solar luminosity, where

$$x = 5\log_{10}(c/H_o) + 5\log_{10} f(q_o, z) - m_B + k_{BOL} \qquad (5)$$

and

$$f(q_o, z) = \{q_o \cdot z + (q_o-1)[(1+2q_o \cdot z)^{1/2} - 1]\}/q_o^2 \qquad (6)$$

with H_o and q_o the standard cosmological parameters, and K_{BOL} the bolometric correction. The form of (5) and (6) derives from Mattig's treatment of a Robertson-Walker metric model.

In practice K_{BOL} presents the problem that QSO's exhibit highly non-thermal spectra, and differences from one object to another are considerable. This has led us to an extensive study of QSO continuum spectrophotometry (see paper by Kidger and Beckman in these proceedings). An overall result is that we need a "bandwidth correction" (a more appropriate term than bolometric correction, since it combines also a k correction) of between 1.5 and 4.0 magnitudes, depending on the class of active nucleus. We are now able to make classifications based on a sample of over 100 objects with comprehensively compiled continua.

Given a shot rate λ, the luminosity of an object is expressed as

$$L = \lambda \eta M_* c^2 \qquad (7)$$

where M_* is the average mass consumed per pulse of radiation, and η is an efficiency factor (see Hills, 1975 for a more detailed treatment). Typically we should expect M_* to be in the range 1 solar mass and η to lie between 0.1 and 0.4; although the present method cannot allow us to separate η and M_*, we do obtain via BL Lac a combined value which is

$$\eta M_* = 9.76 \times 10^{11} \qquad (8)$$

THE HUBBLE DIAGRAM FOR QSOs LEADING TOWARDS q_o

equivalent solar masses per pulse. In making the computation of absolute luminosity, we must relate the shot rate to the observed variance in luminosity ΔL (for which we need to apply equations (4) (5) and (6) to (2)) to obtain

$$\log \lambda = 2 \log \Delta L + \log k \tag{9}$$

We must also ensure that the relativistic correction for redshift is taken into account, via

$$\lambda_{rest} = (1 + z)\lambda \tag{10}$$

where λ_{rest} is the required shot rate in the rest frame of the source, and λ the rate measured on the earth; clearly (7) is then modified to yield

$$L = \lambda_{rest} \, \eta \, M_* \, c^2 \tag{7'}$$

which for values of z out to 3.5 makes this correction very important.

CALCULATION OF q_o.

The equation defining the Hubble diagram is

$$m-M = 5\log_{10}(c/H_o) + 5\log_{10}\{\frac{q_o z + (q_o-1)[(1+2q_o z)^{1/2} -1]}{q_o^2}\} \tag{11}$$

where the apparent and absolute magnitudes are either bolometric or of the same band (e.g blue). The problem presented in plotting

$$m - 5\log_{10} f(q_o, z) + const = M \tag{12}$$

is that although our calibration on BL Lac and on other relatively nearby objects allows M to be deduced from the variability of a QSO over long periods (periods greater than, say 50 pulse times, which may be between 3 years and 20 years depending on the object) m remains subject to the corrections outlined above, i.e. (1) Bandwidth correction (2) Extinction corrections both internal to the source and within the galaxy.

At the present stage of the work we can only produce an illustrative example. Using BL Lac, we insert three values of q_o and derive ηM_* for each. For $q_o = 0.001$, $f(q_o, z) = 0.079$; given $L = 2.7 \times 10^{12} L$ and measuring $\lambda = 2.12$ pulses per year we obtain

$$\eta M_* = 0.06 \, M \text{ per pulse}$$

for our values of q_o we obtain

q_o	0.001	0.5	1
ηM_* (M)	0.086	0.153	0.213

We could assume that $M_* = 1 M_\odot$, in which case $0.086 < \eta < 0.213$, and this agrees well with the theoretical range i.e. $0.057 < \eta < 0.42$ for the mass-radiation conversion factor via purely gravitational infall (Hills, 1975). So far none of our physical assumptions is ruled out on order of magnitude grounds.

Given that we can mark to derive a good bandwidth correction k band, and that the galactic extinction curve can be represented in the form $K_{gal} = \varkappa(b)$ where b is the galactic latitude of the QSO, we can re-express the basic Hubble equation (11) as

$$m_B - M_{BOL} - k_{band} = 5\log_{10}(c/H_0) + 5\log_{10} f(q_0 z) + \varkappa(b) \qquad (13)$$

In order to explore the range of q_0 we plot the regression line corresponding to equation (13) for different values of q_0, and the value which gives unit slope is the preferred outcome. It is also clear that the method does not depend on a numerical value for H_0, which determines only the intercept of the regression line. In this sense our "absolute" calibration via BL Lac is not needed, but has been useful in indicating that the numbers do in fact correspond to realistic models. We aim to bring QSO's within the scope of classical cosmological model tests by this technique.

REFERENCES.

Fahlman, G.G., Ulrych, T.J.: 1976, Ap.J. 201, 277
Fahlman, G.G., Ulrych, T.J.: 1976, Ap.J. 209, 553
Hills, J.G.: 1975, Nature 254, 295
Kidger, M.R., Beckman, J.E.: 1986, Three articles in these proceedings.
Tapia, S., Craine, E.R., Johnson, K.: 1976, Ap.J. 203, 291

LINERS: THE LOW LUMINOSITY END OF AGN

Luc Binette
European Southern Observatory
D-8046 Garching bei München
Federal Republic of Germany

ABSTRACT. Discussion of the evidence for photoionisation as the excitation mechanism in LINERs and its relationship to nuclear activity.

Recent surveys[1,2,3,4] of magnitude-limited samples have shown that most, if not all, "normal" galaxies present some sort of line emission in the nucleus. The significant fraction of galaxies characterised by low excitation lines such as [NII], [OII], [SII] and [OI] has been dubbed LINERs by Heckman[4] while the other major type of spectrum observed is HII region-like with the gas photoionised by young hot stars (Even in this case, the nuclear HII regions, which are intrinsically more luminous[1], can mask an underlying LINER spectrum).

Of special interest is the determination of the excitation mechanism in LINERs which were thought initially to be collisionally (Burbidge & Burbidge[5]) or shock excited (Heckman[4]). The major weakness of such interpretation, as shown by Keel[6], is that it does not reproduce well the correlation existing between different line ratios for a population of LINERs. (Shock models applied to individual objects can also give a very good fit if many parameters are optimised as illustrated in ref. 7 for NGC1052).

When photoionisation is taken as the excitation mechanism, however, these correlations can be qualitatively reproduced by varying only one parameter: U, the ionisation parameter which represents the density ratio of ionising photons to nucleons at the face of the ionised clouds. For instance, Ferland & Netzer[8,9] and Halpern & Steiner[10] showed that by varying U from 10^{-4} to 10^{-2}, one can reproduce the excitation sequence of LINERs up to the higher excitation spectra of Seyferts and QSOs which, they argued, supports the idea of a similar excitation mechanism in LINERs and AGN. Osterbrock & Dahari[11], Pagel[12], Péquignot[13], Filippenko & Halpern[14] and Binette[15] later confirmed that photoionisation accounts better for the spectral signature of LINERs. This is illustrated in Fig. 1 by two diagrams relatively unaffected by reddening or by the geometrical parameters and density of the ionised gas as discussed in ref. 15. (It should be emphasised that the LINERs with a density stratification as shown in ref. 14,20 are by no means typical). Fig. 1 shows that for a sample of emission galaxies not biased towards high luminosity lines (the one depicted is magnitude-limited), the nuclear emission regions have relatively **homogeneous spectral properties**. Inspection of the top diagram suggests that the objects' distribution is characterised by an "upper envelope" towards which the objects tend to concentrate. Filled symbols correspond to objects which fall within 0.17 dex of the empirically defined envelope and the open symbols are assigned to the more distant objects below it. The particular position of the latter can be explained by the optical thinness of the ionised clouds or, alternatively, by the superposition of nuclear HII regions on the LINER phenomenon[15]. These diagrams also indicate that the weak Seyferts discovered by Stauffer[16] in his magnitude-limited survey are an extension to LINERs along the excitation axis defined by OI/OIII.

A combination of shocks and photoionisation is, in principle, possible and has been proposed by Contini & Aldrovandi[17,18] and Binette et al.[19]. However, these models only fit well a subset of LINERs when the photoionisation component is dominant and, therefore, they correspond rather to a second order effect in the context of photoionised calculations.

If photoionisation is the dominant mechanism, what is the nature of the ionising source and what shape does the ionising spectrum take? Most models so far have opted for a non-thermal ionising spectrum by analogy with Seyfert nuclei where the UV is emitted by a single non-stellar object. However, in the few LINERs where a UV excess has been detected[14,20], simple extrapolation of the fitted optical power-law does not yield enough ionising photons to account for the observed line fluxes. This led Filippenko[14,20] to postulate the presence of a "UV bump" at higher energies in order to reconcile photoionisation models with observed fluxes. It is obvious that our knowledge of the UV spectral shape above the Rydberg limit is at best sketchy. In this context, Péquignot's[13] suggestion of a hot accreting disk which would emit as a blackbody (but with a tail at high energies) around 80,000K is quite relevant, since it would explain why the UV component contributes so little to the optical spectra. In the same vein, the photospheric model of 3C 273 by Camenzind & Courvoisier[22] can be scaled down to produce an equivalent UV spectrum.

One must also consider the possibility of hot stars being the source of the ionising continuum. Terlevich & Melnick[21] have suggested that clusters of hot massive stars (labelled Warmers and of the same type as the observed extreme WC or WO Wolf-Rayet stars) might be responsible for the ionisation and excitation of metal rich Giant HII Regions. A varying degree of gas excitation would result from the time evolution of these clusters and yield the low excitation spectra of LINERs. There are, however, weaknesses in this interpretation. For instance, given the short timescales involved[21], if these hot-star clusters are spatially distributed it is hard to imagine how bursts of star formation could operate in phase and give way to a smooth distribution in surface brightness. Alternatively, if these clusters of ionising stars are confined to the nucleus, how does one explain the large extension of the emitting gas in LINERs? One would expect the excited gas to be coextensive with the star forming region, defining a high surface brightness core which would then be limited to the nucleus.

The best evidence for LINERs being at the low luminosity end of AGN activity are the properties they have in common with Seyfert 1's, such as a broad component to the Balmer lines or a significant nuclear X-ray flux; these properties are all observed in a substantial fraction of LINERs[15]. This is illustrated in the diagrams of Fig. 1 where LINERs with broad Hα component show no particular trend compared to the rest of the sample but fit into the same correlation as the others. It appears unlikely that a completely different mechanism is responsible for the broad lines only. If such a mechanism operated, it should also contribute to the other lines and produce its characteristic spectral signature. It is therefore concluded that a common excitation mechanism is at work in most if not all LINERs and that LINERs belong to the same family as AGN but with scaled down properties, as suggested by Netzer[9].

Fig. 1.   Diagrams of [OIII]/(Hα/3) (ordinate is [OIII]/Hβ for models) and [NII]/[OIII] vs [OI]/[OIII]. Measurements are <u>not</u> corrected for reddening, long arrows represent correction for E(B-V)=0.66. Broken lines join multiple measurements of the same object. Continuous lines represent the locus of "integrated" models[15].

REFERENCES

(1) Keel, W.C.: 1983, Ap.J.Suppl. 52, 229. -- (2) Stauffer, J.R.: 1982, Ap.J.Suppl. 50, 517. -- (3) Phillips, M.M., Jenkins, C.J., Dopita, M.A., Sadler, E.M. & Binette, L.: 1985, Astron.J. (submitted). -- (4) Heckman, T.: 1980, A.&A. 87, 152. -- (5) Burbidge, E.M. & Burbidge, G.R.: 1962, Ap.J. 135, 694. -- (6) Keel, W.C.: 1983, Ap.J. 269, 466. -- (7) Fosbury, R.A.E., Mebold, U., Goss, W.M. & Dopita, M.A.: 1978, M.N.R.A.S. 183, 549. -- (8) Ferland, G.J. & Netzer, H.: 1983, Ap.J. 264, 105. -- (9) Netzer, H.: 1983, Proc. 24th Liège Astroph. Coll., Liège, June 21-14, p.398. -- (10) Halpern, J.P. & Steiner, J.E.: 1983, Ap.J. Lett. 269, L37. -- (11) Osterbrock, D.E. & Dahari, O.: 1983, Ap.J. 273, 478. -- (12) Pagel, B.E.J.: 1984, The Structure & Evol. of Normal Gal. Cambridge Univ. Press, p.211. -- (13) Péquignot, D.: 1984, A.&A., 131, 159. -- (14) Filippenko, A.V. & Halpern, J.P.: 1984, Ap.J. 285, 458 & 475. -- (15) Binette, L.: 1985, A.&A., 143, 334. -- (16) Stauffer, J.R.: 1982, Ap.J. 262, 66. -- (17) Contini, M. & Aldrovandi, S.M.V.: 1983, A.&A. 127, 15. -- (18) Aldrovandi, S.M.V. & Contini, M.: 1984, A.&A. 140, 368. -- (19) Binette, L., Dopita, M.A. & Tuohy, I.R.: 1985, Ap.J. (in press). -- (20) Filippenko, A.V.: 1985, Ap.J. 289, 489. -- (21) Terlevich, R. & Melnick, J.: 1985, M.N.R.A.S. 213, 841. -- (22) Camenzind, M. & Courvoisier, T.J.-L.: 1983, Ap.J. Lett. 266, L83.

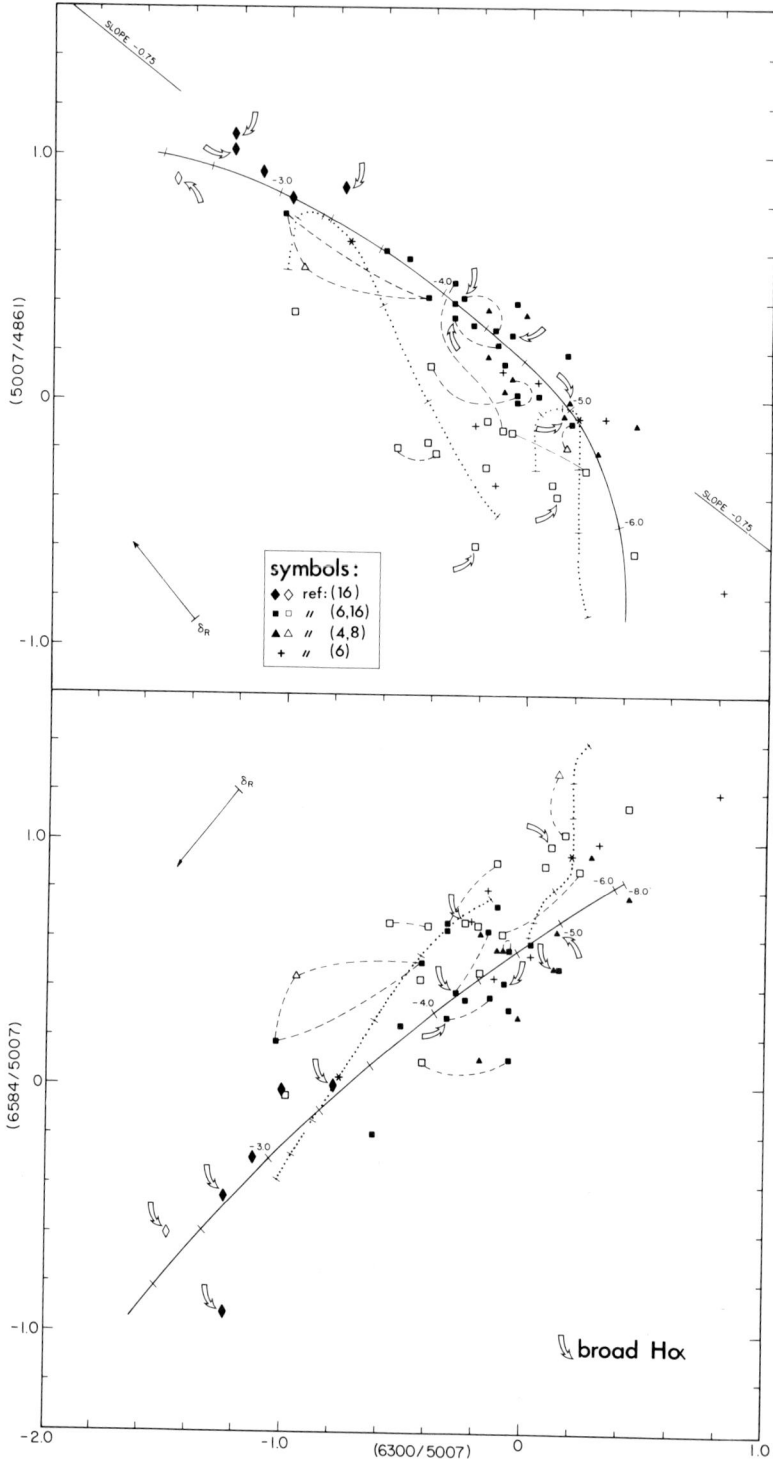

Fig. 1

OBSERVATIONS OF ACTIVE NUCLEI AT CALAR ALTO OBSERVATORY

K. Birkle[1], P. Rafanelli[2], U. Thiele[1]

1) Max-Planck-Institut für Astronomie, Heidelberg (DSAZ)
2) Asiago Astrophysical Observatory, University of Padova

1. INTRODUCTION

CCD images and photographic spectra of Seyfert galaxies have been recently obtained by the authors at the Cassegrain focus of the 2.2m, F/8 telescope of the Calar Alto Observatory (D.S.A.Z.). The main purposes of the observing programme were essentially three :
1) To obtain medium resolution spectra and high quality images of interacting active nuclei in order to study in detail their physical and morphological features and to verify whether nuclear activity can be related to some external disturbances.
2) To monitor some previously observed Seyfert-1 galaxies in order to detect changes in their emission line spectra.
3) To verify the spectral features of some objects remarkable within the Seyfert class for their peculiar line profiles and/or intensity line ratios.
The images have been obtained using the CCD Camera described by Marien and Rauh (1983), equipped with a RCA-CCD detector (type : SID 53 612-XO, sensitivity range: 350-1000 nm, dimensions 512x320 pixel, pixel dimensions 30x30 µ) which covers a sky field of 3x2 minutes of arc at the Cassegrain focus of the 2.2m telescope. Filters at our disposal were : Johnson B and V, OG570, RG780, RG830, GUNN v, g and r.
The spectroscopic observations have been performed using a B.&C. spectrograph in combination with a cooled, two stage, S20 ITT image intensifier. Gratings giving typical dispersions of 96 or 48 A/mm in front of the detector (IIaO plates) were normally used. The plates were calibrated by means of a calibration device, which uses a Lyot filter element and which has been described in detail by Trefzger and Solf (1978). The plates have been scanned at the PDS machine of the Padova Observatory and all data have been reduced using the IHAP programme. We present here in schematic form the preliminary results for three of the more than fifty observed objects, namely for: Mkn 975, NGC 1144, Mkn 1095 (Akn 120).

2. RESULTS
Mkn 975
Mkn 975 is a Seyfert-1 galaxy, the complex structure of which was first-

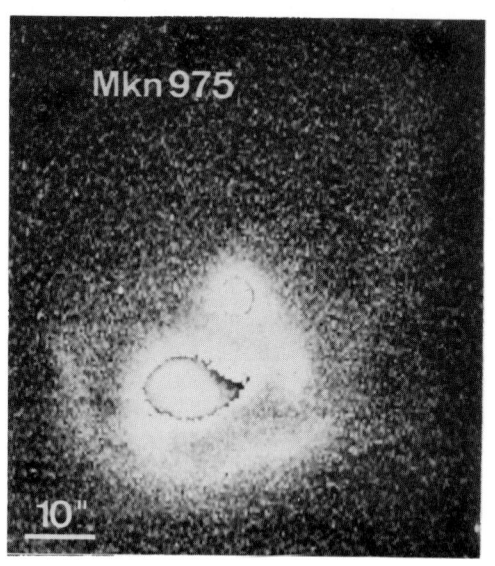 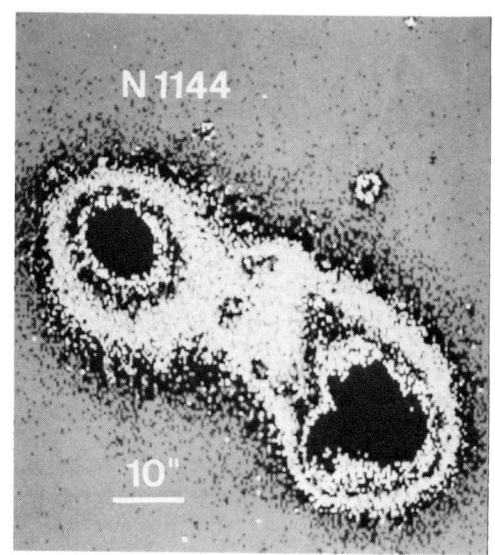

Fig.1 -

V-image of Mkn 975. North is to the top, east to the right.

Fig.2 -

V-image of NGC 1144. North is to the top, east to the right.

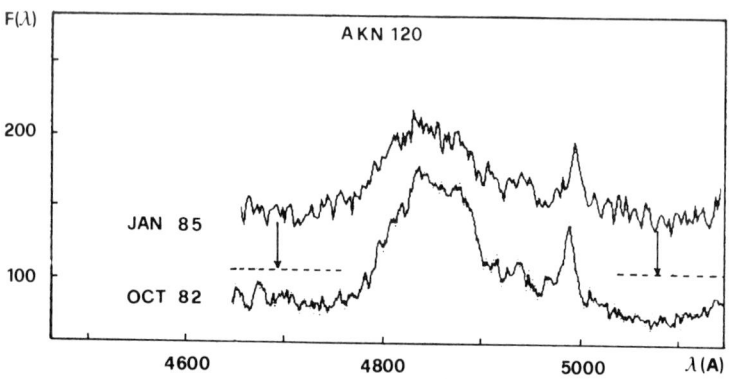

Fig.3 -

Intensity profiles of the H_β and $|OIII|$ lines of Akn 120 at two different dates. The dotted line represents the continuum level of the upper profile. Units in ordinate are 10^{-16} erg cm^{-2} s^{-1} A^{-1}.

ly noticed by its discoverer Markarian, who told also the presence of a 17^m stellar "satellite", 13" to northeast of the Seyfert nucleus. From the POSS prints it is quite doubtful whether this satellite is a foreground star or a galaxy.
Fig. 1 shows a 15^m, V exposure of Mkn 975. In this image the satellite appears embedded in a fuzzy structure, from which two faint oriented spiral arms depart. One extends to the north east direction, the other one to the south east towards a condensation at east of the Seyfert nucleus. A spectrum obtained putting the slit of the spectrograph along the two objects (P.A.=$30°$) does not show any sign of emission or absorption features on the continuum of the "satellite". Conversely the southern object shows a typical Seyfert-1 spectrum with emission lines well concentrated in the nuclear region. On the basis of these observations the satellite can be therefore a foreground or a background galaxy even if the system looks as an interacting pair of galaxies.

NGC 1144

NGC 1144 is a bright Seyfert-2 galaxy, discovered by Huchra et al.(1982). Its complex morphology is shown in Fig.2, where two nearby lying galaxies are clearly visible. Their angular separation is 40", which corresponds to 23 Kpc at the distance of the system d=116 Mpc, derived assuming for the Hubble constant the value H=75 km s^{-1} Mpc^{-1}. The southern component of the system is a spiral galaxy, the spectrum of which shows Seyfert-2 features, the other component is an elliptical galaxy and its spectrum does not show any emission line. In the intermediate region between the two nuclei, a kind of ring made up of a set of bright condensations with typical dimensions 4"-5" (1"=560 pc) is visible. The radiation emitted by these condensations may be of stellar or nebular origin according to their nature. In this case the first assumption is more likely since the spectra of the condensations show a faint continuum only, without traces of nebular emissions.

Akn 120

Akn 120 is a Seyfert-1 galaxy, well known as a highly variable source (e.g. : Peterson et al.,1983). During these last three years we have monitored it in order to reveal variations of the Hβ emission line profile, of its intensity and equivalent width. The flux and equivalent width of the Hβ feature have been measured by identifying the lowest points of the depressions between the FeII λ 4570 blend and Hβ and between Hβ and the FeII $\lambda\lambda$ 5190-5320 blend as continuum and interpolating between these points with a straight line (Peterson et al. 1983). The data at our disposal are two IDS spectra obtained on Oct 19,1982 and Oct 1, 1983 at the 1.5 m ESO telescope at La Silla Observatory and a photographic spectrum obtained at Calar Alto on Jan 1,1985. Tab 1 lists the measured H$_\beta$ equivalent widths and the corresponding H$_\beta$ intensities. The photographic spectrum (in the range around H$_\beta$) has been calibrated in intensity units (erg cm^{-2} s^{-1} A^{-1}) assuming that the flux in the |OIII| $\lambda\lambda$ 4959, 5007 lines is constant over several years. This assumption is used frequently (Phillips, 1978, Osterbrock and Shuder, 1982). It is clear from Tab. 1 and Fig. 3 that the equivalent width of H_β is significantly decreased during 1984 and that this variation is due to the decline of the line intensity and to the rising of the underlying continuum.-

Tab. 1

$W(H_\beta)$ A	$I(H_\beta)$ erg cm^{-2} s^{-1} A^{-1}	Date
140	$11.0 \cdot 10^{-13}$	Oct 19, 1982
130	$12.7 \cdot 10^{-13}$	Oct 1, 1983
90	$9.5 \cdot 10^{-13}$	Jan 1, 1985

REFERENCES

Huchra, J.P., Wyatt, W.F., Davis, M.: 1982, Astron. Journal, 87, 1628
Marien, K.H., Rauh, W.: 1983, Internal report of M.P.I. für Astronomie
Osterbrock, D.E., Shuder, J.M.: 1982, Astrophys. Journal Suppl., 49, 149
Peterson, B.M., Foltz, C.B., Miller, H.R., Wagner, R.M., Crenshaw, D.M., Meyers, K.A., Byard, P.L.: 1983, Astron. Journal, 88, 926
Phillips, M.M.: 1978, Astrophys. Journal Suppl., 38, 187
Trefzger, C., Solf, J.: 1978, Astron. Astrophys, 63, 131

FREQUENCY DEPENDENT POLARIZATION IN BLAZARS

C.-I. Björnsson
NORDITA
Blegdamsvej 17
DK-2100 Copenhagen Ø
Denmark

ABSTRACT. It is argued that the observed properties of the intrinsically frequency-dependent polarization (FDP) in blazars suggest a one- rather than a multi-component source model. Assuming a synchrotron origin of the emitted radiation, a well ordered magnetic field and a sharp break or cutoff in the electron distribution are both obligatory in such models; non-uniform pitch angle distribution or synchrotron losses enhance the resulting FDP. It is emphasized that the existence of such conditions in blazars finds observational support. Furthermore, a specific model is presented in order to illustrate the salient features of one-component models.

1. INTRODUCTION

Although both rare and transient, the evidence for intrinsic frequency-dependent polarization (FDP) in blazars is now secure. The optical continuum emission is thought to be optically thin synchrotron radiation. Since a powerlaw distribution of electron energies give no FDP, most attempts to interpret the observations have been in terms of two (or more) overlapping independent components, each of which with constant polarization. However, there are several observational arguments against such an interpretation.

(i) When the spectral flux distribution deviates from a powerlaw, the spectral index α (defined by $I(\nu) \propto \nu^{-\alpha}$) increases between the infrared and optical regimes (Cruz-Gonzalez and Huchra, 1984). This is just opposite the result expected from overlapping components.

(ii) In most cases, the spectral index does not change during flux variation (e.g. Rieke et al., 1977).

(iii) A minimum in the degree of polarization (P) is expected close to where the position angle (PA) changes most rapidly, implying that both increase and decrease of P with ν should be seen. However, observationally there is a strong tendency for P to increase with ν (Bailey, Hough, and Axon, 1983).

(iv) <u>Time</u> variations in flux does not lend itself, in a straightforward way, to an interpretation in terms of overlapping components (Impey et al., 1982, 1984).

2. PROPERTIES OF ONE-COMPONENT SOURCE MODELS

All of the above observational characteristics suggests a source model where the different components are intimately connected, i.e. a one-component model. In order to produce strong FDP, a one-component source model should have the following properties: (1) a well ordered magnetic field, (2) an electron distribution with a sharp break or cutoff, (3) non-uniform pitch angle (μ) distribution skewed towards smaller values of sin μ. A non-uniform pitch angle distribution will result in any of the following situations: (a) a spatially inhomogeneous electron distribution where the density of electrons correlates inversely with sin μ, (b) a non-uniform distribution of magnetic field direction within the source, (c) an anisotropic electron distribution.

Observational support for a source model with the above suggested properties include: (1) The most direct indication that a well ordered magnetic field exists in blazars is given by the high degree of polarization sometimes observed (e.g. Impey, Brand, and Tapia, 1982). (2) The large spectral curvature seen in several objects (Rieke, Lebofsky, and Wisniewski, 1982) implies both a uniform magnetic field and a sharp high energy break in the electron distribution. (3) The strong magnetic field needed in order to avoid an excessive amount of Compton scattered synchrotron radiation together with the upper limits on the density of a coexisting thermal plasma (from the lack of observed Faraday rotation) imply phase velocities for both Alfvén waves and whistlers close to c. Furthermore, due to the heating by the relativistic electrons, the velocity of sound in a thermal plasma is also expected to be close to c. Thus, isotropization of an initially anisotropic electron distribution by waves propagating in the thermal plasma should be small. Furthermore, due to the short radiative lifetimes for the relativistic electrons, synchrotron losses will produce anisotropic electron distributions skewed towards lower values of sin μ.

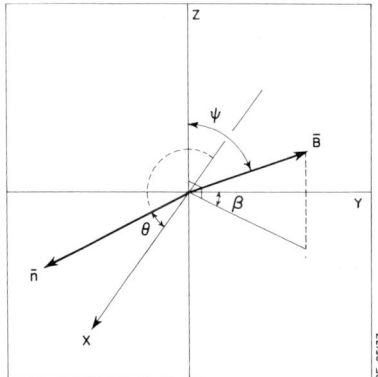

Figure 1. Geometry of the emission region. The direction to the observer n̄ lies in the X-Z plane, and makes an angle θ with the X-axis. The direction of the magnetic field B̄ is specified by the polar angle ψ and azimuthal angle β̄.

Using the coordinate system defined in fig. 1, fig. 2 shows the flux and polarization of a source model where the magnetic field $\bar{B}$ and the spatial distribution of electrons are described jointly by a probability density

$$H(B,\psi,\beta)dBd\psi d\beta \propto \delta(B-B_0)\delta(\psi - \tfrac{\pi}{2})\sin^{-n}\mu\, dBd\psi d\beta \quad -\tfrac{\pi}{2} \leq \beta \leq 0$$

$$= 0 \text{ otherwise.}$$

In the type of source model discussed here, the spectral flux distribution is determined mainly by the distribution of pitch angles (i.e. the magnetic field geometry) and not by electron energies. Thus, strong FDP can occur even without a significant steepening of the spectrum. Furthermore, since FDP is a reflection of the magnetic field geometry, a study of FDP provides a direct route to an understanding of the structure of the innermost regions in blazars and, maybe, also in quasars in general.

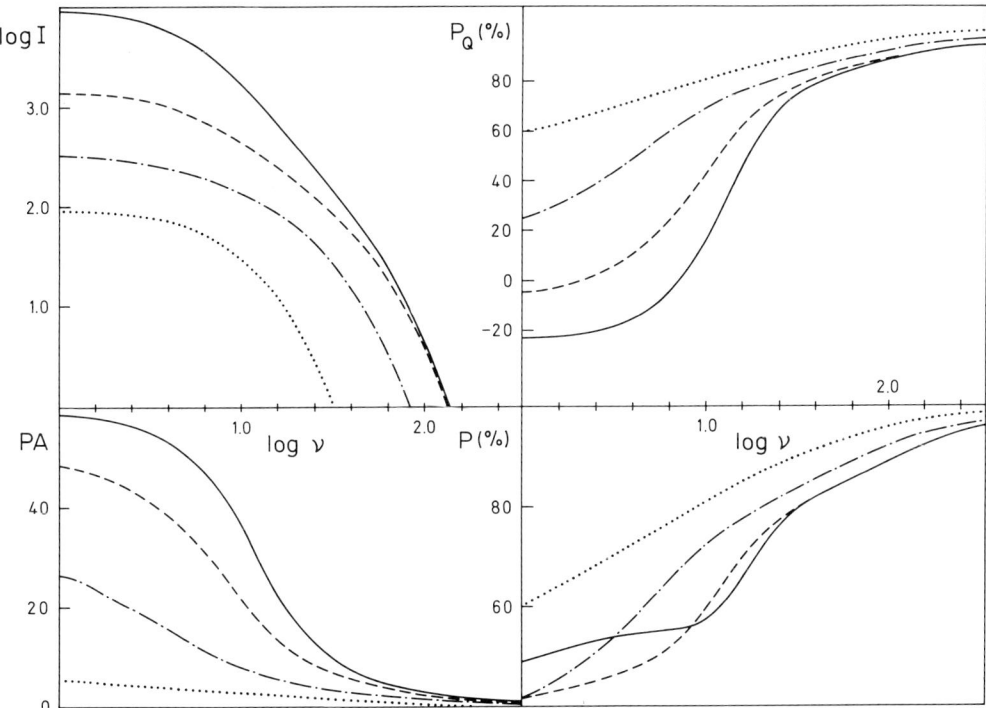

Figure 2. Intensity (I), position angle (PA), degree of polarization (P) and $P_Q \equiv P\cos 2\,PA$ (= P when $-\pi/2 \leq \beta \leq \pi/2$) are shown for different values of n with $\cos\theta = 0.9$. (i) n = 4 (solid line), (ii) n = 3 (dashed line), (iii) n = 2 (dashed-dotted line), (iv) n = 0 (dotted line). The frequency scale is shifted so that, for all n's, maximum intensity occurs at $\nu = 1$.

References

Baily, J., Hough, J.H. and Axon, D.J., 1983, M.N.R.A.S. 203, 339.
Cruz-Gonzalez, I. and Huchra, J.P., 1984, A.J. 89, 441.
Impey, C.D., Brand, P.W.J.L. and Tapia, S., 1982, M.N.R.A.S. 198, 1.
Impey, C.D., Brand, P.W.J.L., Wolstencroft, R.D. and Williams, P.M.,
    M.N.R.A.S. 200, 19; 1984, M.N.R.A.S. 209, 245.
Rieke, G.H., Lebofsky, M.J., Kemp, J.C., Coyne, G.V. and Tapia, S.,
    1977, Ap.J. 218, L37.
Rieke, G.H., Lebofsky, M.J. and Wisniewski, W.Z., 1982, Ap.J. 263, 73.

CCD PHOTOMETRY OF A SAMPLE OF SEYFERT GALAXIES: A PROGRESS REPORT.

C. Bonoli, F. Bortoletto
Osservatorio Astronomico e Istituto di Astronomia
Padova, Italy

F. Bònoli, F. Delpino, V. Zitelli
Osservatorio Astronomico e Istituto di Astronomia
Bologna, Italy

ABSTRACT. A project of surface photometry of a sample of Seyfert galaxies was undertaken with the CCD camera of Asiago Observatory, for the study of the properties both of the nuclei and of the underlying galaxies. Here we present some results from the first observing runs.

The properties of active galaxies have so far been studied mainly by means of spectroscopic investigations of the nuclei. However, a complete photometric study of the whole galaxy would be an important step toward the understanding of their behaviour.
This paper presents some results from an on-going project of CCD photometry of Seyfert galaxies. The final goal of our study is a detailed, calibrated photometric analysis of the images. This surface photometry will provide accurate determination of axial ratios, of nuclear and galactic luminosities and will show whether underlying galaxies actually behave as "normal" galaxies.
The objects so far observed are about 15 in number and are chosen from a homogeneous sample of Seyfert 1 and 1.5 galaxies, that was selected from the first nine Markarian lists in terms of nuclear and total luminosity, in order to avoid any possible source of incompleteness (Cheng et al., 1985).
The images were acquired over several observing runs during last months with the CCD camera attached to the 182cm telescope at Asiago Observatory. The silicon target is 580 x 430 pixels with a scale of 0".28 per pixel. The filter system is the Kitt Peak filter set. The exposure times were about 15 minutes.
A detailed description of the instrument can be found elsewhere (Bortoletto and D'Alessandro, 1985).
The reduction of a typical galaxy image consists of the following procedure. Dark exposures are subtracted from all the frames. Several flat field exposures, obtained observing the sky at dawn each morning, are used to flat-field correct the galaxy image.
Sky background on each frame is determined by making a histogram of all

pixels by data number and locating the peak value by a gaussian fit.
This value is then subtracted from each pixel.
Frames of a selected field in M67 cluster, obtained every night in each
colour, are treated identically with the galaxy pictures and are used
to calibrate the instrument photometry.
Details on the calibration procedure and the photometric accuracy are
given in Bonoli and Bortoletto (1985).
Figures 1a and 1b show a flat-field corrected image of the galaxy Mkn
766 taken in R filter and its isophotal contours respectively; the exposure time was 10 min. To give a rough estimate of the amount of the
signal of Figure 1a, the mean sky background is about 10 A.D. units and
the peak intensity is about 1000.
Figure 2a shows the galaxy Mkn 231, a very interesting object whose highly disturbed morphology suggests to be the remnant of a collisional
process; Figure 2b represents its isophotal map, evidencing some features at low signal level.
B and R isophotal maps of Mkn 79 are shown in Figures 2c and 2d. Figures 2e and 2f represent V and R frames of Mkn 374; the galaxy has two
close interacting nuclei: only the eastern component exhibits a Seyfert
spectrum.

References.
Bonoli C., Bortoletto F., 1985, in preparation.
Bortoletto F., D'Alessandro M., 1985, Rev. Sci. Instr. in press.
Cheng F.Z., Danese L., De Zotti G. and Franceschini A., 1985, Mon. Not.
R. Astr. Soc. 212, 857.

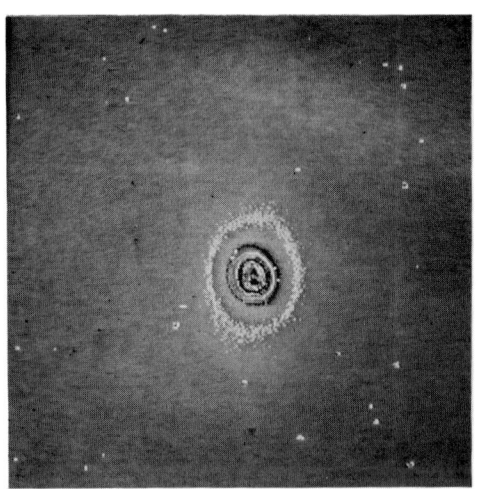
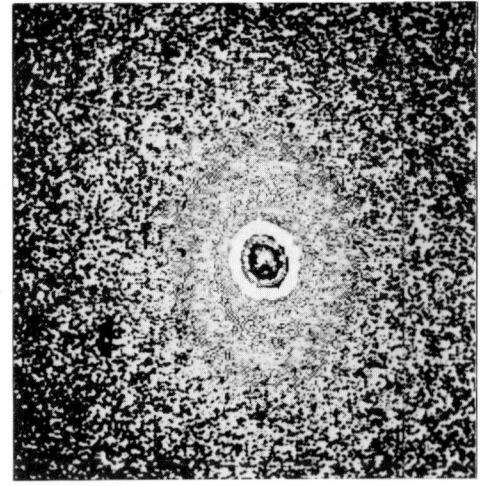

Figure 1 - a) Mkn 766 in R filter and b) its isophotal contours.

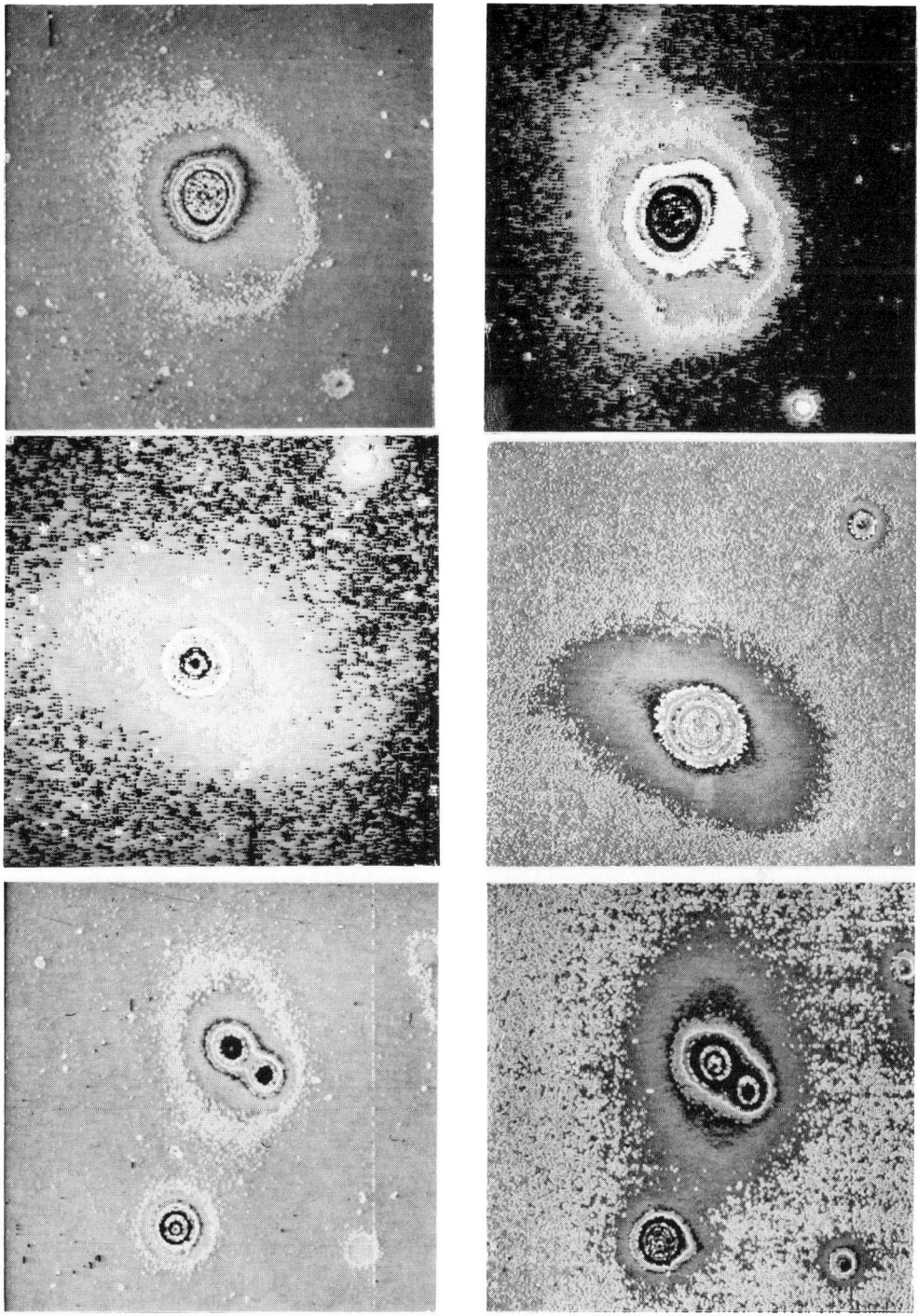

Figure 2 - a) V frame of Mkn 231 and b) its isophotal map; c) and d) B and R isophotal maps of Mkn 79; e) and f) V and R frames of Mkn 374.

A NEW FAINT QSO SURVEY

B.J. Boyle, R. Fong and T. Shanks,
Department of Physics, University of Durham, South Rd.,
Durham DH1 3LE, U.K.
B.A. Peterson,
Mount Stromlo & Siding Spring Observatories, Private Bag,
Woden, Canberra ACT 2606, Australia.

ABSTRACT. We are compiling a large catalogue of QSOs with complete spectroscopic identification. Candidate QSOs are selected from COSMOS machine measurements of UK Schmidt photographic plates using the ultra-violet excess (UVX) technique. These candidates are subsequently observed spectroscopically using the new 40 arc minute fibre optic coupler at the Anglo Australian Telescope (AAT) which allows a maximum of 50 objects to be observed simultaneously. To date we have obtained spectra for 400 (B<21) UVX objects of which approximately 170 are emission line objects. This represents a surface density of UVX QSOs at B<21 of 50 sq. deg^{-1}. This catalogue, which is free from selection effects up to a redshift of 2.2, will be used to study the QSO luminosity function, its evolution with redshift and the clustering of QSOs in the early Universe.

1. INTRODUCTION

It is well known that the vast majority of all low redshift (z<2.2) QSOs, whether selected by optical or radio techniques, exhibit an ultra-violet excess (UVX) with respect to normal galactic stars (see e.g. Veron 1983). It is not surprising, therefore, that many of the major QSO surveys of recent years e.g. Schmidt & Green (1983), Koo and Kron (1982) and Marshall et al. (1984) have searched for QSOs using UVX. Such surveys are limited by (a) the difficulty in producing complete UVX catalogues from photographic plates, and (b) the need for large amounts of telescope time to verify spectro-scopically the QSOs in the UVX sample. Recent developments in fast measuring machines and multi-object spectroscopy, however, now allow large complete QSO catalogues to be obtained without using up prohibitive amounts of telescope time. We describe here how, by using the COSMOS measuring machine and the fibre optic coupler at the Anglo-Australian Telescope (AAT), we are compiling a large UVX QSO catalogue containing, to date, spectroscopic identification for approximately 170 emission line objects.

2. THE UVX METHOD

In preparing our survey we first use COSMOS machine measurements of UK Schmidt Telescope U and B plates to yield positions and magnitudes for approximately 25000 stars in a typical Schmidt 5° x 5° field. Calibration of the stellar magnitudes is then achieved from comparison with faint stellar photoelectric sequences which lie on the measured area of each plate. Using the resulting U-B colours we can then define a sample of stellar images to be UVX, complete to a given B magnitude limit.

These UVX samples were first used in studies of QSO-galaxy clustering. Shanks et al. (1983) found that on a field centred at the South Galactic Pole the UVX sample was significantly anti-clustered with respect to galaxies. Recently the compilation of similar UVX catalogues on a further 7 high galactic latitude Schmidt fields have verified the anti-clustering result (Boyle et al., in preparation). To establish the proportion of QSOs in these UVX samples we carried out an initial survey of the brighter UVX objects using conventional spectroscopic techniques. This survey was limited to B=19 and discovered 13 QSOs from observations of 32 UVX objects in 3 nights on the AAT (Boyle et al. 1985). This 40% success rate agreed with the prediction made by Shanks et al. (1983) to explain the observed level of anti-clustering. Furthermore it proved that significant numbers of QSOs could be identified from COSMOS measurements.

Redshift surveys of QSOs are, of course, extremely important in their own right and so, with the success of the initial survey and the introduction of the fibre optic coupled aperture plate (FOCAP) system (see Gray 1983 and Figure 1) at the AAT, we were encouraged to embark on a major UVX QSO redshift survey using FOCAP. With FOCAP spectra can be obtained for up to 50 objects simultaneously over the full 0.3 sq.deg. field of view at Cassegrain focus of the AAT. We therefore selected samples of UVX objects in 40 arcminute diameter areas from the machine produced catalogues to be spectroscopically surveyed with FOCAP. The survey so far is made up of 10 such areas distributed over 6 UK Schmidt fields. The centres of Schmidt plates are given in table 1. Also listed in table 1 are the magnitude and colour criterion chosen on each field to define the UVX sample. Our magnitude limit of B=21 was decided upon to give approximately 40-50 UVX objects per 40 arcminute field. For all fields we chose our U-B limit to minimise the contamination by galactic stars while ensuring it was red enough to include all the UVX QSOs in our sample.

3. RESULTS

Using the FOCAP system we have, to date, obtained spectra for 400 (B<21) UVX objects in 4 nights. Approximately 150 of these objects are QSOs, a further 20 are narrow line objects. The majority of the other objects are white dwarfs or galactic stars scattered into our UVX sample by photometric errors. Less than 20% of the

QSOs have redshifts which are ambiguously defined, either because only weak emission lines are present in their spectra or they only exhibit one strong line with no other weaker lines being visible. More than 90% of the narrow line objects have accurate redshifts. The spectra for the QSOs and narrow line objects on a typical 40 arcminute area are shown in Figure 2.

We intend to use the survey to determine the luminosity function of QSOs and its evolution with redshift. Since the UVX method is complete up to redshifts of 2.2 our survey is ideally suited for such an analysis. As a preliminary result from the survey, we find that the surface density of UVX QSOs at B<21.0 is approximately 50 sq. deg.$^{-1}$ (see Figure 3). From the other points plotted (all from similar UVX surveys with complete spectroscopic identification) it is clear that the UVX QSO surface density undergoes a dramatic turnover at B=20, indicating that evolution of the QSO luminosity function cannot be parameterised by a pure density evolution law (see e.g. Marshall et al. 1984).

One of the most exciting uses for the survey will be in the analysis of QSO clustering. We intend to use 3 dimensional correlation analyses to look for clusters and superclusters of QSOs, thereby testing the homogeneity of the universe at large redshifts. We have already discovered individual examples of apparent QSO clusters, and we display the spectra for one such cluster in Figure 4. This group of four QSOs, separated by less than 10 arcminutes on the sky, have remarkably similar redshifts (1.95<z<2.03), two of the objects having an identical redshift of 2.003. These two objects have a spatial separation of only $2.1h^{-1}$ Mpc, while all the QSOs in the group are separated by less than $20h^{-1}$ Mpc, ($q_o=0.5$).

The survey will also be able to put strong limits on the space densities of broad absorption line QSOs and BL Lacertae objects, as well detecting other interesting (non QSO) objects such as distant HII regions and RR Lyraes.

4. CONCLUSIONS

Using the combination of the COSMOS measuring machine to produce catalogues of UVX objects and the FOCAP system at the AAT to observe these objects spectroscopically we are producing a large complete QSO catalogue at magnitudes where scant information on QSOs existed previously. So far, we have obtained spectra for approximately 170 emission line objects. When completed, the catalogue will be of fundamental importance in studying the clustering of QSOs and the evolution of the QSO luminosity function.

REFERENCES

Boyle, B.J., Shanks, T., Fong, R., and Clowes, R.G., 1985 MNRAS submitted.
Gray, P.M., 1983, Proc. SPIE, 445, 57.
Koo, D.C., and Kron, R.G., 1982, Astron. and Astrophys., 105, 107.

Marshall, H.L., Avni, Y., Braccesi, A., Huchra, J.P., Tannenbaum, H., Zamorani, G., and Zitelli, V., 1984, Ap. J. 283, 50.
Schmidt, M., and Green, R.F., 1983, Ap. J. 269, 352.
Shanks, T., Fong, R., Green, M.R., Clowes, R.G., and Savage, A., 1983, MNRAS 203, 103.
Veron, P., 1983, 24th Liege Symposium on Astrophysics: QSOs and Gravitational Lenses, P.210, Universite de Liege.

Figure 1. The FOCAP system. One of the aperture plates plugged up with the fibre optic bundle.

TABLE 1

Field	Centre RA	Dec	b^{11}	FOCAP Obs.	Mag Limit	UVX Limit
SGP	0h53	-28°03	-90°	2	B < 21.0	U-B < -0.30
QSF	3h44	-45°	-52°	2	B < 21.0	U-B < -0.20
QNB	10h40	0°	49°	2	B < 21.0	U-B < -0.20
QNY	12h30	0°	62°	2	B < 21.25	U-B < -0.25
QNA	13h40	0°	60°	1	B < 21.25	U-B < -0.20
QSM	22h04	-20°	-55°	1	B < 20.75	U-B < -0.20

A NEW FAINT QSO SURVEY

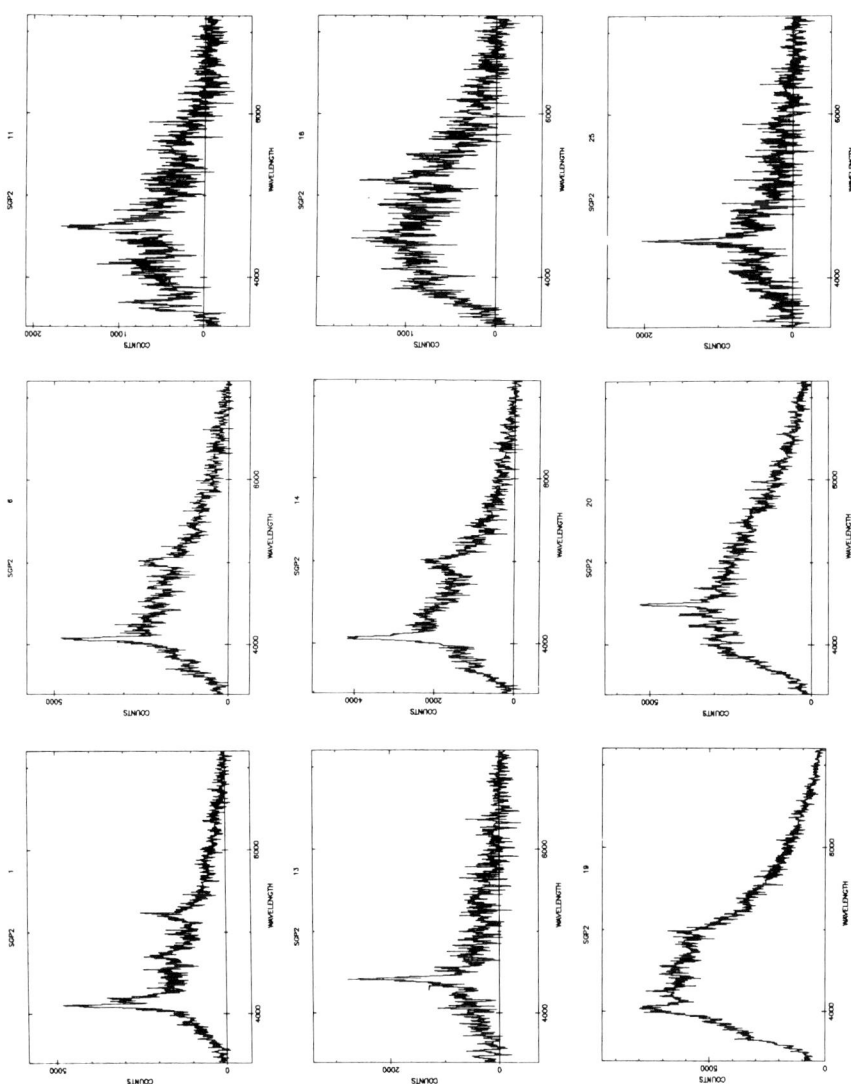

Figure 2. Spectra of 18 emission line objects discovered on one 40 arcminute FOCAP field centred near the SGP. The objects vary in magnitude from B=18.92 (object #47) to B=20.98 (object #11). Redshifts have been determined for all these objects and vary from 0.087 (object #38) to 2.377 (object #1). The integration time for the objects was 9000 secs.

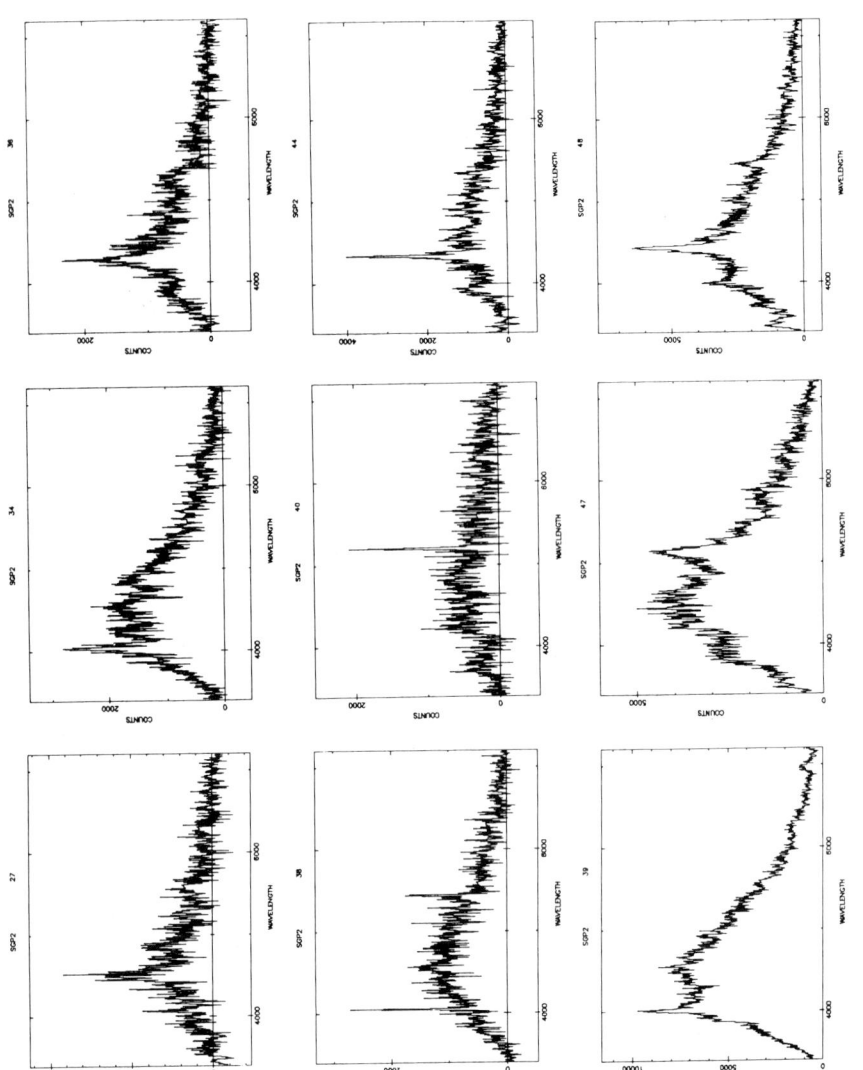

Fig. 2 (cont.)

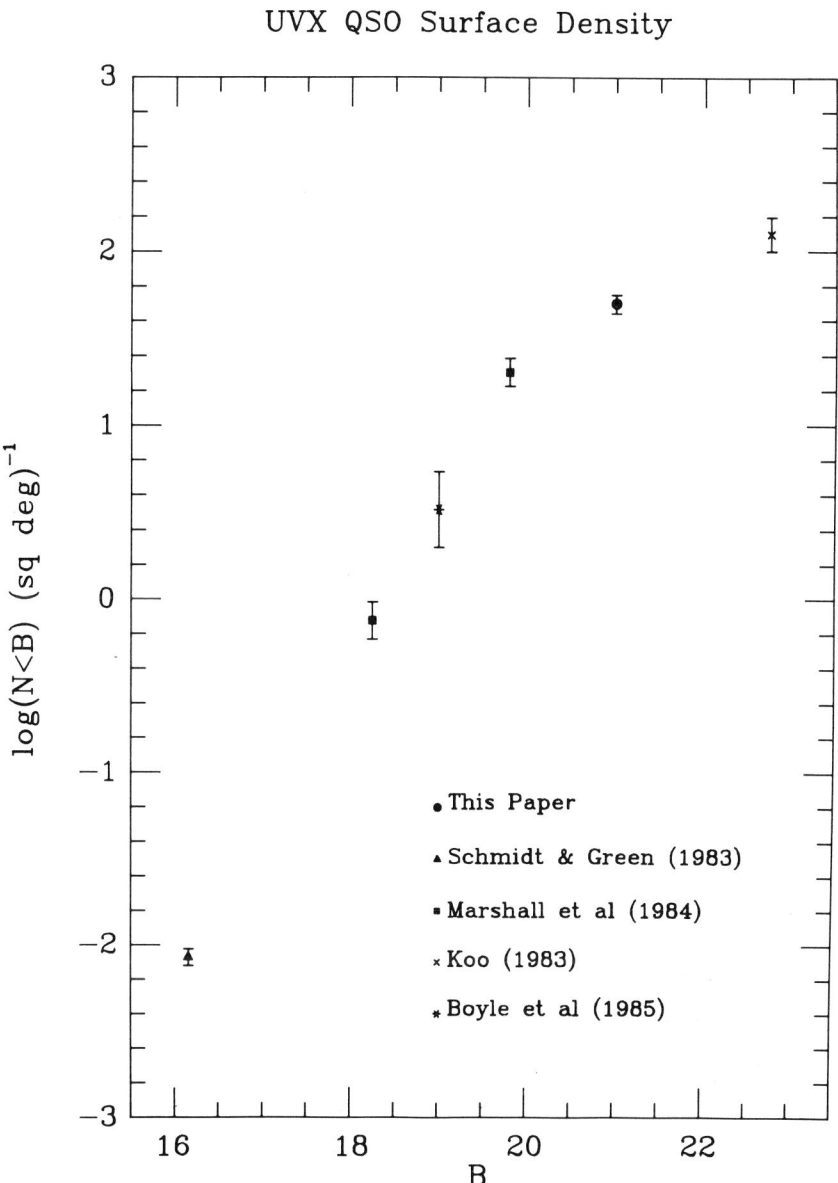

Figure 3. The N(m) Counts for samples of UVX QSOs with complete spectrosopic identification.

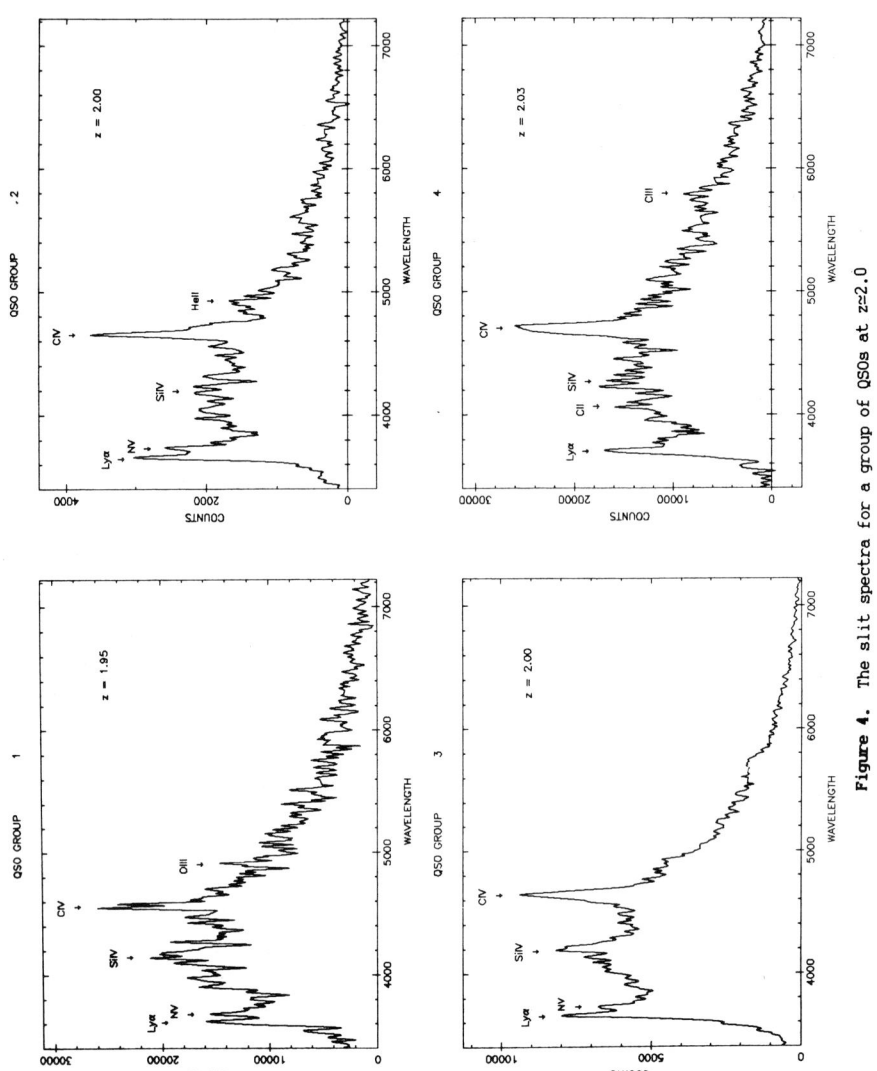

Figure 4. The slit spectra for a group of QSOs at $z \simeq 2.0$

THE RELATIVE FREQUENCY OF COMPACT EXTRAGALACTIC OBJECTS

P. Brosche
Observatorium Hoher List
Universitäts-Sternwarte Bonn
D-5568 Daun
F.R. Germany

ABSTRACT. Central regions of galaxies are seen as the product of a random agglomeration of clouds. The resulting distribution of angular momentum is a Maxwellian one. Small angular momentum is a necessary condition for compact cores. For a given mass in virial equilibrium, a small angular momentum means a fast rotation velocity. We assume the broad line width to be caused mainly by rotation.
A relation between the relative frequency and the angular momentum contrast of the central region is given (relative with respect to all galaxies of the same luminosity). The data for Seyfert 1 galaxies are not in conflict with this relation. The application to quasars seems possible.

1. INTRODUCTION

The observational stimulus for the present contribution was the fact that there is no stringent difference between the gross characteristics of galaxies with and without active nucleus. This could lead to the opinion that all galaxies are sometimes "switched on" and at other times "switched off". Here it is proposed that the difference is genuine and persistent but of random origin.

2. THE UNDERLYING SCENARIO

We proceed from the assumption that the central regions of galaxies grow - as the other parts - from random addition of independently moving entities, say clouds. Then the central limit theorem of probability theory tells us: the angular momentum of the central region will be Maxwellian distributed. The Maxwellian distribution has been found to be consistent with the distribution of the angular momenta of whole galaxies (Brosche 1977). This supports the application to central regions, since they are dynamically isolated in a first order approximation. If P_c is the angular momentum of the central region and P_o its most probable value (for the given class of galaxies), we have for the normalized distribution of $\xi = P_c/P_o$:

$$f(\xi) = (4/\sqrt{\pi}) \, \xi^2 \exp(-\xi^2) \tag{1}$$

For the case $\xi \ll 1$ the following approximation holds for the cumulative frequency

$$F(\xi) = \int_0^\xi f(\xi') d\xi':$$

$$F(\xi) \simeq (4/3\sqrt{\pi}) \xi^3 \tag{2}$$

Objects in mechanical equilibrium with mass M and rotational velocity V possess an angular momentum of the order

$$P \simeq G M^2/V \tag{3}$$

(G = constant of gravity).

For two central regions of equal mass, the one with velocity V_c, the other with average velocity V_o, we have

$$\xi \simeq P_c/P_o \simeq V_o/V_c \tag{4}$$

3. THE MANIFOLD OF GALAXIES

Normal galaxies reveal *two* independent dimensions with respect to the present overall accuracy of the observational parameters (Brosche 1973, Bujarrabal et al. 1981, Mebold and Reif 1981, Whitmore 1984 for spirals; Brosche and Lentes 1982 and 1983, Lentes 1983, Efstathiou and Fall 1984 for ellipticals). So if we fix two independent parameters, say mass and type, everything else if fixed, e.g. also the distribution of mass and angular momentum within a galaxy. Of course, there are random deviations if we go to a level where small numbers come into consideration. Observationally, luminosity may be choosen as variable instead of mass. The large variation in the total absolute magnitude of the galaxies we are dealing with (including the quasars!) justifies the neglect of type in most of our following considerations: It is tacitly assumed that quantities are averaged over type.

4. THE CONNECTION WITH OBSERVATIONS

First, we assume that nearly every compact nucleus (by measure of its angular momentum) is also an active nucleus, whereby the degree of activity could depend on time - smoothly and/or stochastic by the very nature of the addition of 'clouds'.

Furthermore, our most important assumption is the following: the bulk of the matter of the compact nucleus *rotates* in virial equilibrium with velocities given by the width of the broad lines. This assumption is at least a possible one, if not the most plausible one (Osterbrock, 1984). 'Rotation' should be seen in terms of the resulting angular momentum, so it includes the possibility of binary objects (Gaskell, 1983). Now we have to specify V_c and V_o. It looks as if any galaxy - at least from an average luminosity upwards - is 'active' in some sense if only the observations are sensitive enough. Here we refer to a certain lower limit of activity defined by the lower limit of the Seyfert 1 class of galaxies. For such galaxies a distinct lower limit of the observed broad lines width seems to exist (Dibaj, 1983): $v \simeq 1000$ km/s. We identify this with $2V \cdot \sin i$, i being the unknown angle between the nucleus' rotation axis and the vision radius. With $<\sin i> = \pi/4$ we arrive at $V_c \simeq 640$ km/s. The rotation curves of at least the more luminous spiral galaxies are essentially flat; hence it is not important at which radius we read off a $V_o \simeq 200$ km/s. Using these velocities, we obtain $\xi \simeq 0.3$ and a cumulative frequency $F \simeq 2\%$. Such a value seems quite consistent with the relative frequencies of Seyfert 1 galaxies of brighter luminosities in respect to all galaxies (Bhattacharia and Dibaj, 1983). Because of the observational uncertainties and of the hefty ξ-dependence of F this is not a strong argument. On the other hand, the existence of an acceptable (ξ, F)-pair is not a triviality. If, e.g., the bulk of the matter is a black hole with a maximum angular momentum $P = GM^2/c$, this value is so small that the predicted frequencies would fall beyond any observational limit! Therefore this interpretation is *not* consistent with our view.

5. POSSIBLE EXTENSION TOWARDS QUASARS

Let us assume that the activity limit in V_c is rather an universal constant because even quasars show typically V_c from ~ 1000 km/s upwards. The well known correlation between the rotation velocities and the luminosity of galaxies means that V_o is not a constant and therefore ξ *rises with increasing luminosity*. Consequently, we would predict an increasing frequency F with increasing absolute luminosity. We prefer a value $-\partial M_B/\partial \lg V = 5$ for the change of the absolute blue magnitude M_B with rotation velocity V (Bottinelli et al. 1983 and 1984). The step from average galaxies to quasars is about $\Delta M_B = 5^m$. If we use the gradient above for an extrapolation we find $\Delta \lg V \simeq 1$ or $V_o \simeq 2000$ km/s. That would mean that quasars are just the *average* galaxies for their luminosity, at least in terms of the angular momentum of their central region. The predicted F of the order of 50% or 100% are in agreement with the observations (Braccesi et al. 1980, Bhattacharia and Dibaj 1983).

REFERENCES

Bhattacharia, D., Dibaj, E.E., 1983: Astr. Tsirk. No.1263,1
Bottinelli, L., Gouguenheim,L., Paturel,G., de Vaucouleurs, G., 1983: Astron. Astrophys. 118, 4
Bottinelli, L., Gouguenheim,L., Paturel,G., de Vaucouleurs, G., 1984: Astrophys. J. 280,34
Braccesi, A., Zitelli, V., Bònoli, F., Formiggini, L., 1980: Astron. Astrophys. 85, 80
Brosche, P., 1973: Astron. Astrophys. 23, 259
Brosche, P., 1977: Astrophys. & Space Sci. 51, 401
Brosche, P., Lentes, F.Th., 1982: Mitt. Astron. Gesellschaft 55, 116
Brosche, P., Lentes, F.Th., 1983: in *Internal Kinematics and Dynamics of Galaxies*, ed. E. Athanassoula, IAU Symp.100 Reidel, Dordrecht, p. 377
Bujarrabal, V., Guibert, J., Balkowski, C., 1981: Astron. Astrophys. 104, 1
Dibaj, E.A., 1983: Pisma Astron. Zh. 9, 707
Efstathiou, G., Fall. S.M., 1984: Mon. Not. R. Astr. Soc. 206, 453
Gaskell, C.M., 1983: Proc. 24th Liège int. Astrophys. Coll. p. 473
Lentes, F.Th., 1983: in Proc. *Statistical Methods in Astronomy* Symp. Strasbourg, ed. E.F. Rolfe, Noordwijk, European Space Agency SP 201, p. 73
Mebold, U., Reif, K., 1981: Mitt. Astron. Gesellschaft, 51, 143
Osterbrock, D.E., 1984: Q.J.R. Astron. Soc. 25, 1
Whitmore, B.C., 1984: Astrophys. J. 278, 61

AN IDEALIZED MODEL FOR DUST ACCRETION IN ACTIVE GALACTIC NUCLEI.

J. Buitrago and E. Mediavilla
Instituto de Astrofísica de Canarias
University of La Laguna
Tenerife
Spain.

ABSTRACT. With an adequate supply of matter, the inwards drifting of dust particles induced by the Poynting-Robertson effect can lead to the formation of a shell or disk-like structure around the source of radiation. In a steady state, the Poynting-Robertson effect produce a spatial density of dust going like $1/r$ which in turn give rise to a mid-infrared spectrum $F_\nu \propto 1/\nu$. We think that this short report could be of interest for the interpretation of the infrared excess found in many extragalactic sources.

1. INTRODUCTION.

The Poynting-Robertson effect (PRE) has traditionally been studied under the stand-point of the dynamical consequences of the tangential acceleration (opposed to the direction of the motion) induced in the orbits of small bodies around a radiation source (Robertson 1937, Leinert et al. 1983, Buitrago and Mediavilla, 1985). There are, however, other interesting aspects (intimately related to the energy balance) which have been overlooked so far. These aspects are the principal motivation for the present work.

We shall consider "ideal" spherical dust particles, in local thermodynamic equilibrium, which absorb all the electromagnetic radiation incident from a point-like source, part of which is re-emitted isotropically in the rest frame of the particle.

It will be implicitly assumed that the cloud of dust or "PRE-driven" accreting system is optically thin. This seriously restricts the astrophysical situations in which our, rather idealized, model could be applied. But it is essential for a primary understanding of the physical problem at hand.

We make some tentative predictions about the luminosity generated by the PRE emission mechanism taking as reference, for comparison, the standard gas accretion disk model (which in what viscosity and output efficiency concerns is also idealized). We hope that our results, especially the infrared character of the emergent fulx, will help us, and other researches, in future works.

2. ENERGY TRANSFER

The basic conditions configurating the problem have been described by us elsewhere (see Buitrago and Mediavilla 1985). In the special relativistic case, the equation of motion for a dust particle of geometric cross-section A in a static radiation field is given by

$$\frac{dp^\mu}{dt} = A \cdot \rho \cdot (u_\beta \hat{r}^\beta) \left(\hat{r}^\mu - (\hat{r}^\beta u_\beta) u^\mu \right), \qquad (1)$$

where $\hat{r}^\mu$ ia a null vector having its origin in the source of radiation and with the spatial component pointing in the radial direction, u^μ is the 4-velocity of the particle and ρ is the energy density of the radiation field.

Selecting in (1) the time component and denoting the radial and transverse part of the particle's 3-velocity by $v_\shortparallel$ and $v_\perp$ respectively, the net transfer of energy between source and particle can be written

$$\left(\frac{dE}{dt} \right)_T = A\rho c \frac{(1 - v_\shortparallel/c)}{\left(1 - (v_\shortparallel^2/c^2 + v_\perp^2/c^2)\right)} \cdot \left(v_\shortparallel/c - v_\shortparallel^2/c^2 - v_\perp^2/c^2 \right) \qquad (2)$$

3. CIRCULAR ORBITS AND ACCRETION

Let us consider now the case of a shell, or disk shaped configuration, of dust around a luminosity source of mass M. For simplicity, let us consider only circular orbits. Due to the radiation drag, the individual particles lose energy. In the low velocity limit, the net energy transfer can be easily obtained from (2) giving

$$\left(\frac{dE}{dt} \right)_T = - A\rho c \frac{GM}{c^2 r} \qquad (3)$$

Equating (3) to the rate loss of total energy:

$$\frac{d}{dt} \left(-\frac{1}{2} \frac{GMm}{r} \right) = - A \frac{L_{ph}}{4\pi r^2} \frac{GM}{c^2 r} \qquad (4)$$

($\rho c = L_{ph}/4\pi r^2$, where L_{ph} is the luminosity of the source).

From this expression, it is inmediate to obtain the differential equation describing the continuous decrease in orbital radius:

$$\frac{dr}{dt} = - \frac{2\alpha}{r} \qquad (5)$$

(where $\alpha = \frac{AL_{ph}}{4\pi mc^2}$)

In the classical domain, the inward drift velocity (being an effect of second order in v/c) is small. However, provided there is an adequate supply of particles, it is this small effect which is responsible for the formation of a dust accreting system.

The PRE makes particles lose their energy mainly because they lose angular momentum (linear decrease in the classical case, exponential in the general relativistic regime. For details see Buitrago and Mediavilla 1985, op. cit.).

Let us start with a particle, in a circular orbit, far from the source where its total energy is practically mc^2. Since the PRE acts as a small perturbation, we can suppose that the circular character of the orbit is maintained (in a secular way) during the major part of the time in which the particle is drifting inwards towards the source of radiation.

Eventually, the journey ends once the last stable orbit at $r_{LSO} = \frac{6GM}{c^2}$ is reached. In this orbit, we have for the energy,

$$E_{LSO} = mc^2 \left(1 - \frac{2GM}{c^2 r_{LSO}}\right) \left(1 - \frac{1}{\frac{3GM}{c^2 r_{LSO}}}\right)^{1/2} \tag{6}$$

Therefore, the total energy released (radiated) is

$$E_{rad} = mc^2 - E_{LSO} = \eta mc^2 \tag{7}$$

($\eta = 0.057$ in the Schwarzschild Metric).

We thus arrive at the curious result that if we have a continuous supply of dust, in steady state, the expression giving the luminosity generated by the PRE-driven accretion mechanism, is exactly the same as in the case of the standard gas accretion disk model (e.g., Shakura and Dunyaev 1973), namely

$$L_D = \eta \cdot \dot{N}_T \, mc^2 \tag{8}$$

$\dot{N}_T$: total number of particles injected at the outer boundary per unit time.

Essentially, the main difference has its origin in the agent releasing gravitational energy. In the gas model, the luminosity is generated, internally, by viscous forces, while in the dust model we need an external support: the primary radiation source.

At first sight it might appear that dust accretion could be an efficient mechanism for converting gravitational energy in radiation. However, in any realistic astrophysical situation this is far to be the case. The grains cannot approach the source nearer than the distance of sublimation, which, in most cases, is far apart from the region where general relativistic effects are significant and where most of the gravitational energy is released. It could be said that equation (8) correspond to an idealized situation which seems very unlikely of being realized in practice.

To be more specific, for dust, the equivalent of the Eddington limit is given by

$$L_{cr} = 6.67 \times 10^{37} \left(\frac{M}{M_\odot} \right) a.\rho \qquad (9)$$

where a and ρ are the grain radius and density respectively. For $\rho = 2$ g.cm^{-3}, $L_{cr} \simeq L_{Edd} \cdot a$. If $M = 10^8 M_\odot$ and $a \leqslant 1 \mu m$, then $L_{cr} \sim 10^{42}$ erg.s^{-1}.

The last stable orbit is located at $\frac{6GM}{c^2} \simeq 9 \times 10^{13}$ cm, while the sublimation distance ($T \simeq 1800$ K) is 1.14×10^{16} cm, far enough from the relativistic region.

Let us consider now the consequences of the PRE on the radial distribution of dust particles. If we have a spherical unimodal population of dust particles of mass m. In steady state:

$$\dot{N}_T m = m.n(r).4\pi R^2 . \frac{dr}{dt} = \text{const} \qquad (10)$$

(n(r): spatial density).

From the last equation and (5) it is obvious that the spatial density would be

$$n(r) = \left(\dot{N}_T / 8 \pi \alpha \right) . \frac{1}{r} \propto \frac{1}{r} \qquad (11)$$

To single out a specific spatial distribution of dust could be of interest for the interpretation of the infrared spectrum of Seyferts. In the optically thin approximation, it can be shown (e.g. Rees et al. 1969) that for $n(r) \sim 1/r$, the mid-infrared spectrum F_ν goes as $1/\nu$. From the partially completed IRAS survey, there is a sample of more than 20 Seyferts with spectral indices: $-1.25 < \alpha < -0.5$ (Grijp et al. 1985). The turn over position in the spectrum of many AGN is also consistent with a $1/r$ distribution (provided we take into account the outer and inner boundaries of the shell -or disk- and the adequate temperature gradient). Perhaps the action of the PRE mark out the general trend which is forced to change due to transient enviromental effects.

As a final remark, we note that, incidentally, the distance of sublimation that we previously computed is about the same as the corresponding to the outer edge ($r_{out} = 2.10^3 M$) of Supercritical Thick Accretion Disk Models (e.g. Jaroszynski et al. 1980). Although the choice of the outer edge is somewhat arbitrary, thick accretion models predict luminosities in the equatorial plane several orders of magnitude lower than the corresponding to the rotation axis. This could be essential for the eventual existence (sheltered from the bulk of radiation) of dust clouds beyond the outer edge of the disk.

REFERENCES.

Buitrago, J. and Mediavilla, E.: 1985, Astrophys.Space Sci. **109**, 77
Grijp, M.H.P., Milley, G.K., Lub, J. and Jong, T.: 1985, Nature **Vol. 314**, 240
Jaroszynski, M., Abramowicz, M.A., and Paczynski, B.: 1980, Acta Astronómica, **Vol. 30**, 1.
Leinert, C., Röser, S., and Buitrago, J.: 1983, Astron. Astrophys.**118**, 345
Rees, M.J., Silk, J.I., Werner, N.W. and Wickramasinghe, N.C.: 1969, Nature, **Vol. 223**, 778.
Robertson, H.P.: 1937, Monthly Notices Roy. Astron. Soc. **97**, 423
Shakura, N.I., and Sunyaev, R.A.: 1973, Astron. Astrophys. **24**, 337

RAM PRESSURE CONFINEMENT OF THE EMISSION LINE GAS AND THE ASSOCIATED RADIO EMISSION IN AGN'S

T. J. Carroll
Department of Astrophysics
University of Oxford
Oxford, U.K.

ABSTRACT. The inferred high temperature and small size of the broad-line emitting clouds in quasars and Seyfert galaxies requires some confining mechanism to prevent rapid dissipation of those clouds in a time much shorter than their dynamical time scale. Ram pressure confinement by a supersonic wind provides such confinement while avoiding the mass budget and dynamical drag problems of a static confining medium. Reasonable limits on the mass outflow in the wind restricts the size of the high density BLR (broad-line region) to the canonical 1 pc. The limited range in observed ionization parameter also follows given that the mass accretion rate is similar to the mass outflow rate. Flat spectrum ($\alpha \sim 0.5$) radio emission from the BLR is also expected to be produced in the bow shocks around the confined clouds.

The gravitational binding energy per unit volume of a homogeneous spherical gas cloud of density $\rho_c = n_c m_H$ and radius R_c is $3GM\rho_c/5R_c = 4\pi G m_H^2 N_c^2/5$ with N_c a mean nucleonic cloud column density. This is to be compared with the thermal kinetic energy per unit volume $2n_c k T_c$. Photoionization models of the emission line gas (e.g. Kwan and Krolik 1981) give $N_c = 1 \times 10^{23}$ cm^{-2}, $n_c = 4 \times 10^9$ cm^{-3}, and $T_c = 2 \times 10^4$K yielding a ratio of binding to thermal energy of $2 \times 10^{-7} \ll 1$. Since clouds will freely expand on a time scale of $R_c/c_{sc} \simeq (N_c/n_c)(2m_H/kT_c)^{\frac{1}{2}} \sim 1$ year, they cannot be long lived unless a confining medium exists. Although there is no direct observational evidence that clouds must be long lived, the presence of large amounts of low density gas in the BLR due to expansion of the broad line clouds is ruled out by the absence of broad forbidden lines in the spectrum. The presence of a static confining medium in pressure balance with the line emitting clouds as envisaged by Krolik, McKee and Tarter (1981) has problems with mass budget (Perry and Dyson 1985) and with drag restricting the relative velocities of the clouds and confining medium to small values unless the medium is very hot (Weymann et al. 1982). We argue below for ram pressure confinement of the clouds by a supersonically outflowing wind; this mechanism solves the above problems and several other attractive features follow.

It follows from the similarity of the ionization parameters deduced for both the BLR and NLR from photoionization calculations (Carroll and Kwan 1983) that the confining pressure has a radial dependence r^{-2} to high accuracy. A spherically symmetric supersonic wind has ram pressure $\rho_w v_w^2 \propto r^{-2}$. A reasonable temperature for the wind is the Compton temperature $T_w \sim T_{IC} \sim 10^8 K$ (Krolik, McKee and Tarter 1981) with a corresponding sound speed of $c_{sw} \simeq 1300$ km·s^{-1}. The wind velocisty must be greater than this and also greater than the cloud speed for the ionization parameter to remain distance independent. A wind velocity $v_w \gtrsim 2 \times 10^4$ km· satisfies these constraints. Christiansen (1969) has discussed the structure of inertially confined clouds, while Blake's (1972) study of the stability may be inapplicable since the clouds are likely to be insulated by a supersonic evaporative outflow (Krolik, McKee and Tarter 1981).

The mass outflow in the wind is $\dot{M} = 4\pi r^2 \rho_w v_w = 4\pi r^2 (\rho_w v_w^2)/v_w$. The cloud pressure in the BLR is approximately 0.02 dynes·cm^{-2} which must be balanced by the wind ram pressure $\rho_w v_w^2$. Hence, $\dot{M} = 4.0 r_{18}^2/v_{w9}$ M$_\odot$·yr^{-1} with r and v_w in units of 10^{18} cm and 10^9 cm·s^{-1} respectively. It should be noted that the required mass outflow is reduced in inverse proportion to the wind velocity. The required mass loss is also very sensitive to the typical 'size' of the BLR, increasing as r_{BLR}^2. The restricted region of high density emission line gas around the continuum source could then be related to a limitation on the mass supply available to the wind. A BLR much larger than 1 pc would require exorbitant mass loss rates.

The drag associated with a supersonic wind can be very much smaller than that of a static confining medium. The stopping distance for a cloud due to drag is roughly the distance at which a cloud has swept up a column density equivalent to its own: $l_s \sim N_c/n_w$. The confining medium number density required to produce a pressure of 0.02 dynes·cm^{-2} is $n_w = 3.6 \times 10^5/T_{w8}$ cm^{-3} for a static medium with temperature $T_w = 10^8 K$ and $n_w = 1.2 \times 10^4/v_{w9}^2$ cm^{-3} for the wind. This yields a stopping distance of $l_s = 2.8 \times 10^{17} T_{w8}$ cm and $l_s = 8.3 \times 10^{18} v_{w9}^2$ cm respectively. While drag is seen to be significant over the size of the BLR for a static confining medium, its influence is much reduced in the case of a confining wind. Additionally, a stopped cloud will, in the static case, remain at high density. On the other hand, once the relative wind-cloud velocity is reduced to zero, the cloud will expand and reduce its density by a factor equal to the Mach number squared of the flow: $\rho_w v_w^2/\rho_w c_{sw}^2 = M^2 \simeq 240$. Since the emissivity per unit mass for line emission is proportional to n_c, these low density clouds effectively disappear.

An understanding of the limited observed range in ionization parameter $\Xi \equiv L_{ion}/4\pi r^2 n_c kT_c c \sim 0.3 - 2$ (Kwan and Krolik 1981) also follows. Taking the ionizing radiation luminosity to be some fraction ε_r of the accreted rest mass energy, $L_{ion} = \varepsilon_r \dot{M}_{in} c^2$ and $n_c = \dot{M}_{out} v_w/8\pi r^2 kT_c$ gives $\Xi = 2 \varepsilon_r (\dot{M}_{in}/\dot{M}_{out})(c/v_w)$. Given $\varepsilon_r = 0.1$, $v_w = 2 \times 10^4$ km·s^{-1} and equal mass flux in and out yields $\Xi = 3$. Higher wind velocities will reduce this value to a minimum of $\Xi = 0.2$ for $v_w = c$. Lower wind velocities would be less than the BLR cloud velocities thereby leading to differences in the BLR and NLR ionization parameters. Also, if the BLR velocities are determined by the gravity of a central mass (Kwan and Carroll 1982), then the wind, having at least the escape velocity, will naturally have an outflow velocity greater than the maximum velocity of the BLR clouds. An alternative view is to examine the relative efficiencies of accretion energy release and wind luminosity production. If $L_{wind} = \dot{M}_{out} v_w^2/2 = \mathcal{E}_w \dot{M}_{in} c^2$, then $\rightleftharpoons = (v_w/c)(\mathcal{E}_r/\mathcal{E}_w)$ so that $\mathcal{E}_r \gtrsim 3\mathcal{E}_w$ must hold to produce the correct range in $\rightleftharpoons$.

The radio emission associated with shocks around obstacles in a supersonic flow has been discussed by Blandford and Konigl (1979) among others. Roughly half the wind kinetic energy incident upon a cloud will be converted to internal energy behind the bow shock. Some fraction f_r of this will be radiated so that $L_{rad} = L_{wind} f f_r/2$ with f the total cloud covering factor. The mechanical wind luminosity is $L_{wind} = \dot{M}_w v_w^2/2 = 3.2 \times 10^{43} \dot{M}(M_\odot \cdot yr^{-1}) v_{w9}^2$ ergs·s^{-1}. Numerical work by Coleman (1984, 1986) has shown that the spectrum of the emission is a power law of slope 0.75 at Mach 3 decreasing to 0.58 at Mach 7. The Mach number for a $T_w = 10^8$ K and $v_w = 10^4$ km·s^{-1} wind is 7.8 so a fairly flat slope of ~0.5 is expected. The spectrum will break at a frequency $\nu_{BR} \simeq 0.3 \times 2.512 \times 10^{24} B_0^{-3} t_0^{-2}$ Hz with B_0 the preshock magnetic field strength and $t_0 = R_c/c_{sw}$ (Coleman 1984). For $R_c = N_c/n_c = 2.5 \times 10^{13}$ cm, a 10^8 K wind and a magnetic field $B_0 = 0.5$ G ($B_0^2/8\pi = 0.01$ dynes·cm^{-2}), this gives $\nu_{BR} \simeq 1.6 \times 10^{14}$ Hz. Since most of the cloud covering factor is concentrated in the BLR, this radio source is expected to be compact with size $r \sim 10^{18} L_{ion46}$ cm. Such a mechanism can also be responsible for the extended radio emission apparently associated with the forbidden line emission in Seyfert galaxies (Wilson 1981). Although the radio jet does not directly produce the optical line emission, the energy being provided instead by photoionization from the nucleus, the ram pressure will compress the clouds by some factor relative to clouds outside. The compressed clouds will emit more efficiently, per unit mass, by this same compression factor and so will appear brighter than material outside of the jet.

Thanks to Professor D.W. Sciama and the International School for Advanced Studies in Trieste where most of this work was completed.

REFERENCES

Blake, G.M. 1972, Mon.Not.R.astr.Soc., 156, 67.
Blandford, R.D. and Konigl, A. 1979, Astrophys. Letts., 20, 15.
Carroll, T.J. and Kwan, J. 1983, Astrophys.J., 274, 113.
Christiansen, W. 1969, Mon.Not.R.astr.Soc., 145, 327.
Coleman, C.S. 1984. Ph.D. Thesis, Australian National University.
Coleman, C.S. 1986, in preparation.
Krolik, J.H., McKee, C.F. and Tarter, C.B. 1981, Astrophys.J., 249, 422.
Kwan, J. and Carroll, T.J. 1982, Astrophys.J., 261, 25.
Kwan, J. and Krolik, J.H. 1981, Astrophys.J., 250. 478.
Perry, J.J. and Dyson, J.E. 1985, Mon.Not.R.astr.Soc., 213, 665.
Weymann, R.J., Scott, J.S., Schiano, A.V.R. and Christiansen, W.A. 1982, Astrophys.J., 262, 497.
Wilson, A.S. 1981, I.A.U. Symposium No.97, 179.

ULTRAVIOLET SPECTRA OF 3 SEYFERT 1 GALAXIES WITH OPTICAL FE II EMISSION LINES

F.-Z. Cheng[1]*, C.A. Grady[2]*, P.L. Selvelli[3]
[1] Center for Astrophysics, Cambridge, Ma. USA, University of Science and Technology of China
[2] Computer Sciences Corporation - USA
[3] Osservatorio Astronomico di Trieste - ITALY

SUMMARY. UV spectra of 3 Seyfert 1 galaxies, VII Zw 118, I Zw 1, and III Zw 2, which are known to have strong to moderate optical Fe II emissions are presented. The first IUE observations of VII Zw 118 show that its UV spectral index is about 1 as in the optical. The CIV FWHM and FW20%I are broader than those of Lyα. The variability of the UV line profile and/or continuum of I Zw 1 is not significant. Relatively strong emissions in the UV Fe II multiplets are confirmed. III Zw 2 is strongly variable in the UV band. During the period covered by IUE observations its continuum flux changed by a factor of approximately 4.5. The variable CIV flux is correlated with the continuum flux at 1450 $\overset{\circ}{A}$, thus supporting the photoionization model.

IUE observations of Sy I are very important for the interpretation of their radiation mechanisms, structure, evolution, and relation with QSOs. The three objects presented here are believed, on the basis of their optical continuum luminosities, to be intermediate between the classical Seyfert 1 nuclei and QSOs.

While the data of VII Zw 118 come from very recent IUE observations made in March 6, 1985 at GSFC, the data for I Zw 1 and III Zw 2 have been obtained from the IUE databank.

The data reduction for both the archival and observing program spectra are similar. All spectra were reduced using standard IUESIPS software (Turnrose and Thompson, 1984) to the form of slit integrated spectra which were absolutely calibrated using the data of Bohlin and Holm (1980) for the SWP and LWR cameras, and Cassatella and Harris (1982) for the LWP. All spectra were corrected for interstellar extinction using the data of Savage and Mathis (1979). For I Zw 1 correction was made for both the galactic extinction of E(B-V)=0.04 and the extinction estimated for the AGN host galaxy of E(B-V)=0.14

---

* Guest Observer, International Ultraviolet Explorer (IUE) Observatory, operated by Goddard Space Flight Center, National Aeronautics and Space Administration

(Cheng, Danese, and DeZotti 1983). A correction for the galactic extinction component only was made for III Zw 2 and VII Zw 118.

Following the procedure of Wu, Boggess, and Gull (1981) the observed continuum flux at 1450 Å in the Seyfert rest frame was derived by fitting the observed fluxes in continuum windows at 1345, 1450, and 1700 Å with a linear model. Due to uncertainties in the amount of emission line contamination no continuum windows were defined for the long wavelength spectra. A linear extrapolation of the short wavelength spectrum was used instead.

VII Zw 118

Our data are the first ultraviolet observations of VII Zw 118. Figure 1 shows the merged short wavelength spectrum of VII Zw 118. The only lines observed with sufficient signal to noise to permit analysis of the line profile shape are the Lyman alpha and C IV. Blended Si IV+O IV] emission is detected, although with insufficient signal to noise to permit quantitative analysis. The feature observed at the expected position of He II is a blend of two cosmic ray hits. The assumed linear continuum derived from a linear least squares fit to the continuum window data is also shown. Its continuum spectral index is about 1 which is almost the same as in the optical range. The LWP spectrum, where sufficient signal was detected is relatively flat and shows excess emission over that expected for the extrapolation of the linear continuum. Our data was of insufficiently high signal to noise to permit identification of the excess

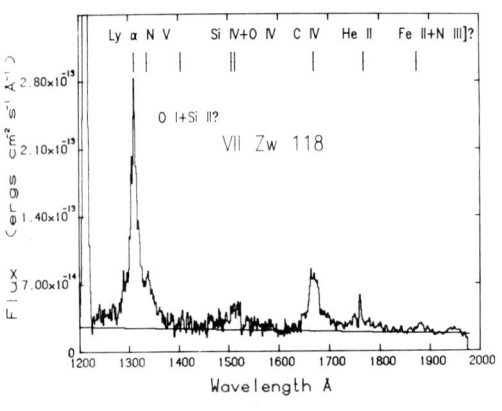

Figure 1. The merged short wavelength spectrum of VII Zw 118.

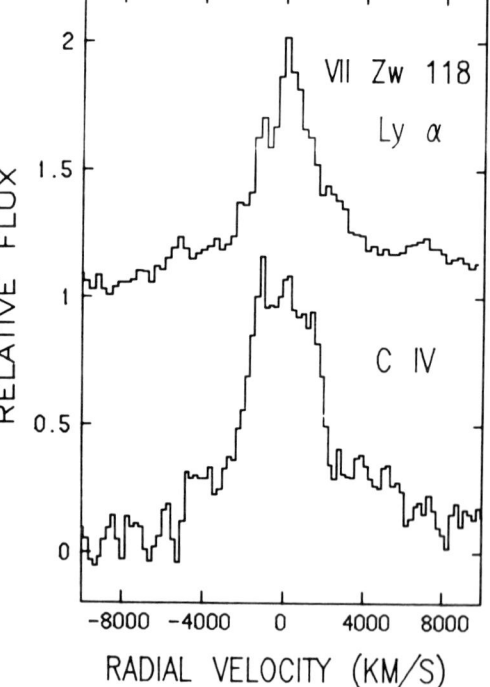

Figure 2. Lyman alpha and CIV line profiles for VII Zw 118.

flux over the linear continuum derived from the SWP data as Fe II emission or emission from a black body source, such as an accretion disk. Optical spectrophotometry for this object is presented in Kunth and Sargent (1979). They have reported a similar rise in flux near 3500 Å in the optical, and the LWP data may represent a continuation of the feature to shorter wavelengths.

Figure 2 presents the Lyman alpha and C IV line profiles with the linear continuum subtracted, and the flux normalized to the flux at the line peak. The data are presented on a radial velocity scale. The zero of the radial velocity scale for each line was determined by visual estimation of the line peak. Lyman alpha is blended with N V below the 20% intensity level in the red wing of the line profile. Inspection of Fig. 2 shows that Lyman alpha is much more sharply peaked than C IV, and that at the 50% intensity level C IV is broader than Lyman alpha.

I Zw 1

The spectra of I Zw 1 and III Zw 2 were inspected and intercompared for significant line profile and/or continuum variation. In the case of I Zw 1 no significant variation in the ultraviolet spectra was observed. All spectra were then co-added to produce a composite spectrum of higher signal to noise.

As noted by Penston, et. al. (1980), the spectrum of I Zw 1 is characterized by strong optical Fe II emission lines, and relatively strong emission in UV multiplets 1, 33, 60, 62, and 63 (Fig. 3). The long wavelength continuum of I Zw 1 can also be fit by a straight line. The measured data are listed in Table 1.

TABLE 1 - I Zw 1 Composite Spectrum

Line	Flux	FWHM
Lyman alpha	$1.18 \times 10^{**}-11$	3650
OI+Si II?	$3.82 \times 10^{**}-13$	
Si IV+OIV	$1.66 \times 10^{**}-12$	
C IV	$1.38 \times 10^{**}-12$	3150
Mg II	$1.39 \times 10^{**}-12$	2160

The following images were included in the composite spectrum:
LWR 1955, SWP 2216, SWP 2333,
SWP 5411, LWR 4673, SWP 5427,
SWP 5497, LWR 4765, LWR 6463,
LWR 6464, SWP 7460, SWP 7484,
LWR 6469, LWR 14626, and SWP 18557. Uncertainties for fluxes and radial velocities are similar to those for VII Zw 118.

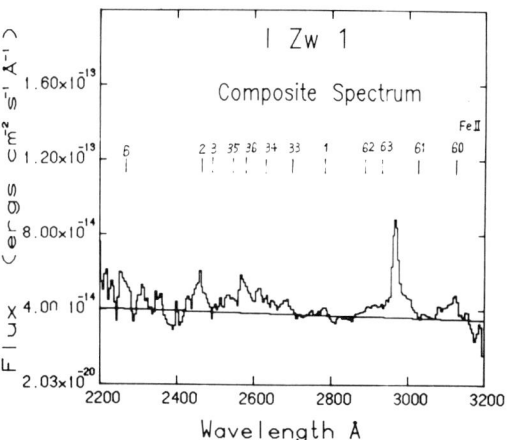

Figure 3. Composite spectrum for I Zw 1. Expanded scale plot of the long wavelength region showing the FeII and MgII emissions.

III Zw 2

III Zw 2 has been observed with IUE several times over the interval 1978-1984. Lyman alpha in many of the spectra was saturated, and MgII is typically of low signal to noise. Our analysis of the emission line spectrum is largely restricted to C IV. During the period covered by the IUE spectra, the continuum flux in III Zw 2 changed by a factor of approximately 4.5. In 1979 the continuum flux increased by a factor of 2.5 in three months. The timescale of the variability limits the size of the BLR to less than 0.1 pc. The C IV flux is also quite variable, and is higly correlated with the continuum flux at 1450Å. Such a correlation strongly supports the hypothesis that photoionization is the dominant source of ionization in the broad line region (BLR).

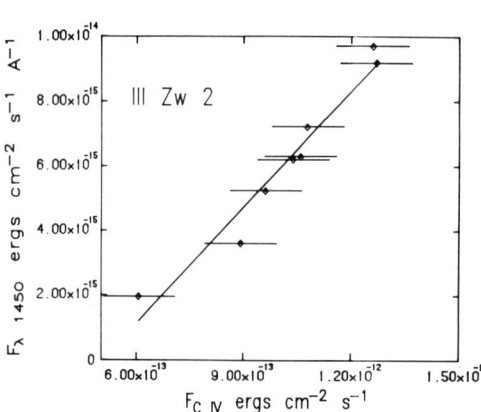

There is also a suggestion that the C IV profile full width at half maximum (FWHM) and possibly also full width at 10% of peak intensity increase with increasing continuum luminosity. Higher signal to noise spectral data will be required to confirm this suggestion. If true, however, such a correlation in line profile width with continuum luminosity suggests as noted by Ulrich, et al. (1984) that radiation pressure may be the dominant source of acceleration for the BLR clouds.

Figure 4. Continuum flux at 1450 A in III Zw 2's rest frame plotted against integrated C IV 1550 A flux.

ACKNOWLEDGMENTS. We wish to express appreciation to GSFC IUEOC for generously allocating the necessary IUE observing time for VII Zw 118. We want to thank Miss L. Abrami and Mr. M. Quartana for their help in the preparation of the manuscript.

REFERENCES

- Bohlin, R.C., and Holm, A.V. 1980, IUE NASA Newsletter 10, 37.
- Cassatella, A., and Harris, A.W. 1982, IUE ESA Newsletter 17, 12.
- Cheng, F.-Z., Danese, L., and DeZotti, G. 1983, MNRAS 204, 13P.
- Kunth, D., and Sargent, W.L.W. 1979, Astron. and Astroph. 76, 50.
- Penston, M.V., et al. 1980, NASA CP-2171, p.757.
- Savage, B.D., and Mathis, J.S. 1979, Ann. Rev. Astr. Astroph. 17, 73.
- Turnrose, B.E., and Thompson, R.W. 1984 IUE IPIM Version 2 (New Software)
- Ulrich, M.H., et al. 1984, Mon. Not. Roy. Astr. Soc. 206, 221.
- Wu, C.-C., Boggess, A., and Gull, T.R. 1981, Ap.J. 247, 449.

DISTRIBUTION OF QUASARS AND THE FORMATION OF LARGE SCALE STRUCTURE IN THE UNIVERSE

Yaoquan Chu & LiZhi Fang
Center for Astrophysics
University of Science & Technology of China
Hefei, Anhui, CHINA

ABSTRACT. Statistical evidence show that the distribution of quasars with $Z < 2$ is different from that of $Z > 2$, the former has large scale clustering while the later has not. All the results are in very good agreement with the developed new scenario of formation of large scale structure in a universe with several components of dark matter.

It has been shown by Fang et. al.(1984) that if the dark matter in the universe contains two components, the dominant consists of weak interaction particles with smaller rest mass (a few ev) and the nondominat consists of more weak interaction particles with larger rest mass (a few hundred ev), the scenario of clustering might be different from both standard isothermal and adiabatic ones. According to this new scenario, there should be two kinds of small scale objects. One is formed due to fragment of large scale objects (supercluster), another is formed before large scale objects form. The distribution of first kind objects should be rather inhomogeneous on the large scale, and the second kind objects should be more homogeneous. Accoding the results obtained by Frenk et. al. (1983), the large scale structure formed only at redshift small than 2 so we expect that the distribution of quasars should difference for $Z < 2$ and $Z > 2$.

Recently several wide field samples of quasars are comparatively available for test the about-mentioned prediction. We test the prediction on the difference distribution of quasars with $Z < 2$ and $Z > 2$, using the Savage-Bolton sample (1979), because which consists of two classes of quasars identified by different methods, i.e. the objective prism technique and the UV-B two colour methods, the redshifts of quasars in this sample spreads on more broad region than that of other surveies. It is then convenient to do the comparision between quasars of $Z<2$ and $Z>2$. Also it already has been investigated on the large scale cluster.(Chu & zhu, 1983). We applied the Nearest Neighbor Test (N.N.T.) to study the distribution of quasars in space. The first step is calculate the distance of each quasar to its nearest neighbor, using the formula as following:

$$D = R(t)\left\{r_j^2\sin^2\theta + \left[r_i(1+r_j^2)^{1/2} - r_j\cos\theta(1+r_i^2)^{1/2}\right]^2\right\}^{1/2} \quad (1)$$

where $R(t)$ is the cosmic scale factor. r_i and r_j are the dimensionless comoving radial coordinates of quasars i and j respectively, the relation between r and Z is:

$$r = \frac{Z(1+0.5Z)}{1+Z} \quad (2)$$

and the angle distance on the sky between the two quasars is θ. Then we compare the mean value of nearest neighbor distance $\langle D_c \rangle$ over the quasars sample to the mean value D_c^* obtained from a radom distribution sample which are deviation from Monte Carlo routine. If clustering exists in quasar sample, $\langle D_c \rangle$ should be smaller than D_c^*. As a measure of the statistical significance, one can define the following fuction

$$\delta = N^{1/2}\frac{\langle D_c \rangle - D_c^*}{\hat{\sigma}} \quad (3)$$

where N is the number of objects. $\hat{\sigma}$ is the standard deviation of D_c^*. The distribution of δ is asymptotically normal with mean 0 and variance being 1. Therefore, $1-P(\delta)$ is the probability that clustering is observed in the sample

The main results, which are showed in Table I. are that an apparent clustering at 95% significant level for the quasars sample of $Z<2$ in both fields of Savage-Bolton sample, i.e. ($02^h, -50°$) and ($22^h, -18°$), and there are no evidence of clustering for $Z>2$.

These results have been supported by other statistical

Table I: Nearest Neighbor Test

Redshift	N	Quasar data $\langle D_< \rangle$ (Mpc)	Monte Carlo data $D_<^*$ (Mpc)	$\hat{\sigma}$	δ	$1-P(\delta)$
\multicolumn{7}{c}{(02^h, $-50°$) field}
$Z < 2$	62	141.7	159.0	79.6	-1.72	97%
$Z > 2$	48	201.0	205.9	13.2	-0.40	66%
\multicolumn{7}{c}{(22^h, $-13°$) field}
$Z < 2$	57	146.7	165.8	77.9	-1.34	97%
$Z > 2$	26	207.1	193.0	75.9	0	--

investigation on the quasars distribution. The clustering of quasar in the sample of Shanks et. al.(1983) has been analysed by 2-dimensional correlation function method. They found that the UVX quasar candidate are clustered. It is well known that the redshifts of quasars identified by UVB color method are smaller than 2. Our conclusion on $Z > 2$ quasar is also the same as that given by Osmer (1981) and Webster (1982). They claimed no evidence for large scale clustering can be found from the CTIO optical quasar sample. Most of the redshifts of quasars in CTIO sample is of $Z > 2$, so it should not be clustering.

The investigation in this paper seems to show that the large scale structure (such as supercluster) formed at the epoch of about $Z \simeq 2$. Before the formation of supercluster some quasars have already come into being. Their distribution, therefore, is distinct from that of galaxies, most of which are probably formed by fragmentation of supercluster.

REFERENCES

Chu, Y. & Zhu, X. 1983. Ap. J., 267, 4.
Fang, L., Li, S.& Xiang, S., 1984, A. Ap. 140, 77.
Frenk, C.S., White, S.D. & Davis, M.,1983, Ap. J., 271, 41.
Osmer, P.S., 1981 Ap. J. 247, 761.
Savage, A. & Bolton, J.G., 1979, M. N., 188, 599
Shanks, T., Fong, R.,1983, M. N., 203, 131.
Webster, A., 1982, M. N., 199, 683.

RELATIVISTIC PARTICLE STREAMS IN AGN

C. S. Coleman
Department of Astrophysics
South Parks Road
Oxford OX1 3RQ
England

ABSTRACT. Under certain conditions a relativistic particle stream is stable against disruption by self-generated turbulence. Such streams could persist and propagate at relativistic velocity in regions of strongly magnetized plasma near the cores of active galactic nuclei (AGN). The anomolously high brightness temperatures and superluminal motions observed in many AGN can be understood in terms of synchrotron radiating streams of electrons and/or positrons, without requiring unrealistic magnetic field configurations or extreme degrees of alignment relative to the observer.

It has been suggested (Camezind, 1984; Coleman, 1984) that the superluminal effects exhibited by many high luminosity AGN result from streaming relativistic particles rather than from bulk plasma flow at high Lorentz factors. In addition to apparently superluminal velocities and anomolously high brightness temperatures, relativistic particle streams offer natural explanations for the circular polarization excess (Valtaoja, 1984) and frequency dependent polarization variations (Björnsson, these procedings) observed in blazars. Both of these effects are difficult to reconcile with models based on isotropic synchrotron emission from a relativistically expanding source.
  Most jet acceleration mechanisms, together with observations of large scale jets, suggest plasma flows with Lorentz factors $\gamma \lesssim 2$. If the active nucleus contains a highly ordered magnetic field, however, a relativistic particle stream may propagate along this field at about the Alfvèn velocity. An upper limit to the field strength at a given distance from the core is provided by the virial theorem (Begelman, Blandford and Rees, 1984), so the Alfvèn velocity at R parsecs from an $M_8 \times 10^8 M_\odot$ active nucleus in a region with number density $n_1 \times 10^1$ cm^{-3} is characterized by a maximum Lorentz factor $\gamma^{(A)} \simeq 62 M_8 n_1^{-\frac{1}{2}} R^{-2}$. Clearly then, by assuming realistic values for field strength and density, one may conclude that Alfvèn velocities can be highly relativistic within a few parsecs of an active galactic nucleus.

At present there exists no observational upper limit to the magnetic field strengths in AGN. It is conceivable, however, that shearing flows in accretion discs may generate fields of magnitude approaching the virial limit (Macdonald and Thorne, 1982). Furthermore, the inclusion of a component of thermal e^+/e^- pairs in the plasma, which may become significant at temperatures well below 10^{10}K, serves to increase the Alfvèn Lorentz factor and consequently to lessen the field strength requirement.

At this point it is necessary to examine the behaviour of relativistic particle streams in the exotic environment of a strongly magnetized plasma ($B^2 > 4\pi\rho c^2$), and in particular to determine if any instability is capable of confining them to propagation speeds significantly below the Alfvèn velocity. There are three instability mechanisms to be examined in this context, and each of these are now discussed separately.

(i) Electrostatic two-stream instability

Achterberg (1981) has shown the electrostatic two-stream instability to be ineffective, except for streams consisting of very relativistic particles with typical Lorentz factor $\gamma_0 \gg (n/n_r)^{1/3}$, where n_r is the number density of stream particles and $n \gg n_r$ by assumption. Scattering then redistributes particle pitch angles α within the narrow cone $\alpha \lesssim \gamma_0^{-1}(n_r/n)^{1/3} \ll 1$. A synchrotron radiating stream in an AGN is therfore virtually unaffected by this instability. (NB: the two-stream instability, being entirely longitudinal, is independent of any homogeneous magnetic field which may be present.)

(ii) Cyclotron wave scattering

The dispersion relations for magneto-hydrodynamic (MHD) waves in a low pressure plasma have been well studied (see eg. Akhiezer et al., Ch.V, 1975). A relativistic particle interacts with an MHD wave propagating at an angle θ to the field direction when the Cherenkov resonance condition $(\omega/k)/\cos(\theta) = c.\cos(\alpha)$ is satisfied. It may be readily demonstrated from the dispersion relations that the LHS of this equation, representing the wave phase velocity along the field direction, is a monotonically increasing function of θ at all frequencies. As a result, the minimum velocity to which streaming particles can be confined by this process is simply the phase velocity of the slowest parallel propagating waves.

Low frequency MHD waves propagate parallel to the field at the Alfvèn velocity, but as ω approaches ω_{Bp}, the proton cyclotron frequency, the wave phase velocity decreases sharply. These slowly propagating cyclotron waves are strongly damped by thermal protons in the background plasma (Akhiezer, Ch.V, 1975). The minimum phase velocity for which cyclotron damping is weak is highly relativistic in a sufficiently hot and strongly magnetized plasma, with Lorentz factor given approximately by:

$$\gamma_m = [3\sqrt{2}\beta_p \omega_{Bp}^2/\omega_P^2]^{1/3}$$

where β_p is the normalized thermal proton velocity and ω_p is the plasma frequency. Relativistic particle streams cannot be confined by cyclotron wave interactions to propagate more slowly than this. For a p^+/e^- plasma with temperature $T_{10} \times 10^{10}$K and the values of density and field strength described above, $\gamma_m \simeq 8.7 T_{10}^{1/6}$, a value which is weakly dependent upon plasma temperature, and typical of what is required to explain the superluminal velocities in most sources.

(iii) High order cyclotron resonances
High order wave/particle resonance scattering occurs when the wave frequency, Doppler shifted to the non-streaming frame, is close to an integer multiple of the particle cyclotron frequency. The phenomenon is a consequence of a particle's Larmor radius becoming comparable with the instability wavelength, and so is much more effective for protons than for positrons or electrons. The strength of order ℓ cyclotron resonance between electrons and Alfvèn waves is governed by the factor $J_\ell^{2/3}(\lambda)$ where J_ℓ denotes the Bessel function of order ℓ and $\lambda = (\omega/\omega_{Bp})\tan(\theta).(\gamma m_e/m_p)\sin(\alpha)$ (Akhiezer et al., Ch.VI, 1975). For $\gamma \leqslant m_p/m_e$, $\lambda << 1$ and all cyclotron resonances are exponentially inhibited. (The $\ell=0$ interaction, effective when $\lambda \to 0$, corresponds to Cherenkov resonance.) Synchrotron radiating streams of positrons and/or electrons are therefore unaffected by high order cyclotron resonance interactions. It is likely, however, that this process effectively prohibits the development of streams of protons or extremely relativistic ($\gamma >> m_p/m_e$) positrons/electrons.

CONCLUSIONS

The superluminal effects and anomolous polarization behaviour found in many AGN may be a consequence of relativistic particle anisotropies (streaming) rather than bulk relativistic flow. Particle streams propagate at high Lorentz factors in a hot, strongly magnetized plasma, and are stable against disruption by any self-generated linear instability mode.

REFERENCES

Achterberg, A. 1981, Astron. Astrophys. 98, 161.
Akhiezer, A.I., Akhiezer, I.A., Polovin, R.V., Sitenko, A.G. and Stepanov, K.N. 1975, 'Plasma Electrodynamics' Vol.I, Pergamon Press, Oxford.
Björnsson, C.I. 1986, 'Structure and Evolution of Active Galactic Nuclei', D.Reidel, Holland. (This volume)
Begelman, M.C., Blandford, R.D. and Rees, M.J. 1984, Rev. Mod. Phys. 56,181.
Camezind, M. 1984, Astron. Astrophys. Lett. 130,L9.
Coleman, C.S. 1984, 'X-ray and UV Emission from Active Galactic Nuclei', p203, ed. Brinkmann, W. and Trumper, J. M.P.E. Report 184.
Macdonald, D. and Thorne, K.S. 1982, Mon. Not. R. astr. Soc., 198,345.
Valtaoja, E. 1984, Astrophys. Spa. Sci., 100,227.

THE BEHAVIOUR OF THE CIV EMISSION LINE STRENGTH IN
VARIABLE SEYFERT I GALAXIES.

L. Colina
Space Astrophysics Division,
SSD-ESTEC, Noordwijk, Holland.

W. Wamsteker
ESA IUE Observatory,Madrid,Spain,
Affiliated with Space Sciences Dept.,

ABSTRACT.The variation of the CIV 1550A line strength in variable Seyfert I galaxies based on low resolution IUE data is discussed. The results suggest that the Baldwin-relation might not be a reliable distance indicator for QSO's.Seyfert I galaxies vary in a way indicating that the emission line gas is matter bounded when in a high state. For F-9 the transition from photon-bounded to matter bounded occurs at $\log(L(1350)) \approx 29.7$ erg/sec/Hz .

1. INTRODUCTION

The "Baldwin relation",which combines the equivalent width of the CIV (1550 A) emission with the continuum Luminosity (Ref.1), is important as a possible distance calibrator for QSO's. Various studies were made, analyzing different QSO samples (Ref.2,3,4,5). These results have also been compared with the behaviour of Seyfert I galaxies, which possibly represent the same phenomena at lower luminosities (Ref.6,7,8). Since all these studies have been made on statistical samples, only indirect arguments can be used to establish the physical reality of this relation . We use here the large variability of individual Seyfert I galaxies in the hope to find a better understanding of this relation. The very large continuum variation, found for the high-luminosity Seyfert I galaxy Fairall 9 (Ref.10,11) is, in this context, especially interesting.

2. RESULTS

In this study we used low resolution spectra of the 4 Seyfert I galaxies NGC 3783, NGC 5548, AKN 120, F-9 and for

NGC 4151 we used the data from Ulrich et al. and Perola et al. (Ref.9,12). All data were obtained with IUE between 1978 and 1985. For all spectra of the first four galaxies we determined, after standard correction for galactic reddening and redshift -assuming $q_0 = +1$ and $H_0 = 50$ Km/sec/Mpc -, the continuum luminosity (L(1350)), the integrated CIV luminosity (L(CIV)) and the CIV equivalent width (E.W.(CIV)).

2.1. Log(E.W.(CIV)) vs Log(L(1350))

In figure 1 we show the results in a log(L(1350)) vs log(E.W.(CIV)) diagram (Baldwin relation). It can be seen that for a single E.W.(CIV) value a wide range of luminosities ($\Delta \log L_\nu \sim 2.0$) is present. This result was also found in a study of a sample of single epoch Seyfert I data by Wu et al. (Ref.6). However, closer inspection of figure 1 shows that each individual galaxy describes in its variation a fairly narrow strip in the log(L(1350)) vs log(E.W.(CIV)) diagram, not unlike a Baldwin-type relation. The results for linear regression through the data points for each galaxy independently are given in table 1. It can

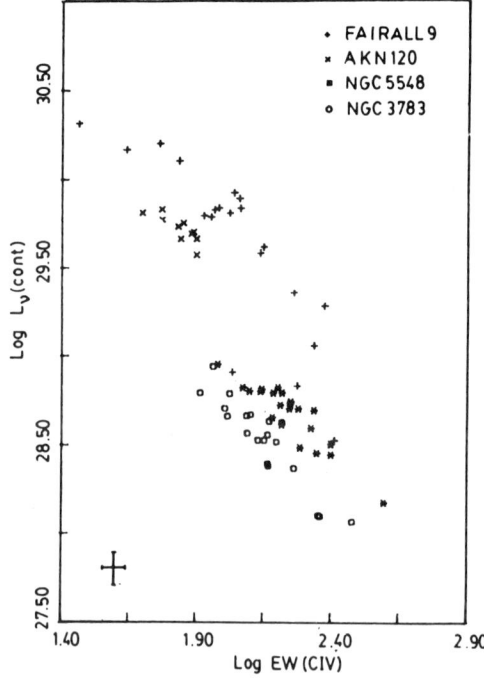

Figure 1. The diagram shows the variations in Log(EW(CIV)) with Log(L(1350)) for 4 Seyfert I galaxies. The 4 galaxies cover a luminosity range of $\Delta L_\nu \sim 100$. Observations made between 1978 and 1985 are marked for each galaxy with different symbols. The considerable variation of the UV continuum flux - measured at 1350 A - is quite clearly illustrated. For each individual galaxy the relation found between Log(EW(CIV)) and Log(L(1350)) (see also table 1) can be easily distinguished.

be seen that highly significant correlations are obtained for all galaxies. The two separate entries for the high luminosity Seyfert I galaxy F-9 will be discussed below. Our choice of the 1350A continuum was made to avoid the difficulty caused by the interrelation of the CIV equivalent width and the 1550A continuum.The use of the 1350A window makes the continuum luminosity an independent variable. This window is nearly completely free of emission lines.

TABLE 1

Log(L(1350))- Log(E.W.(CIV))
(Linear Regression).

Object	Zero Point	Slope	Corr.Coeff.	Comments
NGC 3783	31.92±0.34	-1.57±0.16	-0.85	all data
NGC 5548	31.16±0.36	-1.12±0.16	-0.85	all data
AKN 120	31.44±0.43	-0.96±0.23	-0.83	all data
F-9	31.84±0.26	-1.00±0.14	-0.91	Log(L(1350))>29.7
F-9	33.07±0.55	-1.69±0.27	-0.83	all data

The results in table 1 cover a luminosity range of $\Delta L_\nu > 100$ in L(1350). F-9 is, when brightest, indistinguishable from a low luminosity Q.S.O., while at its faintest it is at the level of the intermediate to faint Seyfert I galaxies. This seems to suggest that one can extrapolate the results of table 1 to QSO luminosities. The implication of this is that the Baldwin-relation is not suitable as a distance indicator for QSO's.

2.2. Log(L(CIV)) vs Log(L(1350))

A more physically meaningful way to represent these data is shown in figure 2 where the same data as given in figure 1, are shown in a Log(L(CIV)) vs Log(L(1350)) diagram. The overall appearance of this diagram strongly suggest a direct relation between the L(CIV) and L(1350) for the Seyfert sample as a whole. Note however that, although the overall trend represents a direct proportionality, for the individual galaxies Log(L(CIV)) varies with Log(L(1350)) with a slope considerably less then unity.The large variations shown by F-9 might present a clue to the interpretation of this diagram.For Log(L(1350))<+29.7 erg sec^{-1} Hz^{-1} F-9 follows quite closely a L(CIV) ~ L(1350) relation, while for Log(L(1350))>+29.7 one finds Log L(CIV) = const = 44.1 erg sec^{-1} (see also results in table 1 for EW (CIV)). Such a behaviour can be most easily understood if for Log(L(1350))<29.7 the line emitting region is photon bounded while for Log(L(1350))>29.7 the line emitting region is matter bounded. Since AKN 120, NGC 5548 and NGC 3783 behave in a matter bounded way at Log(L(1350))<29.7 ergs/sec/Hz,such transition occurs for these galaxies at a lower luminosity. This would imply that the transition luminosity is characteristic for each Seyfert I galaxy. That NGC 3783 varies under matter-bound conditions was also concluded by Wamsteker and Barr (Ref.13) on the basis of the variations seen in the absorption lines. The results of Ulrich et al. (Ref.9) on NGC 4151 on the CIV line strength can be understood under such conditions. Here the variations appear to occur around the transition luminosity which is then at Log(L(1350A))≈27.5 ergs sec^{-1}Hz^{-1}.Details of this work and its further implications on Seyfert I galaxy and QSO modelling will be discussed elswhere (Wamsteker and Colina,in preparation).

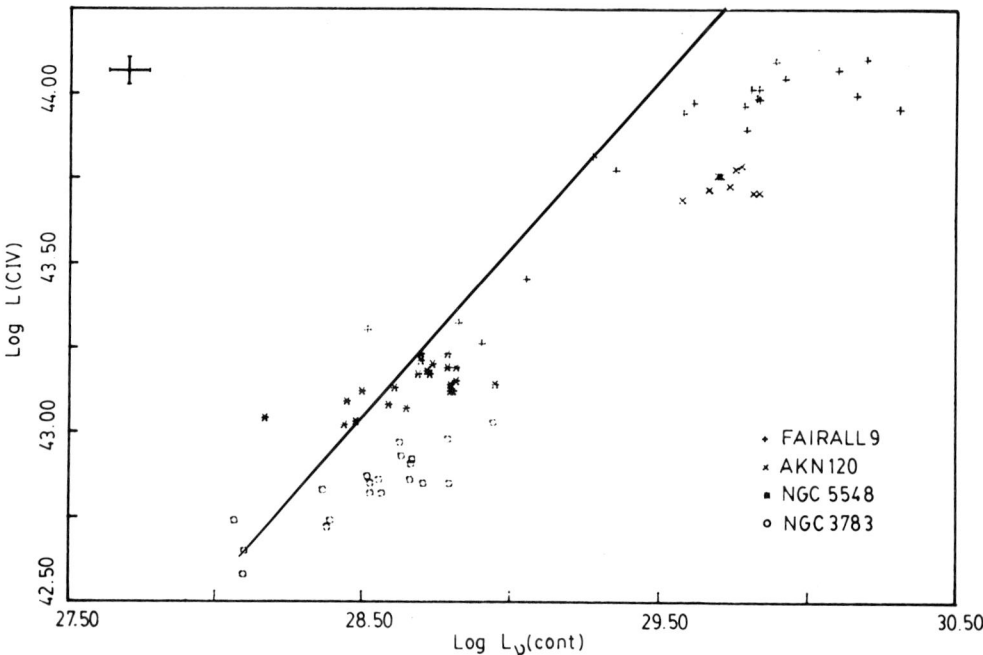

Figure 2. The data shown in this figure are the same as in Figure 1. Only here we show the CIV luminosity rather then the equivalent width. The complete set of data clearly indicates a relation between the UV continuum luminosity and the CIV luminosity. The line drawn is Log(L(CIV))= Log(L(1350))+const. The variation of the individual galaxies suggest that the line emitting regions are matter-bound. Note the large variation in F-9 (+), which indicate at log(L(1350))=29.7 ergs/sec/Hz transition from matter-bound to photon-bound.

3. SUMMARY

IUE results on variable Seyfert galaxies suggest that:

a). The Baldwin-relation most likely is the high-luminosity extension in a luminosity limited sample of objects which behave similar to Seyfert I galaxies, and therefore might not be useful as a distance indicator for QSO's.

b). Seyfert I galaxies represent in their variability a mixture of matter-bound and photon-bound states, where these are separated by what appears to be a critical luminosity which is possibly directly related to the amount of gas available for ionisation in each galaxy.

References:

1. Baldwin, J.A., 1977, Ap.J., 214, 679.
2. Baldwin, J.A., et al., 1978, Nature, 273, 430.
3. Jones, B.J.T., Jones, J.E., 1980, M.N.R.A.S., 193, 537.
4. Osmer, P.S., 1980, Ap.J.Suppl.Ser., 42, 523.
5. Wampler, E.J. et al., 1984, Ap.J., 276, 403.
6. Wu, C.C. et al., 1983, Ap.J., 266, 28.
7. Mushotzky, R., Ferland, G.J., 1984, Ap.J., 278, 558.
8. Netzer, H., 1985, M.N.R.A.S., in press.
9. Ulrich, M.H. et al., 1984, M.N.R.A.S., 206, 221.
10. Wamsteker, W. et al., 1984, ESA SP-218, pg 97.
11. Wamsteker, W. et al., 1985, IUE preprint #4.
12. Perola, C.G. et al., 1982, M.N.R.A.S, 200, 293.
13. Wamsteker, W., Barr, P., 1985, Ap.J., 292, L45.

THE LOW EXCITATION LINES IN QUASARS AND AGN IN THE FRAME WORK OF PHOTOIONIZATION MODELS

S. Collin-Souffrin
Observatoire de Paris-Meudon
92195 Meudon Principal Cedex
France

ABSTRACT. Up to now, the photoionization models, which account well for the intensities of high excitation lines (CIV, CIII$_]^-$, Lα ...) in quasars and AGN, are less successful in explaining the low excitation lines (FeII, MgII, Balmer lines...). In this paper is described a work in which a large number of situations, especially suited for low excitation line emission, are explored, in the framework of photoionization models.

1. INTRODUCTION

Photoionization models* are successful in accounting for intensities of high ionization lines in the spectra of quasars and active galactic nuclei (AGN). They are less convenient when their purpose is to explain the intensity ratios of low excitation lines (FeII, MgII, Balmer and Paschen lines). Especially, the FeII lines are too weak compared to H$_\alpha$, H$_\beta$ and Lα in the photoionization models computed by Kwan and Krolik (1981), Kwan (1984), Wills Netzer and Wills (1984). The idea of a single region producing the whole line spectrum cannot be retained, and one expects at least two regions of different temperatures and densities.

In this communication, I give a summary of a work which will be described in more details in forthcoming papers (Collin-Souffrin and Dumont, 1985, Collin-Souffrin et al. 1985). Its aim is to try to answer this question : are the low excitation lines in quasars and AGN explainable in the frame-work of photoionization models (even in the case of different emission regions) ?

---

* The emissive region in quasars is supposed to be made of many small cloudlets partially covering a central source of continuum.
One calls "photoionization model" the description of the structure of each cloudlet represented by a finite plan-parallel medium which is ionized and heated from one side by a source of continuous radiation (cf for instance Davidson and Netzer, 1979).

In order to tackle this problem, a code has been built, which is mainly designed to study optically thick clouds (up to column densities of 10^{25} cm^{-2} or more) heated by hard X-rays (cf. Collin-Souffrin, Dumont and Joly (1985) for its description) : it computes the ionization and thermal equilibrium of a shell irradiated by a non thermal continuum, solves the radiation transfer in lines and continua, and leads to accurate emergent line intensities even when the optical thickness at the Balmer edge is much larger than unity (this is not the case of previous codes, such as those of Kwan and Krolik (1981) or Kwan (1984)). But it is somewhat difficult to handle, therefore I have built another simplified code, which is sufficient to get an idea of the physical conditions prevailing in the bulk of a region emitting low excitation lines. This simplified version gives estimates of some intense line strengths. It can be used to get "cooling curves" for a thick medium heated by X-rays or by another mechanism, or to determine the dominant process of heating, ionizing or cooling : then approximate analytical expressions can adequatly represent the medium (cf. Collin-Souffrin and Dumont, 1985). Finally, it allows to define the most favorable parameters corresponding to an observed emission spectrum.

In the following section, is given a short description of the simplified computational procedure. A set of results is reported and discussed in the last section.

2. COMPUTATIONAL PROCEDURE

In the simplified computations, the transfer problem is reduced by the use of a mean escape probability formalism for lines and continua, adapted from Faurobert and Frisch (1985) and from Puetter and Levan (1982). A mean distance from the irradiated side is adopted for the X-ray hydrogen and heavy element absorption. Indeed, if a cloud is irradiated from one side by an UV and X-ray continuum - unless the ionization parameter is lower than $\sim 10^{-3}$ - it is divided into an HII zone and an extended HI zone (HI*), these two zones being separated by a narrow transition region. In the the HI* zone, the physical conditions (especially the temperature) are almost constant : it is for this reason that this HI* zone is well represented by an average point. However resonance UV lines (MgII 2800, UV FeII multiplets) are formed in the hot transition region, and are underestimated in this computation, contrary to the H_α or optical FeII multiplets. This difficulty can be overcome for some lines emitted by the HII zone : for instance, L_α can be easily estimated since it depends mainly on the ionization parameter and on the density. In short, this simplified procedure gives reasonably good values only for the average temperature and ionization degree of the HI* zone, and for the line strengths of L_α, H_α and optical FeII multiplets as well as for the integrated emission in the Balmer and Paschen continua.

The atoms are schematized by :
- a 3 -level + continuum for H^0
- a 2 -level + continuum for Mg^+
- a 3 -level + continuum for Fe^+ (as suggested by Wills, Netzer and Wills, 1985).

None of the processes contributing to the statistical equilibrium of the atoms (photoionization, collisional ionization and excitation, charge exchange) is neglected. Ionizations by secondary electrons are taken into account in a simple way. The Compton effect is also introduced.

Thus, the problem reduces to a set of (highly) non linear algebraic equations which are solved by an iteration procedure as functions of the following parameters :
- the total density n_H
- the column density N_H
- the spectral indexes in the optical, UV and X-ray ranges, respectively α_O, α_{UV}, α_X
- the spectral index α_{ox} between 3000 Å and 3 keV
- the ionization parameter U (ratio of the Lyman continuum photon density to the gas density)
- the element abundances
- the high energy X-ray cut-off.

3. SOME RESULTS

The program has been run in a large range of situations involving high densities and high column densities, and very low ionization parameters, which could be particularly suited for low excitation lines. Various values of the spectral index α_{ox} - especially small values which favor the HI* zone with respect to the HII one - were tried.

Table 1 summarizes some of the results. In the left part, it gives the input data : the total density in the HI* zone, the column density, the ionization parameter (which is computed assuming that the density in the HI* zone is about 5 times the density at the head of the HII zone, as is the case if pressure equilibrium is achieved) and α_{ox}. All these runs correspond to $\alpha_O = \alpha_{UV} = \alpha_X = 1$ and to cosmic abundances. In the right part are given some output data. The temperature T_e, the electron density n_e, the populations of the first and second level of hydrogen, n_1 and n_2, and the population of the second level assuming thermodynamic equilibrium between levels 1 and 2 (it is often claimed that there are in thermodynamic equilibrium), the H_α flux in ergs cm^{-2}, (taking into account the contribution of the HII zone), and the intensity ratios, L_α/H_α and FeIIopt/H_α (since FeII is reduced to a 3- level atom, the sum of all the optical multiplets is represented by the line 2 ↔ 3. Provided that the atomic paramaters - collision strengths, oscillators strengths ... - are well chosen, this schematisation is realistic).

Let us first make two remarks concerning this table :

1 - An important H_α emission can be provided by regions of very low ionization parameter (U = 10^{-3}, 10^{-4}) : such regions may correspond to high density clouds (n = 10^{12}) located near the central source, or to relatively low density clouds (n = 10^{10}) located farther from the central source.

2 - Strong departures from LTE between the first secund level of HI

TABLE 1

n	N	U	α_{ox}	T_e	n_e	n_1	n_2	n_{2ET}	$F(H\alpha)$	$\frac{L\alpha}{H\alpha}$	$\frac{FeII_o}{H\alpha}$
1(12)	3(22)	.5(-4)	1.	6056	.70(10)	.99(12)	.90(5)	.13(5)	.16(7)	1.7	.53
	3(24)			5845	.18(10)	1.0(12)	.16(5)	.64(4)	.49(7)	.57	.25
	3(22)	.5(-3)		7622	.45(11)	.95(12)	.72(6)	.69(6)	.26(8)	1.1	.13
	3(24)			6977	.41(11)	.96(12)	.17(6)	.16(6)	.24(8)	1.2	.11
1(11)	3(22)	.5(-4)		5478	.68(9)	.99(11)	.78(4)	.16(3)	.13(6)	2.2	.68
	3(24)			4700	.16(9)	1.0(11)	.12(4)	.46(1)	.30(6)	.90	.68
	3(22)	.5(-3)		6596	.19(10)	.98(11)	.21(5)	.63(4)	.14(7)	2.0	.46
	3(24)			6327	.64(9)	.99(11)	.45(4)	.30(4)	.59(7)	.48	.28
	3(22)	.5(-2)		8000	.15(11)	.85(11)	.13(6)	.13(6)	.26(8)	1.0	.12
	3(24)			7234	.16(11)	.84(11)	.26(5)	.26(5)	.28(8)	1.0	.11
1(10)	3(22)	.5(-4)	.7	6087	.16(9)	.98(10)	.18(4)	.14(3)	.72(5)	.65	.67
	3(24)			4900	.39(8)	.99(10)	.26(3)	.12(1)	.18(6)	.17	.77
	3(22)	.5(-3)		7776*	.26(9)	.87(10)	.88(4)	.96(4)	.90(6)	.43	.31
	3(24)			6400	.12(9)	.58(10)	.51(3)	.36(3)	.26(7)	.11	.55
	3(22)	.5(-2)		8494	.27(10)	.73(10)	.26(5)	.26(5)	.18(8)	.17	.10
	3(24)			7450	.30(10)	.70(10)	.34(4)	.35(4)	.26(8)	.11	.12
	3(23)	.5(-4)	1.	4651	.38(8)	1.0(10)	.24(3)	.36	.16(5)	1.9	.52
	3(23)	.5(-3)		5495	.90(8)	.99(10)	.64(3)	.18(2)	.15(5)	1.9	.76
	3(23)	.5(-2)		6780	.29(9)	.97(10)	.14(4)	.10(4)	.17(7)	1.7	.59
	3(23)	.5(-1)		8510	.75(10)	.25(10)	.92(4)	.73(4)	.42(8)	.69	.11
	3(24)	.5(-4)	1.3	4000	.73(7)	1.0(10)	.52(2)	.59(-2)	.78(4)	3.6	.36
	3(24)	.5(-3)		4726	.24(8)	1.0(10)	.85(2)	.53	.77(5)	3.7	.87
	3(24)	.5(-2)		5800	.98(8)	.99(10)	.15(3)	.54(2)	1.0(6)	2.7	.80
	3(24)	.5(-1)		7397	.35(10)	.65(10)	.32(4)	.29(4)	.30(8)	.90	.11
	3(24)	.5(-4)	1.6	3834	.51(7)	1.0(10)	.32(2)	.16(-2)	.51(4)	5.2	.20
	3(24)	.5(-3)		4493	.17(8)	1.0(10)	.52(2)	.14	.51(5)	5.2	.52
	3(24)	.5(-2)		5430	.66(8)	.99(10)	.84(2)	.14(2)	.61(6)	4.6	.80
	3(24)	.5(-1)		7500	.42(10)	.58(10)	.34(4)	.37(4)	.32(8)	.90	.11

* bad convergence

can be noticed (they are underlined in the Table 1) : these are due to depopulation of the first level by fast secundary electron inonizations.

Concerning now the line strengths, one sees that it is very difficult to account for the observed ratio FeIIopt/Hα $\sim$ 1 in Seyfert 1 nuclei, except in some extreme cases : high density (n $\gtrsim 10^{11}$), low ionization (U < 10^{-3}), flat optical - X-ray index ($\alpha_{ox} \lesssim 1$).

The necessity of a flat optical X-ray spectrum makes however these cases unrealistic. Indeed, although such a small α_{ox} has been observed in a few "red quasars", which show very low excitation spectrum, (cf. Bregman et al., 1985), this is not the case for normal quasars ($\alpha_{ox} \sim 1.2$ in radio loud quasars, and ~ 1.5 in radio quiet ones).

REFERENCES

- Bregman, J.N., Glassgold, A.E., Huggins, P.J., Kinney, A.L., Ap. J. 291, 505, 1985.
- Collin-Souffrin, S. and Dumont, S., in preparation, 1985.
- Collin-Souffrin, S., Dumont, S., Joly, M., Péquignot, D., in preparation, 1985.
- Collin-Souffrin, S., Dumont, S., Joly, M., in preparation, 1985.
- Davidson, K., Netzer, H., Rev. Mod. Phys. 51, 715, 1979.
- Faurobert, M. and Frisch, H., preprint, 1985.
- Kwan, J., Ap. J. 283, 70, 1984.
- Kwan, J. and Krolik, J.H., Ap. J. 250, 478, 1981.
- Wills, B.J., Netzer, H., Wills, D., Ap. J. 288, 94, 1985.

COSMOLOGICAL EVOLUTION OF OPTICALLY SELECTED QSOs

L. Danese, G. De Zotti and A. Franceschini
Istituto di Astronomia
Vicolo dell'Osservatorio 5
I-35122 Padova
Italy

We report on an updated analysis of the cosmological evolution of optically selected QSOs incorporating, in addition to the data sets used in our previous work (Danese, De Zotti & Franceschini, 1985), the samples of Marshall et al. (1984), Koo (1983), Marano et al. (1984), Usher et al. (1983), Mitchell et al. (1984) and Weedman (1985).

Both luminosity dependent density evolution (DE) models and luminosity evolution (LE) models have been tested against:
- Source counts
- Luminosity and redshift distributions
- V'/V'_{max} tests.

Altogether, the data points amount to 58.

The log of the local luminosity function has been expressed as a third order polynomial whose four coefficients have been treated as free parameters. LE models require only one additional free parameter, while four of them have been used for DE models.

RESULTS

We confirm that good fits can be obtained with both classes of models, although the newly added data seem to favor luminosity evolution. For instance, for $q_0 = 0$ we find reduced χ^2 (i.e. χ^2 per degree of freedom) values of 1.25 for LE and of 1.46 for DE. The best fit LE timescale is $\simeq 18\%$ of the Hubble time. Some examples of our results are shown in Figs. 1 - 4.

In general we find that the local luminosity function cannot be represented by a single power law: a reduced $\chi^2 \geq 4$ is found for any model in this case.

Remarkably enough, pure luminosity evolution models predict a local luminosity function of low luminosity ($M_B > -23$) QSOs (Fig. 4) in strikingly good agreement with that of Seyfert 1 nuclei derived by Cheng et al. (1985).

As illustrated by Fig. 1, the surface density of QSOs brighter than B = 19.8 found by Marshall et al. (1984) exceeds by a factor of 2 to 4 that expected on the basis of models satisfactorily fitting the body of data mentioned above, as well as those measured, for the same limiting

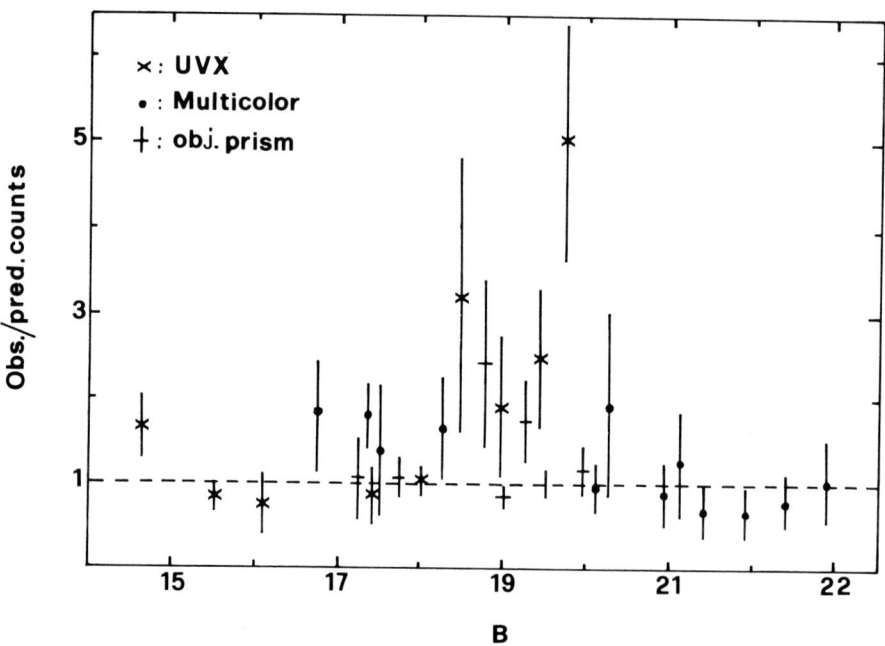

Fig. 1. Ratio of observed to predicted differential counts for a pure luminosity evolution model, $q_o = 0$.

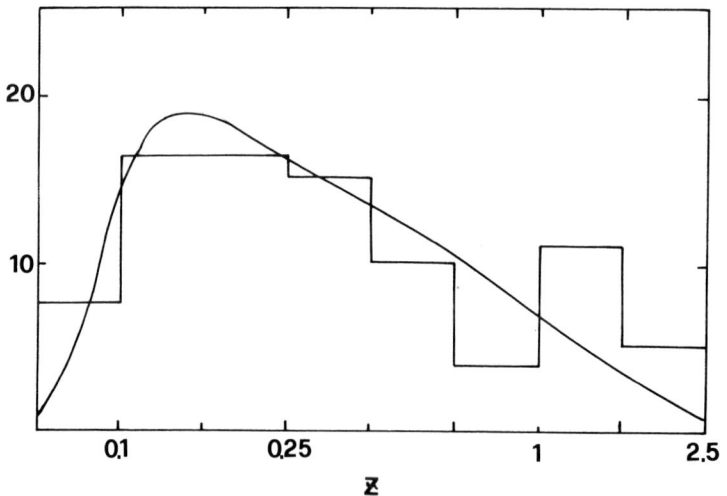

Fig. 2. Pure luminosity evolution ($q_o = 0$) fit to the redshift distribution of the BQS sample (Schmidt & Green, 1983).

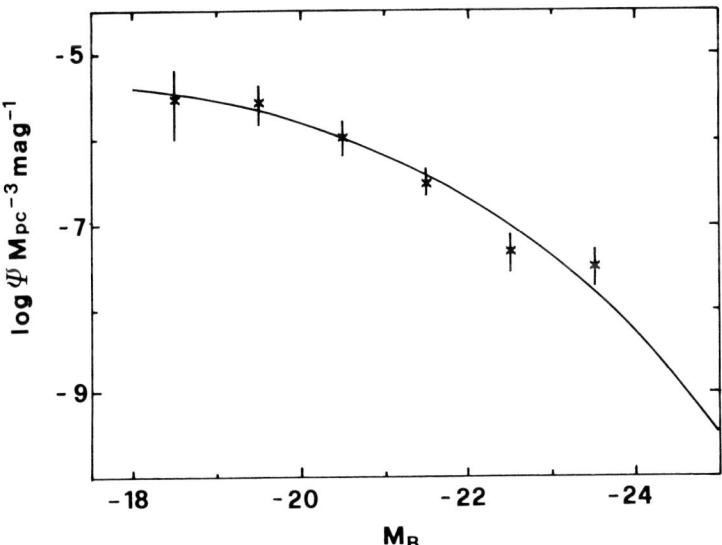

Fig. 3. Comparison between the local luminosity function of low luminosity QSOs predicted by the pure luminosity evolution model and the local luminosity function of Seyfert 1 nuclei (Cheng et al., 1985).

magnitude, by Marano et al. (1984) and Koo (1983). It is thus likely that some earlier calculations, heavily relying on the sample of Marshall et al. (1984), have somewhat overestimated the evolution rate.

It would be interesting to analyze the counting rates of the HEAO 1 A-2 detectors with $1°.5 \times 3°$ field of view in the directions of the areas of the sky apparently yielding discrepant counts. If the mean X-ray spectral index of QSOs is not much steeper than that of Seyfert 1 nuclei, a factor ≥ 2 overdensity for $B \leq 19.8$ should result in an excess flux possibly detectable with the A-2 experiment. HEAO 1 data could then directly test the reality of the relatively large surface density fluctuations suggested by optical counts; combined with the IPC observations of Marshall et al. (1984) they would also supply valuable information on the lumped QSO spectrum.

Figure 4 shows the redshift distributions for faint ($B \leq 22$) QSOs predicted by both DE and LE models, stretched up to $z = 4$. A comparison with the preliminary results of the spectroscopic survey of faint QSOs by Koo (1983) shows that the simple evolution functions considered here entail too many high redshift objects (see also Osmer, 1982). The problem is more serious for DE models which predict that $\simeq 54\%$ of $B \leq 22$ QSOs should lie in the redshift range $2.5 \leq z \leq 4$; in the case of LE models the predicted fraction of high z objects is $\simeq 37\%$. On the other hand a weakening or even a decline of the evolution function at $z \geq 2$ may be expected, in the framework of LE models, from straightforward physical arguments, constraining the active lifetimes (Cavaliere et al.,

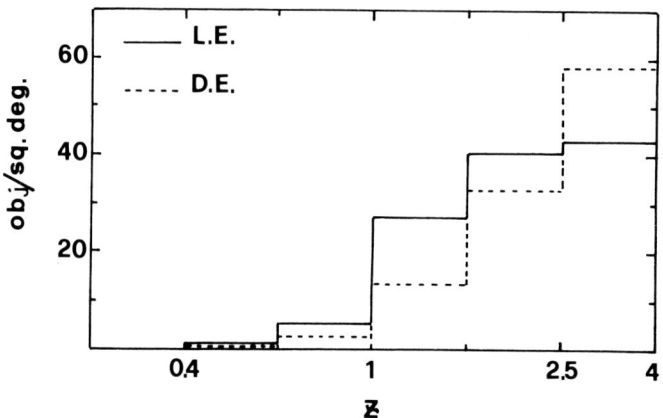

Fig. 4. Redshift distributions for B ≥ 22 predicted by density evolution (dashed line) and luminosity evolution models.

1983, 1984; Setti & Zamorani, 1983), combined with "demographic" studies of active galactic nuclei (Phinney, 1983). Work in this direction is in progress.

References

Cavaliere, A., Giallongo, E., Messina, A., Vagnetti, F.: 1983, Ap. J. 269, 57
Cavaliere, A., Giallongo, E., Vagnetti, F.: 1984, Proc. Symp. "X-ray Astronomy '84", p. 435
Cheng, F.Z., Danese, L., De Zotti, G., Franceschini, A.: 1985, MNRAS 212, 857
Danese, L., De Zotti, G., Franceschini, A.: 1985, Astr. Ap. 143, 277
Koo, D.C.: 1983, Proc. 24th Liège Int. Astrophys. Symp., p. 240
Marano, B., Zamorani, G., Zitelli, V.: 1984, The Messenger, Dec. issue
Marshall, H.L., Avni, Y., Braccesi, A., Huchra, J.P., Tananbaum, H., Zamorani, G., Zitelli, V.: 1984, Ap. J. 283, 50
Mitchell, K.J., Warnock, A., Usher, P.D.: 1984, Ap. J. 287, L3
Osmer, P.S.: 1982, Ap. J. 253, 28
Phinney, E.S.: 1983, Ph. D. thesis, Cambridge University
Schmidt, M., Green, R.F.: 1983, Ap. J. 269, 352
Setti, G., Zamorani, G.: 1983, in COSPAR/IAU Symp. "Advances in High Energy Astrophysics and Cosmology"
Usher, P.D., Green, R.F., Huang, K.L., Warnock, A.: 1983, Proc. 24th Liège Int. Astrophys. Symp., p. 245
Weedman, D.W.: 1985, Ap. J. Suppl. 57, No. 3

THE SMALL WIDE ANGLE TAIL NGC4874 IN THE CENTER OF COMA CLUSTER

L. Feretti and G. Giovannini
Istituto di Radioastronomia, Bologna, ITALY

NGC4874, one of the two dominant galaxies of the Coma cluster, is associated with a radio source of the Wide Angle Tail (WAT) type (Fig. 1), on a very small angular scale (Feretti and Giovannini 1985). The total linear extent, from the core to each tail edge, is ~7 kpc ($H_o = 100$ km s^{-1} Mpc^{-1}), and the radio structure is all contained within the optical boundary of the galaxy (Fig. 2). The total power of the radio source at 1.5 GHz is Log $P_{1.5}$=23.06 W/Hz; the core is very weak: S_5=0.9 mJy, which corresponds to a power Log P_5=20.76 W/Hz. This value is lower than that expected on the basis of the correlation P_{core} - P_{tot} found for radio galaxies (Feretti et al. 1984); however, it is still in agreement with this correlation, due to its large dispersion.

Apart from the size scale, the radio structure is not different from the more extended WAT sources. An unresolved core is detected roughly aligned with two possible jets, which are visible after a gap of radio emission. At about 2.5-3 kpc from the nucleus, the jets bend of about 110° and shade into the extended low brightness tails.

We interpret this bending, which coincides with the transition from jet to tail, as due to ram pressure of the intergalactic medium on the moving galaxy. The radial velocity of the galaxy is v=7180 km/s, which leads to a velocity of Δv=220 km/s with respect to the cluster mean. Therefore this might represent another case of a WAT source associated with a low speed galaxy. Implications of the above interpretations are: 1) the interstellar-intergalactic medium contact surface occurs very close to the nucleus; this is in agreement with a great part of the interstellar medium being swept out from the galaxy (Gisler 1976); 2) the velocity of the jet at the bending point is probably few hundreds km/s, as deduced from momentum balance, assuming the same density for the jet and for the external gas. A low jet velocity at the distance of ~3 kpc from the nucleus can be expected as effect of deceleration due to shocks and entrainment of interstellar material (Bicknell 1984). Also, this low velocity of the jet could be related to the low power of the radio core.

References
Bicknell, G.V.: 1984, Astrophys. J. **286**, 68
Feretti, L., Giovannini, G.: 1985, Astron. Astrophys., in press

Feretti, L., Giovannini, G., Gregorini, L., Parma, P., Zamorani, G.:
    1984, Astron. Astrophys. **139**, 55
Gisler, G.R.: 1976, Astron. Astrophys. **51**, 137

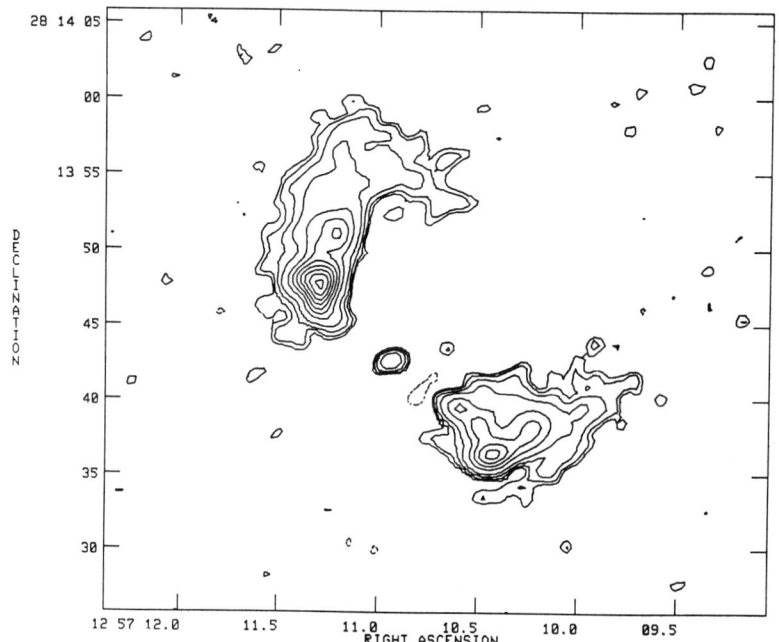

Fig. 1 Contour map at 4.9 GHz (VLA, B array) of NGC 4874. Contours are -0.2, 0.15, 0.2, 0.3, 0.5, 1, 1.5, 2, 2.5, 3, 3.5, 4 mJy/beam.

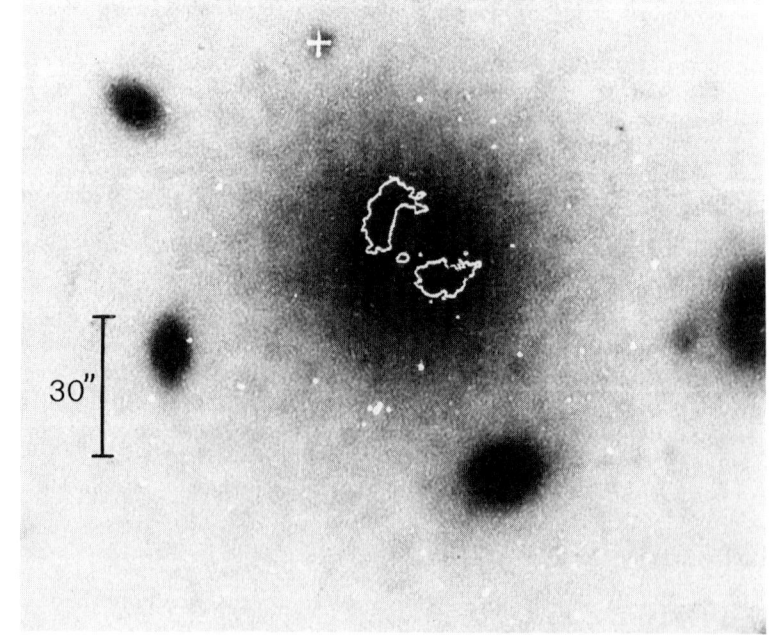

Fig. 2 Lowest contour of 4.9 GHz radio emission superposed to the POSS red print.

COLLIMATED WINDS FROM ACTIVE GALACTIC NUCLEI

A. Ferrari[1][2], E. Trussoni[2], R. Rosner[3], K. Tsinganos[3][4]

(1) Istituto di Fisica Generale dell'Universita', Torino, Italy
(2) Istituto di Cosmogeofisica del CNR, Torino, Italy
(3) Harvard-Smithsonian Center for Astrophysics, Cambridge, USA
(4) Department of Physics, University of Crete, Heraklion, Greece

ABSTRACT. We present a wind-type model for the modulation of jet brightness (knots) on kpc scale. It is shown that the mass distribution of the parent galaxy and the development of unstable modes on the surface of the beam can lead to the periodic formation of internal shocks that can be associated with enhanced particle acceleration. The results are compared specifically with the data of the jet of M87.

1. INTRODUCTION

The origin of extragalactic jets is closely related to the observed activity in AGN: although they may extend for several kpcs, their acceleration and collimation must take place predominantly within a narrow zone of size less than a parsec, where the intense radiation field from an accretion disk may accelerate a supersonic beam (Abramowicz et al. 1980). Following a wind-mode approach, Ferrari et al. (1985) have shown that the radiaton pull and the expansion of the flow at the exit of the accretion funnel lead to the formation of shocks transverse to the propagation direction. Such shocks can be associated with dissipative processes (e.g. particle acceleration, heating) and may be responsible for the activity of AGN. On the other hand they could also arise during the propagation of the flow far away from the nucleus (on kpc scale): in this framework the study of jets (at large scales) can provide a substantial contribution to our knowledge of AGN.

In this perspective we analyze here the propagation of jets outside the zone of acceleration. The model is able to connect the dynamics of jets to processes which can give rise to brightness enhancements. The solutions are presented with reference to the jet of M87, but can be extended to other objects for which quasi-periodic knots are observed and the galactic mass distribution is determined. The comparison can therefore provide estimates for the jet speeds which are consistent with values implied by observations.

Although M87 is not a typical AGN, we shall use it as reference for our model because its jet is well studied. It extends for about 1.5 kpc and displays a sequence of six well-defined knots regularly distributed (average separation ~ 170 pc). Three of them (D-E-F) are relatively weaker and closer to the galactic core, and three (A-B-C) are stronger, more extended and farther away (Nieto and Lelievre 1982, Biretta et al. 1983); the change in regime between the first three and the last three knots is marked by a sharp feature in knot A, which has been interpreted as a strong shock.

Rees (1978) and Blandford and Königl (1980) have discussed how irregularities in the flow velocity and clouds in the plasma can give rise to shocks where particle acceleration and heating produce the observed knots. Ferrari et al. (1983) have instead proposed that the development of long-wavelength Kelvin-Helmholtz modes creates localized compression regions which travel along the jet and at which particle acceleration by interaction with short-wavelength MHD modes occurs. The periodicity of spatial oscillations is linked to the wavelength of the most unstable modes.

We shall adopt this last point of view: namely, in framework of the wind-type model, we assume that linearly-unstable "pinching" modes do grow to form finite-amplitude oscillations, connected with periodic expansions and contractions of the jet cross section. We shall then study the latter effect on the propagation of the beam well outside the nucleus: starting from a jet with constant small opening angle we simplify the picture by assuming a spherically symmetric expansion and a constant amplitude of the perturbations. Then the cross section $A(Z)$ of the perturbed area is given by:

$$A(Z) = [Z + \epsilon_0 Z \sin(\frac{2\pi}{\lambda} Z)]^2 , \qquad (1)$$

where Z is the radial coordinate in units of the base of the beam z_0 (assumed $=100$ pc for the case of M87), ϵ_0 is the amplitude (in units of the beam radius r_0 at z_0) and λ the wavelegth of the oscillation. Concerning the phyisical parameters of the jet we shall assume it supersonic at z_0 and with temperatures $\approx 5 \times 10^6$ - 10^7 K. The most important parameter is the behaviour of the gravitational field through which the jet propagates; again for M87, from Sargent et al. (1978) we can assume a mass distribution in function of the radial distance z (in pc) $m(z) = 1.2 \times 10^8 z^{0.84} M_\odot$ in the range $100 \le z \le 1500$ pc.

2. THE EQUATION FOR THE MACH NUMBER

The conservation laws for mass and momentum fluxes in polytropic flows (index α) along a channel of variable cross-section can be combined into a single equation for the Mach number $M = V/V_s$, where V_s is the sound speed (Ferrari et al. 1985):

$$\frac{M^2-1}{2M^2}\frac{dM^2}{dZ} = (1 + \frac{\alpha-1}{2}M^2)\left[\frac{1}{A(Z)}\frac{dA(Z)}{dZ} - \frac{\alpha+1}{2}\frac{HZ^{n-2}}{h(1+\frac{\alpha-1}{2}M^2)}\right]$$

$$Z = \frac{z}{z_0}, \quad H = \frac{1.9 \times 10^{19} m_{gal,8}}{T_8^{(o)} z_0}, \quad h = \left[\frac{M_o}{A(Z)M}\right]^{g(\alpha)}, \quad g(\alpha) = 2\frac{(\alpha-1)}{(\alpha+1)}; \qquad (2)$$

quantities with subscript or superscript "o" are measured at z_0, while the mass of the galaxy $m_{gal,8}$ and the temperature T_8 are given in units of $10^8 M_\odot$ and 10^6 K respectively.

The solutions of Eq. (2) are critically dependent on the number and nature of its critical points where the flow becomes supersonic, and these in turn depend on the assumed flow geometry. Assuming in Eq.(1) a non-perturbed geometry ($\epsilon_0 = 0$, $A(Z) = Z^2$), we obtain from Eq. (2) only one critical point, whose position is given by the relation:

$$Z_c = F f(\alpha,n); \quad F = \frac{H}{2M_0 g(\alpha)}, \quad f(\alpha,n) = \frac{1}{n_1 - n}, \quad n_1 = \frac{5 - 3\alpha}{\alpha + 1} \leq 1. \quad (3)$$

In general the nature of the critical points is ruled by the index n of the mass distribution: if $n = 0$, i.e. the case of a centrally concentrated mass, Eq. (2) yields the well-known Parker critical point (Parker 1963); for $n < n_1$ or $n > n_1$ the critical point is X-type or O-type respectively. In the former case, we find in fact wind-type critical solutions; in the other case no stationary solution can be found from z_0 to ∞ with a smooth transition between the subsonic and supersonic regimes: we call "trapped" these (unphysical) solutions.

3. APPLICATION TO THE JET OF M87

The main consequence of the development of unstable modes ($\epsilon_0 > 0$) is the possibility of multiple transonic solutions, one of which is continuous, while the others involve shock transitions which occur between pairs of transonic solutions when the Rankine-Hugoniot conditions are fulfilled (Habbal and Tsinganos 1983). Assuming, with reference to M87, $M_0 = 2.5$, $T^{(0)} = 8 \times 10^6$ K, and $\lambda = 170$ pc we have solved Eqs.(2) over a wide range of values of ϵ_0 in isothermal conditions. The results are the following:

(a) For $\epsilon_0 < 0.23$, we find only the continuous transonic solution with oscillations of $M(Z)$. Actually for $\epsilon_0 > 0.15$ multiple critical point appear, but no shocks are allowed with the original continuous solution.

(b) For $\epsilon_0 \geq 0.23$, discontinuous transonic solutions are found as due to the formation

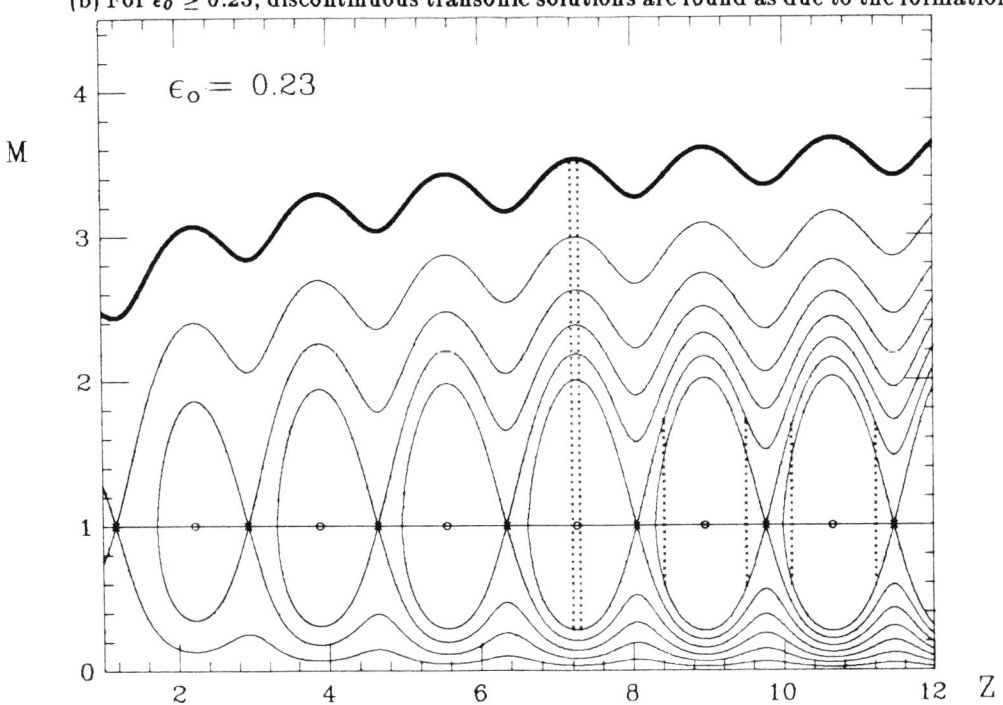

Figure 1. Typical solution topologies for an isothermal flow: the heavy line represents the continuous solution supersonic at the base of the beam, crosses and circles indicate X-type and O-type critical points respectively, dashed vertical lines the location of shocks.

of shocks. These shock transitions occur first between the continuous solution and those associated with the more external critical point, then shift inwards for growing ϵ_o. As a consequence there is a range of ϵ_o for which the inner part of the jet shows oscillations only and the external one has shocks (Fig. 1).

(c) The Mach number at $z \approx 1200$ pc is $M(z = 1200 \text{ pc}) \approx 3.7$ for the continuous transonic solution, and $2 \leq M \leq 3$ for the shocked transonic solutions; this yields velocities $V(z = 1200 \text{ pc}) \approx 1800$ km s^{-1} and $\approx 1000 - 1200$ km s^{-1} respectively.

The positions of knots appear to be associated either with jet compressions (when they are sufficiently strong), or with shock transitions between critical solutions. We use now Fig. 1 to make a comparison with the morphological characteristics of the M87 jet. The shock at ≈ 730 pc gives rise to a sharp transition in the flow pattern; this feature may be associated with knot A, where the jet is thought to slow down drastically. The jet compressions upstream of knot A may be associated with the three inner blobs, while shocks between multiple solutions may generate the outermost knots. These features are not qualitatively affected by different initial velocities of the jet and its temperature (if assumed within observational constraints), while more important is the effect of the gravitational potential well and the polytropic index.

Mass Distribution. The adopted power-law index $n = 0.84$ is close to the critical value n_1 ($n_1 = 1$ in the isothermal case); the solutions pattern is heavily affected only for $n > 0.95$.

Polytropic flows. The assumption of $\alpha > 1$ affects the topologies of solutions because they become ruled by the two parameters ϵ_o and α. For fixed ϵ_o an increase of α leads to the formation of "trapped" solutions, starting from the outermost ones. As a consequence shocks can be more generally distributed along the jet respect to isothermal case, leading to a better fit with observations.

We note that our model is based on a time-independent treatment of the flow, in particular, we have derived only the necessary conditions for the shock formation. Time-dependent computations (Tsinganos et al. 1983) have shown that, at least in the case of the solar wind, shocks do form at the locations expected on the basis of the steady solutions. We intend to test our model by a time-dependent treatment of the flow.

Acknowledgements. Support by the NSF grant AST 83-03522 and by the Italian CNR and MPI are acknowledged.

REFERENCES

Abramowicz, M.A., Calvani, M, and Nobili, L., 1980, *Ap. J.*, **242**, 772
Biretta, J., Owen, M.H., and Hardee, P.E., 1983, *Ap. J. Lett.*, **274**, L27.
Blandford, R.D., and Königl, A., 1980, *Astrophys. Lett.*, **20**, 15.
Ferrari, A., Trussoni, E., and Zaninetti, L., 1983, *Astron. Astrophys.*, **125**, 179.
Ferrari, A., Trussoni, E., Rosner, R., and Tsinganos, K., 1985, *Ap. J.*, **293** in press.
Habbal, S.R., and Tsinganos, K., 1983, *J. Geophys. Res.*, **88**(A3), 1965.
Nieto, J.L., and Lelievre, G., 1982, *Astron. Astrophys.*, **109**, 95.
Parker, E.N., 1963, *Interplanetary Dynamical Processes*, Interscience, New York.
Rees, M.J., 1978, *M.N.R.A.S.*, **184**, 61.
Sargent, W.L.W., Young, P.J., Boksenberg, A., Shortridge, K., Lynds, C.R., and Harwick, F.D.A., 1978, *Ap. J.*, **221**, 731.
Tsinganos, K., Habbal, S.R., and Rosner, R., 1983, in *Solar Wind 5*, M. Neugebauer ed., NASA Science and Technological Information Office, Washington D.C.

COMPARISON BETWEEN OPTICALLY SELECTED AND X-RAY SELECTED ACTIVE GALACTIC NUCLEI

A. Franceschini
Istituto di Astronomia, Vicolo Osservatorio, I-35122 Padova
and
I.M. Gioia[1] and T. Maccacaro[1]
Center for Astrophysics, 60 Garden st., Cambridge MA 02138
[1] Also from Istituto di Radioastronomia, CNR, Bologna, Italy.

ABSTRACT. We compare statistical properties on optically selected and X-ray selected Active Galactic Nuclei by using the available data on the corresponding distributions of the X-ray to optical luminosity ratio. We show that these computations critically depend on the width of these distributions. The use of the observed ones leads to significant inconsistencies between predictions and observations. However, if we assume that these distributions have been "broadened" by external effects, and consequently we use narrower intrinsic widths, we can eliminate most of the discrepancies.

1. INTRODUCTION

The compatibility of properties of X-ray selected and optically selected Active Galactic Nuclei (AGN) was questioned by several authors. The crucial point was to understand whether these two selection techniques are sampling the same parent population or whether different classes of AGN exist. Some evident differences, concerning mainly the shape of the luminosity and redshift distributions and the evolutionary rates, have been explained in part as the result of the established dependence of the X-ray to optical luminosity ratio L_x/L_o (Avni and Tananbaum 1982, Maccacaro and Gioia 1983), or as the result of the X-ray photoelectric absorption and reddening which is probably affecting the low luminosity tail of the X-ray selected samples (Kriss and Canizares 1985). However, several major inconsistencies still exist between the observed properties of the two populations: in particular, it was recently suggested that samples of X-ray AGN may be significantly underpopulated with respect to optical samples (Maccacaro 1984, Setti 1984, Zamorani 1984a).

Using essentially all the statistical information available at the end of 1984, we discuss here the relationship between optically selected and X-ray selected AGN, to seek a possible explanation for these discrepancies. A fuller account will be published elsewhere (Franceschini, Gioia and Maccacaro 1985).

$H_o = 50$ and $q_o = 0$ are used.

2. PREDICTIONS IN THE X-RAY BAND USING DATA ON OPTICALLY SELECTED AGN

Details on the data used on optical and X-ray AGN can be found in Franceschini, Gioia and Maccacaro (1985) and Danese, De Zotti and Franceschini (1985). It is worth to remind here that the only complete sample of X-ray selected AGN at sufficiently faint fluxes is that derived from the Medium Sensitivity Survey (MSS) (Maccacaro, Gioia and Stocke 1984), which is now fully identified in the optical.

Our data on optically selected quasars were confined to objects with blue absolute magnitude $M_B<-23$, to avoid a possible incompleteness with respect to objects of the Seyfert type, but we have used in addition data on the local luminosity function (LLF) of Seyfert galaxies by Cheng et al. (1985). We have also excluded the MSS AGN with $\log L_x<25.5$ (L_x in units of erg/s/Hz at 2 keV), because they may be heavily affected by X-ray absorption and reddening.

Although different parametric models of evolution can successfully describe the data in both wavebands, here we consider only a pure luminosity evolution (PLE) model of the form $L(z)=L(0)\exp(k\tau)$, where τ is the look-back time and k=const, since it seems now to give the best representation of the current data with the minimum number of parameters (see Danese, De Zotti and Franceschini, these Proceedings). Both the optical and X-ray LLFs were represented with 3-rd order polynomial expansions.

If one has some knowledge of the distribution $f(\alpha_{ox})$ of the optical to X-ray energy index $\alpha_{ox}=\log(L_o/L_x)/2.605$ (L_o is the monochromatic optical luminosity at 2500 A), one can easily make predictions for the X-ray number counts, luminosity and redshift distributions, etc., by convolving $f(\alpha_{ox})$ with the best-fit luminosity function of the optically selected AGN and by taking into account the sky coverage of the MSS. We have compared predictions computed in this way with the corresponding X-ray observations and used a χ^2 test to quantify the difference.

As illustrated in Fig.(1), we have found that a very critical parameter which affects this comparison is the standard deviation σ of the distribution of the residuals of α_{ox} with respect to the average dependence on L_o. The use of the observed distribution of α_{ox} with σ=0.2 (Avni and Tananbaum 1982) leads to significant inconsistencies between predictions (dashed lines in Fig.2) and observations (histograms). How-

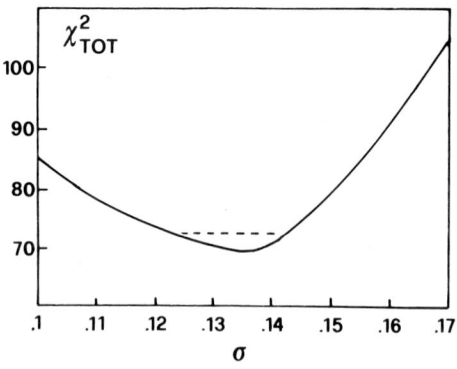

Fig.(1). Total χ^2 from all the optical and X-ray data calculated as a function of the standard deviation σ of the residuals of α_{ox} around its average dependence on L_o. We adopted a gaussian model to represent this distribution. The mean of α_{ox} was kept to the best-fit regression ($\alpha_{ox}=.11 \log L$ -1.9) given by Zamorani 1984b). A best-fit σ≅0.14 is indicated. The formal 90% confidence interval is also shown.

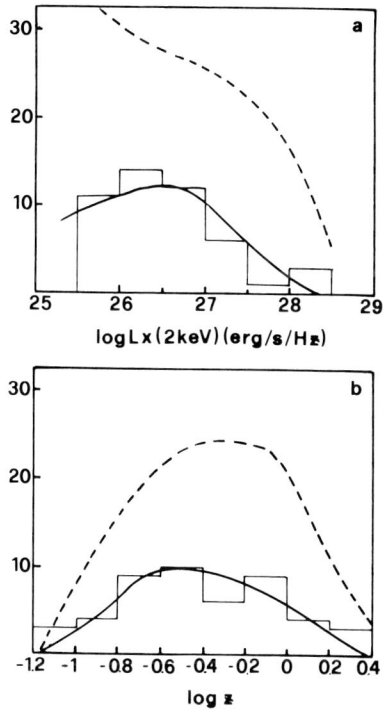

Fig.(2). Comparison between the observed and predicted X-ray luminosity (a) and redshift (b) distributions for the Medium Sensitivity Survey AGN. Dashed lines are predictions from a PLE model describing the optical data and convolved with a gaussian fit to the observed distribution of the residuals of α_{ox} ($\sigma \simeq 0.2$). The continuous lines were calculated with our best-fit $\sigma = 0.14$.

ever, the use of a narrower distribution, with a value of σ equal to the best-fit value indicated in Fig.(1), leads to a substantial improvement of the situation (see continuous lines in Fig.(2)).

The "broadening" of the observed distribution can be interpreted as due to the presence of some "noise" that widens the intrinsic distribution so that values of the standard deviation larger than expected are observed.

3. PREDICTIONS IN THE OPTICAL BAND USING DATA ON X-RAY AGN

Predictions in the optical band using data on X-ray selected AGN plus the distribution of their α_{ox} can be calculated following a line of reasoning similar to that of the previous Section. But the dynamic range in the L_x-z plane covered by data on X-ray AGN is now much smaller. This, combined to a much poorer statistics, results in less tight constraints.

In this case also we find that the sources predicted at optical wave bands by means of the observed α_{ox} distribution of X-ray AGN outnumber those observed by a factor of roughly 3. Again, the use of a smaller standard deviation of the α_{ox} distribution improves the situation. We find a best-fit value $\sigma \simeq 0.11$, while the observed quantity is $\sigma \simeq 0.15$. Fig.(3) shows how our predictions calculated with the best-fit $\sigma = 0.11$ compare with the luminosity distribution of the Bright Quasar Survey AGN (Schmidt and Green 1983).

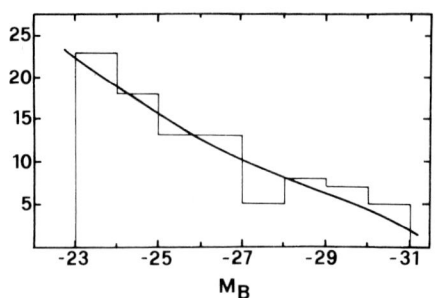

Fig.(3). The optical luminosity distribution of the Bright Quasar Survey AGN. The curve is a prediction according to a PLE model describing data on the X-ray selected AGN convolved with the distribution of their α_{ox} parameter.

4. DISCUSSION AND CONCLUSIONS

In our analysis aimed at the study of the relationship between optically selected and X-ray selected AGN, we have found that the standard deviation σ of the α_{ox} distributions is one of the parameters which affect most these computations. We suggest that the determination of this parameter in both optically and X-ray selected AGN samples suffers from a systematic bias towards high values. Several effects are probably contributing to the broadening of the α_{ox} distributions:
1) Noise due to the intrinsic absorption at optical and X-ray wavelenghts, predominantly in low luminosity sources.
2) Noise due to long term optical and X-ray variability.
3) Errors in the estimated optical and X-ray fluxes.
4) Errors in the conversion from optical and X-ray fluxes to the monochromatic luminosities.

We have found that the combination of these various effects may help explaining the difference between the predicted and observed values of the standard deviation of the α_{ox} distributions.

REFERENCES

Avni,Y. and Tananbaum,H., 1982, Ap.J.(Letters) 262, L17.
Cheng,F.Z., Danese,L., De Zotti,G. and Franceschini,A., 1985, MNRAS 212, 857.
Danese,L., De Zotti,G. and Franceschini,A., 1985, Astr. Ap. 143, 277.
Franceschini,A., Gioia,I.M. and Maccacaro,T., 1985, submitted to Ap.J.
Kriss,G.A. and Canizares,R.G., 1985, submitted to Ap.J.
Maccacaro,T., 1984, in "X-ray and UV emission from Active Galactic Nuclei", eds. W. Brinkman and J. Trumper, pag. 63.
Maccacaro,T. and Gioia,I.M., 1983, IAU Symp. 104, eds. G.O. Abell and G. Chincarini, pag. 7.
Maccacaro,T., Gioia,I.M. and Stocke,J.T., 1984, Ap.J. 283, 486.
Schmidt,M. and Green,R., 1983, Ap.J. 269, 352.
Setti,G., 1984, in "X-ray and UV emission from Active Galactic Nuclei", eds. W. Brinkman and J. Trumper, pag. 243.
Zamorani,G., 1984a, in "X-ray Astronomy 84", eds. M. Oda and R. Giacconi, pag. 419.
Zamorani,G., 1984b, in "VLBI and Compact Radio Sources", eds. R. Fanti, K. Kellermann and G. Setti, pag. 85.

GROUPS AROUND BRIGHT SEYFERT NUCLEI

K.J. Fricke, W. Kollatschny, H. H. Loose
Universitäts-Sternwarte
Geismarlandstr. 11
3400 Göttingen
West Germany

ABSTRACT. We have determined the nuclear galaxy spectra of the members of wide groups (~ 1 Mpc across) around bright Seyfert galaxies. We find a tight relation between the "activity classes" of the nuclei as defined from their emission line spectra and the distance from the central Seyfert galaxy. This relation is characterized by a critical radius.

I. INTRODUCTION

Evidence has been found that a large fraction of quasars (~ 30%) is in a state of interaction with nearby galaxies and an even larger fraction of quasars is connected with groups of ~ 1 Mpc diameter (Hutchings et al. 1984). For Seyfert galaxies, Dahari (1984) found on POSS plates a significant excess of companions indicating a tendency of interactions toward the Seyfert phenomenon.

In an investigation of Arp galaxies, Keel et al. (1985) found enhanced nuclear activity over all Hubble types. This effect does not go away if the well-known correlation of nuclear emission with Hubble type is taken into account. However, these authors also find that ongoing interaction is neither a necessary nor a sufficient condition for nuclear activity. Activity in more distant companions of Seyfert galaxies has first been described by Kollatschny and Fricke (1985) for the group around NGC 4593.

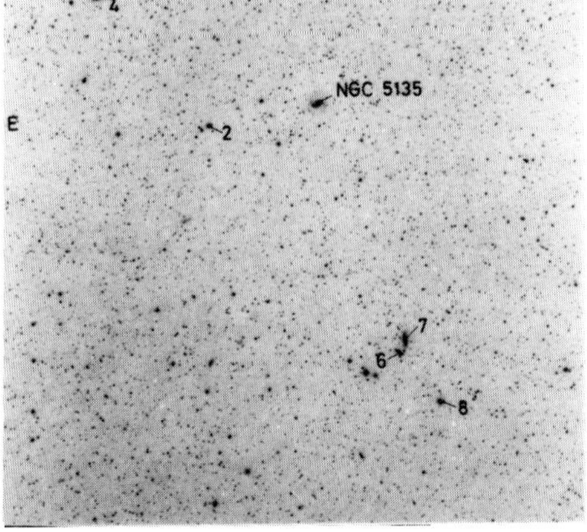

Fig.1: NGC 5135 and its surrounding group

G. Giuricin et al. (eds.), Structure and Evolution of Active Galactic Nuclei, 551–554.

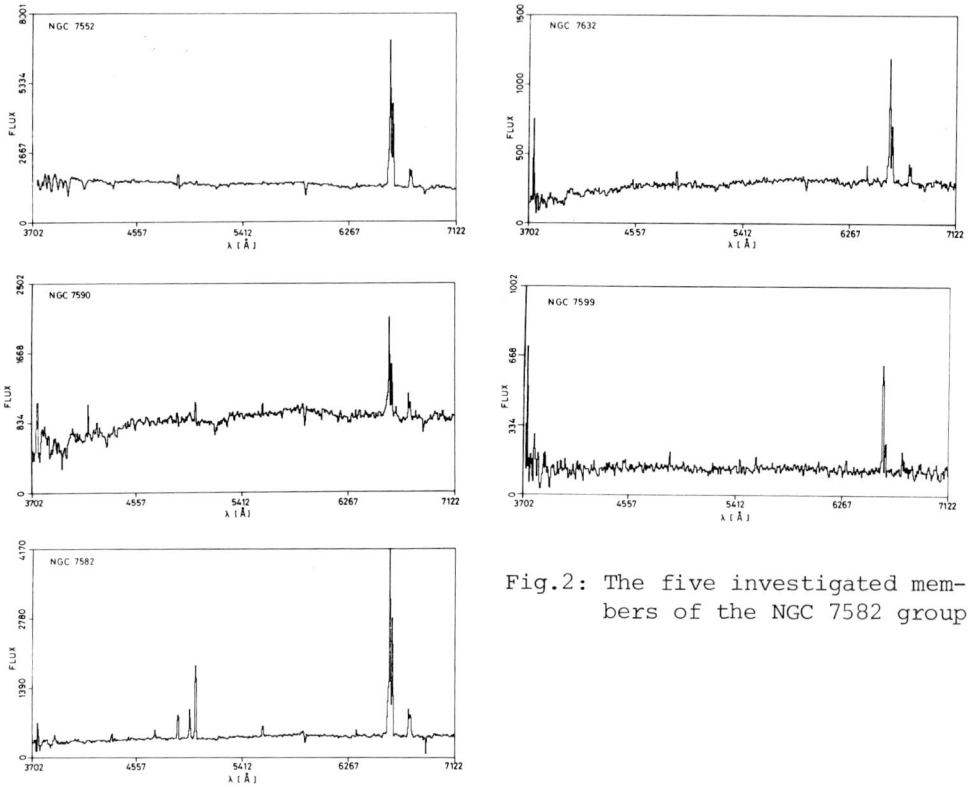

Fig.2: The five investigated members of the NGC 7582 group

II. OBSERVATIONS

We have inspected Palomar and ESO/SRC survey plates and find that Seyfert galaxies - like the low-z quasars - are often located within groups of ~ 1 Mpc size (see, for example, Fig.1). In extension of our work on NGC 4593 we investigated for several groups around Seyfert galaxies from our POSS survey the occurrence and degree of activity in the group members. We observed at ESO spectroscopically nine such groups which are listed in Table 1. Spectra for the NGC 7582 group are given in Fig. 2.

III. RESULTS AND DISCUSSION

The activity state of the nuclei has been characterized in terms of activity classes I to VI as defined in Fig.3. We find enhanced activity in many companions of the central Seyfert galaxy. This is consistent with the theoretical result that interactions are most effective for the feeding of a central source at low encounter velocities and high space densities (Schmutzler and Biermann 1985) typical for groups of galaxies.

In particular we observe that the activity degree of the group members drops sharply beyond a critical radius $R_C \approx 200$ kpc from the Seyfert

Table 1: Groups around Seyfert galaxies

Seyfert	Symbol	R.A.	DEC	z	group members[1]
NGC 4593	○	$12^h37^m01^s$	$-05°04$	0.008	5 (6)
NGC 7582	▽	23 15 38	-42 39	0.005	5
NGC 6221	+	16 48 26	-59 08	0.005	4
NGC 1566	×	04 18 53	-55 03	0.004	7
NGC 2992	*	09 43 18	-14 06	0.007	2
NGC 3783	⊕	11 36 33	-37 28	0.009	3
TOL 1238-364	△	12 38 11	-36 29	0.011	3
NGC 5135	□	13 22 57	-29 34	0.013	8
IC 5063	⊞	20 48 12	-57 15	0.011	2

[1] number of group members with known spectra

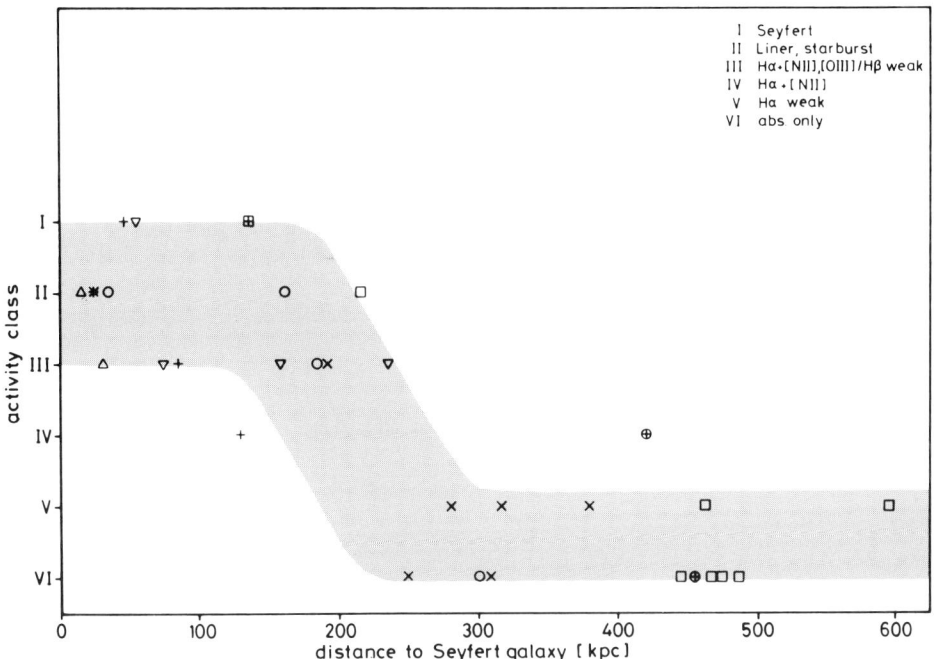

Fig. 3: The activity classes of the group members as a function of their projected distances to the Seyfert galaxies.

galaxy, independent on the morphological type. The critical radius is approximately ten times as large as the tidal radius R_T for a nuclear gas cloud in the gravitational field of a 10^{11} $M_\odot$ galaxy. If the activity of the Seyfert companions is triggered by close encounters with the Seyfert galaxy, which may itself be a result of such encounters, $R_C/R_T \gg 1$ implies that the nuclear activity in the group members lasts much longer than the duration of the close interaction. The relative velocities of the galaxies ($\lesssim 500$ km/s) suggest activity phases of 10^8 - 10^9 years duration. Such activity timescales can be readily explained in terms of our theoretical models for compact star bursts in the nuclei of galaxies (Loose and Fricke 1982). In this model the total duration of a burst or of a sequence of bursts is determined by the details of the energy supplied by massive stars and of the energy losses due to thermal and turbulent dissipation.

Presently we extend this investigation to larger samples of groups with and without Seyfert galaxies.

References

Dahari, O., 1984, *Astron. J.* **89**, 966
Hutchings, J.B., Crampton, D., Campbell, B., Duncan, D., Glendenning, B., 1984, *Astrophys. J. Suppl.* **55**, 319
Keel, W.C., Kennicutt Jr., R.C., Hummel, E., van der Hulst, J.M., 1985, (preprint)
Kollatschny, W., Fricke, K.J., 1985, *Astron. Astrophys.* **143**, 393
Loose, H.H., Fricke, K.J., 1982, Proc. ESO-Workshop *The Most Massive Stars* (d'Odorico, S., Baade, D., Kjär, K., eds.), p. 269
Schmutzler, T., Biermann, P., 1985, *Astron. Astrophys.* (in press)

Acknowledgements

This work has been supported in part by the Deutsche Forschungsgemeinschaft under grants Fr 325/21-1 and Fr 325/15-2.

STATISTICS OF BRIGHT GALAXIES AND CLUSTER MORPHOLOGY

G. Giuricin[1], F. Mardirossian[2], M. Mezzetti[1]
Astronomical Observatory, Via G.B. Tiepolo, 11
I 34131, Trieste, Italy[1]
Department of Astronomy, University of Trieste,
Via G.B. Tiepolo, 11 - I 34131, Trieste, Italy[2]

ABSTRACT

We have investigated the luminosity distributions of the first-ranked, second-ranked, and third-ranked cluster galaxies for different cluster morphological types. Basically, we have found that the two magnitude distributions relative to the cD, B and C, L, F, I clusters (according to Rood and Sastry's classification) differ significantly for the first-ranked members, whereas they are not significantly different for the second- and third-ranked members. This fact points to some differences in the creation mechanism or evolutionary processes relative to the first-ranked members of the cD, B and C, L, F, I clusters.

INTRODUCTION

The small intrinsic scatter in the luminosities of the brightest members of rich clusters of galaxies has been a subject of hot debate in the literature. It has been often argued that these galaxies are special objects and not simply the tail-end of a statistical distribution. Recently Bhavsar and Barrow (1985) have claimed that the distributions of the luminosities of the first-ranked cluster members are not the extreme members of a statistical population (at variance with what is found in the loose groups of galaxies).

In order to cast further light on this point, we have decided to investigate the distributions of the luminosities of the first-, second-, third-ranked cluster galaxies separately for different cluster morphological types (cD, B, C, L, F, I clusters, according to Rood and Sastry's classification). For this purpose we have used the magnitudes given in the list of Hoessel et al. (1980,

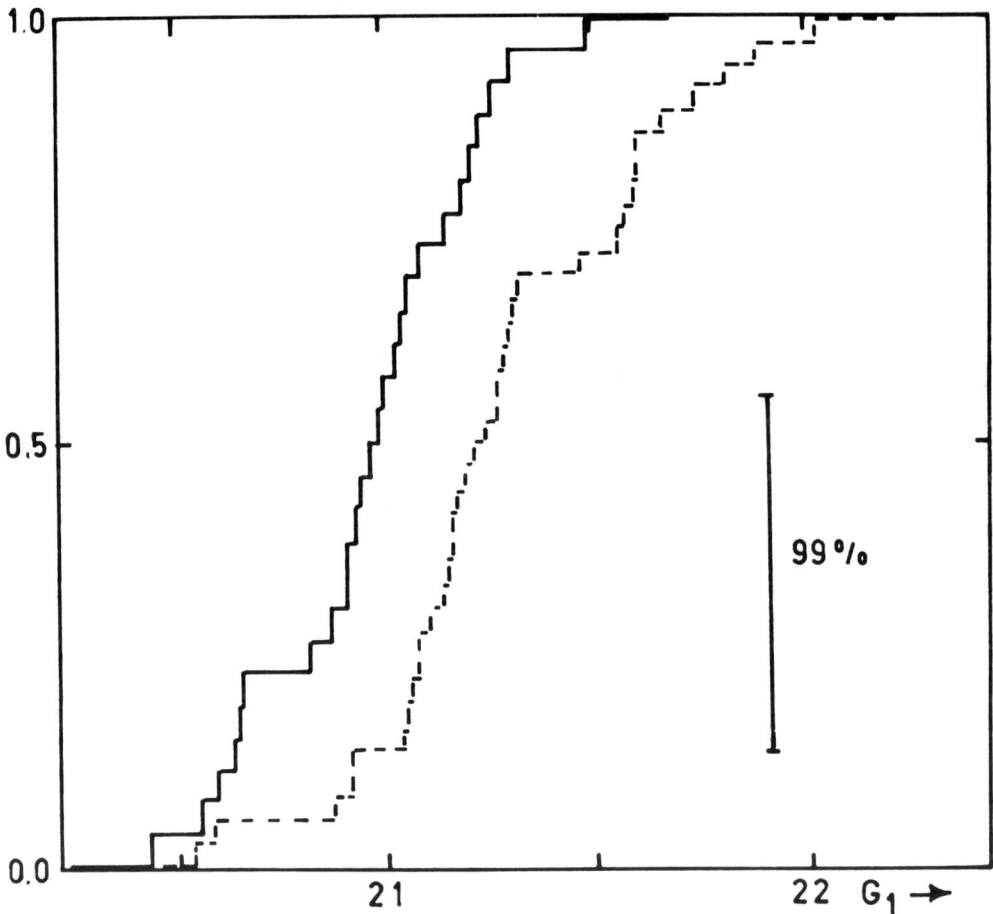

Fig. 1   The lines show the distribution function of the G1 magnitudes (Schneider et al., 1983) of the first-ranked galaxies of the cD,B (solid line) and C,L,F,I clusters (dashed line). The bar represents the separation that the distributions have to reach (at least once) in order not to be similar at the 99% confidence level.

HGT) consisting of 116 first-ranked galaxies in a complete sample of nearby Abell clusters and those of Schneider et al. (1983, SGH), who give values of their magnitudes G1, G2, G3 for 83 intermediate distance Abell clusters.

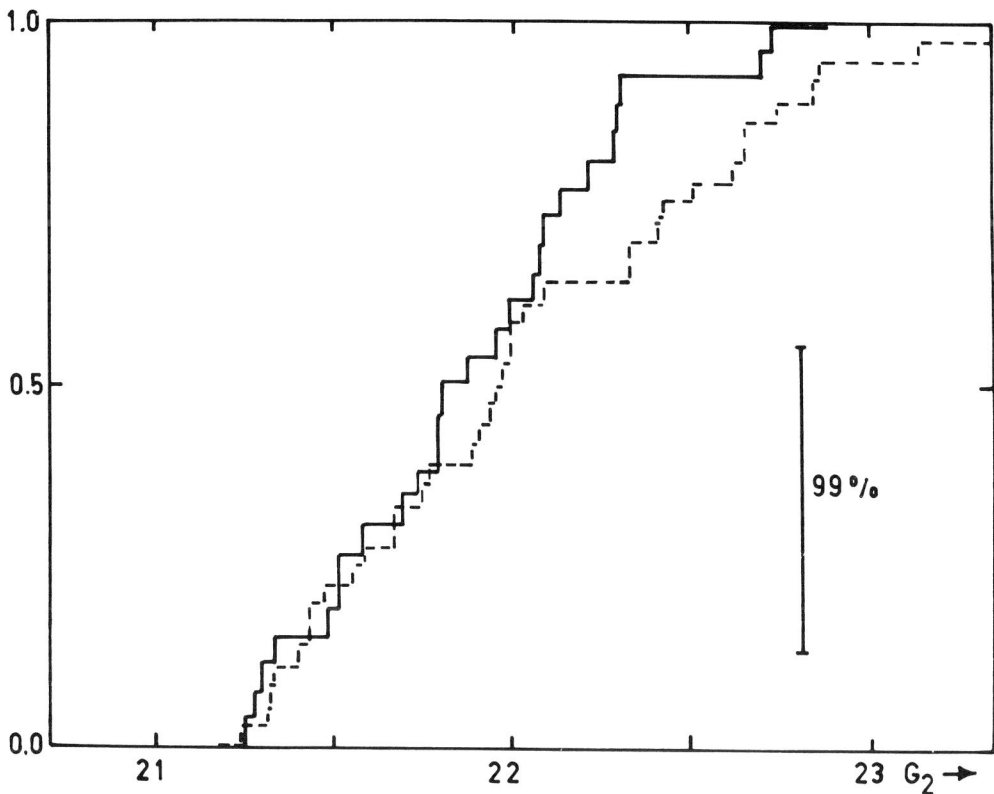

Fig. 2   The same as in Fig. 1, but for the G2 magnitudes of the second-ranked galaxies.

ANALYSIS

Basically, applying a Kolmogorov-Smirnov test on the cumulative distributions of the HGT M1 magnitudes, we have found that the distributions relative to the cD, B, and C, L, F, I clusters differ significantly at the > 99% confidence level. The same holds for the SGH sample of G1 magnitudes (see Fig. 1). On the other hand, there is no significant difference between the two corresponding distributions for the second- and third-ranked cluster galaxies (see Fig.s 2 and 3).

Our findings point to some differences in the creation mechanisms or evolutionary processes relative to the first-ranked members of the cD, B and C, L, F, I clusters.

A more detailed work on this line is in progress.

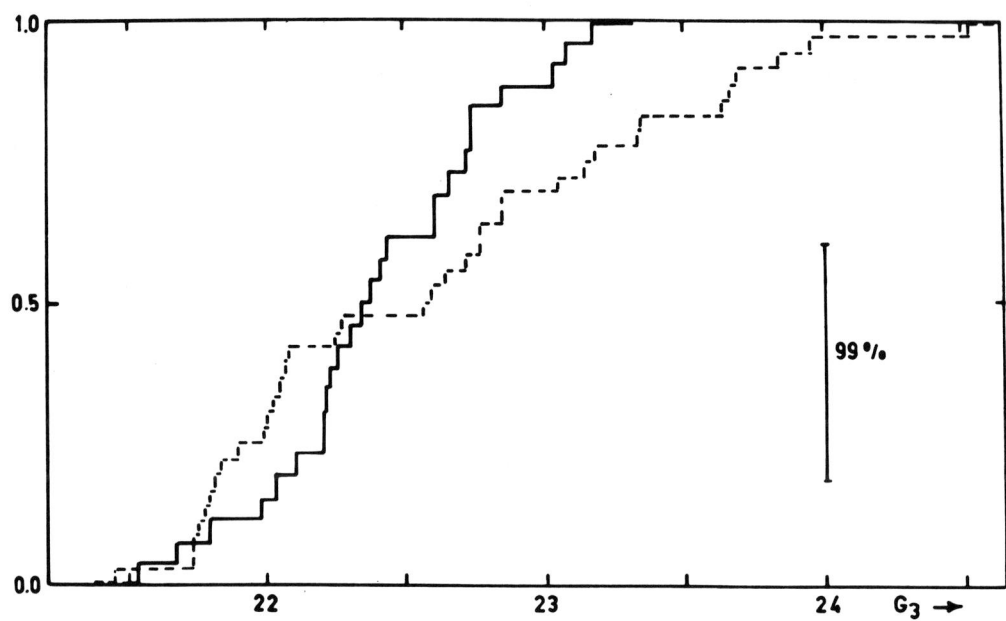

Fig. 3   The same as in Fig. 1, but for the G3 magnitudes of the third-ranked galaxies.

REFERENCES

Bhavsar, S.P. and Barrow, J.D.: 1985, Monthly Notices Roy. Astron. Soc., in press.

Hoessel, J.G., Gunn, J.E., Thuan, T.X.: 1980, Astrophys. J. 241, 486.

Schneider, D.P., Gunn, J.E., Hoessel, J.G.: 1983, Astrophys. J. 264, 337.

FIRST RESULTS FROM A CATALOG OF DATA CONCERNING AGN - FWHM AND FWZI VERSUS L_o FOR CIV AND Hβ

Monique Joly
Observatoire de Paris-Meudon
92195-Meudon Principal Cedex
France

ABSTRACT : From a statistical analysis on a large sample of Seyfert 1 nuclei and quasars, I put into evidence the existence of an homogeneous class of objects defined by a FWHM(Hβ) larger than 1700 km/s. For these AGN, there is a tight correlation between FWZI(Hβ) and the optical luminosity. From this relation one can deduce that the radiation process in low luminosity object occurs far below the Eddington limit with an efficiency factor independant of the mass while at high luminosity the masses involved are very high. It is shown also that in low luminosity objects the CIV line arises in a region with higher velocity dispersion than Hβ.

1. INTRODUCTION

The study of AGN is characterized, up to now, by the accumulation of a lot of disparate data which are still far from being well understood. Once possible way to shed light on them is a statistical approach, in particular the search for correlations between the observational parameters. For this purpose we have built up a catalogue of observational data on QSO's, Seyfert 1 nuclei and broad line radio galaxies from the literature. It is not exhaustive though greatly enlarged compared to previous ones.

I present here the results from a subsample of about 200 objects having known redshifts, apparent magnitudes and full width at zero intensity, FWZI, and at half of maximum, FWHM, of Hβ and CIV 1550.

2. OBSERVATIONAL DATA

Measurements of CIV and Hβ line widths have been collected mainly from published IDS and IPCS observations. In each line, there is a contribution from two distinct regions : the narrow line region (NLR) and the broad line region (BLR), very different in size and density. We are mainly interested here by the BLR, closest to the central source ($\sim 10^{17}$cm diameter). The separation between the two regions is fundamental.

In most cases the resolution power was not sufficient (< 1000) to isolate the NLR contribution and therefore the FWHM measures a line profile including the narrow component. However, some authors have succeeded to divide the broad and the narrow line component and give FWHM related only to the BLR.

We call these data Pure BLR Widths.

3. RELATION BETWEEN THE LINE WIDTHS AND THE OPTICAL LUMINOSITY

3.1. FWHM versus L_o

Figure 1 shows FWH(Hβ) and FWHM (CIV) versus the optical luminosity, L_o, for the whole sample of objects. We have symbolized the Pure BLR widths by a distinct mark on the figure (this only concern Hβ widths, for the CIV 1550 line the narrow and broad components have never been separated). On the whole their distribution is similar to that of the full sample except that below $\sim$ 1700 km/s there is only a small number of objects, all gathered in the low luminosity region ($L_o < 10^{44}$ erg/s). It shows clearly the lack of narrow line AGN's when the BLR component is measured separately. We have therefore divided the full sample into 2 classes :
- class 1 : FWHM(Hβ) < 1700 km/s
- class 2 : FWHM(Hβ) > 1700 km/s

It is obvious that class 2, the distribution of which is very similar to the width distribution of the Pure BLR component, is a population of AGN for which the NLR has a very weak contribution to the line strength.

It is noteworthy that FWHM (CIV) exhibit a lower limit $\sim$ 1700 km/s corresponding exactly to the lower limit of class 2 objects. Moreover, in the common luminosity range, there is a strong similarity between the distribution of FWHM (CIV) and class 2 FWHM(Hβ), which proves that the NLR makes only a weak contribution to the CIV line strength.

3.2. FWZI versus L_o

Figure 2 shows FWZI(Hβ) and FWZI(CIV) versus optical luminosity. The uncertainty on FWZI is mainly due to the uneasy determination of the continuum and in some cases, to a low signal/noise ratio. However we have verified that there is no correlation between the FWZI and the V magnitude. There is a weak correlation between the width of Hβ and L_o.

Now if we consider the class 2 AGN's defined previously there is a strong correlation (figure 3 ; N = 54, r = 0.52, P < 0.001%) of the form : FWZI(Hβ) =10^{-4} $L_o^{0.20\pm0.04}$.

Figure 4 shows FWZI(CIV) and class 2 FWZI(Hβ) versus L_o. Contrary to FWHM, in low luminosity objects the widths distributions are different : the wings of CIV extend farther than those of Hβ. Note the lack of small values of FWZI(CIV), which are always larger than 6000 km/s, and the presence of large values up to 50000 km/s.

A CATALOG OF DATA: AGN-FWHM AND FWZI VERSUS L_o FOR CIV AND Hβ

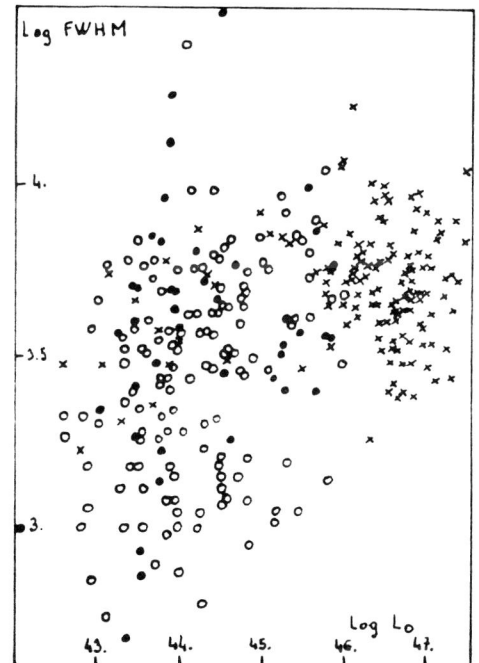

Figure 1. Hβ (open and full circles) and CIV 1550 (crosses) line widths at half magnitude versus optical luminosity. Full circles are for pure BLR measurements.

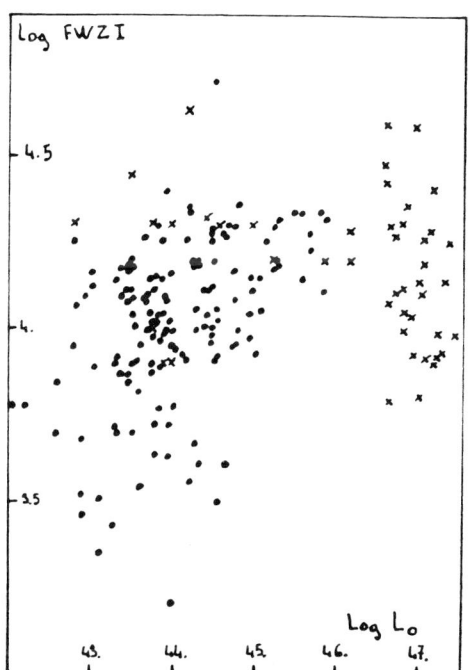

Figure 2. Hβ (circles) and CIV (crosses) line widths at zero intensity versus optical luminosity.

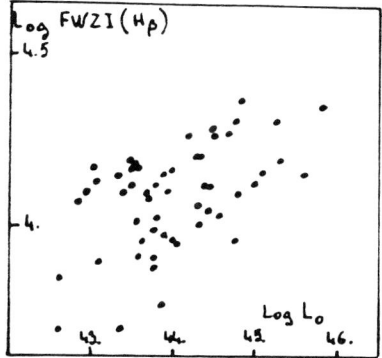

Figure 3. Hβ line widths at zero intensity versus optical luminosity for class 2 objects.

3.3. Interpretation

In the framework of gravitational model of the BLR, the velocity of the clouds is linked to the central mass and the luminosity can be compared to the Eddington luminosity (which is also a fonction of the central mass).

On figure 4 are drawn the line of constant central mass as well as the range $L_O = L_{Edd}$, as a fonction of the physical parameter $U \times n$; n is the density and U in the ionization parameter $U = L/4\pi r^2 n$ (L being the ionizing radiation luminosity, which is about equal to L_O when the spectral index is 1 and r the distance to the central source). U ranges from 10^{-2} in quasars to 10^{-1} in low luminosity Seyfert and n is of the order of 10^{10} cm^{-3} in the BLR. This figure shows that low luminosity AGN's ($L_O < 10^{46}$ ergs/s) radiate far below their Eddington luminosity with an efficiency factor independant of the mass, while for high luminosity AGN's the masses involved are very large ($\sim 10^{11} M_\odot$).

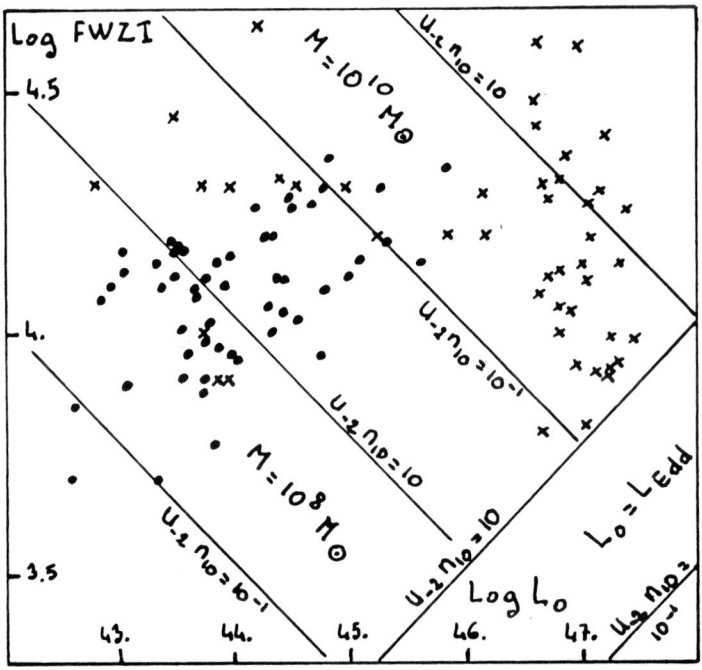

Figure 4. Hβ (class 2 : circles) and CIV (crosses) line widths at zero intensity versus optical luminosity. Straight lines delimit the ranges of masses and the range where $L_O = L_{Edd}$ as predicted by the gravitational model.

4. CONCLUSION

The existence of an homogeneous class AGN's, characterized by a weak contribution of the NLR to the line strength, is put in evidence.

In low luminosity objects the wings of CIV extend farther than those of Hβ but the core of the lines are similar.

An interpretation could be that the BLR extends on a wide scale in Seyfert 1 galaxies contrary to quasars.

In the framework of gravitational models :
- AGN radiate for below their Eddington luminosity ;
- For $L_o < 10^{46}$ ergs/s the efficiency factor is independent of the mass ;
- At high luminosity ($L_o \sim 10^{47}$ ergs/s) the masses involved are large $\sim 10^{11} M_\odot$.

COSMOLOGICAL EVOLUTION IN THE EXTENT OF DOUBLE RADIO GALAXIES

V. K. Kapahi
Tata Institute of Fundamental Research
P.O. Box 1234
Bangalore 560 012
India

ABSTRACT. A comparison of the sizes of radio galaxies of similar radio luminosity at different redshifts is shown to provide strong evidence for size evolution with epoch independent of any luminosity-size relationship.

1. INTRODUCTION

Although quasars and radio galaxies at large redshifts are observed to have a considerably smaller extent associated with their basically double radio structure compared to those at smaller redshifts, it has not been possible to decide unambiguously whether this is a result of cosmological evolution (arising, for instance, from the higher intergalactic density at earlier epochs) or whether it results from a possible inverse correlation between physical size and radio luminosity, because redshift and luminosity are strongly correlated in the existing flux limited source samples. It is important to disentangle the two effects to further our understanding of the formation and subsequent evolution of the double structure of powerful radio sources.

We show here that by combining data from the recent Leiden-Berkeley Deep Survey (LBDS) with data from existing strong source surveys it is now possible to compare the physical sizes of radio galaxies of the same radio luminosity at different redshifts. The results strongly favour an epoch dependence of radio size.

2. THE θ-z RELATION FOR RADIO GALAXIES OF CONSTANT LUMINOSITY

In order to obtain unbiased samples of radio galaxies of a constant luminosity at different epochs we consider the following three complete samples, all selected from surveys made at a frequency of 1.4 GHz.
i) The 36 sources from the BDFL catalogue (Bridle et al. 1972) with $S_{1.4\ GHz} \geq 2$ Jy that are identified with galaxies having $0.075 < z < 0.2$. Radio structures are well known from the literature.

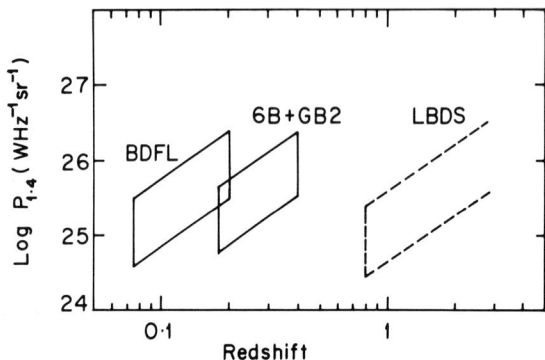

Figure 1. The range of luminosity covered by the three samples.

ii) 48 sources from the GB and GB2 surveys with $S_{1.4} \geq 0.55$ Jy (Machalski & Maslowski 1982) identified with galaxies of m_{pg} between 17.5 and 20, corresponding approximately to the redshift range of 0.18 to 0.4. Optical identifications and VLA structures are known mainly from Machalski and Condon (1985; and references therein).

iii) All the 41 sources in the recent Leiden-Berkely deep radio and optical survey which have $S_{1.4} \geq 10$ mJy and are either identified with galaxies of F magnitude ≥ 22 or are optically unidentified (implying $F \geq 22.75$). In this flux range nearly all the identified galaxies appear to be giant ellipticals with a small dispersion in their absolute magnitudes. A majority of the unidentified sources are also likely to be ellipticals at $z \geq 0.8$ (Windhorst 1984). Angular sizes (θ) or upper limits to them have been derived in the survey from the Westerbork resolution of, about 12" arc.

The relatively narrow range of luminosities covered by the three samples is shown in Figure 1. Most of the galaxies in the LBDS sample could infact be at $z \leq 2$ (Windhorst 1984). Note also that although the structure of many LBDS sources is not fully determined due to the relatively poor resolution, most of them are likely to be standard limb-brightened doubles as their luminosity places them in the class II category of Fanaroff and Riley (1974). It is also worth noting that only one source in the BDFL sample and two sources in the GB samples are known to be of the flat spectrum type ($\alpha < 0.5$). The percentage of such sources in the LBDS sample is also expected to be quite small.

The distributions of θ in the 3 samples are shown in Figure 2 and the corresponding $\theta_{med}(z)$ relation in Figure 3. Although the value of $\theta_m \leq 7"$ arc for the LBDS sources has been plotted somewhat arbitrarily at a median redshift of 1.5, the exact value of z_{med} in this sample is not very crucial for our purpose here. It is clear from Figure 3 that θ_m decreases faster with z than predicted by different world models (including $q_0 = 0$) assuming a constant linear size. This can now only be attributed to cosmological evolution of linear sizes. If the evolution is expressed in the simplest form $\ell \alpha \ell_0 (1+z)^{-n}$, then values of n between 1 and 2 are indicated.

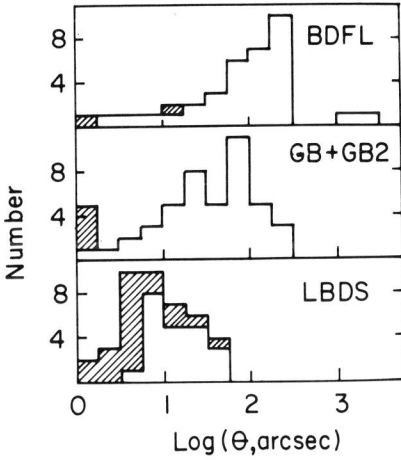

Figure 2. Distributions of angular sizes in the three samples. Shading represents upper limits.

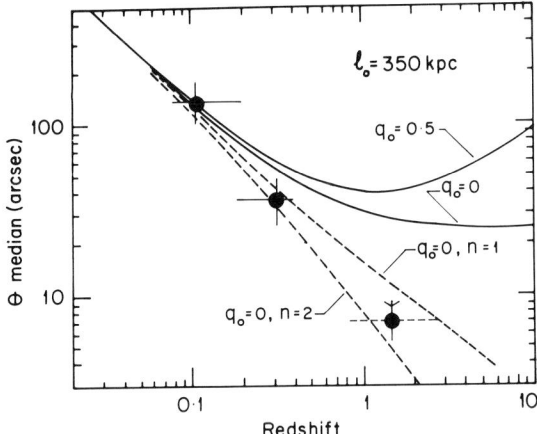

Figure 3. The observed θ-Z relation. Solid and dotted curves show predictions without and with size evolution respectively.

Similar conclusions regarding size evolution have been drawn earlier from an investigation of the θ-S relation for radio sources (eg. Kapahi & Subrahmanya 1982). But it has however been pointed out (eg. Downes 1982; Fielden et al. 1983) that small values of θ_m at low flux levels could arise from a large contribution made be steep-spectrum compact sources that are known to generally have sizes ≲ 30 kpc. This is very unlikely to be the case in the present samples because the median linear sizes of even the resolved sources ($\ell > 50$ kpc) in the BDFL and LBDS samples differ significantly, being about 400 and

140 kpc respectively.

3. CONCLUSION

The new data from the deep radio and optical survey appear to provide strong evidence in favour of an epoch dependence of linear size such as could arise from decreasing intergalactic density with cosmic epoch. Recently it has also been reported by Barthel (1984) that radio quasars at high redshifts appear to have more distorted structure suggestive of a denser medium at earlier epochs. It would be interesting to map with higher angular resolution the LBDS sources considered in our sample in order to compare their structural distortions if any with those in the stronger BDFL sources.

REFERENCES

Barthel P.D., 1984, Ph.D Thesis, University of Leiden.
Bridle A.H., Davis M.M., Fomalont E.B. & Lequeux J. 1972, Astron J. 77, 405.
Downes A.J.B., 1982, in Extragalactic Radio Sources, p.393, eds. Heeschen & Wade, Reidel.
Fanaroff B.L. & Riley J.M., 1974, MNRAS, 167, 31P.
Fielden J. et al., 1983, MNRAS, 204, 289.
Kapahi V.K. & Subrahmanya C.R., 1982, in Extragalactic Radio Sources, p.401, Eds. Heeschen & Wade, Reidel.
Machalski J. & Condon J.J., 1985, Astron J., 90, 5.
Machalski J. & Maslowski J., 1982, Astron J., 87, 1132.
Windhorst R.A., 1984, Ph.D. Thesis, University of Leiden.

THE SPECTRAL INDEX - FLUX DENSITY RELATION AND THE COSMOLOGICAL EVOLUTION OF RADIO SOURCES

V.K. Kapahi[1] and V.K. Kulkarni[2]
[1]T.I.F.R. Centre, Post Box 1234, Bangalore - 560 012, India
[2]Radio Astronomy Centre, Post Box 8, Ootacamund - 643 001, India

ABSTRACT. A careful redetermination of the α-S relation for radio sources selected from metre-wavelengths surveys does not show any significant flattening of α_{median} below $S_{408} \sim 1$ Jy contrary to earlier claims. Constraints on evolutionary models of the luminosity function are briefly discussed.

1. INTRODUCTION

Because of the observed correlation between radio luminosity (P) and spectral index (α) for powerful extended radio galaxies and quasars (eg. Laing & Peacock 1980), distributions of α in metre-wavelength samples at different flux levels can provide useful constraints on models for the cosmological evolution of the radio luminosity function (RLF). Recent work by Gopal-Krishna & Steppe (1982; henceforth GS) and by Steppe & Gopal-Krishna (1984; henceforth SG) shows a significant flattening of α_{median} in samples at $S_{408} \lesssim 1$ Jy. This implies, on the basis of the P-α relation, that P_{median} at $S_{408} \sim 0.1$ Jy is $\gtrsim 30$ times smaller than at $S_{408} \sim 1$ Jy, in disagreement with most successful models of the evolution of RLF that have been proposed to explain the observed source counts. Here we report a redetermination of the α-S relation that does not support the results of GS & SG. We also point out some deficiencies in some of the existing evolutionary models.

2. THE α-S RELATION

In investigating the dependance of α_{med} on flux density it is important that complete and unbiased samples of sources be used, all selected from surveys at the same basic frequency and that the spectral index be always calculated using flux densities at the same set of frequencies. The scaling factors to convert different surveys to a common flux scale must also be accurately known. Some of the samples used by GS & SG do not apparently satisfy one or more of the above requirements. In our determination of the α-S relation we have used only samples selected at a

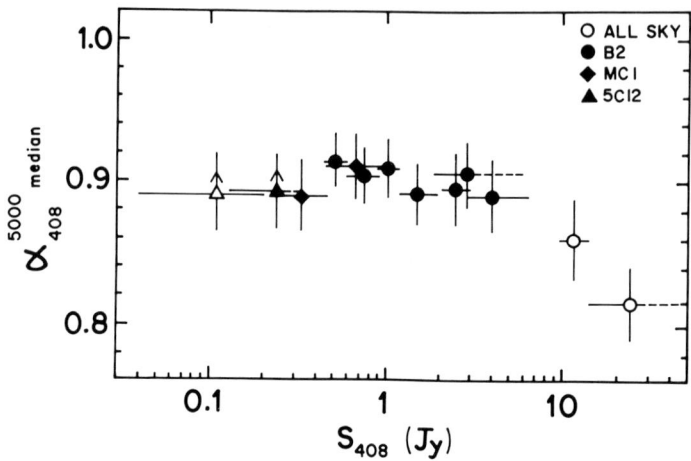

Figure 1. The observed α-S relation. The lowest flux density point refers to (408, 1400) and is from Kulkarni and Mantovani (1985, Astr. Astrophys. in press)

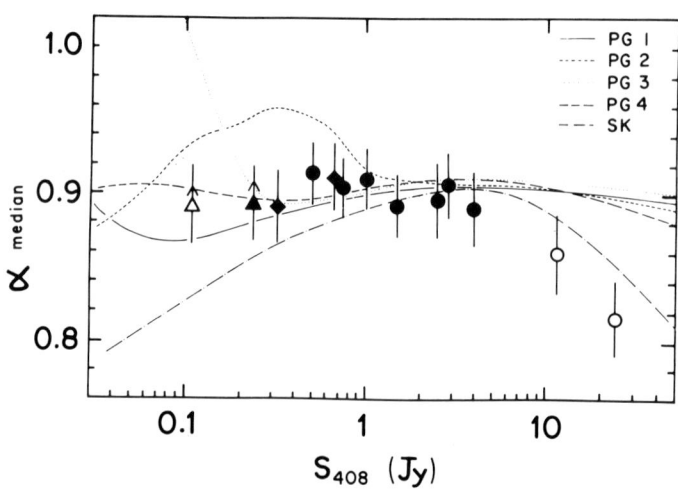

Figure 2. Predictions of different evolutionary models compared with observations.

frequency of 408 MHz for which flux densities near the higher frequency of 5 GHz are also available, (Kapahi & Kulkarni 1985).

The median values of $\alpha(5000-408)$ are shown plotted against S_{med} in Figure 1. It is seen that α_{med} steepens considerably from high to intermediate flux levels (~ 2 Jy) as noted by GS and SG. There is, however, no significant evidence for a flattering of α_{med} below $S \sim 2$ Jy to about $S \sim 100$ mJy, contrary to the earlier claims.

3. DISCUSSION

The α-S relations predicted by the multi-frequency free-form evolutionary models of Peacock & Gull (1981; hereafter PG) are compared in Figure 2 with observations. These models incorporate a P-α relation of the form

$$\alpha = 0.75 \text{ for } P_{408} < 10^{25} \text{ W Hz}^{-1} \text{ sr}^{-1}, \text{ and}$$

$$\alpha = 0.75 + 0.08(\log_{10} P_{408} - 25) \text{ for } P_{408} > 10^{25} \text{ W Hz}^{-1} \text{ sr}^{-1}.$$

Also shown in Figure 2 is the prediction of the parametric model proposed by Subrahmanya & Kapahi (1983; SK). The models appear to fit the data only in a limited range of flux density. The observations clearly provide an important constraint for evolutionary models. This is particularly important at low flux levels (≤ 1 Jy) where the models are based mainly on fitting source counts, with very little information on source redshifts.

It is surprising however that the PG models do not fit the data at the high flux end where there is reasonably good redshift information. The reason for the discrepancy could be that in these models the RLF is specified at 2.7 GHz, and for the steep-spectrum sources ($\alpha > 0.5$) a P-α relation is used to determine the RLF at other frequencies. All sources are assumed to have straight power law spectra. A significant fraction of the strong sources in high-frequency surveys is known to be of the steep-spectrum compact variety (Kapahi 1981; Peacock & Wall 1982). Such sources generally have high radio luminosity and are poorly represented in low-frequency surveys because their spectra show downward curvature at low frequencies. The assumption of a constant spectral index (independent of frequency) therefore results in an overestimation of the number of high luminosity sources at the high flux density end at 408 MHz, thus predicting an α_{med} that is steeper than observed.

Because of the strong correlation between luminosity and redshift in flux limited samples at $S \geq 1$ Jy, it has not really been possible to decide unambiguously whether α correlates with luminosity or with redshift (e.g. Merkelijn-Katgert et al. 1980). Recent work on the evolution of RLF using deep radio and optical data suggests (Windhorst 1984) that the value of Z_{med} at $S_{408} \leq 100$ mJy may be considerably smaller than implied by PG models. The fact that α_{med} does not flatten significantly down to ~ 100 mJy may then imply that spectral index correlates with epoch and not with luminosity.

Kapahi (1985) has recently defined complete samples of radio galaxies selected from surveys at 1.4 GHz that have a constant radio luminosity but are located at different redshifts. Careful measurements of the radio spectra of sources in these samples could help decide whether α correlates primarily with luminosity or with cosmic epoch.

REFERENCES

Gopal-Krishna & Steppe, H. 1982, Astr. Astrophys., 113, 150.
Kapahi, V.K. 1981, Astr. Astrophys. Suppl., 43, 381.
Kapahi, V.K. 1985, MNRAS (in press).
Kapahi, V.K. & Kulkarni, V.K. 1985 (in preparation).
Merkelijn-Katgert, J., Lari, C. & Padrielli, L. 1980, Astr. Astrophys. Suppl. 40, 91.
Peacock J.A. & Gull S.F. 1981, MNRAS, 196, 611.
Peacock J.A. & Wall J. 1982, MNRAS, 198, 843.
Steppe H. & Gopal-Krishna 1984, Astr. Astrophys., 135, 39.
Subrahmanya C.R. & Kapahi V.K., 1983, in Early Evolution of the Universe and Its Present Structure, p47, G.Abell & G.Chincarini (eds.), D. Reidel.
Windhorst R. 1984, Ph.D. Thesis, University of Leiden.

EFFECT OF DYNAMICAL FRICTION ON THE ESCAPE OF A SUPERMASSIVE BLACK HOLE EJECTED FROM THE CENTER OF A GALAXY

Ramesh Chander Kapoor
Indian Institute of Astrophysics
Bangalore 560034
India

ABSTRACT. We have used the impulsive approximation technique to numerically estimate the effect of dynamical friction on the motion of a supermassive black hole (mass $\simeq 10^9 M_\odot$) through a galaxy (mass $\simeq 10^{11} M_\odot$) which has recoiled from the center of the latter as a result of rocket effect (anisotropic emission of gravitational radiation or plasma emission etc.) We find the effect to be minimal for recoil taking place at a velocity larger than that of escape at the center of the galaxy. For recoil velocities less than a certain critical velocity (slightly larger than the central escape velocity), dynamical friction becomes relatively pronounced and damped oscillatory motion of the black hole in the potential well of the galaxy ensues.

INTRODUCTION

Ejection or displacement of matter from the kinematic center of galaxies in the form of compact supermassive bodies (spinar, black hole, white hole, etc) has often been hypothesized to explain certain high energy phenomena such as relativistic and nonrelativistic jets, radio lobes and quasar - galaxy associations etc. (Rees and Saslaw 1975, Kapoor 1976, Shklovsky 1972, 1982, Rees 1982, Narlikar 1984 and references). The ejection of the supermassive object is achieved in various processes by requiring conservation of linear momentum (slingshot/gravitational radiation/plasma emission etc.) which in an extreme case can import an initial recoil velocity $\sim 10^3$ km sec^{-1} or more. In this paper, we deal with one aspect of the problem, viz., tidal influence of the ejected compact supermassive object, most likely a black hole, over the structure of the galaxy using the impulsive approximation technique as developed by Alladin (1965) for the study of interpenetrating collisions of galaxies. The usefulness of the impulsive approximation is discussed by Ahmed and Alladin

(1981).

The observations of interest from this point of view are, for instance, (1) offset location ($\simeq$ 4 kpc) of the nebulosity around 3C 273 from the quasar position (Tyson et al. 1982), (2) disturbances in the inner isophotes of some galaxies pointing in the direction of quasars they are seen near (Arp et al. 1975, Sulentic 1983), (3) the weak Seyfert nucleus-like feature in the outer regions of Mkn 335 (Fricke et al. 1983), (4) displaced location ($\simeq$ 230 pc) of the kinematic center of the rotation curve of NGC 2110 from the active nucleus (Wilson et al. 1983).

Let the black hole, mass M_2, be ejected from the center of the galaxy, mass M_1, at a velocity V_o. For its separation z from the center of the galaxy, let W(z) be their mutual interaction energy and $\Delta U(z)$ the change in the binding energy of the galaxy induced by the black hole. The velocity of the black hole with respect to the center of the galaxy is

$$V(z) = [\frac{2}{\mu}(\tfrac{1}{2}uV_o^2 + W(o) - W(z) - \Delta U(z))]^{\tfrac{1}{2}} \tag{1}$$

The deceleration due to dynamical friction is

$$-f_D = \frac{1}{\mu}\frac{d\Delta U}{dz} \; ; \; u = \frac{M_1 M_2}{M_1 + M_2} \tag{2}$$

The effect of dynamical friction on the motion of the black holes through the galaxy is elicited by a velocity function F(z), defined as

$$F(z) = \frac{V_{esc}(z)}{V(z)} \tag{3}$$

For comparison, we also have the velocity function with dynamical friction neglected

$$F'(z) = \frac{V_{esc}(z)}{V'(z)} \tag{4}$$

where $dV'/dt = -GM_1(z)/z^2$. If $F(z) < 1$ for $0 < z < R_{galaxy}$, the recoil is successful. The limitation of the impulsive approximation is such that it can be considered applicable as long as $F(z) < 1$.

RESULTS AND DISCUSSION

For computations, the galaxy is regarded as a Plummer sphere. Parameters chosen are: $M_1 = 10^{11} M_\odot$, $M_2 = 1.1 \times 10^9 M_\odot$, scale length $\alpha = 2.5$ kpc and

$V_o = 0.60, 0.62, 0.65, 0.70$ and 0.80 (velocities are in units of 1000 km sec^{-1}); note that $V_{esc}(o) = 0.59$. The results are summarized in Fig. 1 for the velocity function and Fig. 2 for dynamical friction as functions of separation of the black hole from the center of the galaxy for various velocities of ejection. The effect of dynamical friction can be seen to be minimal which dwindles as the velocity of ejection is increased to the extent that the dotted and the solid lines overlap for $V_o = 0.80$ in Fig. 1. The motion is reversed in the case of $V_o = 0.60$ since $F(z) > 1$ and soon tends to infinity. The arrows in the $V_o = 0.60$ curves in Fig.1 and 2 imply the iterations becoming inaccurate beyond this point. In such a case, the black hole executes damped oscillatory motion through the galaxy to eventually settle down in its kinematic centre. This aspect of the problem is of astrophysical interest but requires numerical simulations for a correct picture.

The computations suggest that a recoil must take place at a velocity $V_{critical}$, rather than at $V_{esc}(o)$ in order to be a successful one. For the parameters we have chosen, $V_{critical}$ turns out to be $\simeq 1.01\ V_{esc}(o)$. If M_2 is increased, this value would be revised upwards since $\Delta U \sim <(\vec{\Delta V})^2>$; $V_{critical}$ and the force of dynamical friction is slightly increased is we take into account the extended nature of the object, however compact from galactic standards. The object, for the purpose of illustration is taken to be an isothermal sphere of 10^8 stars around a black hole, with a net mass $1.1 \times 10^9 M_\odot$, i.e. a loaded polytrope of Huntley and Saslaw (1975) which is truncated at $\sim$ accretion radius of the black hole. This is reasonable to assume since the black hole that is ejected from the center of the galaxy is likely to be surrounded by a system of accretion disk, gas clouds and a star cluster which can be very large in number if a density cusp has formed prior to ejection and a major portion of it will be carried along because of its being bound to it.

It is apparent that dynamical friction is hardly significant for $M_2 < 10^8 M_\odot$ and $V_o > V_{critical}$. It is interesting to note that the velocity increments for the stars in the galaxy are so small that even an object of $\sim 10^9 M_\odot$ and $V_o \sim V_{critical}$ is unable to pull substantial amounts of galactic material in its wake. Instead, its track would be delineated by a very faoint perturbation in the distribution of stars and gas in its wake, ionizing effects on the interstellar medium, bursts of star formation and hot spots (plasma thrown out at large velocities from the object) giving rise to a blue gradient steepening towards the object. The object would be luminous, $\sim 10^{41}$ erg sec^{-1} if it is just a bare black hole that accretes gas from the interstellar medium while moving through, and $< 10^{44}$ erg sec^{-1} if a star cluster with $< 10^8$ stars in number accompanies it.

The details of the calculations are presented in Kapoor (1985).

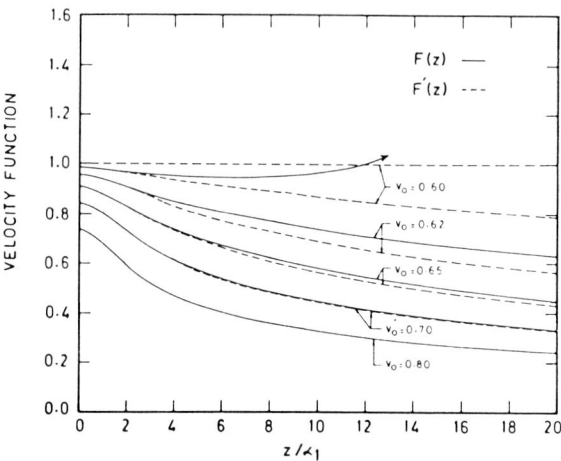

Figure 1. This shows the behaviour of the velocity function as a function of separation z for various velocities of ejection. Upward bend in the $V_o = 0.60$ curve implies a failed recoil. Solid lines take into account the dynamical friction but dotted lines do not.

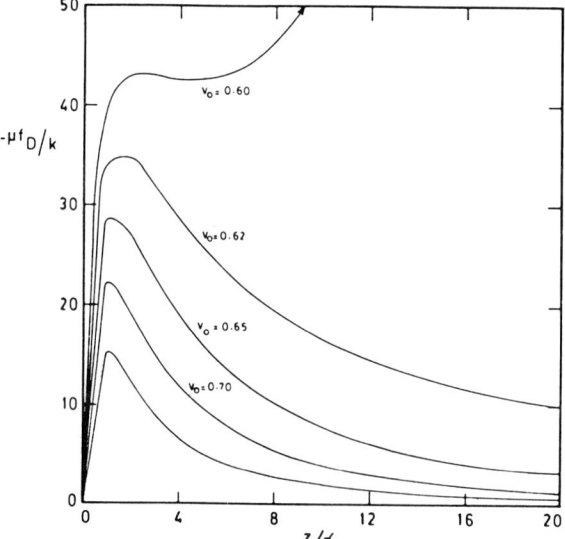

Figure 2. The general effect of the force of dynamical friction on the motion of the object is shown as its separation from the center of the galaxy increases, for various velocities of ejection. The quantity $K = 1.41 \cdot 10^{31}$ dynes.

REFERENCES

Ahmed, F., Alladin, S.M., 1981 Bull. Astron. Soc. India, 9, 40.
Alladin, S.M. 1965, Ap. J. 141, 768
Arp, H.C., Pratt, N.M., Sulentic, J.W., 1975 Ap. J. 199, 565.
Fricke, K.J., Kollatschny, W., Schleicher, H., 1983, A. Ferrari and
    A.G. Pacholczyck (Eds.) Astrophysical Jets (D. Reidel), p. 149
Huntley, J.M., Saslaw, W.C. 1975, Ap. J. 199, 328.
Kapoor, R.C., 1976 Pramana 7, 334.
Narlikar, J.V., 1984, Bull. Astron. Soc. India 12, 1.
Rees, M.J. 1982 in D.S. Heeschen (Ed.) Extragalactic Radio Source
    (D. Reidel), IAU Symp. 97 p. 211.
Rees, M.J., Saslaw, W.C., 1975 M.N.R.A.S. 171, 53.
Shklovsky, I.S. 1972 in D.S. Evans (Ed.) External Galaxies and Quasi-
    Stellar Objects, IAU Symp. 44, 272.
Shklovsky, I.S. 1982 in D.S. Heeschen (Ed.) Extragalactic Radio Sources
    (D. Reidel), IAU Symp. 97, p. 475.
Sulentic, J.W. 1983, Ap. J. 265, L 49
Tyson J.A., Baum, W.A., Kreidl, T. 1982, Ap. J. 257 L1.
Wilson, A.S., Baldwin, J.A., Ward, M.J. and Ulvestad, J.S., 1983, Bull.
    A.A.S. 15, 988.

CLASSIFICATION OF OPTICAL JETS IN GALAXIES

William C. Keel
Kitt Peak National Observatory
National Optical Astronomy Observatories
P. O. Box 26732
Tucson, AZ 85726,   U. S. A.

ABSTRACT. New, deep images of galaxies reported in various catalogs to have optical jets show that most are not jets, but tidal features, polar rings, or superpositions. From those objects that do seem to represent nuclear ejecta, criteria of brightness relative to the parent galaxy, width, and location have been derived, which should produce much purer samples of optical jets. Jet-like features in ESO 0610-23, UGC 3995, and NGC 1598 are described. Finally, a new search of the SRC IIIa-J survey is described, which was intended to incorporate the new criteria and provide a large sample of optically-selected jets for detailed study.

1. INTRODUCTION

   Many galaxies are listed in notes to various catalogs as having apparent optical jets. These objects are of great interest in view of the connections now found between nuclear activity and jet phenomena on parsec to megaparsec scales, especially as possible examples of radio-quiet collimated ejection. Accordingly, these objects have been examined on Sky Survey images, and new images and spectra obtained for the most promising candidates. From these, it is possible to distinguish certain kinds of confusing phenomena from jets, and better define the optical properties of genuine jets.

2. OBSERVATIONS

   Galaxies were taken from the Zwicky (1971), Uppsala (Nilson 1973), ESO-Uppsala (Lauberts 1982), and Michigan-Tololo (Mac Alpine, Lewis, and Smith 1977 et seq.) catalogs. All objects noted therein as having jets or possible jets were examined on the appropriate sky survey fields and those appearing most likely to be jets were observed further.
   Images were obtained using image-tube and video (ISIT) cameras at the KPNO 2.1-m and CCDs at the KPNO and CTIO 4-m telescopes. B and V

frames are available for most objects, and narrow-band Hα frames for a few. All but the image-tube data are suitable for photometry, so that magnitudes and colors of the jet-like features and the parent galaxies could be measured.

Spectra have been obtained for many of the galaxy nuclei, and a few of the apparent jets. The image-dissector scanner systems at the Mt. Lemmon 1.5-m and KPNO 2.1-m, Cryogenic Camera at the KPNO 4-m, and SIT-Vidicon at the CTIO 4-m were employed, with resolution 6-20Å.

In particular, evidence for or against current nonstellar activity in the nuclei was sought, and signs of stars or gas in the apparent jets were a prime goal of the study. In this way, the features in VV144 and I Zw 96 could be securely identified as tidal remnants. Two objects, AM 0207-49A and UGC 3995A, were found to be type 2 Seyfert galaxies.

3. RESULTS

The apparent jets were divided into those probably originating by ejection from the nucleus (of which few were found), and other types such as tidal tails, disrupted companions, polar rings, or photographic artifacts. The classification could be based on direct spectroscopic data (stellar population in the "jet," velocity structure, ionized-gas content), on color and absolute magnitude, or on morphological grounds by comparison with well-established similar objects.

Together with data in the literature on objects such as M87 (de Vaucouleurs and Nieto 1979) and NGC 1097 (Lorre 1978), measurements of the objects observed here could be used to produce a set of criteria for separating jets and false alarms, at least as far as can be done on sky survey plates. These entries may be summarized as follows:
1) Jets have no more than a few percent of the parent galaxies' optical (stellar) luminosity. This helps rule out tidal features, which, even when viewed so as to appear straight, are generally much brighter.
2) Jets are either very asymmetric in intensity or completely one-sided. The most asymmetric set of optical jets reported (and apparently the only one to date) is the N-S pair in NGC 1097, with a total-intensity ratio of ~5:1 at V. This criterion distinguishes edge-on polar rings or bars in faint disks from jets.
3) One-sided fetures along the major axes of highly inclined galaxies are usually spiral arms, appearing asymmetric due to the internal dust distributions, and not jets.
4) Features seen on photographs are not likely to be real objects if they occur very near the plate limit. Chance clumpings of grains superimposed on galaxy halos can appear much like jets.
5) Jets frequently have aspect ratios 10:1 or greater, when their width can be resolved at all from the ground. "Wiggles" in their optically visible portions usually stay within this range; they are straighter and narrower than most competing features.

These proposed criteria do not include the current level of

nuclear activity, and so could in principle be used to identify "fossil" jets from active episodes of now-dormant nuclei.

4. AMBIGUOUS SYSTEMS

Several galaxies in this study show features with some characteristics of jets, but confused or uncertain structure. These systems merit more detailed study.

UGC 3995 is a pair of spirals, very close in projection. The larger seems to show a faint jet or plume to the NW, and has a type 2 Seyfert nucleus. Because of the possibility of confusion with the other galaxy's arms and its faintness, further confirmation of the jet is desirable.

NGC 1598 showed a pair of jet-like features on an enhanced UK Schmidt plate by Hawarden et al. (1979). A smoothed CCD frame (Fig. 1) confirms their presence, but shows additional chaotic stricture nearby, so that it is not clear that they are coherent, distinct features.

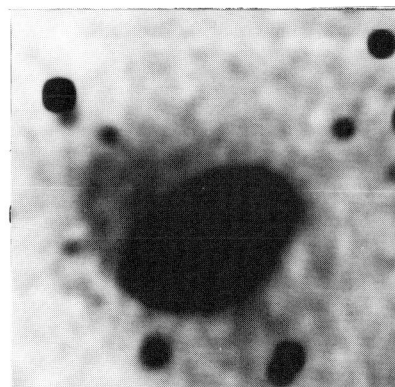

Fig. 1: V-band CTIO 4-m CCD image of NGC 1598, smoothed to show faint jet-like features to the NE and SW, as well as the complex outer disk structure.

ESO 0610-23 (Fig. 2) has an amorphous main body and clumpy "jet." Both are sites of active star formation, rich in H II regions. The inner contours of Fig. 2 show structures revealed by median-filtering the image; dust and asymmetric bright features are present, perhaps indicative of a merger. The total radial-velocity range seen spectroscopically is less than 100 km/s, unusual for merging or interacting systems. The nature of this galaxy remains unclear.

5. THE NEXT STEP: A SURVEY OF THE SOUTHERN SKY FOR OPTICAL JETS

With the experience gained from study of previously identified jet candidates, it was possible to define more precise criteria for the appearance of optical jets on sky-survey plates, and to distinguish them from the tidal features and other false jets which dominate available lists. Using these criteria, the SRC-J survey plates away from the galactic plane have been searched for candidate

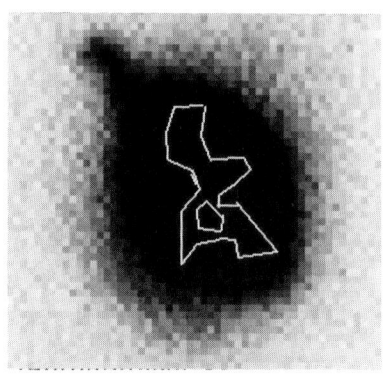

Fig. 2: B-band CTIO 4-m CCD image of ESO 0610-23. The inner contours show structure revealed by median-windowing, with off-center peak and hints of dust lanes. The "jet" to the NE has the spectrum of a group of H II regions.

galaxies which, as well as can be determined, pass these tests. The very deep limiting surface brightness of the SRC J plates makes this material particularly suitable for such a search.

This search has yielded about 100 candidates; most are elliptical or compact in morphology, but a significant fraction are clearly spirals. Among these may be objects similar to NGC 1097. Deep imaging and spectra of these objects should produce useful statistical data on the occurrence of optical jets independent of selection from radio emission or nuclear activity. Details of the survey and candidate objects will appear elsewhere.

The National Optical Astronomy Observatories are operated by the Association of Universities for Research in Astronomy, Inc., under contract with the National Science Foundation.

REFERENCES

de Vaucouleurs, G. and Nieto, J.-L 1979, Ap. J. 231, 364.
Hawarden, T. G., Longmore, A. J., Cannon, R. D. and Allen, D. A. 1979, M.N.R.A.S. 186, 495.
Lauberts, A. 1982, ESO/Uppsala Survey of the ESO(B) Atlas (ESO:München).
Lowe, J. 1978, Ap. J. (Lett.) 223, L99.
MacAlpine, G. M., Lewis, D. W. and Smith, S. B. 1977, Ap. J. Suppl. 35, 302.
Nilson, P. 1973, Uppsala General Catalog of Galaxies (Uppsala: University Press).
Zwicky, F. 1971, Catalog of Selected Compact Galaxies and of Post-Eruptive Galaxies (Zürich: L. Speich).

CERENKOV LINE RADIATION - A New Interpretation of the Emission Lines
of Quasars

T. Kiang *Dunsink Observatory, Dublin, Ireland*
and
J. H. You *University of Science & Technology of China,
Hefei, China*

An electron moving at relativistic speed $\beta = v/c$ in a medium of refractive index n will emit Cerenkov radiation if $\beta n > 1$, and for a given β, this condition may be satisfied over some small wavelength range next to atomic lines, under favourable physical conditions. The radiation so produced will than look like *broad lines*, asymmetrical in shape and characteristically displaced from the atomic line. For this reason, J.H. You called this radiation the "Cerenkov line radiation". The necessary physical conditions are that there should be a sufficient number of the atom in question and of the relativistic electrons.

The first paper on Cerenkov line radiation in English is : J.H.You & F.H.Cheng, *Chinese Physics* 2 (1982) 16-24. A systematic account of the basic theory is given in : J.H.You et al., *MNRAS* 211(1984) 667-677. (Paper I hereafter).

We now indicate the steps that lead to the formula for the total intensity of an optically thick Cerenkov line.

When there are N_e electrons moving at β in a unit volume, the volume emissivity of Cerenkov radiation per unit frequency range is

$$J_\nu = N_e \pi (e^2/c) \nu \beta (1 - \beta^{-2} n^{-2}). \tag{1}$$

We shall always have n and β very close to 1, and the Lorenz factor $\gamma = 1/\sqrt{(1-\beta^2)} \gg 1$, and so, to a high approximation,

$$\beta (1 - \beta^{-2} n^{-2}) = n^2 - 1 - \gamma^{-2}. \tag{2}$$

The spectral variation of n in the neighbourhood of an atomic line of wavelength λ_{ij} is well-known: as we pass λ_{ij}, $n^2 - 1$ rapidly rises from below zero to a peak (usually well within 0.1Å, under all circumstances likely to be encountered), then falls off as the inverse first power of the displacement. In terms of the fractional displacement,

$$u = (\lambda - \lambda_{ij})/\lambda_{ij}, \tag{3}$$

then, we have

$$n^2 - 1 = c_0 u^{-1}, \tag{4}$$

throughout practically the whole range of interest. The constant c_o is

$$c_0 = 4.48(-14) \lambda_{ij}^2 N \tag{5}$$

where N is the number of equivalent classical oscillators with frequency c/λ_{ij},

$$N = N_i f_{ij} + N_j f_{ji} = f_{ij}[N_i - (g_i/g_j)N_j], \tag{6}$$

suffix i always referring to the lower level and suffix j always the upper level, and N, f and g representing the number density, oscillator strength and statistical weight, respectively. Under all normal circumstances, we have $N_i \gg N_j$ and then (6) simplifies into

$$N = f_{ij} N_i \tag{7}$$

For a given γ, Cerenkov radiation is produced in the wavelength range limited by u_{lim}, given by

$$\gamma^{-2} = c_0 \, u_{lim}^{-1}. \tag{8}$$

Substituting (7) in (5), then the result in (4) and (8), then the results in (2) and hence in (1), and multiplying by the factor c/λ_{ij} to get the emissivity per unit interval of u, we have, to the lowest order in u,

$$J_u = N_e c_1 (u^{-1} - u_{lim}^{-1}), \tag{9}$$

where

$$c_1 = 9.74(-22) f_{ij} N_i. \tag{10}$$

A remarkable feature of the Cerenkov line emission now emerges from (9) and (10): the emission is proportional to N_i, the population in the lower level! This feature means that the notorious "Lα/Hβ problem" does not arise if the lines are the "Cerenkov lines". We must next include absorption into our considerations, that is, we must consider the case of large optical thickness. In this case, the emergent intensity per unit interval of u is

$$I(u) = J_u / k \tag{11}$$

where k is the total absorption at the wavelength in question. There are two main contributions to k, the line absorption k_1 and the photo-electric absorption k_2. Again, because the Cerenkov emission depends on the population in the lower level, we need not consider the absorption due to collisional de-excitation.

The line absorption k_1 is proportional to the imaginary part of the complex refractive index, and falls off as u^{-2} throughout practically the whole range of interest:

$$k_1 = c_2 \, u^{-2} \tag{12}$$

where

$$c_2 = 7.47(-25) \lambda_{ij}^2 f_{ij} \Gamma_{ij} N_i \tag{13}$$

Γ_{ij} being the damping constant associated with the line λ_{ij}. The photo-electric absorption k_2 is, to the lowest order in u, independent of u and is sufficiently accurately represented by

$$k_2 = c_3 = 1.04(-2) \lambda_{ij}^3 (N_\ell/\ell^5). \qquad (14)$$

where ℓ is the lowest level of which the photoionization potential is less than the photon energy.

Inserting (9), (12) and (14) in (11) gives the profile of an optically thick Cerenkov line,

$$I(u) = N_e c_1 (u^{-1} - u_{lim}^{-1}) / (c_2 u^{-2} + c_3), \quad 0 < u < u_{lim}. \qquad (15)$$

The total intensity of the Cerenkov line is obtained by integrating $I(u)$ over the range $0 < u < u_{lim}$. The result is

$$I_{tot} = Y \cdot \left(\ln(1 + x^2) - 2[1 - (\arctan x)/x] \right), \qquad (16)$$

where

$$Y = N_e c_1 / 2 c_2 = 4.68(-20) \lambda_{ij}^{-3} (R_i/R_\ell) \ell^5 N_e \qquad (17)$$

and

$$X = u_{lim} / \sqrt{(c_2/c_3)} = 5.29(-3) \cdot (\lambda_{ij}^5 f_{ij} \Gamma_{ij}^{-1})^{\frac{1}{2}} \cdot \gamma^2 N (R_i R_\ell \ell^{-5})^{\frac{1}{2}} \qquad (18)$$

In these expressions we have put $N_i = N R_i$, $N_\ell = N R_\ell$, N being the total number of the atom over all the levels, and R the fraction at the level indicated.

Thus, five unknown physical parameters enter into the expression for the total intensity, N_e, γ referring to the relativistic electrons and N, R_i, R_ℓ referring to the atom in question. However, N_e will be absent from any intensity *ratio* between any two Cerenkov lines. The other four appear in the two combinations $q = R_i/R_\ell$ and

$$p = \gamma^2 N (R_i R_\ell)^{\frac{1}{2}}; \qquad (19)$$

Now, for all the Balmer lines of hydrogen, we have $i = 2$ and $\ell = 3$. Hence the intensity ratio between any two Balmer lines will be independent of q and will be a function of p only. For a selection of values of p between 0 and ∞, our calculated Balmer intensity ratios, calculated according to (16) - (19) are shown in TABLE 1.

TABLE 1

CALCULATED BALMER LINE INTENSITY RATIOS
ACCORDING TO THE *CERENKOV LINE RADIATION* MECHANISM

p	Hα/Hβ	Hγ/Hβ	Hδ/Hβ	Hε/Hβ
0.	46.1	0.116	0.0257	0.0080
5.83(+18)	23.9	0.122	0.0273	0.0085
2.92(+19)	7.42	0.191	0.0498	0.0165
1.75(+20)	3.93	0.338	0.136	0.0616
2.92(+21)	2.92	0.439	0.225	0.127
5.83(+22)	2.63	0.473	0.257	0.154
∞	2.18	0.526	0.308	0.195
Observed	3.7	0.34	0.14	0.05
Classical	2.8	0.50	0.29	0.18

For comparison, we have added two more rows; the row labelled "Observed" shows the median values from samples of 10, 10, 7 and 6 in the quasar data measured by Puetter et al. (1981). Where duplicate measurements have been made by Neugebauer et al. (1979), geometrical means of the two were taken before the median value was decided. The last row labelled "Classical" refers to the calculated values in the "classical" theory of planetary nebulae (the values are taken from Ambarzumian's (1958) text). These values are prototypical of the values given by the current photoionization-cascade models. It has long been known that the observed Balmer line intensities in quasars show a steeper decrease along the series than simple photoionization-cascade models would give. It now appears that the steep gradient is no problem with the Cerenkov mechanism; the observed variation is in fact well-matched by our calculated values for $p = 1.75(+20)$.

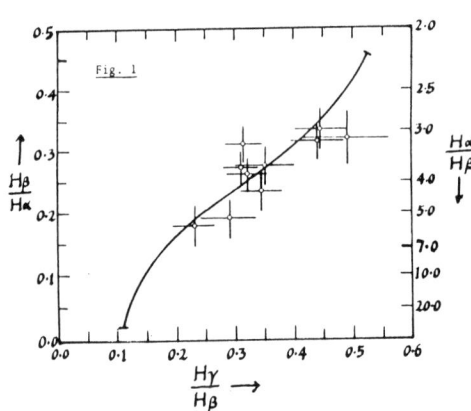

Fig. 1

If we put $\gamma \sim (+3)$, $(R_e R_\ell)^{\frac{1}{2}} \sim (-2)$, then, according to (19), this value of p corresponds to a total density $N \sim (+16)$. This may appear too high, in view of the observed presence of the CIII] 1909 line in the spectrum. However, such a value is not too high from general considerations (the concentration of a galactic mass into physical dimensions measured in light-months), and that it may be that the hydrogen lines are produced by the Cerenkov mechanism in such a high-density environment while the carbon line is produced by the usual mechanism in a more tenuous region.

The present theory of Cerenkov line radiation is highly vulnerable to observational disproof in that it makes many strong predictions. If, for example, we plot our calculated values of Hα/Hβ and Hβ/Hγ one against the other in a "two-ratio diagram", we get a unique curve, with no room for adjustment whatsoever. Then, if the observed lines are predominantly produced by the Cerenkov mechanism, the observed points must fall close to this curve.

Figure 1 illustrates the confrontation of our calculated values with observations in the (Hα/Hβ, Hγ/Hβ)-plot. It is clear that, in this case, the hypothesis that the Cerenkov line radiation mechanism is at work is not disproved by the observations. Of course, it does not prove that the mechanism is at work; but it is an intimation that we should perhaps take up this hypothesis more seriously.

REFERENCES

Ambarzumian V.A. (ed.) "Theoretical Astrophysics", English translation by J.B. Sykes, 1958, Pergamon Press, p.424.
Neugebauer G., Oke J.B., Becklin E.E. & Matthews K., (1979) Ap.J.230:79.
Puetter R.C., Smith H.E.,et al.(1981) Ap. J.243:345.

AN INTERPRETATION OF THE LIGHTCURVE OF BL Lac.

M.R. Kidger and J.E. Beckman
Instituto de Astrofisica de Canarias
Universidad de La Laguna
Tenerife
Spain

ABSTRACT. Extensive monitoring is now available for a good sample of the brighter quasars and BL Lacs. In some cases, lightcurves are well enough sampled to permit stochastic modelling. We present here, a simple model of the lightcurve of the prototype BL Lacertae object, BL Lac in terms of a series of random events, each with rapid rise and exponencial decline. Whilst unable to explain very rapid "optically violent" activity, it seems to provide a satisfactory fit to the long term trends in the lightcurve.

OBSERVATIONS.

Between Pollock et al. (1979), who observed from Rosemary Hill, Florida and Lloyd (1984) who observed from the RGO in England, 630 B-magnitudes have been reported, well spread over the years 1969-1979. Of the observations, 450 are from Pollock et al. (1979) and 180 from Lloyd (1984). Typical errors are about 0.12 magnitudes and 0.06 magnitudes respectively.

THE LIGHTCURVE.

Figure 1 reproduces the lightcurve, in the form of ten days means. Since the weight per point is the reciprocal of the r.m.s. error, the two data sets have an almost equal weight in the final plot. An effect, hinted at by McGimsey et al. (1975) and by Pollock et al. (1980) is a series of declines, each lasting several months and comparitively constant in gradient from decline to decline. Superimposed are more rapid variations, of timescale from tens of days down to days. Figure 2 reproduces a section of the lightcurve where both effects can be clearly seen.

The mean gradient b, is found, for nine observed declines, to be

$$b = -0.00528 \ (\pm 0.00090) \ \text{mag day}^{-1}$$

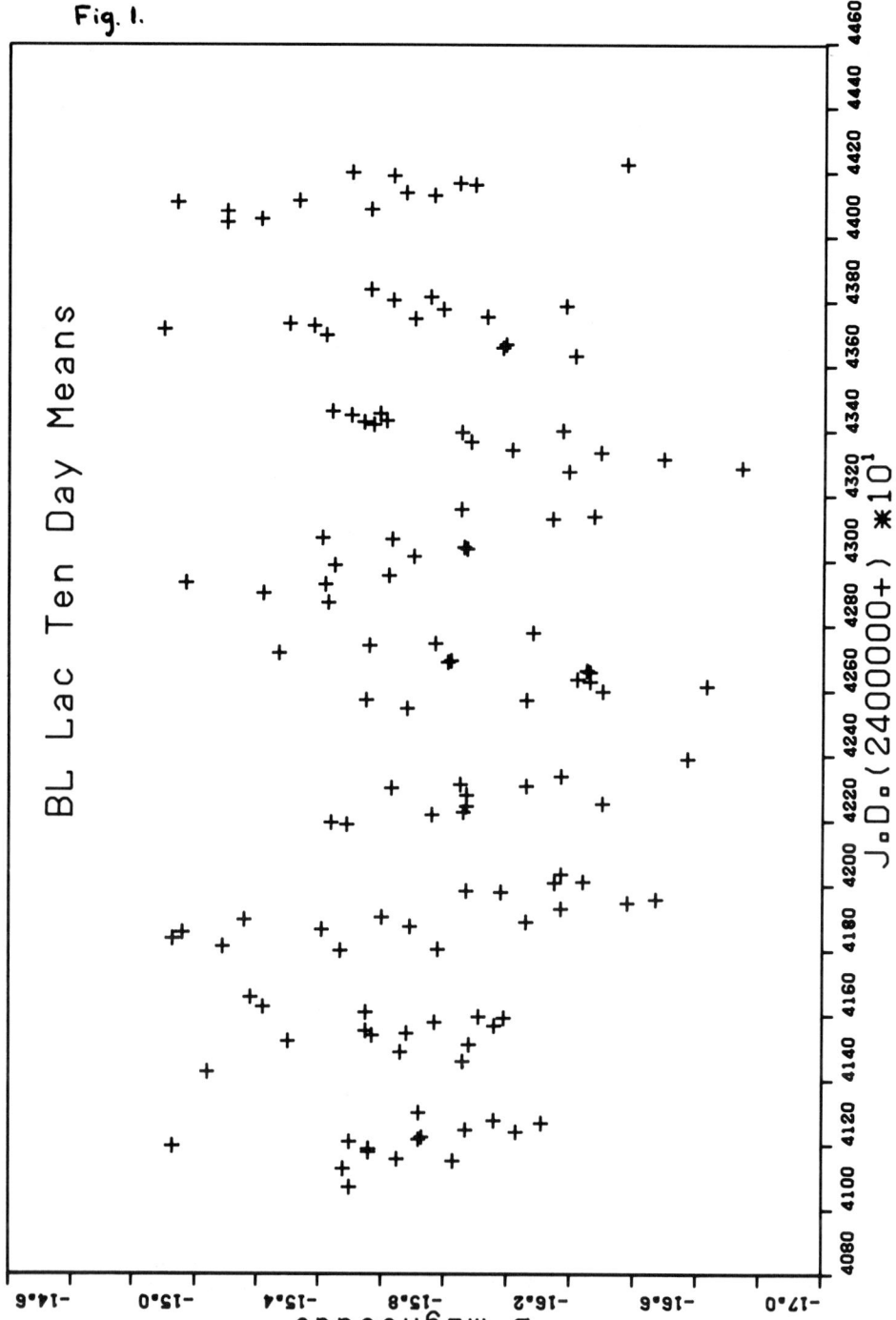

Fig. 1.

An, admittedly simple minded, statistical test, that of dividing the mean by the error, to find the significance assigns this gradient a nominal significance of 6 σ.

Since rises in the lightcurve are steeper and thus shorter, they are more difficult to analyze. However, eight rises give a mean gradient d:

$$d = 0.0120\ (\pm 0.0039)\ \text{mags day}^{-1}$$

which is still better than 3 σ.

If we hypothesize that each rise and subsequent decline corresponds to a shot noise pulse (see, for example Fahlman and Ulrych, 1975, 1976, for theoretical arguments), we can model the curve in terms of shot noise. If each pulse has a similar shape, of intensity:

$$I(t) = A \cdot t^\beta \exp(-\alpha t) \qquad (1)$$

where A is the amplitude, t is the time ordinate and α and β are constants. We can fit a curve of general form

$$m_B = \text{constant} - 1.086 \ln(\Sigma\ [(t-t_i)^{1/2}\ \exp(-0.009(t-t_i))\ ) \qquad (2)$$

to the data points. Where t is the epoch of each observed magnitude m_B and t_i the initiation dates of the shots.

Solving this equation numerically, we find a mean shot rate of 2.12 per year. The error is difficult to assess, since it is necessary to interpolate, by introducing shots which occured during the unobserved 25% of each year, to reduce the residuals of later observations. However, an upper limit is probably ±0.36, as judged from the overall fit.

FOURIER TRANSFORM ANALYSIS.

Fourier transform analysis is very poor at detecting non-sinusoidal periodicities, however we would expect a peak at the shot rate, if this analysis is correct. Our analysis, by the Deeming technique (Deeming, 1975) for non evenly spaced data, does show a small peak, centered at 2.12 year^{-1}, (fig. 3). We further note the existence of the spurious peaks which are created by inhomogeneities of sampling and are produced by true periodicities, or "pseudo-periodicities". Thus we believe that this is a genuine peak in the power spectrum.

SUMMARY.

We believe that we have the first direct evidence for the existance of shot noise in quasar lightcurves and thus a possible means of viewing the central power source. Analysis of four other O.V.V.s, with less well sampled lightcurves, does show similar results, in terms of visibility of shots. An idea as to the power per shot, holds interesting possibilities for calculating directly, the luminosity of quasars and thus correcting the Hubble diagram for luminosity. Research into this possibility is currently underway.

REFERENCES.

Deeming, T.J.: 1975, Astrophys. Space Sci. 36, 137
Fahlman, g.G., and Ulrych, T.J.: 1975, Ap.J., 201, 277
Fahlman, G.G., and Ulrych, T.J.: 1976, Ap.J., 209, 663
Lloyd, C.: 1984, M.N.R.A.S., 209, 697
McGimsey, B.Q., Smith, A.G., Scott, R.L., Leacock, R.J., Edwards, P.L., Hackney, R.L.: 1976, Astron. J., 80, 895
Pollock, J.T., Pica, A.J., Smith, A.G., Leacock, R.J., Edwards, P.L., Scott, R.L.: 1980, Astron. J., 85, 1442

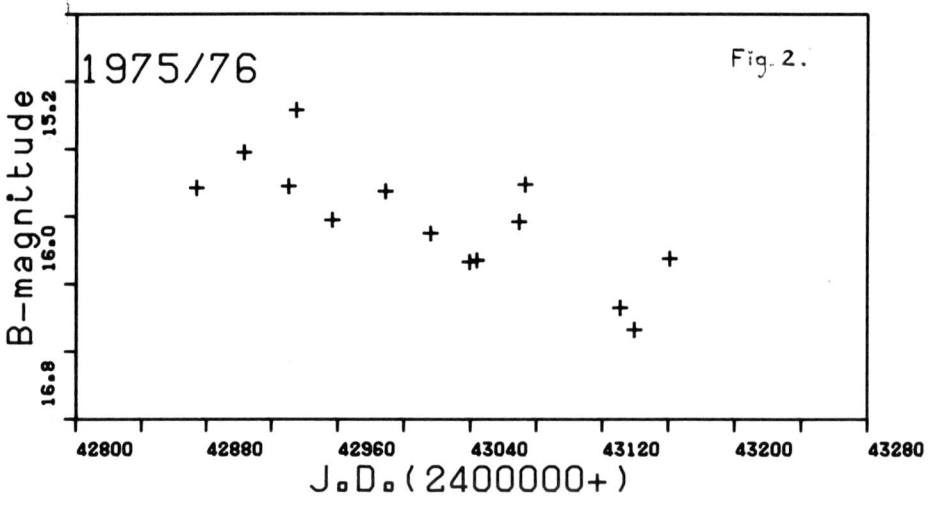

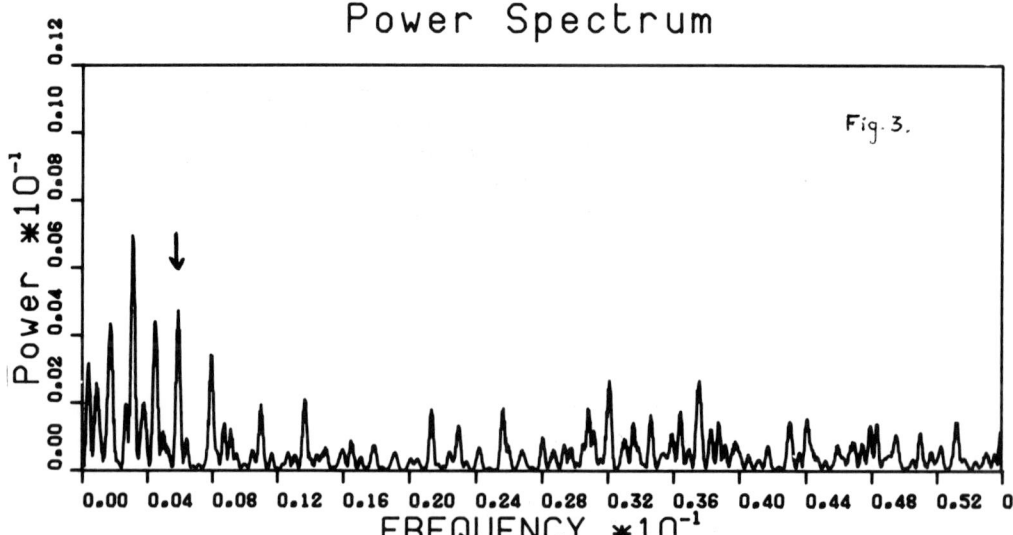

THE I.A.C./Q.M.C. CATALOGUE OF QUASAR MULTIBAND SPECTRA

M.R. Kidger and J.E. Beckman,
Instituto de Astrofísica de Canarias,
Universidad de La Laguna
Tenerife
Spain

ABSTRACT. One of the major problems with using quasars to investigate the Hubble diagram is an uncertainty as to the size of the bolometric correction. Some estimates have suggested that it may be as high as 6 magnitudes (by comparison, except at the two extremes of the temperature distribution, for stars, it is rarely more than a few tenths of a magnitude). To this end, we have compiled data from a wide range of sources, to investigate the behaviour of quasar spectra over the range from radio to X-rays, and hence to be able to calculate this important parameter. Here we present a sample of the compiled spectra, whilst a complete catalogue of around two hundred spectra is in preparation, which will present this work in considerably more detail.

THE CATALOGUE.

Figures 1 to 18 below present spectra for a range of quasars and BL Lac objects. In all but a very few cases, the errors are smaller than the points. The effect of variability are clearly shown in these non-simultaneous spectra as is the difference in range of variability in various bands.

ACKNOWLEDGEMENTS.

We would like to thank Carlos Martínez for his assistance with the computer program used; the staff of the computer centre and "residentes" of the IAC, for advice and assistance and Dolores García and Monica Murphy for considerable assistance with the preparation of this article.

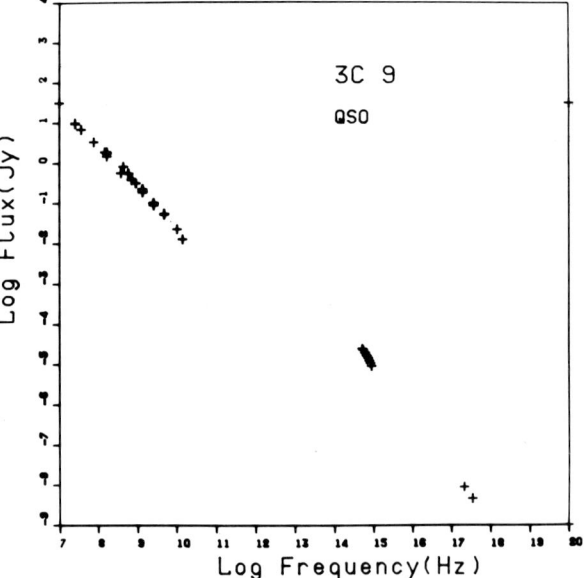

Fig. 1

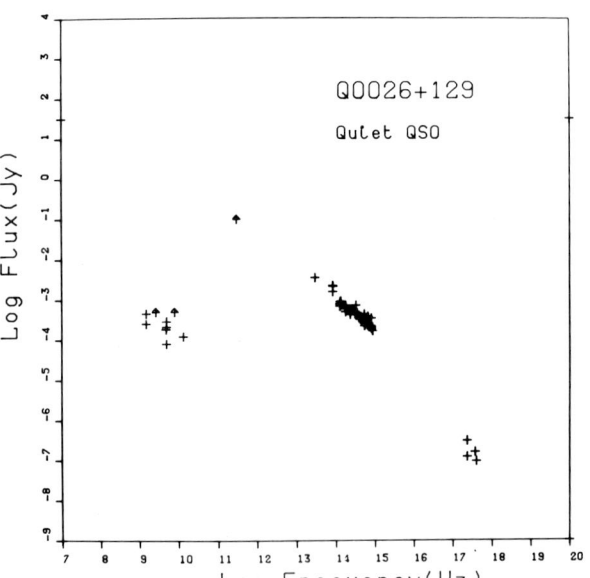

Fig. 2

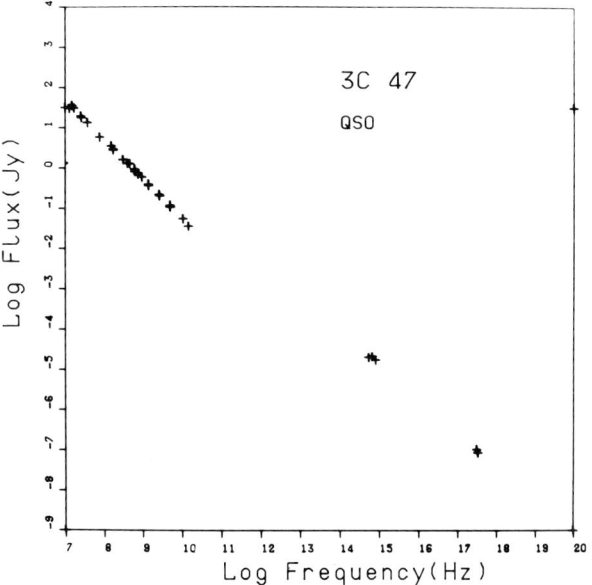

Fig. 3

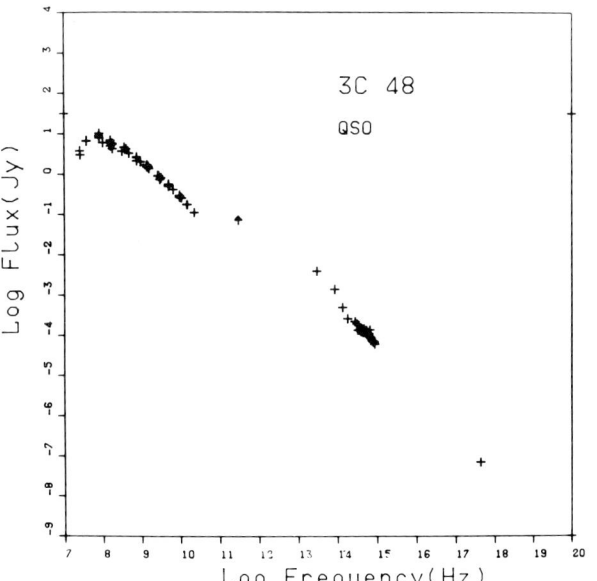

Fig. 4

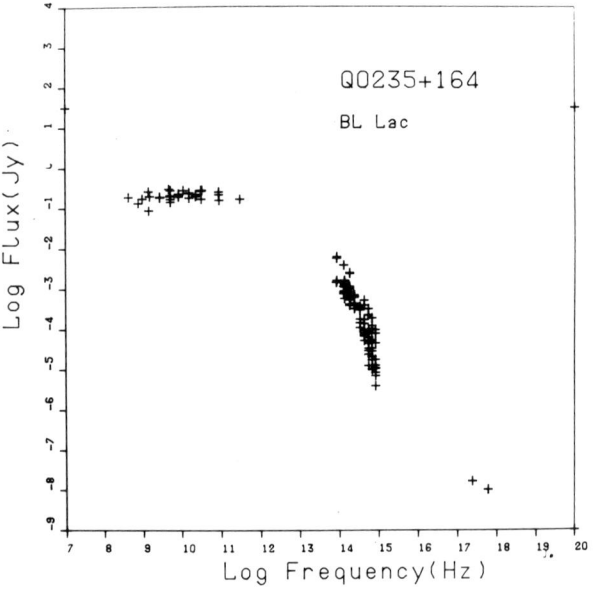

Fig. 5

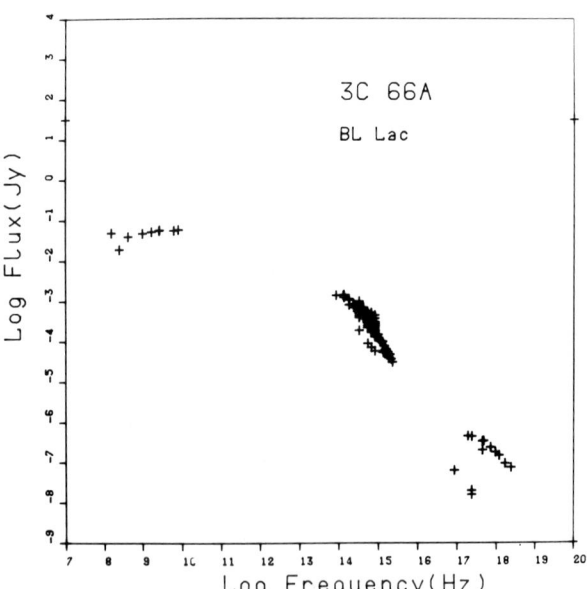

Fig. 6

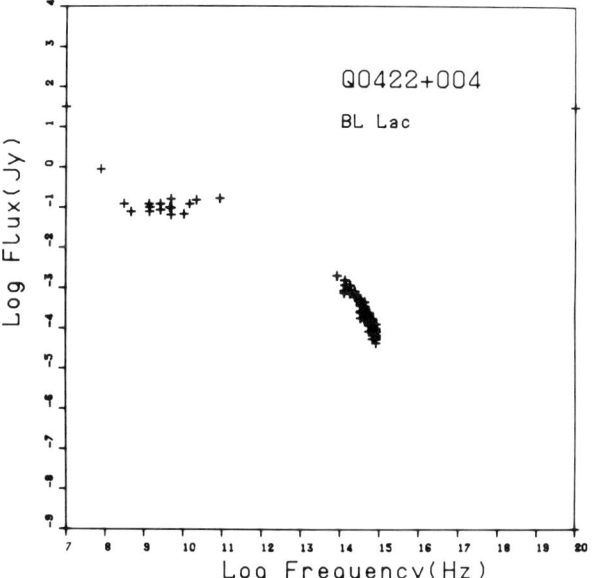

Fig. 7

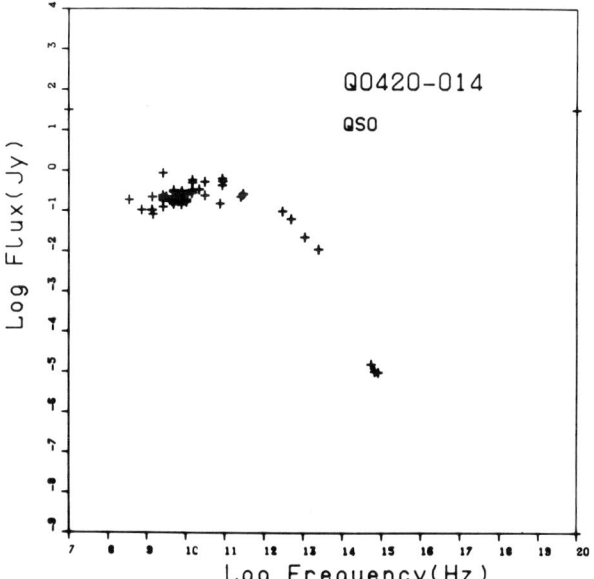

Fig. 8

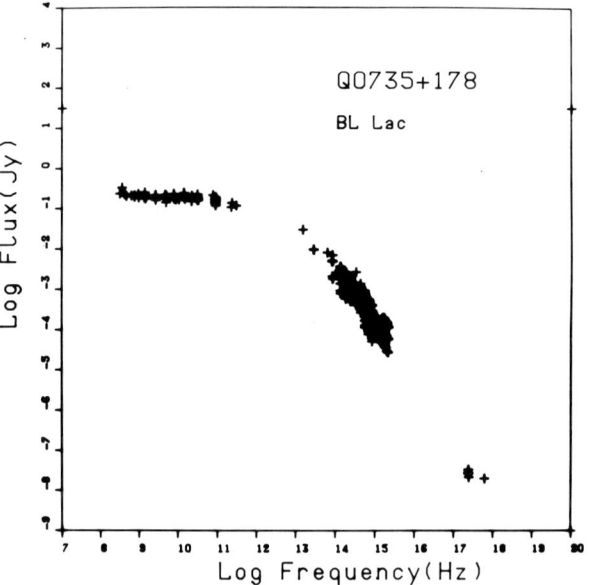

Fig. 9

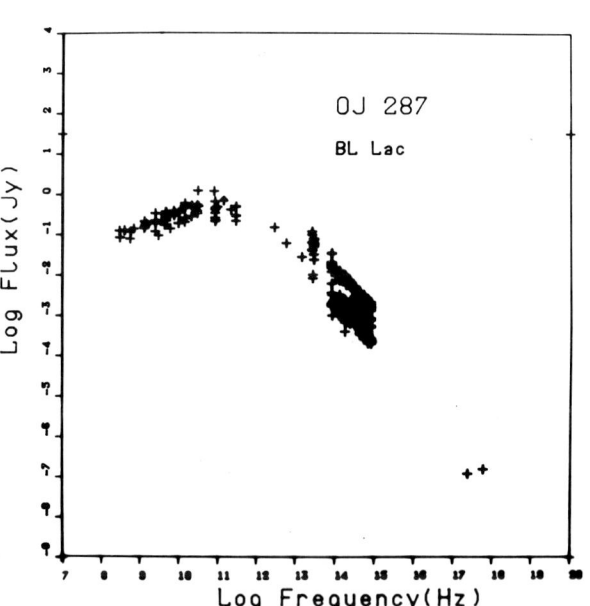

Fig. 10

THE I.A.C./Q.M.C. CATALOGUE OF QUASAR MULTIBAND SPECTRA

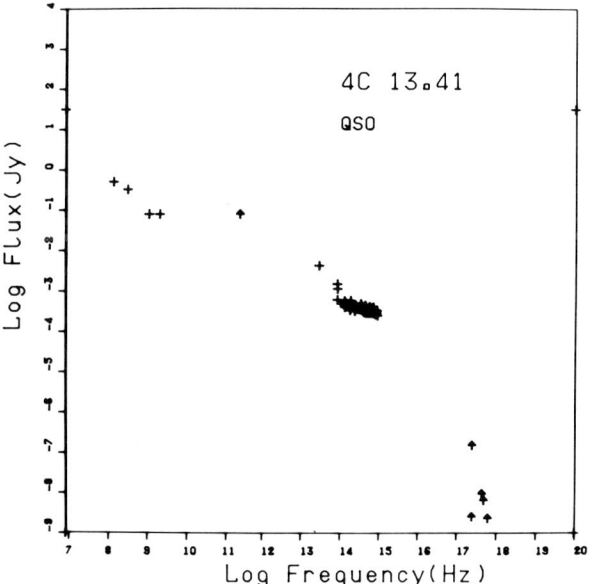

Fig. 11

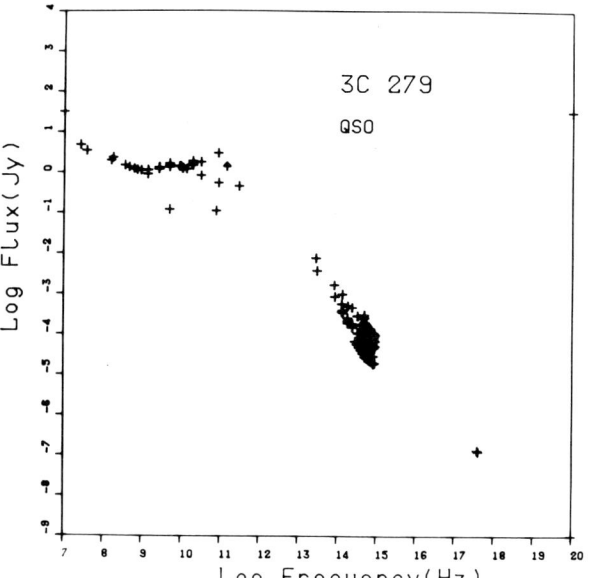

Fig. 12

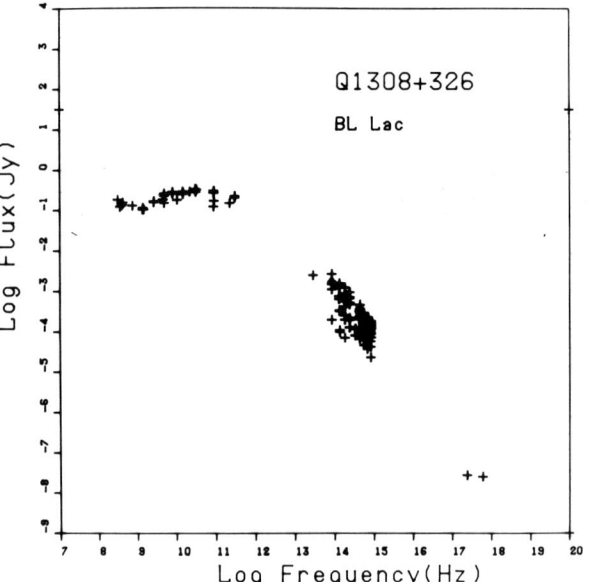

Fig. 13

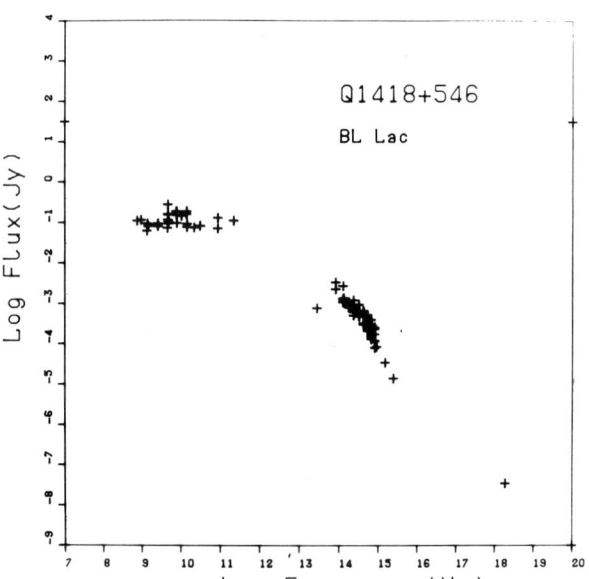

Fig. 14

THE I.A.C./Q.M.C. CATALOGUE OF QUASAR MULTIBAND SPECTRA

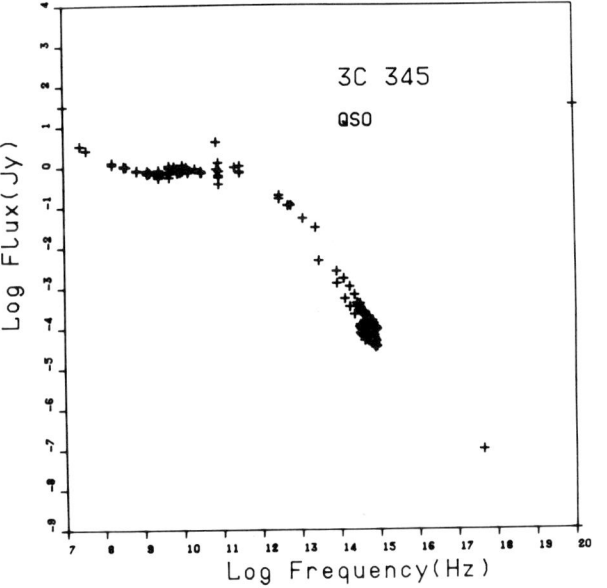

Fig. 15

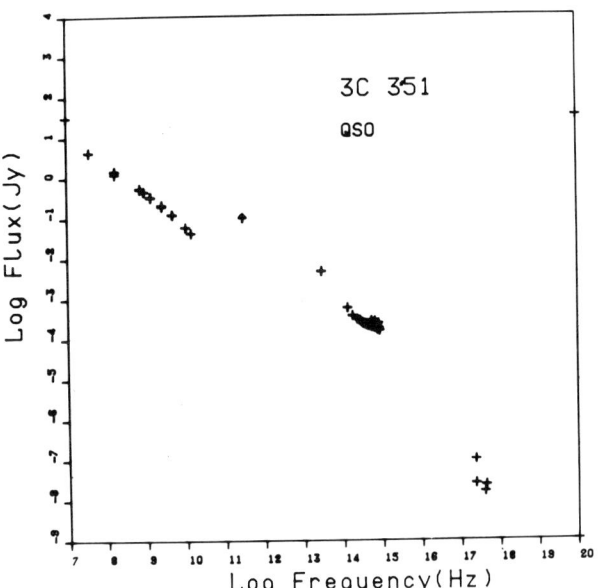

Fig. 16

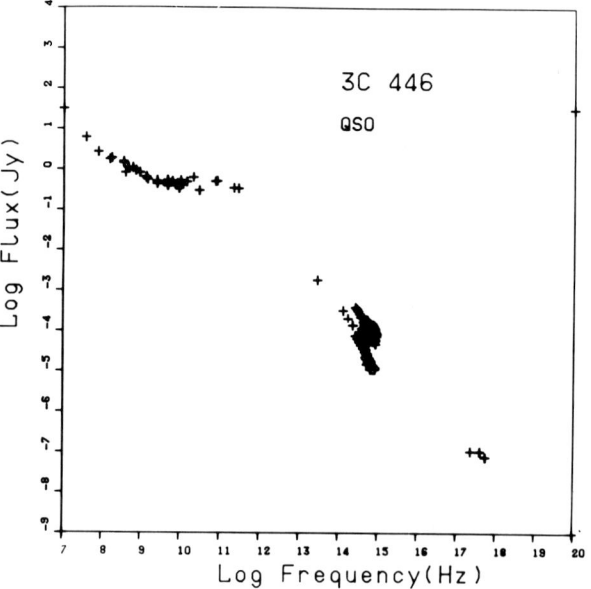

Fig. 17

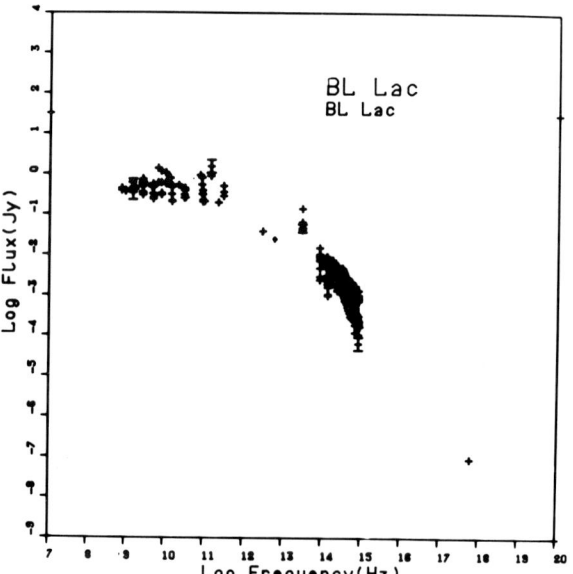

Fig. 18

THE OPTICAL VARIABILITY OF 3C345.

M.R. Kidger and J.E. Beckman,
Instituto de Astrofísica de Canarias
Universidad de La Laguna,
Tenerife,
Spain.

ABSTRACT. The publication by Lloyd (1984) of new and revised data for a sample of active nuclei brought the number of published observations of 3C345 up to over 900, between 1965 and 1980. Since the last major study, by Barbieri et al. (1977), sufficient new data has been produced to merit a complete revision of previous analysis. A compilation of the lightcurve and Fourier Transform analysis are presented here and compared with previous studies. We find no strong evidence for long lasting major periodicities, in contradiction to some other published conclusions.

PUBLISHED DATA.

Seventeen papers have been produced giving numerical magnitudes for 3C345 and three more have presented results in graphical form. For a complete listing see Kidger and Beckman (1985). Of these, the definitive study is still that of Kinman et al. (1968), who managed to combine high temporal coverage and low errors. Their observations comprise nearly a third of the total.

In most cases, the data is presented as B-magnitudes. Where it is not, the correction

$$m_B - m_{pg} = +0.60 \qquad \text{Pollock et al (1979)}$$

has been adopted, although no attempt has been made to use U, or V magnitudes. From 1967 to 1978 there were always a minimum of two monitoring programs being carried out simultaneously, with a maximum of four in 1970-72 and 1975-76.

THE LIGHTCURVE.

Figure 1 reproduces the lightcurve compiled from all the published data. We note a major change in the character of the variability. The first half of the lightcurve shows very large and frequently very rapid

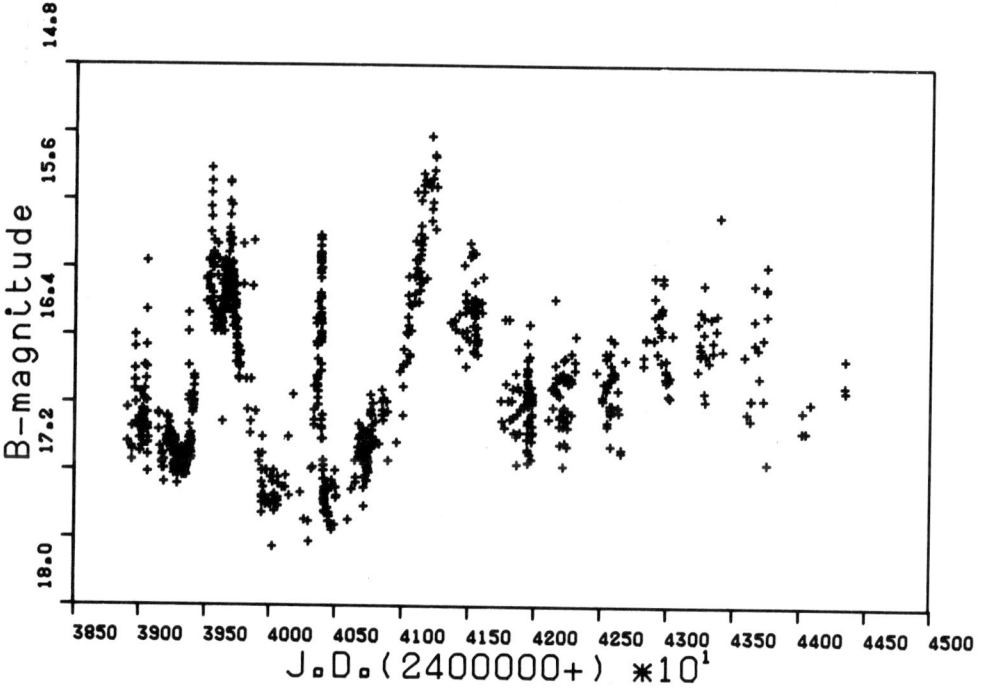

Fig. 1

changes in magnitude. The "spike" near J.D. 2440400 rose from the lowest ever recorded magnitude of 17.62 in seventy days and decayed back to almost the same level in another thirty. However, after the largest recorded peak, which reached magnitude 15.38, at J.D.2441235 and took two years to decline, little in the way of similar activity has been seen. We note a slight upwards trend of about 0.07 magnitudes per year, with superimposed flaring on very short timescales, but no peaks of any great size. In the period covered only by Lloyd (1984) i.e. After 1978, there is a hint of another change in behaviour, more observations of this period would be useful.

PREVIOUS ANALYSIS.

Three major studies, by: Kinman et al. (1968); Smyth and Wolstencroft (1970) and Barbieri et al. (1977) have produced highly contradictory results. As yet no results have been subsequently confirmed by later analysis. Periodicities range from 80.37 and 321.5 day periodicities in flaring (Kinman et al. 1968) to major 800 and 1600 day periodicities (Barbieri et al. 1977). Where extrapolation has been attempted, it has failed to <u>predict</u> the form of the lightcurve with any degree of accuracy, for example, Barbieri et al. (1977) predicted major flares in 1976 and 1980 when the lightcurve is, in fact almost quiescent.

We must conclude that whilst short term periodic, or more probably, pseudo-periodic behaviour may exist, there is no evidence for strong, long-term periodicities over the covered timescale.

NEW ANALYSIS

We used the now popular technique of Deeming (1975) for analysing non-uniformly sampled data. For a practical application of this technique, see Barbieri et al. (1977).

We first defined the spectral window

$$S_N(\nu) = \sum_{i=1}^{N} \cos(2\pi \nu t) \qquad (\ )$$

where there are N observations at epochs t and the discrete Fourier Transform

$$F_N(\nu) = \sum_{i=1}^{N} m_i \cos(2\pi \nu t_i)$$

with m_i the blue magnitude at epoch t_i.

With the new and revised data and an applied $m_B - m_{pg}$ correction, we have almost double the 485 data points of Barbieri et al (1977) and higher temporal resolution.

Figure 2 reproduces the power spectrum found by this analysis.

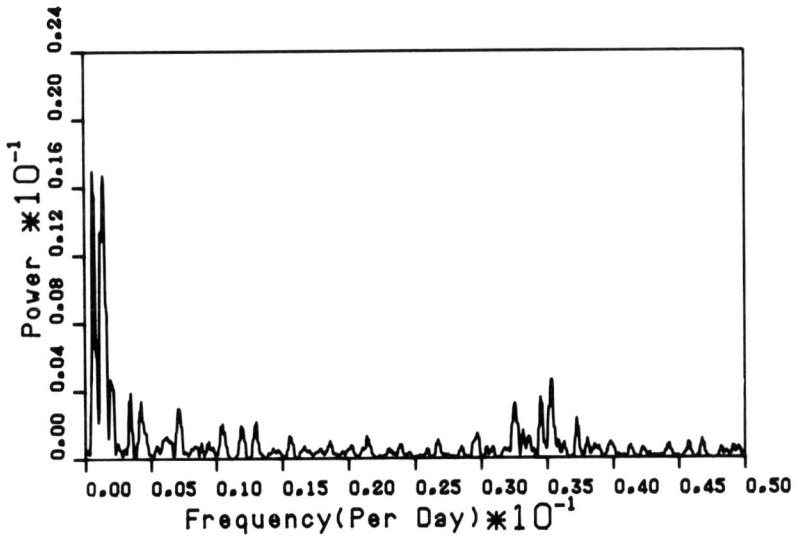

Fig. 2

We do see large peaks near to the 800 and 1600 day peaks of Barbieri et al. (1977) however, these are found, almost certainly to be due to the residuum of the three big outbursts which were pseudo-periodic in form. More interesting are the peaks at

$\nu_1 = 0.00718$ day^{-1}; Period = 140 days
$\nu_2 = 0.05192$ day^{-1} ; Period = 83.9 days
$\nu_3 = 0.01308$ day^{-1} ; Period = 76.5 days

All of these show the "aliasing" effect of true peaks. The first of these peaks duplicates exactly on one of the peaks of Barbieri et al. (1977). The mean of the other two peaks is 80.2 days, close enough to the 80.37 day period of Kinman et al. (1968) to be thought provoking.

CONCLUSIONS.

Elsewhere in these proceedings we present evidence that shot-noise may be directly observable in the lightcurves of some objects. It follows from the nature of shot-noise that a slightly higher shot-rate will produce as highly variable object, but without individual shots being visible (this condition results from many shots overlapping). Simulations suggest that this will occur between about 3 and 5 shots per year. The 140 day peak would correspond to a rate of 2.6 per year and the 80 day double peak to 4.5 per year, these values agree well with the predicted range. We would tentatively suggest that, the peak at 140 day, which shows considerable detail in the wings is typical of the expected output from a shot noise source. We note though, that, as shot-noise is a random process, the 140 day peak has no predictive power and is only an indication of mean behaviour.

REFERENCES.

Deeming, T.J.: 1975, Astrophys. Space. Sci. 36, 137
Kidger, M.R., Beckman, J.E.: 1985 Submitted to Astronomy and Astrophysics

Kinman, T.D., Lamla, E., Ciurla, T., Harlan, E., Wirtanen, G.A.: 1968, Ap.J. 152, 357
Lloyd, C.: 1984, M.N.R.A.S., 209, 697
Pollock, J.T., Pica, A.J., Smith, A.G., Leacock, R.J., Edwards, P.L., Scott, R.L.: 1980, Astron. J., 85, 1442
Smyth, M.J., Wolstencroft, R.D.: 1970, Astrophys. Space Sci. 8, 471

DOUBLE NUCLEUS GALAXIES

W. Kollatschny, K.J. Fricke, J. Hellwig
Universitäts-Sternwarte
Geismarlandstr.11
D-3400 Göttingen
West Germany

ABSTRACT. Double nucleus galaxies may represent late stages of galaxy mergers and may therefore be key objects for understanding galaxy evolution and activity. Here we present spectroscopic data on these objects and their individual nuclei.

I. INTRODUCTION

Quasar activity seems to be strongly connected with close interactions and merging of galaxies (Stockton 1983, Hutchings et al. 1984). By these processes a central source can be effectively fed with low angular momentum gas. We try to explore this basic mechanism within nearby galaxies by detailed spectroscopic studies of double nucleus Markarian galaxies. We define these objects as galaxies with two close compact nuclei at their centers, usually associated with high absolute brightness (see Table 1) and also with the occurrence of tidal arms. These galaxies are quite different in appearance and nature from the clumpy irregular and hot spot galaxies. We consider double nucleus galaxies as late stages of mergers which may resemble - at a lower power level - the close interactions seen in some nearby quasars (Hutchings 1984, Gaskell 1983).

II. OBSERVATIONS

In order to study the properties of double nucleus galaxies we employ optical and UV 2D-spectroscopy and high-resolution optical (Speckle-) and radio (VLA-) imaging together with other data. Detailed optical and UV spectroscopic data on Mkn 266 and other objects have been discussed elsewhere (Kollatschny and Fricke 1984a, b). Radial velocities, apparent and absolute magnitudes, and component separations for the objects studied by us are given in Table 1.

Table 1: Double Nucleus Galaxies

	$v_r [km \cdot s^{-1}]$	m_v	M_v	Separation (")	(kpc)
Mkn 544	7 211	14.5	-20.41	5.9	2.8
Mkn 1027	9 000	14.2	-21.20	10.5	6.1
Mkn 739	8 900	14.1	-21.29	6.6	3.7
Mkn 788	7 200	14.3	-20.61	24.0	11.2
Mkn 789	9 600	14.9	-20.67	4.1	2.5
Mkn 266	8 400	13.4	-21.83	11.9	6.5
Mkn 273	12 000	14.5	-21.50	4.5	3.3
Mkn 463	14 900	14.3	-22.19	4.5	4.3
Mkn 673	10 900	14.5	-21.31	5.3	3.7
Mkn 823	13 850	14.5	-21.83	4.9	4.3
Mkn 480	5 400	14.1	-20.19	8.9	3.1
Mkn 1394	2 400	15.0	-17.53	4.0	0.6
Mkn 296	4 700	15.5	-18.53	10.3	3.2
Mkn 306	5 600	14.5	-19.87	5.8	2.1
Mkn 314	2 100	14.0	-18.24	7.6	1.1
Mkn 930	5 300	14.5	-19.75	4.5	1.6

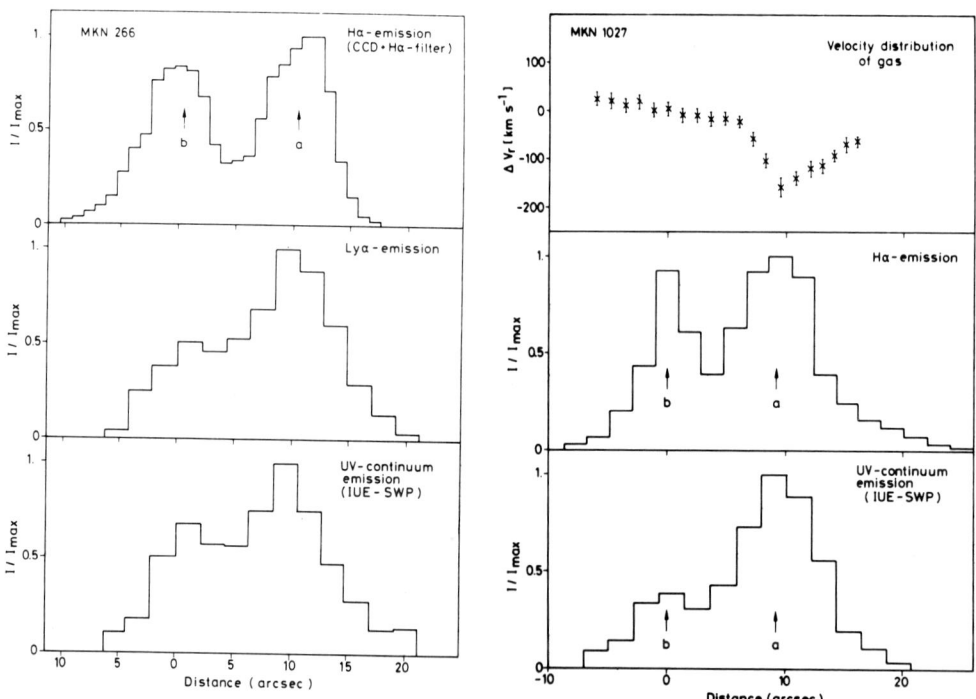

Fig. 1a: Optical and UV tracings of Mkn 266 (left)
  b: The same for Mkn 1027 together with the velocity profile across the nuclei (right)

DOUBLE NUCLEUS GALAXIES

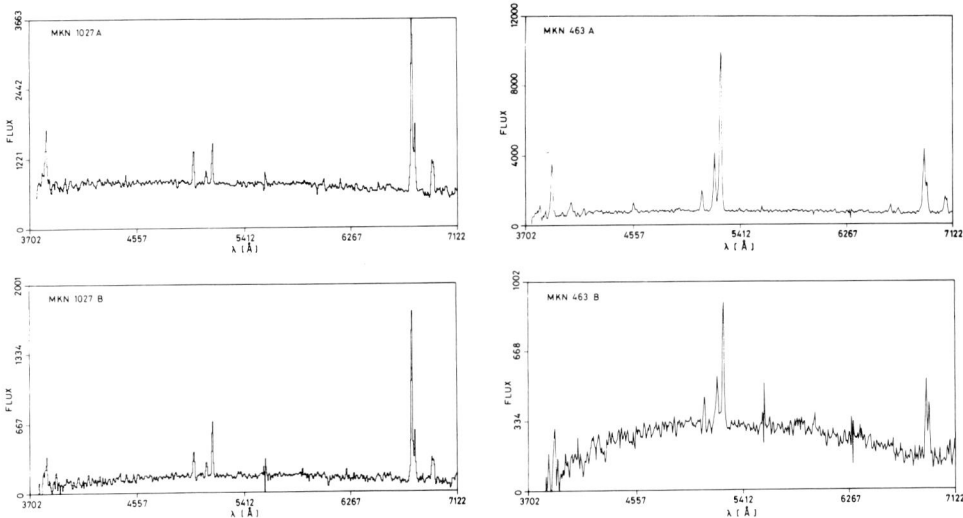

Fig.2: Optical spectra of the double starburst galaxy Mkn 1027 (left) and of the double Seyfert galaxy Mkn 463 (right)

III. RESULTS AND DISCUSSION

In Fig. 1a tracings of Hα, Lyα, and the short wavelength UV-continuum are shown for the double Seyfert galaxy Mkn 266 as an example. Fig. 1b presenting similar data for Mkn 1027 also shows the velocity profile. This profile is fairly complex as may be expected from an ongoing merging process. The optical spectra of the two nuclei of Mkn 1027 show

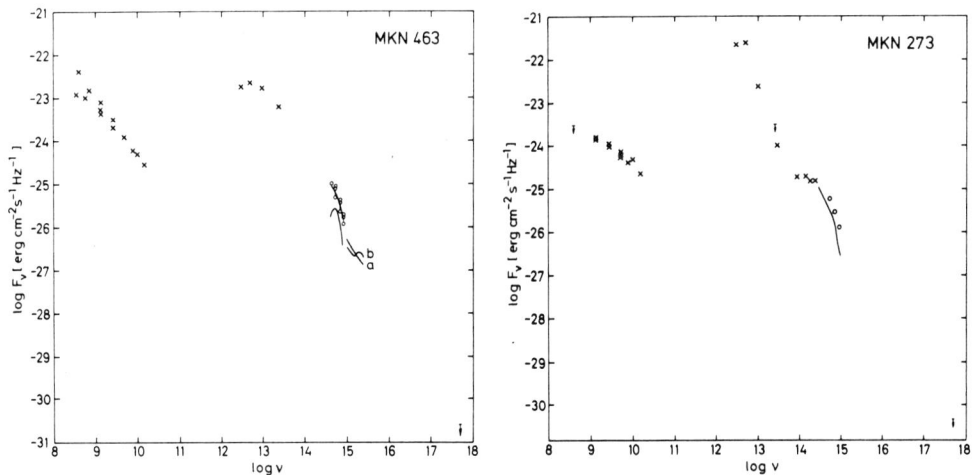

Fig. 3: Radio through X-ray continuum spectra for Mkn 463 and Mkn 273

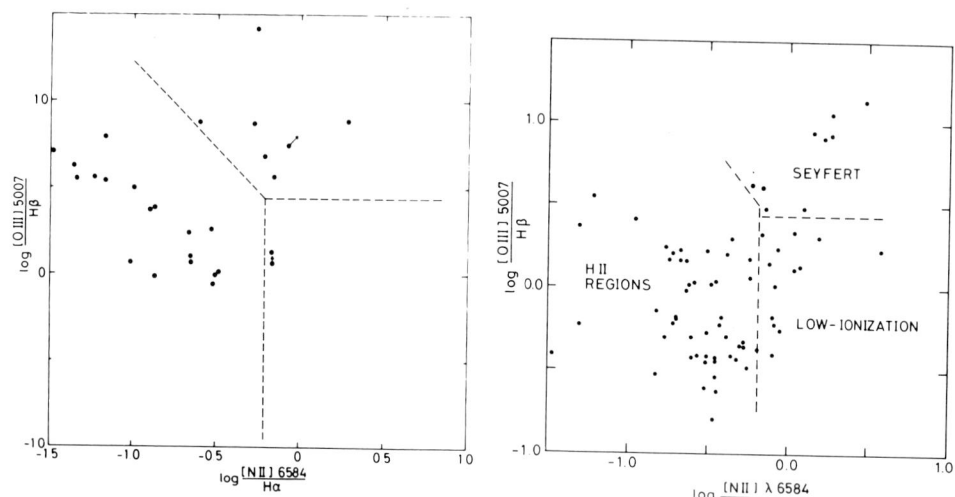

Fig. 4: Observed line ratios of the spectra of the individual nuclei in multiple nucleus galaxies (left); observed line ratios in the spectra of interacting galaxies (Keel et al. 1985) for comparison (right).

starburst character and are depicted in Fig. 2a. Fig. 2b similarly shows the spectra of the double Seyfert Mkn 463. As far as possible we have determined the overall continous spectra of the objects in our sample. Two examples (Mkn 463 and 273) are given in Fig.3. Double nucleus galaxies generally have very pronounced far-infrared emission indicating extreme starburst activity which may be coexistent with nonthermal activity. The ratios of the FIR fluxes at 100 μm and 60 μm are found to be typical for warm dust. According to the criterion by de Grijp et al. (1985) from these ratios most of our objects would be classified as Seyferts. For the components of double nucleus galaxies all combinations of activity types may occur (starburst, Liner, Seyfert 2, Seyfert 1). No connection is found between activity type and separation of the nuclear components. The intrinsically most luminous galaxies in our sample show Seyfert or double Seyfert character. In comparison to Keel et al.'s (1985) sample of interacting galaxies (fig.4) the double nucleus galaxies show on average substantial higher [OIII]/Hβ ratios and a higher percentage of Seyfert nuclei.

References

Gaskell, C.M., 1983, 24th Liège Astrophysical Colloquium on *Quasars and Gravitational Lenses*, p. 473
de Grijp, M.H.K., Miley, G.K., Lub, J., de Jong, T., 1985, *Nature* **314**, 240
Hutchings, J.B., Crampton, D., Campbell, B., Duncan, D., Glendenning, B., 1984, *Astrophys. J. Suppl.* 55, 319
Keel, W.C., Kennicutt, Jr., R.C., Hummel, E., van der Hulst, J.M., 1985, (preprint)

Kollatschny, W., Fricke, K.J., 1984a, *Astron. Astrophys.* **135**, 171
Kollatschny, W., Fricke, K.J., 1984b, Proc. 4th Europ. IUE-Conference, p.91
Stockton, A., 1982, *Astrophys. J.* **257**, 33

Acknowledgements

This work was in part supported by the Deutsche Forschungsgemeinschaft (grants Fr 325/21-1 and Fr 325/15-2). K.J.F. is grateful to Steward Observatory for hospitality and observing time at the 61 inch Mt.Lemmon telescope. He thanks Drs. U. Fink, R. Leach, A. Schultz, and Mr. M. DiSanti for providing instrumentation and help during the observations.

A TEST FOR THE GRAVITATIONAL LENS HYPOTHESIS IN THE
MOST LUMINOUS QUASAR: S5 0014+81.

H. Kühr
Max-Planck-Institut für Astronomie
Königstuhl
D-6900 Heidelberg 1
F.R.G.

ABSTRACT. The most luminous quasar known, S5 0014+81 (z = 3.41), was extensively studied with direct CCD imaging, and with radio interferometric techniques using the VLA as well as MK II and MK III VLBI systems. Optical spectroscopic observations revealed a high column density metal absorption system at z = 1.11, and VLA maps displayed a weak second radio component. We thus had evidence for the possible existence of a gravitational lens, which may have boosted the observed flux of the quasar towards an energy output of about $5*10^{48}$ erg/s. However, results from subsequent VLA and VLBI observations are not compatible with such interpretation, instead the weak radio component is probably associated with a jet-like phenomenon and does not provide evidence for or against a gravitational lens.

1. INTRODUCTION

The flat spectrum radio source S5 0014+81 was found in the MPI 5 GHz S5-Survey (Kühr, et al. 1981) and optically identified on Palomar Observatory Sky Survey plates with a neutral coloured object of 16.5 mag. Several spectrophotometric observations between September 1982 and February 1983 using the Steward Observatory's 90" (2.3 m) and Multiple Mirror telescopes revealed an exceptional quasar. At its emission line redshift of z = 3.41 we calculated its optical/IR luminosity over the observed bandpass (3550 Å - 2.2 μ) to be $3*10^{48}$ erg/s, with H_o = 50 km/s/Mpc and q_o = 0. Thus S5 0014+81 is more powerful than any other known object radiating in the radio through X-ray wavelength range (Kühr, et al. 1983).

For comparison we present in Fig. 1 absolute visual magnitudes for published quasars with z>3, as well as for the most powerful low redshift quasars. We also include three well known variable quasars, and indicate their range in luminosity during recorded outbursts.

Because S5 0014+81 has several times the luminosity of any other quasar yet discovered, we must consider the possibility that its brightness has been enhanced significantly by gravitational lensing of a foreground galaxy. In fact, a high column density metal absorption system (MgII, FeII) is present in the optical spectrum at z = 1.11 and

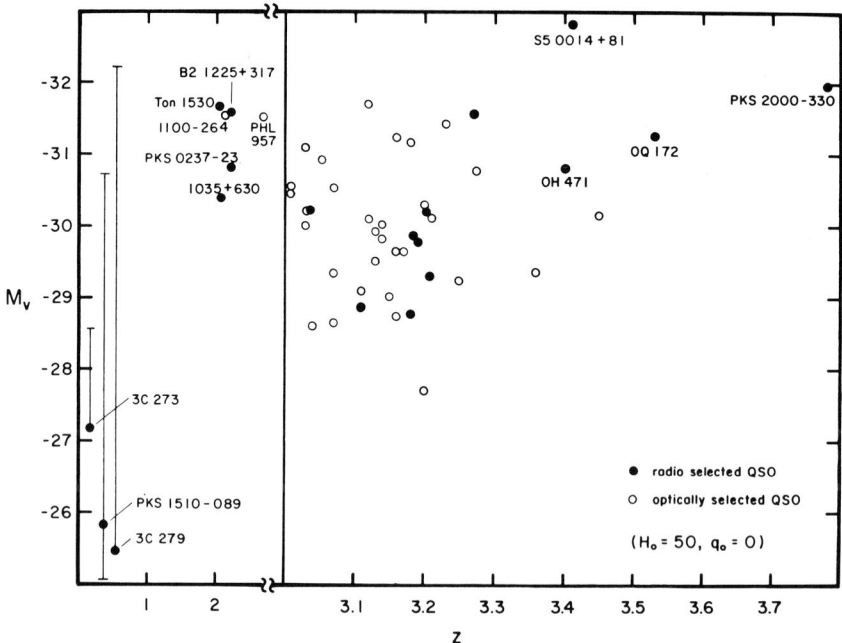

FIGURE 1. Absolute magnitudes for published quasars with z 3 and some powerful objects at lower redshifts. Three OVV quasars are also shown with vertical bars indicating their mangitude range during recorded outbursts, assuming an isotropic emission process.

may account for the absence of UV flux in this object in the short and long wavelengths cameras of the IUE. If this absorber is associated with a galaxy at that redshift, any significant amplification of the image will be accompanied by the presence of multiple images split by 2 arcsec or less.

2. OBSERVATIONAL RESULTS

Deep CCD frames of the quasar were obtained using the Steward Observatory 2.3 m telescope on Kitt Peak and the Max-Planck-Institut für Astronomie 2.2 m telescope on Calar Alto. At seeing conditions of about 1.5 arcsec (FWHM) the image of the quasar remained unresolved, and within 20 arcsec no additional object was detected above our limit of 22.5 mag.

Radio maps of high dynamic range and resolution were obtained with the Very Large Array (VLA). The strong, flat spectrum radio source was detected at 21, 18, 6, 2, and 1.3 cm in the A-array configuration of the telescope and found to be unresolved, with a flux density at 6 cm of 711 ± 2 mJy. In addition, at 6 cm only, a faint second component, located 0.6 arcsec to the south of the quasar, was found with a flux density of 3.3 ± 0.5 mJy. This is shown in Fig. 2.

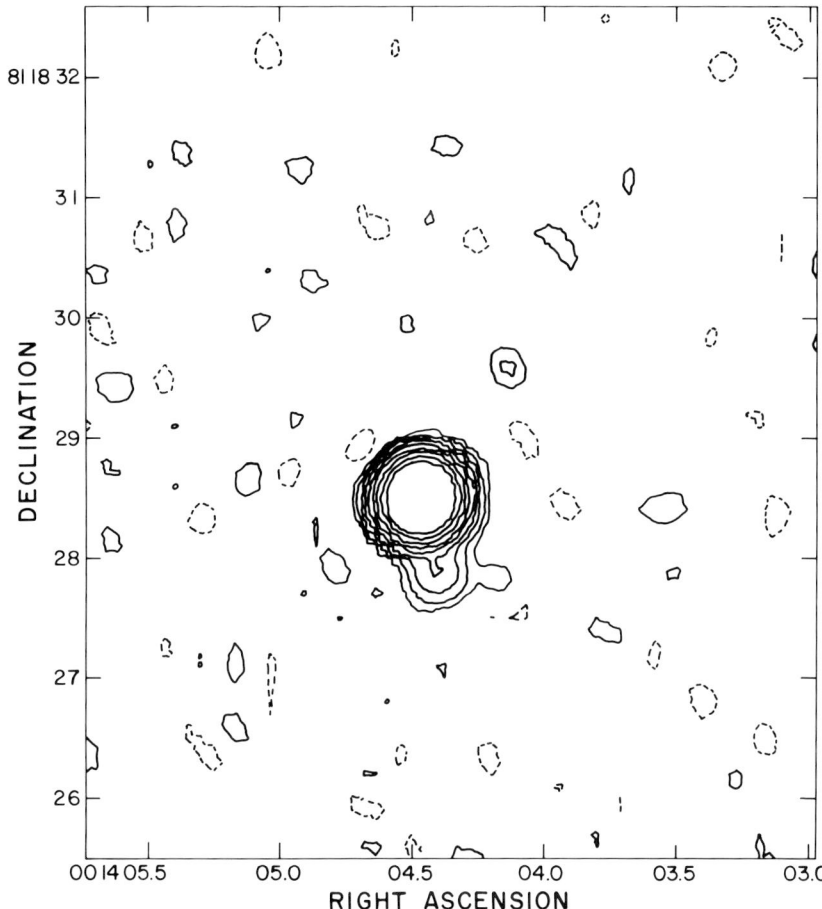

FIGURE 2. The quasar S5 0014+81 observed at 6 cm with the VLA in the A-array configuration. The unresolved flat spectrum radio source with 711 mJy is at the center of the map. In addition a faint second component with 3.3 mJy located 0.6 arcsec to the south of the quasar is displayed.

If the faint component is a lensed image of the quasar the achromatic nature of the lens (Einstein 1936) should leave the radio spectral index unchanged compared with that of the flat spectrum strong component. In subsequent VLA observations at 2 cm in the B-array, however, the faint, second component was not detected and a steep spectral index was derived. Even if we allow for a 10 percent variability of the flux densities at each wavelength, which is the typical variability found from frequent observations during the last five years, and if we further allow for a differential time delay, our upper bounds are still inconsistent with the two components being images of a single quasar.

This was further supported by two experiments, namely MK III and M II VLBI measurements with transatlantic baselines at 6 cm. The former experiment was carried out in order to detect or place upper bounds on the flux density of any compact component at the position of the faint source. We determined the diameter of the main, flat spectrum component to be 0.9 ± 0.1 mas (FWHM). If we assume that both, the weak and strong components are images of a quasar, the law of surface brightness conservation (Misner, Thorne, Wheeler 1973) requires the weak component to be an order of magnitude more compact than the strong component. Thus, for the weak component we would expect the correlated flux density not to be significantly different from the total flux density of 3.3 ± 0.5 mJy, i.e. we would expect the weak source to be essentially unresolved. We did not detect the source. The upper 4 σ (statistical standard deviation) bound for the total flux density of any unresolved source in the area we searched was 1 mJy. This bound is inconsistent with the weak component being a lensed image of the quasar.

The main objective of the MK II experiment was to map the source with milliarcsecond resolution. Only the strong, main component was detected and found to be resolved into a compact component with an extension to the south. Fitting a model of two unresolved gaussian components to the data revealed a flux density of 500 ± 20 mJy for the northern and 140 ± 20 mJy for the southern component. The separation was 0.78 ± 0.02 mas at a position angle of 185.2 ± 1.3 deg.

3. CONCLUSION

The position angles of the VLBI milliarcsecond structure and the weak, steep spectrum VLA component agree to within the errors. This supports the existence of a jet-like phenomenon, extending over 8 kpc (for q_o = 0, H_o = 50 km/s/Mpc) from the central component and detected for the first time in a high redshift quasar.

At the given level of angular resolution and sensitivity the CCD, VLA, and VLBI measurements described do not reveal evidence for lensed multiple images. It is thus likely that S5 0014+81 is indeed intrinsically exceptionally powerful. However, we cannot totally exclude the possibility that a gravitational lens (e.g. a cluster) is present, too weak to produce multiple images, but which nevertheless brightens the source.

REFERENCES

Einstein, A. 1936, Science, 84, 506
Kühr, H., Pauliny-Toth, I. I. K., Witzel, A., and Schmidt, J. 1981, A.J., 86, 854
Kühr, H., Liebert, J. W., Strittmatter, P. A., Schmidt, G. D., Mackay, C. 1983, Ap. J., 275, L 33.
Misner, C. W., Thorne, K. S., and Wheeler, J.A. 1973 (Freeman, San Francisco), p. 589.

THE PROPERTIES OF X-RAY SELECTED BL LAC OBJECTS

T. Maccacaro[1,2], I.M. Gioia[1,2], D. Maccagni[3], and J.T. Stocke[4]
[1] Harvard-Smithsonian Center for Astrophysics, Cambridge, USA.
[2] Istituto di Radioastronomia, Bologna, Italy
[3] Istituto di Fisica Cosmica, Milano, Italy
[4] Steward Observatory, University of Arizona, Tucson, AZ

ABSTRACT

The optical and radio properties of the X-ray selected BL Lacs in the Einstein Medium Sensitivity Survey are presented and compared with those of the HEAO 1 A-2 sample and with those of radio selected BL Lacs. The X-ray selected BL Lacs possess smaller polarized fractions and less violent optical variability than radio selected BL Lacs. These properties are consistent with the substantial starlight fraction seen in the optical spectra of a majority of these objects. This starlight allows a determination of definite redshifts for two of four MSS BL Lacs and a probable redshift for a third. These redshifts are 0.19, 0.297 and 0.59.

As a class, X-ray selected BL Lacs have the largest ratio of X-ray to optical flux of any Active Galactic Nuclei (AGN) yet discovered. We have also found that the number-flux relation for BL Lacs shows a drastic flattening below fluxes of the order of 10^{-12} ergs cm^{-2} s^{-1} and we have interpreted this as evidence that BL Lacs do not share the evolutionary properties of quasars.

The results presented here are summarized from two papers, Maccacaro et al. 1984 and Stocke et al. 1985, to which we refer the reader seeking a detailed discussion of the available radio, optical and X-ray data. For a description of the Einstein Medium Sensitivity Survey see Maccacaro et al. 1982, Stocke et al. 1983, and Gioia et al. 1984.

DISCUSSION

Traditionally, BL Lac objects have been discovered as flat spectrum, highly variable, radio sources. The optical spectra of the counterparts to these sources were found to be featureless, polarized, and variable in total and polarized light. These spectra are also typically very red, sometimes with spectral indexes greater than 3, leading to their detection as strong infrared sources (however a few blue BL Lac objects have been found; e.g. Mkn 421 and 501). At high signal-to-noise or in spectra obtained through an annulus or offset from the nucleus, weak absorption or emission features are often seen. Photometry of the "fuzz" surrounding BL Lac objects, in conjunction with spectroscopy, has led to the general conclusion that BL Lacs inhabit giant elliptical galaxies (Miller, French, and Hawley 1978). Einstein observations (Schwartz and Ku 1983; Maccagni and Tarenghi 1981) found that X-ray emission is a common feature and indeed most BL Lac objects have now been detected in the X-ray band.

Two Markarian objects (421 and 501) were among the first non-radio-selected BL Lac objects and presented somewhat different properties than their radio-selected counterparts. Both objects are clearly the nuclei of giant elliptical galaxies (Ulrich 1978), both possess a smaller fraction of polarized light and exhibit a lower degree of variability than other BL Lac objects. However, they are among the strongest X-ray emitters of their class (Piccinotti et al. 1982).

The Medium Survey BL Lacs all have properties similar to these two Markarian objects. In particular, the optical polarizations of the four MSS BL Lacs are quite low but still consistent with being similar to Mkn 421 and 501. In fact most of the unusual properties of the MSS BL Lacs (low polarization, absence of violent variability, and increased starlight fraction as seen in the optical spectrum) can be reproduced by any object of the Mkn 421 or 501 type at a large enough distance that the observing aperture admits a much larger fraction of starlight contamination.

Moreover, each of these 4 X-ray sources possess at least some of the qualities found in radio selected BL Lacs:

- All but one (1207+397) are a flat spectrum radio source. Two of these sources (1235+632 and 0317+186) show definite variability from only two epochs of observations.
- All but one source (1235+632) is optically polarized although the polarization measurement of 1207+379 is significant only at the 3 σ level. The other two sources exhibit variable optical polarization.
- 1402+043 possesses a featureless optical spectrum, in all the other three cases the stellar absorptions are much weaker than in typical elliptical galaxies.
- all but one source (0317+186) are optically variable at or above the 0.5 magnitude level.

We have compared some of the global properties of the X-ray selected BL Lacs with those of a subset of radio-selected BL Lac objects and with those of X-ray selected AGNs in the MSS. In figure 1 the distribution of α_{ox} (as defined in Tananbaum et al. 1979), monochromatic optical luminosity of these three classes of objects are shown. The X-ray selected BL Lacs are characterized by a lower α_{ox} indexes and while they have X-ray luminosities similar to radio selected BL Lac objects, their optical luminosities are lower and thus comparable to the optical luminosities of X-ray selected AGNs. The number of objects in this analysis is small and, furthermore, radio-selected BL Lacs are highly variable objects at X-ray, optical and radio wavelengths. Both these factors might modify this conclusion, which should therefore be considered as tentative until larger samples can be obtained. This conclusion, however, is consistent with the large starlight fraction seen in several X-ray selected BL Lac objects.

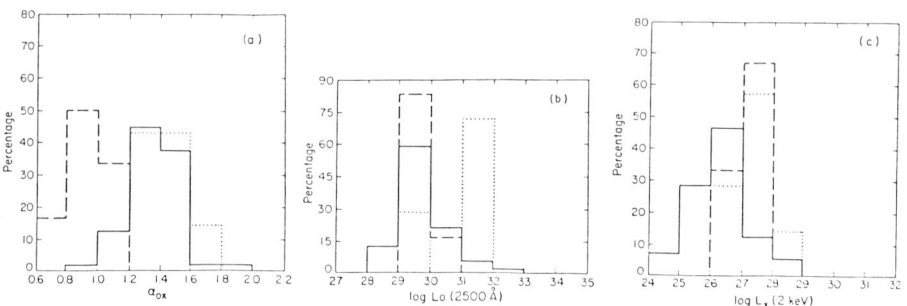

Figure 1. (a) The α_{ox} distribution, (b) monochromatic optical luminosity distribution and (c) monochromatic X-ray luminosity distribution for X-ray selected BL Lac objects (dashed line), X-ray selected MSS AGNs (solid line) and a subset of radio selected BL Lac objects (dotted line) (adapted from Stocke et al. 1985)

There is another indication that X-ray selected BL Lac objects have peculiar global properties. Figure 2 shows a plot of α_{ro} vs. α_{ox} for X-ray selected and radio selected BL Lacs, and for the QSOs observed by the Einstein Observatory. The radio-to-optical spectral index is defined as $\alpha_{ro} = \log(S_{5GHz}/S_{2500A})/5.38$. X-ray selected BL Lacs occupy a specific and separate region of the α_{ro}-α_{ox} plane. With respect to their optical emission, they are weak radiosources but strong X-ray sources with a very small dispersion in α_{ro}. No radio quiet QSO or X-ray selected AGN has such a high ratio of X-ray to optical flux. Furthermore, radio selected BL Lac objects populate the same region as radio-loud quasars. Were this peculiarity shown by X-ray selected BL Lacs only a selection effect, we would expect to find other X-ray selected AGNs in the low α_{ox} region. Instead, the same X-ray selection yields two classes of objects: at the lower end of the X-ray distribution we find AGNs, at the higher end we find featureless objects which we identify as BL Lacs.

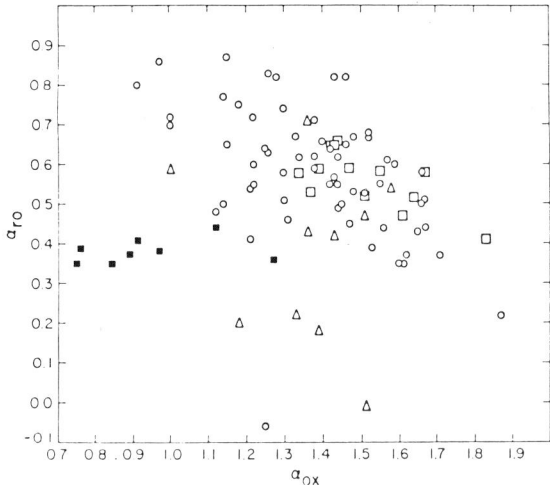

Figure 2. α_{ro} - α_{ox} plot for the X-ray selected BL Lac objects (filled squares), a subset of radio selected BL Lac objects (open squares), the MSS X-ray selected AGNs (triangles) and the QSOs found in Zamorani et al. (1981) (open circles). Radio-quiet QSOs and AGNs fall in the bottom right part of the plot. Note that X-ray selected BL Lac objects all cluster around a α_{ro} value of 0.4 and six out of eight have α_{ox} value smaller than 1.0 (the two objects with α_{ox} 1 are Mkn 501 and 1402+043) (adapted from Stocke et al. 1985).

If these objects are not considered as BL Lacs, then they constitute a new class of AGNs ("X-ray galaxies") defined by strong X-ray emission (compared to optical emission), modest radio emission and absence of strong emission line in the optical spectrum.

In the Medium Sensitivity Survey, sources as faint as 10^{-13} ergs cm^{-2} s^{-1} can be detected. For example, of the 56 AGN detected, 49 have fluxes between 10^{-13} and 10^{-12} and only 7 have fluxes in excess of 10^{-12} ergs cm^{-2} s^{-1}. Quite remarkably the 4 MSS BL Lac objects have all been detected at fluxes larger than 10^{-12}, and there are no detections at smaller fluxes. From the flux distribution of the BL Lacs in the MSS and in the HEAO 1 A-2 all sky X-ray survey of Piccinotti et al. (1982) we have determined that the logN(>S)-logS of these objects shows a significant negative curvature. Within the approximation of two power laws, the logN(>S)-logS is described by a slope $\alpha \sim 1.5$ at fluxes larger than 10^{-12} and by a much flatter slope ($\alpha < \sim 0.8$) at smaller fluxes. Since for a given value of q_o the logN-logS relation is mostly determined by the (local) luminosity function and by its evolution, the present results can and have been used to set constraints on both this functions.

To this end we have considered different slopes γ for the BL Lacs X-ray luminosity function (assumed to be a pure power law) and we have integrated it within a framework of pure luminosity evolution to compute the resulting logN-logS curves. We have found that a marginal agreement with the observed data points and limits can be obtained if the BL Lac X-ray luminosity function has an (integral) slope of 1.25 and no evolution is assumed. This solution, however, is not unique. Although it is not possible to reconcile the data with the assumption that BL Lac objects evolve as much as quasars, we cannot rule out steeper luminosity function and an evolution opposite that required for quasars.

It is worth noting that a slope $\gamma \sim 1.25$ is in agreement with the results of Schwartz and Ku (1983) on the local volume emissivity of BL Lac objects (see also Perez-Fournon and Biermann 1984) and with those of Urry, 1984.

This work has received partial financial support from NASA contract NAS8-30751. JTS acknowledge the continuing support of NASA grant NAG-8442.

REFERENCES

Gioia, I.M., Maccacaro, T., Schild, R.E., Stocke, J.T., Liebert, J.W., Danziger, I.J., Kunth, D., and Lub, J. 1984, *Ap.J.*, **283**, 495.
Maccacaro, T. et al. 1982, *Ap.J.*, **253**, 504.
Maccacaro, T., Gioia, I.M., Maccagni, D., and Stocke, J. 1984, *Ap.J. (Letters)*, **284**, L23.
Maccacgni, D., and Tarenghi, M. 1981, *Ap.J.*, **243**.
Miller, J.S., French, H.B., and Hawley, S.A. 1978a, *Ap.J. (Letters)*, **219**, L85.
Perez-Fournon, I., and Biermann, P. 1984, *Astr.Ap.*, **130**, L13.
Piccinotti, G. et al. 1982, *Ap.J.*, **253**, 485.
Schwartz, D.A., and Ku, W.H.-M. 1983, *Ap.J.*, **266**, 459.
Stocke, J.T., Liebert, J.W., Schmidt, G.D., Gioia, I.M., Maccacaro, T., Schild, R.E., Maccagni, D. and Arp, H.C. 1985, *Ap.J.* submitted.
Stocke, J., Liebert, J., Gioia, I.M., Griffiths, R.E., Maccacaro, T., Danziger, I.J., Kunth, D., and Lub, J. 1983, *Ap.J.*, **273**, 458.
Tananbaum, H. et al., 1979, *Ap.J. (Letters)*, **234**, L9.
Urry, M.H., Kinman, T., Lynds, C., Rieke, G., and Ekers, R. 1975, *Ap.J.*, **198**, 261.

KISO SURVEYS OF ULTRAVIOLET-EXCESS OBJECTS

H. Maehara, T. Noguchi, M. Kondo, and N. Miyauchi-Isobe
Tokyo Astronomical Observatory
Osawa, Mitaka
Tokyo 181
Japan
          and
B. Takase
Kokugakuin University
Higashi, Shibuya
Tokyo 150
Japan

ABSTRACT. Kiso surveys of ultraviolet-excess objects have been carried out with the use of the 105cm Schmidt telescope, in which about 1,200 stellar (KUVs) and 1,100 galaxy (KUGs) images are catalogued from the (U,G,R) triple image plates. Follow-up observations in the optical, infrared, and radio wavelengths reveal the results below in relation to active galaxies;
(1) a few percent of KUVs belong to quasars,
(2) about 80 percent of KUGs possess conspicuous emission lines,
(3) there are some Seyfert-like objects among KUGs.

1. SURVEYS AND DETECTIONS

Surveys of ultraviolet-excess objects have been carried out as programs of the Kiso Schmidt telescope (105/150/330) of the Tokyo Astronomical Observatory. The $6° \times 6°$ field is covered on a 14 inch square plate. Three adjacent images corresponding to U (ultraviolet), G (green), and R (red) bands are exposed on a hypersensitized 103aE plate changing the three color filters; UG1, BPB50, and RG610. The exposure time is so set that the brightness is the same for three images of A0-type stars.

A program is to search for stellar images (KUVs), and the other is to search for galaxies (KUGs). Main survey area is spread along the galactic longitude of $180°$ from the north to the south galactic poles. Other fields are also observed according to some specified programs. The total number of these multi-color-image plates taken is about 140 covering $\sim 4,000$ square degrees. The detection is visually made with a binocular microscope or a TV viewer by means of the enhancement of U images relative to **G** and **R** ones.

The celestial position of the objects is measured using the comparator, and calculated by means of the standard coordinate method with an accuracy of 0."5. Their brightness and the degree of UV-excess are visually estimated from the multi-color-image plates and the Palomar Sky Survey Prints. These values and the finding chart of each object are given in the catalogues.

2. KUVs

A search for stellar objects has been made using the (**U,G,R**) triple image plates. The limiting magnitude of detection ranges from $\sim$ 17 to 18.5 magnitude due to the plate condition. A total of 1,186 objects was detected in 1,400 square degree field, and compiled into the catalogues of Noguchi et al.(1980) and Kondo et al.(1984). The catalogued fields are illustrated in Figure 1. The objects identified in the previous catalogues are 28 quasars, 48 spectroscopic white dwarfs, 36 suspected white dwarfs or high proper-motion stars, and 77 blue objects.

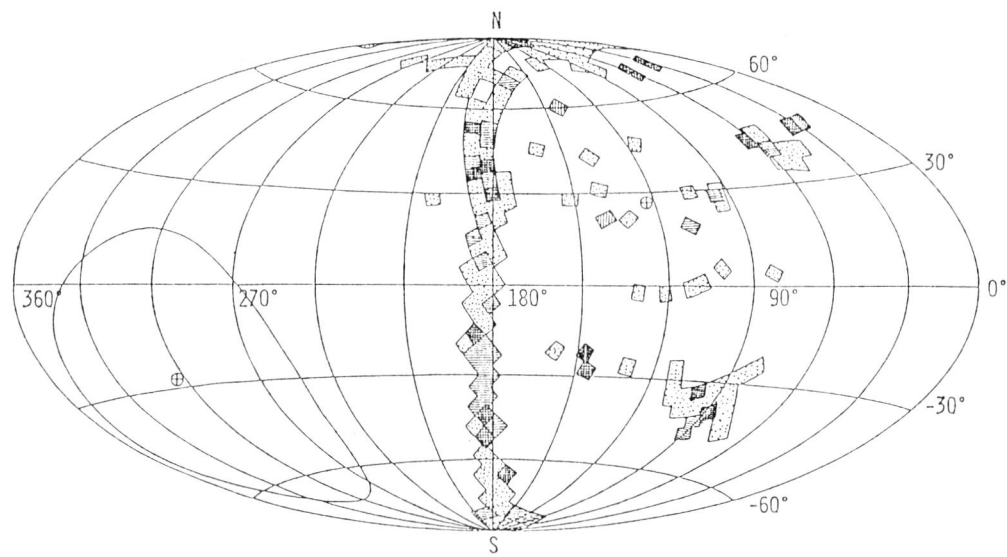

Figure 1. Celestial fields observed and surveyed in this work. Horizontally and vertically hatched areas denote the fields which have been catalogued in the KUV and KUG surveys, respectively. Dotted area denotes the fields observed with multi-color-image plates. The coordinates are galactic ones drawn in the Aitoff's projection.

Proper motions were measured for about 80 objects in the two pilot fields with an accuracy of 0".018 yr^{-1} (Noguchi et al. 1982). It is shown that a fair fraction of the sample objects possesses appreciable amount of proper motions. The space number density of white dwarfs was estimated from this result (Ishida et al.1982) on the assumption that all the KUVs belong to white dwarfs.

A spectroscopic investigation is done with the use of the 188cm telescope of the Okayama Astrophysical Observatory. About 100 spectra have been obtained up to now mainly for bright objects; major constituents belong to white dwarfs of types DA, DB, and DO, or OB subdwarfs. There are candidates of novae, emission-line objects, and quasars among KUV samples. UBV observations are also made with the Okayama 91cm telescope (Kondo et al.1982), and the photographic magnitude and color index are calibrated. In consequence, a few percent of KUVs belong to quasars, though most of them are high-temperature white dwarfs or subdwarfs.

3. KUGs

A search for UV-excess galaxies has been made using the common plates to KUVs and sometimes (**U,R**) double image plates. About 3,000 candidates have been found up to now, and ~ 1,100 objects were listed in Takase and Miyauchi-Isobe (1984, 1985). The faintest galaxies detected is about 17.5 magnitude. The catalogued fields are shown in Figure 1 as well as those of KUVs. The number ratio of KUGs to all galaxies is around 0.25, i.e., one forth of all galaxies shows the UV-excess character. The ratio is higher in the field than in the cluster region (Takase 1980). Markarian galaxies in the common field are also detected as KUGs, and the frequency of the latter is about ten times higher than that of Markarian's. It is mainly because of the fainter limiting magnitude of this survey.

A morphological classification was performed for 142 selected samples (Takase et al.1983). They are classified into the following seven types;
(1) Irregular with clumpy H II regions (Ic)
(2) Irregular with a giant H II region (Ig)
(3) Pair of interacting components (Pi)
(4) Pair of detached components (Pd)
(5) Spiral with knotty arms (Sk)
(6) Spiral with peculiar bar or nucleus (Sp)
(7) Compact (C)
It is suggested that most KUGs belong to irregular or spiral galaxies and not to elliptical ones. There is an obvious correlation between morphological type and the UV-excess degree; the types Ic, Ig, Pi, and Pd tend to have a higher degree, while spiral galaxies have a lower degree.

A high-resolution optical imagery of selected samples shows that some of Ic-type galaxies probably belong to clumpy irregular galaxies, which are in the phase of active star formation (Maehara et al.1985a). About 20 % of all the KUGs are identified as the IRAS sources, and they are abundant with warm dust (Ukita and Karoji 1985). A deep radio observation of 38 KUGs was made with the Nobeyama 45m radio telescope at 10 GHz (Maehara et al.1985b). Their dominant non-thermal radiation suggests frequent supernova explosions.

According to spectroscopic observations of KUGs, about 80 % of all have conspicuous emission lines of Balmer and nebular ones. Most of them are sharp, low-excited emission lines similar to galactic H II regions, and there are strong and broad emission lines characteristic of Seyfert galaxies. A fairly large variation exists in the intensity ratio of the emission lines (Maehara et al.1985c). In consequence, the observational results obtained above reveal the fact that most KUGs have portions of active star formation in large scales, and a small fraction of them have nuclear activities like quasars or Seyfert galaxies.

References

Ishida,K., Mikami,T., Noguchi,T., and Maehara,H. 1982, Publ. Astron. Soc. Japan, 34, 381.
Kondo,M., Noguchi,T.,and Maehara,H. 1984, Ann. Tokyo Astron. Obs. 2nd Ser., 20, 130.
Kondo,M., Watanabe,E., Yutani,M., and Noguchi,T. 1982, Publ. Astron. Soc. Japan, 34, 541.
Maehara,H., Hamabe,M., Takase,B., Bottinelli,L., Gouguenheim,L., and Heidmann,J. 1985a, (to be sumitted to Publ. Astron. Soc. Japan).
Maehara,H., Inoue,M., Takase,B., and Noguchi,T. 1985b, Publ. Astron. Soc Japan, 37, No.2 (in press).
Maehara,H., Noguchi,T., and Takase,B. 1985c, (in preparation).
Noguchi,T., Maehara,H., and Kondo,M. 1980, Ann. Tokyo Astron. Obs. 2nd Ser., 18, 55.
Noguchi,T., Yutani,M., and Maehara,H. 1982, Publ. Astron. Soc. Japan, 34, 407.
Takase,B. 1980, Publ. Astron. Soc. Japan, 32, 605.
Takase,B., and Miyauchi-Isobe,N, 1984, Ann. Tokyo Astron. Obs. 2nd Ser., 19, 595.
Takase,B., and Miyauchi-Isobe,N, 1985, Ann. Tokyo Astron. Obs. 2nd Ser., 20, No.3 (in press).
Takase,B., Noguchi,T., and Maehara,H. 1983, Ann. Tokyo Astron. Obs. 2nd Ser., 19, 440.
Ukita,N., and Karoji,H. 1985, private communication.

STEEP SPECTRUM RADIO SOURCES SHOWING LOW FREQUENCY VARIABILITY

F. Mantovani[1], I. Browne[2], R. Fanti[1], A. Ficarra[1], T. Muxlow[2], L. Padrielli[1], J. Romney[3]
  1. Istituto di Radioastronomia, Bologna, Italy
  2. Nuffield Radio Astronomy Labs., Jodrell Bank, U.K.
  3. N.R.A.O., Charlottesville, U.S.A.

1. INTRODUCTION

The Bologna group has been carrying on a monitoring program at 408 MHz of more than a hundred sources since 1975. Among sources found to be variable or possibly variable (Fanti et al., 1983) a small sample (10) has steep spectrum ($\alpha \sim 0.6$, $S \propto \nu^{-\alpha}$) in the frequency range 0.4-15 GHz.

Sources with steep straight spectral index are usually extended (~ 100 Kpc) with weak central components detected if observed with high dynamical range. The observed variability at low frequency of some of these sources represents a critical point for the mechanisms suggested as explanation of this phenomenon either extrinsic or intrinsic to the sources.

We present here flux density measurements, spectral index behaviour, arcsecond and milliarcsecond observations of some of these sources.

2. DATA

Measured and derived parameters for 4 sources are summarized in the Table 1.

At the given redshifts, all these sources are distant powerful radio sources of comparable luminosity ($P(0.4$ GHz$) \, 10^{28}$ Watts/Hz). Fig.1 shows spectral indices extracted from the compilation of Kühr et al. (1978). They have a constant value $\alpha \sim 0.7$ like the 'normal' power law spectra found in extended sources, indicating that the bulk of the sources are transparent to the radiation. A possible turnover cannot be excluded at frequency lower than 100 MHz.

Light-curves at 0.4 GHz and, for three sources, at 8 GHz taken from Padrielli et al. (in prep.), are shown in Fig. 2. Beside the variation at 0.4 GHz a common feature is the absence of variability at high frequency.

If the variations at low frequency are intrinsic to the source, the observed amplitude and time scale variability would imply the existence of a component with a brightness temperature 10^{14}-10^{16} K. All of them have been mapped with MERLIN at 0.4 GHz or 1.7 GHz, (Fig.3). Their angular size (major axis) ranges from 3 arcsec for 3C99 (0358+004) to 30 arcsec for 3C159 (0621+40).

Both 3C159 and 1422+202 show weak central core and outer symmetrical

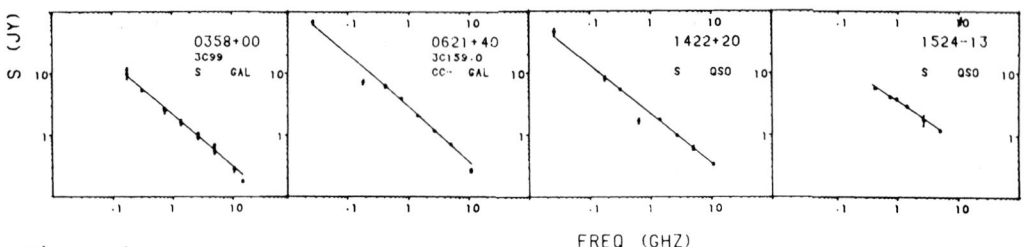

Figure 1

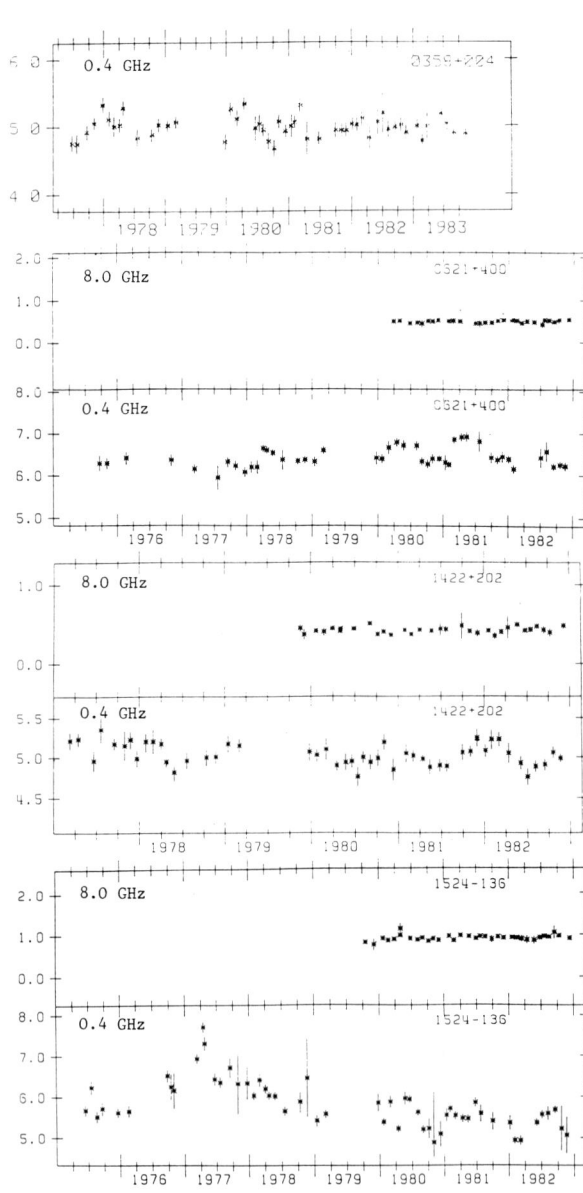

Figure 2

Figure 3

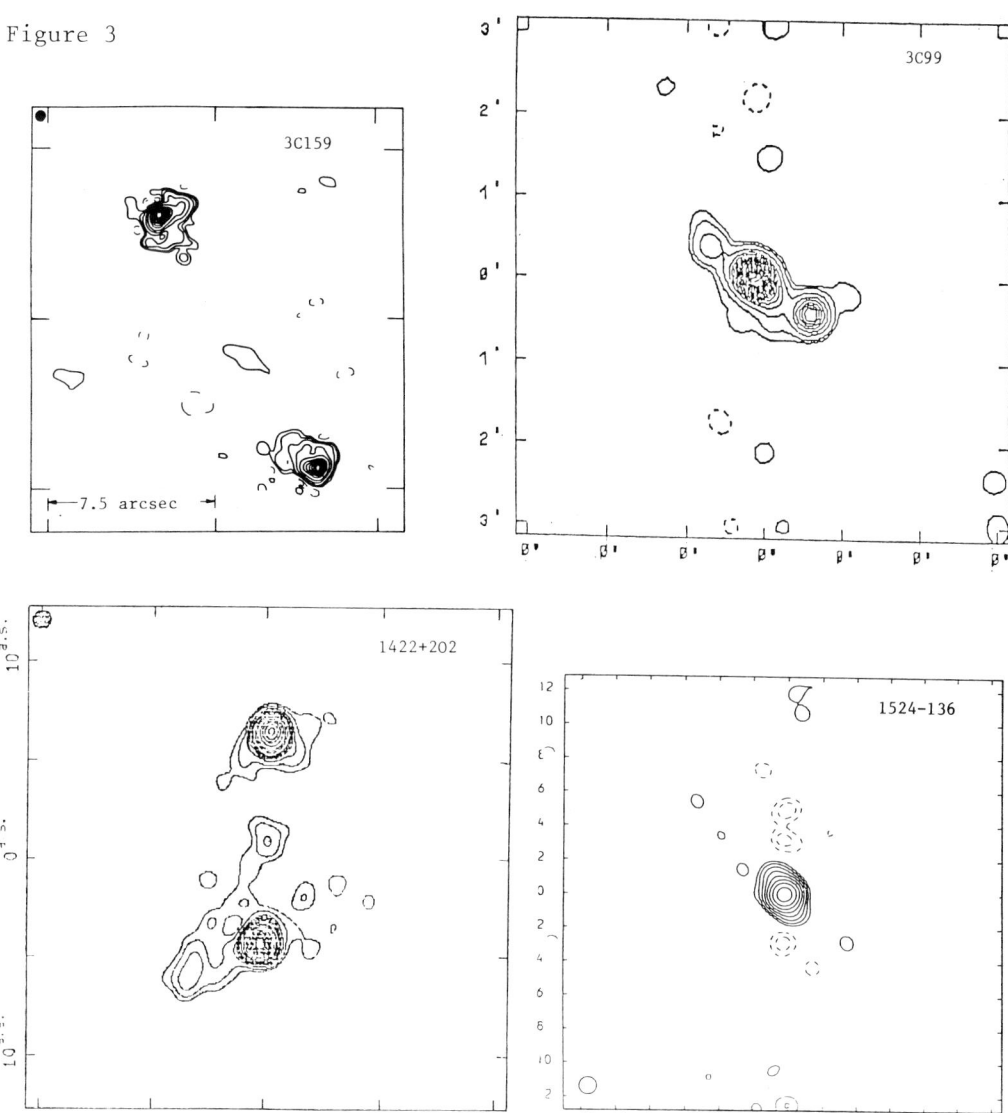

TABLE 1

	3C159	1422+202	3C99	1524−136
Optical Id.	G	QSO	G	QSO
Red-shift	0.483	0.87	0.426	1.687
Lum. (408 MHz) (Watts/Hz)	1.6×10^{28}	0.4×10^{28}	1.0×10^{28}	14.6×10^{28}
Major axis (Kpc)	115	47	7	12

lobes as typically found in extended sources. In particular 3C159 has a twin hot spot in the northern lobe of comparable brightness at 6 cm (Tytler and Browne, 1985). This source (discussed by Browne et al.,1985) has been observed with the VLBI technique using the EVN at 18 cm and no compact components have been detected. On the Effelsberg-Westerbork baseline the limit in sensitivity was 90 mJy.

The source 1422+202 is only marginally detected in the shortest EVN baseline, Effelsberg-Westerbork. With this baseline we got fringes from the two bright hot spots visible in the MERLIN map.

3C99 has a triple structure at the arcsec resolution with a dominant core significantly brighter than outer emission regions. VLBI fringes show evidence of a double structure.

1524-136, which has been the fastest variable found by the Bologna group, has a single slightly resolved component in a MERLIN map at 0.4 GHz. It has also been observed with a VLBI array of 8 telescopes at 18 cm and it has been detected in the transatlantic baselines.

3. DISCUSSION

We are tempted to suggest from this small sample that steep spectrum low frequency variable sources might be subdivided in two groups:
a)'classical' extended double sources;
b) steep spectrum compact sources.

If we believe in their variability at low frequency, sources belonging to group a) represent a critical test for generally accepted mechanisms explaining such a phenomenon, such as bulk relativistic motion along direction near the line of sight and scintillation due to interstellar medium. The former model requires a dominant compact core, absent here, the second interpretation implies compact component, $\leq$ 20 mas in size, clearly in conflict with the radio morphology found for these sources. Members of this group are 3C159 and 1422+202.

Sources in group b) seem to have compact components where the variability is presumed to take place. This structure is typical of core-dominated sources which have flat spectrum core and in the frame of the unified scheme are classical double seen along the line of sight. 3C99 and 1524-136 belong to this group.

So where can we place these sources in such theories?

There is a tendency to believe that compact steep spectrum sources represent an independent class of objects, characterized by: i) physical size of few Kpc; ii) absence of dominant flat spectrum radio core.

Many of them (17 3CR sources) have been monitored monthly by the Bologna group. Tey were selected because of their small angular size and represent a homogeneous but not complete sample. Among them only 3C138 and 3C99 were classified as variable. So low frequency variability is a quite peculiar behaviour for sources belonging to this class, even if at present, data do not unable us to derive results of statistical significance.

REFERENCES

C. Fanti et al., 1983, Astron. Astrophys., 118, 171
H. Kühr et al., 1978, MPIfR preprint No 55
D. Tytler and I. Browne, 1985, M.N.R.A.S. in press
I. Browne et al., 1985, M.N.R.A.S., 213, 945

THE LUMINOSITY FUNCTION OF QUASARS AND LOW LUMINOSITY ACTIVE GALACTIC NUCLEI

Herman L. Marshall
Space Telescope Science Institute
3700 San Martin Drive
Baltimore, MD 21218
U.S.A.

Abstract.

The quasar luminosity function is extended to include the nuclear luminosities of Seyfert 1 galaxies. The model is fitted to the (total) luminosity data for the active galaxies from a complete flux-limited survey. A good fit is obtained with $dN/dL \propto L^{-1.2}$, matching onto the steeper quasar luminosity function near $M_B = -22$. The feature thus produced conforms to the luminosity functions of Koo and Kron using a pure luminosity evolution model derived from brighter quasar data.

1. Data set

The data are taken from a complete sample of 48 Seyfert (1 and 2) galaxies compiled by Huchra (1985) from the CfA redshift survey (Huchra et al. 1983). The sample consists of galaxies selected for $B_{tot} \leq 14.5$ which show broad emission lines. The total, monochromatic luminosities at 2500 Å for the 26 Seyfert 1 galaxies (Fig. 1) were obtained using the formulae of Marshall et al. (1984) and $H_0 = 50$ km s^{-1} Mpc^{-1} and will be quoted in units of $L^* = 10^{30}$ erg s^{-1} Hz^{-1} (corresponding to $M_B = -23.56$). The sample area is 2.7 ster.

2. Method

The method for estimating the parameters of the nuclear luminosity function is described in detail by Marshall (1985). Briefly, the total luminosity, L_T is assumed to result from two components: L_G, the host galaxy luminosity and L_Q, the luminosity of the quasar-like nucleus. No assumption is made regarding the properties of either component, contrasting with the "color-given" and "galaxy-given" method used by Cheng et al. (1985). Instead, it is assumed that the values of L_G are drawn randomly from a portion of the galaxy luminosity function (LF):

$$\psi(L_G) = (\psi_0/L_G^*)(L_G/L_G^*)^{-\alpha} e^{-(L/L_G^*)} \qquad 1)$$

where $\psi_0 = 0.014$ Mpc^{-3}, $L_G^* = 0.086$ ($M_B^* = -20.9$) and $\alpha = 1.3$ (Davis and Huchra 1982). The LF is truncated below $L_{G,0} = 0.01$ ($M_B = -18.6$, cf. Huchra 1985). The nuclear LF estimates of Cheng et al. are also shown in Figure 2. See Cheng et al. for comparisons to total LFs.

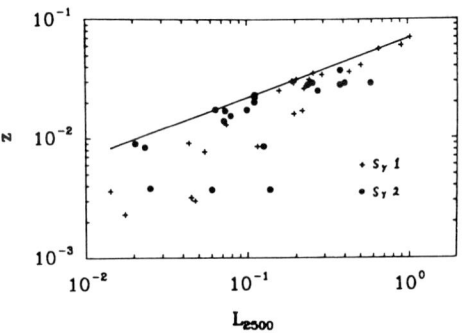

Figure 1. The z vs. total luminosity, L_{2500}, diagram for the Seyfert galaxies in the CfA redshift survey (Huchra 1985). The line represents the magnitude limit, $B_1 = 14.5$.

The quasar LF is assumed to be given by

$$\phi(L_Q) = (\phi_0/L_Q^*)[(L_Q/L_Q^*)^\beta + (L_Q/L_Q^*)^\gamma]^{-1} \qquad 2)$$

and is truncated below $L_{Q,0}$. The parameters ϕ_0 and β can be obtained from the pure luminosity evolution (LE) model F of Marshall (1985a, see Fig. 2): $\beta = 3.6$ and $\phi_0 = \rho_0 (L_Q^*/L^*)^{-\beta+1}$, where $\rho_0 = 22$ Gpc^{-3}. The quantities L_G^* and γ are to be determined. Given the steep slope of the bright end, one expects $\gamma < \beta$, providing a feature comparable to the Cheng et al. LF estimates. For $\gamma < 1$, the LF coverges, otherwise $L_Q > L_{Q,0}$ is required. The value for $L_{Q,0}$ is somewhat arbitrary; $L_{Q,0} = 0.001$ is chosen because the faintest nuclei are known to show $L_Q \lesssim 0.003$.

The most important assumption that must be made concerns the correlation between L_Q and L_G for $L_G > L_{G,0}$, i.e., are more luminous quasars formed in more luminous galaxies? It is assumed here that there is no correlation. The total LF may now be formed:

$$\rho(L_T) = \iint \psi(L_G)\phi(L_Q)\delta(L_T-L_G-L_Q)dL_G dL_Q \Big/ \int \psi(L_G)dL_G \qquad 3)$$

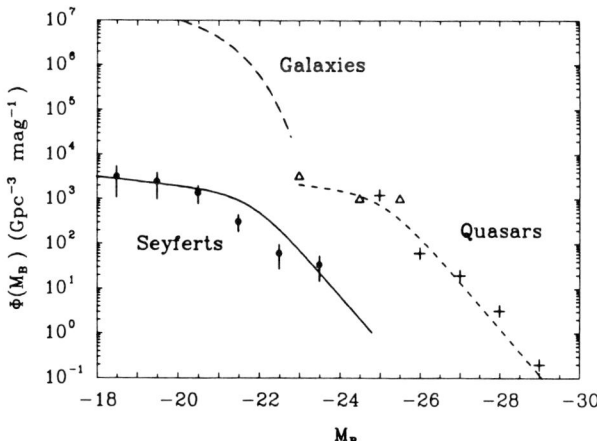

Figure 2. Various luminosity functions (LFs). The galaxy LF is taken from Davis and Huchra (1982); the Sy 1 LF model has parameters $L_Q^*=0.2$ and $\gamma=1.2$; the quasar LF uses the previous Sy 1 LF and the LE model F from Marshall (1985a). The data come from Cheng et al. (1985) - dots, Marshall (1985a) - plus signs; and Koo (these proceedings) - triangles.

A simple likelihood scheme (Marshall 1985b) is then used to determine γ and L_Q^* from the L_T data. A Euclidian universe is assumed because the redshifts are all less than 0.1 - thus the redshifts are unimportant data except for the determination of luminosities and the verification of completeness. Similarly it is assumed that evolutionary effects are negligible.

3. Results

The best fit was obtained with $L_Q^*=0.2(M_B=21.8)$ and $\gamma=1.2$. The uncertainties are large but correlated (Fig. 3). This model was tested against the data using a Kolmogorov-Smirnov test (Marshall 1985b) and is acceptable at the 90% confidence level (p=0.16). The Cheng et al. estimates are consistent with the model except for two points that are low by a factor of three. These differences may be due to the different data sets rather than the different methods.

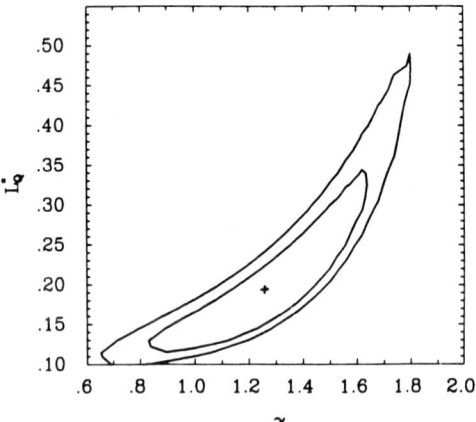

Figure 3. The confidence region for the model parameters, L_Q^* and γ. Note the strong covariances.

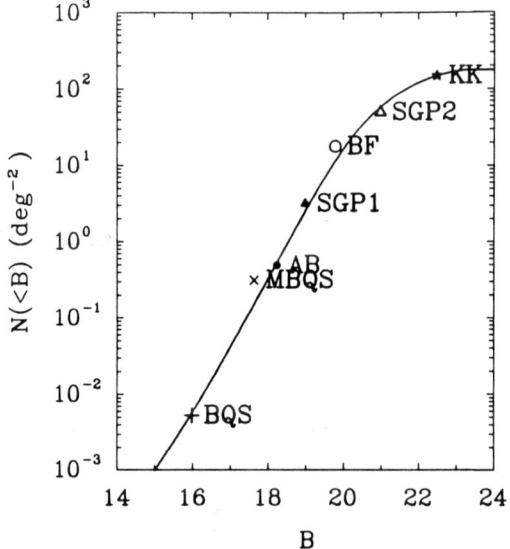

Figure 4. The $N(<B)$ relation. The data are from complete UV excess

surveys: BQS - Schmidt and Green (1983); MBQS-Mitchell, Warnock, and Usher (1984); AB-Marshall et al. (1983) and Braccesi et al. (1980); BF-Marshall et al. (1984); SGP1-Boyle et al. (1985); SGP2-Boyle et al. (these proceedings); KK-Koo and Kron (1985, see also these proceedings). The KK and the SGP2 data should be adjusted downward slightly to eliminate quasars with $z > 2.2$, the integration limit for the $N(<B)$ model. The lines are LE models with $\gamma = 1.2$, $L_Q^* = 0.2$ and $L_Q > 0.001 f(z)$.

A further test reiterates this difference. A power law model was tried using $L_{Q,o} = 0.003$ and a high luminosity truncation at $L_Q = 2$. The slope, 2.1, and the normalization, $\psi = 1250 \text{Gpc}^{-3}$ at $L_Q = 0.05$ ($M_B = -20.3$), were the only fitted parameters. This model doesn't fit at the 95% level and also cannot be extended to high luminosities because too many quasars are predicted at brighter magnitudes. For example, for $\gamma = 2.35$, $\psi_o = 800$ (an extreme of the 90% error region) and no evolution, 0.13 quasars per sq. deg. are predicted with $B < 16$ and $z < 0.5$ (60% have $z < 0.2$), whereas Schmidt and Green (1983) find <0.01 quasars per sq. deg. with $B < 16$ and $z=2$! Thus it is necessary to have a feature in the LF that separates the high and low luminosity portions.

Using the LE model F of Marshall (1985a, $L_Q \alpha [1+z]^{3.5}$) and the LF model from equation (2), then reasonable agreement with the number counts ($z < 2.2$) and LFs of Koo and Kron are obtained (Figs. 2 and 4). The $N(<B)$ prediction can be extended to $B=29$ to demonstrate the usefulness of deep surveys using the Hubble Space Telescope (HST). In particular, if the local LF can be extended to $M_B = -10$, accommodating the active objects found by Fillipenko and Sargent (1985, also these proceedings) and objects with $M_B < -23$ are not eliminated, then these active nuclei would dominate the HST surveys for $B > 26$ and $>1000 \text{deg}^{-2}$ are found with $B<29$. Under the LE hypothesis, the nuclei of these objects are x80 brighter at $z=2.5$, rendering them easier to find relative to the background galaxy light. These objects, should they exist, will provide a key test of the LE hypothesis.

REFERENCES

Boyle, B.J., Shanks, T., Fong, R. and Clowes, R.G. 1985, M.N.R.A.S., submitted.
Braccesi, A., Zitelli, V., Bonoli, F., and Formiggini, L. 1980, Astr. Ap., 85, 80.
Cheng, F., Danese, L. de Zotti, G., and Franceschini, A. 1985, M.N.R.A.S., 212, 857.

Davis, M. and Huchra, J. 1982, Ap. J., 254, 449.
Fillipenko, A. V. and Sargent, W. L. W. 1985, Ap. J. Suppl., 57, 503.
Huchra, J. P., 1985, in preparation.
Huchra, J. P., Davis, M., Latham, D., and Tonry, J. 1983, Ap. J. Suppl., 52, 89.
Koo, D. C. and Kron, R. G. 1985 in preparation.
Marshall, H. L., Tananbaum, H., Zamorani, G., Huchra, J. P., Braccesi, A., and Zitelli, V. 1983, Ap. J., 269, 42.
Marshall, H. L., Avni, Y., Braccesi, A., Huchra, J. P., Tananbaum, H., Zamorani, G., and Zitelli, V. 1984, Ap. J., 283, 50.
Marshall, H. L., 1985a, Ap. J., submitted.
Marshall, H. L. 1985b, in preparation.
Mitchell, K. S., Warnock, A. III, and Usher, P. D. 1984, Ap. J. (Letters), 287, L3.
Schmidt, M. and Green, R. F. 1983, Ap. J., 269, 357.

NEAR INFRARED SURFACE PHOTOMETRY OF NGC1068

Evencio Mediavilla Gradolph and Carlos Sanchez Magro
Instituto de Astrofisica de Canarias
Universidad de La Laguna
Tenerife
Spain

ABSTRACT. Near infrared scans of NGC1068 with 20 and 5 arcsecs apertures have been used to isolate the nuclear emission. The nuclear near infrared emission can be explained by a combination of dust clouds at different positions and temperatures. The same scans have been used for a tentative detection of the ring in the L band.

NGC 1068 has been observed at the near infrared bands J H K L. These observations are especially important in a spectral range where the contrast of the nucleus is maximum and where the size must change in scales well under the 1 arcsec region (Lebofsky et al. 1978 ; Rieke and Lebofsky, 1979) and the extended emission (Telesco et al. 1980 ; Lebofsky et al. 1978; Scoville et al. 1983). Also in this spectral region the spectral index can characterise the kind of emission related to QSO's and active nuclei, an interesting point when there are clear indications that Seyfert 1 and Seyfert 2 are manifestations of similar phenomena (Penston and Perez, 1984).

DISCUSSION

The observations were made with the 1.5m telescope of the I.A.C. (Instituto de Astrofisica de Canarias), at 2400 m altitude and with InSb detector. BS1084 and BS718 were used as calibration stars. On figure 1 and 2 we have represented the calibrated profiles for the different bands and apertures. From these profiles the colours of the galaxy, as well as the integrated magnitudes, can be obtained for different apertures. Our results agree with similar ones published in the literature (Penston et al. 1974; Lebofsky et al. 1978). The values found for the nuclear flux agree with similar data in the literature (Rieke, 1984; Ulrich, 1984). We have tried to fix this steep, spectrum in the near IR by a simple thermal source extinguished by around 15 mag., as indicated by the silicate bump (Lebofsky et al. 1978). The best fit is approximated by a temperature of 650 K. This result indicates the possibility that the near IR emission as well as the far IR emission is generated by dust of decreasing temperature at increasing distances form the nucleus.

The L scans have tentatively detected a ring with a 9σ value relative to the mean. This ring is particularly important because its distance from the nucleus (~16 arcsec) approximately agrees with the results of Telesco et al. (1984) in 20μ, the CO emission (Scoville et al., 1983) and the zones of high excitation in the visible (Beck and Beckwith, 1984). We are probably looking at a ring typical of different types of spiral galaxies. The extinction for the stellar component should be of the order of Av~4mag smaller, but matching the values obtained by the CO column density. This annulus of dust, gas and early type stars may correspond to the cool thermal emission (T~40 K), but at 3.6μ we are probably seeing the hot dust around the recently formed stars. The identification of the detected ring with the CO emission, as well as the knowledge of relative velocities (Rickard et al. 1977), allows an independent mass determination to the observed radius (16 arcsecs on the semimajor axis) $M \simeq 3.10^9 \, M_\odot$

FIGURE 1

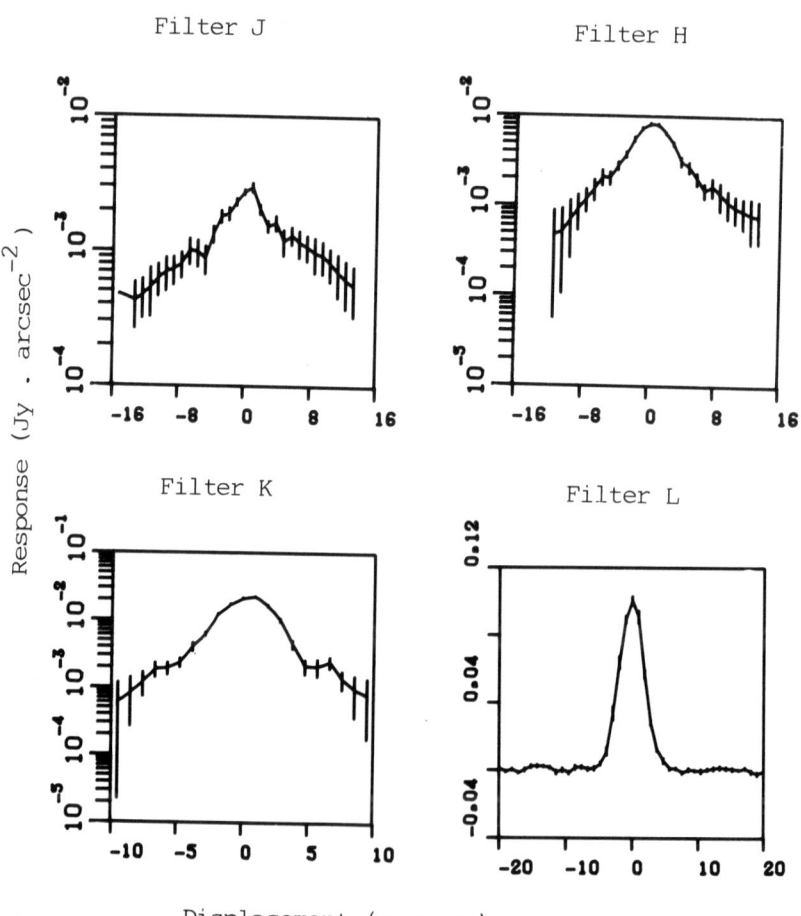

5 arcsecs aperture scans in the N-S direction. Error bars: ± 1σ

FIGURE 2

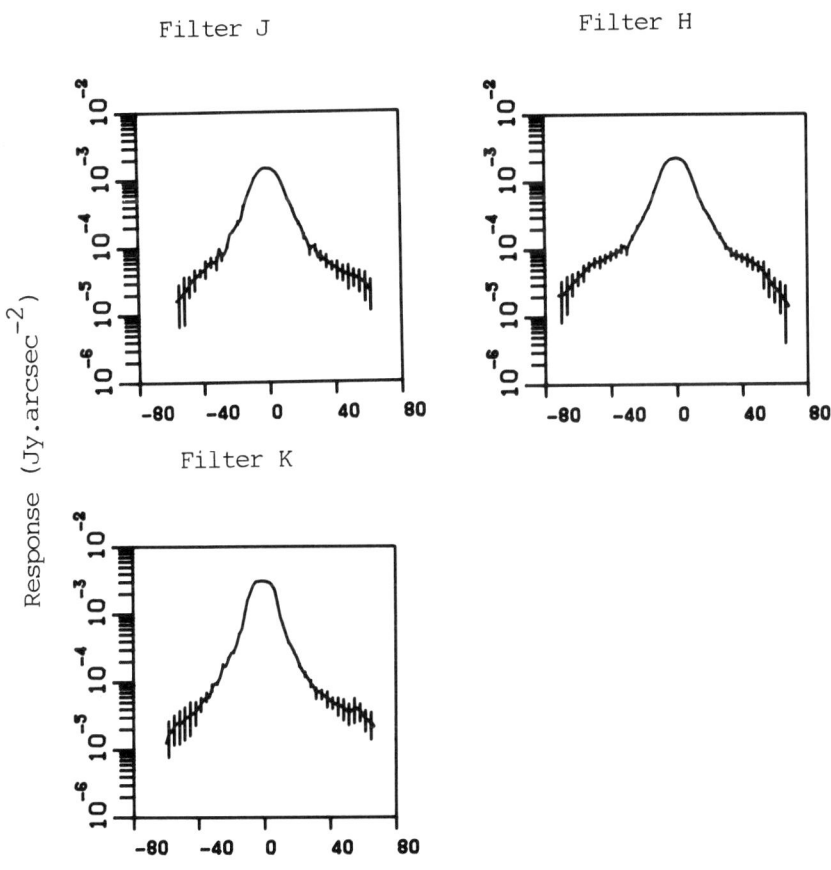

20 arcsecs aperture scans in the N-S direction. Error bars: ±1σ

REFERENCES

Beck, S.C. and Beckwith, S.: 1984, Ap.J., **279**, 563
Lebofsky, M.J., Rieke, G.H., and Kemp, J.C.: 1979, Ap.J. **222**, 95
Penston, M.V., Penson, M.J., Selmes, R.A., Becklin, E.E., and Neugebauer G.: 1974, MNRAS, **169**, 357
Penston, M.V., and Perez, E.: 1984, MNRAS, **211**, Short Communication, 33
Rickard, L.J., Patrick Palmer, Morris, M., Turner, B.E. and Zuckerman, B.: 1977, Ap.J., **213**, 673
Rieke, G.M., and Lebofsky, M.J.: 1979, ARAA, **17**, 477
Rieke, G.H.: 1984, Santa Cruz Workshop on Active Galaxies and QSO's.
Scoville, N.Z., Young, J.S., and Lucy, L.B.: 1983, Ap.J., **270**, 443
Telesco, C.M., Becklin, E.E., and Wynn-Williams, C.G.: 1980, Ap.J., **241**,

L69
Telesco, C.M., Becklin, E.E., Wynn-Williams, C.G., and Harper, D.A.: 1984
Ap.J., **282**, 427
Ulrich, M.H.: 1984, IAU Symposium, **110**, 73

> I would like to pay tribute to my friend and colleague Carlos Sanchez Magro, whose untimely death occurred between the time of the Trieste meeting and the publication of the proceedings. This paper forms the last of his many contributions to the astrophysical literature.
>
> Evencio Mediavilla

ACTIVE GALAXIES IN HIGH DENSITY ENVIRONMENTS

T. K. Menon and Paul Hickson
Department of Geophysics and Astronomy
University of British Columbia
2219 Main Mall
Vancouver, B.C., V6T 1W5
Canada

ABSTRACT. A radio survey of 401 galaxies in small groups with local galaxy space densities ranging from 10^2 to 10^6 Mpc^{-3} has detected 41 sources above a flux limit of 1.5 mJy at 18 cm. These radio sources fall into two catagories according to the optical morphology of the host galaxies. Radio sources in elliptical galaxies occur only in the first-ranked (in optical luminosity) galaxies. They are compact nuclear sources with no detectable extended emission. Those in spirals are more extended and show no preference for optical rank. Our interpretation is that the emission in spiral galaxies is a result of star formation activity, possibly triggered by interactions, and that the sources in elliptical galaxies result from inflow onto a central compact object. This inflow is a direct result of the large scale galaxy environment.

1. INTRODUCTION

It has long been recognized that nuclear activity may be strongly influenced by environmental factors (Miley 1980). Radio surveys of field galaxies (Stocke 1978; Adams, Jensen, and Stocke 1980) and cluster galaxies (Owen 1974; Leir and van den Bergh 1977) show a correlation between the occurrence of radio emission and the local space density of galaxies. A possible explanation is that galactic interactions affect the supply of gas to the radio source through such mechanisms as ram pressure sweeping (Gisler 1976) and mergers (Hausman and Ostriker 1978). Alternatively, the gas supply may result from primordial infall (Gunn and Gott 1972) and may be determined more by initial conditions than by interactions. If dynamical processes are effective in triggering radio emission in galaxies, these effects are expected to be strongest in systems of galaxies with high space density and low velocity dispersion (Hickson, Richstone and Turner 1977). We have therefore surveyed 88 such systems in order to investigate the frequency and morphology of radio emission.

Our sample consists of all compact groups identified by Hickson (1982) which are north of declination -19 deg. The groups were selected from a search of the Palomar Observatory Sky Survey (POSS) red prints for groups satisfying three selection criteria of population, isolation, and mean surface brightness. The catalog and several aspects of the structure of these groups are discussed by Hickson (1982) and Hickson et al. (1984). Details of the observations and analyses are presented elsewhere (Menon and Hickson, 1985). This paper summarizes the observations and the principal results and implications of our work.

2. OBSERVATIONS AND DATA ANALYSIS

Most observations were made at a frequency of 1635 MHz in the C configuration of the Very Large Array (VLA) of the NRAO in April 1983. Detected groups were reobserved in the B-configuration in January 1984. The C-configuration observations were made in the snapshot mode with a bandwidth of 50 Mhz and an integration time of 6 minutes for each group. The integration time for B-configuration observations varied from 30 minutes to 10 minutes and also utilised all four intermediate frequency passbands. The flux density calibration was done using 3C 286 as the primary calibrator and a number of VLA calibrator sources were observed about every 30 minutes. Calibration and mapping was done using standard NRAO programs available at the VLA. By measuring the rms noise in the field of each observed group in a region of the map devoid of any obvious sources we estimate that 1.5 mJy is a conservative uniform detection limit for all groups. It may be possible to improve this limit by further editing and processing of the data. Positions of all sources detected within the optical boundaries of the groups were compared with the optical positions of the member galaxies determined from the POSS prints using the Mann measuring engine at the VLA. The radio positions have an intrinsic accuracy of about 1", the error on the optical positions is about 2" but can be significantly larger for large overexposed galaxies.

3. RESULTS

A total of 41 radio sources above 1.5 mJy were identified with galaxies in the groups. Most sources are located within two or three arcsec of the optical nucleus. The median optical-radio separations are 1.54", and 3.33" for elliptical and spiral galaxies respectively. The larger separations in spiral galaxies suggest that these sources are morphologically different from those in ellipticals.

The most striking result of this survey is shown in figure 1. All elliptical radio galaxies but one are first-ranked in the group in optical luminosity. Only the brightest ellipticals have radio sources, and no group has more than one elliptical radio galaxy. This is not a selection effect as the optical rank distribution of elliptical galaxies in the groups is similar to that of the others, and there is no excess of ellipticals amongst the first-ranked galaxies. This result is highly significant; the probability of this distribution being due to chance is 3×10^{-5}. The effect is not found in spiral radio galaxies.

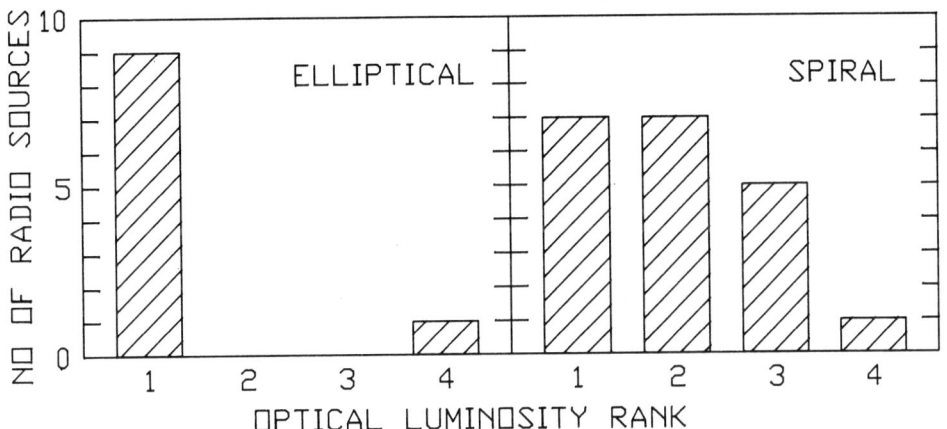

Figure 1. Optical luminosity rank of radio galaxies in compact groups

They show equal preference for all but the faintest galaxies.
 The sources detected here are less powerful than typical rich cluster radio galaxies. The median radio power for sources in elliptical and spiral galaxies is 22.0 and 21.8 (log W/Hz) respectively. Of all the detected galaxies only one (60a) has an extended structure beyond the optical dimensions of the associated galaxy. It is a typical wide-angle head-tail source and extends to almost 70" (60 kpc) while the optical size of the galaxy is only about 25". Of all the 88 groups surveyed in this program, this is the only one known to be in an Abell cluster (A1452). This source is also the most powerful in the sample by an order of magnitude. There are no significant differences between the radio luminosity distributions of elliptical and spiral galaxies. Although this result differs from earlier studies (see Hummel 1980), this can be explained by differences in selection criteria among various samples. The spiral galaxies in our sample have distances similar to those of other galaxy types, so a comparison of the luminosity distributions is meaningful. The luminosity functions for these groups will be discussed in detail in a subsequent paper.

5. DISCUSSION

The probability that the detected sources are background sources not associated with the groups is very small. The number of sources expected to occur by chance were computed from Bridle (1982). At most one or two of the sources could be unrelated to the groups.
 It is evident that the radio sources in spiral galaxies are different from those in elliptical galaxies. In spiral galaxies we are likely seeing emission from supernovae remnants associated with recent star formation. Recent high-resolution data on one such spiral (47a) shows that all the emission is extra-nuclear and related to star formation (Menon and Hickson, in preparation). An inspection of the POSS images shows that the radio galaxies contain twice the number of

interacting galaxies (as indicated by a close and distorted companion) contained in a random sample from the groups. High resolution radio and CCD imaging in progress will shed further light on this problem.

The conspicuous lack of extended radio emission suggests that either these sources do not produce the double radio structure found in rich clusters (Miley, 1980), or that they have not had time to evolve. This suggests that the groups have formed recently, that this triggered the radio activity, and that the groups have short lifetimes. This is consistent with the apparent lack of dynamical evolution in the groups (Hickson, 1982), with the the short dynamical times (Hickson, Richstone and Turner 1977), and with the merging instability exhibited by numerical simulations of galaxy clusters (White 1979, 1982, Aarseth and Fall 1980, Miller 1983, Quinn 1984).

The strong tendency of radio ellipticals to be first-ranked points to the importance of either mass or central location of the first-ranked galaxy to the radio emission mechanism. Even luminous second or third ranked elliptical galaxies do not contain radio sources while a lower luminosity first-ranked galaxy in another group does. This suggests that central location is the important factor and that accretion onto the galaxy may be the dominant mechanism. Regardless of the details, this correlation provides strong evidence that nuclear activity in these galaxies is enhanced by their high density environment.

This work was supported by the Natural Sciences and Engineering Research Council of Canada.

REFERENCES

Aarseth, S. J., and Fall, S. M. 1980, Ap.J., 236, 43.
Adams, M. T., Jensen, E. B., and Stocke, J. T. 1980, A. J., 85, 1010.
Bridle, A. 1982, in Procedings of the NRAO VLA Workshop, ed.
  A. R. Thompson and L. R. D'Addario (Greenbank: NRAO).
Gisler G. R. 1976, Ap.J., 228, 385.
Gunn, J. E., and Gott, J. R. 1972, Ap.J., 176, 1.
Hickson, P. 1982, Ap.J., 255, 382.
Hickson, P. 1984, in Clusters and Groups of Galaxies, ed
  F. Mardirossian, G. Giuricin, M. Mezzetti (Dordrecht: Reidel), p367.
Hickson, P., Richstone, D. O., and Turner, E. L. 1977, Ap.J., 213, 323.
Hausman M., and Ostriker, J. P. 1978, Ap.J., 224, 320.
Hummel, E. 1980, Ph.D. Thesis, University of Groningen.
Leir, A. A., and van den Berg, S. 1977, Ap.J. Suppl., 34, 381.
Menon, T. K., and Hickson, P. 1985, Ap.J., 295, (in press).
Miley, G. 1980, A.R.A.A., 18, 165.
Miller, G. E. 1983, Ap.J., 268, 495.
Owen, F. N. 1974, A. J., 79, 427.
Quinn, P. J. 1984, private communication.
Stocke J. T. 1978, A. J., 83, 348.
White, S. D. M. 1979, M.N.R.A.S., 189, 831.
White, S. D. M. 1982, in IAU SYmposium 100, Internal Kinematics and
   Dynamics of Galaxies, ed. E. Athanassoula (Dordrecht: Reidel), p337.

TOWARDS THE LUMINOSITY FUNCTION OF SEYFERT GALAXY NUCLEI

E.J.A. Meurs
Institute of Astronomy
Madingley Road
Cambridge, CB3 0HA
England

ABSTRACT. The optical and radio luminosity functions of Seyfert galaxies for total emission of the systems are briefly reviewed. A project of CCD-photometry aimed at determining a luminosity function for Seyfert nuclei is described. The merits of integrated and nuclear magnitudes with respect to Seyfert luminosity functions are discussed. Finally, a few remarks on systematic survey incompletenesses are made.

1. INTRODUCTION

Seyfert galaxies are the major type of active galactic nuclei. They are widely considered as objects intermediate between normal galaxies and quasars. Several observed quantities show indeed a continuus range in their values for these objects, for instance luminosities at various wavelengths. A point to check in this regard is how the space density of the Seyferts compares with that for galaxies in general and for quasars. Therefore, luminosity functions for Seyfert galaxies have been studied at optical and radio wavelengths (Meurs and Wilson 1984, hereafter MW). Present efforts are directed towards establishing observational quantities pertaining to the active nuclei proper.

2. SEYFERT LUMINOSITY FUNCTIONS FOR TOTAL EMISSION OF THE SYSTEMS

2.1. Optical luminosity function

A luminosity function (LF) for the Seyfert galaxies in the first nine lists of the Markarian survey has recently been published (MW). At that time, 101 Seyferts of both types (1 and 2) were known in these lists, of which 78 actually went into the LF calculation. Necessary observational data for the galaxies include:
- classification (also type 1 or 2; see also Osterbrock 1986);
- radial velocity (from a list of Seyferts compiled by J. Huchra);
- magnitude (see below);
- radio power at 1415 MHz (from a WSRT survey; see Section 2.2).

For about 2/3 of the sample, magnitudes can be found in Zwicky et al's Catalogue of Galaxies and of Clusters of Galaxies. These are blue integrated magnitudes, denoted here by m_p, which are considered to provide a reasonable measure of the total emission of galaxies for $m_p \geq 14.5$ (Bothun and Schommer 1982) - that is, for the majority of the Markarian Seyfert galaxies. For the sake of some homogeneity in the data, Markarian's estimates have been used for the remainder of the sample, which are somewhat crude but yet on the mean may give a reasonable estimate of m_p for the galaxies. Although a LF for m_p's yields interesting comparisons with LF's for other categories of object (cf. Section 4.1), one surely would like to obtain a LF pertaining to the active nuclei only (Section 3).

The derivation of optical and radio LF's employs standard analysis methods which commonly are applied to this kind of work. The optical LF has been derived using the Vmax method of Schmidt (1968) after allowance for survey incompleteness. It appears that the Seyfert LF joins up with the local LF for optically selected quasars, showing a likely increasing contribution of the active nucleus to the integrated magnitude towards the most luminous Seyferts (cf. MW).

2.2. Radio luminosity function

Also referring to the total emission of the Seyferts is the radio luminosity function (RLF) based on the 1415 MHz data obtained in a survey with the Westerbork telescope (WSRT). The results of this survey were published by de Bruyn and Wilson (1976), Meurs and Wilson (1981), and Wilson and Meurs (1982). The WSRT flux densities do in fact refer to nuclear radio sources (which are dominant in the systems), but cannot resolve very much of the structures of these sources as they have become apparent in VLA studies of some of these objects (e.g. Wilson 1982).

The RLF has been derived by constructing an optical-radio bivariate LF, and combining this with the optical LF. Again, the Seyferts seem to have a position between normal (spiral) galaxies and optically selected quasars (cf. MW).

3. CCD-PHOTOMETRY OF MARKARIAN SEYFERTS

3.1. Observations and analysis

In order to derive a LF for the Seyfert nuclei, a project of CCD-photometry is currently carried out that provides two-dimensional photometric data that can be analysed for nuclear magnitudes. The observations were obtained at the German-Spanish observatory at Calar Alto with the 2.2 m telescope. Most of the Markarian Seyferts that had gone into the LF derivation referred to above (Section 2.1) were observed (75 objects). Limitations in time led to relatively short exposures, frequently with only the Thuan and Gunn (1976) v-filter for which the CCD is not very sensitive anymore. Therefore the recorded images are often restricted to the central regions of the galaxies. For fifteen objects frames in the g- and r-filters were secured as well.

After reduction (see Section 3.2) the final frames are analysed for observational quantities characterising the nuclear and central emission of the systems. Further, a magnitude may be found that suitably describes the increasing incompleteness of the Markarian survey towards fainter magnitudes. Galaxies for which frames in several filters were obtained are to be checked upon dust features.

3.2. Aspects of the CCD-reduction

After subtraction of readout-noise and (negligible) dark current, the frames were flatfielded employing twilight sky exposures. There are characteristic cold columns in the frames, their behaviour being dependent on the count level. This made possible to develop a correction procedure utilizing object-free portions of frames at various count levels, which subsequently has been applied to the frames.

4. DISCUSSION

4.1. Integrated versus nuclear magnitudes

Estimates have been made of the optical LF of the very nuclei of the Seyfert galaxies, particularly the type 1 Seyferts (Véron 1979; Cheng et al. 1985), usually by means of Sandage's colour-given method (Sandage 1973). An attempt to present an overall LF for Seyfert 1 nuclei extending further to higher and lower luminosities is made by Weedman (1986). The result of Cheng et al. is probably the best estimate so far of the nuclear LF, having used a sample comparable to that of MW. Their result is broadly confirmed (Meurs, in preparation) by estimating the nuclear LF using the few nuclear magnitudes directly determined from two-dimensional photometric data that are available at the moment (Yee 1983). This nuclear LF converges with the local LF of optically selected quasars at $M_B \approx -23$, but also with the LF for integrated magnitudes of MW. The magnitudes of such high luminosity Seyferts are thus likely to be dominated by a very bright nucleus. Some of the Seyfert galaxies at about this M_B are at times considered as quasars.

At present, the Cheng et al. nuclear LF provides probably the best comparison of Seyferts with (optically selected) quasars, particularly in order to judge differences in space density between these categories of object for $M_B > -23$. One has to bear in mind that the local LF for quasars that is compared with the Seyfert LF's, is an extrapolation from high redshifts. A convergence of both nuclear LF and LF for integrated magnitudes of the Seyfert 1's is not unexpected in case an active nucleus increasingly dominates the various magnitudes for higher luminosities. How much light of bulges of galaxies is involved both for Seyferts and for quasars, is not yet clear. However, for comparisons of the Seyfert LF with the LF for (normal) field galaxies, as discussed by Osterbrock (1986), one has to compare with the Seyfert LF of MW as the field galaxy LF is given for integrated magnitudes as well. Analogously, comparing a Seyfert LF for integrated magnitudes with a LF for Seyfert nuclei may suggest discrepancies that are not really present (cf.

Weedman 1985).

4.2. Systematic incompletenesses

The Seyfert LF's mentioned above refer essentially to Markarian Seyferts. As discussed more in detail by Osterbrock (1985), there will be several incompleteness-effects introduced by the way these Seyfert galaxies have been found. Known effects include relatively small UV-excess (more easily missed by the Markarian survey but turning up in emission-line surveys) and inclination (nearly edge-on systems not very likely in Markarian survey but turning up in for instance hard X-ray selected samples). As to the latter effect, it is interesting to note that the IRAS-Seyferts (cf. Osterbrock 1986) seem to be predominantly nearly edge-on systems (Meurs, in preparation).

Also, some incompleteness is likely to be introduced at the faint end of the Seyfert LF where the classification of the galaxies may pose some problems as to including the objects in which LF-sample (Seyferts 2, 1.9, 1.8, Liners, NLXG's, etc.); one of these problems being the presence of very weak broad wings in emission line profiles. It is expected that a detailed spectroscopic survey as described by Filippenko (1986) will clarify much in this respect.

REFERENCES

Bothun,G.D. and Schommer,R.A. 1982, Astrophys.J. 255,L23
de Bruyn,A.G. and Wilson,A.S. 1976, Astron.Astrophys. 53,93
Filippenko,A. 1986, this volume.
Cheng Fu-zhen, Danese,L., De Zotti,G., Franceschini,A. 1985,
    Mon.Not.Roy.astr.Soc. 212,857
Meurs,E.J.A. and Wilson,A.S. 1981, Astron.Astrophys.Suppl. 45,99
Meurs,E.J.A. and Wilson,A.S. 1984, Astron.Astrophys. 136,206 (MW)
Osterbrock,D.E. 1986, this volume.
Sandage,A. 1973, Astrophys.J. 180,687
Schmidt,M. 1968, Astrophys.J. 151,393
Thuan,T.X. and Gunn,J.E. 1976, Publ.astr.Soc.Pac. 88,543
Véron,P. 1979, Astron.Astrophys. 78,46
Weedman,D.W. 1986, this volume.
Wilson,A.S. 1982, in: IAU Symp. 97 "Extragalactic Radio Sources", p.179
Wilson,A.S. and Meurs,E.J.A. 1982, Astron.Astrophys.Suppl. 50,217
Yee,H.K.C. 1983, Astrophys.J. 272,473

DISTANCE OF THE FAR GALAXIES AND SCATTERING OF THE PHOTONS IN THE SPACE

M.Missana
Osservatorio Astronomico di Brera
via Brera, n°28
20121 Milano
Italy

ABSTRACT. From the theory of the radiative transfer, including the scattering of the light and the Compton effect, it can be deduced that the intensity of the far astrophysical objects decreases approximatively as $1/R^3$.

INTENSITY AND EFFECTS OF THE LIGHT SCATTERING

The scattering of the photons, with wavelength in the visible region, by the hydrogen atoms of the intergalactic matter, can be rigorously computed with the approximation of the Thomson scattering cross-section and of the simple Compton effect, as shown in the article (Missana,1977).
We indicate with I_o the light intensity of the source, i.e. of the far galaxy at the distance R from the Earth; then the observed intensity $I(R)$, computed with the theory, is given in a first approximation by:

$$I(R) = \frac{I_o}{4 \pi R^2 (1 + R/R_o)} \quad ; \quad (1)$$

$R_o = 2/(\sqrt{3}\, \sigma_t\, N)$ is a constant. With σ_t total Thomson cross-section = $6.7\,10^{-25}$ cm^2 and N number of scattering centres (hydrogen atoms) per unit volume $\sim 10^{-7}$ atoms cm^{-3} (Allen,1973) we have $R_o \cong 6\,10^{12}$ parsecs; R/R_o is usually called optical thickness.
In a better approximation the factor $(1 + R/R_o)$ in the denominator of equation (1) must be replaced by:

$$[1 + R/R_0 + q\ (R/R_0)^2 + ..]\ ,\ \text{with}\ \ 0 < q \ll 1\ .$$

From equation (1) it follows that the intensity of the far galaxies decreases as $1/R^3$ for $R > R_0$.

The equation (1) has been deduced with two rather questionable assumptions . First we assume that the energy balance between incoming and scattered photons , having wavelengths λ' and λ , is given by $\lambda' = \lambda - b\ (1-\cos\vartheta)$ where ϑ is the scattering angle and b is a constant .
That assumption can be accepted for the hydrogen atoms of solar chromosphere but must be discussed in detail for atoms of the intergalactic matter . The second assumption is that the wavelength λ' of the incoming photons is far from a resonance of the scattering atoms . Indeed in the case of resonance scattering one meets great difficulties in the solution of the quantum problem and of the transfer equation . The average expected effect is a remarquable increase of the total scattering cross-section σ_t , and hence a large reduction of the constant R_0 .
In the regions of the visible spectra , where a resonance scattering occurs , the constant R_0 , for what the resonance scattering is concerned , can be smaller than in the solar chromosphere even by a factor 10^{-8} , on the basis of the (few) available measurements in Orion nebula as shown in the article (Missana,1979) . The values of the resonance wavelengths depend of course on the thermodynamic state of the intergalactic matter.

The equation (1) is not the only consequence of the interaction of the photons with the intergalactic gas . Other observable effects are the increase of the halfwidth FWHM and of the red-shift of the spectral lines $d\lambda = \lambda(\text{galaxy}) - \lambda(\text{laboratory})$, which are correlated to the distance of the galaxies . In the case of Thomson scattering and simple Compton effect , far from a resonance , the theoric function $d\lambda = f(\text{FWHM}, R/R_0)$ is given by Table I (Missana,1977) for $R/R_0 = \tau < 20$. Then the value R/R_0 of a given galaxy can be obtained from the study of its red-shift . For instance if the red-shift due to the scattering is about $d\lambda = 50$ A° (10 A° = 1 nm) , as it happens for 4C 48.28 with respect to Ton 616 (Missana,1984) , from the quoted Table I we obtain a difference of optical thickness much larger than 20 . In the case of the two QSO however R/R_0 includes also the contribution of the interstellar matter in which they are embedded . Hence in this example the equation (1) must be adopted in the computation of the intensity of the source , instead of the usual law $I = I_0/(4\pi R^2)$.

Also the shape of the continuum spectrum , and hence the color index , depends on the distance from the Earth , in the present theory . At last the collision of the photons with the intergalactic matter rises its temperature . The exact computation of this effect is not easy ; however from thermodynamical considerations it can be believed that the temperature of the interstellar matter is rised to about 3.2 Kelvin by the interaction with the radiation (Eddington,1926) . In the same approximation the temperature of the intergalactic matter is expected to be smaller than 3.2 K. .

It is worth to notice that all these theoretical results are not in contrast with the present experimental data.

REFERENCES

Allen C.W.:1973,Astrophysical Quantities,The Athlone Press, London,p.293.
Eddington S.A.:1926,The Internal Constitution of the Stars, Cambridge University Press,p.371.
Missana M.:1977,Astrophys.Space Sci.50,409.
Missana M.:1979,Astrophys.Space Sci.61,277.
Missana M.:1984,Atti XXV Riunione Società Astonom.Ital. , Suppl.al Vol.55°-N°1-1984,p.221.

THE OPTICAL LUMINOSITY OF THE INVISIBLE NUCLEUS OF M82.

P. Notni
Zentralinstitut für Astrophysik der AdW der DDR
15 Potsdam
German Democratic Republic

ABSTRACT. A new estimate of the relative luminosity of the hidden nucleus of M82 compared to the disk is given. This estimate is a factor of 10 higher than a previous one; the very blue colour of the scattered halo light can now be explained. An approximate conversion to absolute luminosity shows that this high nuclear luminosity is also compatible with some of the models of Rieke et al (1980).

1. INTRODUCTION. THE DATA.

Violent star formation is going on in the hidden central regions of the irregular galaxy M82 (Rieke et al 1980, references therein). The optical emission of the stars in this region is seen in the blue light scattered in the anomalously dusty halo. Figure 1 shows the data, from which the luminosity of the nuclear star-forming region can be estimated relatively to the disk:
a. The polarization of the scattered light (vectors)
   (Bingham et al, 1976)
b. The colour (reddening-free parameter $Q = U-B - 0.72(B-V)$)
   (isolines of $-Q$; Bronkalla, Notni, Tiersch, 1980).
The more the polarization vectors are turned parallel to the disk with respect to a centrosymmetric pattern, the brighter is the <u>disk</u> relatively to the nucleus. The bluer the scattered light, the brighter is the <u>nucleus</u>.

2. PREVIOUS ESTIMATES OF $I_{nucleus}/I_{disk}$ FROM THESE DATA.

$I_N/I_D = 1:20$ from the polarization pattern (Solinger, Markert, 1975)
$I_N/I_D \approx 1:1$ from the colours ($\max(-Q)=0.8$; Notni, Tiersch, Bronkalla, 1981).
The discrepancy is excessively large.

3. NEW ESTIMATE OF THE RELATIVE NUCLEAR LUMINOSITY

Two possible effects may be the cause of this discrepancy:

1. The centre-to-limb variation of the disk's luminance may be less than assumed by Solinger and Markert. For instance, the interstellar matter may have a considerable scale height; we assume it equal to the scale height of the stars. If the optical depth, τ, rises, the luminance of the disk becomes more and more independent of the viewing angle and the polarization (of scattered disk light allone) in the halo more parallel to the disk.

2. If the light of the central source is polarized parallel to the disk of the galaxy already before scattering ("circumsource polarization"), this polarization likewise turns the directions of the polarization vectors in the desired way (Notni,1985).

Figure 2 presents observations of the polarization pattern (deviations from circular symmetry) compared to model predictions. A comparatively bright nucleus,

$$I_{nucleus}/I_{disk} = 0.7 \pm 0.3$$

is favoured by the curves "1" and "2" (error estimated).

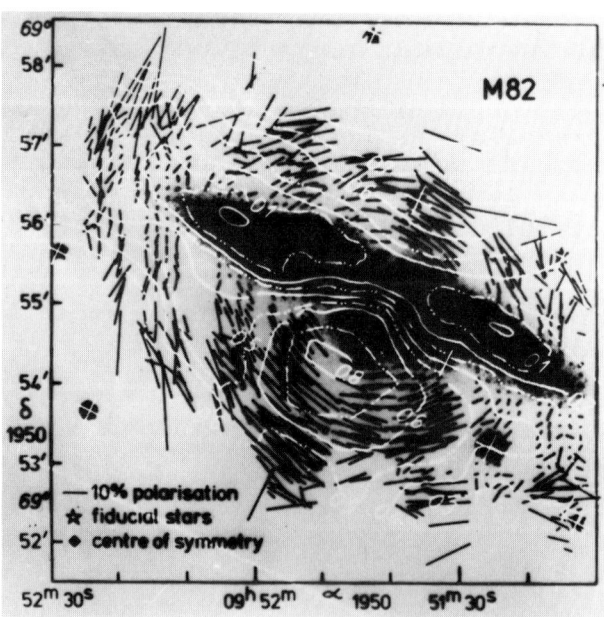

Fig.1. Polarization (vectors) and colours (parameter -Q, isolines) in the halo of M82. The halo light is determined by scattered light which contains a high contribution from the very blue nucleus as indicated by the strongly negative extremal values of Q ($\approx$ -0.8 mag).

Given this bright nucleus, the observed highly negative Q in the halo can also be reproduced (cf.Notni,1985). The contradiction mentioned at the beginning has thus been removed.

4. THE ABSOLUTE LUMINOSITY OF THE NUCLEUS

The relative brightness of the nucleus can approximately be converted to absolute luminosity be relating it to the observed brightness of the disk (we give two estimates, using slightly different values for the corrections at each step):
1. Measured total B-magnitude of the galaxy inside an isophote $B=23^m$, i.e. excluding the outer halo: $B = 9^m.4$ (Bronkalla, Tiersch, Notni, unpublished).
2. Correction to face-on brightness -1.07 (-1.00) (Holmberg 1958: corr = $0.28(\mathrm{cosec}\, i - 1)$, (Notni and Bronkalla 1984: $i = 12°$)

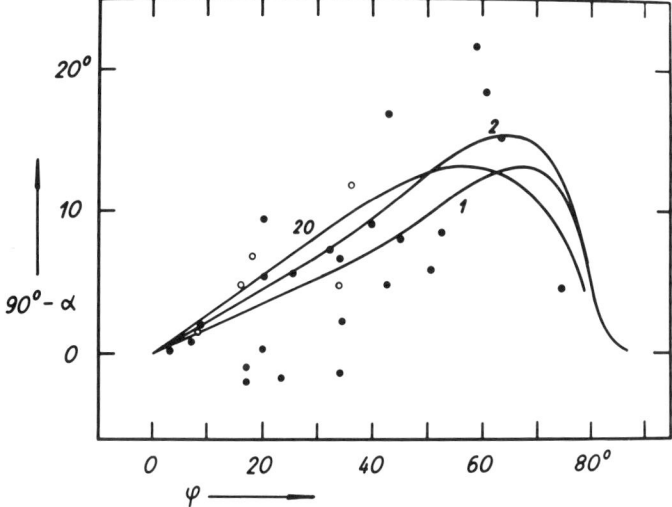

Fig.2. The range of the parameter $I_{nucleus}/I_{disk}$, able to represent the observed variation of the polarization angle, α, with polar distance, φ, in M82

20 - I_N/I_D = 1:20,   original Solinger - Markert model
2 - I_N/I_D = 1:2,   1 - I_N/I_D = 1:1,   our model,
using the following assumptions: <u>a</u>. extinction in the disk distributed as the stars, and $\tau = \overline{0}.92$ (pole-on, $\hat{=}$ 1 mag), <u>b</u>. circumsource polarization parallel to the galactic plane, 2 % near the pole, rising to 4.3% at a polar distance 70°.
Observations: ● Elvius (1962,1969),
              o Visvanathan and Sandage,1972

3. Difference $B_{nuc} - B_{disk}$ +0.40 (+0.75)
   (this paper, $I_N/I_D = 0.7$ (0.5))
4. Extinction pole-on   -3.50 (-3.20)
   (Notni, Tiersch, Bronkalla 1981, $E_{B-V}=0.86$ (0.80))
5 = 1+2+3+4 = blue magnitude of the nuclear region:
                  5.23^m (5.95^m)

  To get the total absolute luminosity of the nuclear region, we apply a crude estimate for the bolometric correction, corresponding to a B1 (B2)-star, of -2.44 (-2.11) and the distance modulus of -27.55 (Tammann and Sandage 1968):   Absolute bolometric magnitude: -24.76^m (-23.71^m) From this the total luminosity is (sun: $M_{bol} = 4.64$)

$$\underline{5.75 \cdot 10^{11}} \; (\underline{2.2 \cdot 10^{11}}) \; L_\odot$$

The difference between the two results illustrates the uncertainties of the procedure. The most uncertain step is perhaps the correction for extinction which assumes that no bluing or reddening occurs in the (last) scattering process.

5. COMPARISON TO MODEL PREDICTIONS

A comparison of this luminosity to model predictions by Rieke et al (1980) for the central star-forming region leads to the following results:

1. Our high luminosity favours young models with few massive stars. Best matched is their model H (age $2.5 \cdot 10^7$ a) not only because of its high predicted luminosity ($1.7 \cdot 10^{11} \; L_\odot$) but also since it has no stars earlier than B0, thus corroborating a low bolometric correction which leads to the more consistent result.
2. Our colours ($Q_{nucleus} \approx -0.9$) favour $\alpha \leq 2$ in the initialmass function $dN(m)/dm \sim m^{-\alpha}$. $\alpha = 3$ would lead to more positive Q.

References
Bingham, R.G., McMullan, D., Pallister, W.S., White, C., Axon, D.J., Scarrott, S.M., 1976, Nature, 259, 463
Bronkalla, W., Notni, P., Tiersch, H.:1980, Astron. Nachr. 301, 217
Elvius, A.:1962, Lowell Obs. Bull. 5, 281; 1969, 7, 117
Holmberg, E.:1958, Meddelande Lund, Ser.II, N°136
Notni, P., Tiersch, H., Bronkalla, W., 1981, Astron. Nachr. 302, 259
Notni, P., Bronkalla, W., 1984, Astron. Nachr. 305, 157
Notni, P., 1985, Astron. Nachr. 306 (in press)
Rieke, G.H., Lebofsky, M.J., Thompson, R.I., Low, F.J., Tokunaga, A.T., 1980, Astrophys. J. 238, 24
Solinger, A.B., Markert, T., 1975, Astrophys. J. 197, 309
Tammann, G.A., Sandage, A.R., 1968, Astrophys. J. 151, 825
Visvanathan, N., Sandage, A., 1972, Astrophys. J. 176, 57

FLUX VARIATIONS AND STRUCTURAL CHANGES IN EXTRAGALACTIC RADIO SOURCES

L. Padrielli[1], M.F. Aller[2], H.D. Aller[2], N. Bartel[3], C. Fanti[1],
R. Fanti[1], A. Ficarra[1], L. Gregorini[1], F. Mantovani[1], L. Matveenko[4],
G.D. Nicolson[5], J.D. Romney[6], K.W. Weiler[7]

    1. Istituto di Radioastronomia CNR, Bologna, Italy
    2. University of Michigan, Ann Arbor, USA
    3. Harvard Smithsonian CFA, Cambridge, USA
    4. Inst. for Space Research Academy, Moscow, USSR
    5. N.I.T.R., Johannesburg, South Africa
    6. N.R.A.O., Charlottesville, USA
    7. National Science Foundation, Washington, DC, USA

1. INTRODUCTION

The flux variability of extragalactic radio sources at decimetric wavelengths (Low Frequency Variability - LFV) is mostly associated with the cores of compact radio sources.

Two categories of models have been proposed to account for this phenomenon: 1) the intrinsic models - among which a general consensus emerges for the interpretation of the flux variations in terms of synchrotron emission of relativistic electrons beamed in a direction close to the line of sight; 2) the extrinsic models - that attribute the LFV to propagation effects through the interstellar medium.

To extend the amount of information available and to test the different theoretical models, multifrequency observations ranging from 0.4 to 14.5 GHz, with a time base of $\sim$10 years have been examined. Also data from two epochs of VLBI observations at 1.7 GHz, with a resolution of 2-3 mas have been studied to search for possible structural changes associated with the flux density variations.

2. CLASSES OF VARIABILITY

Analysis of multifrequency observations at 0.4, 2.3, 4.8, 8.0 and 14.5 GHz of 51 radio sources selected from the Bologna monitoring program (Fanti et al. 1981) allowed us to identify three classes of sources showing LFV and evaluate their occurrence in an unbiased sample of variable sources (Padrielli et al. in prep.).

a) Correlated Broad Band Variability
This class contains sources for which the variability appears to be correlated across the whole radio frequency band. Outbursts occur either quasi-simultaneously at all radio frequencies or delayed in time at lower frequencies, with somewhat reduced amplitude.

In our sample there are 4 good cases of this class (3C 120, 0605-085, 1510-089 and BL Lac) and several possible ones indicating a frequency of occurrence from 10% to 20% in an unbiased sample of LF variables. Two examples (0605-085 and 1510-089) are shown in fig. 1.

For these two sources we obtained 2 epochs of VLBI observations at 1.7 GHz with an array of 8 antennas giving a resolution of 2-3 mas (Romney et al. 1984). The choice of the observing frequency represented a compromise between the low frequency and a resolution sufficient to detect the structural changes expected from models based on relativistic bulk motions along the line of sight. From the time scale of the LF variations between the two epochs we can estimate a maximum angular diameter for the varying component, its brightness temperature, and then a lower limit to the bulk Lorentz factor γ (Fanti et al. 1983)

For the sources 0605-085 and 1510-089 our VLBI observations give evidence of significant structural changes between the two VLBI epochs. A careful study of the fringe visibilities shows that the differences occurring at the longest baselines (b>30 Mλ) can be interpreted in terms of an increase of the angular size of the more compact component. The corresponding expansion rate is in agreement with the γ's derived from the LFV (with the simple assumption of an angle between the line of sight and the ejection direction of the order of $1/\gamma$). The maps are shown in fig.1a and 1b.

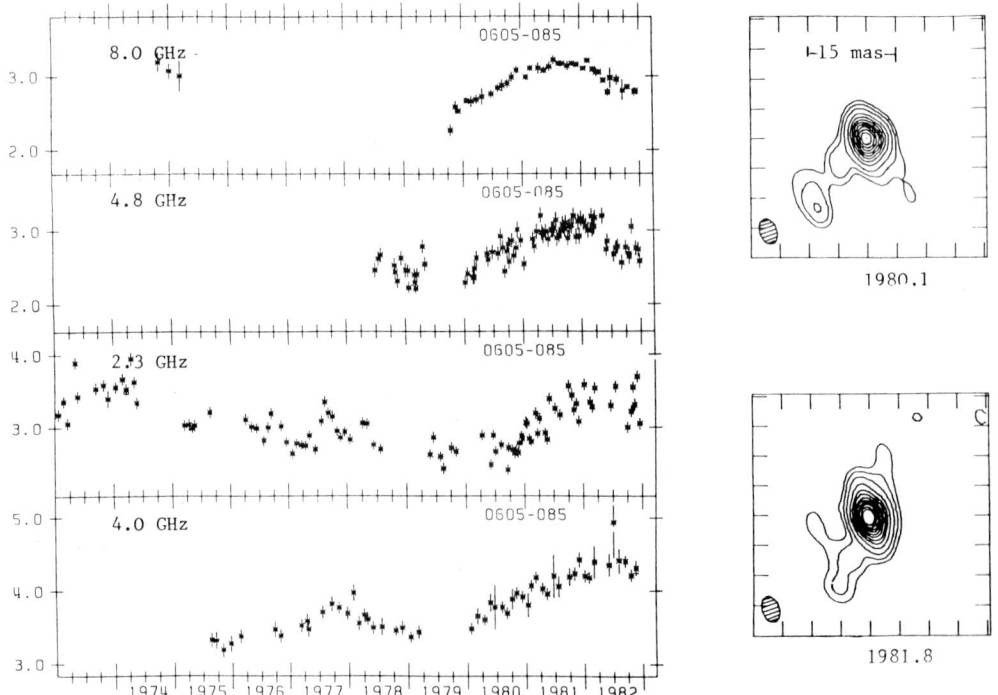

Fig.1a : 0605-085 - Quasar (z=0.5 assumed) - $T_b=10^{13}$K - $\gamma > 3$.
The HPW of the compact structure increased by 0.6+0.2 mas.
Contour Levels : 30, 60, 120, +120 mJy/beam area

b) Variability confined to the Low Frequency range

This class contains sources which are strongly variable at low frequencies ($\nu < 1$ GHz) and weakly at high frequencies. The occurrence rate of this class is 30% in an unbiased sample of LF variables. Fig. 2 shows light-curves of DA406, prototype of the class and studied by several authors (see for references Altschuler et al. 1984). We have 2 epochs of VLBI measurements for only two objects of this class. 0859-140 did not vary in flux density at 0.4 GHz between the two epochs, but 1611+343 (DA406) had a spectacular flux increase at 0.4 GHz. If we interpret this variation in terms of relativistic beaming model, we obtain a lower limit for the Lorentz factor γ of the order of $\sim$12, which leads to an expected expansion of 0.7 mas (for angles of the order of $1/\gamma$ between the beam and the line of sight). Although such a structural change could have been detected, the two VLBI maps do not show significant structural differences between the two epochs (fig.2).

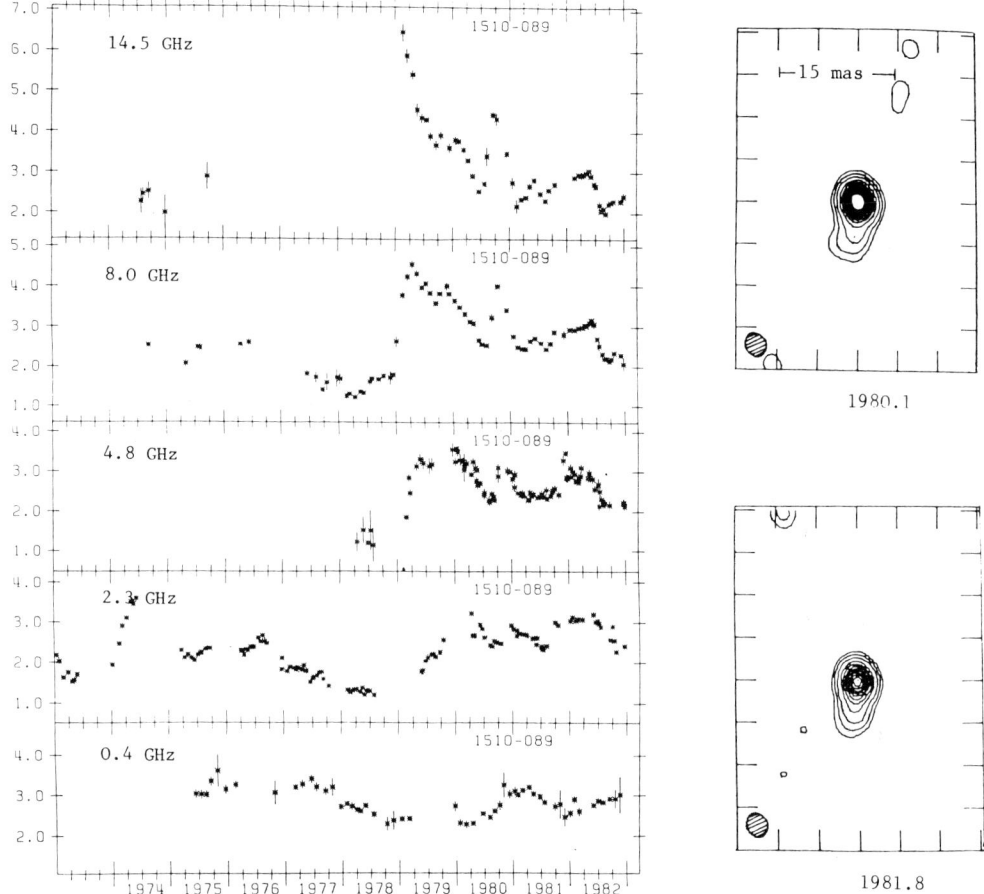

Fig.1b : 1510-089 Quasar (z=0.36) - $T_b = 10^{14}$K - $\gamma > 5$.
In p.a. 0° the HPW of the compact structure increased by 0.9+0.3 mas.
Contour Levels: 35, 70, 140, +140 mJy/beam area

c) Uncorrelated Broad Band Variability

This class consists of sources which have both high ($\nu>5$ GHz) and low ($\nu<1$ GHz) frequency activity which appears to be uncorrelated, with a minimum at intermediate frequencies. The frequency of occurrence of this category of objects in an unbiased sample is between 15% and 30%.

In fig. 3 an example of this class of sources is given (3C 454.3). The VLBI observing frequency of 1.7 GHz lies in the intermediate frequency gap of this source. However in the interval between the first and the second VLBI measurements, the source had a strong burst at 0.4 GHz. The corresponding brightness temperature leads to an extimate of $\gamma>11$. The two VLBI maps (fig. 3) are very similar and the separation between the core (East component) and the jet (West component) which is the region we expect to see at 0.4 GHz, is unchanged. Also other sources in this class do not show the angular expansion that is expected from relativistic beaming models on the basis of the LFV.

3. CONCLUSIONS

We have established the existence of at least 3 classes of LFV. On a preliminary basis, we might interpret the observed different behaviours as the manifestation of two distinct phenomena. Together with the relativistic bulk motion responsible for the high frequency activity sometime extending to the meter wavelengths, another mechanism, probably extrinsic to the source, must coexist to account for the variability in the low frequency range of the sources with uncorrelated LF activity.

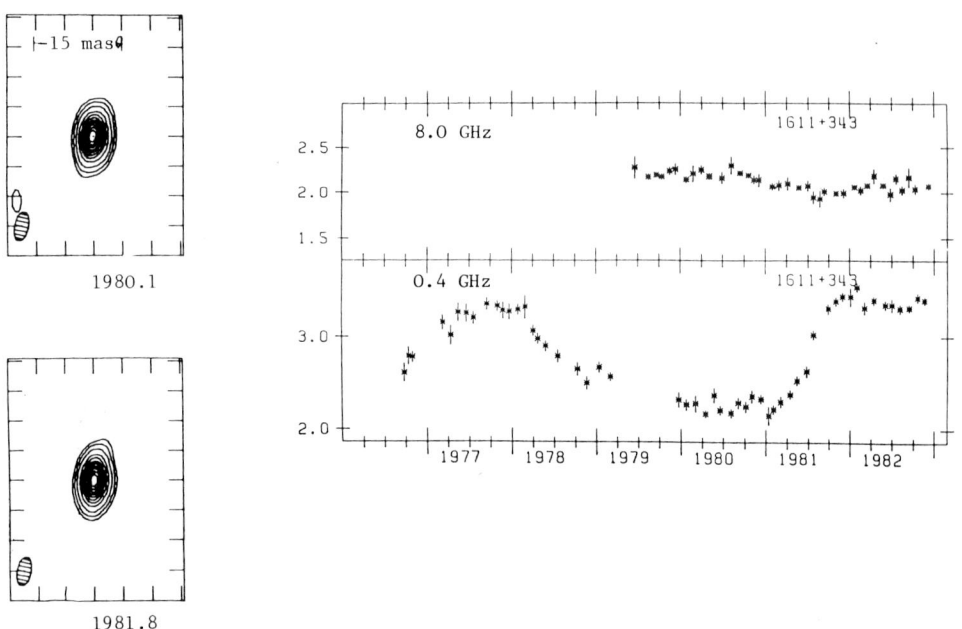

Fig.2 : 1611+343 (DA406) - Quasar (z=1.4) - $T_b = 2 \; 10^{15}$ K - $\gamma > 12$.
No significant variation in fringe amplitudes.
Contour Levels : 45, 90, 180, +180 mJy/beam area

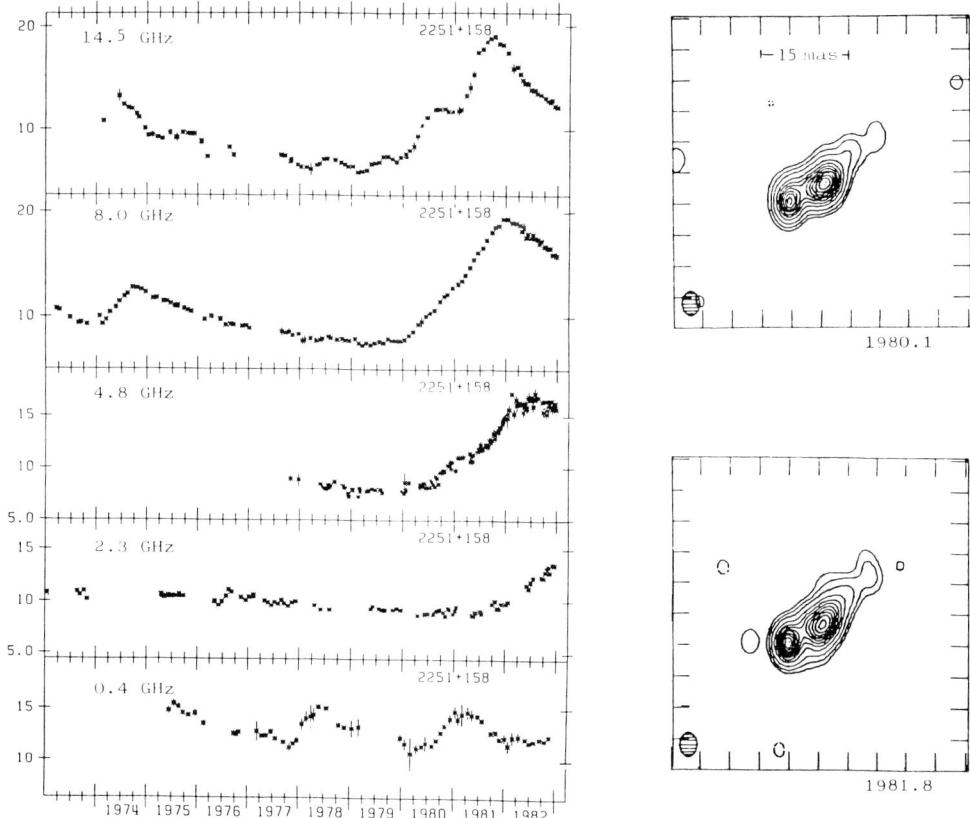

Fig.3: 2251+158 (3C 454.3) - Quasar (Z=0.86) - $T_b=10^{15}$K - $\gamma > 11$.
No significant variation in fringe amplitudes.
Contour levels : 66, 132, 264, +264 mJy/beam area

REFERENCES
Altschuler D.R. et al. 1984, Astronom. J., 89, 1784
Fanti R. et al. 1981, Astron. Astrophys. Suppl. Ser. 45, 61
Fanti C. et al. 1983, Astron. Astrophys. 118, 171
Romney J.D. et al. 1984, Astron. Astrophys. 135, 289

THE MAGNETIC MONOPOLES CONTENT OF GALACTIC NUCLEI, QUASARS, STARS AND PLANETS

Qiuhe Peng [1,2]) Zongyun Li[1]) Deya Wang[3])
[1]) Department of Astronomy, Nanjing University, Nanjing
[2]) Kapteyn Astronomical Laboratory, Groningen
[3]) Purple Mountain Observatory, Nanjing

I. INTRODUCTION

There are two possible sources of Monpoles in celestial bodies: [1]) They may be preserved in the collapse of the objects from original clouds or nebulae. [2]) They may be captured in the time after formation of the objects. The main goal of this work is to estimate the efficiency of the first process, in order to estimate monopole number ratio in the objects, $\xi \equiv N_m/N_B$, where N_m and N_B are the numbers of monopoles and nucleons respectively.

We assume both negative and positive monopoles are separated partly by the stronger local chaotic magnetic field in early universe, so we only consider one kind of the two 'charged' monopoles.

We assume that in the original nebula, $y_n = \frac{G m_B m_m}{g_m^2} \sim 1.9 \times 10^{-25}$ (m_B, m_m and g_m are the mass of nucleon, mass and magnetic charge of monopole respectively, $m_m \sim 10^{16} m_B$, $g_m = 3 \frac{hc}{2e}$[1]), which corresponds to the value for balance between gravitation and Colomb repulsion by the total magnetic charge in a static monopole-concentrated sphare. It is reasonable for $y_n \ll y^0$ ($\lesssim 10^{-20 \pm 1}$) which is the upper limit proposed by Parker[2] and Lazarides[3].

II. PRINCIPLE

How many monopoles fall toward the center of a nebula, when the nebuleus shrinks, depends on the interaction between matter and monopoles. The monopoles which move in medium will exist turbulant electric fields. For neutral medium, the field may cause nearby atoms to be excited or ionized. It consumes kinetic energy of monopoles in a cloud of neutral hydrogen at rate[4] $-\frac{dE_n}{dx} \sim A_n n_H$ erg/cm, $A_n \sim 1.4 \times 10^{-24}$, where n_H(g/cm) is the number density of neutral hydrogen. For the plasma medium, the turbulant electric field excited by moving monopoles creates plasma oscillations which consumes the kinetic energy of monopoles at rate $-\frac{dE_p}{dx} \sim A_p N_e T_e^{-1/2} U_m$ erg/cm, $A_p \sim 5.4 \times 10^{-31}$, where N_e, T_e are the number density and the temperature of electron respectively,

U_m is the velocity of monopole relative to the plasma. When the nebula contracts to some extent, the velocity of monopole, V_m, relative to the stationary frame of reference, is much lower than the infall velocity of mass, $V_m \ll V_R$, and $U_m \sim V_R$.

The most probable distance of travel of monopole relative to matter is $\ell \sim \frac{1}{2} m_v V_m^2 / (-\frac{\alpha \varepsilon}{\alpha X})$, over this distance the kinetic energy of monopole relative to matter is consumed by 'fiction' and infall of monopoles with matter begins. When the nebula contracts to some radius which satisfies relation $\ell \leqslant bR_c$ ($b \simeq (\frac{1}{3} - \frac{1}{5})$), we consider that all monopoles with outward velocity $v \leqslant v_m$ within radius R_c are drawn back by the collapsing mass and fall into the center of nebula at free-fall velocity. Hence number of monopoles in galactic nuclei and stars is equal to that within R_c in original nebula. $N_m \sim N_m^0 (R_c/R_0)^3$, where N_m^0 is total number of monopole in orginal nebula which radius is R_0.

For the galactic nuclei, the contraction of original clouds began in recombination epoch, and the matter in them is plasma. According above mentioned arguments we have $y/y_n \sim 10^2 (\frac{v}{10^{-5}})^{-1} (\beta\Omega) (\frac{V_m}{10^{-4}c})^{-4}$, ($T_e \sim 4 \times 10^3$ K, n0 $\simeq 0.4 \beta\Omega$) where v, which is the ratio of the mass of galactic nucleus to the galaxy, is about 10^{-5}. $\beta (> 1)$ is the ratio of average density of original nebula to that of matter in universe in that era, and Ω is the ratio of average density of matter in universe now to the critical density. And the velocity of monopoles in early universe may be less than the value today, 10^{-3} c[2], for example, $V_m \leqslant 10^{-4}$ c [3].

For stars and planets, it is believed that stars at least population I are formed after formation of the Galaxy and the physical condition in nebulae, which form stars and planets, is similar to today's condition. The matter in nebulae is in neutral state, $T \simeq (10-100)$K, $n \sim (10^4-10^6)$ cm^{-3}, and the velocity of monopoles is about 10^{-3}c. Moreover, it is usually considered that the ratio of mass of star to the nebula, $v' \equiv M/M_0$, is about $(0.3 \sim 10^{-2})$. So we have

$$y/y_n \sim 4.7 \times 10^{-13} (v') b^{3/2} \left(\frac{n_0}{10^5 \text{cm}^{-3}}\right)\left(\frac{M}{M_\odot}\right)^{1/2}\left(\frac{V_m}{10^{-3}c}\right)^{-3}.$$

The ratio corresponding to monopoles captured by objects with mass M, radius R and the age τ is

$$(y/y_n)^{(c)} \sim 2.6 \times 10^{-8} \eta \left[1 + 10^{-6}\left(\frac{10^{-3}c}{V_m}\right)^2 \frac{R_g}{R}\right]\left(\frac{R}{R_\odot}\right)^2\left(\frac{M_\odot}{M}\right)\left(\frac{\phi}{\phi_0}\right)\left(\frac{\tau}{10^{10} \text{ yrs}}\right),$$

where η is the probability of capturing monopole by the object, and R_g is the Schwartzschild radius of the object, and ϕ is the flux of monopoles, ϕ_0 is the Parker limit ($\sim 10^{-16}$ cm^{-2} sec^{-1} sr^{-1})[7], the upper limit of monopole flux in interstellar space is [6.1] $\eta(\frac{\phi}{\phi_0}) \leqslant (10^{-5} - 10^6) \left(\frac{\langle\sigma\beta\rangle}{10^{-27} \text{cm}^2}\right)$, where $\langle\sigma\beta\rangle$ is the average cross section of the catalyzed nucleon decay by monopoles[8]. The second term in the brackets [ ] of expression above, steming from acretion theory, is much smaller than 1 for planets, about 1 for normal stars, and far larger than 1 for compact objects - white dwarfs, neutron stars and galactic nuclei.

III. CONCLUSION

1) The ratio of monopoles to baryons in galactic nuclei and quasars is conserved in the formation process, and may be higher than $Y_n = 1.9 \times 10^{-25}$.
2) The monopole concentration in both normal stars (e.g. the sun) and planets (e.g. the earth) is due mainly to capture after they formed and is much less than Y_n. It is impossible to find monopoles on the surface of planets or at the matter of meteorite.
3) For white dwarfs and neutron stars, the monopoles are due mainly to capture and the ratio is either $Y \ll Y_n$ for $\frac{\langle\sigma\beta\rangle}{10^{-21}\text{cm}^2} \gg 10^{-2}$ or $Y \sim Y_n$ for $\frac{\langle\sigma\beta\rangle}{10^{-27}\text{cm}^3} \ll 10^{-2}$.
4) The monopoles in the centers of the objects must produce radial magnetic field. The detection of the field will place a bound on the cross section of the induced decay of nucleon by monopole.

ACKNOWLEDGEMENTS

The authors should like to sincerely thank Dr. B. Sanders for revising th manuscript in English. They also express thanks to Ms. J. Nunnink and I. Rouwé for typewriting this article. This work is partly supported by the Kapteyn Laboratory.

REFERENCES

(1) Z. Ma and J. Tang, Phys. Letters 126 B (1983) 319
(2) E. Parker, Ap. J. 160 (1970) 383
(3) G. Lazarides et al., Phys. Lett., 100B (1981) 21
(4) E. Bauer, Proc. Camb. Phil. Soc., 47 (1951) 777
(5) X. Li, Scientia Sinica, in press (1985)
(6) E. Kolb et al., Phys. Rev. Lett., 48 (1982) 1146
(7) Q. Peng, Scientia Sinica, in press (1985)
(8) V. Rubakov, JETP. Lett., 33 (1981) 644
    C. Callan, Phys. Rev., D25 (1982) 2141

A MONOPOLE MODEL FOR GALACTIC NUCLEI

Qiuhe Peng [1,2)] Deya Wang[3)] Zongyun Li[1)]
[1)] Department of Astronomy, Nanjing
[2)] Kapteyn Laboratory, Groningen
[3)] Purple Mountain Observatory, Nanjing

I. INTRODUCTION

The monopole may catalyze nucleon decay ([1-4)], $pM \to \pi^0 e^+ M$ (85%) or $\mu^+ \mu^- e^+ M$ (15%). The cross section of this process is either 10^{-26} cm^2 ([1,2)]) or 10^{-36} cm^2 ([5)]). Products of the decay, e^+, π^0, μ^+ and μ^-, transform into photons and some neutrinos, and provides a new energy source in astrophysics.

Monopole content in galactic nuclei and quasars may be higher and reach the saturation value ([6)], $\xi (\equiv N_m/N_B) \sim \xi_n \approx 1.9 \times 10^{-25}$. Due to the induced nucleon decay this is possibly sufficient to supply their huge power of galactic nuclei and quasars and to provide support by radiation pressure against collapse to black holes. The aim of this text is to discuss the possibility for setting up a model of galactic nuclei and quasars not based upon black holes.

II. MODEL

We consider, for simplicity, a spherically symmetric system in a state of static and radiative balance in the frame of Newtonian mechanics, neglecting rotation, convection and the effect made by radial magnetic field produced by monopoles. The balance equation is
$$-\frac{dP(r)}{dr} = \frac{GM(r)}{r^2} \rho(r),$$
where M is total mass within radius r. Pressure P includes gaseous and radiative pressure, $P = P_g + P_r$,
$$P_g(r) = \frac{RT(r)}{\mu} \rho(r), \quad -\frac{dP(r)}{dr} = \frac{k\rho(r)L(r)}{4\pi c^2 r}.$$
The radiative temperature gradient is $-\frac{dT}{dr} = \frac{3}{4a_rc} \frac{k\rho(r)}{T^3} \cdot \frac{L(r)}{4\pi r^2}$, and the luminosity provided by induced decay is $L(r) = \int_0^r 4\pi r^2 \, dr \, \langle \sigma \beta \rangle c N_m(r) N_B(r) \cdot m_B c^2$, where $N_m(r)$ and $N_B(r)$ are the number density of monopoles and nucleons respectively, $c\beta$ is the relative velocity of nucleon to monopole. The opacity, k, is attributed by free-free transition.

Moreover, we take the distribution of density,
$$\rho\genfrac{\{}{\}}{0pt}{}{n}{r} = \rho_c [1 + (\frac{r}{a_n})^2]^{-(2n+1)/2},$$
as a working model for the preliminary

approach, where ρ_c is the central density of the object, $n(= 2, 3, 4, \ldots)$ is the model index, and a_n^2 is much less than R^2 (R is the radius of the object), and a_n can be determined by requirement of consistency. Both n and a_n give the indications of concentration.

We find that the various physical quantities of galactic nuclei in the context of the model can be determined by the luminosity (L) and mass (M) of the object and a parameter ξ ($= \frac{y}{y_n} \cdot \frac{\langle\sigma\beta\rangle}{10^{-27} \text{cm}^2}$). The results in the model are following:

$$L = A_L \eta_n \left(\frac{\rho_c}{1.84 \text{ g/cm}^2}\right) \xi M_8 \quad , \quad A_L \sim 2.1 \times 10^{44} \text{ erg/sec}, \quad \eta_n \sim 1,$$

$$R/R_g = Q_y \xi^{q_\xi} \cdot L_{41}^{-q_L} \cdot M_7^{-q_M} \quad , \quad T_{(R)} = Q_{T_R} \xi^{-S_\xi} \cdot L_{41}^{S_L} \cdot M_7^{-S_M} \text{ (°K)},$$

$$\rho_{(R)} = Q_{\rho_R} \xi^{-m_\xi} \cdot L_{41}^{m_L} \cdot M_7^{-m_M} \text{ g/cm}^3 \quad , \quad \rho_c = Q_{\rho_c} \xi^{-1} \cdot L_{41} \cdot M_7^{-1} \text{ (g/cm}^3\text{)},$$

$$H_{(R)} = Q_H \xi^{-t_\xi} \cdot L_{41}^{t_L} \cdot M_7^{-t_M} \text{ Gauss} \quad , \quad a_n/R = Q_a \cdot \xi^{\nu_\xi} \cdot L_{41}^{-\nu_L} \cdot M_7^{\nu_M},$$

where $L_{41} = L/10^{41}$ erg/sec, $M_7 = M/10^7 M_\odot$, R_g is the Schwartzschild radius of the objects. And

$$q_\xi = \frac{1}{3}\left(\frac{n - 11/8}{n - 5/16}\right), \quad q_L = \frac{1}{3}\left(\frac{n - 19/32}{n - 5/16}\right), \quad q_M = \frac{1}{3}\left(\frac{n + 3/8}{n - 5/16}\right),$$

$$S_\xi = \frac{1}{2} q_\xi, \quad S_L = \frac{1}{2}(q_L + \frac{1}{2}), \quad S_M = \frac{1}{2}(1 - q_M),$$

$$m_\xi = (2n + 1)q_\xi - \frac{1}{3}(n - 1), \quad m_L = (2n + 1)q_L - \frac{2}{3}(n - 1),$$

$$m_M = \frac{2}{3}(n + 2) - (2n + 1)q_M$$

$$t_\xi = 2q_\xi, \quad t_L = 2q_L, \quad t_M = 1 - 2q_M,$$

$$\nu_\xi = \frac{1}{3} - q_\xi, \quad \nu_L = \frac{1}{3} - q_L, \quad \nu_M = q_M - \frac{1}{3}.$$

Table Q_i' values for different n

n	2	3	4	5	6	7	8
Q_y	6.3E2	3.0E.2	2.1E2	1.8E2	1.64E2	1.56E2	1.52E2
Q_{T_R}	2.5E3	3.7E3	4.3E3	4.7E3	4.9E3	5.0E3	5.1E3
Q_{ρ_R}	1.0E-9	3.7E-9	6.2E-9	8.6E-9	9.6E-9	1.1E-8	1.2E-8
Q_{ρ_c}	1.4E-2	1.1E-2	9.7E-3	9.1E-3	8.9E-3	8.6E-3	8.4E-3
Q_H	1.3E1	5.6E1	1.1E2	1.5E2	1.8E2	2.1E2	2.2E2
Q_a	3.7E-2	1.2E-1	2.0E-1	2.8E-1	3.5E-1	4.0E-1	4.5E-1
$\rho_{c,max}$	3.6E4	2.0E4	1.7E4	1.6E4	1.5E4	1.4E4	1.3E4

A MONOPOLE MODEL FOR GALACTIC NUCLEI

$\rho(g_m/cm^3)$

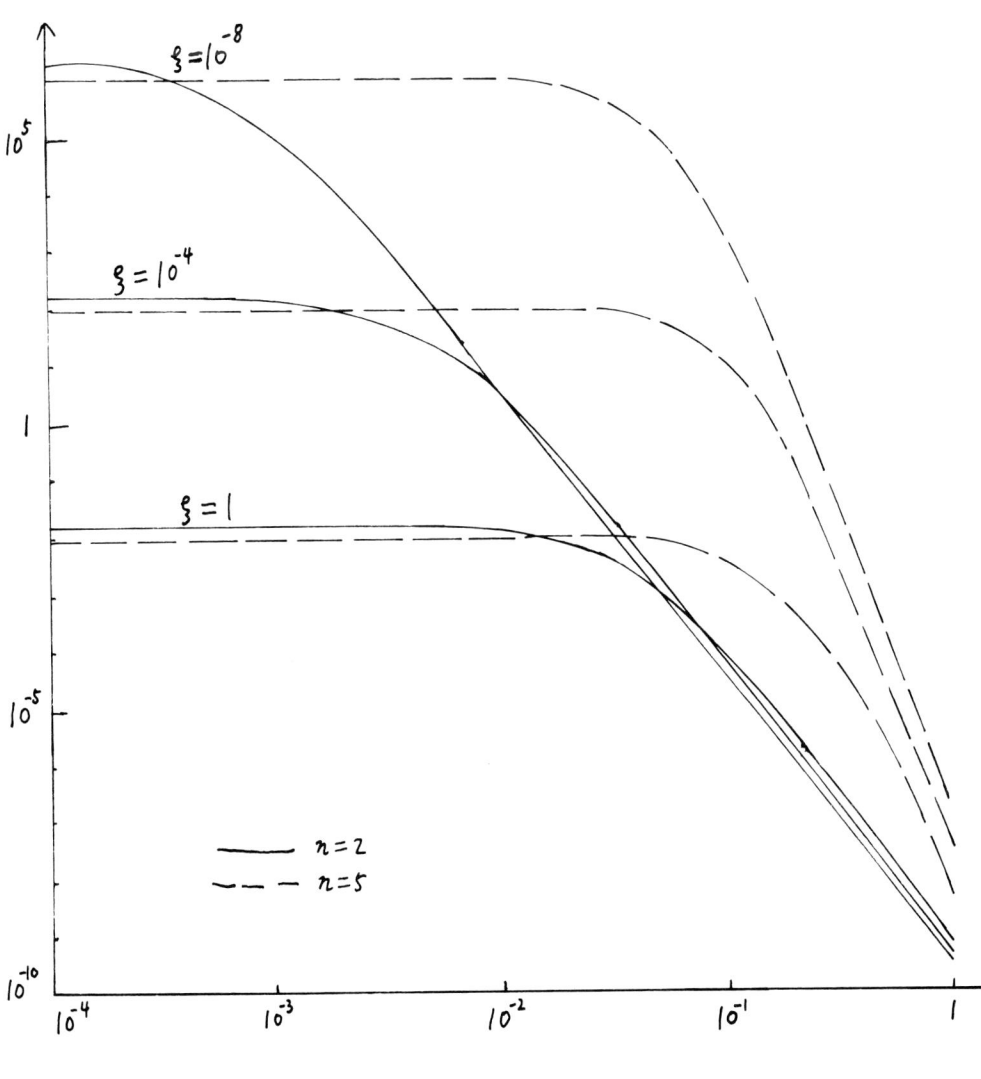

Figure 1. The density distribution

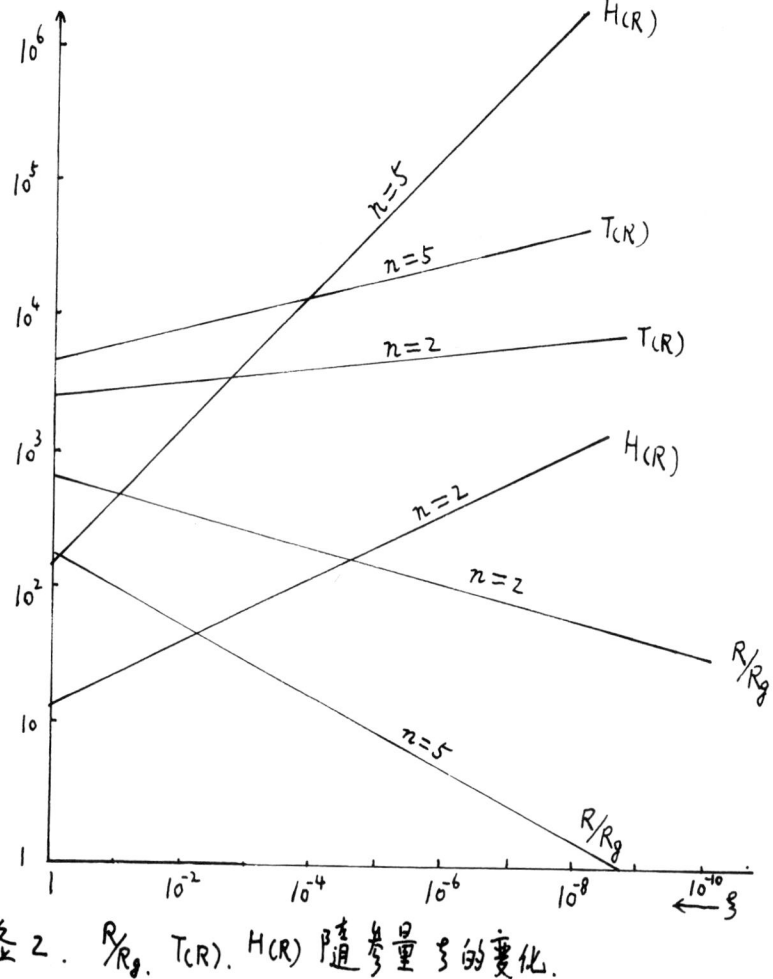

Figure 2. The variances of R/R_g, $T_{(R)}$, $H_{(R)}$ with the parameters ξ.

Models which are not black holes are possible if ρ_c is less than $\rho_{c,max} = F_n M^{-2}$. The values in the last line of the table above is computed for $^8M_G = 3 \times 10^6 M_\odot$. Models which are not black holes correspond $\xi > 10^{-7}$ showed by figure 1. The variations of R/R_g, $T_{(R)}$ and $H_{(R)}$ with parameters ξ (for models n = 2,5) are showed in figure 2.

III CONCLUSION

It's possible for the models of galatic nuclei not to be black holes, $R/R_g > 1$, so long as $\xi > 10^{-7}$. Therefore, we should consider the possibility that the huge power of galactic nuclei and quasars results from the nucleon decay induced by monopoles, and the effect may prevent these objects from collapsing to black holes, given the possibility of the existence of monopoles and the induced decay of nucleons.

ACKNOWLEDGEMENTS

The authors should like to sincerely thank Dr. B. Sanders for revising th manuscript in English. They also express thanks to Ms. J. Nunnink and I. Rouwé for typewriting this article. This work is partly supported by the Kapteyn Laboratory.

REFERENCES

[1] V. Rubakov, JETP. Lett., 33 (1981) 644; Nucl. Phys., B203 (1982) 311; ibid B218 (1983) 240
[2] C. Callan, Phys. Rev., D25 (1982)2141; ibid D26 (1982)2058; Nucl. Phys. B212 (1983)391
[3] Z. Ma and J. Tang, Phys. Lett., 126B (1983)319; Nucl. Phys. B231 (1984) 172; BIHEP-TH-84-14(May 1984)
[4] G. Calucci and G. Vedovato, 9/85/EP(1985)
[5] F. Wilczek, Phys. Rev. lett, 48(1982) 1146
[6] Q. Peng et al. see this symposium

AN EVOLUTIONARY LINK BETWEEN SEYFERT I AND II GALAXIES ?

Enrique Pérez Astronomy Department, University of Sussex
M. V. Penston Royal Greenwich Observatory

ABSTRACT. First spectra of NGC4151 and 3C390.3 from the newly sited Isaac Newton Telescope on the island of La Palma, when compared with old spectra of the same objects taken with the INT at Herstmonceux leads to the conclusion that these two Seyfert galaxies have faded over the years and are adopting Seyfert II-like properties.

NGC 4151

The recent photometric history of this galaxy shows recurrent extended low states, particularly from late 1980. From unpublished IUE and Herstmonceux data, it is known to have started fading again in 1983 October and remained very faint until 1984 July. The INT observations show that it was still very faint during 1984 April. In particular the broad wings of the Balmer lines are weaker than ever previously reported (for example by Antonucci and Cohen 1983). If, as has been recently supposed (Ulrich et al. 1984) the Broad Line Region (BLR) responds to the illumination from the central continuum source, this suggests that the part of the BLR emitting the Balmer lines is at least six light months in radius because the last occasion on which NGC 4151 was known to be bright was in early October 1983. Thus the Balmer line emitting region corresponds closely to the CIII] emitting BLR3 of Ulrich et al. (1984).

In the figure we compare the new data with spectra taken in 1974 with the INT at its previous location at Herstmonceux. The object was then much brighter (in its high state). The new and old spectra have been scaled so that the wavelength scales agree and the narrow emission lines have the same intensity. Note that the new spectrum is flux calibrated, unlike the old one. By straightforward comparison of both spectra, the essentially complete absence of broad component of the Balmer lines can be seen. In 1984 only very weak H wings are present. The spectrum more closely resembles that of a narrow emission line galaxy than that of a Seyfert I. If the quality of the data had been only a little worse it might have been classified as a Seyfert of type II.

3C390.3

This galaxy is known to have been steadily fading over the years since the first plates were taken late last century at the old Royal Observatory at Greenwich. More recently Lloyd (1984) has published a light curve which completes that given by Barr et al. (1980). This shows an overall decrease of about two magnitudes over the past fifteen years. The object is very faint in the INT spectra of 1984, April 23/24.

The figure presents a comparison of spectra taken in 1984 with an INT spectrum taken at Herstmonceux in 1975. Again in this galaxy the broad wings of H and H present in the old data have faded greatly. In this case one might say that the broad lines could now be absent and the galaxy merit classification as a narrow line radio galaxy in contrast with its previous broad line status.

Osterbrock (1978) presented four scans of 3C390.3 covering a period of three years during which the continuum had an overall decrease of about half a magnitude (Lloyd 1984), and which also shows how the H profile varies. Ultraviolet observations (Tadhunter, private comunication) also show a fading of the broad component of L and CIV 1550.

Line variability was also reported by Barr et al. (1980). We do not believe our present data support their idea that long-term continuum variability is also caused by obscuration of the continuum source by a dense cloud of the Broad Line Region. Although the variations of this galaxy are complex, it is apparent that there is a smooth long time-scale continuum variation. It seems to us most likely that the BLR has responded to this smooth decrease. In effect the BLR is caused to fluoresce by the continuum source, as has been invoked to explain the variability in NGC 4151 (Ulrich et al. 1984). In this fluorescence picture, the continuum radiation is emitted more or less isotropically from the nucleus. Mechanisms like cloud absorption or beaming which take place the observers in a favoured direction do not provide a natural explanation of the apparently related line and continuum variations.

One intriguing feature is the fact that these observations of 3C390.3 extend to higher luminosity the class of objects in which the BLR appears to vary in response to the continuum source. It means that a long term study of such systems may enable astronomers to measure the mass of the black holes in these galaxies using the method of Ulrich et al. (1984). Naturally one would expect long time-scales for more luminous objects where the BLR is bigger as appears to be the case in 3C390.3. On the other hand there is mounting evidence that the variations in the true OVV quasars are a byproduct of beaming in which case this method would not work for the most luminous objects.

Discussion

The present optical spectra suggest that transitions from type I to type II Seyfert spectra, or equivalently from broad line to narrow line radio galaxies, can occur on comparatively short timescales. A suitable physical explanation is the switching off of the continuum source, perhaps due to a temporary lack of fuel. This transition is observed as a fading of the continuum followed by a decay in the broad component of the permitted lines, but this feature alone, the broad permitted lines, is not the only difference between the two classes of galaxies.

AN EVOLUTIONARY LINK BETWEEN SEYFERT I AND II GALAXIES?

At other wavelength ranges, there are properties involving large scale phenomena which have to be taken into account. Recently Ulvestad and Wilson (1984) have drawn attention to the greater luminosities and sizes of the radio emitting regions of Seyfert II compared to Seyfert I galaxies. Using this they rule out the suggestion that Seyfert II are obscured Seyfert I (Lawrence and Elvis 1982). It seems to us that the same argument does not exclude a picture in which at least some Seyfert II galaxies are Seyfert Is in which the nuclear continuum source is temporarily off. For example, active galaxies may evolve through a sequence in which they are initially Seyfert I most of the time, later they are equally likely to be observed as either type and finally they behave as Seyfert II nearly all the time. This direction of evolution would be consistent with a growth in size of the radio regions with time, in agreement with the Fig. 2 of Ulvestad and Wilson (1984). This model is tenable provided there is an overlap of radio luminosities of Seyfert I and Seyfert II galaxies.

Other measurements characteristic of extended regions in Seyfert I and II galaxies are the narrow line luminosities. In fact the narrow line regions may be closely related to the radio sources. Indeed some years ago de Bruyn and Wilson (1978) found a correlation between the radio and [OIII] luminosities in Seyfert galaxies. Since then much data has been accumulated and discussed by Yee (1980), Snyder (1981) and, perhaps most interestingly, by Cohen (1983). Some of the discussions in these papers are conducted in terms of equivalent widths or of differences in line luminosity at a fixed continuum luminosity. These discussions would be inappropiate if our picture is correct; rather, one must compare continuum luminosities at fixed line luminosities. Such a comparison is consistent with our proposal since the Seyfert IIs are generally fainter in the continuum than Seyfert Is of the same narrow line luminosities. Our model implies that in any one galaxy the continuum may fade rapidly while the emission from the narrow line region would be essentially constant. Moreover there is a good overlap of narrow line luminosities (see data of Cohen (1983)) between the two classes of Seyfert galaxies. The narrow line luminosities of Seyfert IIs are consistently higher than those in Seyfert Is (as would be expected from the de Bruyn and Wilson (1978) correlation and the differences in radio properties) but is in fact no higher than those in quasars (Figs. 4 of Cohen (1983)).

However it is clear (Figs. 8 of Cohen (1983)) that there are also systematic differences between narrow line ratios. In Seyfert Is the [OIII] to [OII] ratio is systematically higher than in Seyfert IIs; other line ratios are also generally compatible with a higher ionization in Seyfert Is. We suspect this can also be explained in our picture. It is natural that the average spectrum from gas that is more intermittently irradiated by a photoionizing source should display a lower ionization. We recognize, however, a detailed study using a photoionization code is necessary to check this point thoroughly. Another test may be provided by comparing the infrared spectra of Seyfert I and II galaxies where there are known to be

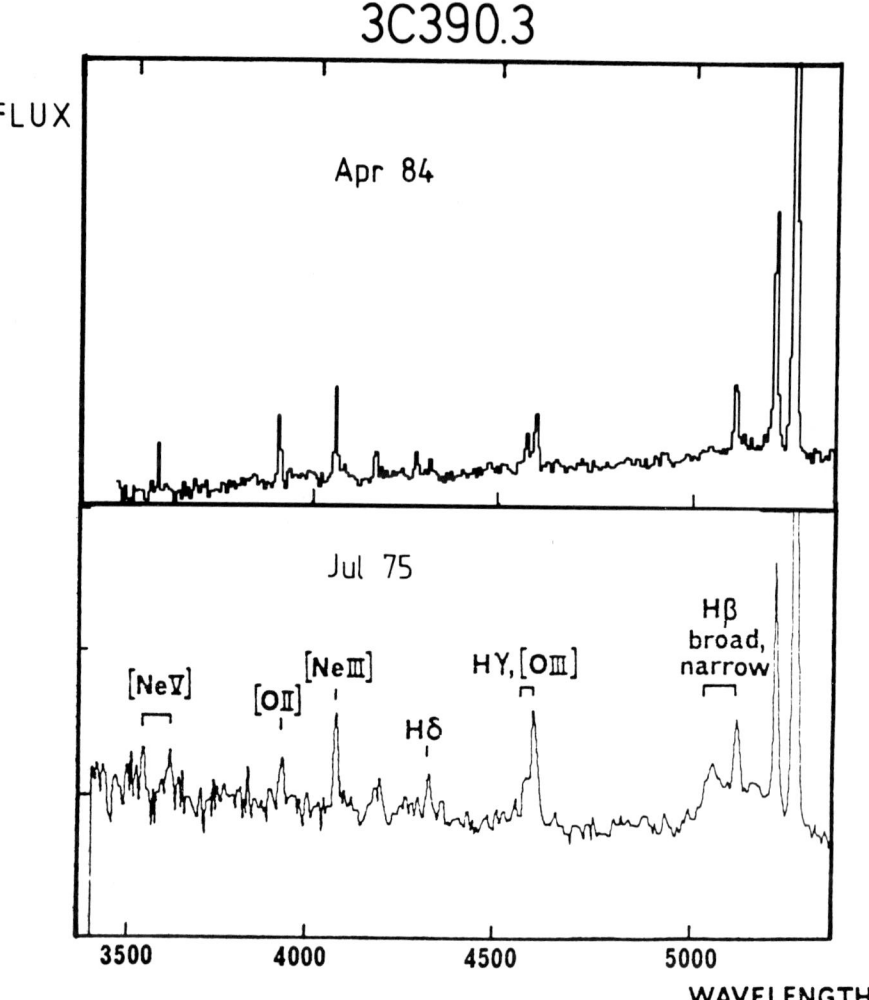

AN EVOLUTIONARY LINK BETWEEN SEYFERT I AND II GALAXIES?

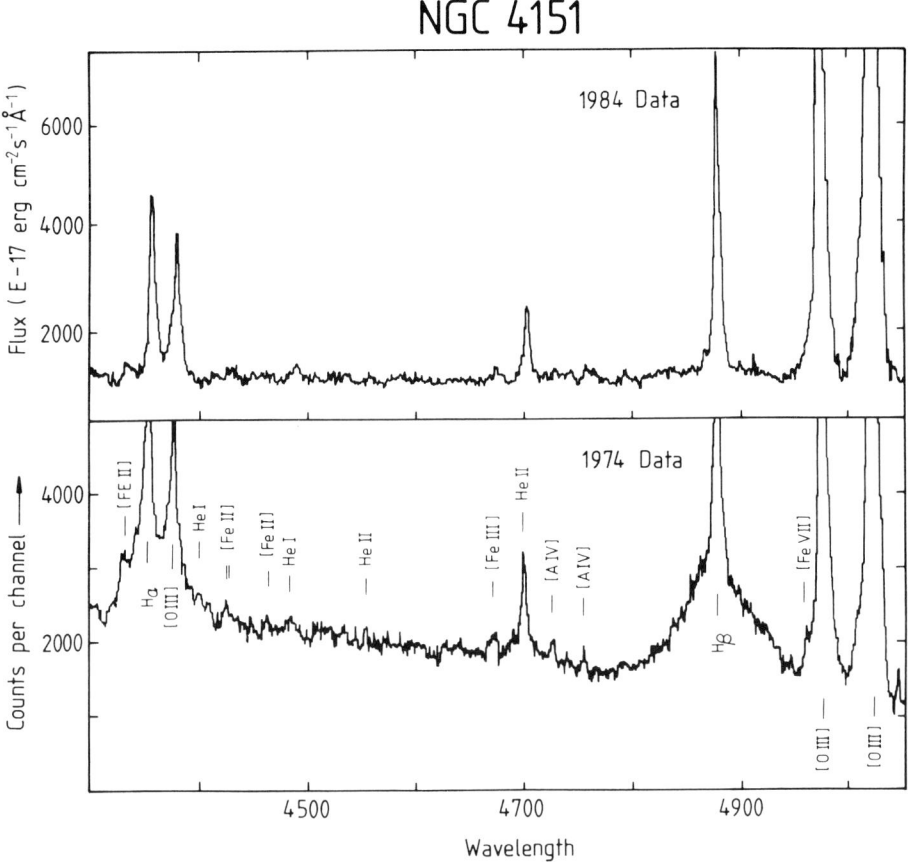

differences between types and some of the infrared components are extended. For example if the properties of large-scale dust components in Seyfert I and II were disjoint, there would be difficulties in maintaining the evolutionary picture we have just outlined. However (Lawrence et al. 1984) our model is not in conflict with the existing data.

Finally, we would also expect cases where Seyfert II galaxies are seen to develop Seyfert I characteristics. It seems there are two cases where this may have happened, namely Mk 6 (Khachikyan, Popov and Egiazaryan 1982) and NGC 1566 (Alloin et al. 1984).

In conclusion, we propose that the source of ionization in Seyfert II galaxies is photoionization by an intermittent source which is now off and that the galaxy develops Seyfert I characteristics when it is on. We believe that the known differences between the extended regions in the two classes of galaxy can be explained by this model if there is an evolution in which the duty cycle (proportion of time active) of the photoionizing source changes along with the properties of the extended region. Further study of this idea is obviously desirable.

References

Lyutyi, V. M., Oknyanskii, V. L., and Chuvaev, K. K., 1985, Sov. Astron. Lett. 10(6)

Penston, M. V., and Pérez, E., 1984, Mon. Not. R. astr. Soc., 211, 33 and references therein.

THE POSSIBLE DETERMINATION OF TWO NEW BL LAC REDSHIFTS

M. Persic and P. Salucci
SISSA-ISAS, Strada Costiera 11, Miramare, Trieste
Osservatorio Astronomico, via Tiepolo 11, Trieste
Italy

ABSTRACT. We present tentative identifications of very faint emission lines in the spectra of the BL Lacertae objects 0754+100 and 1514+197 as given in the atlas of spectra of QSOs and related objects by Wilkes et al.(1983). The proposed identifications, quite common in BL Lac spectra, are discussed.

1. Introduction

Because of the timescale and degree of their variability, BL Lacertae objects can be considered exemplars of the AGN phenomenon at its purest. However, the study of some relevant topics such as their energetics and their evolutionary and population properties are seriously affected by the lack of published redshifts for most of them. Actually, out of the nearly one hundred objects which are currently classified as BL Lacs (Urry, 1984), less than forty have published redshifts. This relatively small number is further reduced if we consider that the determination of some of them is based on absorption lines not originating from the associated galaxy but due to intervening clouds which may be not physically connected to the BL Lac sources themselves.
Waiting for a spectroscopic survey aimed at finding new redshifts, an effort to select some prime candidates to be there included may clearly be worthwhile. According to this perspective, in this work we give tentative, possible identifications of faint emission features in some published spectra of objects still lacking redshift determination.

2. The Analysis of the Spectra

The atlas of QSO spectra published by Wilkes et al.(1983) is particularly suitable for our purposes. It contains 295 spectra (for technical remarks see the atlas), out of which 15 belong to BL Lac objects. Of these, 6 have no published red shifts. On grounds of estimated photometric quality of the spectra we have chosen to study two such objects, 0754+100 and 1514+197. Hereafter the proposed identifications of emission features are reported together with their redshifts (Table 1) and discussed for each of the two spectra, which are reproduced by kind permission of Dr. B. Wilkes and collaborators.

0754+100. The emission feature centred at $\lambda 4685$ has a broad profile similar to the one found in the spectrum of the BL Lac object 1308+326, which was identified as MgII $\lambda 2800$ as well(Miller et al.,1978). We argue that the not perfect agreement between the redshift of this feature and the redshifts of the others is possibly due to the uncertainty in the former's measurement because of the very broad profile of the feature (FWZI is ~300 A). All the other identifications are quite common in BL Lac spectra (Danziger et al.,1978; Miller et al.,1978; Miller and French,1978) (Fig.1).

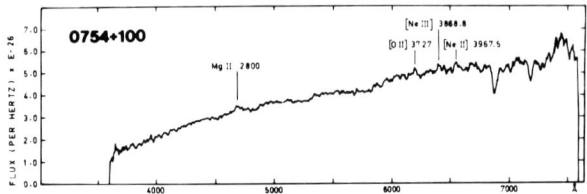

Figure 1. The spectrum of 0754+100 (adapted from Wilkes et al.,1983).

1514+197. The identification of the $\lambda 3950$ feature as CIII] $\lambda 1909$ is consistent with the identification of a rather broad (FWZI is ~200 A) hump centred at $\lambda 5800$ as MgII $\lambda 2800$ at the same redshift and with the expected lack of other easily detectable features at this redshift given the exploitable wavelength coverage ($\lambda\lambda 3700$ through 5900A) of the spectrum and what we know about typical emission-line spectra of AGNs and QSOs. The emission feature at $\lambda 5890$ is a blend of very close together airglow lines (see Meinel et al.,1968) (Fig.2).

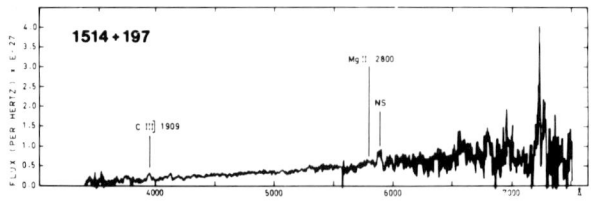

Figure 2. The spectrum of 1514+197 (adapted from Wilkes et al.,1983).

TABLE I

Object	λ_{obs} (Å)	Identification	z
0754+100	4685	MgII 2800	0.67
	6190	[OII] 3727	0.66
	6400	[NeIII] 3868.8	0.65
	6550	[NeIII] 3967.5	0.65
1514+197	3950	CIII] 1908.8	1.07
	5800	MgII 2800	1.07

The identifications of the emission features and their redshifts.

3. Conclusions

The nature of the proposed identifications and their consistency with a single redshift value make these two BL Lac objects 0754+100 and 1514+197 two very promising targets for a future spectroscopic survey devoted to finding new redshifts of QSOs and related objects, in that if the features which have been considered are actually emission lines then their identifications are unique.

Acknowledgements. We thank Dr. B. Wilkes for kind permission of reproducing the spectra and for useful discussions.

References

Danziger, I.J., Fosbury, R.A.E., Goss, W.M.: 1978, in Pittsburgh Conference on BL Lac Objects, Ed. A.M. Wolfe, Univ. of Pittsburgh, Pittsburgh, pennsylvania, 204.
Meinel, A.B., Aveni, A.F., Stockton, M.W.: 1968, Catalog of Emission Lines in Astrophysical Objects, The University of Arizona, Tucson.
Miller, J.S., French, H.B.: 1978, in Pittsburgh Conference on BL Lac Objects, Ed. A.M. Wolfe, Univ. of Pittsburgh, Pittsburgh, Pennsylvania, 228.
Miller, J.S., French, H.B., Hawley, S.A.: 1978, in Pittsburgh Conference on BL Lac Objects, Ed. A.M. Wolfe, Univ. of Pittsburgh, Pittsburgh, Pennsylvania, 176.
Smith, M.E.: 1978, in Pittsburgh Conference on BL Lac Objects, Ed. A.M. Wolfe, Univ. of Pittsburgh, Pittsburgh, Pennsylvania, 211.
Urry, C.M.: 1984, NASA Technical Memorandum 86103.
Wilkes, B.J., Wright, A.E., Jauncey, D.L., Peterson, B.A.: 1983, Proc. Astron. Soc. Australia, Vol. 5, No. 1, 2.

MAGN; MILDLY ACTIVE GALACTIC NUCLEI

Paris Pismis
Instituto de Astrofisica de Canarias,
Universidad de La Laguna,
Tenerife,
Spain,
on leave from

Instituto de Astronomía,
Universidad Nacional Autónoma de Mexico
Apartado Postal 70-264
Ciudad Universitaria
04510 Mexico, D.F.
Mexico

Complexities in the kinematics have been detected in the sixties - in the nuclei of galaxies aside from those defined as Seyferts. The kinematic complexities consist of deviations from the expected circular motion, (generally referred to as not-circular motions) either outward or inward. Some galaxies like M31, NGC 253, NGC 4736 show expansion at their nuclear regions whereas others like NGC 2903, 3351, 3672, 5383 are reported to exhibit inward motions. (References in Pismis, 1979).

At present under the heading Active Galactic Nuclei (AGN) are discussed objects from quasars down to Seyferts 1 and 2. Nuclei active in the radio range, BL Lac objects, Liners, blue compact and Markarian galaxies to name a few, all fall in the group of AGN. However, in my view all galaxies show activity in their nuclei involving the liberation of energy of more or less intensity, thermal and/or non-thermal, and with radial motions and other non-equilibrium manifestations. Activity may be as intense as in a QSO or as mild as in our Galaxy or in M31. The dividing line between the different groups of AGN is becoming fuzzy as observational data increase. I believe that the division line between AGN and those not entering the AGN group is also fuzzy. A classification of activity should therefore be considered only as a working convenience and not taken too hard and fast, the more so since nuclei show variability and may at times pass on to another group; for example NGC 4151, a prototype of a Seyfert 2 galaxy, has varied such that it now is in the class of Seyfert 1!

Despite the vast amount of work, observational as well as theoretical carried out lately, the origin of nuclear activity is unclear. Interesting suggestions abound but the power engine most likely is within the nucleus itself and not due to external effects;

If there are external forces, say encounters with other galaxies advocated by many as causing the onset of activity in the nucleus, the outer more tenuous regions would be far more affected than the nucleus itself. In support of this statement is the case of Seyfert galaxies whose outer regions, the spiral formations, do not seem to show irregularities any more than in a "normal" galaxy. In any case a study of the detailed morphology and the kinematics of many more Seyfert galaxies should give indication as to the relative importance of local vs. external effects.

Although spectroscopic and photometric work on AGN is actively pursued the kinematics and structure of these compact objects are not yet accessible to direct observation. However, it is known that the spectra of QSO's are indistinguishable from those of Seyfert 1 nuclei; it is thus reasonable to expect that the phenomenon of nuclear activity may be similar in both cases though different in their energetics. Thus a closer study of nearby Seyferts might provide valuable hints and constraints to QSO modelling.

I propose the designation MAGN (Mildly active galactic nuclei) for all galactic nuclei the degree of activity of which falls, on the average, below that of Seyfert 2's. One good specimen of MAGN is NGC 253 (Sc) from the nucleus of which outflow is observed to occur within a cone (Gottesman et al., 1976; Ulrich, 1978). Another specimen is NGC 1569 where Pence has found evidence of a past (1-2x10 yrs ago) eruption.

High resolution data have given ample evidence that nuclear velocity fields cannot be explained by rotation alone. Radial motions do seem to exist in a large number of presumably "inactive" galaxies. I shall advocate for the existence of radial motions arising from non-gravitational forces, unspecified as yet. In a recent paper Pişmiş and Moreno have argued that tight nuclear spirals observed in galaxies like NGC 1097, 4314 and 1365 (all three barred) may represent the loci of matter ejected symmetrically from a rotating nucleus. In particular this model, based on ejection, is proposed to explain the nuclear spiral of NGC 4736 for which a detailed velocity field is given by van der Kruit (1976). Adopting reasonable physical parameters, such as the ejection velocity to be 2/3 of that of rotation and a nuclear mass of 3×10^{10} solar masses the computed double loci of the ejecta after a few times 10^7 years has shown a striking similarity to the tight spiral in the nucleus of NGC 4736 (Pişmiş and Moreno, 1984). Moreover the computed outward velocities of the different points along the locus have turned out to be function of direction from the center of the galaxy in agreement with the observations of van der Kruit (1976). NGC 4736 thus qualifies to be a member of the group MAGN. It is interesting to point out that a dependence on direction of outward (or inward) velocities in the nuclear region of NGC 1300 is suggested by Peterson and Huntley (1980) in the discussion of their velocity data. A closer study of the morphology of the nucleus of NGC 1300, unresolved so far, might possibly reveal the existence of a tight spiral. Our model explains also the inward motions of some MAGN; for after reaching the apogalacticon of their elliptical orbits, the ejecta will be returning towards the nucleus. Also both outward and inward motions may coexist in galactic nuclei. In summary I consider the existence of a tight spiral and outflow and inflow velocities signature of MAGN.

It is conceivable that AGN and MAGN are powered by similar mechanisms and that the difference in the manifestations of the activity are due to a difference in the scale of their energetics. For a full account the reader is referred to a review paper by P. Pişmiş to appear in Revista Mexicana Astron. Astrof. Vol. 10, 1985.

In concluding I state once again that activity in galactic nuclei spans a wide range of energetics, the underlying general phenomenon of which may be similar throughout the complete sequence of galactic nuclear manifestations. It is worth calling to attention that as one goes along the sequence of active nuclei from QSO's down, through MAGN, to the least active ones, the nucleus tends to be gradually less compact and the outer regions which are almost undetectable at the QSO's, become relatively more luminous and extended. In support of this statement we mention that in Seyfert 1 galaxies where nuclear activity is more intense than in Seyfert 2's the size of the nucleus is much smaller than a Seyfert 2 nucleus (Brecher, 1977). It should be rewarding to confront this "conjecture" with statistical evidence all along the sequence of AGN and MAGN.

If indeed this conjecture is shown to be valid, a detailed study of the nuclei of nearby active galaxies (MAGN's for example) as regards their morphology, spectral peculiarities and velocity field etc., may provide valuable information in the search of a possible mechanism for the energy production at galactic nuclei.

REFERENCES

Gottesman, S.T., Lucas, R., Weliachew, L., and Wright, M.C.H.: 1976, Ap.J., 204, 699.
Kruit, P.C. van der: 1976, Astron. and Astrophys. 52, 85
Pence, W.D.: 1981, Ap.J. 247, 473
Peterson, C.J., and Huntley, J.M.: 1980, Ap.J., 242, 913
Pişmiş, P.: 1979 in Photometry, Kinematics and Dynamics of Galaxies, Ed. D.S. Evans.
Pişmiş, P.: 1985, Proceeding of the Fourth Latin American Regional Meeting. Rev.Mex. de Astron. y Astrof., **Vol 10**
Pişmiş, P., and Moreno, E.: 1984, Astrofizika 20, 7
Ulrich, M.H.: 1978, Ap.J. 219, 424
Brecher, K.: 1977, in Frontiers of Astrophysics, Chapter 10, p 438, Ed. E.H. Avrett.

ON THE DUPLICITY OF SEYFERT GALAXY NGC 1275 NUCLEUS

I. Pronik and L. Metik
Crimean Astrophysical Observatory, USSR

Seyfert galaxy NGC 1275 has many peculiarities having direct or indirect relation to the activity of its nucleus. This galaxy contains two giant gaseous systems differing in radial velocities by about 3000 km/s. V. Pronik has shown that low-velocity gas system has two kinematical centers located in the nucleus of the galaxy (1979). They differ in radial velocities by 140±20 km/s. The distance between the centers is equal to 1"-2". The line connecting kinematical centers is directed from North to South. The survey of the published data for NGC 1275 galaxy evidences that it is not clear yet how its nucleus is connected with the circumnuclear region (L. Metik, I. Pronik, 1984).

Recently we have investigated nuclear region of NGC 1275 by two-colour photometry method using the negatives obtained in prime focus of 6-m telescope. Six negatives were taken with blue ($\lambda ef \sim 4800 A$) and eight ones with red ($\lambda ef \sim 6650 A$) glass-filters. The image scale on the negatives was 17."5 per mm. The seing was about 1."5. The central region of NGC 1275 obtained with various filters and exposures is shown in Fig. 1. One can see there the nucleus of NGC 1275 labelled as "a", detailes in circumnuclear region "b" and "c" and star "A" - photometrical standard.

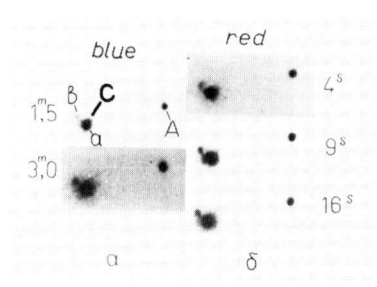

Fig. 1

Fig. 2 shows the map of surface brightness (I) and photometrical densities (D) distribution in the circumnuclear region of NGC 1275. You may find here:

1. The detail "b" stretching towards the S-E region of the

nucleus;

2. The detail "b" and the nucleus "a" are connected by a bar whose direction does not coincide with that of the elongation of the detail "b".

3. The whole regon of interaction of nucleus "a" and detail "b" has a shape of a cone with the top on the detail "b". According to spectral analysis the detail "b" is the compact star cluster

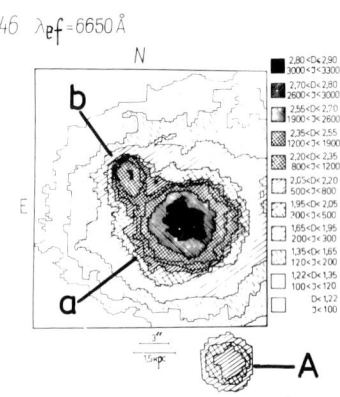

Fig. 2. The photometric map of NGC 1275 circumnucleus region obtained in red light. The figures on the top left – negatives number.

located near the NGC 1275 galaxy (L. Metik, I. Pronik, 1979). The radial velocity of this cluster in respect to the galaxy is 5000 km/s. The spectra showed that near the detail "b" there exists a stream of gas clouds having a wide range of radial velocities: from – 700 km/s to + 4900 km/s in respect to the galaxy. The highest velocity of the gas streem is equal to the velocity of the detail "b", i.e. 5000 km/s.

The brightness in the cone-shaped region of interaction between nucleus "a" and detail "b" is higher than that of the adjacent region. The half-width of passbands of the filters we used equal to ~1000A, therefore we suppose that the cone-shaped region of interaction may consist not only of gas but of stars too.

In Fig. 3 and 4 photometrical maps of the nucleus in two spectral bands are shown. These two figures permit to draw the following conclusions:

1. The nucleus "a" of NGC 1275 is stretched towards the star-formation "b". The elongation is less niticeable in blue light;

2. In the red light the whole circumnuclear region being of 6" (or 3 kpc) is a little stretched towards the detail "b" while

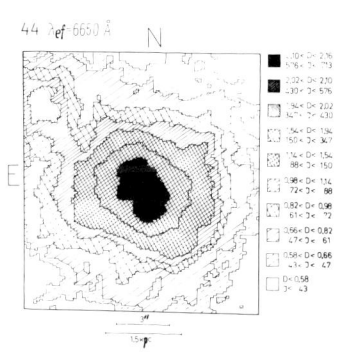
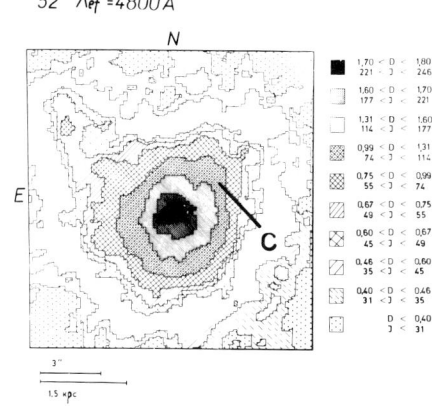

Fig. 3   Fig. 4

in the blue light it is stretched in perpendiculiar direction or towards detail "c".

The elongation of NGC 1275 nucleus is estimated by comparing the photometric profiles of NGC 1275 nucleus and that of "A"-star as shown in Fig. 5. The dimension of photometrical profiles of star

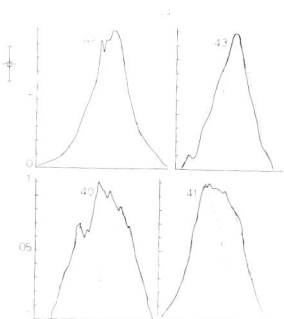

Fig. 5. The photometric tracing made along the line of NGC 1275 nucleus elongation. The dotted lines are the photometric tracing of star "A". The figures correspond to negatives's numbers.

"A" was measured on 8 negatives. At half maximal intensity it equals to $2\rlap{.}''22 \pm 0\rlap{.}''09$. The dimension of photometrical profile of galaxy nucleus in the direction to its elongation was estimated as $1\rlap{.}''5 \pm 0\rlap{.}''06$ or 500 ± 20 pc.

The comparison of integral effect in the frames of photometrical profiles of star "A" and object "b" in two colours on the graphs gives relative intensities $I_b/I_A(6650)=0,28\pm0,02$ and $I_b/I_A(4800)=0,43\pm0,05$. The same comparison of the galaxy's nucleus with "A"-star shows that brightness of the nucleus in both filters has varied by 2 times during 20^m. Such rapid variation evidenced that stretched nucleus cannot be a solid body. We suppose that it consists at least of two Seyfert type nuclei. Dimension of each is less than 1" or 350 pc.

The profiles of the emission lines [OIII], H_β, H_α and [NII] in the radiation spectrum also evidence about the duplicity if NGC 1275 nucleus. The profiles have two components whose shift corresponds to the difference in radial velocities of about 600 km/s (E. Dibay, 1969).

The different direction of elongation of circumnuclear region in red and blue lights indicate that the region of the most active interaction of galaxy nucleus and circumnucleus is shifting with time. About 10 years ago it was located between "a" and "b". As a result of active star-formation at this region the bar "a-b" was originated. Now "a" and "b" objects have been still interacting: bar between them consists not only from stars but from gas too (W.Keel, 1983). But there is a new region of "a" and circumnuclear area that acts. It is located between "a" and "c". It appears that the detail "c" is located on the line of gas accretion from the circumnucleus to the region between two components of the nucleus "a". Indipendent interaction of low-velocity gas with two components of NGC 1275 nucleus can be confirmed by existance of two kinematic centers in this low-velocity gas that were found by V. Pronik (1979).

In the region "a-c" the regular gas-stream is originated which leads to formation of young stars. Fig. 6 shows the energy distribution in the spectrum of "c" - detail (see L. Metik, I. Pronik, 1979). There are two bumps in this distribution. The bump in red region can be associated with the radiation of old red and yellow stars. The bump in the blue region of spectrum can be caused by the presence of young recently formed blue stars.

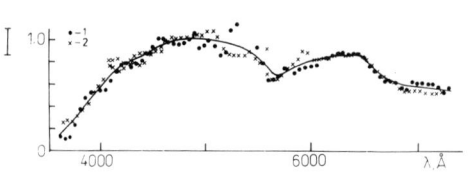

Fig. 6

The gas of NGC 1275 was investigated by many authors but there is no common opinion about its nature. Many hypotheses were put forward e.g.: an explosion in the galaxy nucleus; tidal destruction of satellite galaxy; gas accretion from the intergalactic space and many others. Our data testifys in favour of only two hypotheses. They are as follows:

1. two interacting galaxies,
2. the gas accretion from the intergalactic space. Then, in the first case detail "b" is the nucleus of destroyed galaxy, and in the second one the objects "b" and "c" are groups of stars formed from the gas continuosly flowing between galaxy nucleus and its circumnuclear region. Wirth et al. (1983) supposed that the whole central region of NGC 1275 galaxy is covered by large areas of star-formation explosions and the matter is supplied from the intergalactic space of Perseus cluster. According to our data the most active process of starformation occured in the nearest part of the galaxy nucleus.

REFERENCES

Dibay E.: 1969, Astron. Zh., 46, 725
Keel W.: 1983, Astron. J., 88, 1579
Metik L., Pronik I.: 1979, Astrofizika, 15, 37
Metik L., Pronik I.: 1984, Astrofizika, 21, 233
Pronik V.: 1979, Astrofizika, 15, 51
Wirth A., Kenyon S., Hunter D.: 1983, Astrophys. J., 269, 102

PCD SPECTROSCOPY OF MKN 463 A DOUBLE NUCLEUS SEYFERT-2 GALAXY

P. Rafanelli[1], S. di Serego Alighieri[1,2,+]

1) Osservatorio Astronomico di Padova
2) ST - ECF, Garching bei München
+) Affiliated to the Astrophysics Division
   Space Science Department of E.S.A.

ABSTRACT. PCD spectroscopy of Mkn 463, a double nucleus Seyfert-2 galaxy, has been performed at the 2.2m telescope of the ESO Observatory in order to verify the spatial resolution and the sensitivity of the detecting system. A preliminary analysis of the spectra shows that the asymmetric profiles of the emission lines arising from the Seyfert-2 nucleus are more likely due to dynamical effects rather than to dust extinction.

1. INTRODUCTION

Mkn 463 (m_B=16, z=0.050) is an irregular galaxy, the morphology of which is characterized by a complex outer structure and by a double nucleus elongated in the east-west direction (Adams, 1977, Korovyakovskii et al. 1981). The two nuclei have an angular separation of 4", which corresponds to 3.8 Kpc at the distance of the object d = 199 Mpc, derived assuming for the Hubble constant the value H = 75 Km s^{-1}Mpc^{-1}. A spectroscopic study of the nuclear region of Mkn 463 has been performed some years ago by A.R. Petrosyan et al. (1979) who found that the spectra of both nuclei are characterized by emission features and that the nucleus to the west belongs to the Seyfert-2 class. We have selected this galaxy as test-object for our programme of PCD observations of active galaxies with multiple and interacting nuclei, taking advantage of the fact that the low angular separation of its two nuclei and the relatively wide range of line intensities present in its spectrum give the opportunity to check the spatial resolution and the sensitivity of the system quickly and simultaneously.

2. OBSERVATIONS AND REDUCTION TECHNIQUES

The observations have been performed by one of us (S.S.A.) at La Silla Observatory in February. The ESA-PCD, a two dimensional photon counting system developed as a scientific model for the Faint Object Camera, di Serego Alighieri et al. (1985), was mounted at the B.&C. spectrograph of the 2.2m telescope. Gratings giving dispersions of the order of 89, 44 and 21 A/mm in front of the detector were used. The spectra were taken

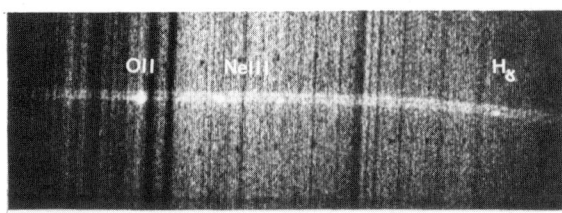

Fig.1 —
PCD spectrum of Mkn 463
Spectral range 3600-4730 A.

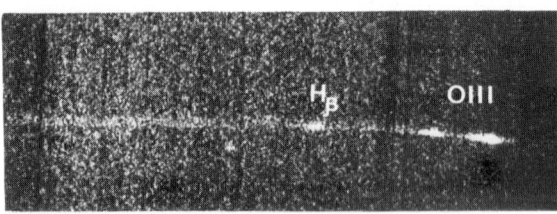

Fig.2 —
PCD spectrum of Mkn 463.
Spectral range 4800-5400 A.

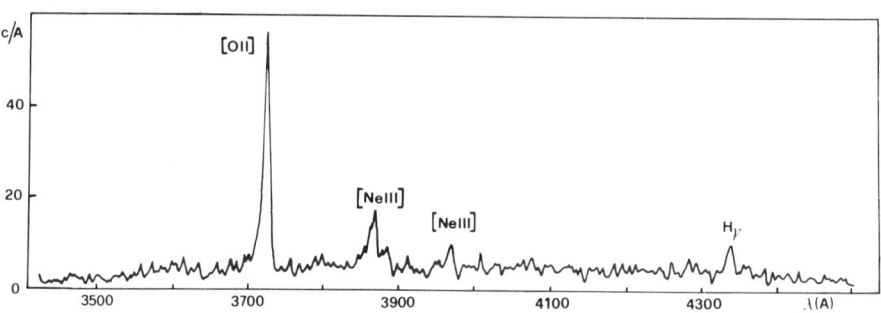

Fig.3 — Intensity tracing of the spectrum of Mkn 463, plotted in the rest system of the object. Units in the ordinates are counts per Angstrom.

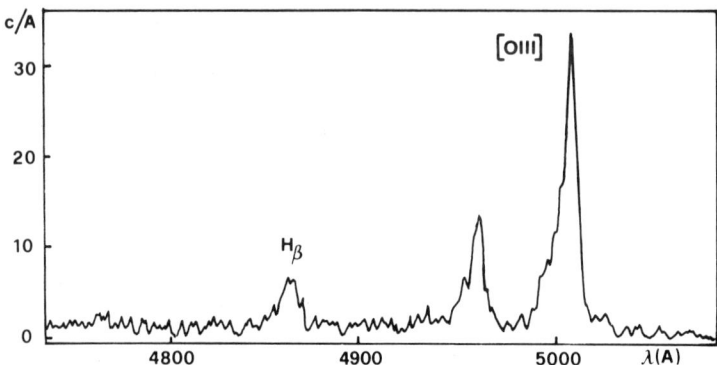

Fig.4 — Intensity tracing of the H_β and $|OIII|$ lines of Mkn 463. Units in the ordinates are counts per Angstrom.

within the 3500-5400 A spectral range, where the sensitivity of the system is highest (see Tab.1 for the quantum efficiency of the detector). Spectra of standard stars were taken each night in order to calibrate the system. Spectra of a He-Ar comparison lamp were obtained after each exposure to evaluate the wavelength scale on the whole image, which has a format of 1024 x 256 pixel. An incandescent lamp with a quartz bulb was employed at the beginning and at the end of the nights to derive the high frequency variations in the sensitivity of the system. The observations have been reduced using the IHAP programme. The procedure is quite long and developes through the following steps:
1 - Correction for curvature of the lines.
2 - Correction for S-distorsion.
3 - Flat-field correction.
4 - Reduction to linear wavelength scale.
5 - Correction for atmospheric extinction.
6 - Reduction to absolute energy units.
Here we present data which have been reduced up to step 4.

Tab. 1 PCD Quantum Efficiency.

λ(A)	Q.E.	λ(A)	Q.E.	λ(A)	Q.E.
3000	25.0 %	4500	14.6 %	6000	2.2 %
3500	23.3 %	5000	9.5 %		
4000	19.7 %	5500	5.3 %		

3. PRELIMINARY RESULTS

The two spectra which we show here have been obtained with two different gratings and cover the spectral ranges 4800-5400 A (D_λ= 21 A/mm, Res.~1A exp. time 40^m) and 3600-4730 A (D_λ= 44 A/mm, Res.~2A, exp. time 30^m) respectively. Fig. 1 and 2 show how the spectra look before being reduced. In Fig. 1 the spectra of both components of Mkn 463 are well separated and clearly visible. It is not the same for the spectrum in Fig.2 where the dispersion is double and the quantum efficiency nearly half (see Tab. 1).
The Seyfert spectrum is dominated by the |OII| $\lambda\lambda$ 3726,3727 doublet (in blend) and by the |OIII| nebular lines, I(|OIII|)/I(H_β) = 7.7, the width of which is nearly twice that of H_β. The other lines visible in the spectrum are |NeIII| $\lambda\lambda$ 3868,3970 and H_γ. H_β and the |OII| $\lambda\lambda$ 3726,3727 |OIII| $\lambda\lambda$ 4959,5007 lines are present in the spectrum of the companion also but they are much fainter and narrower than that arising from the Seyfert nucleus.
The flat-field and sky corrected tracings of the Seyfert spectrum are plotted in Fig. 3 and Fig. 4. It is interesting to note the strongly asymmetrical profile of the |OIII| nebular lines, which show wings extending further to the blue than to the red. Their asymmetry index at 20 % intensity level (A.I. = (B-R)/(B+R)) (A.I.)$_{20}$= 0.33, is similar to that of the other forbidden lines present in the spectrum, see Tab.2, but is completely different from that of H_β. The same consideration is valid for the FWZI and FWHM which is similar for all forbidden lines.

Tab. 2

Line	λ(A)	FWZI (Km/s)	FWHM (Km/s)	(A.I.)$_{20}$
|OII|	3726-27	2500 ± 200	700 ± 50	0.30 ± 0.05
|NeIII|	3868	2400	700	0.33
|OIII|	5007	2300	600	0.33
H$_\beta$	4861	1300	750	0.00

These data suggest that it is quite likely that the asymmetry of the forbidden lines in Mkn 463 is produced by dynamical effects rather than by dust extinction as it would occur in expanding or infalling dusty shells of ionized gas. In this last case dust extinction, which is more efficient at the shortest wavelengths, would produce a |NeIII| λ 3868 line which is narrower and more asymmetric than |OIII| λ 5007, contrary to what we have observed.

In conclusion these first PCD data on Mkn 463 show at a preliminary analysis already interesting results (which will be developed calibrating the spectra) and demonstrate the efficiency of the detector in obtaining medium dispersion spectra of faint sources with relatively short exposure times.-

REFERENCES

Adams, T.F.: 1977, Astrophys. Journal Suppl. , 33, 19
Korovyakovskii, Y.P., Petrosyan, A.R., Saakyan, K.A., Khachikyan, E.E.:
    1981, Astrofizika, 17, 231
Petrosyan, A.R., Saakyan, K.A., Khachikyan, E.E.:1979,Astrofizika,15,373
di Serego Alighieri, S., Perryman, M.A.C., Macchetto, F.: 1985, Astron.
    and Astrophys., in press.

NARROW BAND PHOTOMETRY AND THE REDDENING OF CYG A

Karl D. Rakos and Norbert Fiala
Institut für Astronomie
Türkenschanzstraße 17
A-1180 Wien
Austria

Introduction:

The use of narrow band (FWHM 200 Å) filters in extragalactic research is not new. Our photometric system vz, bz, yz is very similar to the Stroemgren v, b, y passbands. It has many advantages in conection with a CCD camera compared to broad band UBV photometry. These narrow passbands are less affected by emission lines than other filters in use. As example the Syfert galaxy POX 4 (Kunth and Sargent 1983) has for the B filter a total emission line width of a factor four larger than in the v filter calculated for equal spectral width. The filters b and y are not affected by emission lines, except for the very faint He II 4686 Å line and the blend Ar IV and He I 4711 Å. The v bandpass is only affected by H delta.

Narrow band interfrence filters are available for any wavelength region within the sensitivity of a CCD camera, and once calibrated the actual observations can be made in the rest frame of a galaxy cluster in investigation. For this reason our filter system will be always a local system for a chosen value of redshift z, called vz, bz, yz. Small differences (delta z) between the actual z value of the galaxy cluster and the corresponding nominal z value of the filter set should be within a range of +/- 0.03. In such a case a simple K - correction can be applied to the measurements in order to obtain rest frame values. For the purpose of calibration of a filter set it is necessary to have at least one spectrophotometric measurement of a galaxy at the distance of the observed cluster. These disadvantages are more than compensated by the fact that a large number of galaxies can be covered with a few exposures.

Discussion:

There is a very simple relation between color indices (vz - yz) and (bz - yz) for stars of different types. Using the libary of stellar spectra recently published by Jakoby et. al. (1984) and the filter response functions vz, bz and yz (at zero redshift standard Stroemgren v, b and y filter response functions normalized to unity), we have calculated synthetic colors for 22 supergiants well distributed over the spectral types from F4 to K2. There is a significant linear relation

(1a) $(vz - yz) = (2.51 \pm 0.10)(bz - yz) + 0.057 \pm 0.030$

with a correlation coefficient R = 0.985. The relationship for luminossity class III stars is slightly different. We used the spectrophotometric atlas of Gunn and Stryker (1983) to calculate

(1b) $(vz - yz) = (2.56 \pm 0.05)(bz - yz) - 0.205 \pm 0.046$

for 38 stars with a coefficient of correlation R = 0.993. 21 main sequence stars give

(1c) $(vz - yz) = (2.99 \pm 0.13)(bz - yz) - 0.593 \pm 0.099$

with R = 0.983. These relations are shown in Fig. 1. For a metal content very different from the solar value, the lines are slightly shifted to the left or to the right, but the relationships remain linear. In the case of elliptical galaxies, assuming their light is only a linear combination of spectral energy distributions (s.e.d.s) of luminossity class I, III and V stars, we expect the points to fall between the line (1a) and (1c). Of course we may have some amount of reddening within each galaxy. To calculate the direction of reddening we can use the standard values for reddening in the Stroemgren system. For this purpose we used the spectrophotometric atlas of Gunn and Stryker (1983). From this atlas we selected 36 s.e.d.s of stars with known Stroemgren colors, covering the whole spectral sequence. We then convolved the dereddened s.e.d.s with vz, bz and yz response functions and transformed them into magnitudes. The resulting synthetical colors were compared with Stroemgren colors. From this procedure we get a set of transformations

$$(b-y) = (0.961 \pm 0.002)(bz-yz) - 0.039 \pm 0.0004$$

$$m1 = (0.803 \pm 0.030) mz + (0.135 \pm 0.011)(b-y) + 0.267 \pm 0.002.$$

Using the reddening law for Stroemgren photometry (Landold Boernstein, Neue Serie VI 2C, p. 46), we get

(2) $E(vz-yz) = 1.488 E(bz-yz)$, $E(bz-yz) = 3.86 E(mz)$, $AV = 4.13 E(bz-yz)$

It follows: a) that the direction of reddening is as indicated in Fig. 1, and b) that the influence on reddenning to the difference of the color index $mz = (vz - bz) - (bz - yz)$ is very small and can threfore be ignored. Additionaly a correction AV for the visual magnitude of a galaxy can be obtained. The procedure is simple. The actual position of a galaxy in the diagram should be shifted along the reddening line direction at least to the line given by equation (1c).

The difference in (bz - yz) of these two positions represents the minimum intrinsic reddening E(bz - yz) in an extrenal galaxy.

Fortunately, Yee and Oke (1978) published spectrophotometric scans of a sample of elliptical galaxies. They also published the spectral energy distribution of the so called standard elliptical galaxy, a mean of multichannel scans of five first ranked elliptical galaxies. Cyg A, also in this sample, is supposed to be a highly reddend galaxy. We can use this galaxy as a test object. Convolving these spectral scans with vz, bz and yz response functions, we can plot (vz - yz) versus (bz - yz). See Fig. 1. The galaxies are clearly distributed along the line

(3) $\quad (vz-yz) = 2.239 (bz-yz) - 0.046$

with a coefficient of correlation R = 0.954. Cyg A is well shifted in the direction of high reddening. N galaxies in the sample are not plotted in the diagram Fig. 1. Neither N galaxies, nor Cyg A were used to calculate relation (3). They populate a different region in the two color diagrams and can be better seperated from normal elliptical galaxies in the mz versus (vz - yz) diagram. The mean values of color indices for the elliptical galaxies are in (vz - yz) = 0.73 and in (bz - yz) = 0.35. These values are very similar to the values of the standard elliptical galaxy of Yee and Oke (1978), (vz - yz) = 0.69 and (bz - yz) = 0.36.

Of course we have supposed that the light of Cyg A comes from an average elliptical galaxy with average solar chemical composition without strong nonthermal components, and that the standard reddening law can be applied to the dark matter within elliptical galaxies.

Within the sample of elliptical galaxies we used, there are 9 galaxies without any UV radiation excess, the rest of 6 galaxies show UV excess. These groups are well seperated in the diagram. The sandard deviation from the regression line is of the order of 0.06 magnitudes, which is comparable to the errors of the spectrophotomeric scans. The slope of the regression line is very similar to the slope of all three populations of dereddened stars. This suggests that the difference in color of the galaxies is introduced by different stellar composition and not by reddening. Only two galaxies show a larger deviation from the regression line. The simplest interpretation is the galagtic foreground (Cyg A) and the intrinsic reddening within the galaxies (Cyg A and 3C 293). Supposed the regression line (3) for the sample is reddening free (or has small constant amount of reddening), following equation (2) we can derive E (bz - yz) = 0.20 for 3C 293, which is equal to AV = 0.83 or AB = 1.10. The dereddened color indices are (bz - yz) = 0.20 and (vz - yz) = 0.40. In comparison with our sample 3C 293 should have a UV excess and stronger emission lines. The UV excess can be masked by reddening. This is the only galaxiy in our sample, showing moderately strong emission lines without a UV excess.

For Cyg A we found E($bz - yz$) = 0.37, which corresponds to AV = 1.53 or AB = 2.03. The dereddened color of Cyg A is the same to that of 3C 293. Cyg A shows strong emission lines and a UV excess. The first estimate of reddening of Cyg A was discussed by Sandage (1972), who found AV = 0.90 or AB = 1.20. This reddening is significantly lower than the value AV = 2.26 that Osterbrock and Miller (1975) obtained by fitting the observed Balmer decrement in the optical counterpart of Cyg A to that expected for a gas with T = 10^{4} °K and Ne = 10^{4} cm^{-3}. Finally van den Bergh's (1976) photometry of a normal elliptical galaxy near Cyg A yields a galactic foreground reddening of AV = 1.30 or Ab = 1.70. Van den Bergh and Sandage used 19 and 30 arcsec diaphragms for their photometry. Osterbrock and Miller used a 2.7 x 4 arcsec slit, covering the innermost part of Cyg A. Our calculation of reddening is based on the entrance aperture of 10 arcsec diameter of the Yee and Oke (1978) spectrophotometry. The emission in the optical counterpart to Cyg A is concentrated in a region with small dimensions in comparison to the whole galaxy. This means that the Osterbrock and Miller (1975) result needs some aperture correction in the sense that the observed intrinsic reddening is more concentrated to the emission region of the galaxy. The total amount of extinction in the galaxy will therefore be between 1.30 and 2.26, closer to our estimate of 1.53 mag.

References:

Gunn J. E., Oke J. B. 1975, Astrophys. J. 195, 255
Gunn J. E., Stryker L. L. 1983, Astrophys. J. Suppl. 52, 121
Jacoby G. H. et al. 1984, Astrophys. J. Suppl. 56, 257
Osterbgrock D. E., Miller J. S. 1975, Astrophys. J. 197, 535
Sandage A. 1972, Astrophys. J. 178, 25
Van den Bergh S. 1976, Astrophys. J. 210, L63
Yee H. K. C., Oke J. B. 1978, Astropys. J. 226, 753

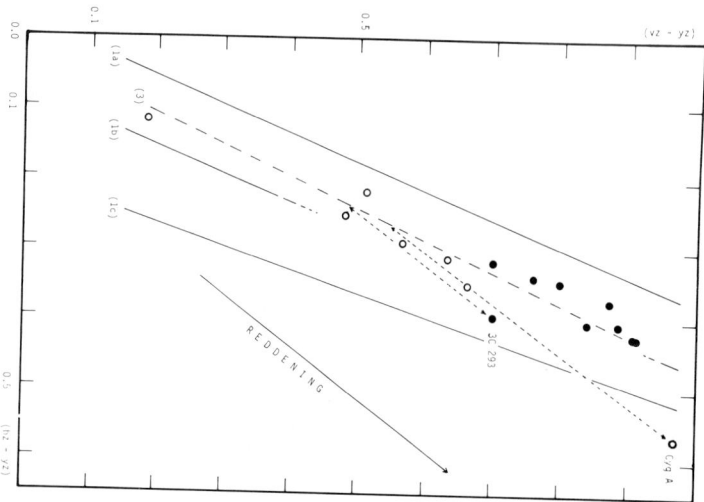

Fig. 1 o galaxies with UV excess, ● galaxies without UV excess

EVOLUTION OF POWERFUL AGN AND THE COLLAPSE OF RICH CLUSTERS

N. Roos
Sterrewacht Leiden
P.O. Box 9513
2300 RA Leiden
The Netherlands

ABSTRACT. The cosmological evolution of active galactic nuclei (AGN) may be described by a luminosity function which depends on redshift and on the (stellar) luminosity or mass of the galaxy. Our mathematical expression for the luminosity function of AGN is based on the "loss-cone filling" model for the fuelling of central black holes in AGN during galaxy mergers. We assume that galactic nuclei containing massive central black holes were formed at $z \simeq 3$ and that the subsequent cosmological evolution of AGN is mainly due to a decrease in the merging rate among galaxies during the subsequent epoch of cluster formation. The radio (and optical) counts of AGN can be reproduced if the merging rate of the most massive galaxies evolves faster than that of the smaller galaxies. At redshifts of about 2.5 the evolution is very similar to pure luminosity evolution and the space density of the most luminous AGN is comparable to the space density of the most massive galaxies. The rapid evolution of powerful AGN is interpreted as an evolution of the merging rate of the brightest galaxies during the collapse and virialization of rich clusters.

1. THE BOLOMETRIC LUMINOSITY FUNCTION.

The tidal disruption rate of stars near a central massive black hole in a galactic nucleus may be enhanced when the star distribution around the hole is perturbed. During the final stage of a merger event this perturbation will be sufficiently large to boost the tidal disruption rate above its stationary level and yield (nonthermal) luminosities as observed in powerful AGN. The "light-curve" of an AGN during such an event can be estimated using simple dynamical arguments. Combining this with an estimate of the merging rate of galaxies with other (mostly smaller) galaxies yields a fractional bolometric luminosity function for galaxies of (stellar) luminosity L_g of the form

$$\phi(L, L_g) = E(L_g) \, t(L, L_g) \, \exp[- E(L_g) \, t(L, L_g)],$$

where ϕ is the fraction of galaxies having bolometric central luminosity L and galactic luminosity L_g, E is the merging rate of galaxies with smaller galaxies (mean mass ratio about 10) and t gives the light curve of an AGN during a single merger event (Roos, 1985a).

The merging rate depends on galaxy luminosity. This property was found in expanding universe simulations. It is a manifestation of a more general mass-neighbour density relation which seems to hold also in the observed clustering properties of galaxies (Roos, 1981). It might explain why (i) the fraction of bright ellipticals among galaxies increases with galaxy luminosity (dwarf ellipticals may have a different origin, see Silk, this volume), (ii) the fraction of bulge galaxies (Sa, SO, E) increases with neighbour density (Dressler, 1980), (iii) the fraction of galaxies that are active increases with galaxy luminosity (Auriemma et al., 1977).

The basic parameters of the model, which we will compare here with radio observations at 1.4GHz, are (i) bolometric over radio luminosity ratio, (ii) central black hole masses, and (iii) merging rate. For the first parameter we choose a value comparable to the ratio estimated for luminous QSO's:

$$\log[\, L/\text{erg s}^{-1}\,] = \log[\, P_{1.4\text{GHz}}/10^{24.5}\text{WHz}^{-1}\,] + 43.4$$

A very good fit to the bivariate radio luminosity function determined by Auriemma et al. (1977) and to the radio luminosity function determined by Windhorst (1985) is obtained using central black hole masses ranging from 10^6 for $M_{B(0)} \simeq -20$ to 10^9 for $M_{B(0)} \simeq -23$. The merging rate is consistent with an estimate from the galaxy correlation functions in position and velocity using the dynamical friction formula. For mergers between bright galaxies of comparable mass it is about $(10^{10}\text{ yr})^{-1}$.

2. COSMOLOGICAL EVOLUTION.

We assume that galaxies and galactic nuclei were formed at about a redshift of 3. During the subsequent epoch of cluster formation the morphological appearance of galaxies may have changed due to galaxy interactions (especially at the high luminosity end of the galaxy luminosity function), but their central structure did not evolve significantly since then. In general the merging rate will evolve relatively slowly, but in rich clusters a rapid evolution on a timescale of a few billion years will take place. Since the most massive galaxies occur mainly in such clusters we expect a rapid evolution for these galaxies. A quantitative estimate of this evolution can be obtained assuming that the density fluctuation spectrum of rich clusters of mass M at the epoch of galaxy formation is given by

$$\delta\rho/\rho \propto M^{-(n+3)/6}$$

where the index n should be in the range $-1.5 < n < -1$ (Gott and Turner, 1977). The mean density and velocity dispersion of galaxies in clusters that are collapsing and virializing at redshift z varies with z as (assuming $q_0 = \frac{1}{2}$)

$$\rho \propto (1+z)^3, \quad \text{and} \quad \sigma \propto (1+z)^{-(1-n)/(2n+6)}$$

The merging rate in clusters varies with density and velocity dispersion as $\rho\sigma^{-3}$ and therefore

$$E \propto (1+z)^{3(n+7)/2(n+3)} \quad \propto (1+z)^{4.5 \text{ to } 5.5}$$

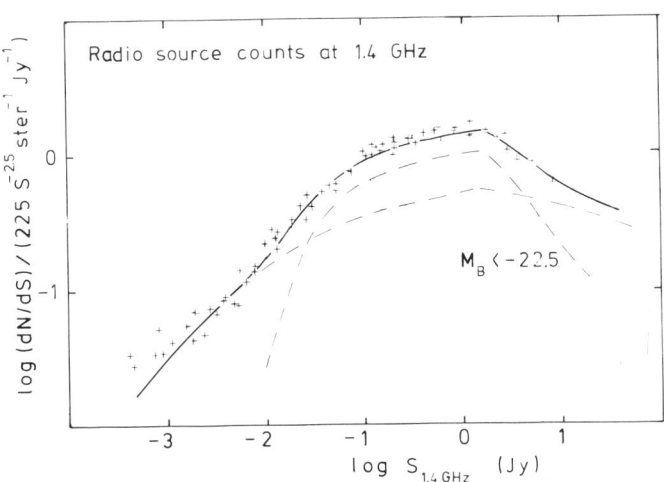

Figure 1 Differential radio source counts at 1.4 GHz. Crosses are the results from several surveys compiled by Windhorst et al. (1985). The model counts are given by the full line. The contribution from the galaxies with $M_{B(0)} < -22.5$ and $M_{B(0)} > -22.5$ are given by the dashed lines.

In figure 1 we compare the radio counts at 1.4 GHz with our model in which the merging rate is given by

$$E(L_g, z) = \begin{cases} E(L_g, 0)(1+z)^5 & \text{for } M_{B(0)} < -22.5 \\ E(L_g, 0) & \text{for } M_{B(0)} > -22.5 \\ 0 & \text{for } z > 2.5 \end{cases}$$

We have adopted $q_0 = \frac{1}{2}$ and $H_0 = 50$ km s^{-1}Mpc^{-1}. Note that only mild evolution is allowed for the smaller galaxies if we do not want to exceed the number counts at low radio flux. The break at about 2 Jy is determined by the Eddington limit we have imposed in the model.

The evolving luminosity function is shown in figure 2. It is interesting to note that (i) the maximum at high luminosities is determined by the (comoving) number density of the rapidly evolving population of massive galaxies, and (ii), we have almost pure luminosity evolution at high luminosities.

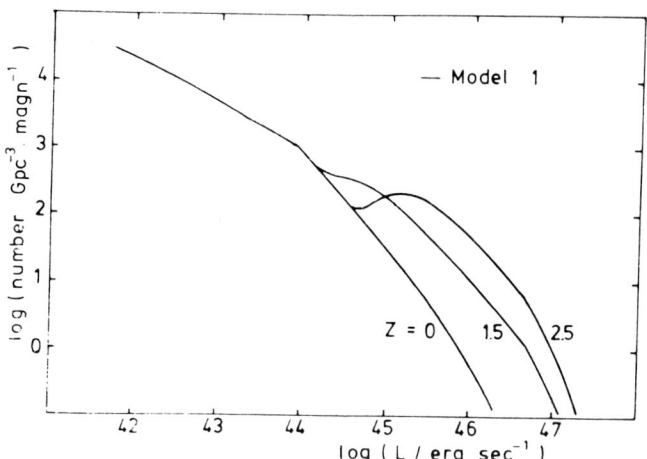

Figure 2 Redshift dependent (bolometric) luminosity function of AGN for the model described above.

We have applied the model also to the optical counts of AGN (Roos, 1985b). A reasonable fit to the high luminosity part of the local luminosity function is obtained using

$$M_{B(0)} \simeq -22.5 - 2.5 \log[\, L/10^{45} \text{erg s}^{-1}\,]$$

The optical counts are reproduced reasonably well using the same parameters as for the radio counts. The main difference is that we have to extent the population of rapidly evolving galaxies to a somewhat lower galactic luminosity or allow a weak evolution of the merging rate for the small galaxies.

The Kerkhoven-Bosscha Fonds is gratefully acknowledged for financial support of my visit to Trieste.

References.

Auriemma, C., Perola, G.C., Ekers, R., Fanti, R., Lari, C., Jaffe, W.J. and Ulrich, M.H. 1977, Astr. Ap. **57**, 41.
Dressler, A. 1980, Ap. J. **236**, 351.
Gott, J.R. and Turner, E.L. 1977, Ap. J. **216**, 357.
Peebles, P.J.E. 1980, *The Large Scale Structure of the Universe* , Princeton, University Press, New Jersey.
Roos, N. 1981, Astr. Ap **95**, 349.
Roos, N. 1985a, Ap. J. **294**, in press.
Roos, N. 1985b, Ap. J. **294**. in press.
Silk, J. 1986, this volume.
Windhorst, R.A., Miley, G.K., Owen, F.N., Kron, R.G. and Koo, D.C. 1985, Ap. J. in press.

LOW-LUMINOSITY ACTIVE NUCLEI IN NEARBY ELLIPTICAL GALAXIES

Elaine M. Sadler
European Southern Observatory
Karl-Schwarzschild-Str. 2
D-8046 Garching bei München
Federal Republic of Germany

ABSTRACT. The ability of an elliptical galaxy to produce a central radio source appears to be determined by both its luminosity and the gas supply in the inner ~ 1 kpc of the galaxy. At least half of all nearby ellipticals contain $10^3 - 10^5$ $M_\odot$ of ionised gas in their central regions (often in association with dust) and show weakly active nuclei characterised by a LINER-like spectrum and radio emission at a level of $10^{20} - 10^{24}$ W/Hz ($10^{37} - 10^{41}$ erg/s).

1. IONISED GAS IN ELLIPTICAL GALAXIES

Elliptical galaxies are in many ways the simplest large stellar systems, consisting (at least in the classical picture) only of a single, fairly homogeneous stellar population without a disk component or the large fraction of gas and young stars seen in spirals. Yet from these basic resources elliptical galaxies can produce powerful radio sources, and so their apparent simplicity makes them good candidates for the study of how active galactic nuclei form and evolve.

A recent survey of Hα/[NII] emission in early-type galaxies (Phillips et al. 1985) has shown that at least 50% of nearby ellipticals have $10^3 - 10^5$ $M_\odot$ of ionised gas in their central regions. The gas is usually confined to the central 1 kpc or less, and has a spectrum characteristic of LINERs (Heckman 1980). Where spatially resolved, the gas is generally in rapid rotation (V $\sim$ 100-300 km/s). There is strong circumstantial evidence, both from the emission-line kinematics and from the association of ionised gas with dust-lanes, which are also common in nearby ellipticals (Sadler and Gerhard 1985), that ionised gas in the centre of elliptical galaxies usually resides in a rotating, kpc-scale disk.

The fraction of galaxies with detected Hα/[NII] emission is independent of galaxy morphology, colour and axial ratio, suggesting that it is not difficult for any early-type galaxy to acquire $\sim 10^4$ $M_\odot$ of gas in the central kpc. Such an amount, if it can be transferred steadily to the innermost few pc, is sufficient to power a radio source of moderate size (Gunn 1979).

It has long been known that radio galaxies tend to show optical emission lines (Schmidt 1965, Disney and Cromwell 1971, Ekers and Ekers 1973, O'Connell and Dressel 1978), and scenarios based on the accretion of gas onto a central massive object (e.g. Rees 1978 and references therein) appear to provide the most plausible model for fuelling AGN, so it is interesting to see whether the presence of a gas supply is in itself a sufficient condition for the formation of a central radio source in an elliptical galaxy.

2. (LOW-LUMINOSITY) RADIO SOURCES

Figure 1 shows the relationship between radio power and absolute magnitude (using H_o = 100 km/s/Mpc) for a magnitude-limited sample of 46 elliptical galaxies observed at 5.0 GHz (6 cm) with the VLA (Sadler, Kotanyi and Jenkins 1985). Twenty-four galaxies (52%) were detected above a level of 0.7 mJy. It is clear from this diagram that there is a correlation of radio power with absolute magnitude, in the sense that the strongest radio sources are found in the brightest galaxies, but there is also a very large spread (up to 3-4 orders of magnitude) in the radio powers of galaxies with similar optical luminosity. The solid line shows how a point at the detection limit of the survey moves with distance; the slope of the radio luminosity - optical luminosity upper envelope is much steeper than this, and so the relationship between radio and optical luminosity is not an artifact of distance effects in a magnitude- and flux-limited sample. Radio sources are both more powerful and more frequent in bright ellipticals than in fainter ones (this is also well-known from earlier radio surveys, e.g. Ekers and Ekers 1973, Auriemma et al. 1977). The large scatter in the observed radio powers suggests that, while the total luminosity of a galaxy limits the maximum power a radio source can achieve, some other parameter determines whether or not a particular galaxy lives up to its potential as a radio emitter.

Thirty-six of the galaxies in figure 1 were included in the Phillips et al. survey, and of these, 21 (58%) show emission lines. If we consider only galaxies brighter than absolute magnitude -19.0 (since fainter galaxies are not well-represented in the sample, and rarely have radio sources), 17/20 (85%) of emission-line galaxies and only 2/10 (20%) of galaxies without emission lines are detected at the VLA. No galaxy with radio power below 10^{20} W/Hz has emission lines, while 78% of those with radio power between $10^{20.5}$ and 10^{22} W/Hz and all galaxies with radio power above 10^{22} W/Hz have emission lines. Since the detection rate of emission lines in the Phillips et al. sample is essentially independent of absolute magnitude, this is not a selection effect related to luminosity but reflects a real difference in the radio properties of elliptical galaxies with and without gas. The difference is significant at a level of about 3σ (5%) in the small sample presented here.

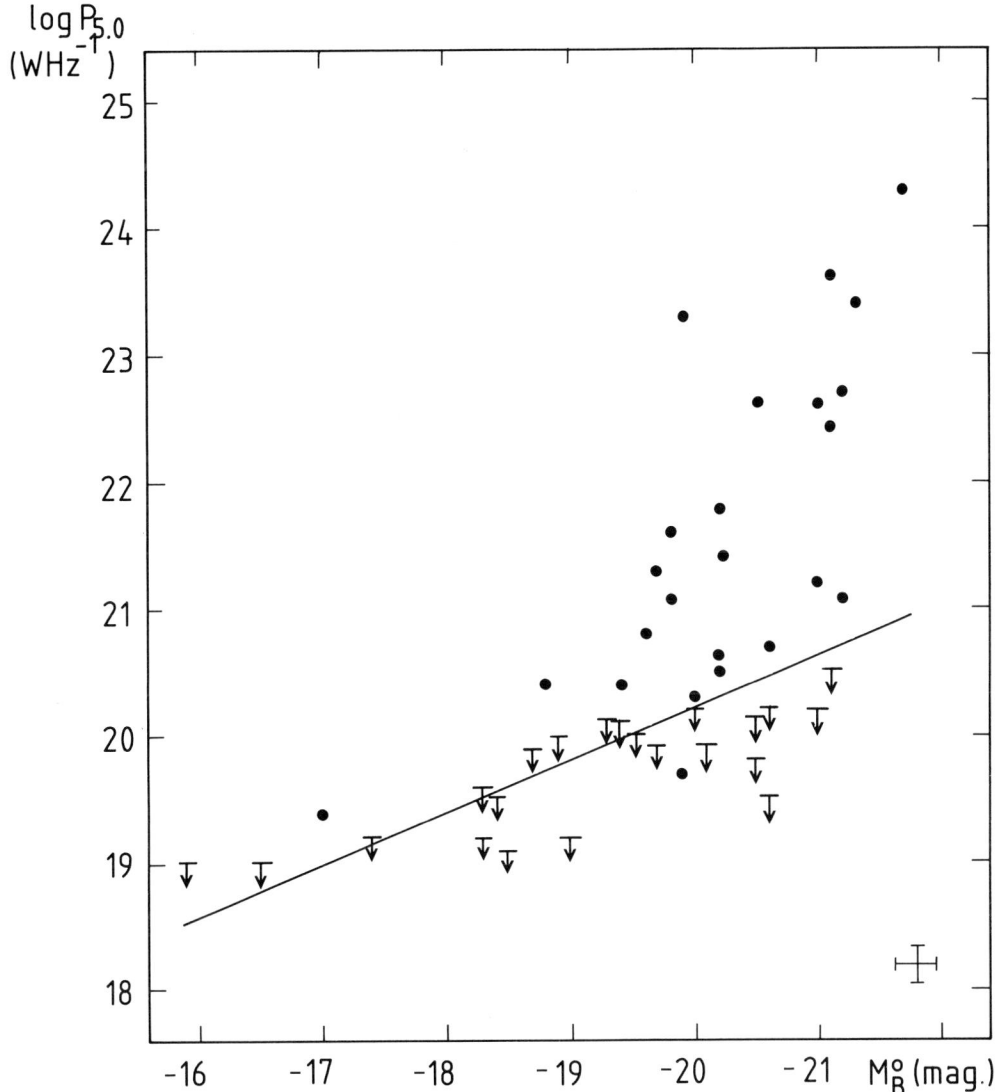

Figure 1: Relationship between absolute magnitude M_B^o and radio power $P_{5.0}$ for a magnitude-limited sample of elliptical galaxies.

3. CONCLUSION

These results suggest that the presence of ionised gas in the centre of an elliptical galaxy (and it seems to be the presence, rather than the amount which is important) is a sufficient condition for the formation of a radio source in an elliptical galaxy brighter than absolute magnitude -19.0 (-5 log h where H_o = 100 h km/s/Mpc). Weak radio emission is nearly ubiquitous in emission-line ellipticals, and since the presence of emission lines does not correlate with other optical properties of the galaxy it is presumably an important parameter in its own right. To the long-standing statement that most radio ellipticals have emission lines, it is now possible to add that most (all?) emission-line ellipticals also have radio sources. Both these statements can be understood naturally in the context of a massive central object fuelled by infalling gas.

The situation for S0 galaxies is less clear. They have fewer, and weaker, radio sources than elliptical galaxies with similar gas content, even when their lower luminosity is accounted for, and there is no obvious correlation between gas content and radio emission. Perhaps S0 galaxies are less able to take advantage of their nuclear gas supply, either because their past history has not led to the formation of a massive central object or because the method of transferring gas from the kpc disk to the central object is less efficient in S0s than in ellipticals.

To summarise, although elliptical galaxies are generally thought of as gas-poor systems, a majority probably contain 10^3 - 10^5 $M_\odot$ of gas in their central regions. Ellipticals with such a gas supply usually have weakly active nuclei, with a LINER emission spectrum and a central radio source of 10^{37} - 10^{41} erg/s. Low-luminosity AGN are therefore a very common feature of nearby elliptical galaxies. Whether they are recent phenomena or the semi-quiescent remnants of former quasars remains an open question.

REFERENCES

Auriemma, C., Perola, G.C., Ekers, R., Fanti, R., Lari, C., Jaffe, W.J. and Ulrich, M.-H. (1977) Astron. Astrophys. 57, 41.
Disney, M.J. and Cromwell, R.H. (1971) Astrophys. J. 164, L35.
Ekers, R.D. and Ekers, J.A. (1973) Astron. Astrophys. 24, 247.
Gunn, J.E. (1979) in "Active Galactic Nuclei" (ed. C. Hazard and S. Mitton). Cambridge: Cambridge University Press, p. 213.
Heckman, T.M. (1980) Astron. Astrophys. 87, 152.
O'Connell, R.W. and Dressel, L.L. (1978) Nature 276, 374.
Phillips, M.M., Jenkins, C.R., Dopita, M.A., Sadler, E.M. and Binette, L. (1985) in preparation.
Rees, M.J. (1978) Physica Scripta 17, 193.
Sadler, E.M. and Gerhard, O.E. (1985) Mon. Not. R. astr. Soc., in press.
Sadler, E.M., Kotanyi, C.G. and Jenkins, C.R. (1985) in preparation.
Schmidt, M. (1965) Astrophys. J. 141, 1.

KINEMATICS IN THE NARROW LINE REGION OF NGC 4151 - EVIDENCE FOR A MERGER ?

Hartmut Schulz
Astronomisches Institut der Ruhr-Universität
Postfach 102148
D-4630 Bochum
Federal Republic of Germany

ABSTRACT. The velocity structure in the narrow line region of the Seyfert galaxy NGC 4151 is studied by means of various long-slit coudé spectra. Spectral line profiles in the NE and SW quadrants display a 2-component structure while those in the NW and SE are narrower and smoother. One component fits to the general rotation pattern of the disk of NGC 4151. The other component may refer to a bipolar radial flow or to gas spiraling inwards within a plane which is perpendicular to the main galactic disk. It is suggested that merging gaseous disks are a common feature of narrow line regions in Seyferts.

1. INTRODUCTION

In general, line profiles from the narrow line region (NLR) of active galaxies display a great variety of profile shapes. This has made it difficult to accept a simple general form of the velocity field of the NLR. More observational insight may be gained from those nearby-Seyferts in which the NLRs can in part be spatially resolved. Ulrich (1973) demonstrated this in a pioneering paper on NGC 4151.
    For this paper we use a new set of long-slit coudé spectra taken at various position angles (PA) and with different plate factors. The spectra were collected at Calar Alto Observatory and have a kinematical resolution between 20 and 40 km/s. The useful spatial resolution is 2 arc sec. The data will be published elsewhere in full detail. It is the purpose of this report to present only the most salient results.

2. OBSERVATIONS

[OIII] $\lambda 5007$ line profiles from a spectrum taken with the slit at PA $35°$ are shown in Fig.1. The basic features are typical for many spectra at PA between $25°$ and $50°$. Inside 4 arc sec

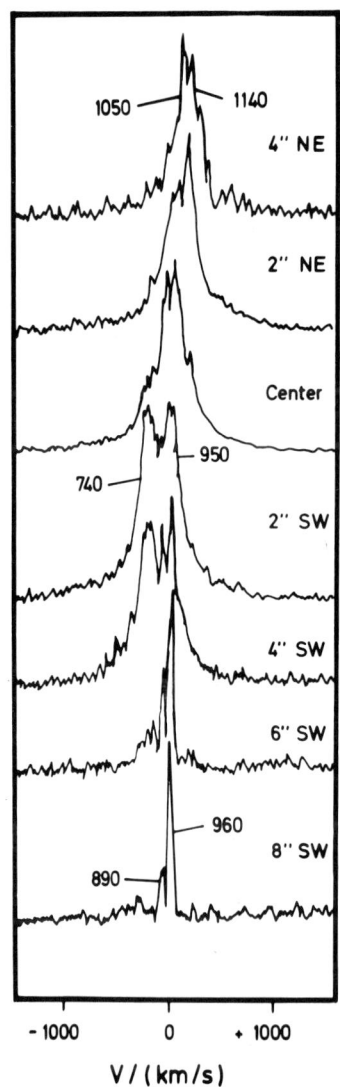

Figure 1. Line profiles of [OIII] λ5007 at PA 35°. Profile components are marked by their typical heliocentric velocities.

from the nucleus (center) the lines are redshifted in the northeast (NE) and blueshifted in the southwest (SW) relative to the nuclear line profile. Essentially no or only small relative shifts are found when the spectrograph slit points to directions nearly perpendicular to these. These findings suggest rotation as the cause of the motions because the major axis of NGC 4151 agrees with the maximum velocity gradient in the NLR. A similar conclusion had already been drawn

by Anderson (1974) on the basis of Ulrich's (1973) data.
However, in light of the accumulated observations since then we suggest a somewhat different interpretation. First we note a significant component structure in the line profiles of Fig.1 which can also been seen in other NE-SW-spectra. Due to the great width of the lines we only give typical approximate ($\pm$ 20 km/s) heliocentric velocities for these components: NE 1050 and 1140 km/s, SW 950 and 740 km/s. Typical FWHM of the components are between 150 and 200 km/s.

In addition, there are two very narrow components (SW 850 and 960 km/s) which are superposed on the line at 4" SW and which appear isolated at 6" and 8" SW. The 850-component is only seen in the spectrum shown in Fig.1 while the 960-component is also visible at other slit-PA pointing NE-SW.

3. INTERPRETATION

The NE-1050 and the SW-950 and SW-960 components have probably to be attributed to normal galactic rotation within the main plane of NGC 4151 due to their consistency with the 21-cm velocity map from Bosma et al. (1977).

The NE-1140 and SW-740 components then appear unlikely to be explained by any motion within the main disk. For example, radial motions would require velocities far in excess of 1000 km/s to be seen that close to the major axis. Moreover, there is no evidence of any radial motion along the direction of the minor axis which would be most sensitive for these. Two other possibilities arise:
1 - There is radial flow - but not confined to the disk. It is clearly possible to devise some kind of bipolar flow to explain the components. In general, radial outflow seems to be favored by several authors to explain spatially unresolved NLR spectra (see e.g. Vrtilek 1983). However, due to the rather sparse evidence for radial motions another possibility also seems attractive:
2 - Gas is spiraling inwards within a tilted plane. Tilted disks have indeed been proposed on theoretical grounds by Tohline and Osterbrock (1982) as an explanation for broad line profiles. References therein, theoretical work by Tubbs (1980) on NGC 5128 and observations of elliptical galaxies (Möllenhoff and Marenbach 1984) show that preferential orbital planes for gaseous disks exist in spheroidal or triaxial stellar systems like elliptical galaxies, prolate spiral bulges or bars.

It is indeed striking that the NLR (or inner oval: see Schulz 1985, his Fig.1) is aligned roughly perpendicular to the major axis of the bar of NGC 4151. This agrees nicely with the predictions of Tohline and Osterbrock (1982).

Thus we suggest that the NE-1140 and SW-740 components may be due to a merging gaseous disk which is just settling in the equatorial plane of the bar at PA $50°$. Since on the sky the main disk of the galaxy appears nearly face-on ($i \lesssim 21°$) we observe the merging disk roughly edge-on. The velocities of the NE-1140 and SW-740 components are indeed typical for gas orbiting with about 200 km/s although their mean is blueshifted relative to the systemic velocity V_s = 999 ± 6 km/s (Schulz 1985).

If this scenario is valid for Seyferts in general it could explain the preference of activity in early type spirals (e.g. galaxies with large bulges) and ellipticals. A study is underway to check whether resolvable NLRs of other galaxies fit into this picture. If true - all Seyferts may be in a late stage of cannibalizing a gas rich dwarf galaxy. The author acknowledges technical support by J.Sis and P. Froelich.

REFERENCES

Anderson,K.S.: 1974, Astrophys.J. 187, 445
Bosma,A., Ekers,R.D., Lequeux,J.: 1977, Astron.Astrophys. 57, 97
Möllenhoff,C., Marenbach,G.: 1984, Mitt.Astr.Ges. 62, 340
Schulz,H.: 1985, Astron.Astrophys. 143, 29
Tohline,J., Osterbrock,D.E.: 1982, Astrophys.J.Lett. 252,L49
Tubbs,A.D.: 1980, Astrophys.J. 241, 969
Ulrich,M.H.: 1973, Astrophys.J. 181, 51
Vrtilek,J.: 1983, Ph.D. thesis, Harvard

T COR B, A "SEYFERT 1 NUCLEUS" AT 1200±100 PARSECS?*

P. L. Selvelli[1], J. Clavel[2], A. Cassatella[2], M. Hack[1]
[1] Astronomical Observatory of Trieste
via Tiepolo 11 - Trieste, Italy
[2] IUE Observatory, ESA**
Apartado 54065 - Madrid, Spain

ABSTRACT. Several symbiotic stars are characterized by UV spectral signatures which are similar, in several aspects, to those observed in the broad-line region of Seyfert galaxies. We suggest to consider the symbiotic stars as a convenient laboratory for the study of the spectroscopic mechanisms which are at work in Seyfert 1 nuclei.

The UV spectral behavior of T Cor B (M3 III + ?) and other symbiotic stars is comparable, in several aspects, to that shown by Seyfert 1 nuclei. Its line spectrum is characterized by the same permitted and semi-forbidden emissions observed in Seyfert 1 nuclei. The strongest emissions of T Cor B are listed in Table 1. The unidentified λ 1594 emissions is present also in the spectrum of NGC 4151 (Fig. 1).

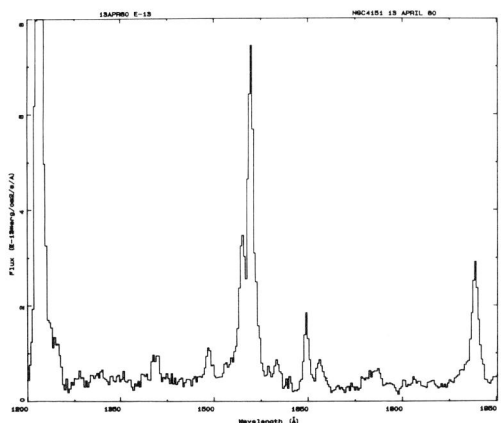

Figure 1. The far UV spectrum of NGC 4151.

* Based on observations by the International Ultraviolet Explorer (IUE) collected at the Villafranca Satellite Tracking Station (VILSPA) of the European Space Agency.
**Affiliated to the Astrophysics Division, Space Sciences Department.

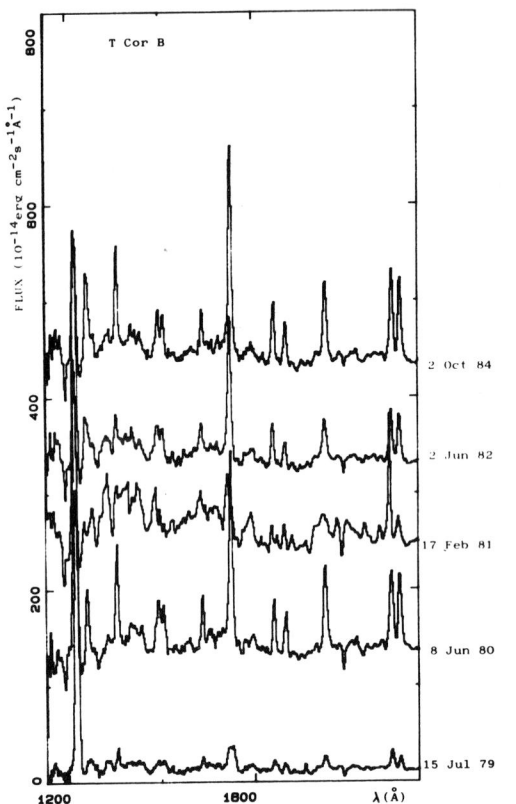

λ(A)	Identification
1240	NV (1)
1285	?
1304	OI (2)
1335	CII (1)
1355	OI (1)
1400	SiIV (1)(+OIV)
1485	NIV
1530	?
1550	CIV (1)
1594	?
1640	HeII (12)
1665	OIII
1750	NIII
1892	SiIII (1)
1908	CIII (0.01)
2330	CII + SiII
2670	AlII (1)
2735	HeII
2800	MgII (1)
2835	OIII (Bowen)
3133	OIII (Bowen)
3188	HeI

Table 1. The Emission Spectrum of T Cor B

Figure 2. UV spectra of T Cor B at different epochs. Successive spectra are offset by 100.

The UV continuum of T Cor B from 1200 to 2000 Å is variable by a factor of up to 10 (Fig. 2) while in the optical there are very small changes. The shape of the continuum over this wavelength range can be represented by a power law $A\lambda^{-\alpha}$ with α ranging from 0.2 to 1.0. When the total flux increases the continuum becomes harder.

Figure 3 clearly shows that in T Cor B the emission-line intensity is also variable by a factor of up to 8 and is positively correlated with the changes in the total flux. This indicates that photoionization is the main energy input mechanism. As a general trend, the excitation also increases when the flux increases.

The ratio SiIII λ1892/CIII λ1908, calculated as a function of n_e, gives log n_e 10.0, a value comparable with the canonical density in the BLR of Seyfert 1 galaxies. A similar value is obtained also from the relative intensities ratio within the NIIIλ1750 multiplet.

The simultaneous presence of ions with a wide range of ionization (from OI to NV) indicates a complex structures of the emitting regions and stratification of the ionization conditions, as found in the best-studied Seyfert 1 galaxies.

The spectral behaviour of T Cor B (lines and continuum) can

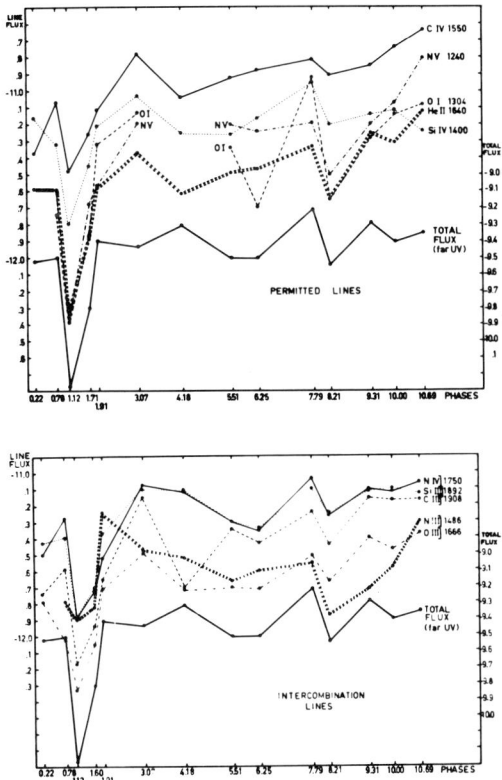

Figure 3. Line and continuum fluxes of T Cor B as a function of the orbital phase. Phase zero was assumed at J.D. 2243700.4.

be interpreted in terms of non-steady-state accretion processes where material is episodically transferred from the Roche-lobe filling giant onto a compact object.

The sole emissions which are outstanding in the near UV and optical spectra of Seyfert' 1, but not clearly evident in T Cor B, are those of FeII. These lines are, however, profusely represented in a recent (23 Jan., 1985) high resolution IUE spectrum of the symbiotic star CH Cyg (M6 III + ?). In previous spectra, resonance fluorescence seemed the excitation mechanism responsible for the FeII emission. In the most recent ones, instead, this mechanism cannot be invoked because there are only pure emissions over the whole IUE spectral range and continuum is very weak. Certainly, another mechanism is at work, though not yet clearly defined (collisional excitation?, charge-exchange reactions?, autoionization of FeI and subsequent decay?). This situation also resembles, in some way, that related to the much debated issue of the excitation mechanism of FeII in Seyfert 1 galaxies and quasars. The strongest FeII emissions in CH Cyg are those of mult. 191, $\lambda 1784$ (Fig. 4), but also other

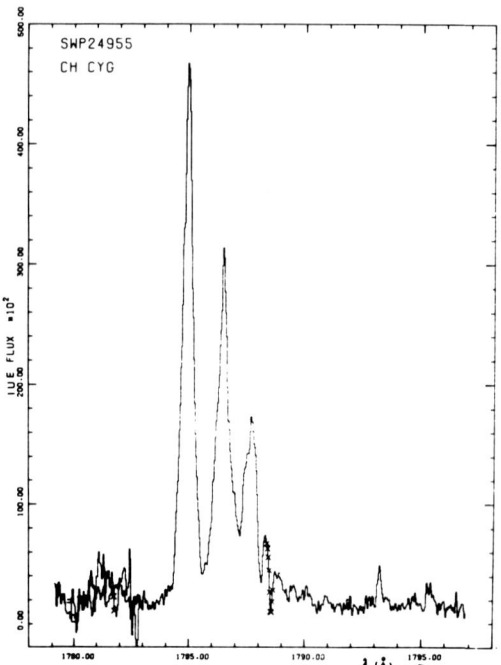

Figure 4. The region of FeII mult. 191 UV in CH Cyg.

multiplets, e.g. 60, 62, 63, 64, 78, etc., to which are due the very broad emission features of FeII in Seyfert galaxies and quasars, are strong in emission. As in Seyfert 1 galaxies, optical depth effects play an important rule in symbiotic stars as well, as can be inferred for CH Cyg from the relative intensities of FeII multiplets with common upper levels, e.g. opt. 42 and UV 3, opt. 6 and UV 36, opt. 2 and UV 2, etc.

From the above-reported analogies and similarities between the behavior of Seyfert 1 galaxies and that of T Cor B and other symbiotic stars, we suggest considering the relatively bright symbiotic stars as a convenient galactic laboratory for the study of the more elusive atomic processes and energy input mechanisms which are at work in the broad-line region of Seyfert galaxies.

ACKNOWLEDGEMENT

We want to thank Miss L. Abrami and Mr. M. Quartana for their help in the preparation of the manuscript.

INTERACTION AND EMISSION MORPHOLOGY

N. A. Sharp
Kitt Peak National Observatory
National Optical Astronomy Observatories[1]
P.O. Box 26732
Tucson, AZ 85726
U.S.A.

ABSTRACT. A simple comparison between broad V-band images and narrow band images tuned to the redshifted wavelength of Hα enables us to make a quick survey of the distribution of emission in galaxies. The appearance of interacting systems can then be contrasted with that of isolated objects. Those systems for which data are presently available seem to split into three classes: (i) nothing happening, despite apparent connections, (ii) emission distributed either smoothly or lumpily throughout, (iii) a central peak of emission in an otherwise smooth profile, for one or more of the galaxies. If this picture is confirmed by further observations, the next stage will be to try to distinguish the dynamical structures and possible gas flow patterns separating the different categories.

1. INTRODUCTION

One of the most interesting places where large scale star formation seems to be occurring is in systems of interacting galaxies. It has been known for some time that such galaxies are bluer than are isolated galaxies, and that they have correlated types and colour indices (the 'Holmberg effect', named after the early discussion by Holmberg, 1958). It was logical to interpret the data in terms of star formation, and this explanation was placed most firmly on a quantitative footing by Larson & Tinsley (1978). Many of the apparent peculiarities of double galaxies have been attributed to star formation induced by the interaction (see, e.g., Sharp & Jones, 1980; Keel et al, 1985). Such studies can only concentrate on systems which appear to be interacting. One difficulty, made clear by the use of N-body 'experiments' and by attempts to reproduce the morphology of observed systems, is the prevalence of 'bridges' and 'tails' which do not connect the galaxies, but which are easily viewed from an orientation which projects a connection from quite widely separated features.

Since young stars are hot, they ionise their surroundings, thus creating strong emission lines, particularly Hα. In fact, careful

[1] Operated by A.U.R.A., Inc. under contract with the N.S.F.

spectrophotometry would enable an estimate to be made of the actual mass in young stars, given an initial mass function. However, it is more important for the present to be able to produce a rapid survey which can distinguish truly interacting (and therefore, subjectively, interesting) systems from sham interactors. The method chosen is to compare the appearance as seen through a narrow-band Hα filter, tuned to the velocity of the system, with a normal broad-band image. Initially, the systems studied were chosen to include at least one elliptical galaxy, and isolated ellipticals were observed for comparison. This is because spirals are full of HII regions and other bothersome disturbances, emitting obfuscatory Hα.

2. OBSERVATIONS AND METHODS

The data discussed here were all obtained with the 2D Photon Counting Array developed by Mt. Stromlo Observatory, used at the f18 Cassegrain focus of the Siding Spring Observatory 1m telescope. The southern systems thus studied will be augmented by CCD observations at the Kitt Peak National Observatory No.1 0.9m telescope when I can get some clear weather. Images were obtained in the V-band and through one of the available set of filters of width about 100Å which was chosen to include the redshifted wavelength of Hα as close as possible to the filter's central transmission.

The images were compared by considering the sky-normalised, sky-subtracted frames. Define g_H, s_H and g_V, s_V to be the flux densities from the galaxy and from the sky, at Hα and V, respectively. After background fitting and subtraction, the available frames are $r_H = g_H/s_H$ and $r_V = g_V/s_V$. Let $s_H = As_V$, where A is a constant which we can determine from the known sky colour. Assume that $g_H = g_V + Bg_V + h$, where B is a function of position which allows for variations in the colour of the stellar population, and h is the emission line flux. Then $h/As_V = r_H - Gr_V$ gives, in slightly odd units, the excess emission due to Hα. From values of V-R for various galaxies ranging around 0.8 to 1.0, the flux ratio $1+B \simeq 0.95-1.14$. For V-R about 0.5 for the Sun (i.e. for the sky), the flux ratio $A \simeq 0.7-0.8$. Therefore $G = (1+B)/A \simeq 1.3$ to 1.6. Without knowing specifically the colour gradients in the galaxies, we cannot know the spatial variation of G. We therefore choose the 'alpha colour' image $r_H - 1.4r_V$, and consider the appearance of the results as contrasted to comparison ordinary galaxies. The essential difference between colour variations and emission regions is the smoothness of the former as compared to the latter.

3. RESULTS

About 20 systems have so far been studied, ranging from isolated ellipticals, through peculiar single galaxies, to interacting doubles and triples. In the 'alpha colour' image, isolated ellipticals have smooth profiles going down from the (defined) zero sky level to a central minimum (most negative point). Even such unusual objects as the spindle

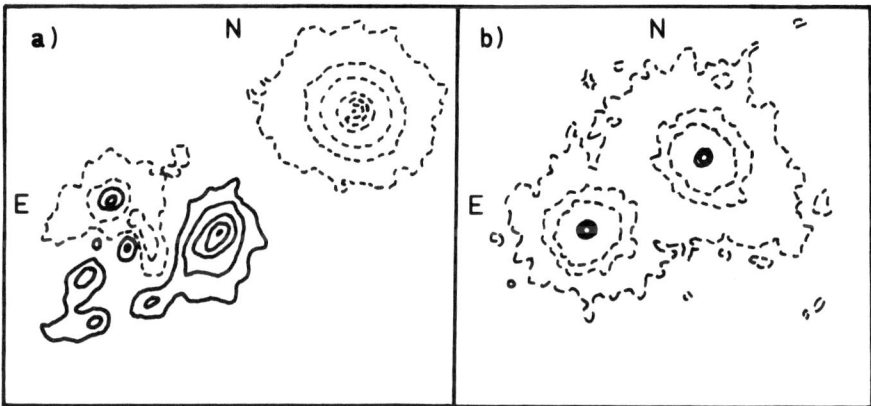

Figure 1. Difference images formed by subtracting a multiple of the V-band frame from the Hα frame. Negative contours are dashed. No activity would correspond to a smooth negative profile dropping to a central minimum.
a) Arp 140. (0^h49^m-7°20') Note the strong, clumpy emission in the spiral component (SE), and the central spike of emission in the elliptical (NW), where the contours turn back.
b) Arp 308. (1^h23^m-1°36') Both galaxies are ellipticals, and both show otherwise smooth contours with very strong central emission spikes.

galaxy NGC4650A (Schechter et al, 1984) show smooth profiles reaching about as low a minimum as do isolated ellipticals. The interacting systems so far seem to divide into three distinct classes: (i) no excess emission, despite close neighbours and apparent bridges and common envelopes. An example is IC2200/IC2200A. (ii) distributed emission, sometimes strong, sometimes weak, sometimes smooth, sometimes lumpy. A particularly strong example is Arp 91 (see Jenkins, 1984). (iii) a strong central spike of emission in an otherwise smooth profile. This last category is not surprising, given the arguments for nuclear activity in interacting systems (e.g. Keel et al, 1985), but this seems to be the first clear morphological support for emission being concentrated at the nucleus. Two examples of this third class are shown in Figure 1, which presents contour plots of the 'alpha colour'. This result is especially interesting, and confirms that gas must be capable of being deposited down into the centre of a galaxy, perhaps in a way induced by interaction.

It is a truism to say that more work is required. Further data are available, but not yet reduced, and more data will, I hope, be obtained soon. It would be particularly useful to be able to perform numerical simulations including both stars and a correct hydrodynamical treatment of the gas.

ACKNOWLEDGMENTS

This project was begun with C.R.Jenkins, who first calculated the image comparison formulae. I am grateful to him, and to W.C.Keel, for many conversations, to M.S.S.S.O. for telescope time, and to K.P.N.O. for time so far lost to weather. Financial support from the Australian National University, and from K.P.N.O./N.O.A.O., has been most welcome.

REFERENCES

Holmberg, E. 1958. Medd.Lunds Astron.Obs.Ser II, No.136, 103.
Jenkins, C.R. 1984. Astrophys.J., 277, 501.
Keel, W.C., Kennicutt, R.C., Hummel, E. & van der Hulst, J.M. 1985. Astron.J., in press.
Larson, R.B. & Tinsley, B.M. 1978. Astrophys.J., 219, 46.
Schechter, P.L., Ulrich, M.-H. & Boksenberg, A. 1984. Astrophys.J., 277, 526.
Sharp, N. & Jones, B. 1980. Nature, 283, 275.

FLOPPY DISCS: A RECIPE FOR THE "OBSERVERS' DREAM" MODEL

Michael D. Smith & Derek J. Raine
Department of Astronomy
University of Leicester
Leicester LE1 7RH
United Kingdom

ABSTRACT. Compton-heating of a disc's photosphere by a hard central X-ray source leads to a back-pressure on the disc surface. This pressure dominates low mass disc (LMD) systems, leading to disc asymmetries and, as a result, large asymmetries in the corona/wind in the nuclear environment. Present knowledge of accretion discs (Rees, these proceedings) indicate that LMD systems may exist in radio galaxies. We propose that the environment controls jet visibility, thus producing 'one-sided' jet structures. Bridle's (1984) "Menu for an All-Purpose Source Model" can then be interpreted.

1. INTRODUCTION

Asymmetries occur in galactic nuclei in a number of contexts. Asymmetric emission lines can be adequately explained by various models involving spherically-symmetric distributions of clouds and dust. Asymmetric radio jets cannot be similarly interpreted with a symmetric configuration (e.g. Bridle 1984) even though relativistic Doppler boosting is expected to play a role in many objects. This leaves two possibilities for intrinsic jet asymmetries:
(a) the twin-jets are physically different from their point of creation;
(b) a reaction to an asymmetric environment occurs on scales < 1 pc..
Here we determine the conditions sufficient to produce a large environmental asymmetry. We shall also consider the effect on jet brightness and attempt to explain the diverse behaviour of extragalactic radio sources as summarised by Bridle (1984).

2. QUALITATIVE OVERVIEW

Following on from the work of Shakura and Sunyaev (1973), Begelman, McKee and Shields (1983, BMS) have shown how a nuclear environment is influenced when material is evaporated off an accretion disc. The heating is provided by a central hard X-ray source. Smith and Raine

(1985) have considered the consequences of a very powerful nucleus
in which strong winds arise. The accretion disc, presumed to stoke
the central engine, is massive. As a result, gravity maintains overall
symmetry (BMS, Appendix) of the disc, corona and wind.

In radio galaxies and radio-loud quasars the accretion disc
may not power the nucleus (Rees, these proceedings). Therefore, the
disc is susceptible to distortion and deviation from the symmetry
plane. A significant asymmetry can be maintained only if the central
source maintains a high pressure gradient across the disc. Moreover,
the disc shape must be such that the pressure gradient pushes material
away from the symmetry plane (balancing the gravitational component
in a steady state solution). Such a configuration, found in the following section, produces naturally a hot extended corona on one-side
of the disc and a low density environment above a relatively cool
photosphere on the other side.

Two identical jets, but oppositely directed, are produced close
to the central engine. One jet enters the low-pressure side. It expands
and attains a high Mach number. This is probably not sufficient for
the jet to avoid some disruption. However, the growth length of fluid
dynamic instabilities will be large. Therefore the internal shocks
which accelerate relativistic particles and keep a jet visible (Norman,
Winkler and Smarr 1984) are unlikely to be present. (Note: we are here
assuming hydrodynamic behaviour for the jet flow).

The jet's expansion rate and Mach number are kept low on the
high pressure side; instabilities grow rapidly. They are not disruptive however (Norman, Winkler and Smarr 1984), but simply transfer
a fraction of the bulk kynetic energy into internal energy via shocks.
In turn, the shocks maintain the low Mach number, promoting the growth
of surface disturbances. This jet may thus become visible.

On larger scales, outside the nucleus, the environment is symmetric:
the radio galaxy properties determined by the bulk motion of the jets'
material are basically identical. Properties determined by the internal
jet structure are not identical since the shocks redistribute the
internal energy across the jet. This should lead to asymmetric hot spot
structures in otherwise symmetric double lobes. Further, the shocks
which propagate down the length of one jet with a fraction of the
jet speed can maintain a high radio surface brightness. Associated
shocks in the ambient medium (Norman, Winkler and Smarr 1984) can
produce optical filamentary structures by initiating thermal instabilities.

Other requirements on an "All-Purpose Source" source model concern
timescale and 'avoidance' behaviour. These depend on the details
of the asymmetric disc which we will now proceed to determine.

3. DISC CHARACTERISTICS

3.1. The Symmetric Disc

Firstly, presume an α-disc in which the central (mid-plane) pressure,
p_c, greatly exceeds the pressure of the surrounding medium, p_s. Then

$$p_c \sim (\dot{M}_a v_\phi^2)/(4\pi^{3/2}\alpha R^2 c_s) \qquad (3.1)$$

using standard disc formulae (Smith and Raine, in preparation), where $\dot{M}_a$ is the mass accretion rate through a radius R (assume a cylindrical coordinate system), v_ϕ is the velocity of rotation $\sim (GM/R)^{\frac{1}{2}}$, c_s is the isothermal sound speed and α is a constant which determines the radial infall velocity through turbulent viscosity: $v_R = \alpha c_s H/R$, where H is the disc scale height given by $H = 2^{\frac{1}{2}}(c_s/v_\phi)R$ (Shakura and Sunyaev 1973). For an opaque disc in thermodynamic equilibrium, and which radiates the dissipated energy locally, we can write $\sigma T^4 \sim \dot{M}_a v_\phi^2/R^2$. This yields

$$c_s \sim 3.10^6 \left(\frac{\dot{M}_a}{1 M_\odot yr^{-1}}\right)^{1/8} \left(\frac{M}{10^8 M_\odot}\right)^{1/8} \left(\frac{10^{18} cm}{R}\right)^{3/8} \text{ cm s}^{-1} \qquad (3.2)$$

where M is the central mass (Note: self-gravity of the disc is negligible in the parameter range discussed below).

A hard X-ray source of luminosity L will heat the outer layers of a flaring disc, forming a corona or wind. The thermal pressure on the disc surface is (BMS)

$$p_s = (L f_o f_1)/(4\pi R^2 \Xi_o' c) \qquad (3.3)$$

where Ξ_o' is the ionisation parameter at the base of the corona and f_o and f_1 represent the degree of X-ray attenuation due to an inner corona and local scattering, respectively. The factor $f_1 \sim 1$ in this case and $f_o > 0.1$ (BMS 1983b). The pressure ratio can be written

$$\frac{p_s}{p_c} = \dot{m}^{-1} \left(\frac{c_s c}{v_\phi^2}\right)\left(\frac{\pi^{\frac{1}{2}}\alpha f_o f_1}{\Xi_o'}\right) \qquad (3.4)$$

where

$$\dot{m} \equiv \dot{M}_a c^2/L. \qquad (3.5)$$

BMS took $\dot{m} \sim 10$ (radiative efficiency of accreting material ~ 0.1) to parameterise the ratio. They found $p_s \ll p_c$ unless R is extremely large (v_ϕ small), where, however, the disc is expected to be optically thin. This appears to imply that a disc remains symmetric and, moreover, the surface pressure has a negligible effect on the location of the flaring surface.

For radio galaxies, however, two factors exist through which equation (3.4) should be adjusted. Firstly, radio galaxies have central luminosities well below the Eddington limit. The mass loss from the disc through a wind, $\dot{M}_w(R < R_o) \propto R$, has then a significant radial dependence. Adjusting $\dot{M}_a$ to account for such a wind, can result in an increase in p_s/p_c as R decreases. However, it can also be shown that $p_s/p_c \propto R^{-\frac{1}{4}}$ is the limiting case in which only a tiny fraction of the accreting material actually reaches the central engine. Therefore we do not expect p_s/p_c to greatly increase and the effect is unimportant except to some specific objects.

Secondly, the nuclei of radio galaxies may be starved of accreting gas (section 2). Then, the disc thickness $H_s < H$ if $p_s > p_c$ i.e. the disc is a slab of approximately constant vertical pressure. Taking $v_R = \alpha c_s H_s/R$ (and using the surface density $\Sigma \sim p_s H_s/c_s^2$) yields

$$H_s^2 \sim \dot{M}_a c_s/(\alpha p_s). \tag{3.6}$$

When $p_s \propto R^{-2}$ and $c_s \propto R^{-3/8}$, this implies that the disc is not flaring. However, the high surface pressure presumes flaring (unless the central source is extended).

We conclude that, approximately, $H_s \propto R$ so that direct radiation approaches at a very low grazing angle to the surface. The local attenuation factor then adjusts p_s. Obviously, f_1 is very sensitive to small changes in H_s/R. For a symmetric disc (3.6) requires

$$f_1 = f_1(R=R_o)(R_o/R)^{3/8} \tag{3.7}$$

Since f_1 decreases with R, the mass lost from the disc is greatly reduced. This is self-consistent with the assumption that an accretion disc actually exists.

3.2. The Asymmetric Disc

A small perturbation to the symmetric disc produces large changes in f_1 and p_s. Moreover, for disturbances of the form $\delta z \propto R^{-n}$ with $n > 0$ the pressure increases on the side perturbed towards the midplane. Therefore, the disc rapidly rises in the inner region until the gravitational component GMz/R^3 balances the pressure gradient. Steady flow solutions in which this equilibrium occurs can be shown to maintain the same form but with different attenuation factors on the top, f_t, and bottom, f_b, of the disc:

$$\ln(f_b/f_t) = (2HH_o v_\phi^2)/(R^2 c_s^2) \tag{3.8}$$

where $z = H_o \propto R$ is the surface closer to the midplane (i.e. the bottom). It follows that

$$f_t \exp(R^{-\frac{1}{4}}) \propto f_b \propto R^{-3/8}. \tag{3.9}$$

We note that $H_o \ll R$ i.e. the <u>disc</u> asymmetry is small.

In the inner region of such a disc the pressure difference across the disc is large. It follows from BMS that the hot extended corona will be highly asymmetric about the midplane. This solution presumes the disc to be optically thick. The vertical optical depth to electron scattering (to ensure that hard X-rays are absorbed near the surface) can be written

$$\tau_{es} = \Sigma \sigma_T/\mu_e \sim \dot{m}(\frac{L}{L_E})(\frac{v_\phi^4}{cc_s^3})^{\frac{1}{2}}(\frac{f_1}{\pi \Xi_o \alpha})^{\frac{1}{2}}, \tag{3.10}$$

on using the above and standard formulae. The optical depth to direct radiation, τ_H, will be a factor $> R/H$ higher. We find

$$\tau_H > (f_1/2\pi\Xi_o)(v_\phi^2/c_s^2)(L/L_E). \tag{3.11}$$

Therefore the optical depth can be large even for $L \ll L_E$, as can the ratio p_s/p_c (Smith and Raine, in preparation).

4. CONCLUSIONS

The asymmetry can be maintained so long as material is continuously injected into the disc at large radii. Very long timescales are thus possible, consistent with observations (Bridle 1984). Moreover, a reformed disc can be asymmetric in either direction. Therefore the model predicts avoidance behaviour rather than flip-flop behaviour. This is consistent with the preferential avoidance of structures in radio galaxies in the sample studied by Rudnick and Edgar (1984).

Extended studies of various aspects of the model are underway. We may thus establish an explanation for the amazing properties of radio galaxies and radio-loud quasars.

ACKNOWLEDGEMENTS

We benefitted greatly from discussions with Robert Laing and information sent by Alan Bridle. The SERC and the Royal Society provided financial assistance.

REFERENCES

Begelman, M.C., McKee, C.F. & Shields, G.A., 1983. Astrophys. J., 271, 70.
Begelman, M.C., McKee, C.F. & Shields, G.A., 1983. Astrophys. J., 271, 89.
Bridle, A.H., 1984. In Physics of Energy Transport in Extragalactic Radio Sources, eds. Bridle, A.H. & Eilek, J.A., NRAO, 135.
Norman, M.L., Winkler, K.-H., Smarr, L., 1984. In Physics of Energy Transport in Extragalactic Radio Sources, eds. Bridle, A.H. & Eilek, J.A., NRAO, 150.
Rudnick, L. & Edgar, B.K., 1984. Astrophys. J., 279, 74.
Shakura, N.I. & Sunyaev, R.A., 1973. Astr. Astrophys., 24, 337.
Smith, M.D. & Raine, D.J., 1985. M.N.R.A.S., 212, 425.

THE LEO INTERGALACTIC NEUTRAL HYDROGEN CLOUD

Yervant Terzian, S. E. Schneider, and E. E. Salpeter
NAIC and Department of Astronomy
Cornell University
Ithaca, New York 14853
U. S. A.

ABSTRACT. The current status of the neutral hydrogen $\lambda 21$ cm radio observations of the intergalactic cloud in Leo are described. These include both HI observations at Arecibo with an angular resolution of ~3 arc min, and VLA D-array observations with an effective resolution of 45 arc sec.

In 1983, we reported the discovery of a large intergalactic neutral hydrogen cloud at a radial velocity of ~960 km/s, in the vicinity of the Leo group of galaxies near M96 (Schneider, et al. 1983). This unexpected detection was made at Arecibo while $\lambda 21$ cm HI calibration measurements were being made on what was supposed to be "empty" sky. Earlier searches to detect intergalactic HI clouds in groups of galaxies have been unsuccessful (e.g. Lo and Sargent 1979), and such observations indicate that intergalactic HI clouds are very rare. However, a few galaxies have shown very extended HI envelopes and streams several times their Holmberg size (e.g. Haynes et al. 1979).

At Arecibo's resolution of ~3 arc min, the main cloud complex appears to be isolated from the surrounding galaxies. The main part of the cloud is about 30 arc min across, and at a distance of 10 Mpc (Leo group of galaxies) this corresponds to ~100 kpc. The column density of HI is ~7 x 10^{19} H cm^{-2}, and for symmetrical cloud models the HI density is ~10^{-3} to 10^{-4} H cm^{-3}.

The observed neutral hydrogen has a mass of ~2 x 10^9 M$_\odot$. The observed HI profiles show a velocity gradient from ~860 to 1040 km/s from one side to the opposite extent of the main cloud, probably resulting from orbital motion about M105 and NGC 3384 (Schneider 1985). Recent optical (Pierce and Tully 1985, Skrutskie et al. 1984), and infrared IRAS (J. R. Houck, private communication, 1985) observations do not show any emission to very low levels.

More recently we have observed the large intergalactic HI cloud with the VLA in its D-array mode (Schneider et al. 1985), and with an effective resolution of 45 arc sec. These observations show that the main HI cloud is resolved and shows a fine structure of a number of prominent clumps. The sizes of these clumps are in the range 40 to 80 arc sec, or 2 to 4 kpc. Each clump contains a few times 10^7 $M_\odot$ of HI mass and has a mean density of 10^{-1} to 10^{-2} HI cm^{-3}. Figure 1 shows the HI brightness distribution obtained with the VLA.

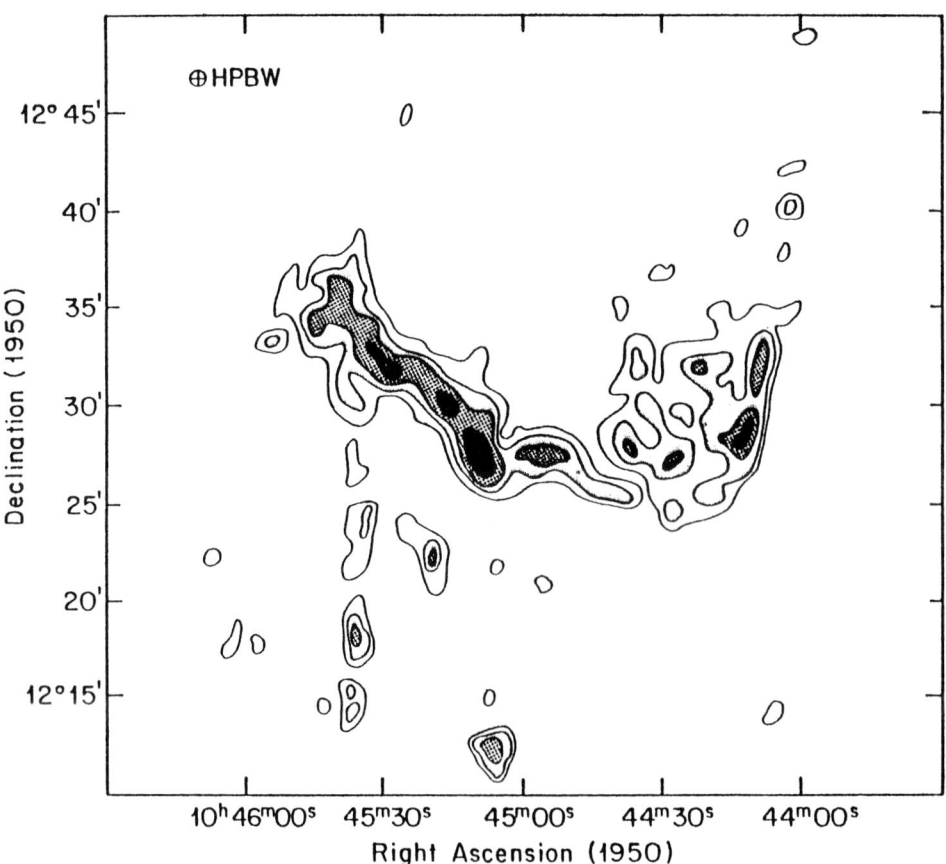

Fig. 1. The neutral hydrogen brightness distribution of the intergalactic cloud in Leo, observed with the VLA in its D-array.

It is possible that the intergalactic cloud in Leo is primordial and it may fragment to form new dwarf galaxies. Such an hypothesis is in accord with the optical and infrared results, showing no evidence of stars or interstellar dust, and it is also in agreement with Schneider's (1985) "ring" model suggesting that the cloud is at least several billion years old.

This work was supported in part by the National Astronomy and Ionosphere Center which is operated by Cornell University for the National Science Foundation, and in part by NSF grant AST 84-15162.

REFERENCES

Haynes, M. P., Giovanelli, R., and Roberts, M. S., 1979, Astroph. J., 229, 83.
Lo, K. Y., and Sargent, W. L., 1979, Astroph. J., 227, 756.
Pierce, M. J., and Tully, R. B., 1985, Astron. J., 90, 450.
Schneider, S. E., Helou, G., Salpeter, E. E., and Terzian, Y., 1983, Astroph. J. Letters, 273, L1.
Schneider, S. E., 1985, Astroph. J. Letters, 288, L33.
Schneider, S. E., Salpeter, E. E., and Terzian, Y., 1985, (preprint).
Skrutskie, M. F., Shure, M. A., and Beckwith, S., 1984, Astroph. J. Letters, 282, L65.

EMISSION LINES INDICATING A UNIVERSAL L/M RATIO FOR THE CENTRAL ENGINE IN AGN ?

Amri Wandel
Astronomy Program
University of Maryland
College Park, MD-20742
U.S.A.

ABSTRACT. Assuming the velocity dispersion of the emission-line clouds is induced by gravity, we calculate the dynamic mass ($M \sim v^2 r /G$) of the central object in a sample of ~ 50 Seyfert 1 nuclei and ~ 40 nearby quasars. Typical values are $10^8 M_\odot$ for Seyferts and $10^9 M_\odot$ for quasars. The dynamic mass is found to be proportional to the continuum luminosity (at the ~ 20σ confidence level), which suggests a universal value of L/L_{Edd}. For the optical band we find $\log <L_c/L_{Edd}> = -2.0 \pm 0.3$. Applying the same technique to the OIII line, one can estimate the dynamic mass within the radius of the narrow-line region (~ 100 pc). For some 50 objects of the sample above, this mass is found to be well correlated with the continuum luminosity and with the dynamic mass estimated from the H_β line.

1. INTRODUCTION

Prominent emission-lines are perhaps the most characteristic feature of quasars and active galactic nuclei. Many of their properties are amazingly similar in different objects, over many orders of magnitude in continuum luminosity. A comparative investigation of these properties and the continuum luminosity may therefore supply valuable clues on the environment that produces the broad lines, and on the unobservable central source.
If the line width is induced by velocity dispersion of the emitting clouds, then the mass of the central object can be estimated from its distance from the clouds together with the velocity dispersion. Previous attempts to deduce the mass associated with the central source in AGN (Dibai 1981a,b,1983; Bassani, Dean and Sembay 1983) do not unveil any particular tendency. Applying a consistent method which determines the distance and velocity from the same spectral feature (the H_β line) to determine the mass, we find the mass to be proportional to the optical continuum luminosity, L_c, and that the ratio L_c/M has a surprisingly low dispersion (Wandel and Yahil 1985). This may indicate that the central source in active nuclei has a characteristic ratio between its luminosity and Eddington luminosity, L_E.

2. DATA

We look for correlations among the continuum luminosity L_c (defined as νf_ν at $\nu = 4000 A$) and the H_β emission line parameters (the luminosity in the line, L_β, and the [Doppler] full width at zero intensity [FWZI], ΔV). Our sample consists of 42 low redshift quasars and 52 Seyfert 1 and broad line radio galaxies. The data are taken from the compilation by Steiner (1981) (corrected for some errors found in that compilation), augmented by the data of Blumenthal et al. (1982). The OIII line widths are taken from Feldman (1982) and Heckman et al. (1984). $H = 50$ kms^{-1}Mpc^{-1} and $q = 0$ are used.

3. THE SIZE OF THE EMISSION LINE REGIONS

An estimate of the size of the BLR, which we take to be comparable to its distance from the central source, is provided by the luminosity in a specific line, e.g., H$_\beta$ (Dibai 1981a,b)

$$r_B = \left(\frac{3 L_\beta}{4\pi f_v j(H_\beta)} \right)^{1/3}, \qquad (1)$$

where L_β and $j(H_\beta) = 1.22 \times 10^{-25} n_c^2$ erg cm^{-3}s^{-1} are the total luminosity and volume emissivity in H$_\beta$ respectively, and f_v is the volume filling factor of the emitting gas. We have used the case B emissivity; in reality $j(H_\beta)$ may be larger, because of collisional exitation, depending on the density, optical depth and degree of ionization (cf. Mathews, Blumenthal and Grandi 1980).

Dibai arbitrarily set the volume filling factor to $f_v = 0.001$. We evaluate it from the covering factor of the clouds, f_a, i.e., the fraction of the sky covered by clouds, as seen from the central source. An exact relation between the two factors requires detailed knowledge of the radial distribution of the sizes and densities of the clouds. To an accuracy of order unity, however, they are related by

$$f_v \approx f_a N(H_\beta) / n_c r_B , \qquad (2)$$

where $N(H_\beta) = 10^{23} N_{23}$ cm^{-2} is the column density of the H$_\beta$-emitting gas (not to be confused with the total column density through the clouds, which may be larger). As the clouds are probably optically thick, $N(H_\beta)$ depends on the ionizing flux and on the shape of the continuum. However, for the lack of a generally recognized model for this dependence, we take N_{23} to be fixed, as discussed below. Substituting equation (2) into equation (1) yields

$$r_B \approx 0.05 \left(\frac{L_{\beta,42}}{n_{c9} f_a N_{23}} \right)^{1/2} pc. \qquad (3)$$

The column density of the line-emitting gas can be estimated from photoionization models (e.g. Netzer 1980; Kwan and Krolik 1981). For lack of their values for the specific objects, we use $n_{c9} = 1$ and $N_{23} = 0.3$. The covering factor has been estimated by a number of authors - e.g. from the ratio of the onizing luminosity to the total luminosity in lines. X-ray data (Mushotzky and Ferland 1984) and IUE data (Kinney et al. 1985) sugest an empirical expression $f_a = (L_c/L_1)^{-0.2}$ for $L_c > L_1$ and $f_a \approx 1$ for $L_c < L_1$, where $L_1 = 2 \times 10^{43}$ ergs^{-1}. For these values eq. (3) yields $r_B \approx 0.1 L_{\beta,42}^{0.6}$ pc.

4. DYNAMICAL INTERPRETATION

The velocity width of the lines is almost certainly due to the bulk velocities of the emitting clouds, since it is greatly in excess of the sound speed in the clouds, but the origin of these velocities is unknown. One model ascribes them to radiation pressure by the central source. While the details may be rather complicated (Mathews 1982), simpler versions make definite predictions of some relations between parameters of the clouds. In the optically thin model of Blumenthal and Mathews (1975, 1979), for example, $\Delta V \propto r_B^{1/2}$ and by equation (3)

$\Delta V \propto L_\beta^{1/4}$, assuming the other variables to be roughly independent of L_β. A regression of the $\log(\Delta V)$ on $\log(L_\beta)$ gives a slope of 0.1±0.02, so the deviation from the predicted 0.25 is significant at the 8σ level. Similar inconsistencies are found if the clouds are optically thick or contained by the ram pressure of the intercloud medium. However, it should be cautioned that these models may be oversimplified, or that other variables have unexpected correlations. See also Gaskell (1985). The evidence presented here is therefore not a repudiation of the model of radiation accelerated clouds.

An alternative picture is that the dispersion velocity of the clouds is induced by gravity. In that case it is expected to be of the order of the free-fall velocity. Hence the mass of the central source can be approximated by

$$M_8 \approx 100 \, r_{B,pc} \, v_9^2 , \qquad (4)$$

where $M = 10^8 M_8 \, M_\odot$ is the mass of the central object and $v = 10^9 v_9$ cm s^{-1} is the three dimensional velocity dispersion. Substituting equation (3) into equation (4), we therefore have

$$M_8 \approx 5 \left(\frac{L_{\beta,42}}{n_c \, _9 f_a \, N_{23}} \right)^{1/2} v_9^2 . \qquad (5)$$

If the H$_\beta$ emissivity is larger than the case B value, the actual mass is lower than the value given by eq. (5). In what follows we identify the three dimensional velocity dispersion with $\Delta V(\text{FWZI})/4 \sim \sqrt{3} \Delta V(\text{HWHM})$.

We find that the mass given by this method is extremely well correlated with the continuum luminosity L_c and with L_β - at the 20σ level (Fig. 1). This correlation is much more significant than the correlation between L_c and ΔV (table 1; cf. eq. [5]), which is only 3-4σ. The linear regression of $\log L_c$ on $\log M$ gives a slope of unity, so that we find the continuum luminosity is proportional to the mass. We find this correlation also in subgroups of the sample: quasars only, Seyferts only, and the coherent quasar sample of Blumenthal et al. (1982). In order to test for flux-dependent selection effects, we repeated the calculations for several bins in continuum flux, with no change in the results.

5. THE NARROW LINE REGION

An analysis similar to that described above may be carried out for the narrow emission lines, which are produced in a much more extended region than the broad lines, typically of the order of a few hundreds pc. The dynamic mass obtained from the narrow lines may therefore sample the mass within a significant part of the galactic nucleus. We use the OIII line, which is the most prominent forbidden line in most AGN, and typical parameter values taken from photoionization models of the OIII region (Kramer and Heckman, private communication): $N = 10^{20}$cm^{-2}, $n = 10^5$cm^{-3} and the emissivity $j(OIII) \approx 10 \, j(H_\beta, case \, B)$. The covering fraction f_a was (somewhat arbitrarily) taken as 0.05. The equations analogous to eqs.(3) and (4) are

$$r_{OIII} \approx 50 \left\{ \frac{L(OIII)_{42}}{n_{c\,5} f_a \, N_{20}} \right\}^{1/2} pc , \qquad (6)$$

and

$$M_8(OIII) \approx 5 \left(\frac{L(OIII)_{42}}{n_{c\,5} f_a \, N_{20}} \right)^{1/2} \left(\frac{v}{300 \text{ km s}^{-1}} \right)^2 . \qquad (7)$$

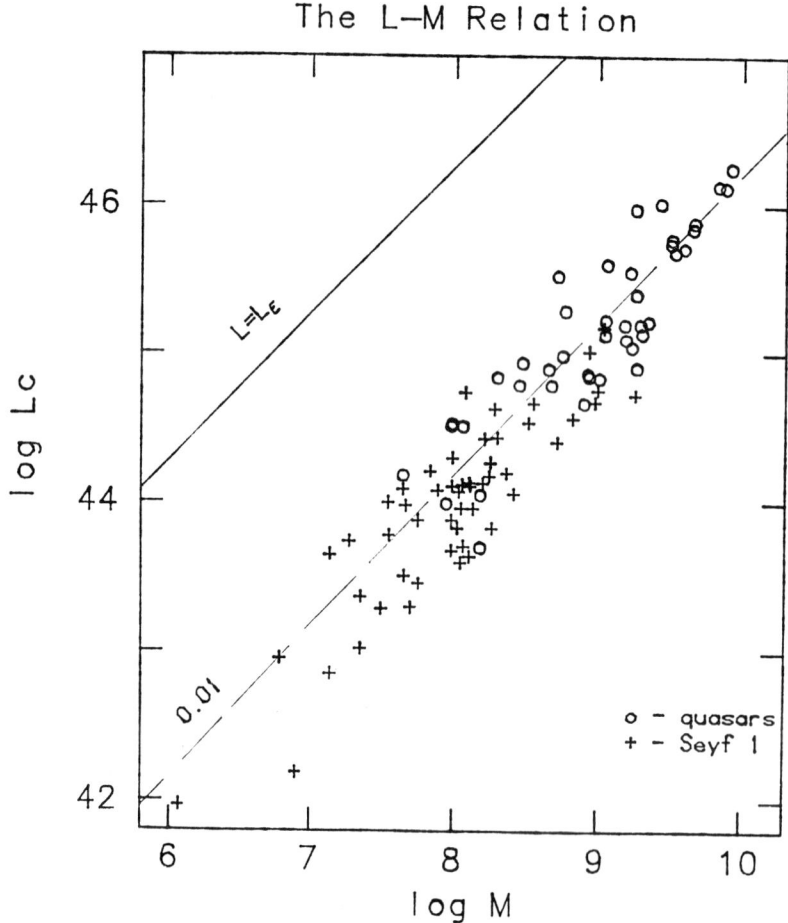

Figure 1: The optical continuum luminosity versus the dynamical (H_β) mass. Crosses mark Seyfert 1 nuclei, while quasars are denoted by circles. The diagonal line marked L_E represents the Eddington limit; The dashed line is the regression of the ordinate on the abscissa.

Table 1.

All entries refer to the logarithms of the variables in the left column. For correlations the table gives the slope of the linear regression (of Y on X) ± its standard deviation. The correlation coefficient is given in parenthesis. Means are given ± the standard deviation.

	Seyfert 1	Quasars	Combined
No. of objects	52	42	94
Correlations (X - Y)			
$L_c - L_\beta$	1.04±.07 (.91)	.98±.07 (.81)	1.11±.04 (.93)
$L_c - \Delta V$	0.13±.03 (.51)	0.12±.03 (.49)	0.09±.02 (.504)
$M^a - L_c$	1.0±.10 (.82)	1.03±.13 (.82)	1.14±.10 (.88)
$M - L_c$	.92±.07 (.88)	.92±.08 (.88)	1.04±.04 (.93)
$M - L_X$	.61±.12 (.64)	.42±.15 (.42)	.71±.08 (.71)
$M(OIII) - L_c$	(52 obj.)		.81±.11 (.74)
$M(H_\beta) - M(OIII)$	(42 obj.)		.75±.11 (.75)
means			
L_c	44.0±.6	45.3±.7	44.7±.9
L_β	42.1±.6	43.6±.6	42.8±1.0
ΔV	4.07±.16	4.14±.15	4.10±.16
M/M_O	7.91±.64	8.97±.69	8.43±.95
L_c/L_E	-2.14±.30	-1.92±.30	-2.04±.32
$M(OIII)$			8.61±.69

M^a is calculated with fixed $f_a = 0.2$. Unmarked M is calculated with $f_a \alpha L_c^{-0.2}$.

In a sample of 52 objects we find the dynamical (OIII) mass to be well correlated with the continuum luminosity (Fig. 2). For 42 of those objects for which the H_β line width was available, this mass is comparable to their H_β dynamical mass, the sample showing a good correlation between the two masses (Fig. 3). It should be stressed, however, that the parameter values for the OIII emission line (especially the emissivity, covering fraction and column density) are less well constrained by observational data than the H_β parameters, and hence the values of the corresponding dynamical mass are less certain.

6. DISCUSSION

The dynamical mass would represent the true mass within the corresponding radius if the velocity dispersion of the line-emitting gas were induced by gravity. This would be the case for three kinematic configurations of the emission-line clouds: 1. radial inflow, 2. Keplerian motion (eventually in parabolic orbits), 3. outflow at a velocity which is of the order of the escape velocity.

As the BLR is typically at a distance of ~ 1 pc from the central source, the dynamical H_β mass is probably dominated by the central compact object. As we have found, this mass is proportional to the continuum luminosity. This proportionality may have a physical meaning beyond a mere correlation. It is convenient to relate the mass to the Eddington luminosity

$$L_E = 4\pi G M m_p c / \sigma_T = 1.3 \times 10^{46} M_8 \text{ erg s}^{-1} . \tag{8}$$

As $L_c/M \propto L/L_E$, it could indicate that all active nuclei have a single universal value of L/L_E.

In order to quantify this hypothesis we have calculated the mean and the standard deviation of $\log(L_c/M)$. If M and L_c were uncorrelated, one would expect $\sigma^2_{L/M,uncorr} = \sigma^2_L + \sigma^2_M = 1.7$, where σ^2_X is the variance of log X. Actually we have $\sigma^2_{L/M} = 0.11$, which strongly supports our hypothesis.

In Fig. 1 we plot $\log(L_c)$ versus $\log(M)$. The correlation is unmistakable. The data points are concentrated along a line of constant L_c/L_E. Numerically we find $\log<L_c/L_E> \sim -2.0\pm0.3$. (The mean is higher for quasars [-1.9] than for Seyferts [-2.1]). It is important to emphasize that this conclusion is quite general, and relies only on the assumption that the velocity dispersion of the clouds is induced by gravity.

Note that L_c is just the optical band; the bolometric luminosity could easily be an order of magnitude larger. Also, the actual masses may be lower by a factor of 2-5 due to higher H_β emissivity and cloud density than used in eq. (5), so that the bolometric luminosity could approach the Eddington limit. (Because of the similarity of their continuum spectra, it is plausible that active nuclei have similar ratios of L_c/L_{bol}, so that L_c could be a good measure of the total luminosity). If, in addition, the logarithmic accretion rate ($\dot{M}/M$) has a preferred value (see e.g. Wandel 1984), then the above evidence would mean that the central power house has a characteristic efficiency, $e = L/\dot{M}c^2$.

An interesting question is whether the value of L/L_E stays constant also in earlier AGNs. Unfortunately, the H_β technique described above can be applied only to relatively low-redshift objects, because for $z > 0.8$ the H_β line is shifted into the IR, which is not observable from the ground. However, at high redshifts ($z > 3$), the H_β line can be observed again in the 2.2 micron. Kühr *etal.* (1985) measured the H_β line of the quasar 5S0014+81 ($z=3.41$), and for their data our technique yields $L_c/L_E \approx 0.3$, a factor of 30 higher than the value for all low-redshift objects. This large factor is contributed by the exceptionally low equivalent width, and by the high luminosity of this quasar. It should be cautioned, however, that evolution in L/L_E is only

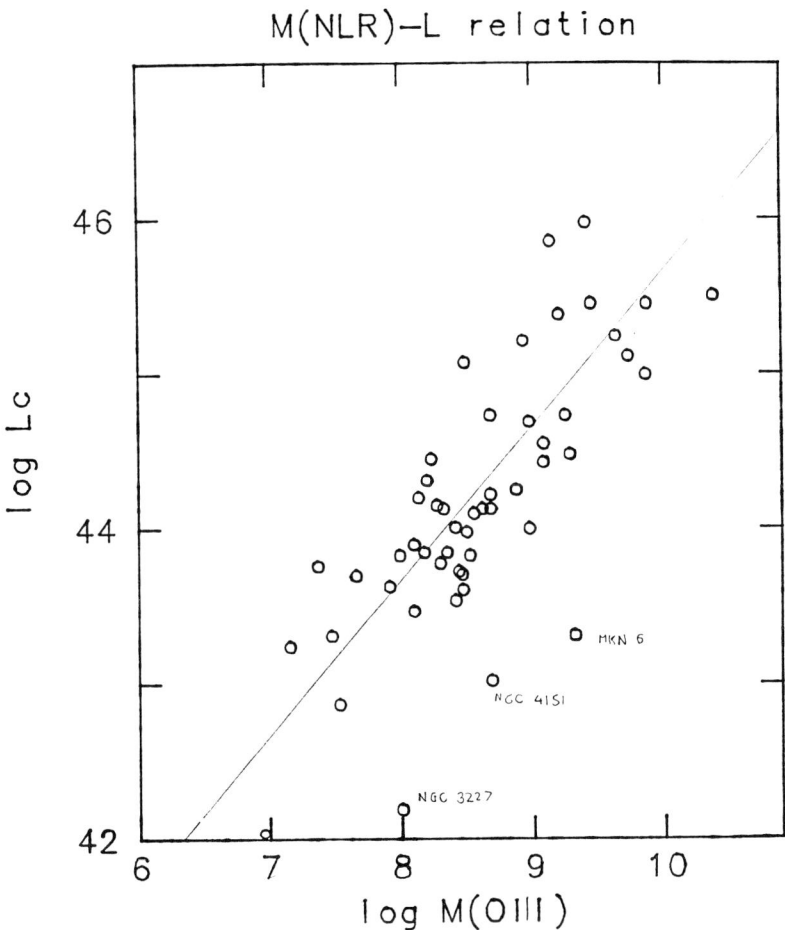

Figure 2: The continuum luminosity vs. the OIII dynamical mass..

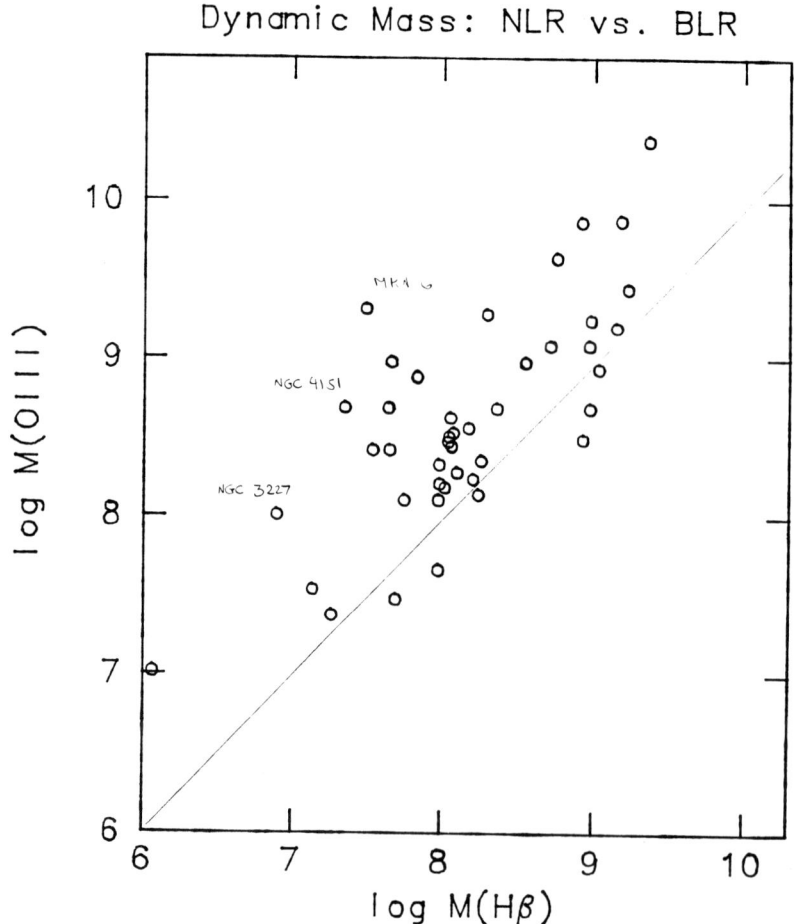

Figure 3: The OIII dynamical mass vs. the H_β dynamical mass.

one explanation, some other interpretations being beamed continuum emission, exeptionally low covering factor, or absorption of the H_β in the IGM.

As we have seen in section 5, the dynamical mass within the OIII region is well correlated with the mass within the BLR, as sampled by the H_β line, and, in most objects, of comparable value (within the uncertainties in the OIII parameters). A possible interpretation of this result is that the mass of the central object dominates the kinematics also at the radius of the NRL. For the masses we find this is not implausible. A correlation between M(OIII) and L_c (and hence also M(H_β)) would also be expected if the NLR formed at a preferred value of continuum flux or ionization parameter (L/r^2 and L/nr^2 respectively), thus enclosing more mass for more luminous objects. In any case, the correlation between the dynamic masses sampled by the BLR and the NLR suggests a common cause. Recently, Wilson and Heckman (1984) found a good correlation between the width of the OIII line and the dispersion of stellar velocities in the central region of the nucleus in a sample of active galaxies (mainly Seyfert 2 and Liners), which does suggest that the OIII emitting gas and the stars are moving in the same gravitational potential.

REFERENCES

Bassani, L., Dean, A.J. and Sembay, S. 1983, *Astron. Astrophys.*, **125**,52.
Blumenthal, G. R, and Mathews, W. G. 1975, *Ap. J.*, **198**, 517.
--------------- 1979, *Ap. J.*, **233**, 479.
Blumenthal, G.R., Keel, W.C., and Miller, J.S. 1982 *Ap. J.*, **257**, 499.
Dibai, E.A. 1981a, *Sov. Astr.*, **24**, 389.
---------- 1981b, *Sov. Astr. Lett.*, **7**, 248.
---------- 1983, *Sov. Astr.*, **28**, 245.
Feldman, F.R., Weedman, D.W, Balzano, V.A., and Ramsey, L.W. 1982, *Ap. J.*, **256** 427.
Gaskell, C.M. 1985, *Ap. J.*, **291**, 112.
Heckman, T.M., Miley, G.K. and Green, R.F. 1984, p. J., **281** ,525.
Kinney, A. L., Huggins, P. J., Bregman, J. N., and Glassgold, A.E. 1985, *Ap. J.* **291**, 128.
Kwan, J., and Krolik, J.H. 1981, *Ap. J.*, **250**, 478.
Kuhr, H., McAlary, C.W., Rudy, R.J., Stittmatter, P.A., and Rieke, G.H. 1984, *Ap. J. Lett.*, **284**, L5.
Mathews, W. G. 1982, *Ap. J.*, **252**, 39.
Mathews, W. G., Blumenthal, G.R. and Grandi, S.A. 1980, *Ap. J.*, **235**, 971.
Mushotzky, R., and Ferland, G.J. 1984, *Ap. J.*, **278**, 558.
Netzer, H. 1980, *Ap. J.*, **236,** 406.
Steiner, J.E. 1981, *Ap. J.*, **250**, 469.
Wandel, A. 1984, *M.N.R.A.S.*, **207**, 861.
Wandel, A. and Yahil, A. 1985, *Ap. J. Lett.* submitted.
Wilson, A.S. and Heckman, T.M. 1984, in the proceedings of the Santa Cruz workshop.

THE EXPLANATION OF THE POSITRON-ELECTRON ANNIHILATION LINE AT OUR GALACTIC CENTER UNDER THE MODEL OF GALACTIC NUCLEI WITH MONOPOLES

Deyu Wang[1]) Qiuhe Peng[2,3]) Zongyun Li[2])
[1]) Purple Mountain Observatory, Nanjing
[2]) Department of Astronomy, Nanjing University, Nanjing
[3]) Kapteyn Astronomical Laboratory, Groningen

I. INTRODUCTION

The observational data on the intense positron-electron annihilation line (AL) at our Galactic Center, detected by ballons([1]) and satellites([2]), may be summarised as follows: the energy of spectral line is 5109.9±0.25 keV with very small gravitational redshift($Z \leq 7 \times 10^{-4}$); the intensity of spectral line is $\sim 6 \times 10^{37}$ erg/sec, corresponding to annihilation of $\sim 10^{43}$ e^+/sec; the width of spectral line is very small, FWMH ≤ 3.2 keV ([3]); and the intensity of spectral line decreases about $\frac{1}{3}$ during 6 Months ([2]). The higher energy photons (continuum) of $E_\gamma \geq 511$ keV are also detected; the integral intensity from $E_\gamma = 500$ keV - 10 MeV is $(8 - 20) \times 10^{37}$ erg/sec([4]), corresponding to about 1/5 the photon number of AL.
We explain these data here, using the model, based on the induced nucleon decay by monopole, $pM \rightarrow \pi^0 e^+ M$ (85%) or $\mu^+\mu^- e^+\mu$ (15%), in the last paper ([5]).

II. MODEL SELECTION

For our Galactic Center, the mass is $M \simeq (3 - 10) \times 10^6 M_\odot$ ([6]), and the luminosity $L \simeq (1 - 3) \times 10^7 L_\odot$ can be estimated from the temperature (less than 5×10^4 K) and the ionization of hydrogen (larger than 10%) in the Sgr A west clouds around the Galactic Center ([7]).
Due to the fact that the observed AL has a very small gravitational redshift and very narrow line width, the positron annihilation processes certainly must occur in extremely low energy region and far way from the Galactic Center, i.e. at $r > 2.2 \times 10^{15}$ cm (for $M = 10^7 M_\odot$). Thus we choose the model n = 3 with $\xi = 1$, $M = 10^7 M_\odot$, $L = 3 \times 10^7 L_\odot$ from our last paper ([5]), then the radius of this object is $R \approx 300 R_g \approx 8.9 \times 10^{14}$ cm, and the position producing AL are outside the surface to the object.

III. THE CASCADE PROCESSES

However, the positrons produced from the catalytic reaction have a relativistic γ of 10^3. It is well known from the theory of positron annihilation [8]: that the positrons with extreme relativistic energy will annihilate with static electron and will produce a forward photon carrying whole collision energy and a backward photon with energy between 250 and 500 keV. The narrow annihilation line observed at the Galactic Center cannot be produced in this way. Therefore, the positron produced from the catalytic reaction and high energy photon from decaying π^0 should undergo a cascade process for multiplication and slowing down within the Galactic Center. This means that photons are emitted mainly through bremsstrahlung by encounters between high energy positrons and electrons, and the photons of $E_\gamma \geqslant 2m_e c^2$ interact mainly with nuclei to produce positron-electron pair.

The longest range of positrons within the Galactic center $(R - R_a)$ is given by $\int_{R_a}^{R} N_B(r) dr = \int_{mc^2}^{E} \frac{dE'}{L(E')}$ where the energy loss rate of positron (mainly through bremsstrahlung) is determined by following formula:

$$-\left(\frac{dE}{dr}\right)_B = n_B(r) L(E) = n_B(r) \int_0^{E-mc^2} E_\gamma \sigma_B(E, E_\gamma) dE_\gamma.$$

According to simple cascade shower theory [9] and considering that positron and photons pass through layers of different thickness inside the Galactic Center, the positron average multiplication number outside the surface of the Galactic nucleus is $(2 - 3) \times 10^2$.

From our model above for the Galactic Center, it can be estimated that the positron number leaving from the Galactic nucleus are $\sim 10^{43} e^+/\text{sec}$ which agrees with observed intensity of AL.

IV. PRODUCING PHOTONS OBSERVED

The positrons will be slowed down after they leave the Galactic nucleus. The effect of the collisional ionizaton will be more and more important in slowing processes when the positron energy decreases. The low energy positron will be annihilated mainly by the following processes: i) direct annihilation: $e^+ + e^- \to \gamma + \gamma$, ii) charge exchange [10]: $e^+ + H \to (e^+, e^-) + p$, iii, radiative recombination $e^+ + e^- \to (e^+, e^-)$, iv) annihilation with boundary electron: $e^+ + H \to 2\gamma + p$, where (e^+, e^-) is a positronium, which could be in two states: triplet state ($^3S_1$) or single state ($^1S_0$). The former decays into three photons with continuous energy between 100 keV and 400 keV: the latter decays into two 511 keV photons. Therefore, the AL with very narrow width observed at the Galactic Center region should be produced by direct annihilation by extremely low energy positrons and $^1S_0$ state positronium decay.

In addition, there is always a number of high energy photons with $E_\gamma > 500$ keV, which are produced directly from the processes inside the Galactic nucleus. This fact is in favour of our model and it is difficult for general black hole models [11].

ACKNOWLEDGEMENTS

The authors should like to sincerely thank Dr. B. Sanders for revising th manuscript in English. They also express thanks to Ms. J. Nunnink and I. Rouwé for typewriting this article. This work is partly supported by the Kapteyn Laboratory.

REFERENCES

[1] W. Johnson, and R. Haymes, Ap. J., 184 (1973) 103 R. Haynes et al., Ap. J., 201 (1975)593
[2] G. Riegler et al., Ap. J., 248 (1981) L13
[3] M. Leventhal, et al., Ap. J., 225 (1978) L11, ibid 240 (1980) 333, ibid 260 (1982) L1
[4] J. Matteson, 'The Galactic Center', AIP Conf. Proc. No 83 (1982) (eds. G.R. Riegler and R.D. Blandford) p. 109-122
[5] Q. Peng et al., see this symposium
[6] J. Lacy et al., Ap. J., 241 (1980) 132
[7] G. Townes et al., Nature 301 (1983) 661; Becklin et al., Ap. J., 258 (1982) 135; I. Gattey, 'The Galactic Center' AIP. Conf. Proc. No 83(1982) p25-32
[8] F. Steckney, Cosmis Gamma Rays. (1971) NASA 59-249
[9] W. Heitter, The Quantum theory of radiation (1954) §38 (Oxford University Press)
[10] R. Drachman et al., Phys. Rev. A14 (1971)100
[11] R. Lingenfelter et al., 'The Galactic Center' AIP Conf. Proc. No 83 (1982) p. 148 - 159)

INDEX OF NAMES

Aarseth, S.J., 59, 640
Abbott, D.C., 98, 110
Abell, G.O., 24, 26, 41, 334, 550, 556, 572
Abramowicz, M.A., 113-114, 117-118, 126, 447, 507, 543, 546
Achterberg, A., 522-523
Adam, G., 42, 272
Adams, M.T., 637, 640
Adams, T.F., 207-208, 689, 692
Ade, P.A.R., 19
Aguirre, C., 271
Aharonian, F.A., 151, 169
Ahmed, F., 577
Akhiezer, A.I., 522-523
Aldrovandi, S.M.V., 476-477
Alladin, S.M., 573, 577
Allen, C.W., 645, 647
Allen, D.A., 77, 582
Aller, H.D., 77, 653
Aller, M.F., 77, 653
Alloin, D., 341, 349, 672
Altamore, A., 275
Altschuler, D.R., 655, 657
Ambartsumian, V.A., 132, 143, 586
Anderson, K.S., 707-708
Andrew, B.H., 87, 90-91
Angel, J.R.P., 64, 77-79, 87, 91, 246, 250
Antonucci, R.R.J., 669
Arakelian, M.A., 258, 269
Arp, H.C., 54-55, 59, 266, 467, 470, 574, 577, 618
Athanassoula, E., 640
Atherton, P.D., 304
Atwood, B., 422, 432-433
Aumann, H.H., 270-271
Auriemma, C., 439, 441, 698, 700, 702, 704
Aveni, A.F., 677
Avni, Y., 6, 8, 226, 237, 242, 250, 334, 350, 397, 399-411, 413-414, 416, 418-419, 494, 540, 547-548, 550, 632
Avrett, E.H., 681
Axford,, W.I., 140, 143
Axon, D.J., 304, 483, 486, 652

Baade, D., 554

Baan, W.A., 461-463
Bahcall, J.N., 6, 8, 422, 433
Bailey, J., 483, 486
Baity, W.A., 169, 271
Balbus, S.A., 108, 110
Baldwin, J.A., 19, 22, 29, 34, 39, 41-42, 60, 110, 195, 198, 208, 210, 370, 433, 525-526, 529, 577
Balick, B., 21, 41, 49, 59, 250, 270, 312-313
Balkowski, C., 502
Balzano, V.A., 196, 208, 210, 216, 225, 735
Barbieri, C., 82, 87, 90-91, 340, 350, 465, 601-603
Bardeen, J., 96, 110, 385, 393
Barr, P., 91, 528-529, 669-670
Barrow, J.D., 555, 558
Bartel, N., 653
Barthel, P.D., 568
Bassani, L., 727, 735
Baud, B., 270
Bauer, E., 661
Baum, W.A., 577
Beck, S.C., 634-635
Becker, R.H., 269-271
Becklin, E.E., 19, 42, 110, 586, 635-636
Beckman, J.E., 471-472, 474, 587, 591, 601, 604
Beckwith, S., 634-635, 725
Begelman, M.C., 130, 134, 142-143, 150, 169, 448, 453, 455, 521, 523, 717, 721
Beichman, C.A., 19, 270-271
Beintema, D.A., 270-271
Bergeron, J., 42, 59, 300, 304, 421-423, 425-426, 429, 433-434
Bertola, F., 176, 190, 299, 302, 304
Bettoni, D., 304
Bhattacharia, D., 501-502
Bhavsar, S.P., 555, 558
Bicknell, G.V., 541
Biermann, P., 77, 248, 251, 303-304, 552, 554, 618
Bignami, G.F., 150, 168-169
Binette, L., 37, 41, 200, 208, 259, 269, 305, 475-477, 704
Bingelli,, 181

Bingham, R.G., 649, 652
Biretta, J., 41, 543, 546
Birkle, K, 479
Bjornsson, C.-I., 483, 521, 523
Blackman, C.P., 26, 41
Blades, J.C., 77, 270, 425-426, 433
Blaes, O.M., 113-114, 119, 121, 123-124, 126
Blake, G.M., 510-511
Blanco, V.M., 269
Bland, J., 302, 304
Blandford, R.D., 75, 77, 108, 110, 114, 124, 126, 143, 150, 152, 163, 166, 169, 246, 250, 448, 455, 511, 521, 523, 544, 546
Bleach, R.D., 272
Blumenthal, G.R., 189-190, 728-729, 735
Boernstein, L., 694
Boggess, A., 226, 514, 516
Boggess, N., 270-271
Bohlin, R.C., 513, 516
Bohuski, T.J., 256, 269
Boisse, P., 422, 425, 433
Boksenberg, A., 42, 143-144, 275, 298, 304, 434, 546, 716
Boldt, E.A., 269-271, 383-385, 387, 389-391, 393-394
Bolton, J.B., 467, 470
Bolton, J.G., 79, 517, 519
Bondi, H., 117, 122, 126
Bonoli, C., 487-488
Bonoli, F., 82, 84-85, 91, 334, 350, 487, 502, 631
Boroson, T.A., 18-19, 21, 41, 49, 59, 299, 304, 311, 313
Bortoletto, F., 487-488
Bosma, A., 707-708
Bothun, G.D., 59, 250, 312-313, 642, 644
Bottinelli, L., 501-502, 622
Bouchet, P., 91
Bowyer, C.S., 271
Bowyer, S., 42, 77
Boyle, B.J., 232, 250, 325, 346, 350, 412, 418, 491-493, 631
Braccesi, A., 85, 91, 224-226, 239, 250, 317, 319, 325, 334, 339, 341, 346-347, 349-350, 365, 370, 403, 405, 412, 419, 494, 502, 540, 631-632
Bracewell, R.N., 91

Brand, P.W.J.L., 484, 486
Branduardi, G., 269, 419
Brecher, K., 681
Bregman, J.N., 16-17, 19, 77, 534-535, 735
Bridle, A., 52, 59, 304, 439, 441, 565, 568, 639-640, 717, 721
Briel, U., 269
Briggs, F.H., 428, 433
Brinkman, W., 270-271, 455, 523, 550
Brinkmann, W., 251
Bromage, G.E., 144, 275, 291
Bronkalla, W., 649, 651-652
Brosche, P., 499-500, 502
Browne, I.W.A., 467, 470, 623, 626
Bruck, H.A., 251
Bruyn,, 671
Bruzual, G., 176-177, 180, 190
Buitrago, J., 503-505, 507
Bujarrabal, V., 500, 502
Burbidge, E.M., 19, 48, 55-56, 59-60, 475, 477
Burbidge, G., 47-48, 50, 54-56, 59, 77, 81, 91, 475, 477
Burke, W.L., 370
Burns, M.L., 151, 169
Butcher, H., 304
Butcher, H.R., 41, 298, 304
Byard, P.L., 482
Byram, E.T., 272

Callan, C., 661, 667
Calucci, G., 667
Calvani, M., 546
Camenzind, M., 476-477, 521, 523
Cameron, A.G.W., 139, 143
Campbell, B., 42, 59, 312-313, 554, 608
Canizares, C., 269
Canizares, C.R., 215, 226, 270, 299, 304, 399, 403, 418
Canizares, R.G., 249-250, 547, 550
Cannon, R.D., 82, 91, 582
Capaccioli, M., 176, 190, 269, 350
Carroll, T.J., 40-41, 509-511
Carswell, R.F., 78, 425, 433-434
Carter, D., 204, 208, 266, 269, 271, 302, 304
Cassatella, A., 513, 516, 709
Castor, J.I., 98, 110

INDEX OF NAMES

Cavaliere, A., 81, 91, 215, 225, 231-232, 241, 246, 250, 333, 361, 385, 393, 412, 418, 450, 455, 539-540
Cavallo, G., 385, 393
Chaffee, F.H., 432-433
Chanan, G.A., 21, 42, 59, 69, 78, 226, 313
Charles, P.A., 22, 42, 195, 210
Cheney, J., 451, 455
Cheng, F.-Z., 22, 41, 195-196, 208, 235, 250, 264, 268-269, 326-327, 330-332, 334, 487-488, 513-514, 516, 537, 539-540, 548, 550, 627-629, 631, 643-644
Cheng, F.H., 583
Chester, T.J., 19
Chiappetti, L., 91
Chincarini, G., 334, 550, 572
Chiu, L.-T.G., 317, 320, 325-326, 334, 340, 350, 370
Christiansen, W., 510-511
Chu, Y., 517-519
Chubb, T.A., 272
Ciardullo, R., 59
Ciurla, T., 604
Clark, G., 269
Clavel, J., 144, 275, 290, 709
Clegg, P.E., 16, 19, 270-271
Clowes, R.G., 340, 350, 467, 469-470, 493-494, 631
Cnuvaev, K.K., 672
Cohen, M.H., 52, 59-60, 81, 91, 108, 110
Cohen, R.D., 669, 671
Cohn, H., 132-133, 139-140, 143
Coleman, C.S., 511, 521, 523
Coleman, G.D., 217, 225
Colgate, S.A., 140, 142-143
Colina, L., 525, 528
Collin,, 429
Collin-Souffrin, S., 531-532, 535
Cominsky, L., 269
Condon, J.J., 18-19, 91, 329, 334, 419, 441, 467, 470, 566, 568
Contini, M., 476-477
Cooke, B.A., 264, 269
Corwin, H.G.Jr., 23, 41, 271
Couch, W.J., 177, 190
Courvoisier, T.J.-L., 476-477
Coyne, G.V., 251, 486
Craine, E.R., 474

Crampton, D., 42, 59, 313, 554, 608
Crane, P.C., 270
Crenshaw, D.M., 482
Cristiani, S., 81, 85, 90-91, 350, 465, 470
Cromwell, R.H., 702, 704
Cruz-Gonzales, I., 78, 483, 486
Cudworth, K.M., 317, 320, 334
Cunniffe, P.E., 298, 304

D'Addario, R., 640
D'Alessandro, M., 487-488
D'Odorico, S., 425-426, 554
Dahari, O., 195, 198-200, 202, 207-208, 210, 241, 250, 475, 477, 551, 554
Danese, L., 41, 208, 226, 232, 250, 269, 334, 412, 418, 488, 514, 516, 537, 540, 548, 550, 631, 644
Danielson, R.E., 142-143
Danziger, I.J., 8, 52, 59, 78, 91, 269, 271, 298-300, 302, 304, 350, 418, 618, 676-677
Davidson, K., 97, 106, 110, 531, 535
Davis, M., 33, 41, 195, 209, 350, 441, 482, 519, 568, 628-629, 632
de Bruyn, A.G., 642, 644
de Grijp, M.H.K., 14, 19, 204, 208, 266-268, 270, 608
de Jong, T., 19, 204, 208, 270, 507, 608
De Robertis, M., 241, 250
De Robertis, M.M., 33, 41, 204, 208
de Vaucouleurs, A., 23, 41, 271
de Vaucouleurs, G., 23, 41, 271, 502, 580, 582
de Young, D., 54, 59
De Zotti, G., 41, 208, 226, 232, 250, 269, 334, 384, 394, 418, 488, 514, 516, 537, 540, 548, 550, 631, 644
Dean, A.J., 727, 735
Deemimg, T.J., 589-590, 603-604
Dehari, O., 49, 59
Dekel, A., 186, 190
Delpino, F., 487
Delvaille, J.P., 271
Demoulin-Ulrich, M.H., 298, 304
Dennefeld, M., 266, 269, 304
Dewitt, B., 455

Dewitt, C., 455
Di Santi, M., 609
di Serego Alighieri, S., 91, 298, 300, 304-306, 689, 692
Dibaj, E., 686-687
Dibaj, E.A., 269, 501-502, 727-728, 735
Dibaj, E.E., 502
Disney, M.J., 262, 264, 269, 304, 702, 704
Djorgovski, S., 299, 305
Dopita, M.A., 41, 304-305, 477, 704
Doroshenko, V.T., 269
Downes, A.J.B., 567-568
Doxsey, R., 78, 270
Drachman, R., 739
Dressel, L.L., 467, 470, 702, 704
Dressler, A., 198-199, 208, 238, 250, 698, 700
Dumont, S., 531-532, 535
Duncan, D., 554, 608
Duncan, M.J., 132, 140, 143
Dyson, J.E., 509, 511

Eastmond, T.S., 24, 26, 41
Eddington, S.A., 647
Edgar, B.K., 721
Edwards, P.L., 590, 604
Efstathiou, G., 500, 502
Eggleton, P.P., 132, 143
Egiazaryan, A.A., 672
Ehlers, J., 470
Eilek, J.A., 721
Einstein, A., 613-614
Ekers, J.A., 702, 704
Ekers, R., 441, 618, 700, 702, 704, 708
Ellis, R.G., 327, 334
Elvis, M., 22, 41, 220, 226, 238, 242, 249-250, 262, 264, 269-270, 272, 384, 389, 394, 418-419, 671
Elvius, A., 144, 275, 651-652
Emerson, J.P., 271
Ennis, D.J., 15, 18-19, 69, 74, 78
Epstein, A., 269
Esipov, V.F., 269
Evans, D.S., 292, 577, 681

Fabbiano, G., 419
Faber, S., 185, 190
Fabian, A.C., 303-304, 383, 394

Fabricant, D., 269
Fackerell, E.D., 143
Fahlman, G.G., 471, 474, 589-590
Fairall, A.P., 258, 262, 269
Fall, S.M., 500, 502, 640
Falomo, R., 78
Fanaroff, B.L., 566, 568
Fang, L., 517, 519
Fanti, C., 91, 623, 626, 653-654, 657
Fanti, R., 59, 82, 87, 91, 441, 550, 623, 653, 657, 700, 704
Faurobert, M., 532, 535
Federici, L., 91
Feigelson, E., 269, 419
Feigelson, E.D., 270
Feldman, F.R., 210, 271, 728, 735
Felten, J.E., 428, 434
Fener, M., 270
Feretti, L., 541-542
Ferland, G.J., 37, 41, 198-199, 208, 475, 477, 529, 728, 735
Ferrari, A., 543-544, 546, 577
Fiala, N., 693
Ficarra, A., 623, 653
Fichtel, C.E., 169
Field, G.B., 104, 110
Fielden, J., 567-568
Filippenko, A.V., 21, 23, 33, 35, 41, 191, 200, 208, 242, 250, 341, 350, 475-477, 631-632, 644
Fink, U., 609
Foltz, C.B., 433, 482
Fomalont, E.B., 441, 568
Fong, R., 191, 250, 350, 418, 491, 493-494, 519, 631
Ford, H., 59
Ford, W.K., 25, 41
Forman, W., 241, 250, 261, 269, 297, 303-304
Formiggini, L., 91, 239, 250, 334, 350, 365, 370, 502, 631
Fosbury, R.A.E., 24, 37, 41-42, 59, 297-300, 302-305, 477, 677
Franceschini, A., 41, 208, 226, 232, 250, 269, 334, 417-418, 488, 537, 540, 547-548, 550, 631, 644
Frank, J., 140, 143
French, H.B., 78, 91, 313, 615, 618, 676-677
Frenk, C.S., 517, 519
Fricke, K.J., 551, 554, 574,

INDEX OF NAMES

577, 605, 609
Fried, J., 18-19, 59, 241, 250, 309, 313-314
Friedman, H., 272
Frisch, H., 532, 535
Froelich, P., 708
Fu-Zhen, C., 219, 226
Fuentes-Williams, T., 207-208

Gallagher, J., 185, 190
Galletta, G., 302, 304
Gandolfi, E., 239, 250, 350, 365, 370
Garcia, D., 591
Garilli, B., 78
Garmire, G.P., 271
Garnier, R., 79
Gaskell, C.M., 78, 207, 209, 239, 250, 370, 501-502, 605, 608, 729, 735
Gaston, B., 4, 8
Gautier, T.N., 271
Gear, W.K., 19
Gehrels, N., 168-169
Gehren, T., 18-19, 49, 59, 310-314
Gerhard, O.E., 701, 704
Ghisellini, G., 63, 76, 78
Giacconi, R., 250, 265, 269-270, 398, 418-419, 550
Giallongo, E., 225, 231-232, 250, 450, 455, 540
Gilden, D., 139, 143
Gillett, F.C., 271
Gioia, I.M., 7-8, 22, 42, 59, 226, 250, 265, 269-271, 305, 350, 418, 547, 550, 615, 618
Giommi, P., 91, 418
Giovanelli, R., 725
Giovannini, G., 541-542
Gisler, G.R., 198, 209, 541-542, 637, 640
Giuricin, G., 250, 555, 640
Glaccum, W., 19
Glass, I.S., 78
Glassgold, A.E., 78, 535, 735
Glendenning, B., 554, 608
Gold, T., 140, 143
Goldreich, P., 114, 125-126
Gondhalekar, P.M., 429, 434
Goodman, J., 143
Gopal-Krishna,, 569, 572
Gorenstein, M., 59
Gorenstein, P., 264, 269
Goss, W.M., 41, 304, 477, 677

Gott, J.R., 637, 640, 698, 700
Gottesman, S.T., 680-681
Gouguenheim, L., 502, 622
Gower, A.C., 42, 110, 313
Grady, C.A., 513
Graham, J.A., 302, 304
Grandi, S.A., 728, 735
Gray, P.M., 492-493
Green, M.R., 494
Green, R.F., 3, 5, 7-8, 21-22, 41-42, 50, 60, 174, 191, 194, 210, 220, 226, 239, 251, 312, 314, 326, 328, 330-331, 334, 341, 350, 361, 363, 365-367, 370, 372, 384, 394, 403, 412, 419, 451-452, 455, 491, 494, 538, 540, 549-550, 631-632, 735
Gregorini, L., 542, 653
Griffiths, R.E., 262, 269-272, 350, 418, 618
Grijp, M.H.P., 506-507
Grindlay, J., 269
Grindlay, J.E., 271
Gruber, D.E., 169, 271
Guibert, J., 502
Gull, S.F., 328, 334, 571-572
Gull, T.R., 226, 514, 516
Gunn, J.E., 23, 42, 198, 208, 226, 312-313, 324, 331, 334, 558, 637, 640, 642, 644, 694, 696, 701, 704
Gursky, H., 269
Gusten, R., 461-463

Habbal, S.R., 545-546
Habing, H.J., 19, 270
Hack, M., 709
Hacking, P., 19
Hackney, R.L., 590
Halpern, J.P., 23, 37, 41, 198-200, 208, 475, 477
Hamabe, M., 622
Hanni, R.S., 448, 455
Hardee, P.E., 546
Harding, A.K., 169
Harlan, E., 604
Harnden, F.R., 269
Haro, G., 467, 470
Harper, D.A., 19, 636
Harris, A.W., 513, 516
Harris, S., 271
Harris, W.E., 341, 350
Hart, M.H., 138, 144
Hartman, R.C., 169

Hartwick, F.D.A., 42, 143, 546
Harvey, G.A., 90-91
Haschick, A.D., 461-463
Hauser, M.G., 182, 191, 271
Hausman, M., 637, 640
Hawarden, T.G., 271, 581-582
Hawkins,, 85
Hawley, J., 114, 126
Hawley, S.A., 78, 615, 618, 677
Haymes, R., 739
Haynes, M.P., 723, 725
Hazard, C., 233, 238, 244, 250, 331, 334, 451, 454-455, 704
Heckman, T.M., 21-24, 26, 30, 32-33, 35, 40-41, 49-50, 59, 191, 199, 209, 241, 250, 253, 258-259, 270, 299, 304, 312-313, 475, 477, 701, 704, 728-729, 735
Heeschen, D.S., 568, 577
Heidmann, J., 622
Heisler, J., 244, 251
Heitter, W., 739
Helfand, D., 78, 269, 399, 418
Hellwig, J., 605
Helou, G., 17, 19, 725
Helpern, J.P., 209
Henry, J.P., 269, 419
Herterich, J., 385, 394
Herterich, K., 168-169
Herzog, A.D., 342, 350
Hewckman, T.M., 78
Hewitt, A., 55, 58-59
Hickson, P., 637-640
Hills, J.G., 140, 143, 471-472, 474
Hoag, A.A., 324, 365, 367, 370
Hodge, P.E., 77
Hoessel, J.G., 555, 558
Holm, A.V., 513, 516
Holmberg, E., 651-652, 713, 716
Holt, S.S., 269-271, 391, 394
Hook, R.N., 304
Horine, E., 59
Houck, J.R., 19, 271, 723
Hough, J.H., 483, 486
Hoyle, F., 56, 59
Huang, K.L., 468, 470, 540
Hubble, E., 216, 225
Huchra, J.P., 22, 33, 41, 50, 55, 59, 78, 194-196, 202, 209, 226, 250, 254, 270, 334, 341, 350, 370, 419, 481-483, 486, 494, 540, 627-629, 632, 641

Huenemoerder, D.H., 216, 226
Huggins, P.J., 535, 735
Hummel, E., 207, 209, 250, 439, 441, 554, 608, 639-640, 716
Hunstead, R.W., 433
Hunter, D., 687
Huntley, J.M., 575, 577, 680-681
Hut, P., 143
Hutchings, J.B., 21, 42, 49, 59, 110, 301, 304, 310-313, 551, 554, 605, 608
Hyland, A.R., 77

Icke, V., 107, 110
Illingworth, G., 342, 350
Impey, C.D., 78, 483-484, 486
Inoue, M., 622
Ipser, J.R., 143
Ishida, K., 621-622

Jacoby, G.H., 696
Jaffe, W., 329, 334, 441, 700, 704
Jakoby, G.H., 693
Jankevics, A., 363, 370
Jaroszynski, M., 114, 126, 506-507
Jauncey, D.L., 91, 677
Jenkins, C.J., 477
Jenkins, C.R., 305, 702, 704, 715-716
Jenkins, E.B., 264, 270
Jenner, D.C., 24, 26, 41
Jennings, R.E., 271
Jensen, E.B., 637, 640
Jensen, K., 271
Jensen, K.A., 271
Johnson, K., 474
Johnson, W., 739
Johnston, K.J., 65, 79
Johnston, M.D., 270
Joly, M., 532, 535, 559
Jones, B., 713, 716
Jones, B.J.T., 334, 529
Jones, C., 241, 250, 269, 297, 303-304
Jones, J.E., 529
Jones, T.W., 54, 59
Julien, P., 269
Junkkarinen, V., 59

Kahn, F.D., 470
Kaiser, N., 455
Kallman, T., 304
Kapahi, V.K., 565, 567-569,

571-572
Kapoor, R.C., 573, 575, 577
Kardashev, N.S., 151, 169
Karoji, H., 622
Katgert, P., 437, 441-442
Katgert-Merkelijn, J.K., 437, 439, 441
Kazanas, D., 151, 169, 389, 394
Kazarian, E.S., 255, 270
Kazarian, M.A., 255, 270
Keel, W.C., 22, 26, 29, 31, 33, 37, 41-42, 52, 59, 199-200, 207, 209, 218-219, 226, 241-242, 249-250, 264, 270, 475, 477, 551, 554, 579, 608, 686-687, 713, 715-716, 735
Kellermann, K., 15, 19, 52, 59, 550
Kellogg, C., 269
Kemp, J.C, 486
Kennicutt, R.C., 207, 209, 250, 554, 608, 716
Kent, S., 59
Kenyon, S., 687
Khachikyan, E.E., 253, 672, 692
Kiang, T., 583
Kidger, M., 471-472, 474, 587, 591, 601, 604
Kinman, T.D., 78, 601-602, 604, 618
Kinney, A.L., 535, 728, 735
Kirillov-Ugryumov, V.G., 169
Kjar, K., 554
Klein, R.I., 98, 110
Knapp, G.R., 298, 304
Koch, D., 269
Kolb, E., 661
Kollatschny, W., 551, 554, 577, 605, 609
Komesaroff, M.M., 78
Kondo, M., 619-622
Konigl, A., 76, 511, 544, 546
Konigl, A.P., 78
Koo, D.C., 175, 177, 183, 191, 224, 226, 232-233, 238, 244, 250, 317, 319-320, 324, 330, 334, 340-341, 344, 346, 350, 362, 412, 418, 438, 441-442, 451, 455, 491, 493, 537, 539-540, 631-632, 700
Korovyakovskii, Y.P., 689, 692
Koski, A.T., 37, 39, 42, 59
Kotanyi, C.G., 702, 704
Kramer,. 729
Kreidl, T., 577

Kriss, G.A., 215, 226, 249-250, 262, 270, 399, 403, 547, 550
Kristian, J., 42, 48, 59, 144, 177, 191, 299, 304, 310, 313
Krolik, J.H., 96, 104, 110, 271, 509-511, 531-532, 535, 728, 735
Kron, R.G., 175, 191, 224, 226, 317, 319-320, 324-326, 333-334, 340, 342, 344, 346, 350, 370, 438, 441-442, 491, 493, 631-632, 700
Kronberg, P.P., 303-304
Krumm, N., 428, 434
Ku, W.H.-M., 79, 399, 418, 618
Kuhr, H., 78, 312-313, 393-394, 611, 614, 623, 626, 732, 735
Kulkarni, V.K., 570-572
Kultarni, V.K., 569
Kumar, S., 114, 126
Kunth, D., 8, 203, 269, 271, 350, 423, 434, 515-516, 618, 693
Kwan, J., 40-41, 279-280, 291, 509-511, 531-532, 535, 728, 735

Lacy, J., 739
Laing, R., 721
Laing, R.A., 569
Lamla, E., 604
Landau, L.D., 122
Landau, R., 69, 78, 450
Landauer, F.P., 42, 144
Landman, U., 412, 418
Lari, C., 441, 572, 700, 704
Larson, R.B., 713, 716
Latham, D., 350, 433, 632
Lauberts, A., 579, 582
Lawrence, A., 220, 226, 262, 264, 270, 671
Lawrence, C.R., 672
Lazarides, G., 659, 661
Leach, R., 609
Leacock, R.J., 590
Lealocock, R.J., 604
Lebofsky, M.J., 268, 271, 484, 633, 635, 652
Ledden, J.E., 78
Leinert, C., 503, 507
Leir, A.A., 640
Leiter, D., 383-385, 387, 389-391, 393-394
Lelievre, G., 543, 546
Lentes, F.Th., 500, 502

Lepp, S., 297, 304
Lequeux, J., 441, 568, 708
LeVan, P.D., 532
Leventhal, M., 150, 169, 739
Lewis, D.W., 202-203, 209, 270, 365, 370, 579, 582
Li, S., 519
Li, X., 661
Li, Z., 659, 663, 737
Liebert, J., 8, 59, 270-271, 305, 313, 350, 418-419, 614, 618
Liebert, J.W., 269
Lifshitz, E.M., 122, 450
Light, E.S., 142-143
Lightman, A.P., 138, 140, 143, 385, 389, 394
Lindblad, P.O., 42, 272
Linfield, R.P., 60
Lingenfelter, R.E., 150, 169, 739
Lipovetskij, V.A., 195, 202, 209, 270
Livio, M., 114, 117-118, 126
Lloyd, C., 587, 590, 601-602, 604, 669-670
Lo, K.Y., 723, 725
Lobofsky, M.J., 486
Locke, J.L., 91
Long, K., 269
Longair, M.S., 251, 451, 455
Longmore, A.J., 271, 582
Lonsdale, C.J., 19
Loose, H.H., 551, 554
Lorre, J.J., 580
Lovelace, R.V.E., 151, 169
Low, F.J., 19, 271, 652
Lowe, J., 582
Lu, J., 113-114, 117-118, 126
Lub, J., 8, 19, 204, 208, 269-271, 350, 507, 608, 618
Lucas, R., 681
Lucy, L.B., 98, 110, 399, 418, 635
Luyten, W.J., 467, 470
Lynden-Bell, D., 132, 143, 357, 370, 471
Lynds, C.R., 42, 143, 546, 618
Lynds, R., 362, 370
Lyutyi, V., 672

Ma, Z., 661, 667
MacAlpine, G.M., 202-203, 208-209, 255, 270, 370, 579, 582

Maccacaro, T., 7-8, 22, 42, 59, 91, 219-220, 226, 246, 248, 250, 264-266, 269-272, 305, 350, 402-403, 413, 418-419, 547-548, 550, 615, 618
Maccagni, D., 250, 269, 615, 618
MacCallum, C.J., 150, 169
Macchetto, F., 298, 300, 305, 692
Macdonald, D., 522-523
Macdonald, D.A., 166, 169, 455
MacGillivary, H.T., 191
Machalski, J., 566, 568
MacKay, C., 313, 614
MacKenty, J.W., 49, 59, 299, 305, 312-314
MacLeod, J.M., 90
Madejski, G.M., 78
Maehara, H., 619, 622
Magro, C. Sanchez, 633
Malin, D.F., 26, 42, 302, 304
Malkan, M.A., 21, 42, 49, 59, 74, 78, 97, 110, 216-217, 226, 310, 313
Mantovani, F., 91, 570, 623, 653
Marano, B., 232, 250, 324-325, 334, 339, 412, 419, 468, 537, 539-540
Maraschi, L., 59, 63-64, 74, 78, 89-91
Marchant, A.B., 132, 143
Mardirossian, F., 250, 455, 555, 640
Mardon, G., 91
Marenbach, G., 707-708
Margon, B., 21, 42, 54, 59, 226, 313
Marien, K.H., 479, 482
Markarian, B.E., 195, 202-203, 209, 254-255, 266, 268, 270, 481, 642
Markert, T., 269, 649-652
Marscher, A.P., 76, 78
Marsden, P.L., 271
Marshall, F.E., 265, 270-271, 383-384, 389, 394
Marshall, H., 412-413, 419
Marshall, H.L., 22, 42, 220, 226, 236, 239, 250, 325, 327-331, 333-334, 341, 346, 350, 363, 365, 370, 400, 402-403, 419, 491, 493-494, 537, 539-540, 627-629, 631-632
Martin, P.G., 78
Martinez, C., 591

INDEX OF NAMES

Maslowski, J., 566, 568
Mason, K.O., 42, 271
Mathews, W.G., 208, 728, 735
Mathez, G., 328
Mathis, J.S., 513, 516
Matteson, J.L., 169, 271, 739
Matthews, K., 19, 586
Matveenko, L., 653
Maza, J., 78
Mc.Kee, C.F., 511
McAlary, C.W., 735
McAlister, H.A., 78
McAlpine, G.M., 365
McCray, R., 304
McGimsey, B.Q., 587, 590
McHardy, I., 78, 269
McHardy, I.M., 261, 270
McKee, C.F., 96, 104, 108, 110, 509-510, 717, 721
McMahon, R., 233, 238, 244, 250, 451, 455
McMillan, S.L.W., 140, 143
McMullan, D., 652
McNutt, D.P., 272
Mebold, U., 41, 304, 477, 500, 502
Medd, W.J., 87, 90-91
Mediavilla, E., 503-505, 507, 633
Meekins, J.F., 272
Meidav, M., 272
Meier, D.L., 330, 334, 453, 455
Meinel, A.B., 676-677
Melnick, J., 30, 42, 476-477
Menon, T.K., 637-640
Merkelijn-Katgert, J., 571-572
Messina, A., 225, 540
Methez, G., 334
Metik, L., 683-684, 686-687
Meurs, E.J.A., 22, 42, 194-196, 209, 219, 226, 256, 268, 270, 440, 442, 641-644
Meyers, K.A., 482
Mezzetti, M., 250, 555, 640
Mijauchi-Isobe, N., 622
Mikami, T., 622
Miley, G., 271, 299, 304
Miley, G.K., 13, 19, 41, 191, 204, 208-209, 270, 442, 608, 637, 640, 700, 735
Miller, G.E., 640
Miller, H.R., 78, 482
Miller, J.S., 37, 42, 78, 91, 208, 253, 271, 615, 618, 676-677, 696, 735

Miller, W.G., 18-19
Milley, G.K., 507
Minkowski,, 174
Misner, C.W., 614
Missana, M., 645-647
Mitchell, K.J., 241, 251, 330, 334, 467, 470, 537, 540
Mitchell, K.S., 631-632
Mitton, S., 704
Miyauchi-Isobe, N., 619
Mollenhoff, C., 707-708
Moncrief, V., 113, 121-122, 126
Moore, R.L., 64, 78-79
Moreno, E., 680-681
Morris, M., 635
Morris, S., 42, 313
Morrison, P., 385, 393
Morton, D.C., 455
Mould, J., 178, 191, 313
Mufson, S.L., 78
Murdoch, H.S., 433
Murphy, M., 591
Murray, S., 269
Murray, S.S., 270
Mushotzky, R.F., 79, 169, 242, 251, 264, 270-272, 384, 388, 394, 529, 728, 735
Muxlow, T., 623

Narayan, R., 114, 125-126
Nardon, G., 90, 350
Narlikar, J.V., 55-56, 59-60, 573, 577
Netzer, H., 37, 41, 82, 84-85, 91, 97, 106, 110, 198-199, 208, 475, 477, 529, 531-532, 535, 728, 735
Neugebauer, G., 11, 13, 15, 17-19, 39, 42, 78, 110, 174, 191, 204, 209, 266, 270-271, 586, 635
Nicolson, G.D., 653
Nieto, J.-L., 543, 546, 580, 582
Nilson, P., 579, 582
Nobili, L., 546
Noguchi, T., 619-622
Noit, I.G., 19
Norman, C., 239, 251
Norman, M.L., 718, 721
Norris, R.P., 461-463
Notni, P., 649-652
Nottale, L., 330, 334
Nousek, J.A., 271
Novick, R., 269
Novikov, I.D., 81, 91, 149, 169

Nugent, J.J., 262, 271
Nulsen, P.E.J., 299, 303-304

O'Connell, D., 143
O'Connell, R.W., 702, 704
O'Dell, S.L., 19, 78-79
Oda, M., 250, 550
Oke, J.B., 18-19, 21, 23,
    41-42, 48-49, 59, 78, 107,
    110, 176, 190, 299, 304, 311,
    586, 695-696
Oknyanskii, V.L., 672
Olnon, F.M., 271
Omizzolo, A., 91, 350
Oort, J., 325, 334
Oort, M.J.A., 437, 442
ORKeel, W.C., 22
Osmer, P.S., 4, 8, 233, 238,
    244, 251, 331, 334, 364-365,
    367-368, 370, 384, 394, 451,
    454-455, 469-470, 519, 529,
    539-540
Osterbrock, D.E., 22, 33, 37,
    39, 41-42, 49, 59, 193,
    195-196, 198-202, 209-210,
    238, 251, 253, 256, 271, 475,
    477, 481-482, 501-502, 641,
    643-644, 670, 696, 707-708
Ostriker, J.P., 106, 110, 244,
    251, 637, 640
Owen, F.N., 78, 442, 637, 640,
    700
Owen, M.H., 546

Pacholczyck, A.G., 577
Paczynski, B., 507
Padovani, P., 251
Padrielli, L., 91, 441, 572,
    623, 653
Page, C.G., 269
Pagel, B.E.J., 475, 477
Pallister, W.S., 652
Palmer, P., 635
Papaloizou, J.C.B., 113-114,
    124-127
Parker, E., 545-546, 659-661
Parma, P., 542
Paturel, G., 502
Pauliny-Toth, I.I.K., 15, 19,
    614
Peacock, J.A., 328, 334, 569,
    571-572
Pearson, T.J., 60
Peebles, P.J.E., 355, 370, 700
Peimbert, M., 22, 42, 200, 210

Pelat, D., 349
Pence, W., 177, 190, 681
Peng, Q., 659, 661, 663, 667,
    737, 739
Penrose, R., 448, 455
Penson, M.J., 635
Penston, M.V., 82, 91, 144,
    272, 275, 304, 515-516, 633,
    635, 669, 672
Pequignot, D., 39, 42, 475-477,
    535
Perez, E., 24, 42, 633, 635,
    669, 672
Perez-Fournon, I., 248, 251, 618
Perley, R., 52, 59
Perola, G.C., 144, 270, 275,
    441, 526, 529, 700, 704
Perry, J.J., 470, 509, 511
Perryman, M.A.C., 298, 300,
    305, 692
Persic, M., 675
Pesch, P., 203, 210, 256, 271
Peters, G., 269
Peterson, B.A., 78, 88, 91,
    250, 350, 418, 432, 434, 467,
    470, 491, 677
Peterson, B.M., 434, 481-482
Peterson, C.J., 680-681
Peterson, L.E., 169, 271
Petre, R., 264, 271
Petrosian, V., 328, 333-334,
    353, 356, 362-363, 370
Petrosyan, A.R., 689, 692
Petteni, M., 144
Petterson, J.A., 96, 110
Pettini, M., 433
Phillips, M.M., 22, 26, 29, 34,
    39, 41-42, 195, 198, 208,
    210, 262, 271, 298, 302,
    304-305, 349, 477, 481-482,
    701-702, 704
Phinney, E.S., 244, 251, 448,
    453, 455, 540
Pica, A.J., 78, 82, 85, 91,
    590, 604
Piccinotti, G., 64, 78, 262,
    271, 615, 617-618
Pierce, M.J., 723, 725
Pineda, F.J., 262, 271
Pismis, P., 679-681
Podurets, M.A., 130, 144
Pogge, R.W., 200, 210
Pollock, J.T., 587, 590, 604
Polnarev, A.G., 169
Polovin, R.V., 523

INDEX OF NAMES

Popov, V.N., 672
Poltasch, S.R., 271
Pounds, K.A., 269-270
Pratt, N.M., 577
Pravdo, S.H., 264-265, 271
Price, R.M., 302, 304
Primack, J., 190
Pringle, J.E., 113-114, 124-127
Pritchet, C., 313
Pronik, I., 683-684, 687
Pronik, V., 686-687
Puetter, R.C., 532, 586
Purgathofer, A.Th., 465, 470
Puschell, J.J., 19, 78
Pye, J.P., 269-270, 419

Quinn, P.J., 640

Radositz, J.V., 19
Rafanelli, P., 479, 689
Raimond, E., 271
Raine, D.J., 717, 719, 721
Rakos, K.D., 693
Ramaty, R., 150, 169
Ramsey, L.W., 210, 735
Rauh, W., 479, 482
Ray, E.C., 140, 143
Readhead, A.C.S., 108, 110
Rees, M.J., 53, 75, 77, 106,
    110, 113, 127, 130-131, 134,
    140, 142-143, 150, 168-169,
    190, 232, 239, 246, 250-251,
    383, 385, 393-394, 447,
    454-455, 506-507, 521, 523,
    544, 546, 573, 577, 702, 704,
    717-718
Reichert, G.A., 31, 42
Reif, K., 500, 502
Reigler, G.R., 169
Reina, C., 264, 271
Reitsema, H.J., 79
Reynolds, S.P., 76, 78
Rice, W., 19
Richstone, D.O., 48, 59, 74,
    78, 637, 640
Rickard, L.J., 634-635
Ricker, G.R., 270, 399, 403, 418
Ricketts, M.J., 269
Riecke, G.H., 78
Riegler, G.R., 271, 739
Rieke, G.H., 268, 271, 483-484,
    486, 618, 649, 652, 735
Rieke, G.M., 633, 635
Riley, J.M., 566, 568
Roberts, M.S., 25, 41, 725

Robertson, H.P., 503, 507
Robertson, J.G., 441
Robinson, I., 143
Robinson, L.B., 19, 60
Robson, E.I., 19
Rolfe, E.F., 502
Romanishin, W., 49, 59
Romano, G., 90-91
Romney, J., 623, 653-654, 657
Rood, H.J., 555
Roos, N., 239, 251, 333, 441,
    697-698, 700
Rose, J.A., 35, 37, 42
Roser, H.-J., 433
Roser, S., 507
Rosner, R., 543, 546
Rothschild, R.E., 150, 169,
    264, 271, 383-384, 389, 394
Rowan-Robinson, M., 11, 19,
    268, 271, 451, 455
Rubakov, V., 661, 667
Rubin, V.C., 25, 41
Ruchti, C.B., 151, 169
Rudnick, L., 721
Rudy, R.J., 261, 271, 735
Ruffini, R., 448, 455
Rusconi, L., 304

Saakyan, K.A., 692
Sadler, E.M., 298, 305, 477,
    701-702, 704
Salpeter, E.E., 81, 91, 356,
    370, 428, 434, 723, 725
Salucci, P., 675
Salvati, M., 82, 91
Sandage, A., 18-19, 23, 25, 42,
    48, 59, 79, 181, 191,
    239-240, 251, 259, 264, 271,
    319, 643-644, 651-652, 696
Sanders, R., 140, 142-143
Sanduleak, N., 203, 210, 256,
    271
Santangelo, P., 250
Sargent, W.L.W., 21-23, 26,
    41-42, 74, 78, 97, 110, 130,
    143, 194, 209, 254, 258,
    270-271, 429, 433-434,
    515-516, 544, 546, 631-632,
    693, 723, 725
Saslaw, W.C., 58-59, 140, 144,
    573, 575, 577
Sastry, G.N., 555
Savage, A., 467, 469-470, 494,
    517, 519
Savage, B.D., 264, 270, 513, 516

Scarrott, S.M., 652
Schechter, P.L., 327, 715-716
Schectman, S.A., 199, 208
Scheuer, P.A.G., 81, 91
Schiano, A.V.R., 511
Schild, R., 305
Schild, R.E., 8, 59, 269, 462-463, 618
Schleicher, H., 577
Schlosman, I., 402, 419
Schmidt, G.D., 313, 614, 618
Schmidt, J., 614
Schmidt, M., 3, 5-8, 22, 42, 74, 78-79, 81, 91, 174, 191, 194, 220, 226, 239, 242, 251, 324, 328, 330-331, 334, 339, 341, 347, 350, 361-363, 365-367, 370, 372, 384, 394, 403, 412, 418-419, 451-452, 455, 491, 494, 538, 540, 549-550, 631-632, 642, 644, 702, 704
Schmutzler, T., 552, 554
Schneider, D., 224, 226, 324, 331, 334, 556, 558
Schneider, S.E., 723-725
Schnopper, H.W., 271
Schommer, R., 313, 642, 644
Schreier, E., 269
Schultz, A., 609
Schulz, H., 705, 707-708
Schutz, B.F., 123-124, 127
Schwartz, D.A., 78-79, 91, 270, 615, 618
Schwarz, J., 270
Schwarzschild, M., 142-143
Sciama, D.W., 511
Scott, J.S., 511
Scott, R.L., 590, 604
Scoville, N.Z., 633-635
Searle, L., 35, 42, 79
Sedmak, G., 304
Seidl, F.G.P., 139, 143
Seielstad, G.A., 60
Selmes, R.A., 635
Selvelli, P.L., 290, 513, 709
Sembay, S., 727, 735
Serlemitsos, P.J., 269-271
Setti, G., 7, 59, 232, 236, 244, 246-247, 251, 540, 547, 550
Seward, F., 269
Seward, F.D., 269
Seyfert, C.K., 81, 91, 253, 271
Shafer, R.A., 246, 251, 271

Shaham, J., 402, 419
Shakura, N.I., 110, 505, 507, 717, 719, 721
Shanks, T., 177, 190-191, 250, 350, 418, 491-494, 519, 631
Shapiro, S.L., 59, 129-130, 132-136, 138-141, 143, 244, 251
Sharp, N.A., 713, 716
Sharples, R.M., 262, 264, 271, 305
Shaver,, 432
Shaviv, G., 95, 106, 110, 402, 419
Sheffer, Y., 82, 84-85, 91
Shields, G.A., 140, 144, 717, 721
Shklovsky, I.S., 573, 577
Shlosman, I., 95, 106, 110
Shortridge, K., 42, 143, 546
Shuder, J.M., 22, 42, 196, 200, 210, 262, 271, 481-482, 671
Shull, J.M., 304
Shure, M.A., 725
Silk, J., 173, 185-186, 189-191, 239, 251, 507, 698, 700
Simon, R.S., 60
Sis, J., 708
Sitenko, A.G., 523
Skrutskie, M.F., 725
Smak, J., 126-127
Smarr, L., 114, 126, 718, 721
Smathers, H.W., 272
Smith, A.G., 82, 85, 91, 590, 604
Smith, B.A., 79
Smith, E., 313
Smith, E.O., 79
Smith, E.P., 59, 250
Smith, G., 59
Smith, H.E., 79, 586
Smith, M.D., 717, 719, 721
Smith, M.E., 677
Smith, M.G., 19, 255, 271, 365, 367, 370, 419, 428, 433-434, 465, 467, 469-470
Smith, S.B., 202, 209, 270, 579, 582
Smyth, M.J., 602, 604
Snijders, M.A.J., 79, 144, 275, 304, 423, 434
Soifer, B.T., 11-13, 17-19, 204, 209, 271
Solf, J., 479, 482

Solinger, A.B., 649-652
Solomon, P.M., 98, 110
Soltan, A., 22, 41, 242, 249-251, 418-419
Spangler, S.R., 78
Spinrad, H., 53, 79, 299, 305, 314
Spitzer, L., 132, 138, 140, 142-144
Sramek, R.A., 210, 365, 367, 370
Stasinska, G., 429, 434
Stauffer, J.R., 22, 31, 33-34, 42, 199-200, 210, 259, 271, 476-477
Steckney, F., 739
Stein, W.A., 19, 78-79
Steiner, J.E., 37, 41, 198-199, 209, 475, 477, 727, 735
Stepanov, K.N., 523
Stepanyan, D.A., 195, 202, 209, 255, 270
Steppe, H., 569, 572
Stern, B.E., 149-150, 169
Stern, R.A., 271
Stevenson, P.R.F., 183, 191
Stewart, G.C., 303-304
Stocke, J., 207-208, 226, 270
Stocke, J.T., 7-8, 22, 42, 55, 59, 69, 79, 250, 265, 269, 271, 297, 305, 341, 350, 418-419, 548, 550, 615-618, 637, 640
Stockman, H.S., 64, 77-78, 87, 91, 246, 250
Stockton, A., 48-50, 59-60, 208, 299, 304-305, 312, 314, 605, 609
Stockton, M.W., 677
Stone, M.E., 140, 144
Stottlemyer, A., 389, 394
Strittmatter, P.A., 56, 79, 130, 144, 313, 419, 614, 735
Stryker, L.L., 694, 696
Subrahmanya, C.R., 567-568, 571-572
Subramanian, K., 56, 60
Suen, W.-M., 455
Sulentic, J.W., 54-55, 60, 470, 574, 577
Sunyaev, R.A., 110, 505, 507, 717, 719, 721
Svensson,, 385
Swings, J.P., 251, 370, 455
Switzer, P., 369
Sykes, J.B., 586

Szalay, A.S., 183, 191, 246, 250
Tadhunter, C.N., 299, 304-305, 670
Tagliaferri, G., 78
Takase, B., 619, 621-622
Tammann, G.A., 23, 42, 181, 259, 271, 652
Tananbaum, H., 7-8, 91, 226, 237, 242, 250, 269, 334, 350, 370, 394, 398-411, 413-414, 416, 418-419, 494, 540, 547-548, 550, 616, 618, 632
Tang, J., 661, 667
Tanzi, E.G., 59, 63, 78, 89-91, 144
Tapia, S., 314, 472, 474, 484, 486
Tarenghi, M., 144, 264, 271, 304, 615, 618
Tarquini, G., 250
Tarter, C.B., 96, 104, 110, 509-511
Taylor, K., 304
Telesco, C.M., 633-636
Terebizh, V.Y., 269
Terlevich, R., 29-30, 34, 41-42, 198, 208, 455, 476-477
Terzian, Y., 723, 725
Teukolsky, S.A., 130, 132-136, 141, 143
Thiele, U., 479
Thompson, A.R., 640
Thompson, D.J., 169
Thompson, I.B., 199, 208
Thompson, R.I., 652
Thompson, R.W., 513, 516
Thorne, K., 386, 394, 522-523, 614
Thorne, K.S., 166, 169
Thorstensen, J.R., 42
Thuan, T.X., 558, 642, 644
Tiersch, H., 649, 651-652
Tinsley, B.M., 713, 716
Toounaga, A.T., 652
Tohline, J., 707-708
Tolman, R., 216, 226
Tonry, J., 350, 632
Toomre, A., 208, 210
Toomre, J., 208, 210
Topka, K., 269
Torres-Peimbert, S., 22, 42, 200, 210
Townes, G., 739
Trefzger, C., 479, 482

Treves, A., 59, 63, 78, 90-91
Tripicco, M.J., 37, 42
Tritton, K.P., 272
Trumper, J., 251, 266, 270-271, 523, 550
Trussoni, E., 543, 546
Tsinganos, K., 543, 545-546
Tubbs, A.D., 707-708
Tucker, W., 297, 303-304, 402, 419
Tully, R.B., 723, 725
Tuohy, I.R., 477
Turner, B.E., 635
Turner, E.L., 298, 304, 637, 640, 698, 700
Turner, M.J.L., 269
Turnrose, B.E., 513, 516
Turnshek, D.A., 434
Tyson, J.A., 574, 577
Tytler, D., 422, 424, 428, 432, 434, 626

Ukita, N., 622
Ulrich, M.H., 24, 42, 52, 60, 79, 130, 144, 275, 277-278, 283-284, 291-292, 441, 516, 526, 528-529, 615, 633, 636, 669-670, 680-681, 704-705, 707-708, 716
Ulrych, T.J., 471, 474, 589-590
Ulvestad, J.S., 110, 671
Ulvestad, M.J., 577
Unwin, S.C., 59-60, 81, 91
Urry, C.M., 79, 246, 251, 675, 677
Urry, M.H., 618
Usher, P.D., 241, 251, 319, 326, 330, 334, 468, 470, 537, 540, 631-632

Vagnetti, F., 225, 231-232, 250, 450, 455, 540
Valtaoja, E., 521, 523
Valtonen, M.J., 59
van Breugel, W., 299, 304
van Breugel, W.J.M., 41, 174, 191
van den Bergh, S., 341, 350, 637, 640, 696
van der Hulst, J.M., 207, 209, 250, 554, 608, 716
van der Kruit, P.C., 680-681
van der Laan, H., 437, 440-442
van Duinen, R., 270
van Heerde, G.M., 437, 442

van Speybroeck, L., 22, 41, 269
Vardanian, V.V., 169
Vedovato, G., 667
Veron, P., 12-13, 15, 17-19, 21-22, 25, 34-35, 42, 85, 253-254, 259, 262, 266-268, 271-272, 325-326, 331-333, 335, 439, 442, 451, 465, 470, 491, 494, 643-644
Veron-Cetty, M.-P., 12-15, 17-19, 25, 42, 85, 254, 259, 266-269, 272, 325-326, 335, 439, 442, 465, 470
Vietri, M., 249, 251
Vignato A., 314
Visvanathan, N., 651-652
Vitello, P.A., 95, 106, 110
Vittorio, N., 250
Vorontsov-Velyaminov, B., 207, 210
Vrtilek, J., 707-708

Wade, C.M., 568
Wagner, R.M., 482
Walker, M., 470
Walker, R.C., 52, 60
Walker, R.G., 271
Wall, J.V., 467, 470, 571-572
Wallace, P.T., 305
Walter, F.M., 271
Wampler, E.J., 18-19, 48, 60, 367, 370, 529
Wamsteker, W., 91, 271, 304, 525, 528-529
Wandel, A., 388, 392-393, 727, 732, 735
Wang, D., 659, 663, 737
Ward, M.J., 26, 41, 77, 262, 264, 272, 577
Wardle, J.F.C., 79
Warnock, A., 241, 251, 326, 330, 334, 470, 540, 631-632
Wasilewski, A.J., 200-201, 204, 210, 254, 256, 268-269, 272
Watanabe, E., 622
Watson, M.G., 269
Webster, A., 519
Weedman, D.W., 21, 42, 195-196, 210, 215-216, 219-223, 225-226, 232, 237, 242, 251, 253, 269, 324, 328, 335, 362, 365, 367, 370, 412, 419, 451, 537, 540, 643-644, 735
Wehinger, P., 19, 59, 311, 313-314

INDEX OF NAMES

Weiler, K.W., 65, 79, 91, 653
Weistrop, D., 79
Weliachew, L., 681
Wellington, K.J., 304
Werner, M., 15, 19, 78
Werner, M.W., 507
Wesselius, P.R., 271
Westerlund, B.E., 79
Westphal, J.A., 42, 144, 191
Weymann, R.J., 419, 421-422, 433-434, 509, 511
Wheeler, J.C., 140, 144, 614
Whelan, J.A.J., 433
White, C., 652
White, N., 388, 394
White, S.D., 239, 251, 519, 640
Whitmore, B.C., 500, 502
Whittle, M., 33, 42, 349
Wickramasinghe, N.C., 507
Wilczek, F., 667
Wilkes, B.J., 91, 242, 250, 394, 418, 675-677
Wilkinson, A., 302, 305
Williams, G.A., 202, 209, 270
Williams, P.J.S., 15, 19
Williams, P.M., 486
Williams, R.E., 130, 144, 434
Wills, B.J., 79, 531-532, 535
Wills, D., 531-532, 535
Wilson, A.S., 22, 26, 41-42, 110, 194-195, 209, 219, 226, 256, 262, 268, 270-272, 304, 440, 442, 511, 574, 577, 641-642, 644, 671
Wilson, C.P., 42, 144
Wilson, J.R., 114, 126
Windhorst, R., 437-438, 440-442, 566, 568, 571-572, 698-700
Winkler, K.-H., 718, 721
Wirtanen, G.A., 604
Wirth, A., 687
Wisniewski, W.Z., 484, 486
Witzel, A., 614
Wlerick, G., 79
Wolfe, A.M., 78, 250, 423, 428, 433-434, 677
Wolstencroft, R.D., 486, 602, 604
Woltjer, L., 7, 232, 246-247, 251
Wood, K.S., 63, 79, 262, 272
Wood, P.A.D., 461, 463
Woods, D.T., 304
Worrall, D.M., 79, 384, 389, 394

Wright, A.E., 91, 677
Wright, M.C.H., 681
Writhe, A.E., 467
Wu, C.-C., 210, 215, 225-226, 514, 516, 526, 529
Wyatt, W.F., 33, 41, 195, 209, 482
Wyckoff, S., 19, 59, 310, 313-314
Wynn-Williams, C.G., 635-636
Wyse, R.F.G., 173, 175, 178, 180, 189, 191

Xiang, S., 519

Yahil, A., 727, 735
Yee, H.K.C., 50, 60, 215, 226, 311-312, 314, 643-644, 671, 695-696
Yentis, D.J., 272
You, J.H., 583
Young, E., 271
Young, J.S., 635
Young, P., 422, 424, 428, 432, 434
Young, P.J., 26, 42, 130, 140, 143-144, 434, 546
Yutani, M., 622

Zambon, M., 91
Zamorani, G., 8, 74, 79, 88, 91, 226, 232, 244, 250-251, 270, 325, 334, 339, 350, 370, 399, 418-419, 494, 540, 542, 547-548, 550, 617, 632
Zaninetti, L., 546
Zdziarski, A.A., 151, 169, 386, 389, 394
Zeilinger, W.W., 302, 304
Zel'dovich, Ya.B., 81, 91, 130, 144
Zemelman, M., 271
Zhu, X., 518-519
Zitelli, V., 91, 226, 232, 250-251, 325, 334, 339, 350, 370, 419, 487, 494, 502, 540, 631-632
Znajek, R.L., 150, 166, 169, 448, 455
Zuckerman, B., 635
Zuiderwijk, E.J., 42, 272
Zwicky, F., 203, 256-258, 272, 579, 582, 642

INDEX OF SUBJECTS

Absorption lines, 422-433
  and LY Forest, 432-433
  cosmological evolution, 424
  degree of ionization, 423, 425-431
  metal-rich, 421
Accretion
  adiabatic, 113-121
  spherical, 121-125
Accretion disks, 114, 717-721
  bel regions formation, 107-109
  clouds formations, 104-106
  line driven winds, 97-103
  stability, 113-114
  thermal instabilities, 103-104
  thick, 113, 125-126
Active galactic nuclei
  dust accretion disks, 503
Active galaxies
  evolution of the gas, 303
  group members
    VLA observations, 637-639
  IRAS observations, 11
  morphology of the gas, 302
  nuclear ionization mechanisms, 300
  origin of the gas, 301
AGNs
  accretion, 447
  accretion on SBH, 96
  bolometric luminosity function, 700
  Central masses, 729
  cosmic X-ray background, 383-384
  data catalogue, 559
  dust accretion disks, 504-505
  electron-positron cauldron, 152-154, 156-157, 162-164, 166-168
  emission line clouds
    confinement, 509-510
    radio emission, 511
  evolution, 697-699
  group members, 239
  hard spectrum formation, 149-152
  high energy e.m. processes, 154-155
  ionized gas, 391-392
  IRAS survey, 266, 268-269
  jets, 96
  line widths vs Lopt, 560-563
  low accretion rate, 242
  luminosity evolution, 631
  luminosity function, 231
  luminosity-mass ratio, 727, 732
  masses, 735
  optical surveys, 254-256, 258-259
  optically selected
    X-ray properties, 548
  photoionization models, 531-533
  relativistic particle streams, 521
    instability, 522-523
  unified model, 449
  X-ray selected, 547
    optical properties, 549
  X-ray surveys, 261-265
  young
    evolution, 386-388
    optical characteristics, 391-393
    physical properties, 384-385
    X-ray spectrum, 389-390
AGNs dust accretion disks, 506

BL Lac
  light curve, 587-588
BL Lac objects, 50-51, 246
  redshifts, 675-677
  spectra, 675-677
  X-ray selected, 615
    spectral indexes distribution, 616
  x-ray selected
    spectral indexes distribution, 617
Black holes
  e.m. energy extraction, 448
Blazars
  comparison with Quasars, 74-75
  emission models, 75-76
  energy distribution, 65, 69
  far UV spectra, 64-65
  luminosity distribution, 71
  models, 484-485
  polarization, 483
  redshift distribution, 71

UV spectra, 63
X-ray spectra, 63

Clouds
  intergalactic
    neutral hydrogen, 723-725
Cluster
  collapse, 697
Cluster morphology
  bright galaxies, 555-557
Compact extragalactic objects
  frequency, 499-501
Cyg A, 695-696

Density function, 6

Elliptical galaxies
  ionized gas, 298
Emission broad line
  NGC 4151, 277-280
Emission line galaxies
  high-dispersion
    objective prism survey, 200-202
  interactions and companions, 207-208
  IRAS observations, 204-205
  low-dispersion
    objective prism survey, 203
Emission line region, 728
Emission lines, 727-729, 732, 735

Faint quasars
  colors, 325-326
  evolution of the LF, 317-318, 330-331
  number counts, 324
  optical sample, 339
  photometry, 319
  proper motions, 320-321
  redshifts, 325-326
  shape of the LF, 330
  spectroscopy, 322-323
  variability, 320-321

Galactic nuclei
  NGC 1275, 683-687
Galaxies
  distance, 645-646
  formation, 179-181
    primordial, 178
  merging, 697-700, 705-708
  nuclei
    megamasers, 461

Gravitational lense
  quasar S5 0014+81, 611
Gravothermal catastrophe, 132

Hubble diagram, 473-474

Infrared active galaxies, 12
Interacting galaxies
  emission morphology, 713-715
Intergalactic hydrogen
  Leo cloud, 723-725

Jets, 52, 717-721
  optical
    survey, 581-582
  optical classification, 579-580
  wind type model, 543
    instabilities, 544-546

LINERs, 23, 29-30
  continuity with Sy 2 galaxies, 200
  continuity with Sy 2 galaxies, 199
  excitation mechanisms, 475-477
  H profiles, 31-32
  ionization mechanisms, 36-39
  line profiles, 35-36
Local luminosity function, 6
Low luminosity active nuclei, 701-704
Luminosity function
  density evolution, 4
  luminosity evolution, 5
  quasars
    evolution, 3

M82
  nuclear luminosity, 649-652
Magnetic monopoles
  in galactic nuclei, 663-667, 737-739
  sources, 659-661
Markarian galaxies
  double nucleus galaxies, 605
Mass density fluctuations, 178
Mildly active galactic nuclei, 679-681
Models
  Fe II OPT/H ratio, 534

Narrow band photometry, 693-696
Narrow emission lines
  NGC 4151, 275

INDEX OF SUBJECTS

variability, 276, 281-290
Narrow emission-line galaxies, 196
Narrow line region, 705-708
Nearby galaxies
  active nuclei, 701-704
  Seyfert nuclei, 21-22
NGC 1068
  IR surface photometry, 633

Optical luminosity function
  evolution, 7

Papaloizou and Pringle instability, 113-118, 121-122, 124-126
Photon scattering, 645-646
Positron-electron annihilation, 737-739
Protogalaxies
  evolution, 184, 187-190
  observations, 175-177
  search for-, 182-184

QSO 3C345
  lightcurve
    shot noise, 604
  optical variability, 601-603
QSOs
  clustering, 454
  continuity with AGNs, 54
  continuity with Radio Galaxies, 53
  continuity with Seyfert galaxies, 48-49
  cosmological evolution, 537
    DB regression, 400-401
  early objects, 244
  heterogeneous samples, 400
  high-redshift
    early universe, 450
    host galaxies, 453
  host galaxies, 452
  identifications
    UVX method, 492-493
  interacting host galaxy, 233
  luminosity evolution, 537-538
  non-cosmological hypothesis, 55-58
  optical evolution rate, 402-404, 406-407
  optically selected
    X-ray LF evolution, 399
    X-ray properties, 416-418
    X-ray quiet, 408-410
  redshift distribution, 539-540
  spectroscopic identification, 491
  X-ray evolution rate, 402-404, 406-407
  X-ray luminosity function, 398-399
  X-ray selected
    cosmological evolution, 412-413
    number counts, 414-415
  X-ray selected samples, 397
Quasar S5 0014+81
  CCD observations, 612-613
  VLA-VLBI observations, 614
Quasars
  bolometric luminosity, 472
  clustering, 518-519
  counts, 221-224
  distribution, 517
  evolution models, 328-329
  evolution of the optical LF, 362-366
  galactic component, 216-217, 299-301
  high-redshift
    host galaxies, 312
  high-redshift cut-off, 367-368, 451
  host galaxies, 309
    spectroscopy, 311
    tidal interactions, 311
  ionized gas, 297
  IRAS observations, 15-16
  light curves, 85, 87-89
    shot noise, 589-590
  light-curves, 471
  low-redshift
    host galaxies, 310
  luminosity evolution, 220-224, 361
  luminosity function, 327, 628
    non-parametric approach, 356-359
  luminosity function evolution, 353
  multiband spectra catalogue, 591
  number evolution, 361
  optical candidates, 468-470
  optical counts, 346
  optical identification, 465
  optical samples, 82, 342-343
    colour selection, 344-345
    completeness, 340-341
    prism selection, 347-349

optical variability, 84
optical versus radio properties, 82
radio loud, 17
   IRAS observations, 18
source function, 360
spectroscopy, 466-467
variability versus redshift, 85
variability
   correlation between wavebands, 90

Radiation mechanisms
   Cerenkov line emission, 583-586
Radio galaxies
   double
      extent evolution, 565-567
   faint
      blue, 439
      red, 439
   ionized gas, 297, 299-300
   wide angle tail, 541
Radio luminosity function
   evolution, 440-441
Radio sources
   a-s relation, 569
   counts, 437-438
   flux variations, 653-657
   free-form evolution models, 571-572
   identification statistics, 437-438
   steep spectrum
      variability, 623
   structural changes, 653-657
Reddening, 694-696

Seyfert 1 galaxies
   Baldwin-relation, 526-529
   continuity with QSOs, 232
   cosmic X-ray background, 232
   local luminosity function, 218-219
   luminosity evolution, 220
   luminosity function, 627
   photoionization, 534
   spectra, 25-26
   UV spectra, 525
      optical Fe II emission, 513-516
Seyfert 2 galaxies
   H profiles, 33-34
   Mkn 463, 689-692

spectroscopy, 689-692
Seyfert galaxies
   CCD photometry, 487-489, 642-644
   connection with quasars, 215
   cosmic X-ray background, 241
   double nucleus galaxies, 607-608
   group members, 551-552
      interactions with galaxies, 553-554
   interacting
      CCD photometry, 479-482
   IRAS observations, 13-14
   luminosity function, 194-196, 641
   megamasers, 462
   members of clusters of galaxies, 198
   NGC 4151, 705-708
   observations, 193
   spectra, 24-25, 669-672, 705-708
   starlight contamination, 27-30
   Sy1-Sy2 transition, 669-672
Star cluster
   numerical relativistic dynamics, 133-134, 138-139
      approximate solution, 140, 142
Star-burst galaxies, 196
Superluminal velocities, 52
Supermassive black holes
   dense star cluster collapse, 129
   ejection from galaxies, 573-575
   formation, 130
Symbiotic stars
   and Seyfert galaxies, 709-712
   T Cor B, 709-712

Tori
   stability, 119-121, 123-125

UV excess objects
   Kiso Surveys, 619-621

X-ray luminosity function
   evolution, 7

INDEX OF OBJECTS

0215+015, 66-67, 426-427, 430
0219+428 (3C 66A), 66-67
0241+622, 49
0317+186, 616
0323+022, 63
0330-380, 454
0358+004, 624
0453-423, 425
0548+322, 63
0605-085, 654
0621+400, 624
0716+71, 64, 66-67
0754+10 (OI090.4), 66-67
0754+100, 675-677
0829+046 (OJ 049), 64, 66-67
0851+202 (OJ 287), 66-67
0859-140, 655
1035+630, 612
1059+730, 49
1100-264, 612
1101+384 (Mkn 421), 63, 66-67
1133+704 (Mkn 180), 66-67
1156+295 (Ton 599), 65-67
1207+379, 616
1207+397, 616
1215+303 (ON 325), 66-67
1218+304, 64, 66-67
1235+632, 616
1308+32, 64
1308+326, 676
1336+135, 454
1402+043, 616-617
1418+54 (OQ 530), 66-67
1422+202, 623-626
1510-089, 64, 654-655
1514+197, 675-677
1514-24 (AP Lib), 64, 66-67
1524-136, 624-626
1611+343 (DA 406), 655-656
1635+119, 309, 313
1641+399, 64
1652+398 (Mkn 501), 64, 66-67
1727+502 (IZ 187), 66-67
1807+698 (3C 371), 66-67
1845+797, 64
2155-302, 64
2200+420 (BL Lac), 66-67
2223+05 (3C 446), 66-67
2237+0305, 50, 55, 57
2251+158 (3C 454.3), 657
2E 0104.2+3153, 55

3A 0557-383, 262-263
3C 9, 592
3C 31, 52
3C 47, 593
3C 48, 11, 14-15, 17-18, 48-49, 54, 593
3C 66A, 52, 57, 594
3C 66B, 52, 57
3C 99 (0358+00), 623-626
3C 120, 52-53, 57, 257, 263, 654
3C 138, 626
3C 159 (0621+40), 623-626
3C 179, 52
3C 232, 423
3C 249.1, 48-49
3C 256, 53
3C 273, 16-17, 49, 52, 58, 81, 168, 476, 612
3C 279, 52, 597, 612
3C 293, 695
3C 295 cluster, 198-199
3C 323.1, 49
3C 345, 16-17, 52, 64, 82, 85, 599, 601
3C 351, 82, 599
3C 371, 51
3C 380, 82
3C 390.3, 64, 257, 669-670
3C 446, 51, 56, 82, 84-85, 87-90, 600
3C 454.3, 52, 656
4C 11.72, 49
4C 13.41, 597
4C 14.82, 309, 313
4C 31.63, 49
4C 37.43, 48-49
4C 39.25, 52
4U 0241+61, 263

Akn 120, 258, 480-481, 525-526, 528
Akn 347, 258
Akn 359, 258
Akn 539, 258
Akn 564, 258
AM 0207-49A, 580
AO 0235+164, 51, 66-67
AP Librae, 51
Arp 91, 715
Arp 140, 715
Arp 220, 12, 18

INDEX OF OBJECTS

Arp 308, 715

B2 1225+317, 612
B2 1308+32, 66-67
BL Lac, 51-53, 472-474, 587-588, 600, 654
BSO 1, 467
BSO 2, 467

Cen A, 168, 302
CH Cyg, 711-712
Coma cluster, 541
Cyg A, 449, 693, 695-696

DA 406, 655
DC 0428-53 cluster, 199

ESO 0610-23, 579, 581-582
ESO 103-G35, 263
ESO 103-G55, 262
ESO 141-G55, 263

F-9, 263, 268, 525-528

Galactic Center, 737-738
GP X 2, 263

H 1613+06, 263-264
H 1814+63, 264
Haro 4, 196-197

I Zw 1, 257, 513, 515
I Zw 26, 257
I Zw 27, 257
I Zw 81, 257
I Zw 92, 257
I Zw 96, 580
I Zw 112, 257
I Zw 121, 257
I Zw 171, 257
I Zw 187, 257
IC 2200, 715
IC 2200A, 715
IC 4329 A, 262-263
IC 4553, 461-462
IC 4553 (Arp 220), 461
IC 5063, 553
II Zw 1, 257
II Zw 14, 257
II Zw 101, 257
II Zw 136, 257
II Zw 171, 257
II Zw 187, 257
III Zw 2, 257, 263, 513-516
III Zw 55, 257

III Zw 77, 257
III Zw 127, 257
IRAS 0428-097, 205-206
IRAS 0450-032, 205-206
IRAS 1249-131, 204-205
IRAS 1319-164, 204
IRAS 1833-654, 204
IV Zw 1, 257
IV Zw 29, 257

Kaz 163, 257

LB 180, 467
LB 2772, 467-468
LB 2811, 468
LB 2813, 468
LB 2828, 467
LB 2837, 467
LB 2844, 467
LB 2845, 467
LB 2847, 468
LB 2850, 467-468
Leo galaxy group, 723

M 31, 142, 679
M 33, 12
M 51, 35
M 67 cluster, 488
M 81, 22, 35-36, 200
M 82, 48, 262, 649-651
M 87, 26-28, 40, 48, 52, 81, 130, 449, 543-546, 580
M 96, 723
M 105, 723
MCG 3-34-084, 423
MCG 8-11-11, 263-264
MCG-5-23-16, 262-264
MCG-6-30-15, 262-263
Mkn 6, 672, 733-734
Mkn 36, 196
Mkn 40, 257
Mkn 78, 262-263
Mkn 79, 263, 488-489
Mkn 176, 257
Mkn 205, 54, 263
Mkn 231, 12, 14-15, 18, 257, 461, 488-489
Mkn 266, 605-607
Mkn 273, 257, 606-608
Mkn 279, 263
Mkn 290, 263
Mkn 296, 606
Mkn 306, 606
Mkn 314, 606
Mkn 315, 257

INDEX OF OBJECTS

Mkn 334, 257
Mkn 335, 263
Mkn 374, 488-489
Mkn 376, 263
Mkn 421, 51, 63-64, 615-616
Mkn 463, 606-608, 689-692
Mkn 477, 257
Mkn 480, 606
Mkn 486, 257
Mkn 501, 51, 64, 615-617
Mkn 507, 263-264
Mkn 509, 263
Mkn 543, 257
Mkn 544, 606
Mkn 590, 263
Mkn 673, 606
Mkn 699, 257
Mkn 739, 606
Mkn 766, 488
Mkn 788, 606
Mkn 789, 606
Mkn 823, 606
Mkn 883, 200
Mkn 926, 263-264
Mkn 930, 606
Mkn 957, 257
Mkn 975, 479-480
Mkn 1014, 14, 49
Mkn 1024, 15
Mkn 1027, 606-607
Mkn 1095 (Akn 120), 479
Mkn 1388, 200
Mkn 1394, 606
MR 2251-178, 36

NGC 253, 679-680
NGC 404, 29-30
NGC 526A, 263
NGC 985, 263
NGC 1052, 32, 302, 475
NGC 1068, 254, 262-263, 462, 633
NGC 1073, 55, 467
NGC 1087, 467
NGC 1087 U1, 467
NGC 1097, 580, 582, 680
NGC 1144, 479-481
NGC 1261, 341
NGC 1275, 48, 64, 254, 683-687
NGC 1300, 680
NGC 1365, 680
NGC 1410, 257
NGC 1566, 263, 553, 672
NGC 1569, 680
NGC 1598, 579, 581
NGC 1808, 259

NGC 2110, 262-263
NGC 2639, 32-33, 200
NGC 2681, 25
NGC 2782, 254
NGC 2841, 29
NGC 2903, 679
NGC 2992, 200, 262-264, 553
NGC 3067, 423
NGC 3077, 254
NGC 3079, 461-462
NGC 3115, 28-29
NGC 3227, 254, 263, 733-734
NGC 3281, 263-264
NGC 3351, 679
NGC 3384, 723
NGC 3516, 254
NGC 3672, 679
NGC 3690, 461
NGC 3783, 263, 525-526, 528, 553
NGC 3842, 55
NGC 3884, 31-32
NGC 3998, 26-27, 35, 200
NGC 4051, 21, 254, 263-265
NGC 4151, 21, 24, 130, 142,
   168, 232, 254, 263, 268,
   275-277, 279, 281, 283, 287,
   290, 292, 391, 526, 528,
   669-670, 679, 705-707, 709,
   733-734
NGC 4192, 24
NGC 4235, 24
NGC 4258, 24, 32, 254
NGC 4278, 24, 32
NGC 4303, 34
NGC 4314, 680
NGC 4319, 54
NGC 4369, 25
NGC 4388, 25, 33
NGC 4579, 200
NGC 4593, 263, 551-553
NGC 4639, 26
NGC 4650A, 715
NGC 4736, 679-680
NGC 4874, 541-542
NGC 5033, 200, 263
NGC 5128, 262, 302, 707
NGC 5135, 553
NGC 5153, 551
NGC 52732, 200
NGC 5383, 679
NGC 5506, 262-264
NGC 5548, 254, 263, 525-526, 528
NGC 5929, 257
NGC 6221, 553
NGC 6814, 254, 263

NGC 6951, 35
NGC 7172, 262-264
NGC 7213, 35, 38-40, 200, 263
NGC 7314, 263
NGC 7469, 263
NGC 7496, 254
NGC 7538, 462
NGC 7582, 262-264, 552-553
NRAO 140, 52

OH 471, 612
OJ 287, 596
OQ 172, 612
OQ 530, 82
OT 546, 51

PG 1351+640, 18
PG 2130+099, 257
PG 2209+184, 257
PGC 0007+10, 257
PHL 658, 82
PHL 957, 612
PHL 1447, 468
PHL 4291, 468
PHL 4295, 468
PHL 4829, 468
PHL 8504, 468
Pictor A, 36, 39
PKS 0237-23, 612
PKS 0241+011, 467
PKS 0245+013, 467
PKS 0256-005, 467
PKS 0300-004, 467
PKS 0323+022, 66-67
PKS 0349-27, 300, 302, 306
PKS 0349-29, 298
PKS 0521-36, 51-52
PKS 0521-365, 66-67
PKS 0537-441, 66-67, 88-90
PKS 0548-322, 66-67
PKS 0634-20, 298, 302, 306
PKS 0735+178, 66-67
PKS 0736+017, 64
PKS 1327-206, 423
PKS 1510-089, 612
PKS 1718-649, 36, 38
PKS 2000-330, 451, 612
PKS 2141+174, 49
PKS 2155-304, 66-67
PKS 2158-380, 300, 303

Q 0026+129, 592
Q 0235+164, 594
Q 0252+0118, 467
Q 0420-014, 595
Q 0422+004, 595
Q 0735+178, 596
Q 1308+326, 598
Q 1418+546, 598

S5 0014+81, 312, 393, 611-614, 732
SBS 1518+593, 257
Sgr A, 737
Sgr B2, 462
SS 433, 54, 275, 290

T Cor B, 290, 709-712
TOL 1238-364, 553
Ton 1530, 612
Ton 1542, 263

U Geminorum, 126
U1, 467
UGC 10683 B, 262-263
UGC 3995, 579, 581
UGC 3995A, 580
UM 283, 196-197
US 3150, 467
US 3472, 467
US 3498, 467
US 3605, 467

V Zw 85, 257
V Zw 317, 257
VII Zw 118, 257, 513-515
VII Zw 490, 257
VII Zw 653, 257
VII Zw 742, 257
VII Zw 838, 257
Virgo A, 26
Virgo Cluster, 26
VV 144, 580

Was 31, 201

Zw 0033+45, 257
Zw 0039+40, 257
Zw 1518+59, 257

LIST OF CONTRIBUTORS

ABRAMOWICZ, M.A., International School for Advance Studies, Strada Costiera 11, 34014 Trieste, Italy
ALLER, H.D., University of Michigan, Ann Arbor, USA
ALLER, M.F., University of Michigan, Ann Arbor, USA
ALTAMORE, A., Istituto Astronomico dell'Universita', Via Lancisi 29, 00161 Roma, Italy
AVNI, Y., Harvard-Smithsonian Center for Astrophysics, Cambridge, Massachusetts 02138, USA and Weizmann Institute of Science Rehovot 76100, Israel
BAAN, W.A., Arecibo-Observatory, Cornell University, P.O.BOX 995, Arecibo, Puerto Rico 00613
BARBIERI, C., Istituto di Astronomia, Universita' di Padova, Vicolo dell'Osservatorio 5, 35100 Padova, Italy
BARTEL, N., Harvard Smithsonian CFA, Cambridge, USA
BECKMAN, J.E., Instituto de Astrofisica de Canarias, Universidad de La Laguna, Tenerife, Spain
BERGERON, J., Institut d'Astrophysique, 98 bis, Boulevard Arago F-75014, Paris, France
BINETTE, L., European Southern Observatory, D-8046 Garching bei Muenchen, Federal Republic of Germany
BIRKLE, K., Marx-Planck-Institut fuer Astronomie, Heidelberg (DSAZ)
BJORNSSON, C.I., NORDITA, Blegdamsvej 17, DK-2100 Copenhagen O, Denmark
BLAES, O.M., International School for Advanced Studies. Strada Costiera 11, 34014 Trieste, Italy
BOYLE, B.J., Department of Physics, University of Durham, South Rd., Durham DH1 3LE, U.K.
BOKSENBERG, A., Royal Greenwich Observatory, Herstmonceux Castle, Hailsham, East Sussex BN27 1RP, U.K.
BOLDT, E., Laboratory for High Energy Astrophysics, NASA/Goddard Space Flight Center, Greenbelt, Maryland 20771, USA
BONOLI, C., Osservatorio Astronomico e Istituto di Astronomia dell'Universita' di Padova, Vicolo dell'Osservatorio 5, 35122 Padova, Italy
BONOLI, F., Osservatorio Astronomico e Istituto di Astronomia, Bologna, Italy
BORTOLETTO, F., Osservatorio Astronomico e Istituto di Astronomia dell'Universita' di Padova, Vicolo dell'Osservatorio 5, 35122 Padova, Italy
BROMAGE, G.E., Astrophysics Group, Rutherford and Appleton

Laboratory, Chilton, Didcot, Oxfordshire OX11 0QX, U.K.
BROSCHE, P., Observatorium Hoher List, Universitats-Sternwarte Bonn, D-5568 Daun, F.R. Germany
BROWNE, I., Nuffield Radio Astronomy Labs., Jodrell Bank, U.K.
BUITRAGO, J., Instituto de Astrofisica de Canarias, University of La Laguna, Tenerife, Spain
BURBIDGE, G., Center for Astrophysics and Space Sciences University of California, San Diego La Jolla, California 92093, and Kitt Peak National Observatory, Tucson, Arizona
CARROLL, T.J., Department of Astrophysics, University of Oxford, Oxford, U.K.
CASSATELLA, A., IUE Observatory, ESA, Apartado 54065, Madrid, Spain
CAVALIERE, A., Astrofisica, Dip. Fisica, II Universita' di Roma, Italy
CHENG, F.Z., Center for Astrophysics, Cambridge, Ma. USA, University of Science and Technology of China
CHU, Y., Center for Astrophysics, University of Science and Technology of China, Hefei, Anhui, China
CLAVEL, J., Observatoire de Meudon, 92190 Meudon, France, and ESA Villafranca Tracking Station, Apdo 54065, Madrid, Spain
COLEMAN, C.S., Department of Astrophysics, South Parks Road, Oxford OX1 3RQ, England
COLINA, L., Space Astrophysics Division, SSD-ESTEC, Noordwijk, Holland
COLLIN-SOUFFRIN, S., Observatoire de Paris-Meudon, 92195 Meudon Principal Cedex, France
CRISTIANI, S., European Southern Observatory, Casilla 19001, Santiago 19, Chile and Istituto di Astronomia, Universita' di Padova, Vicolo dell'Osservatorio 5, 35122 Padova, Italy
DANESE, L., Istituto di Astronomia, Vicolo dell'Osservatorio 5, 35122 Padova, Italy
DELPINO, F., Osservatorio Astronomico e Istituto di Astronomia, Bologna, Italy
DE ZOTTI, G., Istituto di Astronomia, Vicolo dell'Osservatorio 5, 35122 Padova, Italy
di SEREGO ALIGHIERI, S., ST/ECF, Garching bei Muenchen, Federal Republic of Germany - Appliated to the Astrophysical Division Space Science Department of E.S.A.
ELVIUS, A., Stockholm Observatory, 13300 Saltsjobaden, Sweden
FANG, L., Center for Astrophysics, University of Science and Technology of China, Hefei, Anhui, China
FANTI, C., Istituto di Radioastronomia, via Irnerio 46,

LIST OF CONTRIBUTORS

FANTI, R., Istituto di Radioastronomia, via Irnerio 46, 40126 Bologna, Italy
FERETTI, L., Istituto di Radioastronomia, via Irnerio 46, 40126 Bologna, Italy
FERRARI, A., Istituto di Fisica Generale dell'Universita', Torino, Italy
FIALA, N., Institut fuer Astronomie, Turkenschanzstrasse 17, A-1180 Wien, Austria
FILIPPENKO, A.V., Department of Astronomy, University of California Berkeley, CA 94720 USA
FICARRA, A., Istituto di Radioastronomia, via Irnerio 46, 40126 Bologna, Italy
FONG, R., Department of Physics, University of Durham, South Rd., Durham DH1 3LE, U.K.
FOSBURY, R.A.E., Space Telescope European Coordinating Facility, European Southern Observatory, Karl-Schwarzschild-Str. 2, D-8046 Garching bei Muenchen, FRG
FRANCESCHINI, A., Istituto di Astronomia, Vicolo dell'Osservatorio 5, 35122 Padova, Italy
FRIED, J.W., Max-Planck-Institut fuer Astronomie, D-6900 Heidelberg 1, West Germany
FRICKE, K.J., Universitats-Sternwarte, Geismarlandstr. 11, 3400 Goettingen, West Germany
GHISELLINI, G., International School for Advanced Studies, Trieste, Italy
GIALLONGO, E., Istituto Astronomia,dell' Universita di Padova, Vocolo dell'Osservatorio 5, 35122 Padova, Italy
GIOIA, I.M., Center for Astrophysics, 60 Garden st., Cambridge MA 02138, also from Istituto di Radioastronomia, CNR, Via Irnerio 46, 40126 Bologna, Italy
GIOVANNINI, G., Istituto di Radioastronomia, via Irnerio 46, 40126 Bologna, Italy
GIURICIN, G., Astronomical Observatory, via G.B. Tiepolo 11, I 34131 Trieste, Italy
GRADY, C.A., Computer Sciences Corporation, USA
GREGORINI, L., Istituto di Radioastronomia, via Irnerio 46, 40126 Bolo- gna, Italy
HACK, M., Astronomical Observatory of Trieste, via Tiepolo 11, I 34131 Trieste, Italy
HICKSON, P. Department of Geophysics and Astronomy, University of British Columbia, 2219 Main Mall, Vancouver, B.C., V6T 1W5, Canada
HELLWIG, J., Universitats-Sternwarte, Geismarlandstr. 11, 3400 Goettingen, West Germany
JOLY, M., Observatoire de Paris-Meudon, 92195-Meudon Principal Cedex, France
KAPAHI, V.K., Tata Insitute of Fundamental Research, P.O.

Box 1234, Bangalore 560 012, India
KAPOOR, R.C., Indian Institute of Astrophysics, Bangalore 560034, India
KATGERT, P., Leiden Observatory, University of Leiden, The Netherlands
KEEL, W.C., Kitt Peak National Observatory, National Optical Astronomy Observatories, P.O. Box 26732, Tucson, AZ 85726, USA
KIANG T., Dunsink Observatory, Dublin, Ireland
KIDGER, M.R., Instituto de Astrofisica de Canarias, Universidad de La Laguna, Tenerife, Spain
KOLLATSCHNY, W., Universitats-Sternwarte, Geismarlandstr. 11, 3400 Goettingen, West Germany
KONDO, M., Tokyo Astronomical Observatory, Osawa, Mitaka, Tokio 181, Japan
KOO, D.C., Space Telescope Science Institute, 3700 San Martin Drive, Baltimore, Maryland 21218, USA
KUHR, H., Max-Planck-Institut fuer Astronomie, Koenigstuhl, D-6900 Heidelberg 1, F.R.G.
KULKARNI, V.K., Radio Astronomy Centre, Post Box 8, Ootacamund - 643 001, India
LEITER, D., Laboratory for high Energy Astrophysics, NASA/Goddard Space Flight Center, Greenbelt, Maryland 20771, USA
LI, Z., Department of Astronomy, Nanjing University, Nanjing
LOOSE, H.H., Universitats-Sternwarte, Geismarlandstr. 11, 3400 Gottingen, West Germany
LU, J., International School for Advance Studies, Strada Costiera 11, 34014 Trieste, Italy
MACCACARO, T., Center for Astrophysics, 60 Garden st., Cambridge MA 02138 also from Istituto di Radioastronomia, CNR, Via Irnerio 46, 40126 Bologna, Italy
MACCAGNI, D., Istituto di Fisica Cosmica, Milano, Italy
MAEHARA, H., Tokyo Astronomical Observatory, Osawa, Mitaka, Tokyo 181, Japan
MANTOVANI, F., Istituto di Radioastronomia, via Irnerio 46, 40126 Bologna, Italy
MARANO, B., Dipartimento di Astronomia, via Zamboni 33, 40126 Bologna, Italy
MARASCHI, L., Dipartimento di Fisica, Universita' di Milano, Italy
MARDIROSSIAN, F., Department of Astronomy, University of Trieste, via G.B. Tiepolo 11, I 34131, Trieste, Italy
MARSHALL, H.L., Space Telescope Science Institute, 3700 San Martin Drive, Baltimore, MD 21218, USA
MATVEENKO, L., Inst. for Space Research Academy, Moscow, USSR
MEDIAVILLA, E., Instituto de Astrofisica de Canarias, University of La Laguna, Tenerife, Spain

LIST OF CONTRIBUTORS

MENON, T.K., Department of Geophysics and Astronomy, University of British Columbia, 2219 Main Mall, Vancouver, B.C. V6T 1W5, Canada
METIK, L., Crimean Astrophysical Observatory, USSR
MEURS, E.J.A., Institute of Astronomy, Madingley Road, Cambridge, CB3 0HA, England
MEZZETTI, M., Astronomical Observatory, via G.B. Tiepolo 11, I 34131, Trieste, Italy
MIYAUCHI-ISOBE, N., Tokyo Astronomical Observatory, Osawa, Mitaka, Tokyo 181, Japan
MISSANA, M., Osservatorio Astronomico di Brera, via Brera 28, 20121 Milano, Italy
MUXLOW, T., Nuffield Radio Astronomy Labs., Jodrell Bank, U.K.
NEUGEBAUER, G., Division of Physics, Matematics and Astronomy, California Institute of Technology, Pasadena, California 91125, USA
NICOLSON, G.D., N.I.T.R., Johannesburg, South Africa
NOVIKOV, I.D., Space Research Institute Academy of Sciences of the USSR, Moscow 117810, Profsoyuznaja 84/32, USSR
NOGUCHI, T., Tokyo Astronomical Observatory, Osawa, Mitaka, Tokyo 181, Japan
NOTNI, P., Zentralinstitut fuer Astrophysik der AdW der DDR, 15 Potsdam, German Democratic Republic
OORT, M.J.A., Leiden Observatory, University of Leiden, The Netherlands
OSTERBROCK, D.E., Lick Observatory, Board of Studies in Astronomy and Astrophysics, University of California, Santa Cruz, CA 95064
PADRIELLI, L., Istituto di Radioastronomia, via Irnerio 46, 40126 Bologna, Italy
PENG, Q., Department of Astronomy, Nanjing University, Nanjing and Kapteyn Astronomical Laboratory, Groningen
PENSTON, M.V., Royal Greenwich Observatory, Herstmonceux Castle, Hailsham, East Sussex BN27 1RP, U.K.
PEREZ, E., Royal Greenwich Observatory, Herstmonceux Castle, Hailsham, East Sussex BN27 1RP, U.K.
PEROLA, G.C., Istituto Astronomico dell'Universita', Via Lancisi 29, 00161 Roma, Italy
PERSIC, M., SISSA-ISAS, Strada Costiera 11, Miramare, 34100 Trieste and Osservatorio Astronomico, via G.B. Tiepolo 11, I 34131 Trieste, Italy
PETERSON, B.A., Mount Stromlo and Siding Spring Observatories, Private Bag, Woden, Canberra ACT 2606, Australia
PETROSIAN, V., Center for Space Science and Astrophysics, Stanford University, Stanford, CA 94305
PISMIS, P., Instituto de Astrofisica de Canarias, Universidad de La Laguna, Tenerife, Spain - on

leave from - Instituto de Astronomia, Universidad Nacional Autonoma de Mexico, Apartado Postal 70-264, Ciudad Universitaria, 04510 Mexico, D.F., Mexico
PRONIK, I., Crimean Astrophysical Observatory, USSR
RAFANELLI, P., Istituto di Astronomia dell'Universita' di Padova, Vicolo dell'Osservatorio 5, 35122 Padova, Italy
RAINE, D.J., Department of Astronomy, University of Leicester, Leicester LEI 7RH, U.K.
RAKOS, K.D., Institut fuer Astronomie, Turkenschanzstrasse 17, A-1180 Wien, Austria
REES, M.J., Institute of Astronomy, Medingley Road, Cambridge CB 3 OHA, England
ROMNEY, J., N.R.A.O., Charlottesville, USA
ROOS, N., Sterrewacht Leiden, P.O. Box 9513, 2300 RA Leiden, The Netherlands
ROSNER, R., Harvard-Smithsonian Center of Astrophysics, Cambridge, USA
ROWAN-ROBINSON, M., Department of Applied Mathematics, Queen Mary College, Mile End Road, London. El 4NS, England
SADLER, E.M., European Southern Observatory, Karl-Schwarzschild-Str. 2, D-8046 Garching bei Muenchen, Federal Republik of Germany
SALPETER, E.E., NAIC and Department of Astronomy, Cornell University, Ithaca, New York 14853, USA
SALUCCI, P., SISSA-ISAS, Strada Costiera 11, Miramare, 34100 Trieste and Osservatorio Astronomico, via G.B. Tiepolo 11, I 34131 Trieste, Italy
SANCHEZ MAGRO, C., Istituto de Astrofisica de Canarias, Universidad de La Laguna, Tenerife, Spain
SARGENT,W. L.W., Palomar Observatory, California Institute of Technology, Pasadena, CA 91125, USA
SCHMIDT, M., Palomar Observatory, California Institute of Technology, Pasadena, CA 91125, USA
SCHNEIDER, S.E., NAIC and Department of Astronomy, Cornell University, Ithaca, New York 14853, USA
SCHULZ, H., Astronomisches Institut der Ruhr-Universitat, Postfach 102148, D-4630 Bochum, Federal Republic of Germany
SELVELLI, P.L., Osservatorio Astronomico di Trieste, via G.B. Tiepolo 11, I 34131 Trieste, Italy
SHANKS, T., Department of Physics, University of Durham, South Rd., Durham DH1 3LE, U.K.
SHAPIRO, S.L., Cornell University, Center for Radiophysics and Space Research, Space Sciences Building, Ithaca, New York 14853
SHARP, N.A., Kitt Peak National Observatory, National Optical Astronomy Observatories, P.O. Box 26732, Tucson, AZ85726, USA

LIST OF CONTRIBUTORS

SHAVIV, G., Department of Physics, Technion-Israel Institute of Technology, Haifa, Israel 32000
SHLOSMAN, I., Department of Physics, University of Florida, Gainesville, Florida, USA
SILK, J., Department of Astronomy, University of California, Berkeley, CA 94720
SMITH, M.D., Department of Astronomy, University of Leicester, Leicester LEI 7RH, U.K.
SNIJDERS, M.A.J., Royal Greenwich Observatory, Herstmonceux Castle, Hailsham, East Sussex BN27 1RP, U.K.
SOIFER, B.T., Division, of Physics, Matemathics and Astronomy, California Institute of Tecnology, Pasadena, California 91125, USA
STERN, B.E., Institute for Nuclear Research of the Academy of Sciences of the USSR, Moscow 117312, 60-the October Anniversary Prospect, 7a, USSR
STOCKE, J.T., Steward Observatory, University of Arizona, Tucson, AZ
TAKASE, B., Kokugakuin University, Higashi, Shibuya, Tokyo 150, Japan
TANZI, E.G., Istituto di Fisica Cosmica CNR, Milano, Italy
TERZIAN, Y., NAIC and Department of Astronomy, Cornell University, Ithaca, New York 14853, USA
THIELE, U., Max-Planck-Institut fuer Astronomie, Heidelberg (DSAZ)
TREVES, A., Dipartimento di Fisica, Universita' di Milano, Italy
TRUSSONI, E., Istituto di Cosmogeofisica del CNR, Corso Fiume 4, 10133 Torino, Italy
TSINGANOS, K., Department of Physics, University of Crete, Heraklion, Greece
ULRICH, M.H., European Southern Observatory, 8046 Garching bei Muenchen, Federal Republic of Germany
VAGNETTI, F., Istituto Astronomico, I Universita' di Roma, Italy
VAN DER LAAN, H., Leiden Observatory, University of Leiden, The Netherlands
VERON, P., Observatoire de Haute Provence, 04870 Saint Michel l'Observatoire, France
VITELLO, P.A., Science Application Inc., McLean, VA 22102, USA
WAMSTEKER, W., ESA IUE Observatory, P.O.Box 54065, Madrid, Spain, Affiliated with Space Sciences Dept.
WANDEL, A., Astronomy Program, University of Maryland, College Park, MD-20742, USA
WANG, D., Purple Mountain Observatory, Nanjing
WEEDMAN, D.W., Department of Astronomy, The Pennsylvania State University, 525 Davey Laboratory, University Park, PA 16802, USA
WEILER, K.W., National Science Foundation, Washington, DC, USA

WYSE, R.F.G., Department of Astronomy, University of California, Berkeley, CA 94720
YOU, J.H., University of Science and Tecnology of China, Hefei, China
ZAMORANI, G., Istituto di Radioastronomia, via Irnerio 46, 40126 Bologna, Italy
ZITELLI, V., Dipartimento di Astronomia, via Irnerio 46, 40126 Bologna, Italy